PROTEIN FAMILIES

WILEY SERIES ON PROTEIN AND PEPTIDE SCIENCE

Vladimir N. Uversky, Series Editor

Metalloproteomics · Eugene Permyakov

Instrumental Analysis of Intrinsically Disordered Proteins: Assessing Structure and Conformation · Vladimir Uversky and Sonia Longhi

Protein Misfolding Diseases: Current and Emerging Principles and Therapies · Marina Ramirez-Alvarado, Jeffery W. Kelly, and Christopher M. Dobson

Calcium Binding Proteins · Eugene Permyakov and Robert H. Kretsinger

Protein Chaperones and Protection from Neurodegenerative Diseases · Stephan Witt

Transmembrane Dynamics of Lipids · Philippe Devaux and Andreas Herrmann

Flexible Viruses: Structural Disorder in Viral Proteins · Vladimir Uversky and Sonia Longhi

Protein and Peptide Folding, Misfolding, and Non-Folding · Reinhard Schweitzer-Stenner

Protein Oxidation and Aging · Tilman Grune, Betul Catalgol, and Tobias Jung

Protein Families: Relating Protein Sequence, Structure, and Function · Edited by Christine Orengo and Alex Bateman

PROTEIN FAMILIES

Relating Protein Sequence, Structure, and Function

Edited by

CHRISTINE ORENGO
ALEX BATEMAN

WILEY

Published by John Wiley & Sons, Inc., Hoboken, New Jersey
Published simultaneously in Canada

Library of Congress Cataloging-in-Publication Data:

Protein families : relating protein sequence, structure, and function / edited by Christine A. Orengo, Alex Bateman.
pages cm. – (Wiley series in protein and peptide science ; 10)
Includes index.
ISBN 978-0-470-62422-7 (hardback)
1. Proteins. 2. Proteomics. 3. Molecular biology–Data processing. 4. Bioinformatics. I. Orengo, Christine A., 1955– editor of compilation. II. Bateman, Alex, 1972– editor of compilation.
QP551.P695925 2014
572′.6–dc23

2013016212

Printed in the United States of America

10 9 8 7 6 5 4 3 2

CONTENTS

Introduction vii

Contributors xiii

SECTION I. CONCEPTS UNDERLYING PROTEIN FAMILY CLASSIFICATION 1

1 **Automated Sequence-Based Approaches for Identifying Domain Families** 3
Liisa Holm and Andreas Heger

2 **Sequence Classification of Protein Families: Pfam and other Resources** 25
Alex Bateman

3 **Classifying Proteins into Domain Structure Families** 37
Alison Cuff, Alexey Murzin, and Christine Orengo

4 **Structural Annotations of Genomes with Superfamily and Gene3D** 69
Julian Gough, Corin Yeats, and Christine Orengo

5 **Phylogenomic Databases and Orthology Prediction** 99
Kimmen Sjölander

SECTION II. IN-DEPTH REVIEWS OF PROTEIN FAMILIES **125**

6 The Nucleophilic Attack Six-Bladed β-Propeller (N6P) Superfamily **127**
Michael A. Hicks, Alan E. Barber II, and Patricia C. Babbitt

7 Functional Diversity of the HUP Domain Superfamily **159**
Benoit H. Dessailly and Christine Orengo

8 The NAD Binding Domain and the Short-Chain Dehydrogenase/Reductase (SDR) Superfamily **191**
Nicholas Furnham, Gemma L. Holliday, and Janet M. Thornton

9 The Globin Family **207**
Arthur M. Lesk and Juliette T.J. Lecomte

SECTION III. REVIEW OF PROTEIN FAMILIES IN IMPORTANT BIOLOGICAL SYSTEMS **237**

10 Functional Adaptation and Plasticity in Cytoskeletal Protein Domains: Lessons from the Erythrocyte Model **239**
Anthony J. Baines

11 Unusual Species Distribution and Horizontal Transfer of Peptidases **285**
Neil D. Rawlings

12 Deducing Transport Protein Evolution Based on Sequence, Structure, and Function **315**
Steven T. Wakabayashi, Maksim A. Shlykov, Ujjwal Kumar, Vamsee S. Reddy, Ankur Malhotra, Erik L. Clarke, Jonathan S. Chen, Rostislav Castillo, Russell De La Mare, Eric I. Sun, and Milton H. Saier

13 Crispr-CAS Systems and CAS Protein Families **341**
Kira S. Makarova, Daniel H. Haft, and Eugene V. Koonin

14 Families of Sequence-Specific DNA-Binding Domains in Transcription Factors across the Tree of Life **383**
Varodom Charoensawan and Sarah Teichmann

15 Evolution of Eukaryotic Chromatin Proteins and Transcription Factors **421**
L. Aravind, Vivek Anantharaman, Saraswathi Abhiman, and Lakshminarayan M. Iyer

Index **503**

INTRODUCTION

Christine Orengo

Institute of Structural and Molecular Biology, University College London, London, United Kingdom

Alex Bateman

European Bioinformatics Institute, Wellcome Trust Genome Campus, Hinxton, United Kingdom

The protein machine is a triumph of nature that puts any man-made nanotechnology into the deepest shade. Without the myosin motor proteins that drive the actin filaments along the myosin tails in muscle tissue we cannot move. Without the rotating motor protein complex F0/F1 ATPase we cannot generate chemical energy in the form of ATP that is so essential for all life. Every cell in our bodies is a whirring biochemical machine of immense complexity. We are still ignorant of the exact molecular function of many, or perhaps most, of the protein cogs in this machine. To understand all the molecular components of the cell and how they fit together remains one of the greatest challenges for biology.

Charles Darwin had no idea of the molecular complexity that lay in the heart of every cell. However, his theory of evolution by natural selection has given us a framework that allows us to understand how the complexity of the cell and its protein machinery could have arisen from simpler preexisting proteins. By looking at the amino acid sequence of different proteins we can see that nature's major source of innovation is the duplication and subsequent mutation of proteins. The five human hemoglobin genes that share a common function to transport oxygen around the blood have all arisen from a single ancestral gene during the evolution of animals over the last 800 million years. Each of these hemoglobin genes has small differences in sequence and this causes differences in their affinity for oxygen and other properties. The set of proteins that have arisen from a common ancestor through the process of evolution are known as a protein family.

The concept of a protein family as an evolutionary entity has immense implications for understanding biology. Related proteins arising from a common ancestral protein often share a common function. If we can identify a protein in a newly sequenced organism that belongs to the hemoglobin family, then we can infer that its function is likely to be to transport oxygen. Despite having carried out no experiments on this new protein, we can learn something about its function from its amino acid sequence. By carrying out detailed molecular experiments on proteins from a few model organisms, we might hope to understand all proteins in the millions of species on earth.

Our ability to correctly identify proteins that belong to the same family is essential to understanding biology. Our ability to do this has improved immensely over the past 40 years. These improvements have been due to three different factors: (i) improvements in the algorithms and statistics associated with sequence alignment, (ii) the growth in the number of protein sequences, and (iii) the increase in the availability of protein structures.

1 IMPROVEMENTS IN ALGORITHMS FOR SEQUENCE ALIGNMENT

Our ability to see relationships between proteins has been greatly enhanced not just by the wealth of sequence and structures available to us. The sophisticated algorithms and statistics that have been developed allow us to determine which similarities between protein sequence and structures are of true homology and which reflect only chance similarities. While sequence comparison software such as BLAST and Fasta made comparison of sequences accessible, techniques such as profiles, hidden Markov models, and fold recognition gave experts the ability to find relationships between proteins whose common ancestor may have existed more than a billion years ago. Although algorithmic developments that have been extensively covered elsewhere are not the primary focus of this book, we applaud the computational scientists and mathematicians who have given us the tools to unlock the mysteries of the cell's protein machine.

2 THE GROWTH OF PROTEIN SEQUENCES

International genome projects have brought a wealth of diverse protein sequences and this means that in the last 10 years or so there have been significant increases in the number of protein and nucleic acid sequences available. Protein sequence databases now hold more than 20 million sequences. This also gives rise to a large increase in the number of known protein families. For example, automatic classification of protein families suggests that we now have representatives from more than a million families. Protein family classifications such as PhyloFacts or PANTHER (described by Sjolander in Chapter 6), which focus on specific sequence repositories and involve some limited curation, now contain around 93,000 and 71,000 families, respectively.

However, many proteins (nearly 80% in eukaryotes) are multidomain and the million or more protein families currently identified are built up from different combinations of domains. In this sense, domains are the primary building blocks of life and not surprisingly there are far fewer domain families than protein families. Furthermore, there has been a much slower increase in the numbers of domain families—especially over the last 5 years. The most comprehensive domain family resource, Pfam (reviewed by Bateman in Chapter 3) currently identifies nearly 14,000 families. Moreover, many new Pfam families tend to be quite small and species specific, suggesting that we may be close to knowing a significant proportion of the major domain families in nature. With the growth of next generation sequencing, it is likely that we will soon see improved sampling of unusual taxonomic groups and in the next 20 years we are likely to have access to a true sampling of protein space.

Alongside the activities of the international genome sequencing initiatives, worldwide structure genomics consortia have attempted to increase the structural coverage of domain and protein families. Since the structure of a protein is usually much more highly conserved during evolution than the sequence, this data is valuable for detecting remote homologies and has been exploited by resources such as SCOP and CATH to trace far back in evolution and capture universal families common to all kingdoms of life. There appear to be only a few hundred of these, depending on the criteria used to identify them, and some have been extensively duplicated and are highly populated.

By exploiting structural data we see that there are currently less than 3000 domain superfamilies covering nearly 60% of the domain sequences from completed genomes. The term "*superfamily*" denotes a broad grouping of relatives (i.e., including all paralogs and orthologs) even from very divergent species, and remote relatives can have rather different structures and functions within some superfamilies (see, e.g., the HUP superfamily described in Chapter 8). Structural data can also be used to merge domain "families" identified using purely sequence data—for example, Pfam often recognizes "clans" (comprising remotely related Pfam families) in this manner.

The relatively small number of domain superfamilies relative to protein families and the fact that we have nearly classified a complete set of these domain "building blocks" mean that we can begin to understand the assembly of diverse proteins during evolution from different domain combinations and start to derive rules for predicting the likely functional contributions of the domains or how their roles may change in different contexts. This will hopefully allow us to move toward a domain grammar of function that exploits our understanding of the evolutionary changes occurring in different domain families to build a picture of how the complete protein, containing these domains, may function.

The data from some of the structural genomics initiatives adds further support to the hypothesis that we already know a large proportion of all major domain families. For example, the NIH-funded PSI structural genomics initiatives in the States deliberately sought to identify new domain families for which there was no structural data. In their second phase (PSI2: 2005–2010) they primarily focused on new, structurally uncharacterized families in Pfam and related classifications. Powerful HMM–HMM strategies were employed to discard any that were, in fact, distantly related to known families (e.g., in SCOP or CATH) and those remaining were targeted for structure determination. However, despite their lack of sequence similarity to known families, it became increasingly clear as the structures were solved that most of the families were simply divergent relatives of existing families in SCOP or CATH. Only about 20% of them represented completely novel families with novel structures, and many of these novel families were very small, species or subkingdom specific, with less than 100 relatives.

As reported in Chapter 5, some resources (SUPERFAMILY, Gene3D) derive sequence patterns (or HMMs) for domain superfamilies in SCOP and CATH and use these to predict domain relatives in sequences from completed genomes. Their data suggests that the population of superfamilies is very uneven. The trends follow scale-free behavior whereby most superfamilies are rather small, that is, comprising less than 500 relatives while a few (~200) are very large (having >5,000 relatives). This tiny percentage of superfamilies (<5% of all superfamilies) accounts for nearly two thirds of all structural domains classified.

Many are universal and highly promiscuous, combining with multiple other families to give different multidomain combinations. They support a wide range of functions, either by performing a generic role in different protein contexts or by evolving new functions of the domain itself, that is, through residue mutations and structural divergence. For example, changes in the nature and location of catalytic residues in the active site have been observed. Structural variations can alter the active site geometry to enable binding of different substrates and/or reshape surface features promoting changes in domain or protein interaction partners.

As the sequence and structure data grows—and especially as structural genomics initiatives target new families—the mechanisms by which domains change during evolution will become clearer as also the extent to which they fuse with different partners to give new proteins. However, the coverage of current classifications and the insights already derived from them motivated us to compile this book now, both to convey some of the current knowledge and to present some fascinating examples of the role families play in creating the rich diversity of life we see around us and study as biologists.

3 MOTIVATION FOR THE BOOK

The idea that we may now have accumulated knowledge on all the major protein domain families is borne out by the fact that a large proportion (between 70% and 90%) of domain sequences from most completed genomes can be classified in curated domain families in Pfam. In addition, the technologies for recognizing distant relatives of existing families and confidently assigning new families have matured over the last decade with powerful strategies such as profile–profile comparisons identifying incredibly distant and divergent relatives, some of which may have undergone significant structural changes as well.

Protein and domain family classifications are becoming increasingly and routinely used to annotate newly sequenced proteins, for example, from meta-genome studies or completely sequenced genomes. So a review of protein families—how to identify them and what the analyses of these families tells us about the evolution of the proteins and their impact on the phenotypic repertoire of the organisms they are found in—seemed both timely and valuable for biologists wishing to use these resources to infer functions for their proteins of interest.

There are now many protein, domain, and motif classification resources, some very comprehensive (e.g., Pfam or SCOP) and others only focusing on specific families (e.g., related to a disease or a particular functional activity) or biological processes (e.g., kinases). In order to give a flavor of the technologies used for finding families and the insights they bring, we decided to divide the book into three sections. The first covers strategies for identifying and characterizing the families. Since we felt that it would be unrealistic to capture in a single book the different technologies and data exploited and presented by all family classifications, we invited contributions from authors of the larger scale, more comprehensive resources who could provide overviews of the challenges and strategies related to their own types of classification. We decided to organize the book into three sections. The first section titled "Concepts Underlying Protein Family Classification" of this book reviews the major strategies for identifying homologous proteins and classifying them into families. In the second section titled "In-Depth Reviews of Protein Families" of this book, there is a collection of reviews on some fascinating superfamilies for which we have substantial amounts of data (sequences, structures, and functions) allowing us to trace the emergence of functionally diverse relatives and providing structural insights into the mechanisms modifying their functions. Chapters in the third section titled "Review of Protein Families in Important Biological Systems" review groups of families associated with a particular biological theme (e.g., the protein families involved in the cytoskeleton, reviewed by Baines and coauthors).

We would like to thank all of the authors who contributed to this book. We have been delighted that so many experts from the world over were able to devote their time to create this collection of knowledge. We believe that this work will be useful for student and group leaders alike and hope that you enjoy reading the book as much as we have.

CONTRIBUTORS

Saraswathi Abhiman, National Institutes of Health, National Center for Biotechnology Information, National Library of Medicine, Bethesda, MD, USA

Vivek Anantharaman, National Institutes of Health, National Center for Biotechnology Information, National Library of Medicine, Bethesda, MD, USA

L. Aravind, National Institutes of Health, National Center for Biotechnology Information, National Library of Medicine, Bethesda, MD, USA

Patricia C. Babbitt, Department of Biopharmaceutical Sciences, UCSF Mission Bay, San Francisco, CA, USA

Anthony J. Baines, School of Biosciences, University of Kent, Canterbury, UK

Alan E. Barber II, Department of Biopharmaceutical Sciences, UCSF Mission Bay, San Francisco, CA, USA

Alex Bateman, European Bioinformatics Institute, Wellcome Trust Genome Campus, Hinxton, UK

Rostislav Castillo, Division of Biological Sciences, University of California, San Diego, La Jolla, CA, USA

Varodom Charoensawan, Department of Biochemistry, Mahidol University, Bangkok, Thailand, Integrative Computational BioScience (ICBS) Center, Mahidol University, Bangkok, Thailand

Jonathan S. Chen, Division of Biological Sciences, University of California, San Diego, La Jolla, CA, USA

Erik L. Clarke, Division of Biological Sciences, University of California, San Diego, La Jolla, CA, USA

Alison Cuff, Institute of Structural and Molecular Biology, University College London, London, UK

Benoit H. Dessailly, National Institute of Biomedical Innovation, Osaka, Japan

Nicholas Furnham, European Bioinformatics Institute, Wellcome Trust Genome Campus, Cambridge, UK

Julian Gough, Department of Computer Science, University of Bristol, Bristol, UK

Daniel H. Haft, J Craig Venter Institute, Rockville, MD, USA

Andreas Heger, Department of Physiology, Anatomy and Genetics, MRC CGAT/Functional Genomics Unit, University of Oxford, Oxford, OX, UK

Michael A. Hicks, Department of Biopharmaceutical Sciences, UCSF Mission Bay, San Francisco, CA, USA

Gemma L. Holliday, European Bioinformatics Institute, Wellcome Trust Genome Campus, Cambridge, UK

Liisa Holm, Department of Biological and Environmental Sciences, Institute of Biotechnology, University of Helsinki, Helsinki, Finland

Lakshminarayan M. Iyer, National Institutes of Health National Center for Biotechnology Information, National Library of Medicine, Bethesda, MD, USA

Eugene V. Koonin, National Center for Biotechnology Information, National Library of Medicine, National Institutes of Health, Bethesda, MD 20894, USA

Ujjwal Kumar, Division of Biological Sciences, University of California, San Diego, La Jolla, CA, USA

Juliette T.J. Lecomte, Department of Biophysics, Johns Hopkins University, Baltimore, MD, USA

Arthur M. Lesk, Department of Biochemistry and Molecular Biology, Huck Institute for Genomics, Proteomics and Bioinformatics, The Pennsylvania State University, University Park, PA, USA

Kira S. Makarova, National Center for Biotechnology Information, National Library of Medicine, National Institutes of Health, Bethesda, MD, USA

Ankur Malhotra, Division of Biological Sciences, University of California, San Diego, La Jolla, CA, USA

Russell de la Mare, Division of Biological Sciences, University of California, San Diego, La Jolla, CA, USA

Alexey Murzin, MRC Laboratory of Molecular Biology, Cambridge, UK

Christine Orengo, Institute of Structural and Molecular Biology, University College London, London, UK

Neil D. Rawlings, Wellcome Trust Sanger Institute, Wellcome Trust Genome Campus, Hinxton, Cambridgeshire, United Kingdom

Vamsee S. Reddy, Division of Biological Sciences, University of California, San Diego, La Jolla, CA, USA

Milton H. Saier, Division of Biological Sciences, University of California, San Diego, La Jolla, CA, USA

Maksim A. Shlykov, Division of Biological Sciences, University of California, San Diego, La Jolla, CA, USA

Kimmen Sjölander, Plant & Microbial Biology, Bioengineering, Berkeley, CA, USA

Eric I. Sun, Division of Biological Sciences, University of California, San Diego, La Jolla, CA, USA

Sarah Teichmann, MRC Laboratory of Molecular Biology, Cambridge, UK

Janet M. Thornton, European Bioinformatics Institute, Wellcome Trust Genome Campus, Cambridge, UK

Steven T. Wakabayashi, Division of Biological Sciences, University of California, San Diego, La Jolla, CA, USA

Corin Yeats, Institute of Structural and Molecular Biology, University College London, London, UK

SECTION I

CONCEPTS UNDERLYING PROTEIN FAMILY CLASSIFICATION

1

AUTOMATED SEQUENCE-BASED APPROACHES FOR IDENTIFYING DOMAIN FAMILIES

LIISA HOLM

Department of Biological and Environmental Sciences, Institute of Biotechnology, University of Helsinki, Helsinki, Finland

ANDREAS HEGER

Department of Physiology, Anatomy and Genetics, MRC CGAT/Functional Genomics Unit, University of Oxford, Oxford, UK

CHAPTER SUMMARY

Proteins are made up of one or more protein domains. The identification of these domains and classification into domain families gives a comprehensive overview of the known protein universe and helps in the determination of both fold and function of newly discovered proteins. A multitude of automated methods for recognizing domain boundaries and making domain family assignments have been developed over the last 20 years. This chapter gives a historical overview of some of these methods and then goes on to discuss one of them, automatic domain delineation algorithm (ADDA), in detail. ADDA uses pair-wise sequence comparisons to define protein families, now captured in Pfam-B. The advantages of using ADDA are discussed along with the improvements that need to be made, for example, to distinguish cysteine-rich domain families from otherwise similar cysteine free protein families. Finally, the challenges that this field still faces, such as the need for more

Protein Families: Relating Protein Sequence, Structure, and Function, First Edition.
Edited by Christine Orengo and Alex Bateman.

powerful computational resources and better sensitivity in detecting remote homologous, together with new directions for research have been reviewed.

1.1 INTRODUCTION

Domains are the building blocks of proteins. The identification of domain families yields a compact description of the protein universe and helps the assignment of fold and function to newly sequenced proteins. Domain family classification must solve two intimately linked problems: sequences have to be cut into segments (domains), and these segments have to be unified into domain families. On the one hand, the delineation of domain boundaries is straightforward, if all members of a domain family have been identified. On the other hand, domain boundaries are needed to identify family membership correctly. Over the years, a multitude of fully automated procedures for protein sequence clustering have been derived. Most methods cluster a sequence space graph that represents similarity relationships detected by all versus all sequence comparison. The approaches differ in the choice of algorithm and the way to avoid the effects of domain chaining, spurious similarities and partial detection of homology. Here, we review the variety of methods and describe one of them, ADDA, the current source of Pfam-B, in detail.

1.2 MOTIVATION BEHIND AUTOMATED CLASSIFICATION

The genomic era in molecular biology has brought on a rapidly widening gap between the amount of sequence data and first-hand experimental characterization of proteins. Fortunately, the theory of evolution provides a simple solution to the computational assignment of protein structure and function to uncharacterized sequences: functional and structural information can be transferred between homologous proteins. Homologs carry the memory of common ancestry in their amino acid sequences as a result of functional constraints that have persisted through successive generations. Today, sequence similarity searching is still the most widely used tool to predict the function or structure of anonymous gene products that come out of genome sequencing projects.

Grouping proteins into families is useful in two ways. First, it leads to more sensitive detection of new members and improved discrimination against spurious hits based on the essential conserved features in a family as expressed by profiles (position-specific scoring matrices or PSSMs) (Gribskov et al., 1987), (Hidden Markov Models) HMMs (Eddy, 1998), or patterns (Sigrist et al., 2002). Second, having established family membership, the query sequence can be placed in the context of the evolutionary tree of the family for accurate functional inference. It is also easier to spot inconsistent second-hand annotations in the tree context.

Taken by its colloquial meaning the concept of a family seems deceptively simple: members of a family are related by common descent. Thus, protein sequences derived from a common ancestor by speciation and gene duplication fall naturally into families. This is distinct from orthology (Fitch, 1970), in which only sequences related through speciation are considered.

The multidomain architecture of proteins complicates matters. Domains are the building blocks of proteins and correspond to compact three-dimensional (3-D) structures that fold individually (Wetlaufer, 1973). Genomic events such as gene fusion (Sali, 1999) and genome rearrangement cause domains to recombine creating multidomain proteins with components deriving from many domain families (Doolittle and Bork, 1993). As a result, if we go sufficiently far back in time, segments in a protein sequence might derive from different ancestral sequences.

Thus, the meaning of the term family varies with context. Classifications of domains strive toward maximal unification of all homologous sequence segments. In the context of clustering complete protein chains, the term family is usually combined with some notion of functional conservation. In particular, the rise of genomics and the availability of many complete genomes have shifted the focus toward grouping proteins, which perform equivalent biological functions in many organisms. The desired classifications are more fine-grained, and domain composition is seen as a cue to specific gene function.

The usage of the terms family and superfamily (unified family) also are not uniform and can represent different levels of the functional hierarchy. The protein information resource (PIR) definition of superfamilies (Dayhoff et al., 1983) is conservative in terms of sequence identity, while structure-based classifications unify remote homologs whose structural and functional features suggest a common evolutionary origin despite very low sequence identities (Holm and Sander, 1998b; Andreeva et al., 2008; Cuff et al., 2011).

Historically, domain families have been identified one by one and based on similarities to individual proteins under study by individual scientists. The process starts from the compilation of a multiple alignment of similar sequences. Methods for finding similar sequences and the thresholds deemed safe to infer homology from similarity differ between different sequence classifications. In order to deal with the rapid growth of sequence databases, semiautomated approaches extrapolate manually created descriptions of families to all sequences. Libraries of profile models have been generated around sets of particular interest, such as all known structures (Dodge et al., 1998; Schäffer et al., 1999; Teichmann et al., 2000; Gough et al., 2001; Yeats et al., 2010; Marchler-Bauer et al., 2011) and large families (Letunic et al., 2009; Finn et al., 2010). The coverage of these databases has increased rapidly. For example, Pfam 25.0 (Finn et al., 2010) contains 12,273, HMMs that cover about 77% of all sequences (54% of all amino acid residues) in Uniprot release 2010_05 (The Uniprot Consortium, 2011). Semiautomated approaches currently provide the most useful tools for biologists interested in the domain composition of protein sequences.

Fully automated approaches to define protein sequence families have attracted considerable attention. Fully automated methods have the benefit of achieving full coverage and internal consistency. Furthermore, a global clustering can yield novel discoveries and scientifically provide new insights into the evolution of the protein universe.

Current methods cluster a sequence similarity graph based on the all-against-all comparison of protein sequences. Graph properties are used to infer the boundaries of clusters of homologous proteins or domains. In the next section, we describe the sequence similarity graph and several clustering methods. We then describe one method, ADDA, in more detail.

1.3 CLUSTERING THE SEQUENCE SPACE GRAPH

All-against-all comparison of protein sequences, using traditional database search tool such as BlastP (Altschul et al., 1997) or Fasta (Pearson and Lipman, 1988), yields a view of the geometry of protein space. Neighbor lists of each sequence induce a representation of protein space as a graph whose vertices (nodes) are the sequences. If there was a perfect correspondence between sequence similarity and homology, then groups of homologous sequences would be easily identifiable as maximal cliques in the sequence space graph. In reality, the situation is less fortunate, or more complicated, in three ways.

Firstly, only parts of two similar sequences may be related by homology. This leads to the phenomenon known as domain chaining. For example, a sequence with two domains A and B will share similarity with any protein that contains domain A and any protein that contains domain B. It is not necessary that domains A and B co-occur in the neighbors. Thus, the sequence space graph is nontransitive: a sequence that is related to sequence X and Y does not imply that sequences X and Y are related. This holds true even in the case of perfect homology detection as multidomain proteins are members of multiple, overlapping maximal cliques representing the domain families (Fig. 1.1a).

Secondly, there are spurious similarities between nonhomologous sequences. Composition bias is a major, but not exhaustive, source of spurious similarities. Spurious similarities may have quite good *e*-values (Fig. 1.1b).

Thirdly, not all homologous relationships are detected as statistically significant. Models of sequence evolution are based on comparing position-specific target distributions of amino acid frequencies to a background distribution—the sharper the target distribution, the higher the information content. It is important to understand that the *p*-values or *e*-values (scaled for database size, Chapter 4) returned by profile models indicate the risk of false positives and are quiet on false negatives. In other words, sequence similarity is not a condition for homology. Mutations leave a continuous trace in sequence space, but mutational paths can be long and divergent. Structure comparisons back up the notion of domain families forming elongated clusters in sequence space (Fig. 1.1c). While two sequences at opposite ends in the elongated cluster might not share enough sequence similarity to infer homology, homology might be established by following the trace through intermediate sequences in sequence space (Park et al., 1997).

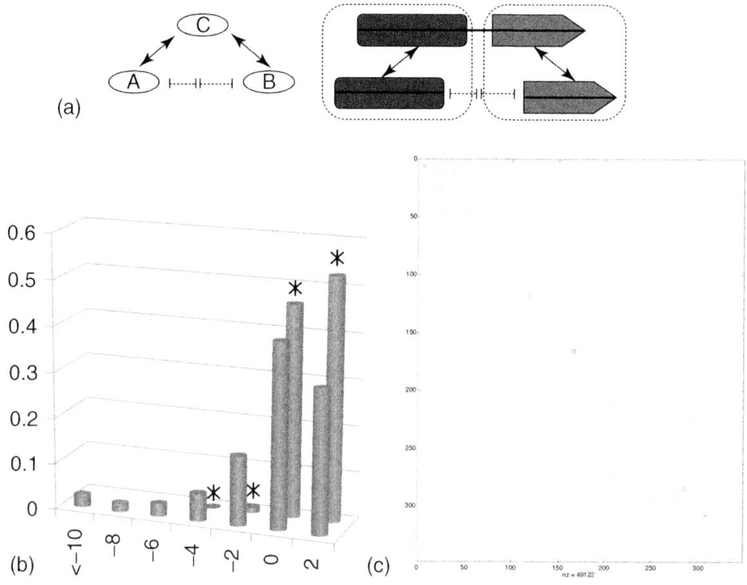

Figure 1.1 (a) The sequence space graph is not transitive due to domain chaining. Two sequences A and B need not be homologous (broken link) even though they share homology (arrows) with a third sequence C. C is a multidomain protein (right) with membership in two domain families (dashed boxes). (b) Overlap between the Blast *e*-value distributions of homologous and nonhomologous sequence pairs. The *x*-axis is the log10 of the *e*-value, and the *y*-axis is the frequency of pairs. All domain sequences from Astral40 were compared against each other. For each query, the *e*-value to the nearest neighbor from the same SCOP superfamily (homologous) and to the nearest neighbor from a different SCOP class (unrelated, marked with asterisk) was recorded. 2097 query domains had a match in both categories. (c) PSI-Blast adjacency matrix for a set of amidohydrolases (PFAM clan CL0034). Dots indicate that sequences are detected by iterative profile searching starting from one query protein. Note the asymmetry and incompleteness of remote homolog detection. Mid-gray squares on the diagonal denote known structures, which confirm the superfamily.

Owing to spurious similarities and domain chaining, the majority of sequences belong to one huge connected component at biologically interesting levels of similarity. Graph clustering leads to the identification of domain families but has to account for noise (missing and false edges).

Over the years, a multitude of fully automated procedures for protein sequence clustering have been derived and are described below. Some have been derived to make sense of BLAST results leaving the structure of the sequence graph intact and allowing to browse the graph of sequence similarities at different levels of granularity (Krause and Vingron, 1998; Yona et al., 1999). Others, partially motivated by structural genomics initiatives, segment sequences in domains and attempt remote homolog identification (Gouzy et al., 1999;

Heger and Holm, 2003). A third set of methods aim to group orthologous and in-paralogous proteins for functional inference by taking into account sequence space topology (Tatusov et al., 1997) and/or reweighting the graph (Enright et al., 2002; Joseph and Durand, 2009).

Objectives between different methods vary. In our opinion, a meaningful evolutionary classification must be based on domains to account for not only speciation and duplication events but also recombination and genomic rearrangements. These domains can exist in different protein contexts. For example, Pawson proposed a model for the functional divergence within domain families based on different protein contexts (Jin et al., 2009). Another school is concerned with comparative proteomics (Li et al., 2003). Here the goal is to map functionally equivalent gene products between species. These studies are usually restricted to the proteomes of a restricted set of species.

Domain family classifications must solve two problems: sequences have to be cut into domains (domain cutting/splitting) and these domains have to be classified into families (clustering/unification). These two problems are intimately linked. On the one hand, delineation of domain boundaries is relatively straightforward, if all members of a domain family have been identified. On the other hand, domain boundaries are needed to assign class memberships correctly.

Methods for domain classification differ in how they try to separate these two problems. In the sequence clustering field, domain cutting has been performed either before or after unification. Cutting before unification has the advantage that subsequent clustering is straightforward, as sequence segments will belong only to a single cluster. However, the signal on which to base cutting is weak as sequence alignments are only an unreliable guide toward domain boundaries (see Section 4.2). Cutting after unification is popular, because the availability of family context permits splitting based on recombination events (mobile modules). However, the data structures in this approach are complex as homologous segments are combined without knowledge of domain boundaries.

Unification before cutting is popular as generic graph clustering algorithms can be employed. However, the clustering is complicated because sequence similarities are not metric distances in a well-behaved space. In these algorithms, the relationship between two protein sequences is encapsulated in a single value, which confounds the degree of sequence similarity with the effects of domain chaining. Approaches differ in the choice of algorithm and the way to avoid the effects of domain chaining.

The simplest clustering approach is hierarchical clustering, where edge weights are the *e*-values for sequence similarity. Single linkage is popular because it is easy to compute and parallelizable (Olson, 1995). However, single-linkage clustering is highly sensitive to domain chaining and is misguided by false positives, which occur even at stringent *e*-values. Average linkage is computationally more expensive (Loewenstein et al., 2008) but yields better separation of protein sequences with different domain combinations (e.g., A vs AB vs B).

Another type of approach modifies the sequence space graph before clustering. For example, putative instances of domain chaining can be removed (Enright and

Ouzounis, 2000) to decompose the graph before clustering. Rescoring similarities based on various types of neighborhood correlation (Song et al., 2008; Jin et al., 2009; Joseph and Durand, 2009) can strengthen edges between homologous sequences and down-weight spurious edges, thus enhancing the clique of the graph.

Graph clustering approaches can be rule-based (cluster of orthologous groups (COGs) or flow-based (minimum cut of a graph) or simulate dynamic processes on a graph (Markov cluster algorithm (MCL), super paramagnetic clustering (SPC), special clustering of protein sequence (SCPS)). These and other applications are described in the next section.

1.4 HISTORICAL OVERVIEW OF SEQUENCE CLUSTERING ALGORITHMS

The field of automated domain family prediction has a history of 20 years dating back to the development of fast sequence database searches (Pearson and Lipman, 1988; Altschul et al., 1990). In the following, we give a historical overview of ingenious ideas to illustrate the breadth of the field.

SYSTERS (Krause and Vingron, 1998) avoids the problem of domain chaining by using a very high threshold. Systers finds connected components by single-linkage clustering. In a perfect cluster, every member is a neighbor of every other member (the cluster is fully connected, i.e., a clique). A nested cluster is a proper subset of another set. Maximal clusters are not contained in any other set. A pair of overlapping maximal clusters has common members and unique members. In Systers, casualties of domain chaining are found in the overlapping clusters. SYSTERS was later extended to use the idea of minimum cut to identify subfamilies (Krause et al., 2005).

ProtoMap (Yona et al., 1999) performs a hierarchical clustering, varying the threshold of statistical significance, stepwise, from very high (10–100) to quite permissive. At each step, the algorithm is applied on the classes of the previous classification, to obtain the next one, at the more permissive threshold. Connections between clusters that are not strongly connected are rejected while clusters that are strongly connected get merged. The criteria for merging were optimised empirically. Rejected connections may be genuine although distant homologies.

PRODOM (Gouzy et al., 1999; Bru et al., 2005) sorts a list of (nonfragmentary) protein sequences by decreasing size. The shortest sequence is taken as a complete, single domain protein and all instances of it are removed by database searching in the list of larger protein sequences. Left-over fragments are entered into the database and the process is continued until the list is empty.

COGs (Tatusov et al., 1997) analyze complete genomes to construct a directed graph of nearest-neighbor relationships between species. The graph is first scanned for cliques of at least three bidirectional nearest-neighbor sequences, leading to dynamic thresholding of the sequence similarity graph. Cliques that share an edge are further merged to form a COG.

GENERAGE (Enright and Ouzounis, 2000) checks in the adjacency matrix of the sequence space graph if transitivity holds for each triplet of connected sequences. If it does not, the linking protein is flagged as a potential multidomain protein and excluded from single-linkage clustering. It is added to two or more clusters at a later stage.

TribeMCL (Enright et al., 2002) is based on the MCL algorithm (Van Dongen, 2000), a generic graph clustering algorithm. The MCL algorithm is based on the insight that there will be many possible paths between vertices within the same cluster, while there will be only few between vertices in different clusters. The MCL algorithm exploits this insight by simulating stochastic flow in networks. Edges between tightly linked clusters are upweighted while spurious edges between clusters are downweighted until the graph falls into distinct clusters. A scaling parameter governs the granularity of the resulting clusters.

CluSTr (Kriventseva et al., 2001; Petryszak et al., 2005) performs a Monte-Carlo simulation in order to replace similarity scores in the similarity matrix with a statistical measure of significance of each pair-wise comparison. Hierarchical clusters are then created using single linkage.

ProClust (Bolten et al., 2001; Pipenbacher et al., 2002) avoids domain chaining by an asymmetric distance measure. Clusters are formed by strongly connected components and further unified using family hidden Markov Models.

ProtoNet (Sasson et al., 2003; Loewenstein et al., 2008) implements a memory-constrained version of average linkage clustering that permits its application to large-scale data sets. The resultant tree is not cut, although nodes in the tree are annotated with respect to their "purity" compared to the keywords and annotations of the sequences grouped.

CHOP (Liu and Rost, 2004) cuts proteins from entirely sequenced organisms beginning from very reliable experimental information (protein data bank (PDB)), proceeding to expert annotations of domain-like regions (Pfam-A), and completing through cuts based on termini of native protein ends. It was estimated that about 20–40% of the fragments that CHOP generates are likely to contain more than one domain.

SPC (Tetko et al., 2005) applies the SPC (Getz et al., 2000) algorithm to the sequence space graph. Each node is assigned a spin vector that encodes the cluster label. A short-range coupling function propagates spin gates to other nodes nearby. Spin–spin correlations, as a function of a temperature parameter, indicate the stability of the clustering. The advantages of SPC include that the number of clusters is determined by the algorithm itself, it is stable against noise, it generates a hierarchy, and it is able to identify nonspherical clusters.

SCPS (Paccanaro et al., 2006; Nepusz et al., 2010) applies spectral clustering to the sequence space graph. The algorithm partitions the graph into clusters by analyzing the eigenvectors and eigenvalues of a matrix, which is derived from the similarity matrix. Similar to MCL, it studies the random walk of a particle on the graph, but the focus is particularly where the particle spends most of its time before reaching the stationary distribution.

EVEREST (Portugaly et al., 2006) breaks up sequences into segments containing putative domains based on pair-wise sequence alignments. Segments are clustered, multiply aligned and summarized as HMMs. A set of known protein domain families are used to train a classifier that separates domain HMMs from spurious ones. The HMMs are then used to rebuild the collection of sequence segments and the process is iterated. A final step selects the highest quality HMMs amongst overlapping and competing HMMs.

CLUSS (Kelil et al., 2007) is an alignment-free method. Instead of using all-versus-all sequence alignment to create the sequence space graph, the graph is derived from an all-versus-all sequence comparison using the collection of shared identical subsequences between a sequence pair as similarity measure. Sequences are then grouped by single linkage and the resultant tree is cut to yield tight clusters.

FORCE (Wittkop et al., 2007) applies ideas from force-based graph layout algorithms: tightly interconnected vertices should be grouped in a 2-D representation. After transformation, sequences are clustered using single linkage based on their location on the 2-D plane. See also (Rahmann et al., 2007).

MACHOS (Wong and Ragan, 2008) analyses blocks of common neighbors sets in a multiple sequence alignment. Segmentation is done at the boundaries between blocks. The resultant segments are then clustered using the MCL algorithm.

Joseph and Durand (2009) rescore edges in a similarity graph based on local graph structure. Cliques are recovered using neighborhood correlation.

Yang et al. (2010) apply affinity propagation (Frey and Dueck, 2007) to the sequence space graph.

Most of the older approaches above are not actively maintained as the annotations provided by semiautomated methods have proved popular with biologists. Incrementally updating the sequence space graph requires considerable commitment and investment, although precomputed data sources exist (Rattei et al., 2006; Heger et al., 2008). Additionally, the rapid growth of sequence databases requires methods to scale well. Also, as more and more sequences are added through automated gene prediction pipelines from whole genome or meta-genome sequencing projects, the likelihood of fragments and gene prediction artifacts has increased. To our knowledge, only ADDA (Section 4) remains in production use and is routinely applied to the set of all known protein sequences. Full-length clustering methods are applied to smaller datasets, for example, in the context of defining groups of orthologous sequences for a limited set of completely sequenced genomes, which is an active area of research (Li et al., 2003; Fulton et al., 2006; Kristensen et al., 2010; Flicek et al., 2011).

1.5 RELATED METHODS

Alternatives to the sequence space graph have been developed. Some tools arrange the results of homology searches to facilitate the visual identification of domains (Guan and Du, 1998). Repeated domains in a sequence can be found by alignment of a sequence to itself (Heringa and Argos, 1993; Pellegrini et al., 1999; Heger and Holm, 2000). Methods to define a set of representative sequences (Holm and Sander, 1998a; Park et al., 2000; Li et al., 2001, 2002) provide a coarse clustering and are often used as a preprocessing step to reduce the size of the sequence set to be clustered. Very ambitious approaches attempted a global classification on residue level (Heger and Holm, 2001; Heger et al., 2007).

Numerous approaches have attempted to predict domain boundaries from a protein sequence alone. The base line is given by Wheelan et al. (2000), who showed that the distribution of observed domain lengths and segment numbers per sequence is able to predict with surprising success, the most likely domain decomposition for a single sequence based entirely on its length. A multitude of machine learning methods have been used to identify sequence features that are associated with domain boundaries, including neural networks (Nagarajan and Yona, 2004; Sim et al., 2005; Cheng et al., 2006; Ye et al., 2008), support vector methods (Sikder and Zomaya, 2006; Chen et al., 2010) and general regression (Yoo et al., 2008). Features used in these methods are predicted domain linker regions based on their amino acid composition (Galzitskaya and Melnik, 2003; Suyama and Ohara, 2003; Dumontier et al., 2005), predicted secondary structure elements (Marsden et al., 2002), and predicted relative solvent accessibilities (Cheng et al., 2006; Sikder and Zomaya, 2006). Another set of methods applies structural domain assignments on predicted 3D structures or contact maps (George and Heringa, 2002; Rigden, 2002; Kim et al., 2005). Most of these methods have been trained and evaluated on protein sequences with known structure and a limited number of domains. They are expected to struggle with long sequences and complex domain architectures.

1.6 QUALITY ASSESSMENT

There is a need for a systematic evolutionary classification of all protein sequences, and several systematic, global clusterings have been proposed. Quality control is a key issue. Can carefully designed automatic, algorithmic approaches match the quality or improve the consistency of manually curated collections?

Comparison between family classifications is not straightforward because of their different definitions, scopes, and purposes. Databases often chosen as reference are PFAM and structural classification of proteins (SCOP), although they might not always be appropriate. Measures of cluster correspondence and their particularities of computation differ between every study.

To our knowledge, no large-scale independent evaluation has been performed. The task is formidable as published results are derived from different input data. After mapping to a common sequence set the question remains if observed differences are due to data or method. Implementations, if obtainable, might prove to be not portable. The situation is better for generic graph clustering methods as they use the same data structure (Yang and Zhang, 2008). Recently, Chen et al. (2007) applied latent class analysis (LCA) to compare three methods to group orthologs.

To conclude, in sequence analysis, there is the fundamental problem that statistical significance does not guarantee a biologically significant relationship. If the problem is too complex to formalize, manual curation by experts is the only solution.

1.7 ADDA—THE AUTOMATIC DOMAIN DELINEATION ALGORITHM

In this section, we provide an overview of ADDA (Heger and Holm, 2003), the algorithm behind the current definition of Pfam-B families (Finn et al., 2010).

1.7.1 High Level Overview

ADDA is a method to define protein sequence domain families based on pairwise sequence alignment information alone. Its objective is maximal unification: each domain family should contain all homologous domain sequences and no analogous domain sequences, that is, domain sequences that are not related by evolutionary descent.

ADDA explicitly models the noise in the sequence databases using a "block model" of multiple alignments. The block model incorporates noise due to sequence fragments and either truncated or spurious alignments.

ADDA separates the confounding problems of domain delineation and family unification by approaching each one in turn. Firstly, multidomain proteins are split into separate domains. A global optimization involving all sequences ensures that domain boundaries are placed consistently. Secondly, after domain decomposition, domains are clustered into families based on sequence similarity.

1.7.2 Domain Decomposition

ADDA's model is conceptually straightforward. In an ideal world, alignments would begin and stop exactly at domain boundaries, if no two proteins shared the same domain combination in the same order. In this ideal world, a multiple alignment built from a sequence database search with a multidomain protein exhibited a block structure (Fig. 1.2a and c) as a result of its domain composition.

In the real world, the block structure is confused by various types of noise (Fig. 1.2b and d):

Multidomain Proteins. Aligning adjacent domains in two protein sequences results in a single alignment. In this case, one alignment represents the recurrence of more than one domain and thus is longer than a single domain and the aligned segment has to be split.

Motifs and Fragments. Local alignments tend to be truncated if the sequences are distant homologs. Here, one alignment represents the recurrence of a partial domain resulting in residues not covered by the alignment. Similarly, fragments cause alignments to end before domain boundaries.

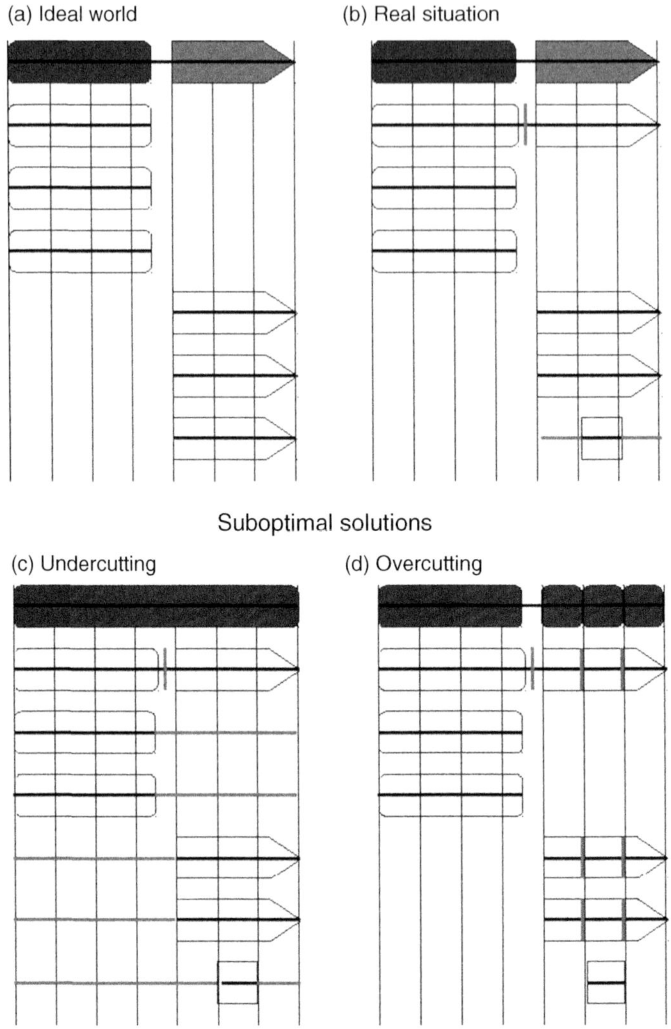

Figure 1.2 Block structure of multiple alignments, in an ideal case where alignments cover full domains and multidomain proteins, have no two domains in the same order. There are seven sequences in this universe. The multiple alignment of the multidomain protein is produced by piling up pair-wise alignments and shows a clear block structure (a) where the domain structure of the query is immediately obvious. In the real situation, multidomain proteins and alignment fragments cause deviations in the block structure (b). Alignments between multidomain proteins have to be split. At the same time, alignments to a motif or fragment do not cover all residues in a domain. The thick gray vertical bars indicate penalties in the objective function for alignments spanning multiple domains or not covering domains. Bottom: Suboptimal domain assignments increase penalties in the objective function. Not splitting the multidomain protein incurs extra penalties through alignments not covering complete domains (c). Oversplitting adds penalties for alignments extending beyond domains (d).

Homologous Overextension. Local alignments extending a few residues beyond domain boundaries if domains are flanked by regions of sufficient similarity.

Spurious Alignments. Nonhomologous regions can be aligned, sometimes giving significant scores. The alignments might match anywhere on the sequence and thus give misleading information about domain length or location.

ADDA models noise due to multidomain proteins, motif alignments, fragments, and spurious links. It defines an objective function that quantifies the deviation from the ideal block structure for a given partition of sequences into domains. The objective function includes probabilistically defined and empirically derived penalties for alignments that extend over a complete domain and alignments that span multiple domains. Conceptually, this approach is related to a minimum message/description length (Wallace and Boulton, 1968; Rissanen, 1978) formulation of the problem finding the best partition of protein sequences into domains that best encodes the observed pair-wise alignment information.

The objective function is optimized globally, that is, simultaneously for all proteins in the sequence set. The global view allows identification of joined alignments due to multidomain proteins and truncated alignments due to motifs and fragments (Fig. 1.3). The optimization step includes evidence from all sequences and can thus balance between cutting too little (based on unresolved multidomain proteins) and cutting too much (leading to fragmented sequences due to cutting at every alignment end) (Fig. 1.4).

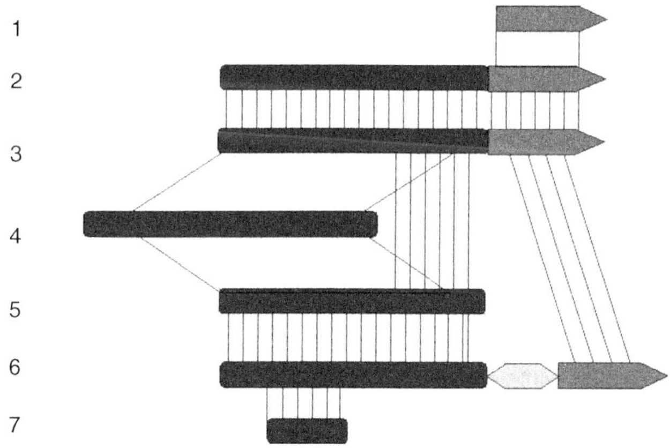

Figure 1.3 A global view corrects for motifs, fragments, and domain chaining. Seven sequences (horizontal bars) are shown with alignments between them (thin lines). Sequence pair 3,5 only aligns in a short conserved motif. Linking sequence 4 and sequences 2 and 6 from subfamilies indicate that the domain is larger than the motif. Sequence 7 is a fragment, but the truncated alignment is compensated for by the alignment between sequences 5 and 6. Sequence domains in different contexts resolve multidomain protein sequences 2, 3, and 6.

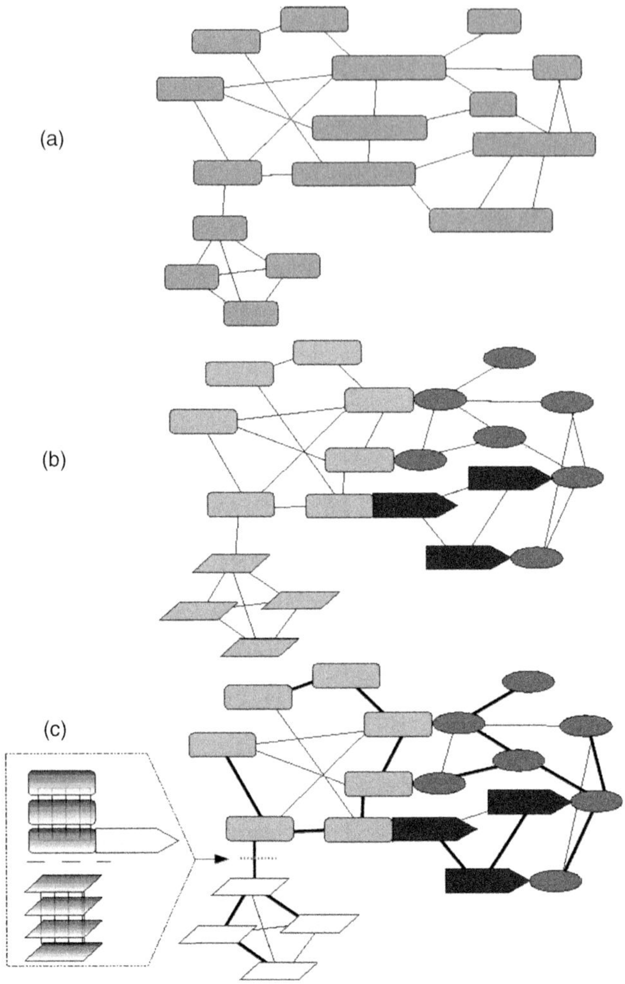

Figure 1.4 Family unification is simplified by the knowledge of domain boundaries. (a) In this toy universe of four domain families and 16 sequences the unstructured sequence alignment graph suggests a single cluster due to domain chaining and spurious alignments. (b) Domain boundaries decompose the sequence alignment graph into a domain alignment graph. Individual components contain domain families as family unification due to domain chaining is resolved. However, spurious links remain and might link unrelated domain families (see bottom left). (c) Spurious links are removed by profile–profile alignment using the immediate neighborhood of the sequences compared. The final clustering yields four clusters (shown as various shades of gray) containing individual domain families.

1.7.3 Family Unification

Once sequences are correctly split into domains, problems posed by domain chaining and sequence fragments disappear and sequences can be simply grouped by sequence similarity.

ADDA assumes that protein sequences of a given family fluctuate around a stable point in sequence space given constant evolutionary constraints ("punctuated equilibria" (Eldredge and Gould, 1997)). If the latter change, for example, if an enzyme starts working on a new substrate, new variants derived from the family will move to a new location in sequence space: a new subfamily has been created. Consecutive changes leave a footprint in sequence space that allows walking from any subfamily to any other either directly, if similarity is within the detection range of sequence profile models, or via a sequence of intermediate steps.

With ADDA, we follow this footprint of a protein domain family in sequence space. Evolutionarily related domains are assumed to occupy continuous neighborhoods. Unrelated domain families should be demarcated by a sharp boundary with dissimilar sequence patterns on either side. Unification proceeds by domain walking between closest neighbors, where each step is checked by pair-wise profile–profile comparison between the adjacent domains. Rejected steps (edges of the sequence space graph) result in domain family boundaries.

1.7.4 Parameterizing

ADDA requires few parameters and these can be learned from data. Parameters of the objective function are estimated from an existing benchmark domain decomposition (SCOP (Murzin et al., 1995)) superimposed on the current alignment graph. Similarly, the alignment score threshold separating homologous from nonhomologous alignments is determined using SCOP.

1.7.5 ADDA Implementation

While ADDA attempts to be rigorous in its approach, the actual implementation requires some trade-offs. For example, the space of all possible domain partitions is too large to enumerate exhaustively. Hence, the objective function is optimized partially, hierarchically, locally, and iteratively.

Partially. Not all possible domain decomposition of a protein sequence are examined, but only those that are suggested by alignment ends.

Hierarchically. Protein sequences are split recursively into the two parts that provide the largest number of nonoverlapping alignments. Splitting stops once the objective function does not increase.

Locally. The objective function is evaluated for each protein sequence and its local neighborhood separately.

Iteratively. As domain boundaries in one protein sequence inform on protein boundaries on another protein sequence, the optimization is run until there is no improvement of objective function summed over all neighborhoods.

This optimization strategy will not guarantee that the final result is a global optimum.

Furthermore, not all pair-wise alignments between domains are tested, but a single-linkage clustering is employed. The clustering is performed using a metric that is an empirical combination of the *e*-value of an alignment and how well it corresponds to the domain boundaries.

Fig. 1.5 summarizes the individual steps in ADDA.

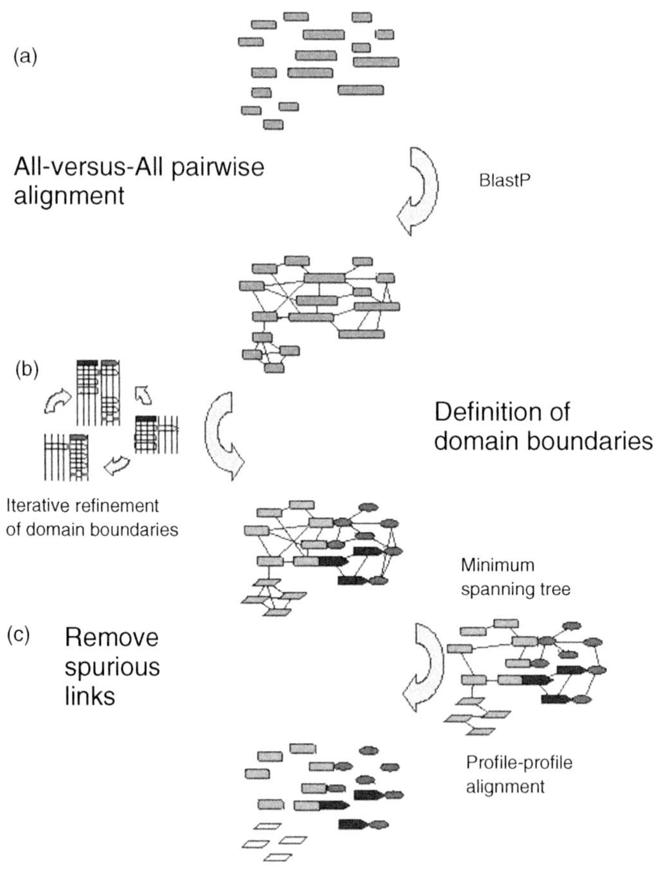

Figure 1.5 Overview over the steps in the ADDA algorithm. (a) Compute pair-wise alignments with BlastP. (b) Refine domain boundaries via an iterative process optimizing the objective function. (c) Arrange domains in a minimum spanning tree and remove putative spurious links using profile–profile alignment.

1.8 RESULTS

ADDA's objective is to achieve a meaningful decomposition of protein domain families. Globally, tests have shown that the decomposition is largely successful (Heger and Holm, 2003; Wong and Ragan, 2008). Pfam-B is based on ADDA since release 23.0 of August 2008 (domain families that overlap with Pfam-A are removed from Pfam-B). Since then, 593 new ADDA families have been promoted to Pfam-A (releases 24.0 and 25.0), amounting to 27% of the recent growth of Pfam-A. The new families contributed by ADDA are heavily enriched in domains of unknown function (68%). This shows the utility of automatic domain family classification in charting the still unknown regions of protein space.

ADDA is efficient enough to be applied on the full set of known protein sequences and is sufficiently robust to not require the removal of fragments or mispredictions. Nevertheless, certain types of domain families, such as cysteine-rich domains, present a challenge. Cysteines are relatively rare amino acids and their presence, conservation, and location in a sequence is highly informative in distinguishing family members from nonfamily members lacking cysteines. However, because of their importance, cysteines mask other features that could be used to discriminate between cysteine-rich families. As a result, cysteine-rich families are poorly resolved.

While ADDA achieves a good global decomposition, there is low level contamination of protein domain families with members of other protein domain families. These are often a consequence of incomplete splits where ADDA failed to separate two adjacent domains. However, their effect through domain chaining is limited. The current implementation of ADDA leaves room for improvement. Several of the heuristic shortcuts could be replaced by more rigorous evaluation of the objective function. Domain boundaries might be improved by including sequence property information, for example, to identify domain linkers. Finally, a full minimum message/description length formulation that includes family unification would provide more rigor to ADDA's model.

1.9 CONCLUSIONS

Global organization of protein sequences into domain families is needed to direct functional and structural genomics and to reap the harvest of these initiatives. The benefits from a description of all protein domain families are more sensitive detection by profile searches, faster search times against a smaller database (profile library), and improved consistency in function and structure assignment.

The field offers a number of challenging computational problems. Sequence search methods fail to detect remote homologs consistently and complex domain architectures complicate the application of generic clustering algorithms. The sheer size of the sequence space graph, approximately one Terabyte, stretches the capacity of common hardware configurations offered by supercomputer centers.

With semiautomated approaches increasing their coverage of abundant domains, the need for fully automated domain family detection methods has somewhat diminished. Current efforts are now concentrating on grouping orthologous full-length protein chains for functional inference.

REFERENCES

Altschul, S.F., Gish, W., Miller, W., Myers, E.W., and Lipman, D.J. (1990) Basic local alignment search tool. J Mol Biol, 215, 403–410.

Altschul, S.F., Madden, T.L., Schäffer, A.A., Zhang, J., Zhang, Z., Miller, W., and Lipman, D.J. (1997) Gapped BLAST and PSI-BLAST: a new generation of protein database search programs. Nucleic Acids Res, 25, 3389–3402.

Andreeva, A., Howorth, D., Chandonia, J.-M., Brenner, S.E., Hubbard, T.J.P., Chothia, C., and Murzin, A.G. (2008) Data growth and its impact on the SCOP database: new developments. Nucleic Acids Res, 36, D419–D425.

Bolten, E., Schliep, A., Schneckener, S., Schomburg, D., and Schrader, R. (2001) Clustering protein sequences—structure prediction by transitive homology. Bioinformatics, 17, 935–941.

Bru, C., Courcelle, E., Carrère, S., Beausse, Y., Dalmar, S., and Kahn, D. (2005) The ProDom database of protein domain families: more emphasis on 3D. Nucleic Acids Res, 33, D212–D215.

Chen, F., Mackey, A.J., Vermunt, J.K., and Roos, D.S. (2007) Assessing performance of orthology detection strategies applied to eukaryotic genomes. PLoS One, 2, e383.

Chen, P., Liu, C., Burge, L., Li, J., Mohammad, M., Southerland, W., Gloster, C., and Wang, B. (2010) DomSVR: domain boundary prediction with support vector regression from sequence information alone. Amino Acids, 39, 713–726.

Cheng, J., Sweredoski, M.J., and Baldi, P. (2006) DOMpro: protein domain prediction using profiles, secondary structure, relative solvent accessibility, and recursive neural networks. Data Min Knowl Discov, 13, 1–10.

Cuff, A.L., Sillitoe, I., Lewis, T., Clegg, A.B., Rentzsch, R., Furnham, N., Pellegrini-Calace, M., Jones, D., Thornton, J., and Orengo, C.A. (2011) Extending CATH: increasing coverage of the protein structure universe and linking structure with function. Nucleic Acids Res, 39, D420–D426.

Dayhoff, M.O., Barker, W.C., and Hunt, L.T. (1983) Establishing homologies in protein sequences. Methods Enzymol, 91, 524–545.

Dodge, C., Schneider, R., and Sander, C. (1998) The HSSP database of protein structure-sequence alignments and family profiles. Nucleic Acids Res, 26, 313–315.

Doolittle, R.F. and Bork, P. (1993) Evolutionarily mobile modules in proteins. Sci Am, 269, 50–56.

Dumontier, M., Yao, R., Feldman, H.J., and Hogue, C.W.V. (2005) Armadillo: domain boundary prediction by amino acid composition. J Mol Biol, 350, 1061–1073.

Eddy, S.R. (1998) Profile hidden Markov models. Bioinformatics, 14, 755–763.

Eldredge, N. and Gould, S.J. (1997) On punctuated equilibria. Science, 276, 338–341.

Enright, A.J., Van Dongen, S., and Ouzounis, C.A. (2002) An efficient algorithm for large-scale detection of protein families. Nucleic Acids Res, 30, 1575–1584.

Enright, A.J. and Ouzounis, C.A. (2000) GeneRAGE: a robust algorithm for sequence clustering and domain detection. Bioinformatics, 16, 451–457.

Finn, R.D., Mistry, J., Tate, J., Coggill, P., Heger, A., Pollington, J.E., Gavin, O.L., Gunasekaran, P., Ceric, G., Forslund, K. et al. (2010) The Pfam protein families database. Nucleic Acids Res, 38, D211–D222.

Fitch, W.M. (1970) Distinguishing homologous from analogous proteins. Syst Zool, 19, 99–113.

Flicek, P., Amode, M.R., Barrell, D., Beal, K., Brent, S., Chen, Y., Clapham, P., Coates, G., Fairley, S., Fitzgerald, S. et al. (2011) Ensembl 2011. Nucleic Acids Res, 39, D800–D806.

Frey, B.J. and Dueck, D. (2007) Clustering by passing messages between data points. Science, 315, 972–976.

Fulton, D.L., Li, Y.Y., Laird, M.R., Horsman, B.G.S., Roche, F.M., and Brinkman, F.S.L. (2006) Improving the specificity of high-throughput ortholog prediction. BMC Bioinformatics, 7, 270.

Galzitskaya, O.V. and Melnik, B.S. (2003) Prediction of protein domain boundaries from sequence alone. Protein Sci, 12, 696–701.

George, R.A. and Heringa, J. (2002) SnapDRAGON: a method to delineate protein structural domains from sequence data. J Mol Biol, 316, 839–851.

Getz, G., Levine, E., and Domany, E. (2000) Coupled two-way clustering analysis of gene microarray data. Proc Natl Acad Sci USA, 97, 12079–12084.

Gough, J., Karplus, K., Hughey, R., and Chothia, C. (2001) Assignment of homology to genome sequences using a library of hidden Markov models that represent all proteins of known structure. J Mol Biol, 313, 903–919.

Gouzy, J., Corpet, F., and Kahn, D. (1999) Whole genome protein domain analysis using a new method for domain clustering. Comput Chem, 23, 333–340.

Gribskov, M., McLachlan, A.D., and Eisenberg, D. (1987) Profile analysis: detection of distantly related proteins. Proc Natl Acad Sci USA, 84, 4355–4358.

Guan, X. and Du, L. (1998) Domain identification by clustering sequence alignments. Bioinformatics, 14, 783–788.

Heger, A. and Holm, L. (2000) Rapid automatic detection and alignment of repeats in protein sequences. Proteins Struct Funct Bioinform, 41, 224–237.

Heger, A. and Holm, L. (2001) Picasso: generating a covering set of protein family profiles. Bioinformatics, 17, 272–279.

Heger, A. and Holm, L. (2003) Exhaustive enumeration of protein domain families. J Mol Biol, 328, 749–767.

Heger, A., Korpelainen, E., Hupponen, T., Mattila, K., Ollikainen, V., and Holm, L. (2008) PairsDB atlas of protein sequence space. Nucleic Acids Res, 36, D276–D280.

Heger, A., Mallick, S., Wilton, C., and Holm, L. (2007) The global trace graph, a novel paradigm for searching protein sequence databases. Bioinformatics, 23, 2361–2367.

Heringa, J. and Argos, P. (1993) A method to recognize distant repeats in protein sequences. Proteins, 17, 391–341.

Holm, L. and Sander, C. (1998a) Removing near-neighbour redundancy from large protein sequence collections. Bioinformatics, 14, 423–429.

Holm, L. and Sander, C. (1998b) Touring protein fold space with Dali/FSSP. Nucleic Acids Res, 26, 316–319.

Jin, J., Xie, X., Chen, C., Park, J.G., Stark, C., James, D.A., Olhovsky, M., Linding, R., Mao, Y., and Pawson, T. (2009) Eukaryotic protein domains as functional units of cellular evolution. Sci Signal, 2, ra76.

Joseph, J.M. and Durand, D. (2009) Family classification without domain chaining. Bioinformatics, 25, i45–i53.

Kelil, A., Wang, S., Brzezinski, R., and Fleury, A. (2007) CLUSS: clustering of protein sequences based on a new similarity measure. BMC Bioinformatics, 8, 286.

Kim, D.E., Chivian, D., Malmström, L., and Baker, D. (2005) Automated prediction of domain boundaries in CASP6 targets using Ginzu and RosettaDOM. Proteins, 61(7 Suppl), 193–200.

Krause, A., Stoye, J., and Vingron, M. (2005) Large scale hierarchical clustering of protein sequences. BMC Bioinformatics, 6, 15.

Krause, A. and Vingron, M. (1998) A set-theoretic approach to database searching and clustering. Bioinformatics, 14, 430–438.

Kristensen, D.M., Kannan, L., Coleman, M.K., Wolf, Y.I., Sorokin, A., Koonin, E.V., and Mushegian, A. (2010) A low-polynomial algorithm for assembling clusters of orthologous groups from intergenomic symmetric best matches. Bioinformatics, 26, 1481–1487.

Kriventseva, E.V., Fleischmann, W., Zdobnov, E.M., and Apweiler, R. (2001) CluSTr: a database of clusters of SWISS-PROT+TrEMBL proteins. Nucleic Acids Res, 29, 33–36.

Letunic, I., Doerks, T., and Bork, P. (2009) SMART 6: recent updates and new developments. Nucleic Acids Res, 37, D229–D232.

Li, L., Stoeckert, C.J., and Roos, D.S. (2003) OrthoMCL: identification of ortholog groups for eukaryotic genomes. Genome Res, 13, 2178–2189.

Li, W., Jaroszewski, L., and Godzik, A. (2001) Clustering of highly homologous sequences to reduce the size of large protein databases. Bioinformatics, 17, 282–283.

Li, W., Jaroszewski, L., and Godzik, A. (2002) Tolerating some redundancy significantly speeds up clustering of large protein databases. Bioinformatics, 18, 77–82.

Liu, J. and Rost, B. (2004) CHOP: parsing proteins into structural domains. Nucleic Acids Res, 32, W569–W571.

Loewenstein, Y., Portugaly, E., Fromer, M., and Linial, M. (2008) Efficient algorithms for accurate hierarchical clustering of huge datasets: tackling the entire protein space. Bioinformatics, 24, i41–i49.

Marchler-Bauer, A., Lu, S., Anderson, J.B., Chitsaz, F., Derbyshire, M.K., DeWeese-Scott, C., Fong, J.H., Geer, L.Y., Geer, R.C., Gonzales, N.R. et al. (2011) CDD: a conserved domain database for the functional annotation of proteins. Nucleic Acids Res, 39, D225–D229.

Marsden, R.L., McGuffin, L.J., and Jones, D.T. (2002) Rapid protein domain assignment from amino acid sequence using predicted secondary structure. Protein Sci, 11, 2814–2824.

Murzin, A.G., Brenner, S.E., Hubbard, T., and Chothia, C. (1995) SCOP: a structural classification of proteins database for the investigation of sequences and structures. J Mol Biol, 247, 536–540.

Nagarajan, N. and Yona, G. (2004) Automatic prediction of protein domains from sequence information using a hybrid learning system. Bioinformatics, 20, 1335–1360.

Nepusz, T., Sasidharan, R., and Paccanaro, A. (2010) SCPS: a fast implementation of a spectral method for detecting protein families on a genome-wide scale. BMC Bioinformatics, 11, 120.

Olson, C.F. (1995) Parallel algorithms for hierarchical clustering. Parallel Comput, 21, 1313–1325.

Paccanaro, A., Casbon, J.A., and Saqi, M.A.S. (2006) Spectral clustering of protein sequences. Nucleic Acids Res, 34, 1571–1580.

Park, J., Holm, L., Heger, A., and Chothia, C. (2000) RSDB: representative protein sequence databases have high information content. Bioinformatics, 16, 458–464.

Park, J., Teichmann, S.A., Hubbard, T., and Chothia, C. (1997) Intermediate sequences increase the detection of homology between sequences. J Mol Biol, 273, 349–354.

Pearson, W.R. and Lipman, D.J. (1988) Improved tools for biological sequence comparison. Proc Natl Acad Sci USA, 85, 2444–2448.

Pellegrini, M., Marcotte, E.M., and Yeates, T.O. (1999) A fast algorithm for genome-wide analysis of proteins with repeated sequences. Proteins, 35, 440–446.

Petryszak, R., Kretschmann, E., Wieser, D., and Apweiler, R. (2005) The predictive power of the CluSTr database. Bioinformatics, 21, 3604–3609.

Pipenbacher, P., Schliep, A., Schneckener, S., Schönhuth, A., Schomburg, D., and Schrader, R. (2002) ProClust: improved clustering of protein sequences with an extended graph-based approach. Bioinformatics, 18(2 Suppl), S182–S191.

Portugaly, E., Harel, A., Linial, N., and Linial, M. (2006) EVEREST: automatic identification and classification of protein domains in all protein sequences. BMC Bioinformatics, 7, 277.

Rahmann, S., Wittkop, T., Baumbach, J., Martin, M., Truss, A., and Böcker, S. (2007) Exact and heuristic algorithms for weighted cluster editing. Comput Syst Bioinform Conf, 6, 391–401.

Rattei, T., Arnold, R., Tischler, P., Lindner, D., Stümpflen, V., and Mewes, H.W. (2006) SIMAP: the similarity matrix of proteins. Nucleic Acids Res, 34, D252–D256.

Rigden, D.J. (2002) Use of covariance analysis for the prediction of structural domain boundaries from multiple protein sequence alignments. Protein Eng, 15, 65–77.

Rissanen, J. (1978) Modeling by shortest data description. Automatica, 14, 465–471.

Sali, A. (1999) Functional links between proteins. Nature, 402(23), 25–26.

Sasson, O., Vaaknin, A., Fleischer, H., Portugaly, E., Bilu, Y., Linial, N., and Linial, M. (2003) ProtoNet: hierarchical classification of the protein space. Nucleic Acids Res, 31, 348–352.

Schäffer, A.A., Wolf, Y.I., Ponting, C.P., Koonin, E.V., Aravind, L., and Altschul, S.F. (1999) IMPALA: matching a protein sequence against a collection of PSI-BLAST-constructed position-specific score matrices. Bioinformatics, 15, 1000–1011.

Sigrist, C.J.A., Cerutti, L., Hulo, N., Gattiker, A., Falquet, L., Pagni, M., Bairoch, A., and Bucher, P. (2002) PROSITE: a documented database using patterns and profiles as motif descriptors. Brief Bioinform, 3, 265–274.

Sikder, A.R. and Zomaya, A.Y. (2006) Improving the performance of DomainDiscovery of protein domain boundary assignment using inter-domain linker index. BMC Bioinformatics, 7(Suppl 5), S6.

Sim, J., Kim, S.-Y., and Lee, J. (2005) PPRODO: prediction of protein domain boundaries using neural networks. Proteins, 59, 627–632.

Song, N., Joseph, J.M., Davis, G.B., and Durand, D. (2008) Sequence similarity network reveals common ancestry of multidomain proteins. PLoS Comput Biol, 4, e1000063.

Suyama, M. and Ohara, O. (2003) DomCut: prediction of inter-domain linker regions in amino acid sequences. Bioinformatics, 19, 673–674.

Tatusov, R.L., Koonin, E.V., and Lipman, D.J. (1997) A genomic perspective on protein families. Science, 278, 631–637.

Teichmann, S.A., Chothia, C., Church, G.M., and Park, J. (2000) Fast assignment of protein structures to sequences using the intermediate sequence library PDB-ISL. Bioinformatics, 16, 117–124.

Tetko, I.V., Facius, A., Ruepp, A., and Mewes, H.-W. (2005) Super paramagnetic clustering of protein sequences. BMC Bioinformatics, 6, 82.

The UniProt Consortium (2011) Ongoing and future developments at the Universal Protein Resource. Nucleic Acids Res, 39, D214–D219.

Van Dongen (2000). Graph clustering by flow simulation. PhD Thesis. University of Utrecht, The Netherlands.

Wallace, C.S. and Boulton, D.M. (1968) An Information Measure for Classification. Comput J, 11, 185–194.

Wetlaufer, D.B. (1973) Nucleation, rapid folding, and globular intrachain regions in proteins. Proc Natl Acad Sci USA, 70, 697–701.

Wheelan, S.J., Marchler-Bauer, A., and Bryant, S.H. (2000) Domain size distributions can predict domain boundaries. Bioinformatics, 16, 613–618.

Wittkop, T., Baumbach, J., Lobo, F.P., and Rahmann, S. (2007) Large scale clustering of protein sequences with FORCE -a layout based heuristic for weighted cluster editing. BMC Bioinformatics, 8, 396.

Wong, S. and Ragan, M.A. (2008) MACHOS: Markov clusters of homologous subsequences. Bioinformatics, 24, i77–i85.

Yang, F., Zhu, Q., Tang, D., and Zhao, M. (2010) Using affinity propagation combined post-processing to cluster protein sequences. Protein Pept Lett, 17, 681–689.

Yang, K. and Zhang, L. (2008) Performance comparison of gene family clustering methods with expert curated gene family data set in Arabidopsis thaliana. Planta, 228, 439–447.

Ye, L., Liu, T., Wu, Z., and Zhou, R. (2008) Sequence-based protein domain boundary prediction using BP neural network with various property profiles. Proteins, 71, 300–307.

Yeats, C., Redfern, O.C., and Orengo, C. (2010) A fast and automated solution for accurately resolving protein domain architectures. Bioinformatics, 26, 745–751.

Yona, G., Linial, N., and Linial, M. (1999) ProtoMap: automatic classification of protein sequences, a hierarchy of protein families, and local maps of the protein space. Proteins, 37, 360–378.

Yoo, P.D., Sikder, A.R., Zhou, B.B., and Zomaya, A.Y. (2008) Improved general regression network for protein domain boundary prediction. BMC Bioinformatics, 9(1 Suppl), S12.

2

SEQUENCE CLASSIFICATION OF PROTEIN FAMILIES: Pfam AND OTHER RESOURCES

ALEX BATEMAN

European Molecular Biology Laboratory, European Bioinformatics Institute, Wellcome Trust Genome Campus, Hinxton, UK

CHAPTER SUMMARY

Pfam is a popular database collection of protein families and domains. This chapter briefly outlines how the resource is made, as well as some of its recent developments. In addition, some of the philosophical aspects that makes Pfam different from other similar collections are outlined. This chapter also makes some comparisons to other similar resources such as InterPro, Simple Modular Architecture Research Tool (SMART), PROSITE profiles, TIGRFAMs, and Conserved Domain Database (CDD).

2.1 INTRODUCTION

This chapter describes the classification of protein sequences using sequence-based profile methods such as hidden Markov models (HMMs). In contrast to the previous chapter (see Chapter 1) where automatic methods were used to provide a comprehensive classification, the clusters of proteins described here are created in a more manual manner. This in general leads to higher quality protein family clusters, but means that the classifications are generally incomplete.

Protein Families: Relating Protein Sequence, Structure, and Function, First Edition.
Edited by Christine Orengo and Alex Bateman.

One of the first classifications of protein sequence available was Prosite (Sigrist et al., 2010). This database used short regular expressions or sequence motifs to identify related sets of protein sequences. Other early sequence classifications were the PRINTS database (Attwood et al., 2003) and the BLOCKS database (Henikoff et al., 1999). Although, still used today these resources have been largely superseded owing to their lack of comprehensiveness.

The Pfam database was created 16 years ago (Sonnhammer et al., 1997) and has grown to become one of the standard tools for analysis of genomes. Pfam is primarily a collection of protein domains as well as protein families. The Pfam database and its developments are well documented in the annual database collections published by Nucleic Acids Research (Bateman et al., 2002; Bateman et al., 1999; Bateman et al., 2000; Bateman et al., 2004; Finn et al., 2006; Finn et al., 2010; Finn et al., 2008; Punta et al., 2012; Sonnhammer et al., 1998). In this chapter the underlying philosophy that has led to the success of Pfam and other related resources is discussed. Related resources such as SMART (Letunic et al., 2012), PROSITE profiles (Sigrist et al., 2010), CDD (Marchler-Bauer et al., 2011) and TIGRFAMs (Selengut et al., 2007) share a similar approach that will be described later.

2.2 Pfam

Release 26.0 of Pfam contains 13,672 curated entries. These entries are known as Pfam-A and represent the high quality portion of the database. Pfam-A contains all the well known and published domains such as PH and immunoglobulin domains. To maintain a good coverage of known sequences Pfam-A is supplemented by an automatically generated set called Pfam-B. Pfam-B is derived from the Automatic Domain Decomposition Algorithm (ADDA) database (see Chapter 1) and contains 460,125 protein families of lower quality. These families tend to contain uncharacterized proteins in addition to less well conserved portions of characterized proteins.

For each Pfam-A family there are four key pieces of data stored. These are shown in Fig. 2.1 that illustrates the flow of work in building a Pfam-A family. Firstly, a *SEED* alignment is created. This alignment contains a set of representative members of the family. In general, the SEED alignment is designed to be as small as possible while capturing the diversity in the family. Seventy six percent of families have 50 sequences or less, and 98% have 200 sequences of less in the SEED alignment. The sequences in the SEED alignment are usually aligned using either the MUSCLE (Edgar, 2004) or the MAFFT (Katoh et al., 2002) alignment tools with some minor modifications to ensure that the boundaries of the alignment are flush at the amino and carboxyl termini. In general, the SEED alignments do not change or only change slightly between Pfam releases.

The SEED alignment is used to construct a profile-HMM, which is a probabilistic representation of the SEED alignment, which can be used to find members of the family in new protein sequences. The collection of Pfam-A profile-HMMs is a valuable resource that is widely used to annotate domains and families in newly sequenced genomes.

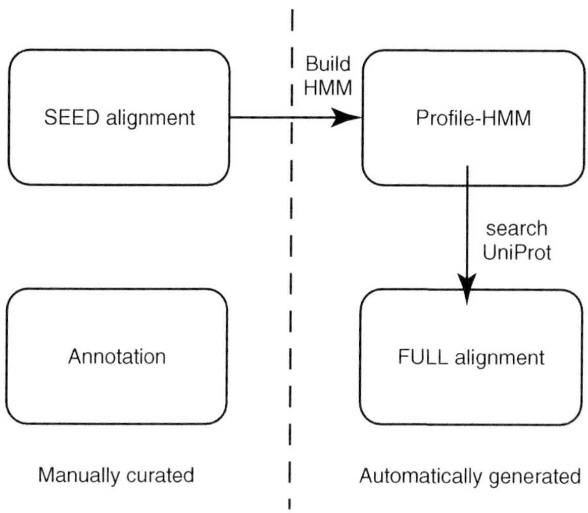

Figure 2.1 The Pfam workflow from SEED alignment to FULL alignment.

The Pfam-A profile-HMMs are used to search the UniProtKB sequence database to identify members of each family. The Pfam biocurators choose a threshold for each profile-HMM such that no known false matches to the model are included. This manual approach allows Pfam to have a very low rate of false matches compared to automated clustering approaches (see Chapter 1). All the matches in UniProtKB that are above the gathering threshold are aligned against the profile-HMM to generate a *FULL* alignment. FULL alignments can be very large, for example the ABC transporter family contains over 163,000 sequences. Unlike the compact SEED alignments, 18% of FULL alignments in Pfam-A contain over 1000 sequences. Given the rapid growth in the protein sequence databases, it seems likely that large families will become the normal situation.

So why have two different types of alignments? The workflow described above allows Pfam to be easily updated and maintained. Storing the SEED alignments means that new sets of profile-HMMs can be quickly generated when new improved versions of the HMMER software becomes available. Over the years, there have been several large improvements in the profile-HMM software. The second aspect of the Pfam workflow that enhances maintainability is that the profile-HMM collection can be searched against new sequence databases as they become available and new homologs incorporated in the FULL alignment.

2.3 SMART, PROSITE PROFILES, CDD AND TIGRFAMs

Pfam was the first database to introduce the concept of a stable SEED alignment for constructing profiles. This approach was also adopted by several other related collections of protein family profiles. Probably the most closely similar resources to Pfam are SMART and TIGRFAMs. Both of these databases also use the same profile construction and search software, but each has a different focus. SMART has a much smaller number (1009) of very high quality families. SMART has mainly focused on extracellular domains, signaling domains, and transcriptional regulatory protein domains. TIGRFAMs that is maintained at the J. Craig Venter Institute is a collection of bacterial families. A particularly noteworthy feature of TIGRFAMs is the careful curation of thresholds aimed not to maximize the number of sequences matched, but to define groups of proteins with the same function. This is because TIGRFAMs is primarily used for automatic function annotation of newly sequenced bacterial genomes. PROSITE profiles such as SMART focuses on delivering a relatively small number (1034) of high quality families using the generalized profile software (Bucher and Bairoch, 1994) to identify homologs. Finally, the CDD is a collection of protein domains maintained at the National Center for Biotechnology Information. CDD carefully crafts SEED alignments based on the superposition of protein structures. Structural features are annotated to these alignments allowing for detailed annotation of individual residues in protein sequences. A summary of family resources is presented in Table 2.1 below.

2.4 PHILOSOPHY OF Pfam

Although the curated SEED alignment concept was a critical advance in creating protein family databases there is still a lot of confusion about what a SEED alignment is. From the perspective of Pfam the SEED alignment is simply a set of representative sequences that guide the construction of the profile-HMM. We are careful to remove sequences that look as though they are mis-predicted fragments of genomes. However, there are currently no efforts made to include sequence from model organisms or those that have structures in preference to

TABLE 2.1 Popular Protein Family Collections

Database Name	Num. Entries (Release)	Access URL
Pfam	13,672 (26.0)	http://pfam.sanger.ac.uk/
PROSITE Profiles	1,034 (20.81)	http://prosite.expasy.org/
PROSITE Patterns	1,308 (20.81)	http://prosite.expasy.org/
PRINTS	2,156 (42.0)	http://www.bioinf.manchester.ac.uk/dbbrowser/PRINTS/index.php
SMART	1,009 (7.0)	http://smart.embl-heidelberg.de/
CDD	42,693 (3.06)	http://www.ncbi.nlm.nih.gov/Structure/cdd/cdd.shtml
TIGRFAMs	4,102 (11.0)	http://www.jcvi.org/cgi-bin/tigrfams/index.cgi
InterPro	23,236 (37.0)	http://www.ebi.ac.uk/interpro/

other sequences. This contrasts to the approach used by SMART and CDD to more carefully select the sequences found in the SEED alignments. To keep the SEED alignment to a small size, if many sequences share more than 80% sequence identity then only one representative sequence is kept in the alignment. For very diverse families we may reduce that identity threshold, for example to 60%, to keep the alignment to a reasonable size. This allows our curators and users to easily view the sequence alignment on screen.

In the early history of Pfam, the tools used to make multiple sequence alignments were somewhat unreliable and an experienced biologist could improve the alignments through manual editing. Some alignments were manually modified in the past but now multiple sequence alignment software is sufficiently accurate and there is little to be gained by manual alignment.

In resources such as SUPERFAMILY and Gene3D described in a subsequent chapter (see Chapter 4) the SEED alignment concept has been replaced with SEED sequence. These are taken from the well-curated structural classifications SCOP and CATH respectively. By using iterative sequence searching, these resources can achieve comparatively high coverage to Pfam and they benefit from new sequences being deposited in the sequence database that may relate otherwise distant proteins together. In Pfam these sequences must be included in the SEED alignment before they contribute to the profile-HMM to find new distant homologs.

A major motivation for the growth and development of Pfam is the concept of increasing the fraction of known sequences that fall into families. We call the fraction of sequences from UniProtKB that fall into Pfam-A families the *sequence coverage* and the fraction of all of the amino acids from UniProtKB that fall into Pfam-A families the *residue coverage*. For Pfam release 26.0 (April, 2010) the sequence coverage was 79.42% and the residue coverage was 57.68%. This means that the majority of proteins fall into a known family. But how long will it take to complete the Pfam classification? The likely answer is that it is not possible ever to attain 100% coverage of sequences or residues. There appear to be many sequences called *ORFans* that appear to show no relationship to any other sequence and therefore do not fall into known families. These account for about 5% of most bacterial genomes.

Ever since the inception of Pfam there has been a philosophical debate about whether to focus on the quality or quantity of protein families. The second line from the first paper describing Pfam puts this tension center stage, "One of the main challenges when constructing such a database is to simultaneously satisfy the conflicting demands of completeness on the one hand and quality of alignment and domain definitions on the other." This is still just as true today as it was then. Is it better to construct 100 families per year that are as near to perfect as possible when 1000 slightly poorer quality families could be created in the same amount of time? Pfam has tended to err on the side of quantity and tried to maintain quality through extensive automatic quality control measures. This perhaps contrasts with SMART and PROSITE profiles that have focussed on quality.

The most important quality control measure in Pfam is the overlap criterion, which states that, "Not even a single residue of a protein may belong to two different families." This is useful because whenever a new family is built one can ask whether it overlaps with any existing family. If it does then there are only a few possibilities:

i. The overlapping sequence is a false positive match in the new family
ii. The overlapping sequence is a false positive match in the pre-existing family
iii. The two families are related to each other

In cases (1) or (2), resolving the overlap improves the quality of the Pfam classification by removing incorrect matches from a family. If the families are found to be related to each other then we can merge the two families to make one greatly improved single family. However, for the most divergent protein families such as the Rossmann domain (see Chapter 8) or the HUP superfamily (see Chapter 7) it is not possible to build a single profile-HMM that is able to capture all members of the family. This has presented a problem historically and so in 2005 the concept of a clan was introduced (Finn et al., 2006) A clan is a collection of Pfam-A families that are known to be evolutionarily related to each other. Within this group of families the overlap criterion is not enforced. So, for example, for the Rossmann domain there are multiple related Pfam families that overlap specificity to some extent. These partially overlapping families provide coverage of different parts of the diverse superfamily, which allows us to maximize the overall coverage of the family. There are currently over 500 clans in Pfam.

2.5 HMMER3 AND Jackhmmer

Pfam was the first protein family database to use profile-HMMs. The profile-HMM formalism for proteins was introduced by Anders Krogh and colleagues in 1994 and this put profiles onto a more rigorous statistical footing. Pfam uses the HMMER package implementation of profile-HMMs written by Sean Eddy. A new version of the software package, HMMER3, was recently released. Firstly, this version has improved sensitivity over previous versions through the use of forward scores rather than viterbi scores. Owing to a range of prefilters, HMMER3 is 100 times faster than the older HMMER2 software and is close to the speed of BLAST for sequence comparison. This has greatly widened the use of the package and allows labs with only modest compute resources to carry out large scale analyses with the Pfam profile-HMM library.

The HMMER3 package includes a new program called *Jackhmmer*. This program works in a similar way to PSI-BLAST (Altschul et al., 1997) taking a single query sequence and then iteratively searching a sequence database, finding progressively more homologs to the query sequence. We have found Jackhmmer to be very useful for rapidly creating new Pfam-A families.

2.6 SOURCES OF NEW FAMILIES

To increase coverage, the Pfam biocurators have continued to build new families. The major sources for new families in past years have been Pfam-B and other related resources. Pfam-B, the automatically generated supplement, has been used to identify large clusters of proteins that are absent from Pfam-A. For example, the Pfam family for the C-terminal domain of the ATP-synthase alpha and beta chains (Pfam accession PF00306) was constructed from Pfam-B family number 15 from Pfam release 1.0. This family now has over 12,000 members. 6022 Pfam-A families or just under half of all Pfam families have been derived from Pfam-B families.

Over time, we have included families on the basis of entries in other family databases to help increase Pfam's completeness while reducing duplication of effort. In the early years, PROSITE was an excellent source of new families with it's high quality annotation. PROSITE patterns were generally quite short, matching an active site or other conserved sequence region, and so more extended sequence alignments were created for Pfam. To date, 978 Pfam families have been derived from PROSITE entries. The COGs (Clusters of Orthologous Groups) resource (Tatusov et al., 2001) was also a valuable source of families with 774 new Pfam entries, and more recently the Eukaryotic databases Eukaryotic Clusters of Orthologous Groups KOGs was used to provide 213 new Pfam families. TIGRFAMs (Haft et al., 2001) is a resource for annotation of microbial genomes using profile-HMMs much like Pfam. 268 Pfam families were derived from this database. However, it is worth noting that while TIGRFAMs aims to create families that can be confidently assigned a specific function, Pfam tries to match the largest number of homologs. So the actual memberships of the Pfam family may be significantly expanded over that found in TIGRFAMs or any of the other databases used as a source of new families. Finally a smaller number of families 91 and 52 were derived from the more specialist databases SMART (Schultz et al., 1998) and PRINTS (Attwood et al., 2003), respectively.

As mentioned previously, databases such as SUPERFAMILY and Gene3D use sequences from protein structures to iteratively create families. Because protein structures are a very valuable resource for understanding biology we have tried to maintain a very high coverage of proteins in the PDB. This coverage is around 95% (Sammut et al., 2008), which is high compared to the coverage of UniProtKB. By taking any sequences from PDB that have no hit to a Pfam family we have created 907 new Pfam families.

Although PDB coverage is very high, we still do not have complete coverage of even bacterial genomes. In Pfam release 24.0 about 80% of the *Clostridium difficile* proteome was matched by Pfam-A. To better understand the nature of the missing 20% of sequences we took each one and carried out an iterative Jackhmmer search against the UniProtKB sequence database. We found that about half of the proteins not covered by Pfam-A built large alignments (>100 matches) with Jackhmmer, with most of these overlapping with existing families. This suggests that there are many false negative hits for Pfam families. To try and incorporate new families into Pfam that match these missing (false negative) sequences we used Jackhmmer searches from unmatched genome proteins. The results of these searches have been used to construct new families to be added into existing Pfam clans. To date 776 new Pfam families have been created. These have increased the sequence coverage of the *C. difficile* proteome to 87%.

2.7 ANNOTATION OF FAMILIES

Many aspects of Pfam-A family construction have been automated to make it as simple as possible to create high quality protein families. A major bottleneck remains in creating a useful description of the family. In general, the aim of the Pfam family annotation has not been to provide a full review of what is known. To fully research each family and the functions of the proteins it contains would take a curator several weeks. Therefore the approach taken is to identify a small number of references and information that can provide the interested researcher with leads to find out more. In cases where a family has already been documented in another database such as PROSITE or InterPro, we simply provide a link to that resource's annotation to save duplication of effort.

Many of the most recent Pfam-A families added to Pfam are Domains of Unknown Function or DUFs as we call them. About 25% of all new families are in the DUF class (Bateman et al., 2010). It is often difficult to find papers that describe proteins belonging to these new families. Sometimes that is because no one has ever carried out any experimental work on the family. Sometimes though, the papers describing the family are not located because the authors have called the protein by a different name from that used in the sequence databases. Once one starts looking for synonyms of proteins in a family one has to be extremely careful not to retrieve papers for unrelated proteins that have the same name. One might expect that literature text mining tools would be of great help for this, but in our experience these tools are good at finding papers for well characterised families, but poor at relating papers to more obscure protein families. Often we rely on a protein sequence or an alignment in the paper to be sure that the manuscript is describing the protein in our new family.

Pfam is part of the InterPro consortium that integrates data from 12 databases of protein families and domains (Hunter et al, 2009). One of the great strengths of this resource is its annotation. For many entries there is a highly detailed description of the family that surpasses what one can find in any of the individual databases. Yet, even the InterPro annotation falls short of what a user might really want, which would be a full review article for the family. The biocurators in each of these databases know about a lot of different families, but cannot be experts for all of them. What we would ideally like is for the annotation of each family to be contributed by molecular biology experts. This process of gathering information from the wider experimental community is known as community annotation. The PROSITE database has for many years had nominated experts for many of its families whom users could contact to get more information. Pfam itself allowed users to submit new annotation and references via a web-based form. However, contributions through this route were rather small in number and often lacking in detail. This had led us, along with others having similar experiences, to write off community annotation.

The possibility of community annotation resurfaced with the growing success of the online encyclopedia Wikipedia. In 2008, we decided to use Wikipedia as the annotation source for the Rfam database of RNA families (Daub et al., 2008). Although there were only a handful of articles about noncoding RNAs in Wikipedia at the time, we created 600 new articles on the basis of the existing Rfam annotation. These Wikipedia articles were then imported back into Rfam on a nightly basis. If noncoding RNA experts wished to update the Rfam annotation then they only needed to edit the corresponding Wikipedia article. Since that time, the articles in Wikipedia that Rfam imports have been edited over 9000 times by more than 1000 unique users (Gardner et al., 2011). This experience gave us confidence that a similar approach could work for Pfam.

In Pfam release 26.0 we added links to approximately 1500 existing Wikipedia articles about protein families in addition to creating a new article titled, "DUF" that we link the 3000 DUF articles to. These articles are displayed on the Pfam-A family page as shown in Fig. 2.2. Since that release we are getting a steady stream of edits to these articles in addition to suggestions for new articles to link each Pfam-A to. Unlike Rfam, Pfam does not yet have an article linked to every family. We are currently in the process of partially automating the process

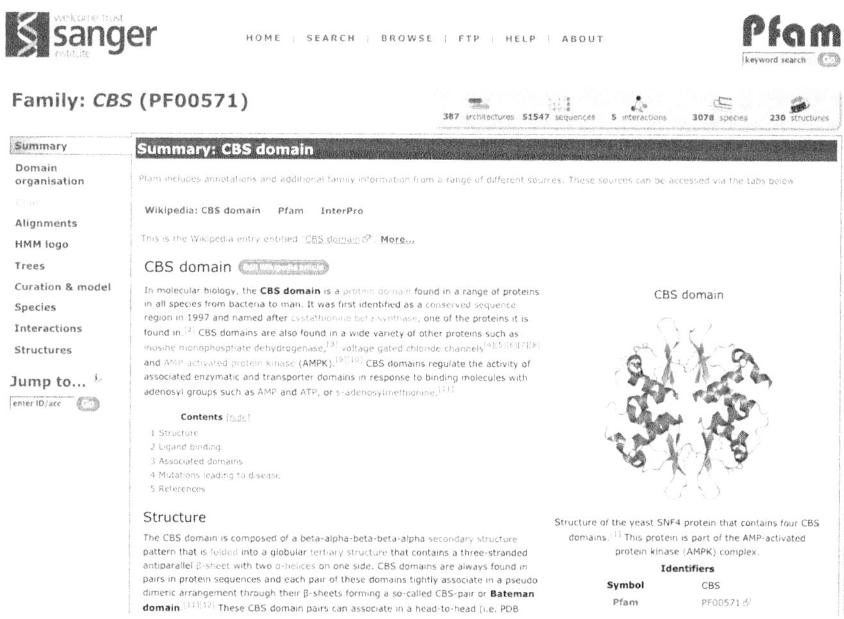

Figure 2.2 A screenshot of a Pfam-A family page illustrating the Wikipedia derived summary of the CBS domain.

to create new Wikipedia articles from Pfam and InterPro content, and hope to rapidly increase the Wikipedia coverage of Pfam families.

2.8 THE InterPro COLLECTION

As mentioned, InterPro provides a single resource that aggregates information from 12 different protein family resources. This can greatly simplify analyses for biologists who may only need to visit InterPro to get an overview of their proteins rather than visiting numerous websites. An example of this integrated view is shown in Fig. 2.3. InterPro also provides the InterProScan software for simultaneously searching against all its member databases.

2.9 THE FUTURE OF SEQUENCE CLASSIFICATION

As a classification of protein domains, Pfam is probably the most comprehensive resource, yet it is still far from complete. Even by aggregating all the major protein family databases our understanding of protein space is still incomplete. There probably still exist many thousands of relatively small families that have restricted phylogenetic spread. However, if we really wish to understand biology in its fullness these families need to be identified and characterized. However, as

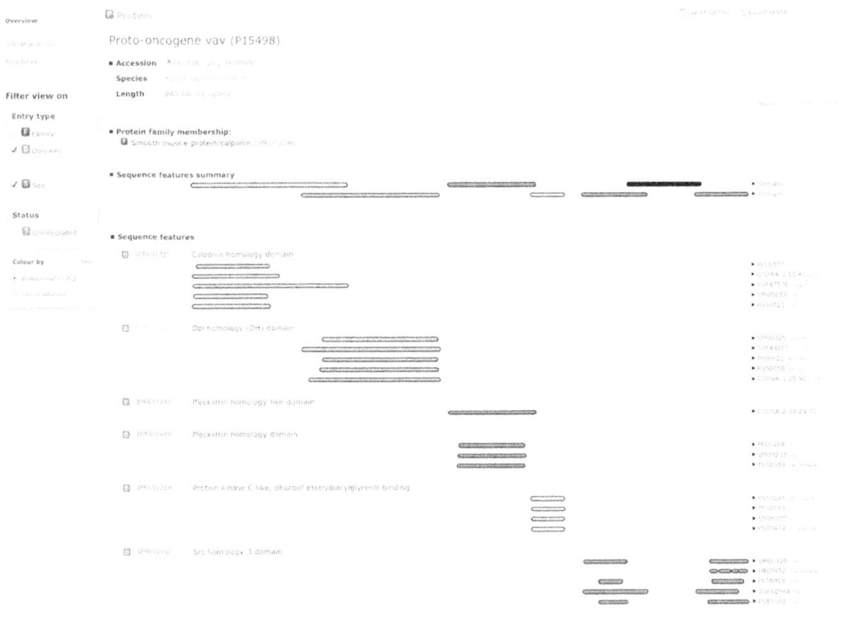

Figure 2.3 A screenshot of the InterPro aggregated view of a protein sequence.

the number of sequences in the databases grows so too does the apparent number of protein families. Kunin et al. (2003) found a linear increase in the number of families over time (Kunin et al., 2003). Therefore, while the community still finds the Pfam and related classifications useful for its research, we must strive toward the goal of a complete and accurate classification of all proteins. Even with improved technologies for finding sequence similarities it could still be many decades before we have a complete classification for protein sequences.

REFERENCES

Altschul, S.F., Madden, T.L., Schaffer, A.A., Zhang, J., Zhang, Z., Miller, W., and Lipman, D.J. (1997) Gapped BLAST and PSI-BLAST: a new generation of protein database search programs. Nucleic Acids Res, 25, 3389–3402.

Attwood, T.K., Bradley, P., Flower, D.R., Gaulton, A., Maudling, N., Mitchell, A.L., Moulton, G., Nordle, A., Paine, K., Taylor, P. et al. (2003) PRINTS and its automatic supplement, prePRINTS. Nucleic Acids Res, 31, 400–402.

Bateman, A., Birney, E., Cerruti, L., Durbin, R., Etwiller, L., Eddy, S.R., Griffiths-Jones, S., Howe, K.L., Marshall, M., and Sonnhammer, E.L. (2002) The Pfam protein families database. Nucleic Acids Res, 30, 276–280.

Bateman, A., Birney, E., Durbin, R., Eddy, S.R., Finn, R.D., and Sonnhammer, E.L. (1999) Pfam 3.1: 1313 multiple alignments and profile HMMs match the majority of proteins. Nucleic Acids Res, 27, 260–262.

Bateman, A., Birney, E., Durbin, R., Eddy, S.R., Howe, K.L., and Sonnhammer, E.L. (2000) The Pfam protein families database. Nucleic Acids Res, 28, 263–266.

Bateman, A., Coggill, P., and Finn, R.D. (2010) DUFs: families in search of function. Acta Crystallogr, 66, 1148–1152.

Bateman, A., Coin, L., Durbin, R., Finn, R.D., Hollich, V., Griffiths-Jones, S., Khanna, A., Marshall, M., Moxon, S., Sonnhammer, E.L. et al. (2004) The Pfam protein families database. Nucleic Acids Res, 32, D138–D141.

Bucher, P. and Bairoch, A. (1994) A generalized profile syntax for biomolecular sequence motifs and its function in automatic sequence interpretation. Proc Int Conf Intell Syst Mol Biol, 2, 53–61.

Daub, J., Gardner, P.P., Tate, J., Ramskold, D., Manske, M., Scott, W.G., Weinberg, Z., Griffiths-Jones, S., and Bateman, A. (2008) The RNA WikiProject: community annotation of RNA families. RNA, 14, 2462–2464.

Edgar, R.C. (2004) MUSCLE: multiple sequence alignment with high accuracy and high throughput. Nucleic Acids Res, 32, 1792–1797.

Finn, R.D., Mistry, J., Schuster-Bockler, B., Griffiths-Jones, S., Hollich, V., Lassmann, T., Moxon, S., Marshall, M., Khanna, A., Durbin, R. et al. (2006) Pfam: clans, web tools and services. Nucleic Acids Res, 34, D247–D251.

Finn, R.D., Mistry, J., Tate, J., Coggill, P., Heger, A., Pollington, J.E., Gavin, O.L., Gunasekaran, P., Ceric, G., Forslund, K. et al. (2010) The Pfam protein families database. Nucleic Acids Res, 38, D211–D222.

Finn, R.D., Tate, J., Mistry, J., Coggill, P.C., Sammut, S.J., Hotz, H.R., Ceric, G., Forslund, K., Eddy, S.R., Sonnhammer, E.L. et al. (2008) The Pfam protein families database. Nucleic Acids Res, 36, D281–D288.

Gardner, P.P., Daub, J., Tate, J., Moore, B.L., Osuch, I.H., Griffiths-Jones, S., Finn, R.D., Nawrocki, E.P., Kolbe, D.L., Eddy, S.R. et al. (2011) Rfam: Wikipedia, clans and the "decimal" release. Nucleic Acids Res, 39, D141–D145.

Haft, D.H., Loftus, B.J., Richardson, D.L., Yang, F., Eisen, J.A., Paulsen, I.T., and White, O. (2001) TIGRFAMs: a protein family resource for the functional identification of proteins. Nucleic Acids Res, 29, 41–43.

Henikoff, J.G., Henikoff, S., and Pietrokovski, S. (1999) New features of the Blocks Database servers. Nucleic Acids Res, 27, 226–228.

Hunter, S., Apweiler, R., Attwood, T.K., Bairoch, A., Bateman, A., Binns, D., Bork, P., Das, U., Daugherty, L., Duquenne, L. et al. (2009) InterPro: the integrative protein signature database. Nucleic Acids Res, 37, D211–D215.

Katoh, K., Misawa, K., Kuma, K., and Miyata, T. (2002) MAFFT: a novel method for rapid multiple sequence alignment based on fast Fourier transform. Nucleic Acids Res, 30, 3059–3066.

Kunin, V., Cases, I., Enright, A.J., de Lorenzo, V., and Ouzounis, C.A. (2003) Myriads of protein families, and still counting. Genome Biol, 4, 401.

Letunic, I., Doerks, T., and Bork, P. (2012) SMART 7: recent updates to the protein domain annotation resource. Nucleic Acids Res, 40, D302–D305.

Marchler-Bauer, A., Lu, S., Anderson, J.B., Chitsaz, F., Derbyshire, M.K., DeWeese-Scott, C., Fong, J.H., Geer, L.Y., Geer, R.C., Gonzales, N.R. et al. (2011) CDD: a Conserved Domain Database for the functional annotation of proteins. Nucleic Acids Res, 39, D225–D229.

Punta, M., Coggill, P.C., Eberhardt, R.Y., Mistry, J., Tate, J., Boursnell, C., Pang, N., Forslund, K., Ceric, G., Clements, J. et al. (2012) The Pfam protein families database. Nucleic Acids Res, 40, D290–D301.

Sammut, S.J., Finn, R.D., and Bateman, A. (2008) Pfam 10 years on: 10,000 families and still growing. Brief Bioinform, 9, 210–219.

Schultz, J., Milpetz, F., Bork, P., and Ponting, C.P. (1998) SMART, a simple modular architecture research tool: identification of signaling domains. Proc Natl Acad Sci USA, 95, 5857–5864.

Selengut, J.D., Haft, D.H., Davidsen, T., Ganapathy, A., Gwinn-Giglio, M., Nelson, W.C., Richter, A.R., and White, O. (2007) TIGRFAMs and Genome Properties: tools for the assignment of molecular function and biological process in prokaryotic genomes. Nucleic Acids Res, 35, D260–D264.

Sigrist, C.J., Cerutti, L., de Castro, E., Langendijk-Genevaux, P.S., Bulliard, V., Bairoch, A., and Hulo, N. (2010) PROSITE, a protein domain database for functional characterization and annotation. Nucleic Acids Res, 38, D161–D166.

Sonnhammer, E.L., Eddy, S.R., Birney, E., Bateman, A., and Durbin, R. (1998) Pfam: multiple sequence alignments and HMM-profiles of protein domains. Nucleic Acids Res, 26, 320–322.

Sonnhammer, E.L.L., Eddy, S.R., and Durbin, R. (1997) Pfam: A comprehensive database of protein domain families based on seed alignments. Proteins, 28, 405–420.

Tatusov, R.L., Natale, D.A., Garkavtsev, I.V., Tatusova, T.A., Shankavaram, U.T., Rao, B.S., Kiryutin, B., Galperin, M.Y., Fedorova, N.D., and Koonin, E.V. (2001) The COG database: new developments in phylogenetic classification of proteins from complete genomes. Nucleic Acids Res, 29, 22–28.

3

CLASSIFYING PROTEINS INTO DOMAIN STRUCTURE FAMILIES

ALISON CUFF

Institute of Structural and Molecular Biology, University College London, London, UK

ALEXEY MURZIN

MRC Laboratory of Molecular Biology, Cambridge, UK

CHRISTINE ORENGO

Institute of Structural and Molecular Biology, University College London, London, UK

CHAPTER SUMMARY

This chapter reviews domain classifications with particular focus on the two most comprehensive protein structure classification resources—structural classification of proteins (SCOP) and class, architecture, topology, homology (CATH).

General approaches to protein structure comparison, as well as methods used for recognizing domain boundaries are discussed. Both CATH and SCOP are in active development and still keep pace with the exponential increase in the number of protein structures being deposited in the Protein Data Bank (PDB). The classification hierarchies of both databases are reviewed and differences in the philosophies of the resources are discussed. The challenges that arise when classifying protein domain structures into superfamilies and the caveats that should be employed when assigning domains to the same fold group are explored.

Protein Families: Relating Protein Sequence, Structure, and Function, First Edition.
Edited by Christine Orengo and Alex Bateman.

We consider the types of insights that can be gained into protein evolution from investigating diverse structural superfamilies. We also discuss the similarities and differences between the two resources and conclude with an introduction to the collaboration between CATH and SCOP, to provide consensus classifications via a new web portal, Genome3D.

3.1 INTRODUCTION

The emergence of X-ray crystallographic techniques in the 1970s and the analyses of the increasing numbers of protein structures deposited in the PDB (Bernstein et al., 1977) revealed deep evolutionary relationships between proteins, previously thought to be unrelated from their sequence compositions. The 3D structures also confirmed that proteins could be decomposed into constituent globular entities (domains) that often occurred in different multidomain contexts.

Chothia and Lesk were the first to quantitatively demonstrate the extent to which the structures of homologous domains (e.g., the globins (Lesk and Chothia, 1980) and immunoglobulins (Lesk and Chothia, 1982)) were conserved during evolution (see Chapter 9 for overview of the globin superfamily). As more data has accrued, these observations have been confirmed by studies of other domain families, in which relatives with very diverse sequence identities and often diverse functions were found to adopt similar 3D structures.

The now iconic plot of Chothia and Lesk (1986) capturing the relationship between protein sequence similarity and structural similarity of homologs revealed that even in the midnight zone where homologs share less than 15% sequence identity there could still exist significant structural similarity. These very distant evolutionary relationships, suggested by structural similarity, were confirmed by other experimental data, for example, conservation of molecular functions, key functional residues or similarities in some unusual or distinctive aspects of the chemistry.

Recently, more comprehensive studies of divergence in sequence and structure in all known protein superfamilies, suggest that at least 40% of residues in a domain, usually in the core, adopt structural conformations conserved throughout the superfamily (Dessailly et al., 2010, see for example Fig. 3.1).

As it became clear that structural data could play a valuable role in detecting evolutionarily related domains, a number of domain structure classifications emerged (e.g., 3DEE (Siddiqui et al., 2001), SCOP (Murzin et al., 1995), CATH (Orengo et al., 1997), COMPASS (Sadreyev and Grishin, 2003), (Dali Domain Dictionary) DDD (Holm and Sander, 1998)) of which the most comprehensive (e.g., SCOP and CATH) still manage to keep pace with the exponentially increasing numbers of structures being deposited in the PDB, both by traditional structural biology groups and increasingly by the international structural genomics initiatives (Todd et al., 2005; Chandonia and Brenner, 2006; Dessailly et al., 2010).

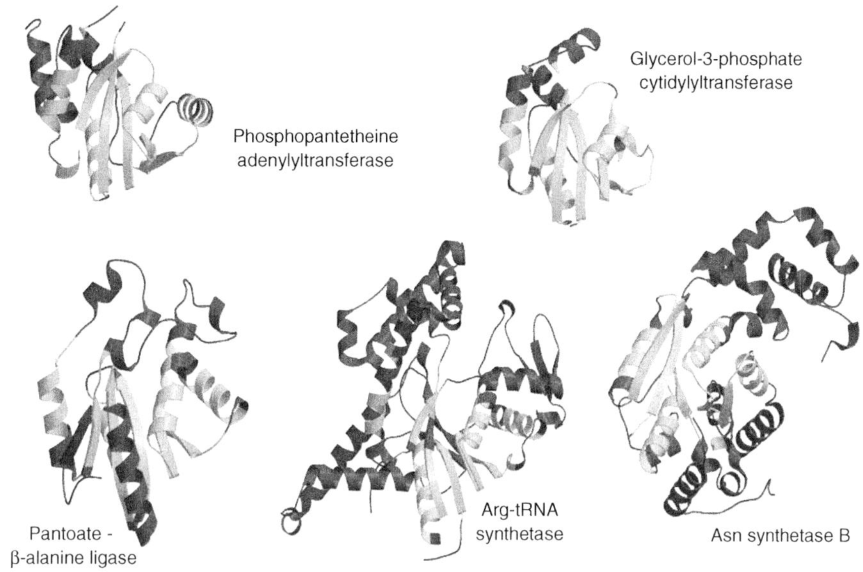

Figure 3.1 Example domains from the HUP superfamily (CATH code 3.40.50.620) showing conserved structural core regions (light gray) and embellishments (dark gray) present in this family. This family is particularly structurally and functionally diverse, but a structural core is still evident.

Both SCOP and CATH classify structures at the domain level as this was deemed to be the primary unit of evolution. For all kingdoms in life, a large proportion of the proteins in a genome comprise multiple domains, ranging from 60% in prokaryotes to as high as 80% in eukaryotes (Teichmann et al., 1998). These numbers were originally determined through prediction of domains in sequences from a few hundred completed genomes and the proportions remain similar today, despite the 10-fold increase in the number of organisms now available for analysis.

The value of classifying proteins at the domain level was further endorsed by analyses of the domain predictions in completed genomes, which confirmed the extent to which domains were being duplicated and combined in diverse combinations. There are currently less than twenty thousand domain families classified (including SCOP, CATH and Pfam); yet, more than a million single and multidomain protein families (Lee et al., 2005) and capturing families at the domain level provide insights into how new proteins emerge during evolution, by domain fusions and also how changes in the domains can alter protein functions.

The definition of protein domain can be based on whether we are considering the functional role of some region of the protein or a compact structural entity within the protein. For example, a structure-based definition, as often employed in SCOP and CATH, largely identifies regions of the protein sequence, which fold into compact and independent globular structures. Domain boundaries rarely occur in the middle of secondary structure elements or near hydrophobic clusters and there are usually more residue contacts occurring within the domain than between domains. A function-based classification may group two such structural domains if they are always associated together in their functional role. Evolutionary history of proteins may also affect the domain definition. For example, in SCOP, a multidomain protein resulting from a probable gene duplication event may be split into structurally similar units rather than compact domains.

Although protein domain structures are much more highly conserved during evolution than their sequences, structural changes can occur, usually outside the conserved cores of the domains. The success of the structural classifications in capturing such extensive information on domain families was therefore partly due to the emergence of powerful new algorithms for comparing protein structures, which emerged in the late 1980s and early 1990s (e.g., DALI (Holm and Sander, 1994), COMPARER (Sali and Blundell, 1990), (Sequential Structure Alignment Program) SSAP (Taylor and Orengo, 1989), (Structural Alignment of Multiple Proteins) STAMP (Russell and Barton, 1992), see also below) and which exploited sophisticated strategies for handling the significant insertions/deletion of residues (indels), which occurred in some highly divergent families and that can seriously affect protein alignment.

While this chapter focuses on structure-based domain classifications, Chapter 2 describes other comprehensive domain classifications (e.g., Pfam (Punta et al., 2012), (Simple Modular Architecture Research Tool) SMART (Letunic et al., 2012), (High-quality Automated and Manual Annotation of microbial Proteomes) HAMAP (Lima et al., 2009), and PRINTS (Attwood et al., 2003), largely derived using sequence data alone and which often place more weight on grouping domains according to their functional similarity. The most populated of these have been integrated in InterPro, a comprehensive resource hosted by the European Bioinformatics Institute (EBI) (Hunter et al., 2011) (see Chapter 2 for further details).

Where data is available, the structure-based domain classifications (e.g., SCOP and CATH) do also consider the functional role of the protein in which the domain occurs, and provide functional information to complement the structural classification.

Here we review the strategies employed by the two most comprehensive classification resources, SCOP and CATH, and consider the challenges faced and the caveats that should be considered when assigning a protein domain to a domain structure superfamily. We also describe some of the insights into protein evolution that can be gleaned by considering the structural data relating to very diverse superfamilies. While, the most commonly used domain structure classifications are the SCOP and CATH resources, the PDB and conserved domain database (CDD) (Marchler-Bauer et al., 2011) do provide structural neighborhood information on proteins, which can include identification of homologies, but these are not explicit classifications as in SCOP and CATH.

CATH and SCOP currently capture domains from at least 80% or more of the proteins deposited in the PDB. In addition to these large resources, other more specialized resources exist such as the membrane structures classification

TABLE 3.1 Specialized Protein Databases

Name of Resource	Website (URL)	Type of Information
Conserved Domain Database (CDD)	http://www.ncbi.nlm.nih.gov/Structure/cdd/cdd.shtml	Annotated multiple sequence alignment models for domains and protein structures
CAMPS 2.0	http://webclu.bio.wzw.tum.de/CAMPS2.0/	Classification of membrane proteins
Periodic Table of Coiled Coils (CC+ database)	http://coiledcoils.chm.bris.ac.uk/ccplus/search/periodic_table	Classification of coiled-coil structures according to size and complexity of structures
Thiamine Diphosphate Dependent Enzyme Engineering Database (TEED)	http://www.teed.uni-stuttgart.de/	Database providing information on the sequence and structure of Thiamine diphosphate dependent proteins
Structure-Function Linkage Database (SFLD)	http://sfld.rbvi.ucsf.edu	Classification of mechanistically diverse enzyme superfamilies
HOMSTRAD	http://www-cryst.bioc.cam.ac.uk/-homstrad/,	Database of protein structure alignments for homologous families

of the Frishman group (Neumann et al., 2011). Other resources have been listed in Table 3.1.

3.2 THE CLASSIFICATION HIERARCHIES ADOPTED BY SCOP AND CATH

As mentioned above, multiple resources attempt to identify protein domain families. As the structure-based resources exploit 3D structure data to detect much more remote relationships, the groups of homologs they identify are generally referred to as *superfamilies*, as they capture relatives with a broad range of features (i.e., very diverse sequences and often diverse functions). In contrast, the purely sequence-based resources tend to group closer homologs more likely to be less diverged and having similar functions. These groups are generally referred to as *families*.

Furthermore, while most sequence-based resources of domain families discussed in Chapter 2 concentrate on evolutionary relationships, the structure-based classifications, SCOP and CATH, organize the domains in additional ways. In adddition to capturing evolutionary relationships at the superfamily level, these resources then further organize the superfamilies into fold groups, architectures, and structural classes (see Fig. 3.2). The fold group clusters

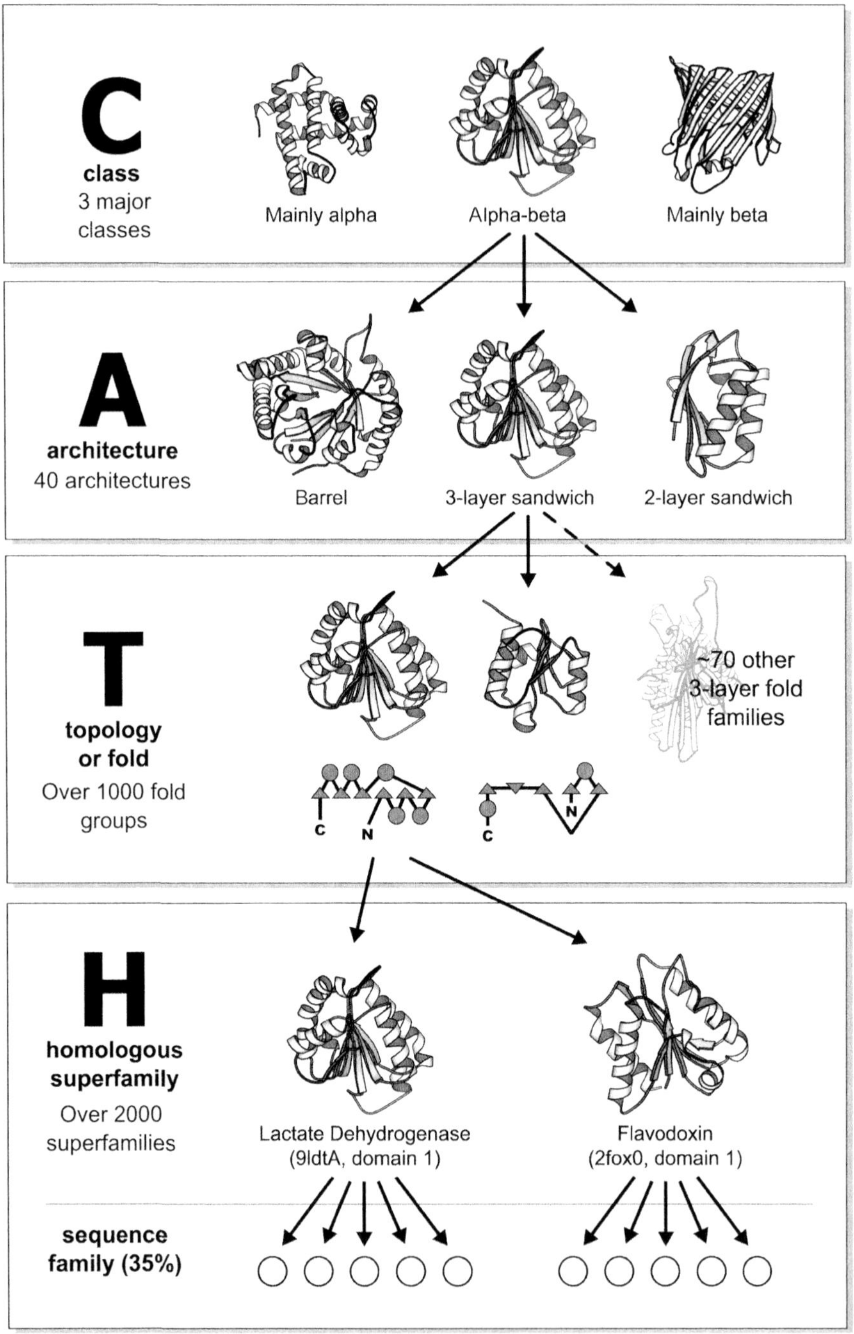

Figure 3.2 The CATH hierarchy.

together superfamilies deemed to have 3D structures adopting similar topologies, that is, they have a significant proportion of equivalent secondary structures (i.e., superimposable in 3D) connected in the same order in the polypeptide chain. Above the level of the fold group, CATH also formally recognizes the architecture or shape adopted in 3D: that is, the arrangement of secondary structure elements in 3D, regardless of their connectivity. SCOP does not explicitly adopt an architecture level but does provide descriptions of regular or distinct architectures for some individual superfamilies.

Finally, the broadest level in both SCOP and CATH is the protein class, which captures the composition in terms of secondary structure (i.e., predominantly α-helical or β-strands or a mixture of both). Figure 3.2 shows the levels employed in the CATH classification.

3.3 CHALLENGES IN IDENTIFYING DOMAINS IN PROTEINS

Generally, it is much easier to recognize individual domains by viewing or analyzing the 3D structure than by inspecting the protein sequence. In most multidomain proteins, domains are separated by small linking regions, whose characteristics can be difficult to recognize from 1D sequence information alone but which are often unstructured in 3D, making it easy to visually identify the distinct, compact, globular structural regions associated with the protein domains.

However, while it is relatively easy to visually recognize the constituent domains of two domain multidomain proteins, for larger, more complex proteins with three or more domains, it can often be difficult to distinguish the different domains. Furthermore, to keep pace with the increases in data in the PDB, it is now necessary to employ automated approaches to suggest domain boundaries, which can then be validated by visual inspection.

These automated approaches are mostly based on the 3D structure. However, some recent approaches also exploit sequence-based methods by using sensitive search protocols (e.g., Hidden Markov Models (HMMs)) to identify sequence regions related to domains of known structure (see below).

3.3.1 Automatic Structure-Based Approaches for Domain Identification

Over the last three decades many algorithms have been designed to identify constituent domains in multidomain proteins using 3D-structure data. These range from *ab initio* methods based on structural characteristics, through to approaches that exploit homology to detect domain regions, which are structurally similar to previously classified domain structures.

3.3.2 *Ab-Initio* Methods Used for Automatic Recognition of Domain Boundaries in the CATH Classification

Ab-initio methods were pioneered in the 1980s by the elegant protein unfolding units (PUU) algorithm of Holm and Sander (1994) which is still one of the most accurate methods available. In this approach, domains are treated as separate units of mass connected by springs and changes in inertia following cuts in the polypeptide chain are calculated and used to determine the optimal residue positions for separating the domains. More widely used approaches perform analyses of residue contact density and endeavor to cut the chain in positions, which cause minimum changes to the contact densities of the separated domains. Some successful strategies also look for hydrophobic clusters. The DhcL method (Koczyk and Berezovsky, 2008) employs a Van der Waals model of domains to predict boundaries.

Holland and Bourne (2006) have undertaken a comprehensive review of the different approaches used and the dConsensus website associated with the PDB presents the domain predictions provided by the most successful of these (Alden et al., 2010).

For the CATH classification, domains are identified using a consensus approach, which combines three orthogonal methods of domain identification. The PUU method, from Holm using spring-based analyses to cut the protein chain, the Detective algorithm of Swindells and Thornton (Swindells, 1995) that searches for large hydrophobic clusters at the core of a putative domain and the Domain Maker (DOMAK) algorithm of Barton et al. (Siddiqui and Barton, 1995), which assesses changes in contact density for different cuts of the chain. As with many bioinformatics prediction protocols, a consensus approach seeking agreement between the algorithms, improves the accuracy considerably. For example, initial benchmarking using a hand curated set of CATH domain assignments demonstrated that both accuracy and sensitivity could be increased by combining assignments from all the three algorithms (Jones et al., 1998). Unfortunately, only a relatively small proportion of domain boundaries can be accurately assigned in this way. Analyses by the Bourne group (Alden et al., 2010) showed agreement between multiple methods for only 63% of their benchmark set.

In SCOP domain boundary assignment was initially largely done by manual inspection of the protein structures and has evolved over the time to take into account sequence defined and functionally defined domains. In the beginning, if a protein contained at least one recurrent structural domain, its other domains were also classified as potentially independent evolutionary units, even if they were unique. More recently, these additional domains are only classified separately if there is sequence or structural evidence for the domain's independent origin, otherwise they are appended to the principle domain

3.3.3 Homology-Based Methods Exploiting the Recurrence of Domains in Different Multidomain Contexts

As mentioned already, the completion of many prokaryotic and eukaryotic genomes confirmed earlier speculations on the duplication of domains during evolution (Jones et al., 1998; Lee et al., 2003). Analyses of thousands of completed genomes have shown more conclusively the extent to which domains have been duplicated, shuffled within genomes and fused to other domains giving rise to different multidomain protein architectures. More generally, multidomain proteins "grow" by accretion of new domains at the N or C termini of the protein with a relatively small percentage of domains being inserted elsewhere in the protein (Björklund et al., 2005, 2006). For many of the most highly populated domain superfamilies, domain relatives are found in a diverse range of multidomain contexts (e.g., see the different domain contexts observed for relatives within the HUP domain superfamily, Chapter 7).

Several sequence and structure-based methods for domain identification exploit this concept of recurrence to look for constituent domains in newly sequenced or newly structurally characterized proteins. Both SCOP and CATH classifications exploit domain recurrence and use structure comparison algorithms to recognize previously classified domain families in newly determined multidomain structures. In fact, as mentioned already, SCOP only classifies constituent domains into separate domain families where they have been observed in different multidomain contexts. SCOP validates previously reported structural similarities, for example,, found by the DALI structure comparison algorithm, but it mainly relies on its human expertise to find other occurrences of a domain in more difficult cases. CATH uses the CATHEDRAL method (Redfern et al., 2007), which combines two in-house structure comparison methods (GRATH (Harrison et al., 2003), and SSAP (Taylor and Orengo, 1989)) as described below.

Some sensitive sequence-based approaches (e.g., profile–profile or HMM–HMM) can also help in identifying recurrent domains. However, while these methods can be good at confirming the existence of the domain in a new multidomain context, especially for a domain that is structurally diverse from any previously classified, the boundaries may be less well assigned. For SCOP, sequences from new structures are scanned against Pfam models to identify close matches for domain regions, which can aid boundary assignment while CATH performs HMM scans against a library of CATH HMMs from classified domains (CATH–HMMscan).

A major challenge for methods exploiting domain recurrence is the structural divergence that has been observed to occur in some domain families. For at least 100 domain superfamilies in CATH (and these are the most highly populated) at least twofold and up to fivefold variation has been observed in the sizes of domains (Reeves et al., 2006, see Fig. 3.3 for an example). While relatively few superfamilies (<5% of CATH) have been found to exhibit this extreme divergence, these superfamilies are very large accounting for nearly 60% of domain sequences in UniProt and completed genomes (Cuff et al., 2009a, 2009b, see Fig. 3.4).

Because of this problem and the fact that currently less than 10% of the relatives in these very large superfamilies have structures solved (see discussion of the structural genomics initiatives below), manual inspection of the protein structure and validation is a key step in reliable domain boundary identification, especially for proteins that are remote from any currently classified in the structural resources. For both SCOP and CATH, domain boundaries are manually validated for all remote homologs (i.e., less than 80% sequence identity

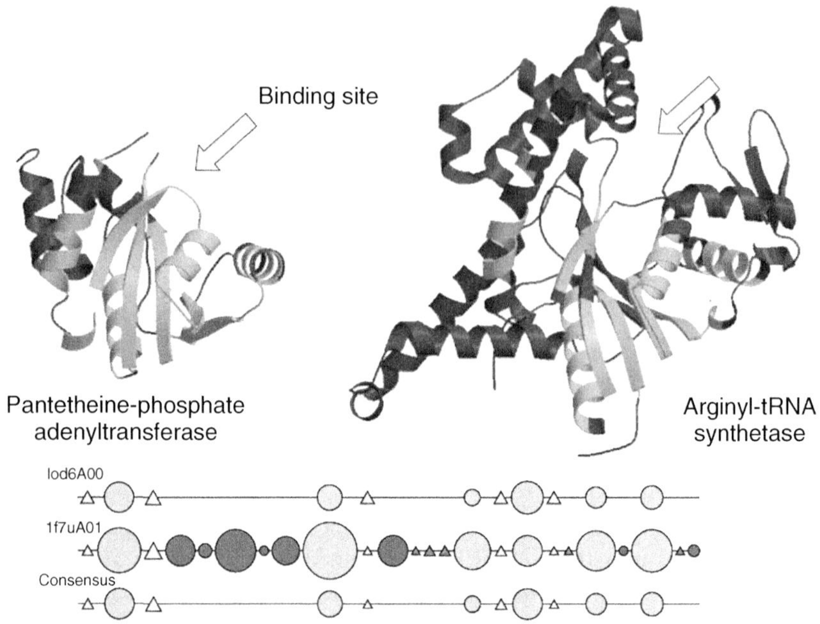

Figure 3.3 Two domains from the structurally diverse HUP superfamily, in which a significant difference in the size of the domains can be seen. The conserved structural core for both domains are highlighted in light gray. The schematic below the Molscript images is a 2DSEC plot displaying the differences in secondary structure content of the two domains above; the circles representing beta-sheet structures and the triangles alpha-helical, the sizes of both corresponding to the size of the secondary structure element. Common helices and strands are in light gray while inserted helices and strands are in dark gray.

to classified relatives). For CATH, this follows prediction of domain boundaries using the range of structure-based and sequence-based automated methods mentioned already above (i.e., DOMAK, Detective, PUU, DBConsensus, CATHEDRAL, CATH–CATH–HMMscan) and the results are presented on a web page to guide the curator in final manual refinement of the boundaries.

Following domain boundary identification, the next step is to determine whether the domains in a newly determined protein structure can be assigned to existing superfamilies in the classification. Clearly, where domain boundaries have been identified using methods exploiting domain recurrence, a preliminary assignment to a particular superfamily has already been made.

3.4 STRUCTURE-BASED APPROACHES FOR IDENTIFYING RELATED FOLDS AND HOMOLOGS

Both SCOP and CATH use structure comparison algorithms to help recognize homologous domains. As mentioned already these methods benefit from the

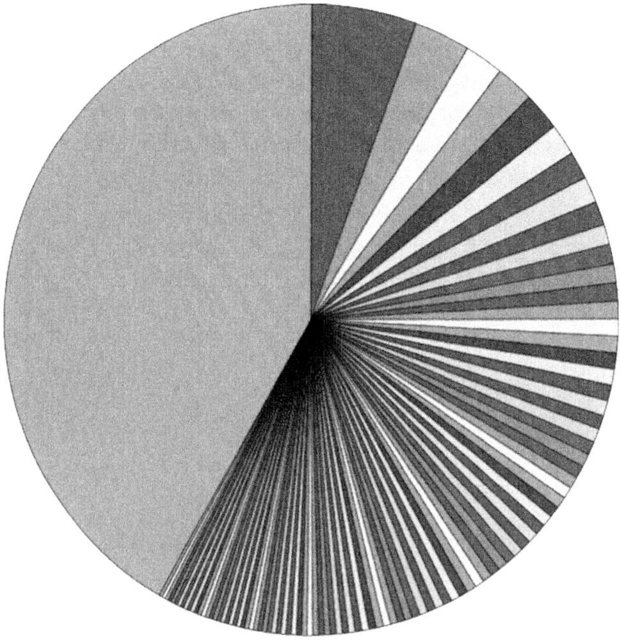

Figure 3.4 Pie chart displaying the number of domain sequences in the most highly populated superfamilies in Uniprot and completed sequences. Each segment corresponds to one of each of these superfamilies. It can be seen that the most highly populated 100 superfamilies correspond to approximately 60% of domain sequences.

significant structure conservation across relatives in a domain superfamily. As mentioned already, residues in the core of the domain structure are usually highly conserved across a superfamily and this core can comprise 40% or more of the residues in the domain.

For relatives sharing 30% or more sequence identity (number of identical residues as a percentage of the number of residues in the smallest domain) there is likely to be significant conservation in the global structure. This level of sequence identity is often used as a threshold for identifying relatives with sufficient structural similarity to use as a template structure for 3D modeling of a structurally uncharacterized sequence. A significant proportion of structures deposited in the PDB do comprise domains sharing at least 30% sequence identity with a domain already classified in CATH.

Below the level of 30% sequence identity (e.g., in the so-called "twilight zone" or even lower the "midnight zone" (<15% sequence identity)) the Chothia and Lesk plot of sequence versus structure similarity reveals that considerable structural divergence can occur between two homologous domains.

In fact, recent analyses of Grishin and others (Krishna and Grishin, 2005; Friedberg and Godzik, 2005; Sippl et al., 2008) have revealed examples of homologous relatives where the structure has changed dramatically making it challenging for even the most robust structure comparison methods to recognize any structural similarity with which to infer an evolutionary relationship.

Studies based on diverse families classified in CATH have shown that for some extremely divergent homologs, where the structure has changed significantly, relatives can sometimes be better identified using highly sensitive sequence-based HMM–HMM comparisons than structure-based methods (Reid et al., 2007). However, these algorithms are computationally expensive and for the vast majority of homologs (>99%) structure-based methods will be faster and a more accurate way of identifying homologs to a newly solved protein structure.

As a large proportion of residues in the domain cores are structurally conserved between homologous domains, even when no detectable level of sequence similarity remains, this relatively large "signal" or "structural fingerprint" can be used to derive superfamily 3D profiles, in an analogous manner to sequence profiles or HMMs used to capture the most conserved positions in domain sequences (Redfern et al., 2009).

3.5 APPROACHES TO STRUCTURE COMPARISON

There are two main approaches used in comparing protein structures based on whether internal distances within each protein are compared between proteins (intra-residue methods) or whether the proteins are superposed (inter-residue methods). For both types of approaches optimization strategies (e.g., dynamic programming, simulated annealing) are employed to handle significant insertions or deletions of residues (indels) in very distant structural homologs. It is outside the scope of this chapter to review all the structure classification methods, or even the major philosophies (see Koehl, 2006 for a review). However, representative approaches of both types will be briefly discussed as differences in these methods can contribute to disagreements in the relationships identified by the different resources. As SCOP does not rely on any particular automated method or methods, we contrast approaches adopted by CATH with the DALI algorithm, which assists SCOP indirectly.

3.6 THE DALI ALGORITHM

Phillips introduced the first intra-residue comparison method, in 1970 (Phillips, 1970). This compares the patterns of internal residue contacts, visualized in distance plots, between pairs of protein structures. Clearly, proteins with similar 3D arrangements of secondary structures display similar contact patterns in the distant plot. However, in more distantly related proteins, significant structural differences between the relatives resulting from indels outside the structural core, can obscure similarities in the distance plots and various optimization strategies have been used to cope with these differences and identify equivalent residues.

While there have been many automated implementations of intra-residue comparison (see Brown et al., 1996; Kolodny et al., 2005 for reviews). Perhaps the most powerful and widely used is the DALI algorithm developed by Holm and Sander in 1994. This generates contact maps for all hexa-peptides in a given structure and then uses powerful optimization strategies to compare contact maps between structures and identify corresponding fragments of proteins, which can be successively concatenated to generate information on global structural similarity (Holm and Sander, 1994).

DALI has been widely used by structural biologists to search for structural similarities for newly solved protein structures and its findings have usually been included in the published reports and have often facilitated manual classification of the new structures. The algorithm was developed by Holm and Sander (1994) and initially exploited to generate a clustering of all domain structures in the PDB, allowing structural relationships between domains to be automatically identified for all homologous and nonhomologous domains. This led to the identification of so-called "attractor" motifs—structural motifs commonly recurring in domains with different global structures. Recent benchmarking by different independent groups (Redfern et al., 2007; Kolodny et al., 2005) has shown that DALI remains one of the most accurate methods available.

3.7 THE SSAP ALGORITHM USED FOR FOLD RECOGNITION IN CATH

The SSAP algorithm, developed by Taylor and Orengo (1989), which have been used to classify domain structures in CATH also exploits intra-residue data by comparing residue structural views between proteins. For a given residue, the "view" corresponds to the set of vectors from the C_β atom of that residue to the C_β atoms of all other residues in the protein. Dynamic programming is used to handle residue insertions and deletions and align views between residue pairs in the different domains and information on probable residue equivalences is accumulated and then analyzed by a final dynamic programming analysis to give the most probable alignment. Benchmarking using a gold standard set of domain structures, that is, domains for which both SCOP and CATH agree on manually curated boundaries and homologue classification, has shown comparable performance to DALI and other popular algorithms (e.g., STRUCTAL (Subbiah et al., 1993; Kolodny et al., 2005)).

3.8 FAST APPROXIMATE METHODS USED TO RECOGNIZE FOLDS IN CATH

Residue-based structure comparison can be considerably more computationally expensive than sequence analysis, as algorithms are necessarily searching for similarities in 3D space. One effective approach to speed up the comparisons is to simply compare secondary structure elements between the query structures. There are at least 10-fold fewer of these than residues and it is possible to compare relatively few features, for example, midpoint distances, pair-wise orientations in 3D.

Although the analyses of Chothia and Lesk and others (Lesk and Chothia, 1980, 1982) have shown that secondary structure orientations can vary by up to 50° between relatives, secondary structure orientations in the core of the protein structure are usually quite well conserved and methods comparing secondary structures between domains have been found to reliably detect the correct fold in searches across all superfamilies.

For rapid fold recognition of domains in CATH, approaches such as the GRATH (Harrison et al., 2003) and CATHEDRAL (Redfern et al., 2007) algorithms compare graphical representations of the protein structures. In these methods, secondary structures become the nodes in a 2D graph and the edges connecting these nodes are labeled with information on the distances between the secondary structures and their relative orientations in 3D. Algorithms, such as the Bron–Kerbosch algorithm (Bron and Kerbosch, 1973) can be used to search for common subgraphs between the proteins, representing structurally related domains. These fast secondary structure-based methods can be up to 1000 times faster than residue-based approaches.

3.9 MEASURING STRUCTURAL SIMILARITY

While many algorithms generate a residue alignment between the proteins based on the structure comparison and may return a specific score related to the features they are comparing, it is common to calculate the "distance" between the structures following superposition in 3D. Therefore, the residue alignment generated by the algorithm is used to guide the superposition of the structures.

Rossmann and Argos (1976) developed the first implementation of a rigid body superposition method, in 1976. Their approach attempts to superpose two protein structures by first translating the center of gravity of one structure onto the other in 3D. Subsequently one structure is rotated relative to the other to minimize the distances between the equivalent residues in the two structures. Many structure comparison methods return a list of equivalent positions between domains to enable their superposition using a Rossmann and Argos like approach.

In practice, the root mean squared deviation (RMSD), calculated as in Equation 3.1 below, is commonly used.

$$\mathrm{RMSD} = \sqrt{\frac{1}{N}\sum_{i=1}^{N}\delta_i^2} \tag{3.1}$$

In order to assess the global structural similarity between two proteins, it is important to consider the number of residues which can be superposed within a certain distance threshold, for example, 3Å. This should then be considered as a proportion of the total residues in the complete structure. For example, superposition of only 20% of residues with a distance threshold less than 3 Å would suggest that the correspondence between the structures only related to a common structural motif, for example an alpha–beta plait motif comprising three secondary structures.

A normalized RMSD is calculated for classification in CATH. This takes into account of the proportions of the two structures being superposed. For example, multiplying the RMSD by the number of residues in the larger structure and dividing by the number of residues superposed between the structures.

A more rigorous approach to assessing the similarity between structural domains is to contrast the score returned by the method for a particular domain against the scores returned by comparison of that domain against a large set of random unrelated domains so that a *Z*-score or *P*-value can be calculated. Several authors have suggested using this approach (Levitt and Gerstein, 1998; Kolodny et al., 2005) but unlike large sequence scans, structure comparison is complicated by the high recurrence of relatively few secondary and supersecondary motifs in different fold contexts, giving rise to a random score distribution that makes the calculation of probability values more problematic.

In addition to the challenges faced in handling the indels occurring during evolution between homologs, further problems in structure comparisons can arise from induced conformational changes. For example, it is possible to miss similarities in protein structures if there are changes induced by ligand binding or other effectors.

Algorithms such as DALI and SSAP that divide the structures into fragments before comparing them or use dynamic programming strategies for aligning equivalent regions tend to be more successful in recognizing local similarities between the proteins, which remain despite these global changes, than rigid body methods that attempt to superpose the structures.

3.10 MULTIPLE STRUCTURE ALIGNMENT

In CATH, multiple structural alignments are generated for reasonably structurally coherent groups of relatives (i.e., superposable within 9 Å normalized RMSD), using the CORA algorithm (Orengo, 1999), which applies a modified version of the double dynamic programming algorithm developed for SSAP. Multiple alignments are displayed on the web site and a topology drawing program (2DSEC (Reeves et al., 2006)) can be used to visualize the equivalent secondary structures in the multiple alignment and identify inserted secondary structures (see Fig. 3.3 for an example).

In SCOP, multiple structural alignments are manually created for selected superfamilies. They are based on so-called superfamily characters, the conserved structural details usually associated with the conserved functional sites. These are effectively given higher weights in superposition than other similar features. The SCOP-related resource, SISYPHUS (Andreeva et al., 2007) presents information on nontrivial structural alignments between remote homologs and pairs of structures sharing rare or interesting structural features. These alignments are difficult to obtain automatically and considerable manual intervention is required.

3.11 CLASSIFICATION PROTOCOLS

Although there are currently almost 80,000 protein structures in the PDB there is considerable redundancy in the data and as homologs with high levels of sequence identity (e.g., $\geq$90% sequence identity) are highly likely to have very similar structures and functions, it is sensible to cluster these relatives and select representatives before submitting them to any classification protocol. For example, using this threshold for redundancy we currently see a 2.5-fold reduction in the data from 79,851 proteins to 31,132 representatives.

Subsequently, classification follows four major steps (i) identification of domain boundaries, (ii) comparison of domains to previously classified structures, using structure and/or sequence-based strategies to assign domains to an existing superfamily. In some cases, further evidence, for example, functional similarity may be required to confirm homology, and (iii) for those domains not assigned to a homologous superfamily investigation of structural similarity to classified fold groups or identification of novel folds. Domains that can be assigned to a fold group but not to an existing superfamily then seed novel superfamilies. (iv) Finally, for any domains possessing novel folds the general shape or architecture of the domain and residue composition (e.g., alpha, beta, or alpha–beta) will be considered and used to further categorize the fold group. This is done to populate an explicit architecture assignment in CATH. While SCOP often describes the architecture adopted by the structure, for example, β-barrel, there is no formal level in the classification. In SCOP and CATH, all steps involve manual curation, especially for identifying novel folds and novel superfamilies (step 3).

3.11.1 Domain Boundary Assignment

Domain boundary recognition algorithms have already been described above and most approaches (including consensus methods that combine the results of multiple algorithms) achieve less than 80% accuracy and therefore curation is still necessary. It is important to minimize errors at this stage as they will be propagated by the domain prediction resources (e.g., SUPERFAMILY (Wilson et al., 2009), Gene3D (Lees et al., 2012), GenTHREADER (Jones, 1999)) which use domain sequences from the structures in SCOP and CATH to build HMM libraries for structural annotation of genome sequences (see Chapter 4).

CATH and SCOP use somewhat different criteria for recognizing domains, in that SCOP only recognizes domains that have been seen in more than one evolutionary context (i.e., multidomain context) while CATH will additionally divide multidomain structures if there are constituent globular regions that appear well packed and able to fold semi-independently. In some cases this has resulted in division of proteins (e.g., some kinases) in which the constituent "structural domains" have never been observed in isolation and which always function together. For example, the active site may appear at the domain interface and there may be active site residues contributed by both domains. The rationale behind this approach in CATH was to collect information that might help to elucidate domain folding and identify evolutionary or structural characteristics that might impact on this.

3.11.2 Superfamily Identification

Usually, both SCOP and CATH, classifications proceeds from the superfamily up through the fold assignment, to the structural class. That is, sequence-based protocols are first applied to recognize very closely related domains already classified in CATH. Pair-wise sequence search protocols are employed first (e.g., BLAST, Smith–Waterman (SSEARCH)) and subsequently the more computationally expensive HMM and HMM–HMM-based methods (e.g., HHblits (Remmert et al., 2011), COMPASS (Sadreyev and Grishin, 2003), (Profile Comparer) PRC (Madera, 2008)) are used to identify more remote homologs.

Structure comparison methods are then used (i.e., DALI, CATHEDRAL, SSAP) to recognize more distant homologs. Again statistically significant similarity is required and extensive manual curation may be required in the case of remote homologs. For classification into an existing homologous superfamily two out of three of the following criteria are usually sought (i) significant sequence similarity, (ii) significant structure similarity, and (iii) evidence of functional similarity, for example, similar molecular function annotations in the enzyme commission (EC) or gene ontology (GO) resources or conservation of catalytic residue or cofactor binding residues in the active site. Similarity in rare structural motifs can also be an indication of homology.

A distant homology between different domain families may well be overlooked at the time of structure determination or initial classification. This could be due to the very different sizes of their representative structures, the absence at that time of related proteins with similar folds, missing evidence of their functional similarity, etc. As new evidence is discovered, the families may be reclassified into the same superfamily. For example, the structurally and functionally characterized *N*-deoxyribosyltransferase and Adenosine diphosphate ADP-ribosyl cyclase-like families had been initially classified in SCOP into different folds, but eventually unified in the same superfamily a decade later (Andreeva et al., 2008) (Fig. 3.5).

As mentioned already, although structure tends to be much more highly conserved during evolution than sequence, there are examples of distant homologs being missed by structure comparison protocols because extensive divergence in structure has occurred. These very remote relationships can sometimes be detected via powerful HMM–HMM methods, which are sensitive enough to pick up sequence similarity in these sites (see Reid et al., 2007 for some examples). Although these relatives share similar domain cores, structural divergence outside

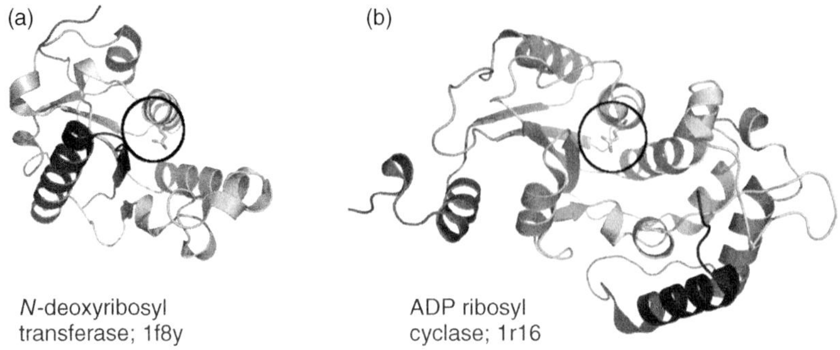

Figure 3.5 Functional evidence for distant homology of two proteins with moderate structural similarity. (a) *N*-deoxyribosyl transferase and (b) Adenosine diphosphate (ADP)-ribosyl cyclase share a similar catalytic mechanism, in which a catalytic glutamate residue (shown as circled stick) forms a covalent intermediate with the (deoxy)ribosyl group of the respective substrate. This invariant glutamate residue occupies the equivalent site in the common structural core of these enzymes that otherwise is relatively small and shared by many functionally unrelated proteins of a Flavodoxin-like fold.

this core is very dramatic and so that the structural similarity score falls below the thresholds used for homologue detection.

Sequence-based protocols are capable of recognizing a surprisingly high proportion of these very remote homologs. In CATH a recent machine learning protocol (CATHsvm (Cuff et al., 2009b)) combines information on sequence, structure, and functional similarity to increase the detection of these homologs. Information from the different detection methods (e.g., CATH–HMMscan, SSAP, CATHsvm) is presented to the curator via a web-based classification table, which summarizes all the data and provides links to relevant publications for further validation.

3.11.3 Fold Group Identification

In both SCOP and CATH, once a domain has been assigned to a particular superfamily, other descriptors in the classification (e.g., fold group, architecture, protein class) are usually automatically inherited. For domains that cannot be assigned to any existing superfamily the structural similarity to all representative domains in the classification is examined to determine whether the domain can be assigned to an existing fold group.

However, in both SCOP and CATH the fold group has become a rather broad structural "bin" in that although the relatives within a fold group should have similar topologies there are no widely accepted quantitative thresholds that can be applied to automatically assign domain structures to the same fold group. When the structural classifications first emerged and contained relatively few structures (i.e., ~3000 nonredundant domains) it was relatively easy to assign domains to very distinct fold groups. However, as the classifications have expanded and some superfamilies have encompassed more and more structurally diverse homologs, some differing quite considerably in their structural characteristics outside the core, the fold groups to which they belong have also, necessarily broadened.

In CATH, the fold group is now assigned based on structural similarities in the cores of the domains. In other words new structures that cannot be assigned to any existing superfamily but show significant structural similarity to the core regions of one or more relatives in a classified fold group, are assigned to that fold group.

A significant proportion of newly determined structures can now be assigned to an existing fold group (e.g., >98% in the last three releases of CATH). However, recent analyses of the classifications have shown that some fold groups often have quite large structural motifs in common with domains in other fold groups (Cuff et al., 2009a, 2009b). This is largely due to the existence of recurrent supersecondary motifs in domains within a particular structural class.

In some cases, fold groups have been reclassified or split to redress this problem and ensure that all relatives share significant degrees of structural similarity. CATH also provides structural neighborhood information, generated by scanning all new domain structures against nonredundant representatives from each CATH superfamily, to highlight similarities between fold groups.

In this sense the fold group can be considered as a more subjective level in the classification than the superfamily, grouping together domains that have similar topologies in their core for the purposes of identifying stable arrangements of secondary structures. Comparison of the SCOP and CATH classifications (see Hadley and Jones (1999), Csaba et al. (2009), and below) have shown more substantial differences in fold assignments than in classifications at the superfamily level, because of the somewhat arbitrary thresholds used to define fold similarities.

3.11.4 Handling Fold Divergence in Homologous Superfamilies

Extreme structural divergence between very remote homologs in some superfamilies mean that they no longer share the same fold—even by the loose definition of fold described above. Figure 3.6 shows relatives from the P-loop hydrolase superfamily where the structural overlap between some domains (i.e., the number of residues superposing below an RMSD of 3Å) is less than 30% of the larger domain. Even more dramatic changes in 'fold' have been observed (see Krishna and Grishin, 2005).

However, while the fold group is clearly a more subjective assignment, given the lack of quantitative criteria in identifying fold similarities, there are clearly some advantages to retaining some description of fold group. Many of the superfamilies do adopt very discrete folds and so the identification of these "fold" bins can help in understanding folding characteristics of proteins. In addition, this further organization of superfamilies makes it easier to identify those regions of structure space where proteins cluster by virtue of common folding motifs. Furthermore, by using a simple numbering scheme to label each level in the

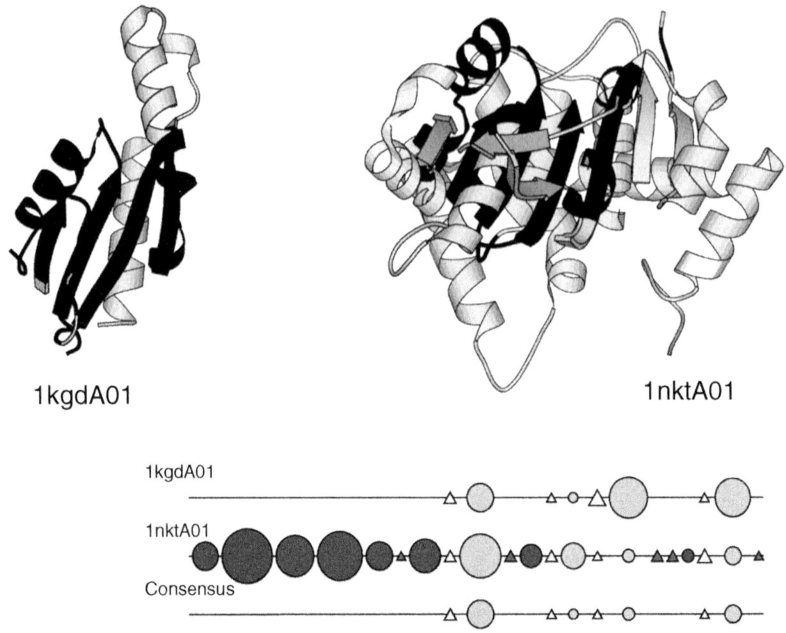

Figure 3.6 Structural diversity of two P-loop nucleotide hydrolase domains. The conserved structural core is highlighted in black. Gray regions indicate structural domains specific to that particular domain. The corresponding 2DSEC plot clearly shows the secondary structures common to both domains.

hierarchy it is very easy to extract relatives from within a particular fold group or class for structural analyses.

Both SCOP and CATH attempt to address the problems of fold variation within a superfamily and overlapping fold groups by providing information on the structural neighbors to each domain. In CATH this means reporting structural matches to domains, which are above the level of structural similarity deemed significant (SSAP score of 70 or normalized RMSD less than 3.5 Å over 60% of the domain). This identifies structural overlaps between domains in different fold groups. In SCOP there are plans to segregate structural and probable evolutionary relationships that would make fold an attribute of a given evolutionary lineage rather than a hierarchical level. Whether the fold group should be dropped completely and structure space more appropriately be viewed as a continuum is still a moot point, though it has been proposed by several researchers (Sippl et al., 2008; Kolodny et al., 2006) and is discussed further below.

3.11.5 Subclassification of Homologous Superfamilies into Functional Families

As the classifications became more highly populated with diverse relatives, and as more functional data became available from public resources such as EC,

(Kyoto Encyclopedia of Genes and Genomes) KEGG, and GO it became clear that structural divergence could be accompanied by significant divergence in function, during evolution. Many reviews describe the functional changes occurring in particular superfamilies and several global studies have reported the general trends observed and the structural mechanisms by which functional changes can be mediated. While loss or change in function can arise from mutations in key functional residues, for example, catalytic residues in the active site, changes can also be brought about by more dramatic changes in the structure.

For example, structural embellishments comprising secondary structure elements commonly occur in loops connecting core secondary structures, but come together in 3D and can change the geometry around the active site altering binding specificities (Dessailly et al., 2010). Structural embellishments can also change surface features enabling different protein interactions and assemblies (see Chapter 7 for details of structural changes in the HUP superfamily). These changes are most frequently observed between paralogous domains (i.e., relatives arising through domain duplication) although some recent studies have reported functional divergence between orthologous (i.e., relatives arising through speciation), too (Studer et al., 2008; Studer and Robinson-Rechavi, 2009).

In SCOP, relatives within very populous and divergent superfamilies are further classified into families according to functional similarity as reported in the literature and publicly available functional annotation databases (e.g., Swiss-Prot, EC, GO, KEGG, etc). This analysis is done manually and the extent of functional subclassification varies between superfamilies depending on the information available. Division of smaller SCOP superfamilies into families are usually sequence based. Families are further subclassified into proteins of distinct known functions. Interestingly, homologues from different species (orthologs from different organisms) normally share greater structural similarity of their active sites than family members of different functions (paralogs) do, even if their overall sequence similarity may well be less than average for the family.

In CATH automated protocols are employed to subclassify superfamilies by using all the sequence information available for a particular superfamily. Sequence relatives for each CATH superfamily are identified in UniProt and other sequence repositories using HMM protocols (see Chapter 4 for more details). This expands the size of the classification 150-fold and can significantly increase the amount of experimental functional information available in some superfamilies.

Once sequence relatives have been assigned to each superfamily, functional families are identified by performing agglomerative clustering of domain relatives according to their sequence similarity (Lee et al., 2010). Relatives are first clustered on pair-wise sequence similarity (i.e., 80% sequence identity and above). Subsequently, multiple sequence alignments of each sequence cluster are used to derive HMMs and HMM–HMM comparisons used to recognize clusters having statistically significant similarity. Experimental functional annotations from GO (preferably molecular function), where available, are also used to guide the clustering and terminate the process before functionally variable clusters are merged (Rentzsch and Orengo, 2013).

Identifying robust criteria for obtaining pure functional families (termed *FunFams*) is not trivial and cautious thresholds are currently used that often result in the generation of multiple subfamilies having highly similar functions. The FunFam protocol attempts to recognize those domain relatives performing similar functional roles in parent sequences, with diverse GO terms, as these should be kept within the same cluster, but there are considerable challenges associated with this. While assessment of the purity of the FunFams for function assignment, in a recent Critical Assessment of protein Function Annotation (CAFA) experiment (Radivojac et al., 2013), demonstrated that they are competitive with other purely homology-based approaches, it was clear that much progress is still needed in function prediction and no methods have solved this problem yet.

Information on functional families (FunFams) is shown in dedicated web pages for each superfamily on the CATH website. This includes GO molecular functions for relatives in each FunFam and visualizations of representative structures highlighted with conserved residue sites, identified by the entropy-based Scorecons algorithm (Valdar, 2002) (see Fig. 3.7 for an example). Relatives are also linked through to information on the protein networks they participate in, presented on associated Gene3D web pages (see chapter 4 for more details).

3.12 POPULATION OF THE HIERARCHY

Table 3.2 shows the number of different entries at each level of the SCOP and CATH classifications (note that SCOP does not have a formal architecture level and so no numbers are cited for SCOP architecture, although all the most highly populated architectures in CATH have similar labels—for example, beta sandwich—in SCOP).

CATH identifies 40 different architectures, of which the five most highly populated account for nearly 50% of domain sequences in CATH, are shown in Fig. 3.8. There are currently more than 2500 (1900) domain superfamilies in CATH (SCOP) and more than 1000 fold groups in both hierarchies. However, Fig. 3.4 shows that these groupings are not evenly populated and follow a power law as described in several reviews of the data (Qian et al., 2001; Chothia et al., 2003). This means that the majority of domain superfamilies comprise few relatives while the top 5% (i.e., ~100 superfamilies) are very highly populated accounting for nearly 60% of the sequences in CATH. Of the most highly populated fold groups (Rossmann, TIM barrel, immunoglobulin, OB fold, globin, Jelly roll, trefoil) the Rossmann, TIM and immunoglobulin folds account for over 20% of all domain sequences and have been the "top" largest fold groups as early analyses of CATH and SCOP in the 1990s identified these "superfolds" (Orengo et al., 1994).

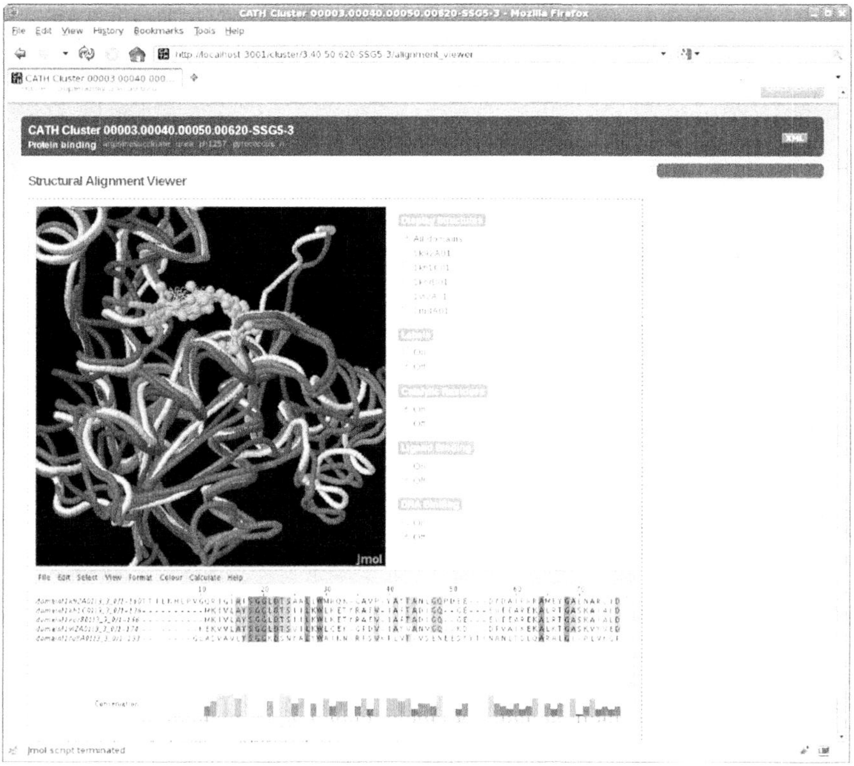

Figure 3.7 Screenshot showing, for a particular functional family in CATH, superpositions of representative structures, annotated with catalytic residues the multiple structure alignment giving information of residue conservation.

3.13 COMPARISONS BETWEEN SCOP AND CATH

In order to compare resources it is first necessary to decide on thresholds for agreement. Recent analyses conducted by the developers of SCOP and CATH used a range of thresholds to infer agreement in domain assignments and superfamily classifications. Using relatively conservative thresholds (60% of the domain residues should overlap to identify a common domain and 60% of nonredundant relatives (at 35% sequence identity) should overlap between two superfamilies or two-fold groups, then there is agreement between the two resources for nearly 70% of the superfamilies. There is much less agreement at the fold level but this is not surprising given the lack of any quantitative or evolutionary definition of fold similarity. Perhaps more surprising is the lack of agreement between domain boundary assignments in the two resources, which probably stems from arbitrary decisions to leave or remove long linker regions between domains or to only chop domains found in multiple contexts.

TABLE 3.2 Statistics from the CATH (v3.4) and SCOP (v1.75) Classifications

CATH Hierarchy	Number of CATH Domain Representatives	Number of SCOP Domain Representatives	SCOP Hierarchy
Class	4	7	Class
Architecture	40		
Topology or fold	1282	1195	Fold
Homologous superfamily	2549	1962	Superfamily
Functional families (all predicted sequences)	47039	3902	Family (structures only)
Total number of domains	152920	110800	Total number of domains of known structure

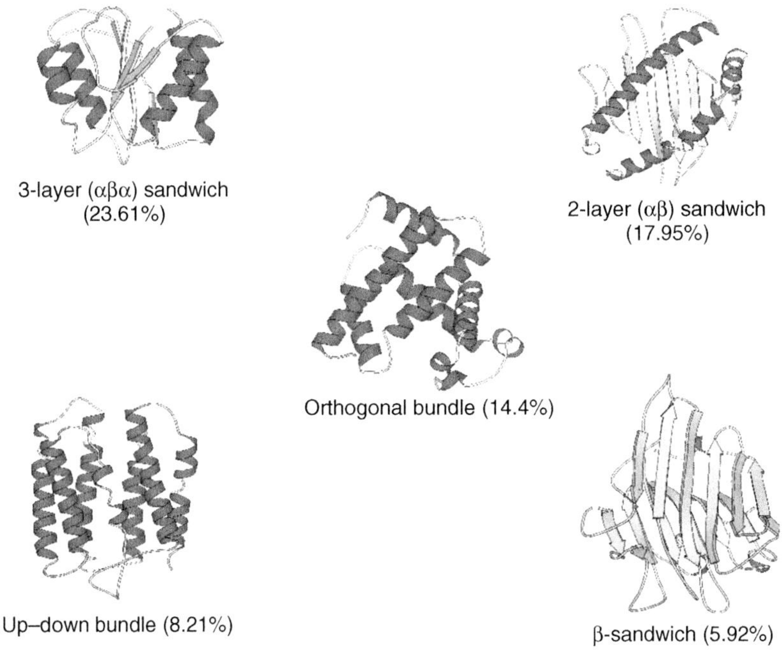

Figure 3.8 The most highly populated architectures in CATH. The percentages correspond to the proportion of structures predicted to share this architecture in the genomes.

These analyses largely agree with earlier ones (Hadley and Jones, 1999) that have analyzed the different types of disagreement between the resources and more recent approaches (Csaba et al., 2009). Csaba and colleagues employed a graph-based SCOP–CATH mapping method and a higher overlap threshold of 80% and found that around two thirds of the superfamilies in SCOP could be mapped to an equivalent homologous superfamily in CATH.

A recent collaboration between SCOP and CATH aims to regularly identify similarities and differences between the resources and present these on a common portal, GENOME3D, which provides links to both and which also links to structure predictions for the two resources generated by SUPERFAMILY, Gene3D, PHYRE and the Genomic Threading Database (GTD) for some key model organisms.

3.14 HIERARCHICAL CLASSIFICATIONS VERSUS STRUCTURAL CONTINUUM

There has been considerable speculation on the value of employing "hierarchical" domain classifications. Certainly there are problems in meaningfully assigning all domains to strict class, architecture and fold levels as these levels are not defined by clear, quantitative criteria and the assignments are very subjective giving ambiguous assignments in the two resources (Hadley and Jones, 1999). The superfamily level has the clearest biological meaning as objective, quantitative and statistical criteria are used to identify homologous relationships. The level of errors associated with superfamily assignments are unlikely to be significantly different from sequence-based domain resources (i.e., Pfam, PRINTS, SMART) and may even be less with regard to the capture of distant relationships as the structural information clearly helps in tracing further back in evolution.

However, the fold group is clearly problematic and has become increasingly so as faster structure comparison methods enabled relationships between all structures to be explored. Sippl, Honig, and others have given compelling reasons for discounting this level as there are many structural overlaps between fold groups arising from similarities in large supersecondary motifs—particularly where these motifs constitute a significant proportion of the domain (see example in Fig. 3.9).

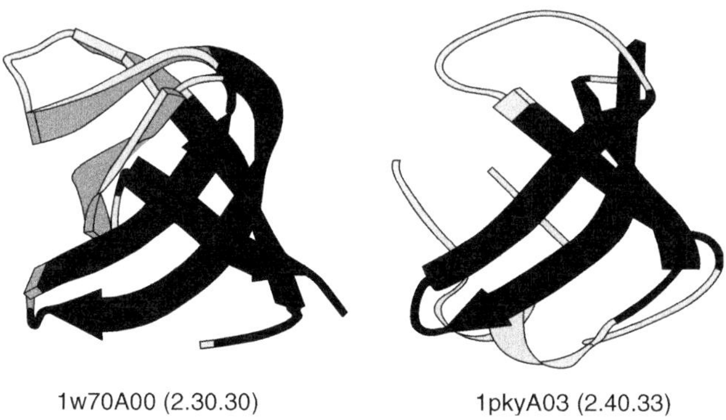

Figure 3.9 Structural overlap involving two domains, one possessing a β-roll and the other a β-barrel architecture. The region of structural overlap is shown in black and is 65%.

However, recent analyses (Cuff et al., 2009a, 2009b) showed that many superfamilies in CATH (86%) have folds that are shared by all relatives in the superfamily and are distinct from any other fold in CATH. The problem largely arises from some highly populated superfamilies which have diverged significantly in structure during evolution and, as mentioned already above, in which some relatives no longer share similar folds over their whole structures. However, they do retain significant fold similarity if only the cores of the domains are considered (see Fig. 3.1). Furthermore, those superfamilies and fold groups overlapping because of common motifs constitute less than 10% of the superfamilies and fold groups in CATH. However, they do contain at least half of all nonredundant CATH entries. For these regions of fold space there is more of a structural continuum linking fold groups but other parts of fold space comprise more discrete islands (see Fig. 3.10).

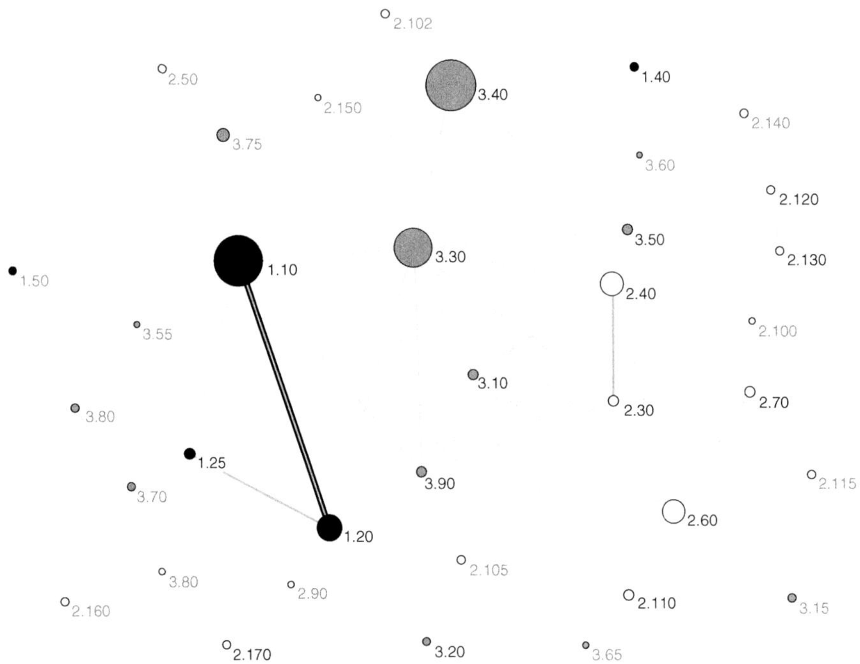

Figure 3.10 Plot showing the extent of structural overlap between different CATH architectures. Each point represents a CATH architecture code. The thickness of the lines represents the number of overlapping superfamilies between the architectures. The size of the circles correspond to the number of sequence clusters (at 35% sequence identity) in that architecture. Those architectures shown to overlap with at least one other in the CATH database are labeled as follows: 1.10 = α-orthogonal, 1.20 = α-up–down bundle, 1.25 = α-horseshoe, 2.30 = β-roll, 2.40 = β-barrel, 2.60 = β-sandwich, 2.70 = distorted β sandwich, 2.120 = β-6-propellor, 2.130 = β-7-propellor, 3.10 = $\alpha\beta$-roll, 3.30 = 2-layer $\alpha\beta$-sandwich, 3.40 = 3-layer($\alpha\beta\alpha$) sandwich, 3.50 = 3-layer ($\beta\beta\alpha$) sandwich, 3.70 = $\alpha\beta$-box, 3.80 = $\alpha\beta$-horseshoe, and 3.90 = $\alpha\beta$ complex.

3.15 WEBSITES

Both SCOP and CATH websites allow users to search the resources with query PDB codes or UniProt sequences. In addition, CATH provides the CATHEDRAL web server (Redfern et al., 2007) that allows users to search with PDB coordinates for query structures and returns a list of significant structural matches. The classifications can be browsed to see representative structures from each superfamily and fold group.

In CATH, structures are linked through to PDBsum (Laskowski, 2009) which shows a summary of derived information and multiple links (e.g., to functional resources such as GO, Pfam) for a given PDB entry. Information on domain sequence relatives for CATH superfamilies is now integrated from Gene3D (see Chapter 4) to provide more experimental annotations on protein functions and multiple sequence alignments. As mentioned already this functional information is also provided as dedicated FunFam pages (see Fig. 3.7).

For highly populated enzyme superfamilies in CATH there is a link through to the FunTree analysis (see Fig. 3.11 for example). FunTree is a new resource, developed in a collaboration between the Thornton and Orengo groups (Furnham et al., 2012). It uses multiple structure alignments from CATH and domain sequence relatives from Gene3D to build a phylogenetic tree that displays the emergence of different functions in the superfamily during evolution. Relatives are annotated with functional information on cofactors and substrates bound and on the chemistry performed by the different relatives in the superfamily. FunTree also provides information on the multidomain contexts of the different relatives.

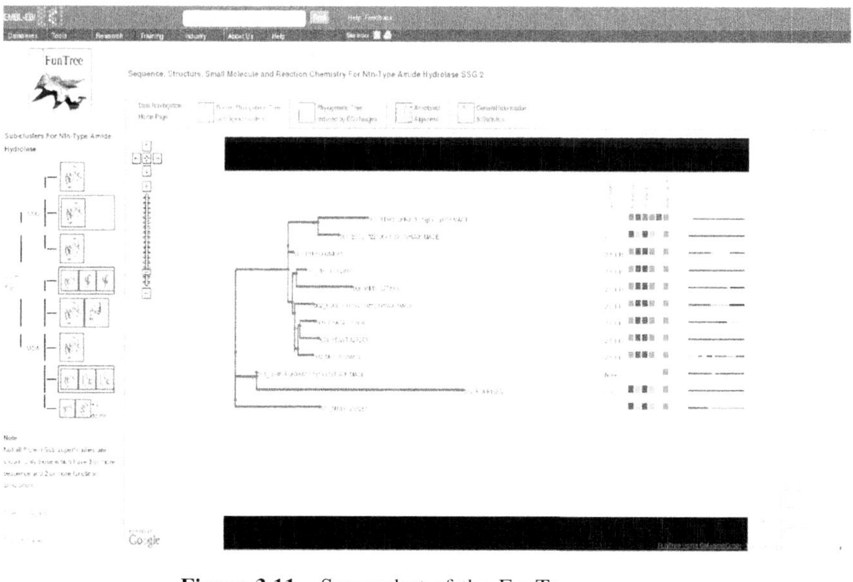

Figure 3.11 Screenshot of the FunTree resource.

REFERENCES

Alden, K. et al. (2010) dConsensus: a tool for displaying domain assignments by multiple structure-based algorithms and for construction of a consensus assignment. BMC Bioinformatics, 11, 310.

Alexandrov, N. and Shindyalov, I. (2003) PDP: protein domain parser. Bioinformatics, 19, 429–430.

Altman, R. and Gerstein, M. (1994) Finding an average core structure: application to the globins. Proc Second IntConf Intell Sys Mol Biol, 2, 19–27 Menlo Park (CA): AAAI Press.

Andreeva, A., Prlić, A., Hubbard, T.J., and Murzin, A.G. (2007) SISYPHUS–structural alignments for proteins with non-trivial relationships. Nucleic Acids Res, 35(Database issue), D253–D259.

Andreeva, A., Howorth, D., Chandonia, J.-M., Brenner, S.E., Hubbard, T.J.P., Chothia, C., and Murzin, A.G. (2008) Data growth and its impact on the SCOP database: new developments. Nucleic Acids Res, 36, D419–D425.

Attwood, T.K et al. (2003) "PRINTS and its automatic supplement, prePRINTS." Nucleic Acids Res 31(1), 400–402.

Bernstein, F.C, Koetzle, T.F, Williams, G.J, Meyer Jr., E.E, Brice, M.D, Rodgers, J.R, Kennard, O, Shimanouchi, T, Tasumi, M (1977) The Protein Data Bank: a computer-based archival file for macromolecular structures. J Mol Biol, 112: 535.

Björklund, A.K. et al. (2006) Expansion of Protein Domain Repeats. PLoS Comput Biol, 2(8), e114.

Björklund A.K. et al. (2005) Domain rearrangements in protein evolution. J Mol Biol 353(4): 911–923

Bron, C. and Kerbosch, J. (1973) Algorithm 457: finding all cliques of an undirected graph. Commun ACM, 16, 575–579.

Brown, N.P., Orengo, C.A., and Taylor, W.R. (1996) A protein structure comparison methodology. Comput Chem, 20(3), 359–380.

Chandonia, J.M. and Brenner, S.E. (2006) The impact of structural genomics: expectations and outcomes. Science, 311, 347–351.

Csaba, G., Birzele, F., and Zimmer, R. (2009) Systematic comparison of SCOP and CATH: a new gold standard for protein structure analysis. BMC Struct Biol, 17(9), 23.

Chothia, C. and Lesk, A.M. (1986) The relation between the divergence of sequence and structure in proteins. EMBO J, 5, 823–826.

Chothia, C., Gough, J., Vogel, C., and Teichmann, S.A. (2003) Evolution of the Protein Repertoire. Science, 300, 1701–1703.

Cuff, A. et al. (2009a) The CATH Hierarchy Revisited—Structural Divergence in Domain Superfamilies and the Continuity of Fold Space. Structure, 17, 1051–1062.

Cuff, A.L., Sillitoe, I., Lewis, T., Redfern, O.C., Garratt, R., Thornton, J., and Orengo, C.A. (2009b) The CATH classification revisited–architectures reviewed and new ways to characterize structural divergence in superfamilies. Nucleic Acids Res, 37(Database issue), D310–D314.

Dessailly, B.H., Redfern, O.C., Cuff, A.L., and Orengo, C.A. (2010) Detailed analysis of function divergence in a large and diverse domain superfamily: toward a refined protocol of function classification. Structure, 18(11), 1522–1535.

Finn R.D. et al. (2011)., HMMER web server: interactive sequence similarity searching. Nucleic Acids Res Web Server Issue 39:W29–W37.

Friedberg, I. and Godzik, A. (2005) Fragnostic: walking through protein structure space. Nucleic Acids Res, 33, W249–W251.

Furnham, N., Sillitoe, I., Holliday, G.L., Cuff, A.L., Laskowski, R.A. et al. (2012) Exploring the evolution of novel enzyme functions within structurally defined protein superfamilies. PLoS Comput Biol, 8(3), e1002403.

Guo, J., Xu, D., Kim, D., and Xu, Y. (2003) Improving the performance of DomainParser for structural domain partition using neural network. Nucleic Acids Res, 31, 944–952.

Hadley, C. and Jones, D.T. (1999) A systematic comparison of protein structure classifications: SCOP, CATH and FSSP. Structure, 7(9), 1099–1112.

Harrison A et al. (2003) Recognizing the fold of a protein structure. Bioinformatics 19(14):1748–1759.

Holland, T.A., Veretnik, S., Shindyalov, I.N., and Bourne, P.E. (2006) Partitioning protein structures into domains: why is it so difficult? J Mol Biol, 361(3), 562–590.

Holm, L. and Sander, C. (1998) Dictionary of recurrent domains in protein structures. Proteins, 33, 88–96.

Holm, L. and Sander, C. (1994) Parser for protein folding units. Proteins Struct Funct Genet, 19, 256–268.

Hunter, S. et al. (2011) InterPro in 2011: new developments in the family and domain prediction database. Nucleic Acids Res, 40(Database issue), D306–D312.

Jones, D. et al. (1998) Domain assignment for protein structures using a consensus approach: Characterisation and analysis. Protein Sci, 7, 233–242.

Jones, D.T. (1999) GenTHREADER: an efficient and reliable protein fold recognition method for genomic sequences. J Mol Biol, 287, 797–815.

Koczyk, G. and Berezovsky, I.N. (2008) Domain Hierarchy and closed Loops (DHcL): a server for exploring hierarchy of protein domain structure. Nucleic Acids Res, 36(Web Server issue), W239–W245.

Koehl, P (2006) Protein Structure Classification. Chapter 1, Rev Comput Chem. John Wiley and Sons.

Kolodny, R., Koehl, P., and Levitt, M. (2005) Comprehensive evaluation of protein structure alignment methods: scoring by geometric measures. J Mol Biol, 346(4), 1173–1188.

Kolodny, R., Petrey, D., and Honig, B. (2006) Protein structure comparison: implications for the nature of 'fold space', and structure and function prediction. Curr Opin Struct Biol, 16(3), 393–398.

Krishna, S.S. and Grishin, N.V. (2005) Structural drift: a possible path to protein fold change. Bioinformatics, 21, 1308–1310.

Laskowski, R.A. (2009) PDBsum new things. Nucleic Acids Res, 37, D355–D359.

Lee, D. et al. (2003) A structural perspective on genome evolution. Curr Opin Struct Biol, 13, 359–369.

Lee, D., Grant, A., Marsden, R.L., and Orengo, C. (2005) Identification and distribution of protein families in 120 completed genomes using Gene3D. Proteins, 59(3), 603–615.

Lees, J., Yeats, C., Perkins, J., Sillitoe, I., Rentzch, R., Dessailly, B.H., and Orengo, C. (2012) Gene3D: a domain-based resource for comparative genomics, functional annotation and protein network analysis. Nucleic Acids Res, 40(D1), D465–D471.

Lesk, A.M and Chothia, C 1980a How different amino acid sequences determine similar protein structures: The structure and evolutionary dynamics of the globins. J Mol Biol, 136 225–270.

Lesk, A.M. and Chothia, C. (1982) Evolution of proteins formed by beta-sheets. II The core of the immunoglobulin domains. J Mol Biol, 160, 325–342.

Letunic, I., Doerks, T., and Bork, P. (2012) SMART 7: recent updates to the protein domain annotation resource. Nucleic Acids Res, (Database issue), D302–D305.

Levitt, M and Gerstein, M , (1998) "A unified statistical framework for sequence comparison and structure comparison" 95. 11, 5913–5920.

Lima, T. et al. (2009) HAMAP: a database of completely sequenced microbial proteome sets and manually curated microbial protein families in UniProtKB/Swiss-Prot. Nucleic Acids Res, 37(Database issue), D471–D478.

Madera, M. (2008) Profile Comparer: a program for scoring and aligning profile hidden Markov models. Bioinformatics, 24(22), 2630–2631.

Madej, T., Gibrat, J.F., and Bryant, S.H. (1995) Threading a database of protein cores. Proteins, 23(3), 356–369.

Manning, G., Plowman, G.D., Hunter, T., and Sudarsanam, S. (2002) Evolution of protein kinase signaling from yeast to man. Trends Biochem Sci, 27(10), 514–520.

Marchler-Bauer, A. et al. (2011) CDD: a Conserved Domain Database for the functional annotation of proteins. Nucleic Acids Res, 39(Database issue), 225–229.

Mizuguchi, K., Deane, C.M., Blundell, T.L., and Overington, J.P. (1998) HOMSTRAD: a database of protein structure alignments for homologous families. Protein Sci, 7(11), 2469–2471.

Moutevelis, E. and Woolfson, D.N. (2009) A periodic table of coiled-coil protein structures. J Mol Biol, 385(3), 726–732.

Murzin, A.G., Brenner, S.E., Hubbard, T., and Chothia, C. (1995) SCOP: a structural classification of proteins database for the investigation of sequences and structures. J Mol Biol, 247, 536–540.

Neumann S. et al. (2011)., Camps 2.0: exploring the sequence and structure space of prokaryotic, eukaryotic, and viral membrane proteins. Proteins Struct Funct Bioinformatics, 80, 839–857.

Orengo, C.A., Jones, D.T., and Thornton, J. (1994) Protein superfamilles and domain superfolds. Nature, 372, 631–634.

Orengo, C.A., Michie, A.D., Jones, S., Jones, D.T., Swindells, M.B., and Thornton, J.M. (1997) CATH - a hierarchic classification of protein domain structures. Structure, 5, 1093–1108.

Orengo, C.A. (1999) CORA–topological fingerprints for protein structural families. Protein Sci, 8(4), 699–715.

Pegg SC, Brown SD, Ojha S, Seffernick J, Meng EC, Morris JH, Chang PJ, Huang CC, Ferrin TE, Babbitt PC (2006) Leveraging enzyme structure-function relationships for functional inference and experimental design: the structure-function linkage database. Biochemistry 45(8):2545–2555

Phillips, D.C. (1970) The development of crystallographic enzymology. Biochem Soc Symp Volume 30, 11–28

Punta, M. et al. (2012) The Pfam protein families database. Nucleic Acids Res, 40(Database issue), D290–D230.

Qian, J., Luscombe, N.M., and Gerstein, M. (2001) Protein family and fold occurrence in genomes: power-law behaviour and evolutionary model. J Mol Biol, 313, 673–681.

Radivojac, P. et al. (2013) A large scale evaluation of computational protein function prediction. Nature Methods, 10, 221–227.

Redfern O.C. et al. (2007) CATHEDRAL: a fast and effective algorithm to predict folds and domain boundaries from multidomain protein structures. PLoS Comput Biol 3(11):e232.

Redfern, O. et al. (2009) FLORA: A novel method to predict protein function from structure in diverse superfamilies. PLos Comput Biol, 5(8).

Reeves, G.A., Dallman, T.J., Redfern, O.C., Akpor, A., and Orengo, C.A. (2006) Structural diversity of domain superfamilies in the CATH database. J Mol Biol, 360(3), 725–741.

Reid, A.J., Yeats, C., and Orengo, C.A. (2007) Methods of remote homology detection can be combined to increase coverage by 10% in the midnight zone. Bioinformatics, 23(18), 2353–2360.

Remmert, M., Biegert, A., Hauser, A., and Söding, J. (2011) HHblits: lightning-fast iterative protein sequence searching by HMM-HMM alignment. Nat Methods, 9(2), 173–175.

Rentzsch, R. and Orengo, C. (2013) A. Protein function prediction using domain families. BMC Bioinformatics, 14 (Suppl 3), S5.

Rossmann, M.G. and Argos, P. (1976) Exploring structural homology of proteins. J Mol Biol, 105, 75–96.

Russell, R.B. and Barton, G.J. (1992) Multiple protein sequence alignment from tertiary structure comparison. Proteins Struct Funct Genet, 14, 309–323.

Sadreyev, R. and Grishin, N. (2003) COMPASS: a tool for comparison of multiple protein alignments with assessment of statistical significance. J Mol Biol, 326, 317–336.

Sali, A. and Blundell, T.L. (1990) Definition of general topological equivalence in protein structures. A procedure involving comparison of properties and relationships through simulated annealing and dynamic programming. J Mol Biol, 212(2), 403–428.

Siddiqui, A.S. and Barton, G.J. (1995) Continuous and discontinous domains: an algorithm for the automatic generation of reliable protein domain definitions. Protein Sci, 4, 872–884.

Siddiqui, A.S., Dengler, U., and Barton, G.J. (2001) 3Dee: a database of protein structural domains. Bioinformatics, 17, 200–201.

Sippl, M.J., Suhrer, S.J., Gruber, M., and Wiederstein, M. (2008b) A discrete view on fold space. Bioinformatics, 24(6), 870–871.

Studer, R.A., Penel, S., Duret, L., and Robinson-Rechavi, M. (2008) Pervasive positive selection on duplicated and nonduplicated vertebrate protein coding genes. Genome Res, 18(9), 1393–1402.

Studer, R.A. and Robinson-Rechavi, M. (2009) How confident can we be that orthologs are similar, but paralogs differ? Trends Genet, 25(5), 210–216.

Subbiah, S., Laurents, D.V., and Levitt, M. (1993) Structural similarity of DNA-binding domains of bacteriophage repressors and the globin core. Curr Biol, 3, 141–148.

Swindells, M.B. (1995) A procedure for detecting structural domains in proteins. Protein Sci, 4, 103–112.

Taylor, W.R. and Orengo, C.A. (1989) Protein structure alignment. J Mol Biol, 208, 1–22.

Teichmann, S.A., Park, J., and Chothia, C. (1998) Structural assignments to the Mycoplasma genitalium proteins show extensive gene duplications and domain rearrangements. Proc Natl Acad Sci USA, 95, 14658–14663.

Todd, A.E., Marsden, R.L., Thornton, J.M., and Orengo, C.A. (2005) Progress of structural genomics initiatives: an analysis of solved target structures. J Mol Biol, 348, 1235–1260.

Valdar, W.S. (2002) Scoring residue conservation. Proteins, 48(2), 227–241.

Widmann M et al. (2010) The Thiamine diphosphate dependent Enzyme Engineering Database: A tool for the systematic analysis of sequence and structure relations. BMC Biochem. 11 : 9

Wilson, D., Pethica, R., Zhou, Y., Talbot, C., Vogel, C., Madera, M., Chothia, C., and Gough, J. (2009) SUPERFAMILY–sophisticated comparative genomics, data mining, visualization and phylogeny. Nucleic Acids Res, 37(Database issue), D380–D386.

Zhou, H., Xue, B., and Zhou, Y. (2007) DDOMAIN: Dividing structures into domains using a normalized domain-domain interaction profile. Protein Sci, 2007(16), 947–955.

4

STRUCTURAL ANNOTATIONS OF GENOMES WITH SUPERFAMILY AND Gene3D

JULIAN GOUGH
Department of Computer Science, University of Bristol, Bristol, UK

CORIN YEATS AND CHRISTINE ORENGO
Institute of Structural and Molecular Biology, University College London, London, UK

CHAPTER SUMMARY

The two major protein classifications are Class, Architecture, Topology, Homology (CATH) and Structural Classification Of Proteins (SCOP), which take 3D atomic resolution structures from the Protein Data Bank (PDB), divide them into one or more domains and group them into a multi-level classification including, for example, families and superfamilies. As only a small minority of known protein sequences (less than 0.001%) have even part of their structure solved, both CATH and SCOP have sister resources, Gene3D, and SUPERFAMILY respectively; these databases detect and classify protein domains in sequences of unknown structure using hidden Markov models (HMMs).

Both resources are based on the same methodological principles despite independent development, however there are also significant technical differences. They both use the structure-based domain classifications to create multiple alignments of homologous sequences of known and unknown structure, representing each superfamily. These alignments are used to build

Protein Families: Relating Protein Sequence, Structure, and Function, First Edition.
Edited by Christine Orengo and Alex Bateman.

the HMMs, which in turn can be used to detect and classify CATH/SCOP domains in completely sequenced genomes and other sequence collections such as UniProt. The websites contain precomputed results for tens of millions of sequences with an interface for browsing, searching, comparing and visualizing the data individually or in a genomic/phylogenetic context.

4.1 INTRODUCTION

The two major structural domain classifications are SCOP (Murzin et al., 1995) and CATH (Orengo et al., 1997) and other notable structural resources are described in Chapter 3. This chapter is primarily about SUPERFAMILY (Gough, 2002) and Gene3D (Buchan et al., 2002), which provide protein sequence domain annotations as companion resources to SCOP and CATH, respectively. The two resources have been independently developed, however the two have some major elements in common. Although they share some concepts, the independently developed technical solutions to similar problems are often distinct, so the duplication of effort and the independent search for solutions has been a productive engine driving the technology forward, as well as allowing for increased confidence from agreement between the two resources.

The main thing that sets SUPERFAMILY and Gene3D apart from other sequence family databases is that they are both entirely based on a structural classification, and furthermore that this classification is inherited (from SCOP and CATH). Most other domain databases, such as Pfam (Bateman et al., 2000) as described in Chapter 2 use a sequence-based classification instead of structure where the derivation of their own classification is part of the process. In summary, the principle goal of SUPERFAMILY and Gene3D is the identification of domains of known structure in amino acid sequences of proteins of unknown structure and the mapping of a structural classification onto the domains identified in the sequence.

4.2 THE IMPORTANCE OF BEING HIGH THROUGHPUT

An essential quality of this process is the ability to carry out the domain assignments in a high throughput manner. This is the ability not only to annotate whole genomes, but all genomes and other large scale sequence collections. Also of great importance is to update the annotations as new genomes are sequenced and previously sequenced genomes have their protein sequence sets updated and structural families are defined or expanded. The exponential growth in protein sequence data requires that any annotation system scales effectively as well. This enables large scale studies of protein evolution and detection of evolutionary trends regarding domain duplication and the emergence of diverse multi-domain architectures in different kingdoms of life. This quality sets the two structural annotation resources discussed in this chapter apart from the bulk of "structure prediction" methods of the kind that compete in the biennial Critical Assessment of Structure Prediction (CASP6) competition. These include threading methods such as GenThreader (Jones, 1999), Phyre (Kelley and Sternberg, 2009) and Robetta (Chivian et al., 2003).

SUPERFAMILY and Gene3D share some technology with many of the methods that compete successfully in CASP (most notably HMMs—see technology section for more on HMMs), where the focus is on a best guess prediction for a limited number of targets, often including input from a human expert. However, the principle focus for SUPERFAMILY/Gene3D is on high confidence annotations in large datasets (>60,000,000 sequences, ~5 orders of magnitude larger than that required for CASP). Despite this, SUPERFAMILY did take part as a fully-automatic method in the 2002 round of CASP for the purpose of demonstrating where the high throughput methods lie with respect to the best state-of-the-art predictors. Both the CASP-winning technology and the high throughput technology has come a long way since 2002 but the result at that time was that SUPERFAMILY performed overall somewhere in the middle-of-the-road and was able to produce the best prediction for at least one target. So there exist other methods that are superior in performance (Zhang et al., 2005), but conversely, the SUPERFAMILY/Gene3D approach is close in accuracy for all genomes to what is possible on small numbers of sequence targets.

4.3 THE USE OF STRUCTURAL INFORMATION

A philosophical point that is often raised is whether SUPERFAMILY and Gene3D use structural information when assigning domains to protein sequences. The short answer is "No, not directly"; they work by taking sequences of known structure and trying to detect sequence homology with genome (or other) sequences. The long answer is that, subtly, structural information is used via the classification. The structural classification allows the grouping of related sequences (and hence the sequence profiles used for homology detection) together to represent a single structural domain type ("family"). A particular advantage of using the structural classifications is that structural similarity is almost always easier to detect than sequence similarity. Furthermore the structure-based domain classification can be used for validation and quality control of sequence profiles, indirectly contributing to the performance. From an abstract point of view, the whole procedure is trained on the structural classification.

4.4 APPLICATIONS

The two resources take as input two sources of primary data: 3D structures via the classification system SCOP/CATH and the amino acid protein sequences from genome sequencing projects and large-scale repositories. From the inputs the two resources output secondary data: domain assignments in genomes and their classification. This secondary data is not limited in its applications but, based on evidence from citations, we can see that it is most commonly used by the scientific community in a handful of ways:

- In its simplest form, an annotation to a single protein sequence can be useful to somebody whose research is focused on that single protein of interest, be it for structural studies or molecular biology/biochemistry, for example, the MET(Gherardi et al., 2003) protein.
- On a larger scale, genome (or other) annotation projects routinely use the assignments, for example, in mouse (Okazaki et al., 2002), in combination with other methods such as BLAST to annotate their sequences.
- Looking beyond a single genome is perhaps where the assignments begin to show their power. Studying groups of assignments across many genomes is informative for investigations motivated by the goal of understanding the sequences containing domains belonging to, for example, the same structural fold in CATH, the same evolutionary superfamily in SCOP, or the same functional family (where this corresponds to a CATH/SCOP classification), etc. There have been many studies of this type, which use SUPERFAMILY/Gene3D, for example, globins (Vinogradov et al., 2005), and in fact this is replacing the historically more common (but less consistent or repeatable) situation where each study would attempt to do genome assignments themselves.
- The assignments have also been used for comparative genomics studies where all assignments for a set of genomes are compared across genomes (Abroi and Gough, 2011; Marsden et al., 2006).
- More recently, the assignments have been used, primarily as extensions to the resources themselves by the authors, for function prediction (Lees et al., 2012; de Lima Morais et al., 2011); this and other tools directly connected to the resources are discussed later.

What the authors had in mind when the resources were conceived were global studies across all domains and all genomes, because sequence in the context of organisms is essential to our understanding of protein structure and evolution. The authors have successfully applied their resources in this way to investigate structural evolution (Chothia et al., 2003), and to identify superfamilies that would have been present in the last universal common ancestor (Ranea et al., 2006), as have several others.

4.5 HISTORY

The brief history of the resources is as follows. The structure of Myoglobin was solved in 1958 (Kendrew et al., 1958) and the PDB (Berman et al., 2000) was established in 1971 with 7 structures. The Haemophilus influenza genome

(Fleischmann et al., 1995) was first sequenced in 1995, which was the same year that the SCOP and CATH classifications were publicly launched via the web and presented at the same conference, by which time the PDB had 3091 entries. SCOP was published in 1995 and CATH was published in 1997 by which time there were over 5000 structures available. In the year 2000, a CATH PSI-blast (Altschul et al., 1997) server was established (Pearl et al., 2000) and the HMM-based SUPERFAMILY server also appeared (Gough et al., 2000), published again in 2001 with 50 genomes (Gough et al., 2001). Gene3D was published (Buchan et al., 2002) in 2002 containing 36 genomes and using PSI-blast, then subsequently with HMMs in 2005 including 120 genomes (Lee et al., 2005). In 2005/2006 SUPERFAMILY/Gene3D joined the InterPro (Mulder et al., 2007) consortium, an umbrella resource linking to other nonstructural domain databases. At the time of writing in 2011 SUPERFAMILY contains 1774 cellular genomes and 119 meta-genomes. Gene3D principally focuses on Ensembl Genomes (Kersey et al., 2012) for comparative genomics, which currently contains 361 genomes, although this number is increasing rapidly now that it is expanding beyond its initial remit of vertebrate and model organism genome annotation. In addition it provides nonchromosomal sequences and extended genome annotations, for example, operon data. Both also contain genomes imported from the UniProt (Apweiler et al., 2004) "complete proteomes sets," which currently contains 1639 cellular genomes.

4.6 TECHNOLOGY

As mentioned in the introduction one of the most important underlying technologies used by the resources is the hidden Markov model (Barrett et al., 1997; Durbin et al., 1998; Eddy, 1996; Nielsen and Krogh, 1998) (HMM). To understand why HMMs are at the core of SUPERFAMILY and Gene3D we must first understand the challenges of the assignment problem and then what HMMs are, and why they are the most powerful methods that can be applied in this context. As explained, the most basic function of the SUPERFAMILY and Gene3D resources is to assign domains of known structure as defined by SCOP/CATH to protein sequences. This is achieved via the concept of homology. As protein domains descending from a common ancestor will share the same structure, and as this evolutionary descent takes place by a series of single mutations, there may be a detectable similarity in proteins of the same structure, that is, homology. Thus if similarity (homology) can be detected between a genome sequence and a sequence of a known structural domain, then it can be inferred that it has the structure of that domain. A detailed comparison of sequence homology detection methods can be seen in the benchmark papers by Park et al. (1998), Madera and Gough (2002), and Reid et al. (2007). However they can be summarized into three categories: pair-wise sequence comparison (Altschul et al., 1990; Biegert and Soeding, 2009; Smith and Waterman, 1981), sequence-profile comparison (Altschul et al., 1997; Durbin et al., 1998; Johnson et al., 2010; Karplus et al., 1998) and finally profile–profile (Madera, 2008; Söding, 2005) comparison software. The application and theory behind these methods are described in more detail in Chapters 5 and 1.

4.7 HIDDEN MARKOV MODELS

Pair-wise (sequence–sequence) comparison methods are fast and conceptually easy and thus have historically been popular. Pair-wise is most appropriate when searching for the single most similar match or closely related sequences, usually not using the transitivity of sequence homology. Sequence homology is transitive, that is, if there is detectable similarity when A is homologous to B and B is homologous to C, we can infer homology between A and C, even in the absence of any detectable similarity. There is a common pitfall with using transitivity called the "domain problem." Full length multi-domain sequences may make a match to other multi-domain sequences only in the region of one common domain, thus breaking transitivity via matches to other domains not present in the original sequence; this is one of the most common causes of mis-annotation of protein sequences. Domain-based classifications such as SCOP and CATH however lend themselves perfectly when making use of the power of transitivity, using profile methods, while also providing a means for minimizing this effect. With profile methods, related sequences are grouped together and represented by a single profile for the group; a sequence can be compared to the profile to detect similarity that implies shared homology to the group, in this case a domain family classified in SCOP/CATH. Figure 4.1 shows that profile methods can detect much more distant evolutionary relationships than pair-wise methods, and that HMMs perform the best. An additional advantage is that the computational expense scales roughly linearly with the number of sequences and the comparatively slower growing number of models; this contrasts with pairwise-based approaches, which scale quadratically with the number of sequences. Thus, both SUPERFAMILY and Gene3D consist of a library of HMMs representing all the domains in the classification. Profile–profile methods perform better still, and in principle the HMMs in the two resources can be used with profile–profile methods if you have a profile to search against the library. In practice this is not yet used for annotating all genomes because the step of converting even a single proteome to its representative set of profiles is computationally expensive.

4.8 BUILDING MODELS

A "profile" is a representation of a multiple sequence alignment (Fig. 4.2), in the case of HMMs using a rigorous statistical framework. To arrive at a multiple sequence alignment for a structural domain, an iterative search is conducted against a large database to populate the alignment with divergent sequences. Starting with the sequence of a domain of known structure, Gene3D uses Jackhmmer and SUPERFAMILY uses the automated SAM iterative

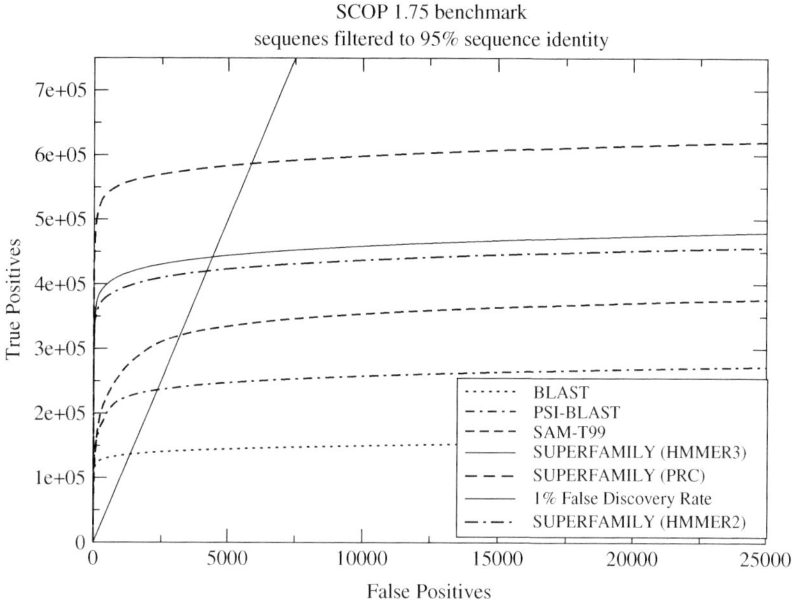

Figure 4.1 This is a plot of a "leave-one-out" benchmark comparing the sensitivity of sequence comparison methods. This was carried out by scoring the sequences of domains from SCOP filtered to 95% sequence identity against each other. A hit is classed as "true" if it belongs to the same SCOP superfamily and "false" if it belongs to a different SCOP fold. True/false hits are plotted cumulatively in descending order of confidence score. Thus it can be seen that the pair-wise method (BLAST) performs much less well than profile methods (PSI-BLAST and SAM-T99). It can also be seen that the curated SUPERFAMILY models perform better than those built by SAM-T99, and that on those models HMMER3 Johnson et al., 2010 scoring is superior to HMMER2 (Eddy, 1998) (same models as the SUPERFAMILY curve), and that the profile–profile method (PRC) performs better than the sequence-profile methods; the PRC curve was calculated using curated SUPERFAMILY models versus automatically-generated SAM-T99 models.

"target" procedure (Barrett et al., 1997), which starts using a pair-wise sequence comparison to collect closely related homologs from UniProt (Apweiler et al., 2004) to construct an initial alignment. The first model built from this alignment is then searched back against UniProt, and as the HMM is more sensitive than the pair-wise comparison, it finds new homologs that are added to the alignment. A new model is built and the process iterates a fixed number of times (various sequence filtering steps have been omitted from this description). The multiple sequence alignment from the ultimate round is used to build the HMM, which goes into the library, in the case of SUPERFAMILY using SAM, and in the case of Gene3D using HMMER3. Both then use the recently released HMMER3 for the scoring of genome sequences against the library of models. HMMER3 is slightly more sensitive and two orders of magnitude faster than the previous HMMER2 version or SAM scoring. Notable

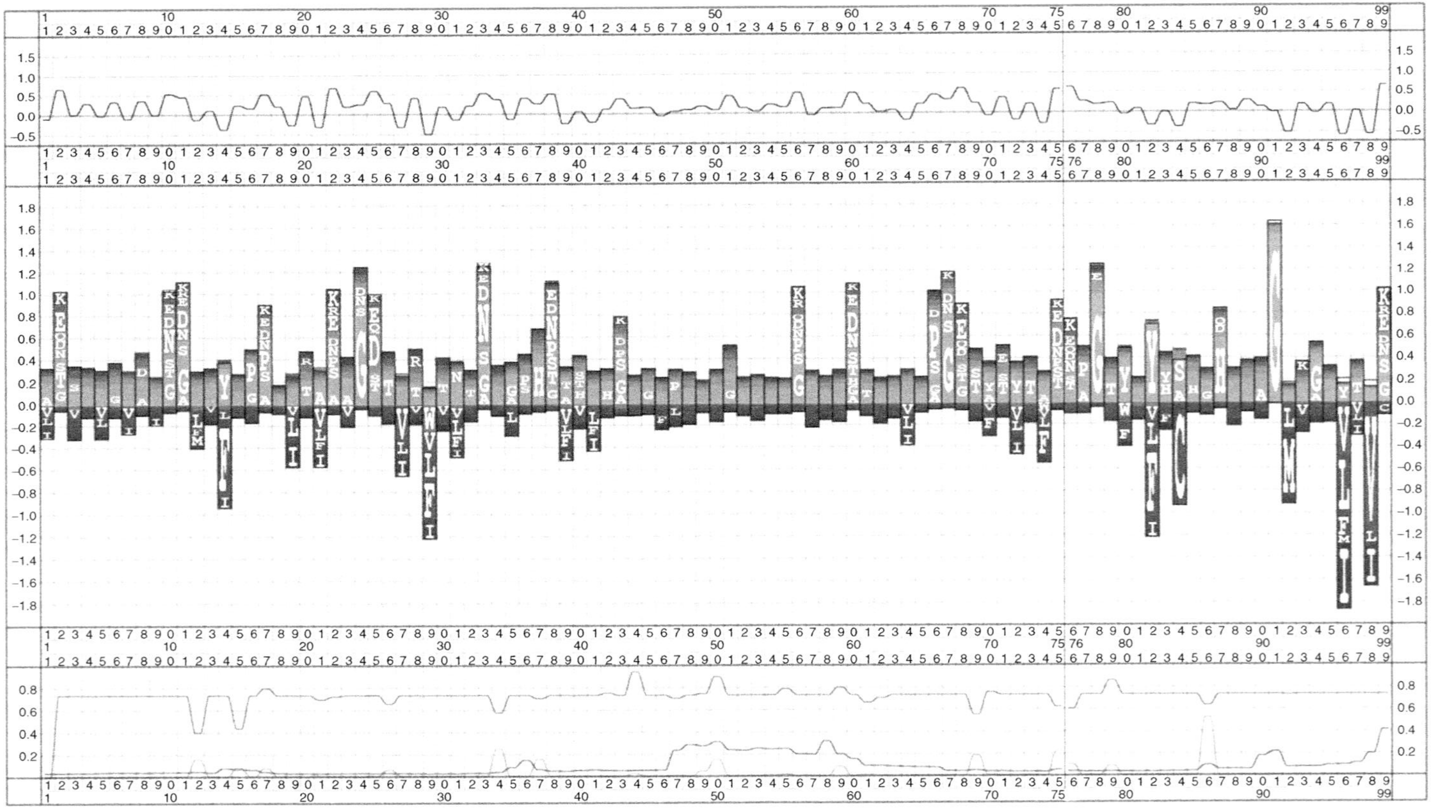

Figure 4.2 A graphical representation of a hidden Markov model representing the Cupredoxin superfamily. The top plot is the average hydrophobicity, the middle plot shows match emission probabilities with the size of each column proportional to the difference to the background distribution and columns partitioned into amino acids based on the ratio of their probabilities. The bottom plot gives the probability of insertions, insert extensions, and deletions.

alternatives to the SAM iterative procedure are PSI-blast (Schaffer et al., 2001) and the newer HMMER3-based "Jackhmmer" (Johnson et al., 2010). PSI-blast is less sensitive and, with the advent of HMMER3, is no faster than Jackhmmer. Actually in the case of SUPERFAMILY, models are also built with these other iterative procedures and combined, although primarily models resulting from the SAM "target" procedures (with various parameters) are used.

4.9 DOMAIN ANNOTATIONS

Domains are assigned by SUPERFAMILY and Gene3D at the superfamily level ("H(omology)-level in CATH"). The two resources both differ from other profile databases in that they both use multiple models to represent a single superfamily. The fundamental reason for this is that superfamilies can be extremely diverse, both in terms of sequence variation and structural composition, and with potentially threefold variation in domain length (i.e., number of residues). So while each model may match common sets of domains, they will also each match a different set of outliers. A collection of HMMs, iterated by the automatic procedures seeded using various different domain sequences belonging to the same structural superfamily as a starting "seed," more flexibly models the superfamily than any one single HMM could. This does produce the complication that for any sequence search several models may match the same region, providing different estimates of the domain boundaries.

As a single assignment is more useful than a list of multiple matches, both resources independently developed sophisticated techniques for processing the list of matches into a consolidated picture of domains for a protein. To solve the problem of resolving conflicting domain boundaries between overlapping prediction results, Gene3D uses the DomainFinder3 (Yeats et al., 2010) technique, which uses a network representation combined with heaviest-weighted clique finding to identify the "best" representative subset of matches, while the SUPERFAMILY technique uses statistics on observed domain length distributions (Madera et al., 2004). For any given input sequence, the final output is a coherent string of domains or domain "architecture." An architecture is an evolutionarily/biologically relevant object that can be useful when analyzing genomes and is defined as the unique classification and ordering of domains on the amino acid sequence. In both databases architectures are stored as strings of domains and displayed on the website graphically as in Fig. 4.3.

4.10 HIGH THROUGHPUT COMPUTATION

The challenge of carrying out sequence searches for all completely sequenced genomes versus the HMM libraries and producing architectures, alignments, and so on, is a significant part of the work required to maintain these two resources. In the past, specialized computer hardware was required to run SUPERFAMILY in

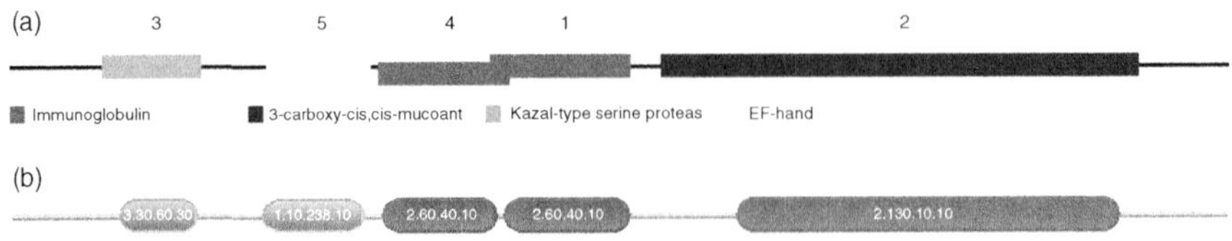

Figure 4.3 The SCOP and CATH domain architectures for Follistatin-related protein 4 (UniProt accession Q6MZW2) from SUPERFAMILY (a) and Gene3D (b) websites. The assignments (and choice of coloring) are independently generated by both resources.

the form of field programmable gate arrays (Luethy and Hoover, 2004) (FPGAs) processing the inner loop of the dynamic programming for HMM scoring for large numbers of sequences in parallel. Gene3D is run on a sizeable computer cluster, but originally had to use PSI-blast because it was faster than HMMs, then HMMER2 that is faster than the SAM scoring used at the time by SUPERFAMILY. Now both resources use HMMER3 for scoring, which is a huge technical advance as it runs two orders of magnitude faster than HMMER2, primarily through the use of a fast, accurate heuristic filtering algorithm that reduces the search space substantially. In the face of accelerating genome sequencing and the so-called next-generation sequencing machines, the advance in HMMER3 was critical for continuing the massive annotation task faced by SUPERFAMILY and Gene3D. In addition to completely sequenced genomes next-generation sequencing is producing a rapidly expanding need for processing other sequence sets such as meta-genomes of environmental samples. To cope with this explosion of sequence, SUPERFAMILY is now run entirely using commercial cloud computing, which is more flexible with varying demand and promises to be more scalable for the future.

4.11 DEVELOPMENT OF NEW BIOINFORMATICS ALGORITHMS

As touched upon in the introduction, the landslide of data is not only a challenge that must be addressed at the level of computer hardware and infrastructure, but also is at the heart of all design decisions for the resources. However good a newly developed technology is, if it does not scale well with the increasing rate of sequencing, it will make a limited contribution and soon become obsolete. A good example of this at work is the SUPERFAMILY sub-classification procedure for families (Gough, 2006; Wilson et al., 2007). As the name suggests, SUPERFAMILY assigns domains into superfamilies, which are subdivided in the SCOP hierarchy into families with complete participation, that is, every domain belongs to a family and every family belongs to a superfamily (even if some superfamilies only have one family). A standard approach to family annotation would entail a task of equal magnitude and difficulty to that at the superfamily level, doubling the human and computational resources required. To add this additional level to the annotation while retaining sustainability of the resource required a new algorithm to be developed that would not need to be updated as new structures are solved and would not increase the computational load on processing new genomes as they are sequenced. The resulting algorithm operates on domains already assigned to a superfamily as a conditional subclassification problem, which reduces the combinatorial search space by approximately three orders of magnitude. Furthermore, the hybrid pair-wise/profile method (required because families span a range of close and distant homologies) for family subclassification reuses the alignments generated by the original search procedure when the superfamily assignments were generated; the calculation coming at virtually no additional cost. As a by-product the closest structure is accurately predicted, something that the deliberately broad HMMs are very poor at.

Similarly, Gene3D has recently added subfamily assignments that consider the function of domains as well as their structure. CATH does not subclassify domains into families based on functional information, as is done in SCOP. Instead, functional subclassification is performed on all the sequence relatives within a superfamily in Gene3D. An agglomerative clustering protocol has been developed that iteratively merges clusters of relatives using profile–profile comparisons. Clustering stops when certain criteria on functional consistency are no longer met (Lee et al., 2010). The profile-based clustering approach, together with restriction of computationally intensive steps to just those sequences with well evidenced functional annotations, keeps the compute demands reasonable. Furthermore, the resulting profiles can be used to functionally classify new sequences, such as for a metagenomic project. The alignments of the functionally similar domains also highlight key functional residues, providing insights into the effects of sequence variants.

Another area where the two resources provided complementary approaches to the same problem is in the phylogenetic profiling of domains for co-occurrence and hence functional association. Gene3D have developed the Phylo Tuner (Ranea et al., 2007) method for inferring interacting genes from their phyletic patterns. A phyletic pattern is simply the distribution across genomes for a given object ignoring the relationships between genomes, in this case multi-gene families, including the copy number in each genome. The SUPERFAMILY equivalent (Wilson et al., 2009) uses a scoring function for phyletic similarity using all 1774 genomes (see Fig. 4.4); in comparison Phylo-Tuner has been designed for use with hierarchical clusters—originally by clustering a superfamily at ten levels of sequence identity from 30% to 100%, but currently being adapted for the use with the new functional families. Clusters with low information content are removed and the Euclidian distances of the phyletic profiles compared. As some superfamilies are mono-functional while others are multi-functional, using hierarchical profiles increases the chance of capturing the relationship at biologically significant grouping and hence the strongest coevolutionary signal. A variant of the maximum-weighted clique process from DomainFinder is used to select the best representative set of relationships between subfamilies.

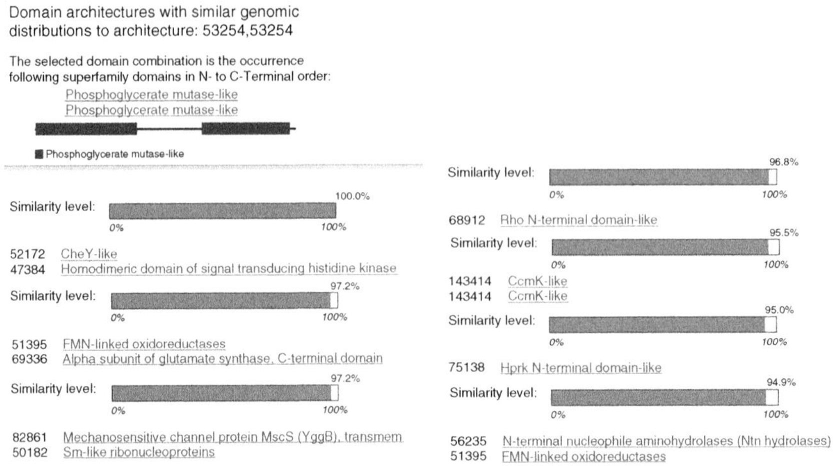

Figure 4.4 An example of phylogenetic pattern matching for a domain architecture on the SUPERFAMILY website. The architecture is two phosphoglycerate mutase-like domains in tandem (shown in dark gray) and the resulting architectures that share a similar distribution and copy number across genomes are listed below in decreasing order of similarity (mid gray score bars). In this example single domain and two-domain architectures are identified as the most similar but they can be any length.

4.12 GENOMES

At the time of writing, Gene3D and SUPERFAMILY provide annotations for close to 2000 genome assemblies (see Table 4.1 for a summary species). The coverage of individual genomes can vary widely within each resource, although 100% coverage would never be expected. Often much of the remaining sequence can be accounted for by nondomain features such as transmembrane helices and disordered regions. For instance, in SUPERFAMILY over 70% of the sequences in vertebrate genomes have at least one domain assigned and ~45% of all amino acids are covered by domains. For other organisms it is closer to 65% coverage of sequences and 57.5% coverage of amino acids because the average number of domains per protein is less. This covers the majority of globular (ordered) protein sequence and is sufficient to draw general conclusions about genome-wide trends and do cross-genome comparisons. It should be noted that for Eukaryotic genomes the quality of the assembly affects the annotation.

The coverage could be increased by including domain families that do not have a structural representative, for example, the coverage of the human genome by Pfam (Bateman et al., 2000) is 72% of sequences and 41% of amino acids, and 65% and 43% respectively by SUPERFAMILY, and 62% and 34% by Gene3D. The difference in domain and residue coverage by SUPERFAMILY and Gene3D partly reflects a difference between SCOP and CATH. CATH domain definitions

TABLE 4.1 The Number of Genomes and Strains Currently Annotated by Both Resources

	UniProt/ENSEMBL (Gene3D and SUPERFAMILY)	Individual genomes (SUPERFAMILY)
Cellular genomes	1639	1770
Animals	81	109
Fungi	67	115
Plants	12	35
Protists	32	40
Total eukaryotic secies	192	301
Total eukaryotic strains	224	357
Total prokaryotic species	1007	958
Total prokaryotic strains	1415	1413
Viral genomes	1016	2473

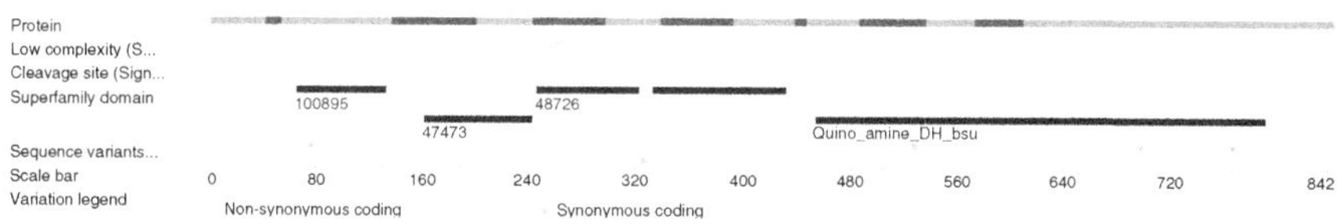

Figure 4.5 The ENSEMBL genome browser view for ENSP00000265342 (same protein as in Fig. 4.3) showing the live SUPERFAMILY and Gene3D DAS tracks.

aim to precisely delineate the structural unit, while SCOP aims to define domains as minimum evolutionary units, so a structure classed as a single domain by SCOP may be classed as multiple domains by CATH, with intra-domain linkers excluded from the definitions (Fig. 4.5).

As an example, combining Pfam and Gene3D annotations using a variant of the DomainFinder process gives the combined residue coverage of 49%, together with sequence coverage of 80%. Gene3D provides two sets of proteome assemblies—Ensembl Genomes and the UniProt complete genomes set. SUPERFAMILY provides the individual genomes from many sources (including ENSEMBL) in addition to UniProt-derived protein sets.

Although the two resources have potentially less coverage than databases such as Pfam that include families with no known structural representative, they do provide a more comprehensive coverage of those superfamilies of known structure through the recognition of more distant homologues. A fundamental issue in genome-wide comparative genomics studies is determining whether a family is absent or sufficiently diverged to be unrecognizable, placing an emphasis on completeness in homologue recognition over coverage.

4.13 *e*-VALUE SCORES

A word of caution on how to use *e*-values: An *e*-value is similar to false discovery rate (FDR), and is an estimation of the number of *errors per query* (see the biological sequence analysis book (Durbin et al., 1998) for further reading). What this means is that for a given query, if you choose a cut-off *e*-value of, for example, 0.01, then you would expect 0.01 errors with an *e*-value lower than your cut-off. Using a cut-off of 0.01 you will have an error in your results one time in every 100 queries, which is a tolerable rate of error for most research investigations. The crucial part of the *e*-value definition is the "per query" part, and this is particularly true when using SUPERFAMILY and Gene3D for whole genome analysis. The *e*-values change depending on the query. If you search two databases you will get two times the errors; this will produce the same number of errors as searching one database that has twice as many sequences. Therefore the *e*-value calculation depends on the database size, and the same sequence hit will have a different *e*-value if it is the result of a larger or smaller total query. The *e*-values in SUPERFAMILY and Gene3D are calculated as a single sequence searched against a model library; hence, if you count the number of errors in a complete genome containing 1000 sequences, it will be 1000 times the *e*-value cut-off you choose. The cut-off displayed by Gene3D and SUPERFAMILY is less than 0.0001), which means there will be one error for 10 genomes of size 1000, or for example, five errors for a 50,000 sequence genome. Approximately, half of the potentially false hits will be ignored because they conflict with a stronger, true assignment. However, as the remaining errors may stand out because they do not make biological sense, it is crucial to understand how to interpret the *e*-values when data mining; if you are looking for something unusual in thousands of genomes totalling millions of sequences you will find hundreds of errors using the default cut-off in the databases, several of that will appear biologically unusual.

4.14 OTHER SEQUENCE SETS

In addition to completely sequenced genomes, major sequence databases such as UniProt (both resources) and the National Center for Biotechnology Information (NCBI)'s RefSeq (Pruitt et al., 2012) (Gene3D) are included. SUPERFAMILY also annotates 119 meta-genomes, all plasmid sequences from the complete bacterial genome set, a handful of early draft genomes, some structural genomics targets (Gough, 2002), sequences from the PDB and sets of pseudogenes for 10 animals. The sets are regularly updated as new genomes are sequenced and existing ones are rereleased, in some cases automatically. For example, in SUPERFAMILY the immediate processing of updates originating from the PDB (weekly) and UniProt (monthly) are automated. Both SUPERFAMILY and

Gene3D also update the model libraries periodically, following the major releases of SCOP (roughly biennial) and CATH (annual).

4.15 DATA ACCESS

In this age of information on biology no resource should stand alone. Integration and communication between databases is essential to the advancement of the field. There are several ways in which SUPERFAMILY and Gene3D interplay with various research communities (Wang et al., 2008). As the number of applications and methods of using sequence annotation data have grown so have the resources developed novel methods of distributing it. Both resources host servers that provide websites—which in themselves have grown in utility and complexity—and simple download sites, as well as several forms of web service for use by external websites and client software (Yeats et al., 2011).

The simplest form of distribution is the provision of precalculated annotations, such as for individual genomes and large-scale sequence databases, and portable forms of the software used to generate these annotations. A more sophisticated approach for accessing the resources is the provided web services. An interesting example is the distributed antenna system (DAS) (Dowell et al., 2001). DAS defines a format for a compressed XML stream that is generated on request by a client program. In the case of Ensembl the client is actually the Ensembl website (Hubbard et al., 2002), which is able to access resources from a large number of registered distributors and display them interactively through a single portal to researchers. Both resources also include more generalized web services that provide programmatic access to hosted sequence searches and precalculated data.

Beyond technological approaches both resources are closely involved with other sequence annotation databases at a research level. For instance, as mentioned in the historical paragraph in the introduction, both resources are members of the InterPro consortium (Hunter et al., 2011), which describes itself as "... an integrated database of predictive protein 'signatures' used for the classification and automatic annotation of proteins and genomes." In essence, InterPro includes relationships between signatures (e.g., families/superfamilies/motifs) from member resources, producing a set of hierarchically integrated signatures with hand-written functional information. Typically, the superfamilies of Gene3D and SUPERFAMILY provide the broadest signatures of the InterPro hierarchy, while other resources provide more specific functional subfamilies. InterPro provides consolidated annotations through their umbrella website, as well as packaging all the member resources' search pipelines into a single bundle called *InterProScan*. The InterPro website provides HTML links to member databases and reciprocally SUPERFAMILY and Gene3D link into InterPro creating a navigable web of connections enabling users to jump between and cross-reference results from SUPERFAMILY, Gene3D, and the other member databases (Table 4.2).

TABLE 4.2 The coverage and domain statistics for the human genome assignments to ENSEMBL release 59_37d from (a) SUPERFAMILY, (b) Gene3D and (c) for Gene3D the supplementary "unclassified domain" (UNC) HMMs also included. Unclassified domains in CATH are structures that have been decomposed into their constitutive domains but not yet evolutionarily classified. Hence it is not possible to provide meaningful family sizes and combinations in the "Gene3D + unclassified" column. The combined assignments are provided by the Gene3D website

	SUPERFAMILY	Gene3D	Gene3D +UNC	
Sequences	79063	79063	79063	Total number of proteins
	51209	49036	51888	Have at least one assignment
	443	443	443	Average length of a protein sequence
	43	34	38	Percentage of amino acid coverage
Domains	102889	111862	122060	Total number of domains identified
	1106	1281	–	Number of unique superfamilies
	1377	3877	–	Number of unique families
	296	107.2	TBD	Average domain length
	93	87	–	The average superfamily size
Domain combinations	1694	3029	–	The number of unique domain pairs
	7260	4342	–	The number of unique architectures

4.16 ANALYSIS TOOLS

4.16.1 Providing an Analysis Platform

In this section we expand on the role of the websites in making the data more easily accessible, and provide an analysis platform for biologists. Although the websites were mentioned briefly, they serve a role beyond data distribution, and also provide visual platforms for carrying out domain-based analyses of sequence data, both on a small and a large scale.

The primary function of the websites is to present precalculated domain assignments with integrated information about species and function descriptions from external resources, along with links to related resources and built-in analysis tools. Domain architectures are shown using interactive graphics, along with access to downloaded text-based descriptions. Function descriptions include molecular, enzymatic, and pathway information (Gene Ontology (GO) (Ashburner et al.,

2000), Enzyme Classification (EC (Webb, 1992), GenomeRNAi (Gilsdorf et al., 2010), Kyoto Encyclopedia of Genes and Genomes (KEGG) (Kanehisa et al., 2012), along with other sequence annotations and variant or phenotype information (e.g., UniProt (Apweiler et al., 2004), Online Mendelian Inheritance in Man (OMIM) (Hamosh et al., 2005)).

Aggregated information is displayed for superfamilies, functional subfamilies, pathways, and genomes. Significant effort is put in by both resources in providing informative interactive interfaces (e.g., Fig. 4.6) for this collected information. The integration of these diverse annotations allows researchers to quickly gather up the key information about a protein and how this may relate to the structural domains found in them.

Sequence alignments underpin domain families, and access to useful representations is provided by both resources (discussed further in "Sequence Alignments" below). The first analysis tool any sequence analysis database must provide is a facility for annotating a researcher's amino acid and/or nucleotide sequences. The view of the results includes similar information as that displayed for precalculated annotations, including links to relevant information in the websites and to sequence alignments. Both sites also provide further feature prediction and sequence alignment tools.

The recognition and display of structural domain assignments provides the core of the Gene3D and SUPERFAMILY platforms. The next layer of functionality lies with aggregating the proteins by domain assignments, or domain architectures, genomes, function, and pathway. Examples of provided tools include: the distribution of domain and protein families amongst different taxonomic lineages, and means of viewing over-represented families (discussed further below in "Domain Families" and "Comparative Genomics"). The inclusion of extra annotations from external resources greatly improves the utility of these tools; for instance, it becomes easy to see what functions are associated with a domain architecture that is over-represented in a pathogenic bacterium.

In summary, Gene3D and SUPERFAMILY each provide an integrated service for domain homologue recognition and annotation, using cutting edge discovery software and providing a comprehensive and flexible set of tools for the access, distribution and analysis of the resulting data.

4.16.2 Sequence Alignments

The sequence alignments are derived from the HMMs; therefore, in addition to the pair-wise alignment of the input sequence and the sequence of the domain of known structure both aligned to the model, multiple sequence alignments are also possible by aligning all sequences to the same HMM. SUPERFAMILY provides a webpage where alignments are dynamically generated by selecting sequence sets, for example, all members of a SCOP family or superfamily, genome sequence hits, or user uploaded sequences. Multiple alignments of predicted sequences in Gene3D can be viewed by linking through to the relevant superfamily in CATH,

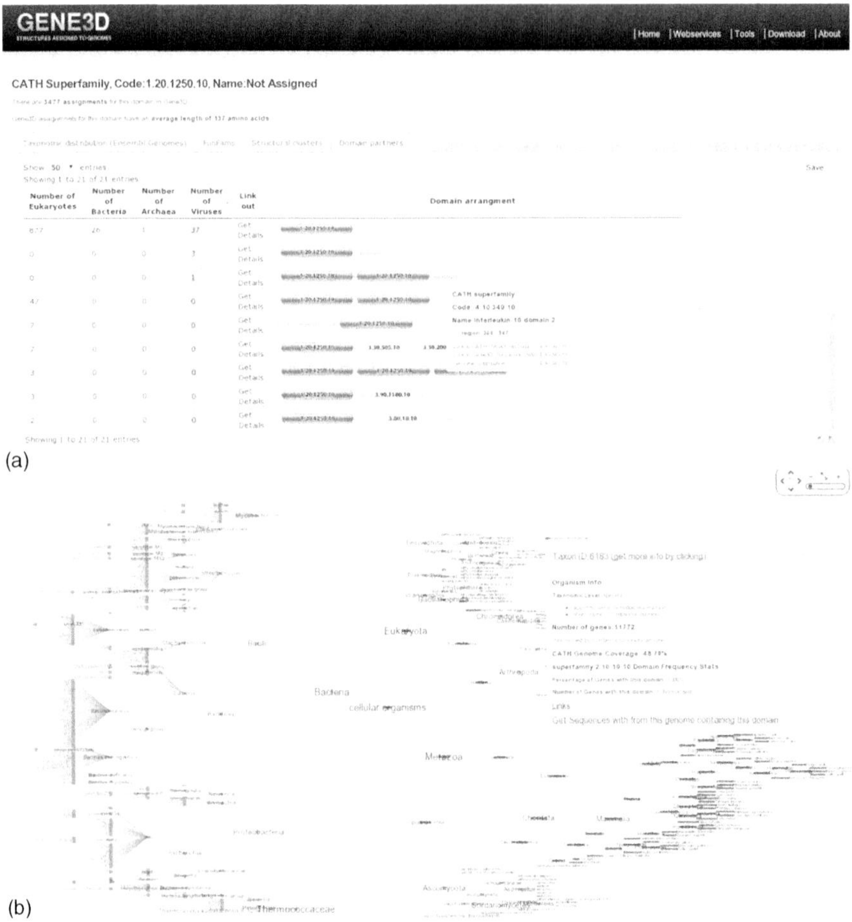

Figure 4.6 (a) Gene3D superfamily page shows functional and structural summaries for a superfamily along with the MDA's that are found for a CATH superfamily and its counts in various branches of life. Links provide the ability to retrieve individual genes. (b) Ensembl genome distribution of CATH superfamily 2.10.10.10. Having a higher proportion of genes that contain the superfamily is indicated by the increased width of the node and the change in shade. (c). The TaxViz visual representation of the phyletic pattern of protein-kinase like superfamily domains in 1734 genomes. The size of the circle is proportional to the average number of domains per organism in that phylogenetic clade, for example, 566 per metazoan genome and 1200 per plant genome. The bottom left shows that Eucalyptus has the most with 3720 domains and 42 genomes only have a single copy of the domain. The inner circle illustrates the relative abundance in the three superkingdoms, and clicking on any circle plots a new image displaying the relative abundance at lower taxonomic levels. (d) Presence–absence pattern of the nuclear receptor ligand-binding domain across the eukaryotic species tree of life. The left panel illustrates the overview of the eukaryotic tree, with a branch (edge) highlighted in green if the domain can be found in all genomes under the clade attached to the branch. The right panel is the zoomed-in version of the kingdom viridiplantae (plants), which further contains two clades, embryophytes (land plants), and chlorophyta (green algae). (*See insert for color representation of the figure*)

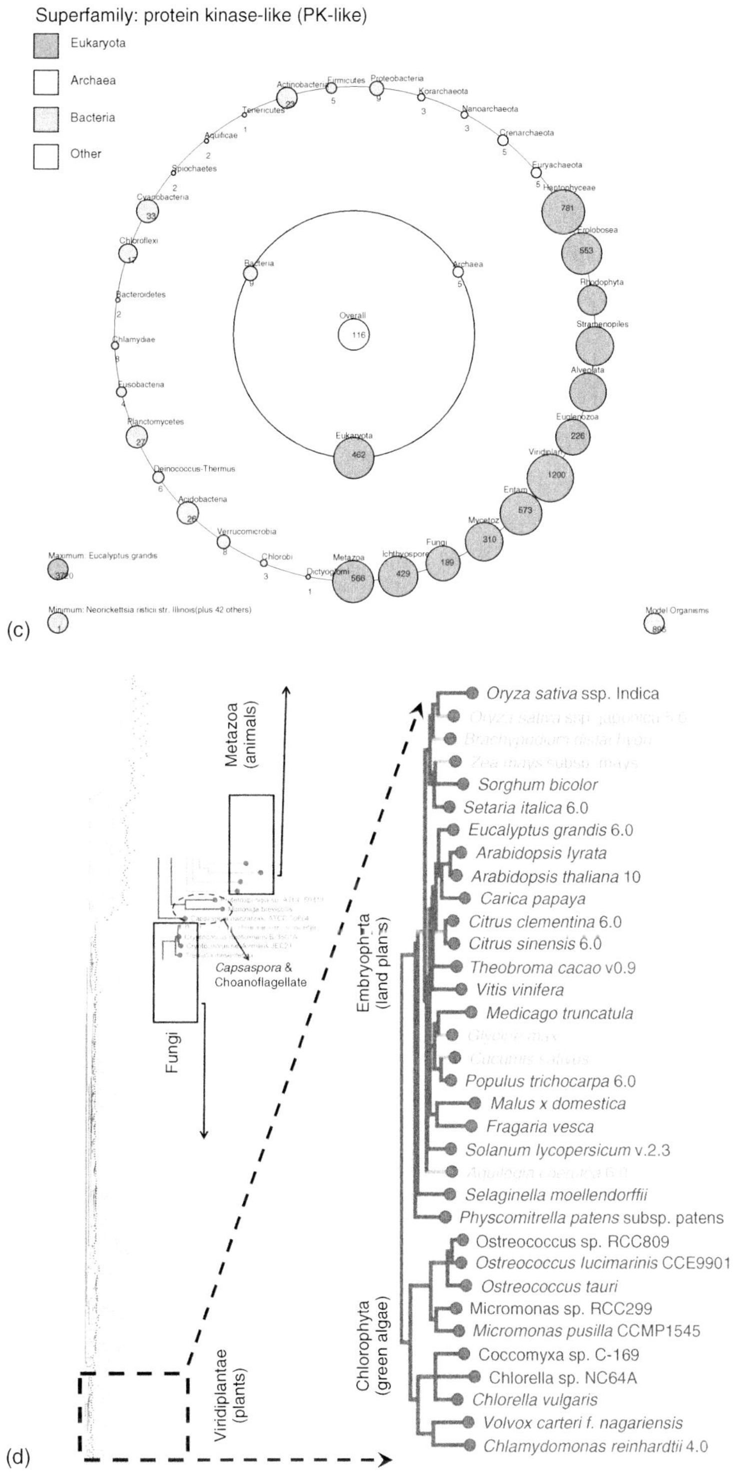

Figure 4.6 *(Continued)*

where they are presented for each functional subfamily together with a representative structure for that subfamily. There is also an option to view multiple structure alignments for a given superfamily. Conserved residue positions identified using the Scorecons (Valdar, 2002) algorithm are highlighted on the sequence alignment and on the structure, as are any known catalytic residues captured in the Catalytic Site Atlas CSA (Porter et al., 2004).

4.16.3 Domain Families

The basic analysis tools discussed above allow a user to investigate individual sequences and with the multiple sequence alignments begin to look at whole families. SUPERFAMILY and Gene3D have global analysis tools on their websites specifically for genome-wide investigations of families, superfamilies, and architectures. Previously mentioned at the end of the technology section are the tools in SUPERFAMILY and Gene3D for matching domains/architectures with a similar phyletic profile. In SUPERFAMILY this is available, for example, by clicking on the architecture image (Fig. 4.3) of a sequence and following the link "Other domain architectures with similar genomic distribution." CATH superfamily distributions and taxonomic distributions of MDA's can be viewed in Gene3D via the superfamily summary page (Fig. 4.6a) Superfamily distributions can also be viewed more graphically using an interactive network representation (generated using Cytoscape-Web (Lopes et al., 2010)), allowing zooming or clicking on nodes for more information (Fig. 4.6b).

Within SUPERFAMILY, in addition to listing the genomes in which a domain/architecture occurs and searching for similar phyletic patterns, there are two ways to visualize the distribution across genomes. The tab "Taxonomic Distribution" displays the TaxViz image (Fig. 4.6c) giving an overview of the relative representation in different kingdoms and organisms. On the same page is a link to "Plot the distribution on a phylogenetic tree" that will show, using vector graphics (Pethica et al., 2010), an interactive species tree of organisms with the membership of the domain/architecture in question highlighted (Fig. 4.6d).

4.16.4 Comparative Genomics

The previous paragraph describes ways of investigating and visualizing domains and architectures and visualizing their distribution across genomes. There are also tools that work across all domains/architectures for a given genome or set of genomes, that is, comparative genomics tools. These are excellent for identifying proteins, which are responsible for specialization or specifically relate to the unique phenotype of an organism. In SUPERFAMILY this information can be obtained by selecting the "unusual Superfamilies" and "Domain Pairs" options. A ranked list of superfamilies for a given genome with respect to all other genomes in the same superkingdom, is shown with the most outstandingly abundant or unique at the top of the list and those underrepresented or missing at the bottom.

The "Domain pairs" option shows a page listing all of the superfamily pairs, which are observed together in at least one architecture from a sequence of the genome in question. The results can also be visualized as a network of interacting superfamilies as in Fig. 4.7a.

Gene3D provides a graphical means of comparing genomes in order to display enriched domains and domain compositions. As an example Fig. 4.7b shows over-represented superfamilies from a comparison between two species of staphylococcus. We see for example the penicillin binding protein associated domain is enriched in *Staphylococcus haemolyticus* along with genes containing both the glycosidases (3.20.20.80) and Golgi alpha-mannosidase II (2.60.40.1180) CATH superfamilies.

4.16.5 Additional Data

As mentioned in a previous section Gene3D imports and displays external data, for example, Pfam and SUPERFAMILY, but perhaps most complimentary to all of these are the predictions of disorder by DisoPred (Jones and Ward, 2003). SUPERFAMILY and Gene3D also generate additional forms of data internally, for example, for coiled-coil oligomeric states (Rackham et al., 2010).

Both SUPERFAMILY and Gene3D include GO and EC data. In the case of SUPERFAMILY domain-centric GO and EC mappings to superfamilies, families, architectures, and supra-domains (Vogel et al., 2004) (pairs, triplets, and longer combinations of domains recurring in different architectures) are derived from primary GO sources using a method described in the 2011 database paper (de Lima Morais et al., 2011).

The functional terms associated with SUPERFAMILY/Gene3D objects, when mapped to sequences, allow the inference of function on a genome-wide scale. In SUPERFAMILY enrichment of functional terms can be analysed (through the "GO," "EC" and "Phenotype" links under the "Ontologies" heading in the navigation bar on the top left of every web page) and plotted on a species tree in a way similar to that described for domains/architectures above, and there are links to interactively explore the evolution of a function over time and across lineages. Using the same methodology as for GO and EC, SUPERFAMILY also provides at the time of writing eight types of domain-centric phenotype ontology including human, mammalian, plant, anatomical, etc.

Gene3D also provides additional functional annotation in the form of mapping to KEGG pathways (Kanehisa et al., 2012), links to microarray expression data, genetic disease information from OMIM, Drugbank (Knox et al., 2011), GenomeRNAi, and interaction data from the main interaction databases. The interaction data is displayed along with individual protein feature tracks and higher level summaries where the CATH superfamily statistics of the network can be viewed (Fig. 4.7b). Any sequence set in Gene3D can also be filtered using linear motifs and regular expressions, such as those found at the eukaryotic linear motifs (ELM) (Dinkel et al., 2012) resource, which can be useful for interpreting interaction pages for example.

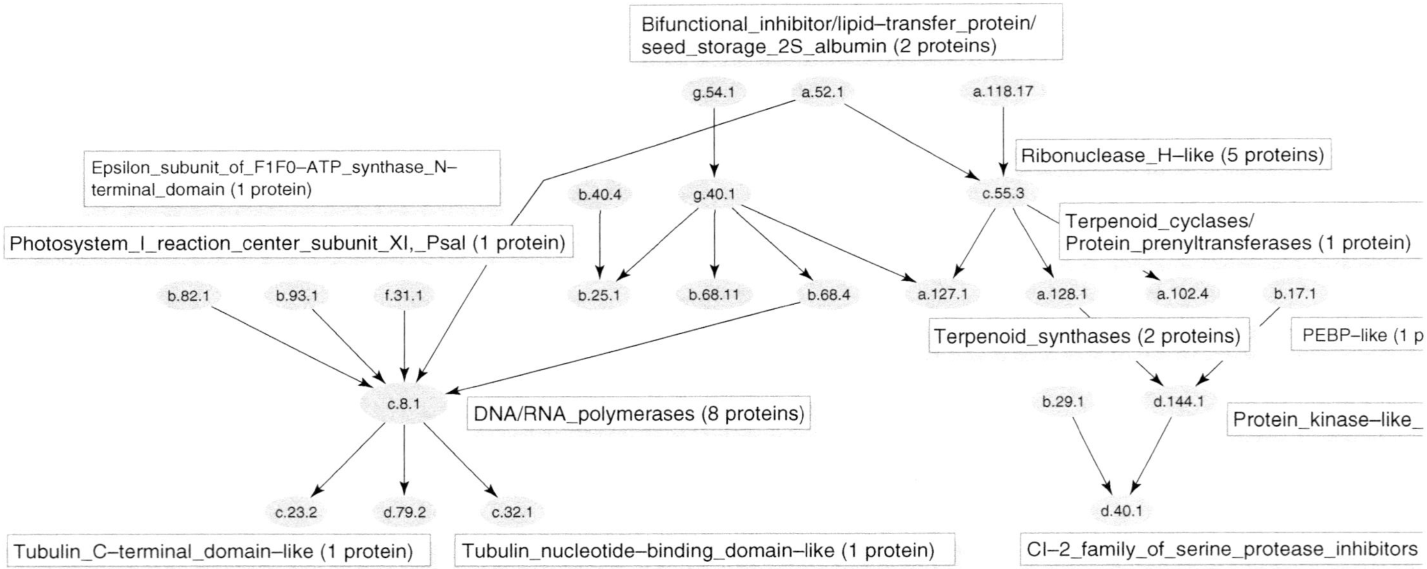

Figure 4.7 (a) From the SUPERFAMILY website, the main part of the network of domain pairs that are unique to the Cocoa genome and not seen in any other genome. If there is a protein in Cocoa that contains a pair of domains, which do not co-occur in any other genome then the two superfamilies are drawn as nodes and an edge connects them. The annotation labels appear on mouse-over in the browser, but several are shown on the figure for information. (b) From the Gene3D website, nodes represent CATH superfamilies, and links (edges) between CATH superfamilies represent genes with domains from both these superfamilies. For edges connecting up hexagons, the superfamilies co-occur in same gene more in *Staphylococcus haemolyticus* JCSC1435 than in *Staphylococcus epidermidis* RP62A and vice versa for edges connecting up diamonds. The edge width corresponds to the absolute value of the log ratio of superfamily pair counts, i.e. thicker means there is a bigger difference for this superfamily composition between the two species. Hexagon indicates the number of genes with this superfamily is higher in *S. haemolyticus* JCSC1435 than in *S. epidermidis* RP62A and vice versa for Diamond. The hexagon linked to the square indicates that this combination of domains occurs in both genomes but it more common in *S. haemolyticus*. A user can click on a node or edge to see more information and retrieve the relevant genes. The node size corresponds to absolute value of log ratio of the number of genes with this superfamily between the two species.

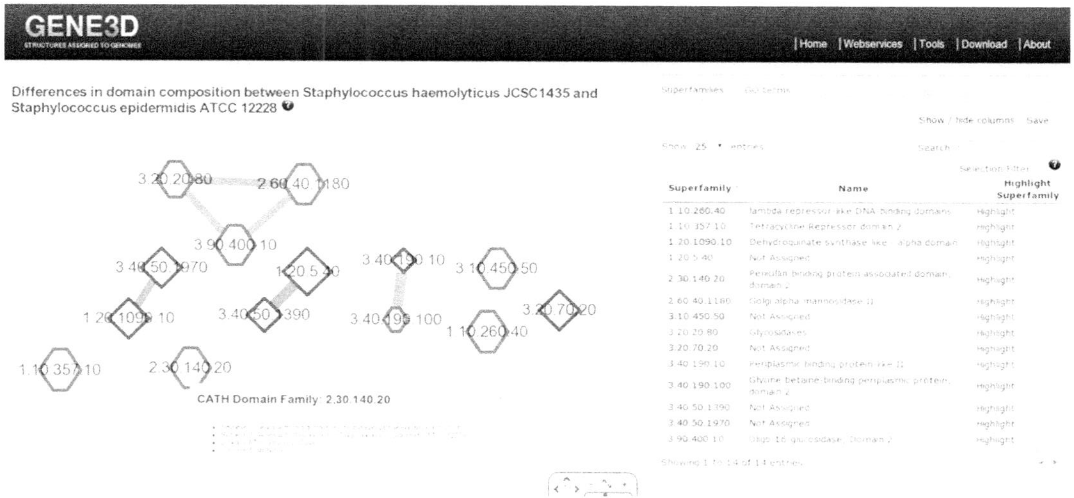

Figure 4.7 (Continued)

4.16.6 The Reference Species Tree of (Sequenced) Life

This section and the previous section include mention of the visualization of domains, families, functions, etc. highlighted on a phylogenetic species tree. One of the most important additional data outputs from SUPERFAMILY is the reference species tree for all completely sequenced genomes. For it to be possible to plot evolutionary characters on a tree, first a reference tree is required, and SUPERFAMILY provides the most accurate complete fully-resolved species tree of sequenced organisms. The tree is built using maximum likelihood phylogenetics (Stamatakis, 2006) taking as input superfamilies, families, architectures, and supra-domains as evolutionary characters, whilst constrained to the NCBI taxonomy (Wheeler et al., 2001). The resulting reference species tree is provided to the community as a structured query language (SQL) table and as a Newick format file. The uses of a binary reference species tree of sequenced genomes for research go far beyond the visualization application already described. One example of its use is for the reconstruction of ancestral nodes in superfamily; using Dollo parsimony the repertoire of characters in ancestral genomes is reconstructed for the whole eukaryote lineage and presented within SUPERFAMILY. Near the top of each genome web page for any eukaryote under "Taxonomy," the taxonomic classification has links at levels that correspond to nodes in the tree of sequenced genomes that display the same (browse-able) web page and data as for a genome, but for the reconstructed ancestor.

4.17 CONCLUSION

The solving of protein structures, combined with the advent of powerful profile-based homologue recognition technologies, has allowed the exploration of structure in protein sequences being produced at a vast rate by modern "*-omics*" technologies. In turn the sequence information allows the detailed analysis of conservation and evolution of protein structures, which in turn deepens our understating of how structure determines function. SUPERFAMILY and Gene3D provide platforms for this kind of study, based on the SCOP and CATH structural classifications respectively. Through using profiles for homology recognition the platforms are designed for future sustainability with the steady increase in the production of structure data and explosion of sequence data. As the quantity of both these primary data increases, the quality and coverage of information provided by SUPERFAMILY and Gene3D improves, gradually getting closer to ultimate completeness. In addition to the resulting information, which is freely available for download, additional tools for analysis and comparison aid users in exploiting it.

REFERENCES

Abroi, A. and Gough, J. (2011) Are viruses a source of new protein folds for organisms? – Virosphere structure space and evolution. Bioessays, 33(8), 626–635.

Altschul, S.F., Gish, W., Miller, W., Myers, E.W., and Lipman, D.J. (1990) Basic local alignment search tool. J Mol Biol, 215(3), 403–410.

Altschul, S.F., Madden, T.L., Schaffer, A.A., Zhang, J., Zhang, Z., Miller, W., and Lipman, D.J. (1997) Gapped {BLAST} and {PSI-BLAST}: a new generation of protein database search programs. Nucleic Acids Res, 25(17), 3389–3402.

Apweiler, R., Bairoch, A., Wu, C.H., Barker, W.C., Boeckmann, B., Ferro, S., Gasteiger, E., Huang, H., Lopez, R., Magrane, M., Martin, M.J., Natale, D.A., O'Donovan, C., Redaschi, N., and Yeh, L.L. (2004) UniProt: the Universal Protein knowledgebase. Nucleic Acids Res, 32, D115–D119.

Ashburner, M., Ball, C.A., Blake, J.A., Botstein, D., Butler, H., Cherry, J.M., Davis, A.P., Dolinski, K., Dwight, S.S., Eppig, J.T., Harris, M.A., Hill, D.P., Issel-Tarver, L., Kasarskis, A., Lewis, S., Matese, J.C., Richardson, J.E., Ringwald, M., and Rubin, G. (2000) Gene ontology: tool for the unification of biology. The gene ontology consortium. Nat Genet, 25(1), 25–29.

Barrett, C., Hughey, R., and Karplus, K. (1997) Scoring hidden Markov models. CABIOS, 13(2), 199–216.

Bateman, A., Birney, E., Durbin, R., Eddy, S.R., Howe, K.L., and Sonnhammer, E.L. (2000) The Pfam protein families database. Nucleic Acids Res, 28(1), 263–266.

Berman, H.M., Westbrook, J., Feng, Z., Gilliland, G., Bhat, T.N., Weissig, H., Shindyalov, I.N. et al. (2000) The Protein Data Bank. Nucleic Acids Res, 28(1), 235–242.

Biegert, A. and Soeding, J. (2009) Sequence context-specific profiles for homology searching. Proc Natl Acad Sci USA, 106(10), 3770–3775.

Buchan, D.W., Shepherd, A.J., Lee, D., Pearl, F.M., Rison, S.C., Thornton, J.M., and Orengo, C.A. (2002) Gene3D: structural assignment for whole genes and genomes using the CATH domain structure database. Genome Res, 12(3), 503–514.

Chivian, D., Kim, D.E., Malmström, L., Bradley, P., Robertson, T., Murphy, P., Strauss, C.E.M. et al. (2003) Automated prediction of CASP-5 structures using the Robetta server. Proteins, 53(6 Suppl), 524–533.

Chothia, C., Gough, J., Vogel, C., and Teichmann, S.A. (2003) Evolution of the protein repertoire. Science, 300(5626), 1701–1703.

de Lima Morais, D., Fang, H., Rackham, O.J.L., Wilson, D., Pethica, R., Chothia, C., and Gough, J. (2011) SUPERFAMILY 1.75 including a domain-centric gene ontology method. Nucleic Acids Res, 39, D427–D434.

Dinkel, H., Michael, S., Weatheritt, R.J., Davey, N.E., Van Roey, K., Altenberg, B., Toedt, G. et al. (2012) ELM–the database of eukaryotic linear motifs. Nucleic Acids Res, 40(Database issue), D242–D251.

Dowell, R.D., Jokerst, R.M., Day, A., Eddy, S.R., and Stein, L. (2001) The distributed annotation system. BMC Bioinformatics, 2(1), 7.

Durbin, R., Eddy, S.R., Krogh, A., and Mitchison, G. (1998) Biological Sequence Analysis, Cambridge University press, Cambridge.

Eddy, S.R. (1996) Hidden Markov models. Curr Opin Struct Biol, 6(3), 361–365.

Eddy, S.R. (1998) Profile hidden Markov models. Bioinformatics, 14, 755–763.

Fleischmann, R., Adams, M., White, O., Clayton, R., Kirkness, E., Kerlavage, A., Bult, C., Tomb, J., Dougherty, B., and Merrick, J. (1995) Whole-genome random sequencing and assembly of Haemophilus influenzae Rd. Science, 269(5223), 496–512.

Gherardi, E., Youles, M.E., Miguel, R.N., Blundell, T.L., Iamele, L., Gough, J., Bandyopadhyay, A. et al. (2003) Functional map and domain structure of MET, the product of the c-met protooncogene and receptor for hepatocyte growth factor/scatter factor. Proc Natl Acad Sci USA, 100(21), 12039–12044.

Gilsdorf, M., Horn, T., Arziman, Z., Pelz, O., Kiner, E., and Boutros, M. (2010) GenomeRNAi: a database for cell-based RNAi phenotypes. 2009 update. Nucleic Acids Res, 38(Database issue), D448–D452.

Gough, J. (2002) The SUPERFAMILY database in structural genomics. Acta Crystallogr D: Biol Crystallogr, 58, 1897–1900 International Union of Crystallography.

Gough, J., Chothia, C., Karplus, K., Barrett, C., and Hughey, R. (2000) Optimal hidden Markov models for all sequences of known structure, in Currents in Computational Biology (eds S. Miyano, R. Shamir, and T. Takagi), Universal Academy Press INC., Tokyo, pp. 124–125.

Gough, J. (2006) Genomic scale sub-family assignment of protein domains. Nucleic Acids Res, 34(13), 3625–3633. doi: 10.1093/nar/gkl484

Gough, J., Karplus, K., Hughey, R., and Chothia, C. (2001) Assignment of homology to genome sequences using a library of hidden Markov models that represent all proteins of known structure. J Mol Biol, 313(4), 903–919. doi: 10.1006/jmbi.2001.5080

Hamosh, A., Scott, A.F., Amberger, J.S., Bocchini, C.A., and McKusick, V.A. (2005) Online Mendelian Inheritance in Man (OMIM), a knowledgebase of human genes and genetic disorders. Nucleic acids research, 33(Database issue), D514–D517.

Hubbard, T., Barker, D., Birney, E., Cameron, G., Chen, Y., Clark, L., Cox, T. et al. (2002) The Ensembl genome database project. Nucleic Acids Res, 30, 38–41.

Hunter, S., Jones, P., Mitchell, A., Apweiler, R., Attwood, T.K., Bateman, A., Bernard, T. et al. (2011) InterPro in 2011: new developments in the family and domain prediction database. Nucleic Acids Res, 40(D1), D306–D312. doi: 10.1093/nar/gkr948

Johnson, L.S., Eddy, S.R., and Portugaly, E. (2010) Hidden Markov model speed heuristic and iterative HMM search procedure. BMC Bioinformatics, 11, 431.

Jones, D.T. (1999) {GenTHREADER}: an efficient and reliable protein fold recognition method for genomic sequences. J Mol Biol, 287(4), 797–815.

Jones, D.T. and Ward, J.J. (2003) Prediction of disordered regions in proteins from position specific score matrices. Proteins, 53(6 Suppl), 573–578.

Kanehisa, M., Goto, S., Sato, Y., Furumichi, M., and Tanabe, M. (2012) KEGG for integration and interpretation of large-scale molecular data sets. Nucleic acids research, 40(Database issue), D109–D114.

Karplus, K., Barrett, C., and Hughey, R. (1998) Hidden Markov models for detecting remote protein homologies. Bioinformatics, 14(10), 846–856.

Kelley, L.A. and Sternberg, M.J.E. (2009) Protein structure prediction on the Web: a case study using the Phyre server. Nat Protoc, 4(3), 363–371.

Kendrew, J.C., Bodo, G., Dintzis, H.M., Parrish, R.G., Wyckoff, H., and Phillips, D.C. (1958) A three-dimensional model of the myoglobin molecule obtained by x-ray analysis. Nature, 181(4610), 662–666.

Kersey, P.J., Staines, D.M., Lawson, D., Kulesha, E., Derwent, P., Humphrey, J.C., Hughes, D.S.T. et al. (2012) Ensembl Genomes: an integrative resource for genome-scale data from non-vertebrate species. Nucleic acids research, 40(Database issue), D91–D97.

Knox, C., Law, V., Jewison, T., Liu, P., Ly, S., Frolkis, A., Pon, A. et al. (2011) DrugBank 3.0: a comprehensive resource for "omics" research on drugs. Nucleic Acids Res, 39(Database issue), D1035–D1041.

Lee, D.A., Rentzsch, R., and Orengo, C. (2010) GeMMA: functional subfamily classification within superfamilies of predicted protein structural domains. Nucleic Acids Res, 38(3), 720–737.

Lee, D., Grant, A., Marsden, R., and Orengo, C. (2005) Identification and distribution of protein families in 120 completed genomes using Gene3D. Proteins, 59, 603–615.

Lees, J., Yeats, C., Perkins, J., Sillitoe, I., Rentzsch, R., Dessailly, B.H., and Orengo, C. (2012) Gene3D: a domain-based resource for comparative genomics, functional annotation and protein network analysis. Nucleic acids research, 40(Database issue), D465–D471.

Lopes, C.T., Franz, M., Kazi, F., Donaldson, S.L., Morris, Q., and Bader, G.D. (2010) Cytoscape web: an interactive web-based network browser. Bioinformatics (Oxford, England), 26(18), 2347–2348.

Luethy, R. and Hoover, C. (2004) Hardware and software systems for accelerating common bioinformatics sequence analysis algorithms. Drug Discov Today, 2(1), 12–17.

Madera, M. (2008) Profile comparer: a program for scoring and aligning profile hidden Markov models. Bioinformatics, 24(22), 2630–2631.

Madera, M. and Gough, J. (2002) A comparison of profile hidden Markov model procedures for remote homology detection. Nucleic Acids Res, 30(19), 4321–4328.

Madera, M., Vogel, C., Kummerfeld, S.K., Chothia, C., and Gough, J. (2004) The SUPERFAMILY database in 2004: additions and improvements. Nucleic Acids Res, 32(10), 235–239.

Marsden, R.L., Lee, D., Maibaum, M., Yeats, C., and Orengo, C.A. (2006) Comprehensive genome analysis of 203 genomes provides structural genomics with new insights into protein family space. Nucleic Acids Res, 34(3), 1066–1080.

Mulder, N. J., Apweiler, R., Attwood, T. K., Bairoch, A., Bateman, A., Binns, D., Bork, P., et al. (2007). New developments in the InterPro database. Nucleic Acids Res, 35(Database issue), D224–D228. 10.1093/nar/gkl841

Murzin, A.G., Brenner, S.E., Hubbard, T., and Chothia, C. (1995) SCOP: a structural classification of proteins database for the investigation of sequences and structures. J Mol Biol, 247(4), 536–540.

Nielsen, H. and Krogh, A. (1998) Prediction of signal peptides and signal anchors by a hidden Markov model. Proc Int Conf Intell Syst Mol Biol, 6, 122–130.

Okazaki, Y. et al. (2002) Analysis of the mouse transcriptome based on functional annotation of 60,770 full-length cDNAs. Nature, 420(6915), 563–573.

Orengo, C.A., Michie, A.D., Jones, S., Jones, D.T., Swindells, M.B., and Thornton, J.M. (1997) CATH–a hierarchic classification of protein domain structures. Structure, 5(8), 1093–1108.

Park, J., Karplus, K., Barrett, C., Hughey, R., Haussler, D., Hubbard, T., and Chothia, C. (1998) Sequence comparisons using multiple sequences detect three times as many remote homologues as pairwise methods. J Mol Biol, 284(4), 1201–1210.

Pearl, F.M., Lee, D., Bray, J.E., Sillitoe, I., Todd, A.E., Harrison, A.P., Thornton, J.M., and Orengo, C.A. (2000) Assigning genomic sequences to CATH. Nucleic Acids Res., 28(1), 277–282.

Pethica, R., Barker, G., Kovacs, T., and Gough, J. (2010) TreeVector: scalable, interactive, phylogenetic trees for the web. PLoS One, 5(1), e8934.

Porter, C.T., Bartlett, G.J., and Thornton, J.M. (2004) The Catalytic Site Atlas: a resource of catalytic sites and residues identified in enzymes using structural data. Nucleic Acids Res, 32(Database issue), D129–D133.

Pruitt, K.D., Tatusova, T., Brown, G.R., and Maglott, D.R. (2012) NCBI Reference Sequences (RefSeq): current status, new features and genome annotation policy. Nucleic Acids Res, 40(Database issue), D130–D135.

Rackham, O., Madera, M., Armstrong, C.T., Vincent, T.L., Wolfson, D.N., and Gough, J. (2010) The evolution and structure prediction of coiled coils across all genomes. J Mol Biol, 403(3), 480–493.

Ranea, J.A.G., Sillero, A., Thornton, J.M., and Orengo, C.A. (2006) Protein superfamily evolution and the last universal common ancestor (LUCA). J Mol Evol, 63(4), 513–525.

Ranea, J.A., Yeats, C., Grant, A., and Orengo, C.A. (2007) Predicting protein function with hierarchical phylogenetic profiles: the Gene3D Phylo-Tuner method applied to eukaryotic genomes. PLoS Comput Biol, 3(11), e237.

Reid, A.J., Yeats, C., and Orengo, C.A. (2007) Methods of remote homology detection can be combined to increase coverage by 10% in the midnight zone. Bioinformatics (Oxford, England), 23(18), 2353–2560.

Schaffer, A.A., Aravind, L., Madden, T.L., Shavirin, S., Spouge, J.L., Wolf, Y.I., Koonin, E.V. et al. (2001) Improving the accuracy of PSI-BLAST protein database searches with composition-based statistics and other refinements. Nucleic Acids Res, 29(14), 2994–3005.

Smith, T.F. and Waterman, M.S. (1981) Identification of common molecular subsequences. J Mol Biol, 147(1), 195–197.

Stamatakis, A. (2006) RAxML-VI-HPC: maximum likelihood-based phylogenetic analyses with thousands of taxa and mixed models. Bioinformatics, 22(21), 2688–2690.

Söding, J. (2005) Protein homology detection by HMM–HMM comparison. Bioinformatics, 21(7), 951–960.

Valdar, W.S.J. (2002) Scoring residue conservation. Proteins, 48(2), 227–241.

Vinogradov, S.N., Hoogewijs, D., Bailly, X., Arrendondo-Peter, R., Guertin, M., Gough, J., Dewilde, S. et al. (2005) Three globin lineages belonging to two structural classes in genomes from the three kingdoms of life. Proc Natl Acad Sci USA, 102(32), 11385–11389.

Vogel, C., Berzuini, C., Bashton, M., Gough, J., and Teichmann, S.A. (2004) Supra-domains: evolutionary units larger than single protein domains. J Mol Biol, 336, 809–823.

Wang, L., Tao, J., Kunze, M., Castellanos, A.C., Kramer, D., and Karl, W. (2008) Scientific cloud computing: early definition and experience, in High Performance Computing and Communications, IEEE Computer Society, Washington, DC, pp. 825–830.

Webb, E.C. (1992) Enzyme Nomenclature 1992: Recommendations of the Nomenclature Committee of the International Union of Biochemistry and Molecular Biology on the Nomenclature and Classification of Enzymes, Published for the International Union of Biochemistry and Molecular Biology by Academic Press, San Diego.

Wheeler, D.L., Church, D.M., Lash, A.E., Liepe, D.D., Madden, T.L., Pontius, J.U., Schuler, G.D., Schriml, L.M., Tatusova, T.A., Wagner, L., and Rapp, B.A. (2001) Database resources of the National Center for Biotechnology Information. Nucleic Acids Res, 29(1), 11–16.

Wilson, D., Madera, M., Vogel, C., Chothia, C., and Gough, J. (2007) The SUPERFAMILY database in 2007: families and functions. Nucleic Acids Res, 35, D308–D313.

Wilson, D., Pethica, R., Zhou, Y., Talbot, C., Vogel, C., Madera, M., Chothia, C., and Gough, J. (2009) SUPERFAMILY-sophisticated comparative genomics, data mining, visualization and phylogeny. Nucleic Acids Res, 37, D380–D386.

Yeats, C., Lees, J., Carter, P., Sillitoe, I., and Orengo, C. (2011) The Gene3D Web Services: a platform for identifying, annotating and comparing structural domains in protein sequences. Nucleic acids research, 39(Web Server issue), W546–W550.

Yeats, C., Redfern, O.C., and Orengo, C. (2010) A fast and automated solution for accurately resolving protein domain architectures. Bioinformatics, 26(6), 745–751.

Zhang, Y., Arakaki, A.K., and Skolnick, J. (2005) TASSER: an automated method for the prediction of protein tertiary structures in CASP6. Proteins, 61, 91–98.

5

PHYLOGENOMIC DATABASES AND ORTHOLOGY PREDICTION

Kimmen Sjölander

Plant & Microbial Biology, Bioengineering, Berkeley, CA, USA

CHAPTER SUMMARY

This chapter discusses the major phylogenomic databases providing precalculated gene family trees and/or orthology relationships, reviews the methods used to identify orthologs, and explains the main orthology subtypes. The mechanisms by which gene families develop novel functions are reviewed, and a protocol for a structural phylogenomic prediction of function, including information from protein structure, is presented. Finally, a case study to illustrate the pros and cons of reconstructing phylogenies for individual domains versus whole-domain architectures is included.

5.1 THE EVOLUTION OF NOVEL FUNCTIONS AND STRUCTURES IN GENE FAMILIES

Dobzhansky's famous saying "Nothing in biology makes sense except in the light of evolution" is as relevant to understanding the diversity of life forms as it is to the interpretation of genes and the proteins they encode (Dobzhansky, 1973). What genes *do*—their biochemical functions, the biological processes and pathways in which they participate, the conditions under which they are expressed, intermolecular interactions and cellular localization—can be determined experimentally or (more commonly) predicted using bioinformatics

Protein Families: Relating Protein Sequence, Structure, and Function, First Edition.
Edited by Christine Orengo and Alex Bateman.

methods. Many bioinformatics methods make use of statistical modeling techniques to integrate heterogeneous and often noisy data over families of related genes. Without evolution as an underlying structure, we cannot hope to interpret these data, and Dobzhansky's proclamation might be accurately truncated to "nothing in biology makes sense." But bring in evolution, and things start to fall into place; a picture starts to form.

The connection between gene family evolution and protein structure and function is profound, and these relationships are exploited by bioinformatics methods. Gene families develop novel functions through a variety of evolutionary processes, including point mutation, insertion and deletion, gene duplication and loss (Ohno, 1970), and gene fusion and fission events (Kummerfeld and Teichmann, 2005). Because of the redundancy of the genetic code, some nucleic acid changes result in silent substitutions at the amino acid level with no impact on function. Nonsynonymous mutations (those that change the amino acid) will be tolerated easily at some positions, but disastrous at others, such as catalytic sites, and quickly removed from a population. Some positions in a protein's three-dimensional (3-D) structure allow significant variation across a family while others do not: insertions or deletions at buried positions can disrupt a protein's structural stability, while those in exposed loop regions typically have fewer intramolecular constraints, allowing loop regions and flexible linker regions between domains to vary in length across homologs. Bioinformatics methods make use of these evolutionary conservation patterns to identify the boundaries of structural domains and predict catalytic residues, specificity determinants, and other types of functionally important positions. Gene duplication events can be localized on a phylogenetic tree using gene tree–species tree reconciliation; rapid rates of functional diversification along one (or both) lineages following duplication will be reflected by long branch lengths in a reconstructed phylogeny.

Proteins are composed of independently folding globular *domains*, each of which contributes to the protein's function in the cell; the ordered series of these domains is a protein's *multidomain architecture*. Gene fusion and fission events and other sources of domain shuffling can result in dramatic changes in a protein's molecular function and the processes and pathways in which it participates.

For all these reasons, evolutionary reconstruction provides the natural framework for protein functional annotation. Phylogenetic information has been shown to improve the accuracy of a wide range of bioinformatics methods, including ortholog identification, multiple sequence alignment (MSA), profile and hidden Markov model (HMM) construction, protein structure prediction, and prediction of ligand-binding and enzyme-active-site residues. In fact, it is difficult to identify a single bioinformatics task that is *not* assisted through the use of evolution.

The combination of evolutionary reconstruction with information from protein structure is powerful. This approach, which we call *structural phylogenomics* (Sjölander, 2010), provides a computational and intellectual scaffold on which experimental and annotation data can be hung, to allow a nuanced view of the different structures and functions explored by a gene family—the molecular equivalent of Darwin's *endless forms most beautiful and most wonderful*.

5.2 HOMOLOGS, ORTHOLOGS, PARALOGS, AND OTHER EVOLUTIONARY TERMS

Homologs are genes related by a common ancestor. This common ancestry may be detectable at the sequence level, for example, using pairwise sequence comparison methods such as BLAST (Altschul et al., 1997), or may require advanced bioinformatics approaches designed to detect remote homologs. Comparison of protein structures (discussed in Chapter 3) can reveal evolutionary relationships below the *Twilight Zone* of detectable sequence similarity (roughly 25% pairwise sequence identity). Homology can be partial (e.g., along a single domain) or based on agreement on a common multidomain architecture (Addou et al., 2009; Krishnamurthy et al., 2007). Sequence identity is often used to infer homology; however, this general rule of thumb is less robust than statistical measures, which include amino acid substitution scores and penalties for insertion and deletion. BLAST *E-values* (expectation values) represent the expected number of matches with a particular score in a database of the size searched. *E*-values <0.001 are generally considered sufficient to infer homology, although some bioinformatics systems will require more significant *E*-values as a basis of functional inference. In general, homology detection methods that use pairwise sequence comparison (such as BLAST) are less sensitive than those that use multiple sequences; methods in the latter class include intermediate sequence search, hidden Markov model methods (Krogh et al., 1994), and iterated profile searches (such as PSI-BLAST) (see Park et al., (1998) for the results of a benchmark comparison of these methods).

Also refer to Chapter 1 for more discussion on the methods for comparison and clustering of sequences.

The term *ortholog* was first proposed by Fitch (1970) to differentiate genes related by speciation (*orthologs*) from those related by duplication (*paralogs*). Gene duplication events can provide a release from evolutionary constraints: following duplication, genes can evolve new functions (*neo-functionalization*) or partition the ancestral function (*sub-functionalization*) (Ohno, 1970). Some of these copies may prove to be dysfunctional and subsequently abandoned (becoming *pseudogenes* or lost entirely), but successes will be retained.

Zmasek and Eddy (2002) introduced two terms: *super-ortholog* and *ultra-paralog*. Two genes are each other's super-orthologs if and only if every node on a path joining them in a gene tree corresponds to a speciation event. Super-orthology is a very restrictive definition, but has the advantage of being transitive (by contrast, the standard definition of orthology does not imply transitivity). Ultraparalogs are defined as genes joined by a path on a phylogenetic tree where each ancestral node corresponds to a duplication event with no intervening speciation events (i.e., they must be found in the same species, and no genes outside that species can be in the same subtree). Sonnhammer and Koonin introduced additional terms, including *inparalog* (paralogs in the same species) and *outparalog* (paralogs in different species) (Sonnhammer and Koonin, 2002). Selected orthology types are illustrated in Fig. 5.1.

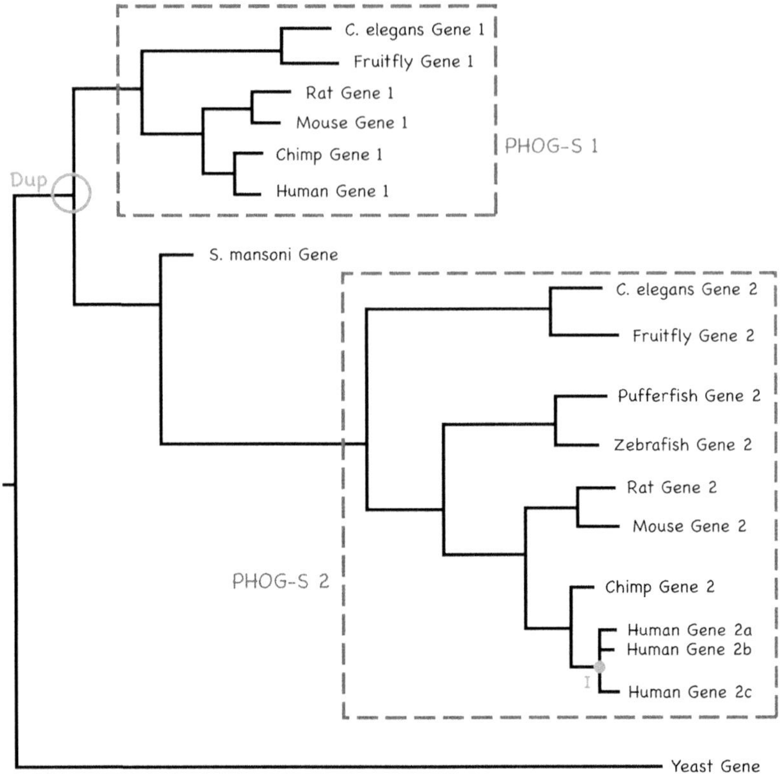

Figure 5.1 Orthology and paralogy subtypes and the PHOG algorithm. We include this toy example of gene family evolution to illustrate the main orthology subtypes and how the PHOG algorithm uses tree distances and topology jointly to infer orthologs. "Dup" indicates a duplication event in the animal lineage, and 'I' represents a group of predicted inparalogs. The two boxed subtrees (PHOG-S 1 and PHOG-S 2) correspond to super-orthology groups; PHOG-S 2 includes a possible inparalogous subtree with human genes 2a, 2b, and 2c. Standard phylogenetic orthology prediction protocols consider only the tree topology, including the *Schistosoma mansoni* gene in an orthology group with the Gene 2 clade. However, PHOG uses both tree distance and topology to enhance orthology identification precision; since the tree distances between the *S. mansoni* gene and genes in PHOG-S 1 are smaller than those between it and genes in PHOG-S 2, it is excluded from PHOG-S 2. This toy example also illustrates the nontransitivity of the standard definition of orthology, which requires only that the most recent common ancestor of two genes correspond to a speciation event. By this definition, the yeast gene is orthologous to Mouse Gene 1 and Mouse Gene 2, and to Rat Gene 1 and Rat Gene 2, and to all of the other sequences in the tree. However, Mouse Gene 1 is clearly not orthologous to Rat Gene 2 (they are paralogs, since they are related by gene duplication). *Source*: Reprinted by permission of Oxford University Press.

5.3 THE STANDARD FUNCTIONAL ANNOTATION PROTOCOL

The standard protocol to functional annotation uses an approach called ***function prediction by homology***, *annotation transfer*, or *transitive annotation*. This approach is widely used to provide predictions of function for genes in newly sequenced genomes. The most common pipeline employs BLAST, a heuristic approach to rapidly scan large sequence databases for homologous sequences; if a match can be found with significant similarity, the annotation of the match is transferred to the query. This procedure is based on the assumption that significant sequence similarity implies an evolutionary relationship and that evolution conserves function, from which it follows that the function of the query and its homolog should agree.

This assumption is largely correct—homology **can** be inferred on the basis of sequence similarity (although the converse is not true; lack of detectable sequence similarity does not mean two proteins are unrelated[1])—but there are two fundamental flaws in the logic. First, evolution is not simply conservative; it is innovative, using mechanisms such as gene duplication, domain shuffling, and changes at key positions to allow functional and structural diversification across homologs. Because of this, two genes (or proteins) can have significant BLAST scores and yet very different functions and domain architectures (Brenner, 1999; Galperin and Koonin, 1998; Green and Karp, 2005; Jones et al., 2007; Mi et al., 2003; Sjölander, 2004).

The second problem is that existing functional annotation errors can be propagated by an annotation transfer protocol (Gilks et al., 2002). Misannotations are actually quite common: a recent study shows that functional annotation error rates in the major sequence databases range from 5% to 63% (Schnoes et al., 2009). Problematically, the provenance of a functional annotation is rarely included by annotation transfer protocols, so that errors are difficult to both detect and correct. The problem is exacerbated by the persistence of annotation errors in sequence databases: while some resources (e.g., SwissProt) will modify a protein's functional annotation if evidence of error is provided, GenBank will not; only the individual who contributes the original annotation is allowed to change it, and there is little incentive for modifying those annotations.

Given these problems, it is reasonable to ask why such an error-prone annotation protocol continues to be widely used. Would it not make more sense to only transfer annotations from orthologs sharing the same multidomain architecture and for which there is experimental support for the annotation being transferred? Unfortunately, such restrictions would result in very sparse functional annotations for three main reasons: (i) fewer than 1% of proteins have any experimental support for their functional annotations (du Plessis et al., 2011); (ii) most ortholog identification protocols are designed for application to whole genomes only; and

[1]Structure comparisons have documented many cases where sequence similarity is below the detectable range but evidence for a common ancestor is convincing; see Chapter 3.

(iii) gene model errors and sequence fragments (and other issues) make it difficult to specify a protein's actual multidomain architecture.

An additional complication is that ortholog detection and structural phylogenomic approaches are often difficult to automate, computationally expensive, or dependent on advanced bioinformatics expertise, limiting their large-scale application. Orthology and phylogenomic databases including precalculated orthology relationships and phylogenetic trees that can be used in a functional annotation pipeline are discussed later in this chapter.

5.4 ORTHOLOGY IDENTIFICATION METHODS AND DATABASES

As noted earlier, orthologs are genes related by speciation events from a common ancestor, and are presumed to share a common function. Ortholog identification is central to many tasks in bioinformatics, including reconstructing species phylogenies (Delsuc et al., 2005), protein functional annotation (Eisen, 1998), predicting protein–protein interaction (Matthews et al., 2001), and phylogenetic profile construction (Pellegrini et al., 1999). Note that the standard definition of orthology allows lineage-specific expansion with potential functional divergence subsequent to the speciation event. For a more specific inference of function and for phylogenetic reconstruction of species relationships, the more selective super-orthology relationships may be required.

There are two main approaches to orthology identification: *pairwise* (or *graph-based*) and *phylogenetic tree-based* (Gabaldon et al., 2009; Kuzniar et al., 2008) methods. Graph-based methods use pairwise comparisons between whole genomes to construct a graph in which nodes (vertices) correspond to genes and edges (connections between vertices) are weighted by sequence similarity; graph-clustering algorithms can then be applied to infer orthology. Examples of graph-based methods include Orthologous MAtrix (OMA) (Roth et al., 2008), Clusters of Orthologous Groups (COGs) (Tatusov et al., 2003), InParanoid (O'Brien et al., 2005), OrthoMCL (Li et al., 2003), eggNOG (Jensen et al., 2008), ClusTr (Petryszak et al., 2005), ProtoNet (Kaplan et al., 2005), and Systers (Meinel et al., 2005).

Chapter 1 contains additional details on clustering resources.

5.5 CHALLENGES IN PHYLOGENETIC METHODS OF ORTHOLOG IDENTIFICATION

Phylogenetic (or phylogenomic) methods of orthology prediction construct and then analyze gene trees to separate orthologs from paralogs. Most phylogenetic methods use a process called *tree reconciliation*: reconciling a gene tree and a trusted species tree to identify putative duplication and gene loss events. Tree reconciliation can be challenging, for two primary reasons. First, a fully resolved and reliable species tree may be unavailable (e.g., in cases of horizontal gene transfer between microbes). Second, incongruence between the gene tree and species tree (Hahn, 2007) can prevent the application of tree reconciliation (see Dufayard et al., (2005) for an elegant solution to this problem).

Incongruence between gene trees and species trees is quite common in protein families; it can stem from incomplete lineage sorting (e.g., Pollard et al., 2006) or horizontal gene transfer, or from errors in the inferred phylogenetic tree. Species trees can also be inherently ambiguous (see e.g., the UniProt taxonomy). Errors in gene tree topology can occur in the coarse branching order (e.g., between paralogous groups) higher in the tree, or in the fine branching order between closely related taxa (species). Both types of error are quite common in protein family phylogenies because of limited phylogenetic signal stemming from a small number of sites (MSA columns) used as input to the phylogenetic reconstruction (Saitou and Nei, 1986); the latter type of tree topology error can cause incongruence between the species tree and the inferred gene tree. While phylogenetic methods for species phylogeny estimation can exploit information from many genes (e.g., by constructing a gene matrix), individual proteins have at most a few hundred sites, and protein functional or structural domains may be quite short. These issues are compounded by the use of amino acid data instead of nucleic acid data for closely related taxa and by the inclusion of sequence fragments in a phylogenetic analysis.

Phylogenetic methods making use of a gene tree but not requiring reconciliation with a species tree include LOFT (van der Heijden et al., 2007), PhylomeDB (Huerta-Cepas et al., 2008), COCO-CL (Jothi et al., 2006), and PhyloFacts Orthology Group (PHOG) (Datta et al., 2009).

Advantages of phylogenetic methods include the ability to discriminate between standard orthologs and super-orthologs (Zmasek and Eddy, 2002) and to include sequences from partly sequenced genomes. Disadvantages of phylogenetic methods include the difficulty in automating a phylogenetic tree-construction pipeline (including technical challenges involved in gene tree–species tree reconciliation), the computational cost, and dependence on human expertise.

5.6 EVALUATING ORTHOLOG IDENTIFICATION METHODS

Orthology identification methods have been evaluated in a variety of ways, including comparing predicted orthologs against those based on manually curated and reconciled phylogenetic trees, and with respect to consistency in functional annotations across putative orthologs (Altenhoff and Dessimoz, 2009; Gabaldon et al., 2009). Both approaches can be criticized: orthology prediction based on phylogenetic trees will perform well on the first benchmark, and homology-based orthology prediction will tend to have high consistency in functional annotation since most functional annotations are transferred by homology.

5.7 ORTHOLOGY DATABASES

We include a brief discussion of three orthology databases: OrthoMCL, InParanoid, and PHOG. Each of these three resources has been shown to provide highly accurate predictions of orthology, albeit using different methods and targeting different taxonomic groups. OrthoMCL and InParanoid are selected to represent methods that use graphical analyses of pairwise comparisons between whole genomes, and the PHOG database is selected as a representative of phylogenetic ortholog prediction. Both OrthoMCL and InParanoid have focused on eukaryotic genomes, while PHOG targets both microbial and eukaryotic genomes.

InParanoid (O'Brien et al., 2005) and MultiParanoid (Alexeyenko et al., 2006) are graph-based orthology databases from the Sonnhammer group. InParanoid is designed for the usage case where a biologist is primarily interested in identifying orthologs across two species, whereas MultiParanoid is designed for cases where multiple species are being compared. InParanoid has been tuned to have a low overall error rate (both false positives and negatives), and Multiparanoid is intended for the case when the selected species are roughly evolutionarily equidistant from each other. InParanoid version 7 provides orthology relationships between 100 fully sequenced eukaryotic species plus *Escherichia coli*, organizing the 1.3 million proteins in these genomes into 42.7 million ortholog pairs (Ostlund et al., 2010).

The OrthoMCL algorithm (Chen et al., 2006) extends other BLAST-based methods (such as COG/KOG) by (i) recognizing the concept of reciprocal "better" hits, that is, collapsing inparalogs that show greater similarity within species than between species; (ii) introducing a normalization step prior to clustering that (using the simplifying assumption that all species are equidistant) minimizes the impact of differences in codon bias, evolutionary rates, and other issues to improve discrimination between orthologs and paralogs; and (iii) employing Markov clustering as a rapid and automated strategy for clustering. OrthoMCL was designed to facilitate ortholog identification in eukaryotes, where frequent gene duplication produces multiple (in and out) paralogs. OrthoMCL does not explicitly address other problems associated with eukaryotic genomes, including (i) the problem of incomplete genomes attributable to sequence assembly issues and the difficulty of accurate gene finding in multi-exon genes, (ii) multiple isoforms, and (iii) the relatively high frequency of multidomain proteins. Performance metrics based on assessing the consistency of the Enzyme Commission (EC) numbers, phyletic pattern analysis, and a latent class analysis indicate excellent performance on a reference set of proteomes. OrthoMCL5 incorporates complete genome sequences for 15 archaeal species, 36 bacteria, and 98 eukaryotes.

Berkeley PHOG (for PhyloFacts Orthology Group) (Datta et al., 2009) is a phylogenetic method for orthology prediction that makes use of precalculated trees for protein families and domains in the PhyloFacts database (Krishnamurthy et al., 2006). PHOG differs from most phylogenetic orthology prediction methods in three ways. First, PHOG does not use gene tree–species tree reconciliation;

instead it uses a taxon-counting approach to identify putative duplication events. Second, PHOG uses tree distances (the sum over tree edge lengths between leaves in the tree) to identify subtrees for which the pairwise evolutionary relationships between species *within* the subtree are closer than to the same species *outside* that subtree; this is designed to improve the precision of orthology identification. Third, PHOG uses a tree-distance threshold to expand super-orthology groups to include additional homologs; this threshold is a parameter of the algorithm and can be adjusted to increase the overall recall of ortholog detection and to target specific taxonomic distances. The PHOG database includes orthologs predicted for bacterial, archaeal, and eukaryotic genomes and can be queried by UniProt accession or by BLAST search using input protein sequences in FASTA format. The PHOG algorithm is illustrated in Fig. 5.1.

5.8 PHYLOGENOMIC DATABASES

Phylogenomic databases provide gene family trees (and often orthologs predicted on the basis of these trees). All gene family trees are estimated from MSAs of proteins clustered on the basis of some criteria: most resources attempt to group sequences into families agreeing at the overall multidomain architecture, either using automated software tools (e.g., the FlowerPower algorithm) or through a combination of bioinformatics software and manual curation. Some databases are restricted to proteins from whole genomes, while others include sequences from partly sequenced genomes. Many databases focus on a particular taxonomic lineage (e.g., microbial, or animal genomes), while others have no effective restrictions.

Table 5.1 presents the major phylogenomic databases. The main differences between these lie in (i) taxonomic representation (e.g., TreeFam aims at coverage of animal gene families, Panther primarily targets eukaryotic gene families, and PhylomeDB and PhyloFacts target gene families across the Tree of Life); (ii) homology clustering methods (e.g., all resources listed attempt to cluster proteins into common domain architecture classes, while PhyloFacts also provides trees for Pfam domains); (iii) balance between automation and manual curation (e.g., Panther identifies protein subfamilies manually while PhyloFacts uses the SCI-PHY (Subfamily Classification in Phylogenomics) algorithm (Brown et al, 2007)); and (iv) ways of accessing the data.

We present additional details on three phylogenomic databases: PhyloFacts, PhylomeDB, and Panther.

5.9 PhyloFacts

The PhyloFacts Phylogenomic Encyclopedia (Krishnamurthy et al., 2006) is hosted by the Berkeley Phylogenomics Group (available at http://phylogenomics.berkeley.edu/). PhyloFacts originated in the UC Santa Cruz

TABLE 5.1 Phylogenomic and Orthology Databases

Database	URL	Notes
eggNOG	http://eggnog.embl.de/	PWO; primarily bacteria
Ensembl Compara	http://www.ensembl.org/info/docs/compara	Trees. all Ensembl genomes
HOGENOM	http://pbil.univ-lyon1.fr/databases/hogenom/	Trees. primarily bacteria
InParanoid	http://inparanoid.sbc.su.se/	PWO; ToL
MultiParanoid	http://multiparanoid.cgb.ki.se/	CO; ToL
OMA	http://www.cbrg.ethz.ch/research/orthologous	CO; ToL
OrthoMCL-DB	http://www.orthomcl.org/	CO; primarily Eukaryotic
Panther	http://www.pantherdb.org/	Trees; 16 Prokaryotes, 32 Eukaryotes
PhyloFacts/PHOG	http://phylogenomics.berkeley.edu/	Trees; ToL
PhylomeDB	http://phylomedb.org/	Trees; ToL
TreeFam	http://www.treefam.org/	Trees; primarily animal genomes

The "Notes" column includes a description of the taxonomic distribution: ToL indicates a database targets a broad (and roughly equal) representation of orthology groups across the Tree of Life; those primarily focusing on specific taxonomic groups are indicated. All resources listed provide precalculated orthologs, including pairwise orthologs only (PWO), clusters of orthologs (CO), or gene trees (Trees) from which orthologs are identified. Unless indicated explicitly, databases restrict homologs to species with whole genomes or to a small set of target species. The primary difference with this general rule is PhyloFacts/PHOG, which includes homologs/orthologs from the UniProt protein sequence database including both whole and partly sequenced genomes.

participation in the Second CASP (Critical Assessment for Protein Structure Prediction) experiment, in which a library of HMMs was constructed to predict protein 3-D structure domains (Karplus et al., 1997). That original structural focus was expanded to include protein functional subclassification using the precursor to the SCI-PHY algorithm (Sjölander, 1998) and formed the core of the Panther system employed at Celera Genomics in the functional annotation of the human genome (Venter et al., 2001).

PhyloFacts is an encyclopedia of protein families across the Tree of Life, and is organized around the concept of protein family "books." Each book contains a collection of primary and derived data: a homology cluster, MSA, phylogenetic tree, predicted orthologs, Gene Ontology (GO) annotations, Kyoto Encyclopedia of Genes and Genomes (KEGG) pathway associations, homologous structures, and so on. The PhyloFacts pipeline is displayed in Fig. 5.2. Additional details on the database construction pipeline and associated web servers are provided in Glanville et al., (2007), Krishnamurthy et al., (2006), and in Section 5.16.

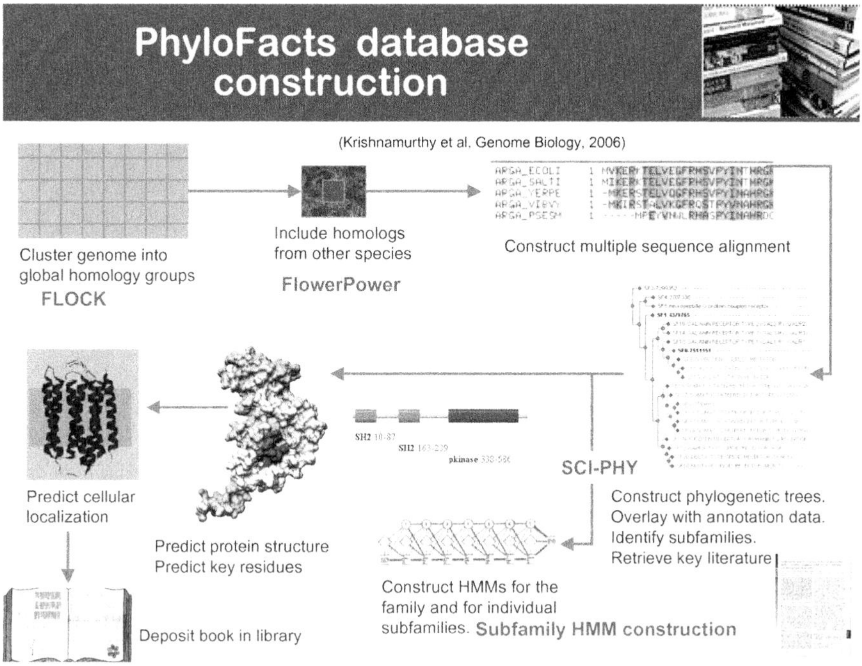

Figure 5.2 The PhyloFacts Phylogenomic Encyclopedia construction pipeline. The input to the pipeline is a genome (or set of genomes), which we cluster into homologous groups using the FLOCK (Full-Length Overlap Clustering of Kindred) tool. One representative is selected from each group and used as input to FlowerPower to gather additional homologs from the UniProt database. A multiple sequence alignment for the FlowerPower cluster is constructed using MAFFT, followed by phylogenetic tree construction, subfamily identification using Subfamily Classification in Phylogenomics (SCI-PHY), orthology prediction using PHOG, identification of homologous Pfam domains and of transmembrane helices using Phobius, prediction of homologous PDB structures using the hidden Markov model (HMM) methods, and retrieval of experimental and annotation from various resources.

PhyloFacts includes phylogenetic trees for protein functional and structural domains and for whole proteins clustered into multidomain architectures using the FlowerPower algorithm (see Fig. 5.3). While most phylogenomic resources restrict sequences to whole genomes or are biased toward specific taxonomic groups, PhyloFacts has no restrictions: sequences are drawn from across the Tree of Life from the UniProt sequence database. Experimental and annotation data are retrieved from a variety of resources, including UniProt, the GO, KEGG, and EC. A number of ways of accessing PhyloFacts families are provided, including sequence accession and HMM-based classification of protein sequences in FASTA format. Subfamilies are identified using the SCI-PHY algorithm (Brown et al., 2007). Orthology relationships are predicted using the PHOG algorithm,

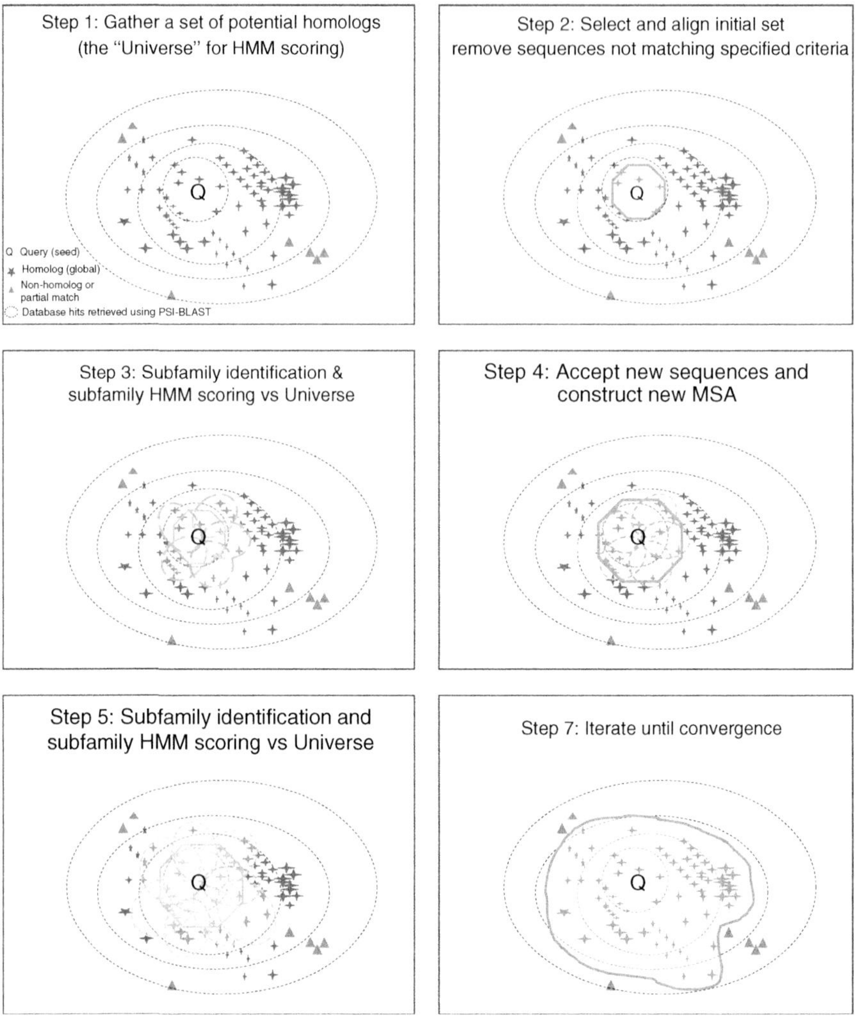

Figure 5.3 FlowerPower algorithm. FlowerPower is an iterated homology clustering tool similar to PSI-BLAST, but using subfamily hidden Markov models to retrieve homologs, and allows users to specify the type of domain architecture similarity required across homologs. The default parameterization is designed to retrieve sequences having identical multidomain architectures. In each iteration, sequences are scored against all the HMMs created in the previous round, and sequences whose HMM scores meet a predefined significance threshold are aligned to the top-scoring HMM. That alignment is examined, and the candidate is either accepted or rejected. The new alignment is analyzed to find subfamilies using the SCI-PHY algorithm, and subfamily HMMs are constructed. The process is iterated until convergence criteria are met (number of iterations, a maximum number of sequences has been selected, or no new sequences meet the user specifications). (*See insert for color representation of the figure.*)

based on analysis of trees in PhyloFacts, and can be viewed using the PHOG web server (Datta et al., 2009).

PhyloFacts also provides stand-alone web servers for the INTREPID functional site identification algorithm (Sankararaman et al., 2009), an automated phylogenomic pipeline called *PhyloBuilder* (Glanville et al., 2007), and a web server for the SATCHMO-JS simultaneous tree construction and MSA program (Hagopian et al., 2010).

5.10 SUBFAMILY CLASSIFICATION IN PhyloFacts

PhyloFacts includes two mechanisms to classify sequences to functional subfamilies: sequences included in a PhyloFacts family (e.g., by FlowerPower) are divided into subfamilies using SCI-PHY, while sequences submitted through the web site are assigned to these precalculated subfamilies using subfamily HMMs.

The SCI-PHY (Brown et al., 2007) *de novo* subfamily identifies subfamilies using a hierarchical (agglomerative) clustering protocol from an MSA: a symmetrized form of relative entropy between profiles estimated for subtrees is used to determine the branching order of the tree, and an encoding cost (based on Dirichlet mixture densities (Sjölander et al., 1996)) is used to cut the tree into subtrees defining the subfamily decomposition. This approach does not explore all possible cuts of a tree, but is computationally efficient, allowing application to very large datasets (thousands of sequences). Given a decomposition of an MSA into subfamilies, we then construct a set of subfamily HMMs; these subfamily HMMs are then used to classify user-submitted sequences to SCI-PHY subfamilies. A logistic regression approach is used to separate sequences that can be classified to SCI-PHY subfamilies and those that represent possible novel subtypes.

There is no mathematically precise definition of a functional subfamily; it depends on the context. For instance, the GPCRDB (G-protein coupled receptor database) and NucleaRDB databases classify sequences at varying levels of functional specificity, and the EC uses four fields to subdivide enzymes into a hierarchy of increasingly specific functional similarities. In general, the term "functional subfamily" is meant to imply agreement at a fairly precise level of function, for example, agreeing at all four fields of the EC number.

De novo clustering or classification methods can be roughly divided into hierarchical (typically, agglomerative) algorithms and those that attempt to divide a group into a predefined number of classes (e.g., k-means, where k is the number of classes). However, the correct number of classes may not be known *a priori*, making a k-means approach unappealing for many applications. Protein families can be organized into phylogenetic trees, but exploring all possible ways of cutting a tree into an arbitrary number of subtrees is exponential in the number of leaves in the tree.

5.11 PhyloFacts 3.0

Earlier versions of PhyloFacts focused primarily on eukaryotic genomes; PhyloFacts 3.0 has been expanded to include significant coverage of microbial genomes. As of November 29, 2011, PhyloFacts 3.0.2 includes over 7 million proteins from across the Tree of Life drawn from the UniProt database. Proteins are clustered into 81,797 families, of which 25,446 represent Pfam domains and 56,351 represent multidomain architectures. PhyloFacts 3.0.2 is available at http://makana.berkeley.edu/phylofacts/. Protein family data, including HMMs, phylogenetic trees, and MSAs for PhyloFacts families, can be downloaded from http://makana.berkeley.edu/phylofacts/downloads/.

5.12 PhylomeDB

PhylomeDB is a database for genome-wide collections of gene phylogenies, or *phylomes*. PhylomeDB has more than 400,000 individual gene trees and alignments and can be explored online or programmatically to inquire about evolutionary relationships for individual genes or at a genomic scale. PhylomeDB stores phylomes for most model organisms but also for nonmodel species from the three domains of life. A particularly useful application of PhylomeDB is its integrated phylogeny-based prediction of orthology and paralogy relationships (Gabaldon, 2008). This utility has been extended through a consistency-based approach, called *MetaPhOrs*, which provides orthology and paralogy predictions for hundreds of genomes based on gene trees stored in PhylomeDB and other public repositories (Pryszcz et al., 2011). MetaPhOrs orthology and paralogy predictions come with a confidence score based on the number of trees observed and their level of congruence.

5.13 PANTHER

PANTHER (Protein Analysis Through Evolutionary Relationships) is a resource for protein families and manually curated subfamilies (Mi et al., 2010). The current release contains 6594 trees covering approximately 400,000 protein-coding genes from 48 organisms (16 prokaryotes and 32 eukaryotes). Among phylogenomic resources, PANTHER has the greatest emphasis on manual curation: functional subfamilies are defined manually and are annotated by the GO Consortium as part of the Reference Genome Project. Panther data can be accessed through the web (http://pantherdb.org) and by FTP (ftp://ftp.pantherdb.org/).

5.14 STRUCTURAL PHYLOGENOMICS: IMPROVED FUNCTIONAL ANNOTATION THROUGH INTEGRATION OF INFORMATION FROM STRUCTURE AND EVOLUTION

The term "phylogenomics" was coined by Eisen (1998) to describe a protocol to predict protein function in an evolutionary context. Additional discussion of phylogenomic prediction of function is available in (Sjölander, 2004). (The term *phylogenomics* is also used to refer to reconstructing the evolution of species using multiple genes, for example, by a concatenated gene matrix approach (Delsuc et al., 2005).)

Structural phylogenomic prediction of gene function includes information about protein multidomain architecture and information from 3-D structure to improve the precision and recall of functional annotation (Sjölander, 2010). Functional annotation precision can be improved by using knowledge (or prediction) of a protein's domain architecture, for example, to restrict orthology calls to proteins sharing a common domain architecture. Protein structure prediction methods can be used to obtain clues about function (see Chapter 3). Finally, phylogenies can be estimated for individual structural domains separately (Sjölander et al., 2011) as shown in Fig. 5.4.

5.15 SPECIFIC ISSUES IN PHYLOGENOMIC PIPELINES

A phylogenomic analysis is described below, including a brief discussion of issues involved in constructing phylogenies and predicting orthologs based on individual domains (Sjölander et al., 2011). For each task in the pipeline, we discuss how we address this stage in constructing the PhyloFacts resource.

Step 1. Dataset selection (Clustering homologs). The first decision involves whether to restrict homologs to those that agree at the multidomain architecture or to construct a phylogeny for a single domain, allowing divergence outside of this domain. Homology clustering software suitable for domain-based phylogenies includes BLAST, PSI-BLAST, and JackHMMER (Johnson et al., 2010); program outputs simply need to be interpreted correctly to extract subsequences matching the homologous domain. The FlowerPower tool (Krishnamurthy et al., 2007) is designed specifically to cluster sequences into common multidomain architectures, although alternative methods that filter homology clusters to remove sequences that align only partially (based on pairwise BLAST comparisons) have been developed.

Another critical decision in this first stage is whether to restrict homologs included in a phylogenetic analysis, for example, to species with sequenced genomes. Dataset restrictions can facilitate some analyses, but have two limitations: first, sparse taxon sampling has been shown to contribute to errors in

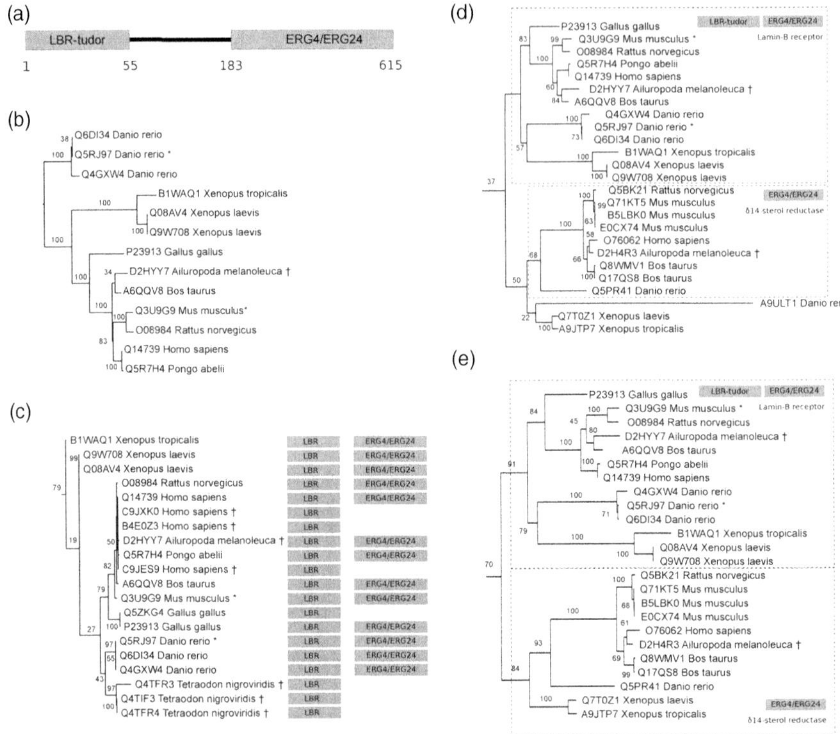

Figure 5.4 Phylogenetic analysis of human Lamin-B receptor. This figure illustrates the advantages and specific challenges involved in phylogenetic reconstruction using different domains of a protein's multidomain architecture. The query sequence used as the starting point for homology clustering for each tree displayed is UniProt sequence Q14739. Orthologs selected by TreeFam in mouse and zebrafish (*Danio rerio*) are indicated with an asterisk. Sequence fragments are marked with a dagger. (a) Pfam domain architecture for Q14739. (b) Maximum likelihood (ML) tree of proteins sharing the same domain architecture identified using FlowerPower. (c) ML tree of proteins aligning to the N-terminal LBR_tudor domain; a subtree of the full tree is shown, restricted to the vertebrate lineage. Pfam domains found for the full-length amino acid sequences are displayed at right. (d) and (e) ML trees of sequences matching the C-terminal ERG4/ERG24 domain (restricted as in (c) to the vertebrate lineage) constructed using RAxML (d) and FastTree (e), respectively. Super-orthology groups are boxed with dashed lines, and sequences within each super-orthology group have identical domain architectures and functions. In both (d) and (e), the upper subtree contains the human Lamin-B receptor and orthologs; sequences in the lower subtree are missing the N-terminal LBR-tudor domain. Note that zebrafish protein A9ULT1 included by RAxML (albeit with low bootstrap support) was excluded by FastTree, allowing predicted super-orthologs in the lower subtree of (e) to expand to include the two *Xenopus* sequences. Reprinted by permission of Oxford University Press.

tree topologies (Hillis et al., 2003; Zwickl and Hillis, 2002); second, restrictions on sequences included in an analysis may limit the experimental and annotation data available to a functional inference.

Finally, some superfamilies are too large to include all members in a single phylogeny (e.g., the Pfam 7TM_1 family of G-protein coupled receptors); in these cases, it may be necessary to represent the family by a *set covering* of smaller, overlapping subsets.

The PhyloFacts database includes phylogenetic trees for both functional and structural domains and for sequences agreeing along their multidomain architecture. We target some genomes for thorough coverage; these are used as seeds for clustering homologs using FlowerPower (default parameters are used for multidomain architecture clustering; for domain-based phylogenies, we use parameters designed for *global* alignment—global to the query and local to the database hit). Homologs are retrieved from UniProt, and are not restricted to whole (sequenced) genomes or to particular taxonomic groups. We use a set-covering strategy to model large protein families; individual proteins can belong to many different trees, with one (or more) trees for each individual Pfam domain and one (or more) for the multidomain architecture. For each Pfam domain and for the multidomain architecture, we designate one PhyloFacts "book" that is "best" for each sequence (based on the data available for the family). Some PhyloFacts families span the Tree of Life, while others are restricted to a small set of species. We reduce redundancy (overlap between families) using various techniques, including preclustering sequences in input genomes and culling families with significant overlap during postprocessing.

Step 2. MSA construction and alignment masking. Top-ranked MSA construction methods include MAFFT (Katoh and Toh, 2008), MUSCLE (Edgar, 2004), SATCHMO-JS (Hagopian et al., 2010), ProbCons (Do et al., 2005), T-Coffee (Notredame et al., 2000), and ClustalW (Thompson et al., 2002). MAFFT and MUSCLE are commonly used in constructing phylogenomic databases because of their scalability to large datasets and excellent performance on benchmark datasets used to evaluate alignment accuracy.

MSAs are typically masked prior to phylogenetic tree construction. A fundamental assumption of a phylogenetic reconstruction is *positional homology*: that is, characters observed at a site are related by descent from a common ancestral character. Alignment masking is commonly used to remove columns where this assumption appears questionable, such as those containing large numbers of gap characters or regions with high sequence variability. However, standard MSA masking protocols that are appropriate for phylogenomic reconstruction of species phylogenies (i.e., concatenated gene matrix approaches, containing potentially thousands of sites) may not be appropriate for application to a single protein family containing at most a few hundred sites. In particular, if the homology cluster includes sequences with extreme sequence and structural divergence, standard masking protocols will remove a large fraction of positions, leaving very few sites for a phylogenetic reconstruction.

In PhyloFacts, we use the MAFFT software to construct MSAs, and mask positions with >70% gaps.

Step 3. Phylogenetic tree estimation using the masked MSA as input. Numerous phylogenetic reconstruction software packages exist, ranging from methods that are fast and can handle very large datasets (e.g., QuickTree (Howe et al., 2002) and FastTree (Price et al., 2010)) to those that are more computationally demanding (e.g., RAxML (Stamatakis et al., 2005) and MrBayes (Huelsenbeck and Ronquist, 2001)). The PHYLIP resource, maintained by Joe Felsenstein at the University of Washington (http://evolution.genetics.washington.edu/phylip.html), provides numerous software tools for phylogenetic tree construction and visualization. In PhyloFacts, we use FastTree and RAxML to construct maximum likelihood (ML) trees.

Step 4. Overlaying the tree with experimental and annotation data. A primary source of functional annotation is provided for many species through the GO (Ashburner et al., 2000); GO annotations include evidence codes denoting whether the annotation has experimental support (e.g., evidence code IDA, for *inferred from direct assay*) or was derived computationally (e.g., evidence code IEA, for *inferred from electronic annotation*). The SwissProt database (a subset of the UniProt resource, at http://www.uniprot.org/) is a leading resource of high accuracy manually curated and detailed annotations; independent analysis shows SwissProt annotations to be highly precise (Schnoes et al., 2009).

PhyloFacts includes data from the GO, UniProt (SwissProt and TrEMBL), the Protein Data Bank (PDB), EC, KEGG, and other resources. These data are used to provide summary annotations for families, subfamilies, and orthology groups, and are overlaid on protein family trees for manual examination using the PhyloScope tree viewer (see Fig. 5.5).

Step 5. Analyzing the tree to distinguish between orthologs and paralogs. As noted earlier, most phylogenetic methods for ortholog identification use gene tree–species tree reconciliation to identify putative duplication and loss events. Automated methods are required for large-scale phylogenetic inference of orthologs (e.g., to construct phylogenomic databases); however, individual protein families can be analyzed manually, particularly when the family is relatively small. We recommend restricting predicted orthologs to those found in subtrees with high bootstrap support. Subtrees that are restricted to sequences from a single taxon (species) can be treated as putative inparalogs (or ultraparalogs) and included in an orthology group (i.e., using a similar protocol as that employed by OrthoMCL and PHOG); in many cases, these will correspond to alternative isoforms or gene model errors for the same gene (see, e.g., Fig. 5.4). Identifying these cases is relatively straightforward when sequences are drawn from whole genomes, in which case the proteins can be mapped to their underlying genomic locus using translated BLAST.

Software tools for tree reconciliation include RAP (Dufayard et al., 2005), EnsemblCompara (Vilella et al., 2009), RIO (Zmasek and Eddy, 2002), Orthostrapper (Storm and Sonnhammer, 2002), and NOTUNG (Chen et al., 2000). Many tree reconciliation tools require the gene tree to agree with the trusted species

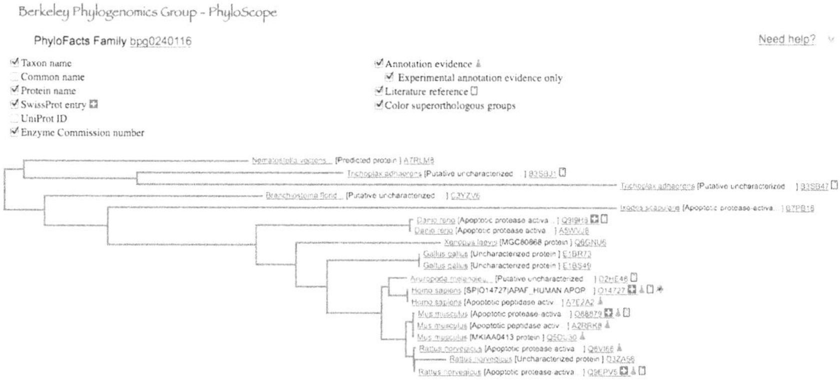

Figure 5.5 PhyloScope display of a protein family in PhyloFacts. Shown here is the PhyloFacts family tree for human apoptotic protease-activating factor 1 (SwissProt sequence identifier APAF_HUMAN), displaying a tree for proteins that align globally to APAF_HUMAN. Icons at the leaves display various types of data available for members of the family. For instance, APAF_HUMAN has several associated crystal structures in the Protein Data Bank (PDB); these are indicated by an icon resembling a crystal structure. Similarly, published papers are available for several members, indicated by page icons. SwissProt sequences are indicated by the Swiss flag, and experimental data supporting Gene Ontology annotations are indicated by a flask. Mousing over an icon reveals summary data, and clicking on an icon will bring up a webpage from the corresponding resource. PhyloScope can display trees with up to a few hundred members (larger trees are displayed using the Archaeopteryx software).

phylogeny; however, incongruence between the gene tree and species tree is common in protein family phylogenies, and is particularly problematic for phylogenies for short domains (Sjölander et al., 2011). Other complications in tree reconciliation include the presence of alternative isoforms and duplicate entries of the same gene: these can be labeled (incorrectly) as inparalogs by tree reconciliation software. If sequences included in the phylogeny have been restricted to those from whole genomes, examining the genomic locus for these sequences can separate these from true inparalogs.

A final complicating factor in this stage is that automated tree reconciliation software tools are designed to identify which internal nodes in a tree correspond to speciation and which to duplication events. This is insufficient for application to phylogenies estimated for individual domains found in different multidomain architectures. For instance, in a gene tree restricted to orthologs, internal nodes of the tree will correspond to speciation events, and internal nodes of protein *superfamily* phylogenies (including orthologs and paralogs) will represent speciation and duplication events. However, in cases where a phylogeny is estimated for a structural domain found in different multidomain architectures, internal nodes of the tree may also represent gene fusion and fission events. If the leaves of the tree are overlaid with the predicted multidomain architecture for the full-length proteins from which the domain was extracted, branch points on the tree corresponding to domain architecture rearrangements can be identified manually. Note, however, that it may not be possible to predict the multidomain architecture precisely for sequences with underlying gene model errors or for which one or more domains vary significantly from the family consensus (in which case they may not be detected by an HMM derived for that domain).

PhyloFacts uses the PHOG algorithm to identify orthologs, using precomputed trees based on both Pfam domains and whole proteins (multidomain architecture clustering). The PHOG algorithm does not use tree reconciliation to identify orthologs, but uses a taxon-counting approach and a combination of tree distance and topology to identify orthologs.

Step 6. Inferring the Function(s) of Individual Sequences in the Family Based on their Placement in the Annotated Tree. Because super-orthology is a transitive relationship, any experimentally supported functions or manually derived annotations (e.g., from a resource such as SwissProt) can often be safely transferred across all the sequences in that subtree. For sequences matching the standard definition of orthology, but not super-orthology, additional analyses will be needed to determine whether functions may have diverged since the most recent common ancestor. However, the sparsity of experimentally supported annotations (and of manually curated proteins) often necessitates the use of more tentative functional assignments, perhaps based on a *subtree neighbor* outside an orthology group (Zmasek and Eddy, 2002). Storing the provenance of the annotation (i.e., the accession of the sequence whose annotation is being transferred and the evidence for the type of orthology relationship used as the basis for annotation transfer) is highly recommended.

5.16 IMPROVING FUNCTIONAL INFERENCE USING INFORMATION FROM PROTEIN STRUCTURE

The protocol outlined here has focused on the identification of orthologs as the basis for functional annotation. However, even among orthologs, functions can diverge, and paralogs can occasionally have greater functional similarity than orthologs (see, e.g., Studer and Robinson-Rechavi (2009)). Functional changes between orthologs are particularly likely if species divergence is high or if mutations have occurred at critical positions.

Including information from protein structure can assist in avoiding functional inference errors. First, we recommend restricting functional inferences to orthologs sharing the same multidomain architecture. Second, residues known to be essential for function (such as catalytic sites in enzymes) should be examined at this stage. For instance, proteins predicted to be serine proteases ought to agree at the catalytic triad; information about these sites is stored in the Catalytic Site Atlas (Porter et al., 2004). Evolutionary conservation patterns can be used alone (e.g., using INTREPID (Sankararaman and Sjölander, 2008)) or in combination with information from protein 3-D structure (e.g., using Discern (Sankararaman et al., 2010)) to predict enzyme active sites and other functionally important positions. These predictions of functional sites can be compared across the family; strong disagreements with members of functional groups whose annotations were to be transferred should be taken into account before transferring an annotation.

5.17 EXAMPLE CASE STUDIES

Example 1: Functional annotation error stemming from annotation transfer based on partial homology. Plant receptor-like proteins (RLPs) and receptor-like protein kinases (RLKs) have homologous extracellular leucine-rich-repeat (LRR) regions followed by a transmembrane (TM) helix, but RLKs have a C-terminal cytoplasmic kinase domain following their TM helix while RLPs have only a short cytoplasmic tail (Fritz-Laylin et al., 2005). *Oryza sativa* protein Q851L1_ORYSJ from the UniProt database is annotated as a "putative LRR receptor-like protein kinase" (i.e., an RLK). However, Pfam analysis shows only LRRs and no kinase domain. While false negatives are not uncommon for some Pfam domains (e.g., TIR domains), kinase domains have a distinctive amino acid signature that makes them difficult to miss (and there is insufficient room after the last detected LRR for a kinase domain). Assuming the gene model is correct, this rice protein is more likely to be an RLP than an RLK. The top match in *Arabidopsis thaliana* is a *bona fide* RLK, suggesting that this sequence is likely to have been annotated by an annotation transfer protocol with an RLK.

Example 2: Pros and cons of domain-based phylogenies versus those based on global similarity. In this analysis (reported separately in Sjölander et al., (2011)), and shown in Fig. 5.4), we constructed phylogenies for homologs to human Lamin-B receptor (UniProt sequence Q14739). The human Lamin-B receptor has an N-terminal LBR_tudor domain 55 amino acids in length, and a C-terminal ERG4_ERG24 domain 430 amino acids in length. We constructed three phylogenies. In the first, we required homologs to agree over the entire multidomain architecture; in the other two, we required only that homologs contained one of the two Pfam domains (allowing sequences to differ outside that domain). From each phylogenetic tree, orthologs were predicted using the PHOG algorithm.

As expected, all orthologs found in the phylogeny based on agreement at the multidomain architecture level are also found in phylogenies for individual PFAM domains. However, the less restrictive clustering criteria used in the Pfam trees results in the inclusion of novel sequences. For example, the N-terminal LBR_tudor domain phylogeny includes additional proteins from chicken (*Gallus gallus*), human, and pufferfish (*Tetraodon nigroviridis*). Detailed analyses (not shown) reveals that the "novel" human and chicken sequences are proper subsequences of full-length proteins included in all three trees (i.e., they correspond to incorrect gene models for the same genes). By contrast, the pufferfish ortholog is found only in the LBR_tudor tree, illustrating the utility of a domain-based phylogeny and ortholog identification: if the clustering had been restricted to sequences aligning globally, the pufferfish gene would not have been detected.

On the other hand, this case study also illustrates the potential problems with using short domains to estimate phylogenetic trees. For instance, the phylogenetic placement of the mouse ortholog (Q3U9G9) is incorrect in the tree estimated from the N-terminal LBR_tudor domain, but is correct in the phylogenies derived from the global MSA and the C-terminal ERG4_ERG24 domain, both of which were estimated using many more sites. This incorrect placement can be attributed to the reduced information available in the short MSA used as the basis of the phylogenetic reconstruction, and is compounded by the use of amino acid instead of nucleic acid data.

Another potential problem with a domain-based phylogenetic reconstruction is the presence of "promiscuous" domains (i.e., domains found in many combinations). However, analysis of the PFAM domains of full-length proteins included in a phylogeny shows a close correspondence between proximity in the phylogenetic tree and agreement at domain architecture, allowing the inference of overall domain architecture for sequence fragments and increasing the reliability of domain-based orthology prediction.

5.18 REVIEW OF KEY POINTS

in this chapter, we have reviewed the mechanisms by which gene families evolve novel functions and structures and the different bioinformatics methods used to predict gene function. A small number of bioinformatics databases providing pre calculated phylogenetic trees and orthology relationships have been presented, along with a simple protocol that can be used to cluster proteins into families and identify orthologs.

REFERENCES

Addou, S., Rentzsch, R., Lee, D., and Orengo, C.A. (2009) Domain-based and family-specific sequence identity thresholds increase the levels of reliable protein function transfer. J Mol Biol, 387, 416–430.

Alexeyenko, A., Tamas, I., Liu, G., and Sonnhammer, E.L. (2006) Automatic clustering of orthologs and inparalogs shared by multiple proteomes. Bioinformatics, 22, e9–15.

Altenhoff, A.M. and Dessimoz, C. (2009) Phylogenetic and functional assessment of orthologs inference projects and methods. PLoS Comp Biol, 5, e1000262.

Altschul, S.F., Madden, T.L., Schaffer, A.A., Zhang, J., Zhang, Z., Miller, W., and Lipman, D.J. (1997) Gapped BLAST and PSI-BLAST: a new generation of protein database search programs. Nucleic Acids Res, 25, 3389–3402.

Ashburner, M., Ball, C.A., Blake, J.A., Botstein, D., Butler, H., Cherry, J.M., Davis, A.P., Dolinski, K., Dwight, S.S., Eppig, J.T. et al. (2000) Gene ontology: tool for the unification of biology. The Gene Ontology Consortium. Nat Genet, 25, 25–29.

Brenner, S.E. (1999) Errors in genome annotation. Trends Genet, 15, 132–133.

Brown, D.P., Krishnamurthy, N., and Sjölander, K. (2007) Automated protein subfamily identification and classification. PLoS Comput Biol, 3, e160.

Chen, F., Mackey, A.J., Stoeckert, C.J. Jr., and Roos, D.S. (2006) OrthoMCL-DB: querying a comprehensive multi-species collection of ortholog groups. Nucleic Acids Res, 34, D363–D368.

Chen, K., Durand, D., and Farach-Colton, M. (2000) NOTUNG: a program for dating gene duplications and optimizing gene family trees. J Comput Biol, 7, 429–447.

Datta, R.S., Meacham, C., Samad, B., Neyer, C., and Sjölander, K. (2009) Berkeley PHOG: PhyloFacts orthology group prediction web server. Nucleic Acids Res, 37(Web Server issue), W84–W89.

Delsuc, F., Brinkmann, H., and Philippe, H. (2005) Phylogenomics and the reconstruction of the tree of life. Nat Rev, 6, 361–375.

Do, C.B., Mahabhashyam, M.S.P., Brudno, M., and Batzoglou, S. (2005) ProbCons: probabilistic consistency-based multiple sequence alignment. Genome Res, 15, 330–340.

Dobzhansky, C.T. (1973) Nothing in biology makes sense except in the light of evolution. Am Biol Teach, 35, 125–129.

du Plessis, L., Skunca, N., and Dessimoz, C. (2011) The what, where, how and why of gene ontology–a primer for bioinformaticians. Brief Bioinform, 12(6), 723–735.

Dufayard, J.F., Duret, L., Penel, S., Gouy, M., Rechenmann, F., and Perriere, G. (2005) Tree pattern matching in phylogenetic trees: automatic search for orthologs or paralogs in homologous gene sequence databases. Bioinformatics, 21, 2596–2603.

Edgar, R.C. (2004) Muscle: multiple sequence alignment with high accuracy and high throughput. Nucleic Acids Res, 32, 1792–1797.

Eisen, J.A. (1998) Phylogenomics: improving functional predictions for uncharacterized genes by evolutionary analysis. Genome Res, 8, 163–167.

Fitch, W.M. (1970) Distinguishing homologous from analogous proteins. Syst Zool, 19, 99–113.

Fritz-Laylin, L.K., Krishnamurthy, N., Tor, M., Sjölander, K.V., and Jones, J.D. (2005) Phylogenomic analysis of the receptor-like proteins of rice and Arabidopsis. Plant Physiol, 138, 611–623.

Gabaldon, T. (2008) Large-scale assignment of orthology: back to phylogenetics? Genome Biol, 9, 235.

Gabaldon, T., Dessimoz, C., Huxley-Jones, J., Vilella, A.J., Sonnhammer, E.L., and Lewis, S. (2009) Joining forces in the quest for orthologs. Genome Biol, 10, 403.

Galperin, M.Y. and Koonin, E.V. (1998) Sources of systematic error in functional annotation of genomes: domain rearrangement, non-orthologous gene displacement and operon disruption. In Silico Biol, 1, 55–67.

Gilks, W.R., Audit, B., De Angelis, D., Tsoka, S., and Ouzounis, C.A. (2002) Modeling the percolation of annotation errors in a database of protein sequences. Bioinformatics, 18, 1641–1649.

Glanville, J.G., Kirshner, D., Krishnamurthy, N., and Sjölander, K. (2007) Berkeley phylogenomics group web servers: resources for structural phylogenomic analysis. Nucleic Acids Res, 35, W27–W32.

Green, M.L. and Karp, P.D. (2005) Genome annotation errors in pathway databases due to semantic ambiguity in partial EC numbers. Nucleic Acids Res, 33, 4035–4039.

Hagopian, R., Davidson, J.R., Datta, R.S., Samad, B., Jarvis, G.R., and Sjölander, K. (2010) SATCHMO-JS: a webserver for simultaneous protein multiple sequence alignment and phylogenetic tree construction. Nucleic Acids Res, 38, W29–W34.

Hahn, M.W. (2007) Bias in phylogenetic tree reconciliation methods: implications for vertebrate genome evolution. Genome Biol, 8, R141.

Hillis, D.M., Pollock, D.D., McGuire, J.A., and Zwickl, D.J. (2003) Is sparse taxon sampling a problem for phylogenetic inference? Syst Biol, 52, 124–126.

Howe, K., Bateman, A., and Durbin, R. (2002) QuickTree: building huge neighbour-joining trees of protein sequences. Bioinformatics, 18, 1546–1547.

Huelsenbeck, J.P. and Ronquist, F. (2001) MRBAYES: Bayesian inference of phylogenetic trees. Bioinformatics, 17, 754–755.

Huerta-Cepas, J., Bueno, A., Dopazo, J., and Gabaldon, T. (2008) PhylomeDB: a database for genome-wide collections of gene phylogenies. Nucleic Acids Res, 36, D491–D496.

Jensen, L.J., Julien, P., Kuhn, M., von Mering, C., Muller, J., Doerks, T., and Bork, P. (2008) eggNOG: automated construction and annotation of orthologous groups of genes. Nucleic Acids Res, 36, D250–D254.

Johnson, L.S., Eddy, S.R., and Portugaly, E. (2010) Hidden Markov model speed heuristic and iterative HMM search procedure. BMC Bioinformatics, 11, 431.

Jones, C.E., Brown, A.L., and Baumann, U. (2007) Estimating the annotation error rate of curated GO database sequence annotations. BMC Bioinformatics, 8, 170.

Jothi, R., Zotenko, E., Tasneem, A., and Przytycka, T.M. (2006) COCO-CL: hierarchical clustering of homology relations based on evolutionary correlations. Bioinformatics, 22, 779–788.

Kaplan, N., Sasson, O., Inbar, U., Friedlich, M., Fromer, M., Fleischer, H., Portugaly, E., Linial, N., and Linial, M. (2005) ProtoNet 4.0: a hierarchical classification of one million protein sequences. Nucleic Acids Res, 33, D216–D218.

Karplus, K., Sjölander, K., Barrett, C., Cline, M., Haussler, D., Hughey, R., Holm, L., and Sander, C. (1997) Predicting protein structure using hidden Markov models. Proteins, 1, 134–139.

Katoh, K. and Toh, H. (2008) Recent developments in the MAFFT multiple sequence alignment program. Brief Bioinform, 9, 286–298.

Krishnamurthy, N., Brown, D., and Sjölander, K. (2007) FlowerPower: clustering proteins into domain architecture classes for phylogenomic inference of protein function. BMC Evol Biol, 7(1 Suppl), S12.

Krishnamurthy, N., Brown, D.P., Kirshner, D., and Sjölander, K. (2006) PhyloFacts: an online structural phylogenomic encyclopedia for protein functional and structural classification. Genome Biol, 7, R83.

Krogh, A., Brown, M., Mian, I.S., Sjölander, K., and Haussler, D. (1994) Hidden Markov models in computational biology. Applications to protein modeling. J Mol Biol, 235, 1501–1531.

Kummerfeld, S.K. and Teichmann, S.A. (2005) Relative rates of gene fusion and fission in multi-domain proteins. Trends Genet, 21, 25–30.

Kuzniar, A., van Ham, R.C., Pongor, S., and Leunissen, J.A. (2008) The quest for orthologs: finding the corresponding gene across genomes. Trends Genet, 24, 539–551.

Li, L., Stoeckert, C.J. Jr., and Roos, D.S. (2003) OrthoMCL: identification of ortholog groups for eukaryotic genomes. Genome Res, 13, 2178–2189.

Matthews, L.R., Vaglio, P., Reboul, J., Ge, H., Davis, B.P., Garrels, J., Vincent, S., and Vidal, M. (2001) Identification of potential interaction networks using sequence-based searches for conserved protein-protein interactions or "interologs". Genome Res, 11, 2120–2126.

Meinel, T., Krause, A., Luz, H., Vingron, M., and Staub, E. (2005) The SYSTERS Protein Family Database in 2005. Nucleic Acids Res, 33, D226–D229.

Mi, H., Dong, Q., Muruganujan, A., Gaudet, P., Lewis, S., and Thomas, P.D. (2010) PANTHER version 7: improved phylogenetic trees, orthologs and collaboration with the Gene Ontology Consortium. Nucleic Acids Res, 38, D204–D210.

Mi, H., Vandergriff, J., Campbell, M., Narechania, A., Majoros, W., Lewis, S., Thomas, P.D., and Ashburner, M. (2003) Assessment of genome-wide protein function classification for Drosophila melanogaster. Genome Res, 13, 2118–2128.

Notredame, C., Higgins, D.G., and Heringa, J. (2000) T-Coffee: a novel method for fast and accurate multiple sequence alignment. J Mol Biol, 302, 205–217.

O'Brien, K.P., Remm, M., and Sonnhammer, E.L. (2005) Inparanoid: a comprehensive database of eukaryotic orthologs. Nucleic Acids Res, 33, D476–D480.

Ohno, S. (1970) Evolution by gene duplication, Springer-Verlag, New York.

Ostlund, G., Schmitt, T., Forslund, K., Kostler, T., Messina, D.N., Roopra, S., Frings, O., and Sonnhammer, E.L. (2010) InParanoid 7: new algorithms and tools for eukaryotic orthology analysis. Nucleic Acids Res, 38, D196–D203.

Park, J., Karplus, K., Barrett, C., Hughey, R., Haussler, D., Hubbard, T., and Chothia, C. (1998) Sequence comparisons using multiple sequences detect three times as many remote homologues as pairwise methods. J Mol Biol, 284, 1201–1210.

Pellegrini, M., Marcotte, E.M., Thompson, M.J., Eisenberg, D., and Yeates, T.O. (1999) Assigning protein functions by comparative genome analysis: protein phylogenetic profiles. Proc Natl Acad Sci USA, 96, 4285–4288.

Petryszak, R., Kretschmann, E., Wieser, D., and Apweiler, R. (2005) The predictive power of the CluSTr database. Bioinformatics, 21, 3604–3609.

Pollard, D.A., Iyer, V.N., Moses, A.M., and Eisen, M.B. (2006) Widespread discordance of gene trees with species tree in Drosophila: evidence for incomplete lineage sorting. PLoS Genet, 2, e173.

Porter, C.T., Bartlett, G.J., and Thornton, J.M. (2004) The Catalytic Site Atlas: a resource of catalytic sites and residues identified in enzymes using structural data. Nucleic Acids Res, 32, D129–D133.

Price, M.N., Dehal, P.S., and Arkin, A.P. (2010) FastTree 2–approximately maximum-likelihood trees for large alignments. PLoS One, 5, e9490.

Pryszcz, L.P., Huerta-Cepas, J., and Gabaldon, T. (2011) MetaPhOrs: orthology and paralogy predictions from multiple phylogenetic evidence using a consistency-based confidence score. Nucleic Acids Res, 39(5), e32.

Roth, A.C., Gonnet, G.H., and Dessimoz, C. (2008) Algorithm of OMA for large-scale orthology inference. BMC Bioinformatics, 9, 518.

Saitou, N. and Nei, M. (1986) The number of nucleotides required to determine the branching order of three species, with special reference to the human-chimpanzee-gorilla divergence. J Mol Evol, 24, 189–204.

Sankararaman, S., Kolaczkowski, B., and Sjölander, K. (2009) INTREPID: a web server for prediction of functionally important residues by evolutionary analysis. Nucleic Acids Res, 37, W390–W395.

Sankararaman, S., Sha, F., Kirsch, J.F., Jordan, M.I., and Sjölander, K. (2010) Active site prediction using evolutionary and structural information. Bioinformatics, 26, 617–624.

Sankararaman, S. and Sjölander, K. (2008) INTREPID–information-theoretic TREe traversal for Protein functional site identification. Bioinformatics, 24, 2445–2452.

Schnoes, A.M., Brown, S.D., Dodevski, I., and Babbitt, P.C. (2009) Annotation error in public databases: misannotation of molecular function in enzyme superfamilies. PLoS Comput Biol, 5, e1000605.

Sjölander, K. (1998) Phylogenetic inference in protein superfamilies: analysis of SH2 domains, in Proceedings, International Conference on Intelligent Systems for Molecular Biology, Vol. ISMB 6, pp. 165–174.

Sjölander, K. (2004) Phylogenomic inference of protein molecular function: advances and challenges. Bioinformatics, 20, 170–179.

Sjölander, K. (2010) Getting started in structural phylogenomics. PLoS Comput Biol, 6(1), e1000621.

Sjölander, K., Datta, R.S., Shen, Y., and Shoffner, G.M. (2011) Ortholog identification in the presence of domain architecture rearrangement. Brief Bioinform, 12(5), 413–422.

Sjölander, K., Karplus, K., Brown, M., Hughey, R., Krogh, A., Mian, I.S., and Haussler, D. (1996) Dirichlet mixtures: a method for improved detection of weak but significant protein sequence homology. Comput Appl Biosci, 12, 327–345.

Sonnhammer, E.L. and Koonin, E.V. (2002) Orthology, paralogy and proposed classification for paralog subtypes. Trends Genet, 18, 619–620.

Stamatakis, A., Ludwig, T., and Meier, H. (2005) RAxML-III: a fast program for maximum likelihood-based inference of large phylogenetic trees. Bioinformatics, 21, 456–463.

Storm, C.E. and Sonnhammer, E.L. (2002) Automated ortholog inference from phylogenetic trees and calculation of orthology reliability. Bioinformatics, 18, 92–99.

Studer, R.A. and Robinson-Rechavi, M. (2009) How confident can we be that orthologs are similar, but paralogs differ? Trends Genet, 25, 210–216.

Tatusov, R.L., Fedorova, N.D., Jackson, J.D., Jacobs, A.R., Kiryutin, B., Koonin, E.V., Krylov, D.M., Mazumder, R., Mekhedov, S.L., Nikolskaya, A.N. et al. (2003) The COG database: an updated version includes eukaryotes. BMC Bioinformatics, 4, 41.

Thompson, J.D., Gibson, T.J., and Higgins, D.G. (2002) Multiple sequence alignment using ClustalW and ClustalX. Curr Protoc Bioinformatics Chapter 2, Unit 2.3.

van der Heijden, R.T., Snel, B., van Noort, V., and Huynen, M.A. (2007) Orthology prediction at scalable resolution by phylogenetic tree analysis. BMC Bioinformatics, 8, 83.

Venter, J.C., Adams, M.D., Myers, E.W., Li, P.W., Mural, R.J., Sutton, G.G., Smith, H.O., Yandell, M., Evans, C.A., Holt, R.A. et al. (2001) The sequence of the human genome. Science, 291, 1304–1351.

Vilella, A.J., Severin, J., Ureta-Vidal, A., Heng, L., Durbin, R., and Birney, E. (2009) EnsemblCompara GeneTrees: complete, duplication-aware phylogenetic trees in vertebrates. Genome Res, 19, 327–335.

Zmasek, C.M. and Eddy, S.R. (2002) RIO: analyzing proteomes by automated phylogenomics using resampled inference of orthologs. BMC Bioinformatics, 3, 14.

Zwickl, D.J. and Hillis, D.M. (2002) Increased taxon sampling greatly reduces phylogenetic error. Syst Biol, 51, 588–598.

SECTION II

IN-DEPTH REVIEWS OF PROTEIN FAMILIES

6

THE NUCLEOPHILIC ATTACK SIX-BLADED β-PROPELLER (N6P) SUPERFAMILY

Michael A. Hicks, Alan E. Barber II, and Patricia C. Babbitt

Department of Bioengineering and Therapeutic Sciences, UCSF Mission Bay, San Francisco, CA, USA

CHAPTER SUMMARY

Functionally diverse enzyme superfamilies (SFs) are defined in this chapter as groups of evolutionarily related proteins that conserve constellations of functionally important active site residues but whose substrates, products, and even overall reactions can be substantially different. Several of these SFs are classified in a publically available resource, the Structure–Function Linkage Database (SFLD), which links protein sequences and structures with their chemical reactions. The classification of some protein families into SFs, however, can be especially difficult when complex variations in sequence, combined with an incomplete knowledge of the determinants of function, prevent a full understanding of the sequence–structure–function relationship. The nucleophilic attack six-bladed β-propeller (N6P) SF provides an interesting model for evaluating these complexities based both on the limited number of functionally characterized proteins in the SF and on the great diversity of its sequences and functions. This chapter describes the SF and its classification in the SFLD and other protein databases. Additionally, we describe the N6P SF in the context of other proteins sharing the β-propeller fold and

Protein Families: Relating Protein Sequence, Structure, and Function, First Edition.
Edited by Christine Orengo and Alex Bateman.

highlight the challenges in defining accurate functional boundaries between structurally similar groups. Finally, we discuss how the SF context provides a strategy for selecting protein targets for experimental characterization that can be especially informative for functional inference.

6.1 INTRODUCTION

By August 2013, the UniProtKB/TrEMBL protein database contained more than 41 million sequences and was rapidly increasing in size, with the rate of genome sequencing continuing to increase as technological advances drive down sequencing costs. Metagenomics and other new sequencing efforts also continue to ramp up, promising to multiply the number of available protein sequences many times over. As these databases grow, a decreasing proportion of these proteins can be experimentally characterized, necessitating assignment of molecular function almost exclusively using computational methods. To meet these challenges, annotation strategies have become more sophisticated, benefiting especially from the use of multiple orthogonal methods to improve the accuracy of functional prediction (reviewed in Frishman, 2007; Loewenstein et al., 2009; Rentzsch and Orengo, 2009). Accompanying these, new resources have been developed to classify a large proportion of the known protein universe of evolutionarily related proteins into groupings so that what is known about their sequence as well as structural and functional relationships can be used as a starting point for inference of their functional properties (see Chapters 2, 4, and 5).

Additional resources target specific classes of proteins for exploration and classification of structure–function relationships. This chapter introduces an important subset of enzyme SFs that we describe as "functionally diverse" and for which data and information is accessible using one of these specialized resources, the SFLD (Pegg et al., 2006) (http://sfld.rbvi.ucsf.edu/). The SFLD defines properties of enzyme SFs with a focus on mapping information about their molecular functions and mechanisms to similarity-based classification of their sequences, structures, and active site features. The SFLD also provides tools for investigating the evolution of the many different reaction families these SFs contain and predicting the reaction and substrate specificities of new members discovered in genome and other sequencing projects.

The next section of this chapter provides some basic background important for understanding our analysis of one SFLD SF, namely the N6P SF. First, we discuss the concept of functionally diverse enzyme SFs using as an example the well-curated enolase SF (Babbitt et al., 1996; Gerlt et al., 2012; Gerlt et al., 2005) and show how the study of these types of SFs can be used to provide clues about how new enzyme reactions evolve from a conserved structural scaffold and active site features. This is followed by a brief summary of the organization of the SFLD and the tools and information it contains, accompanied by a description of how the SF context can be used to improve our ability to predict the molecular functions of enzymes, annotate new members of an SF as they are added to the public databases, and identify those that have been misannotated. Finally, to describe our large-scale analysis of structure–function relationships across the 2500 proteins of the N6P SF, we introduce protein similarity networks, a relatively new application of network technology to examine and illustrate broad functional trends in protein SFs from the context of sequence and structural similarity (Atkinson et al., 2009).

We focus in the body of this chapter on the N6P SF. This SF is much less well understood than the enolase SF and illustrates how large-scale computational analysis to define structure–function relationships can aid in predicting the reaction specificity for enzymes of unknown function (called "unknowns" in the rest of this chapter). The N6P SF currently contains approximately 2500 sequences (Hicks et al., 2011) and represents a subset of greater than 17,000 proteins we analyzed within an even larger six-bladed β-propeller fold class. (These >17,000 proteins represent the closest matches outside of the N6P SF but do not include some SFs considered by other resources to belong to the fold class, such as the neuraminidases.) In this chapter, we describe in some detail how the SF context allows us to make inferences about how differences in the sequence and structural properties of the N6P SF map to differences in their functional properties. This SF was chosen because of the importance of its known reactions in many types of biological processes, including the synthesis of natural products that are used as precursors for the development of drugs targeting a broad group of human diseases (Ma et al., 2006). Because of the complexity of their structure–function relationships and the relative lack of experimentally or structurally characterized members, this SF also exemplifies especially well some of the challenges associated with classification and annotation of new enzymes as they are discovered in genome projects.

An online classification of the N6P SF as described in this chapter is available at the SFLD, along with interactive versions of sequence similarity networks such as those shown in this chapter. Additional tools provided by the SFLD enable users to classify their sequences of interest into subgroups and families within the SF.

6.2 BACKGROUND, RESOURCES, AND TOOLS IMPORTANT FOR UNDERSTANDING THIS CHAPTER

6.2.1 Functionally Diverse Enzyme Superfamilies

Functionally diverse enzyme SFs are sets of evolutionarily related proteins (also called *divergent proteins*) that conserve constellations of functionally important active site residues but whose substrates, products, and even overall reactions can be substantially different (Babbitt and Gerlt, 1997; Gerlt and Babbitt, 2001). Within each SF, we define as a family the set of sequences, typically from multiple organisms, that catalyze the same reaction in the same way. Between these two levels, subgroups are defined as needed to describe subsets of sequences that cluster together on the basis of sequence and structural features. Using this hierarchy, the SFLD can link sequence, structure, and functional properties in an especially useful way: that is, it allows us to associate these properties to enzymes at both the SF and family levels. Figure 6.1 illustrates the SFLD hierarchy for the enolase SF. This classification hierarchy also allows us to address problems of "over-annotation" to a specific function that cannot be validated from available information, thereby avoiding misannotation of reaction specificity. For example, enzymes that can only be assigned to an SF on the basis of structural and active site motifs conserved among all of its members but not to a specific family are annotated only as belonging to that SF. This allows us to associate functional properties of an enzyme only with the specific aspects of its active site that enable those properties. As a result, this classification scheme provides a blueprint for the inference of functional properties that is more reliable for functional annotation than simple approaches based on overall sequence similarity (Schnoes et al., 2009). Many other approaches use related principles, some of which are described in other chapters in this book (Chapters 7 and 8).

In functionally diverse enzyme SFs, divergent evolution of such broadly different chemical reactions can be described by the "chemistry-constrained" model of enzyme evolution. In this model, the ancestral structural scaffold can be "re-engineered" by nature for a variety of functions by reusing a conserved active site that mediates a fundamental chemical capability such as a partial reaction.

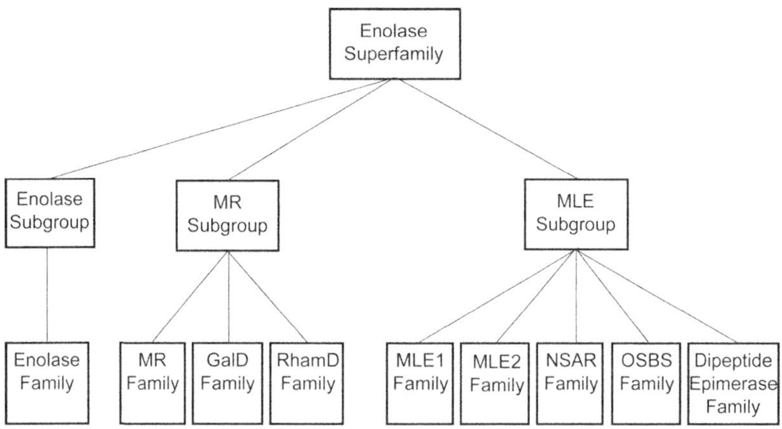

Figure 6.1 Classification hierarchy for some representative subgroups and families of the enolase SF. At the SF level, all members share a subset of active site residues required to initiate the partial reaction they all have in common. Subgroups designate subsets of the SF for each of which an additional set of active site residues are conserved, distinguishing the subgroups from each other. The family level describes an isofunctional group of proteins that catalyze the same reaction in the same way. Typically, each is distinguished by a family-specific set of residues associated with reaction and substrate specificity. Although different families can evolve at different rates, the sequence identity within families is generally greater than between families.

Built on this foundation, variations in substrate binding and overall chemistry evolve (Babbitt and Gerlt, 1997; Gerlt and Babbitt, 2001; Glasner et al., 2007). Ultimately, the new overall reactions that result must be consonant with both the constraints imposed by the SF's common chemical capability and the nature of the reactions that a new substrate can undergo. While this model has been useful for developing a hierarchical classification of linked structure–function properties in enzymes, we note that evolutionary relationships within an SF, especially for very divergent members, can be difficult to define, as we show below for the N6P SF.

As of August 2013, the enolase SF comprised nearly 23,000 sequences (up from ~8000 just two years earlier) and 22 different experimentally characterized reactions. A large number of these sequences are still of unknown function despite two decades of work by many groups to understand their structure-function relationships. Six of the experimentally characterized enolase SF functions are shown in Fig. 6.2a. Remarkably, all of these different reactions are mediated by highly

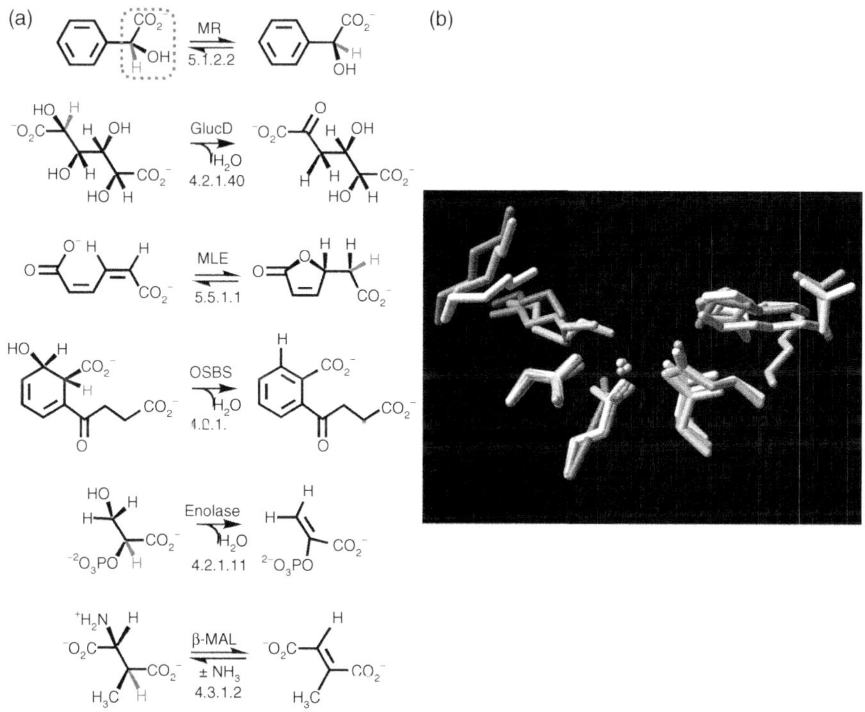

Figure 6.2 Structure–function mapping for some members of the enolase SF. (a) Chemical reactions catalyzed by six enolase SF members. In the top reaction, the dotted box designates the ligand substructure associated with a proton abstraction step. This proton abstraction step is used to initiate the reaction. (b) Superposition of conserved active site residues of these six proteins, each representing one of the reactions shown in (a). The conserved position of the divalent metal ions required for each reaction is also shown.

similar overall structures and active sites (Fig. 6.2b). This finding can be rationalized by the observation that these active site similarities are associated with a partial reaction common to all members of the SF, for example, abstraction of a proton on a carbon alpha to a carboxylate group (as shown in the common substructure box in Fig. 6.2a), leading to the formation of a common type of enolate anion intermediate. Although some of their pairwise sequence identities are less than 10%, all sequences assigned to the SF show conservation of this proton abstraction machinery and metal-binding ligands.

Formation of the intermediate in the enolase and other functionally diverse SFs is typically followed by an additional partial reaction(s) that differ in each enzyme family, leading to a range of different products. Thus, even for the thousands of enolase SF sequences for which we cannot assign a reaction specificity, we can reasonably infer that all will catalyze the common partial reaction and that their substrates will contain the substructure moiety associated with the proton abstraction step (Fig. 6.2a). Thus, the SF context provides a rules-based approach for inference of the SF's common functional capability of all of its members. Further, the nature of this step restricts the search space for functional inference of their reaction and substrate specificities.

Members of functionally diverse SFs are difficult to annotate and easy to misannotate because they all look alike because of the conserved sequence and structural and functional characteristics they share (Fig. 6.2). Since they have diverged to catalyze different reactions, families within these SFs can be distinguished by additional sequence and structural features that define their reaction and substrate specificities. But because the different families in an SF often evolve at different rates and exhibit low sequence similarity both within and between families, their specificity-conferring features are often challenging to identify (see, e.g., Glasner et al., 2006; Nguyen et al., 2008). As a result, misannotation levels for these families can be alarmingly high (Schnoes et al., 2009).

An estimate of the number of functionally diverse enzyme SFs in nature can be made using PDBSprotEC (Martin, 2004) and SCOP (Andreeva et al., 2008) to map the enzyme nomenclature (EC) (International Union of Biochemistry and Molecular Biology: Nomenclature Committee and Webb, 1992) subsubclasses (reaction types) to structural SFs. Defining as functionally diverse those SFs containing two or more families different at the third digit of the EC class (representing reaction specificity), over 250 SFs are predicted (Almonacid and Babbitt, 2011). Although currently comprising about a third of known enzyme SFs, this count is likely an under-representation, as it only takes into account those enzymes for which EC numbers have been identified and for which the reaction specificity is known. Thus, not only are structure–function relationships in functionally diverse enzyme SFs complex, making them especially challenging for understanding the roles of these proteins in many different applications, but the significant proportion of the enzyme universe they represent also underscores the importance of addressing these challenges.

6.2.2 Structure–Function Linkage Database (SFLD)

The SFLD is a publicly available resource that links evolutionarily related sequences and structures from functionally diverse SFs to their chemical reactions. It is unique in linking the overall and partial reactions (or other chemical capabilities) common to all members of a SF, subgroup, or family with conserved sequence and structural elements that enable them. The SFLD ontology and schema were designed to represent enzyme structure–function linkage at these multiple levels of granularity to allow annotation of a function common to SFs even if family-level reaction and substrate specificity could not be predicted (Pegg et al., 2005). This schema also allows us to address problems of "over-annotation" to a specific function that cannot be validated from available information. The SFLD principally targets SFs that are believed to have evolved according to the chemistry-constrained model of evolution described above, as these types of SFs present special challenges for inference of reaction and substrate specificity. In tying together a theoretical model that uniquely links structure and function in each of the SFs it contains, the SFLD provides a level of organization and highly curated information that allows users to leverage the data for biological insights (Pegg et al., 2006). A wide range of browse and search tools are available to aid users in accessing SFLD information, along with visualization tools providing interactive access to such data as curated structures, chemical reactions, multiple alignments, and similarity networks. The SFLD also includes orthogonal information such as operon context and links to several other information resources devoted to classification and analysis of enzymes.

6.2.3 Introduction to Protein Similarity Networks

The sequences in the SFs targeted by the SFLD typically number in the thousands, so that even using redundance filtering it is not possible to create a global picture of their sequence/structure relationships using phylogenetic trees or multiple alignments, as these are slow to generate and realistically cannot be used with thousands of diverse sequences (or hundreds of structures). Moreover, for their structure–function mappings, many different visualizations of each of their constituent subgroups are often needed, requiring a method for querying the data that is very fast to construct, easy to update on a regular basis, and able to capture large amounts of information in a single view. To meet these requirements, we have begun to use sequence- and structure-similarity networks.

For a particular SF, many networks can be generated in a few days using several types of similarity metrics, for example, based on protein sequence or structure, or ligand or reaction similarity, and viewed using various clustering algorithms, layouts, and thresholds. In capturing all connections among proteins better than a user-specified threshold, networks provide information not available from alignments or trees (Atkinson et al., 2009). Especially useful, similarity networks can be linked to functional properties by overlaying the networks with different types of biological and chemical information, providing a simple and fast way to investigate functional trends from the context of sequence or structural similarity. For example, by painting these networks with the known reaction specificities of previously characterized families, it is easy to identify clusters of known function, sometimes allowing us to hypothesize functional properties of unknowns that cluster nearby.

Although networks have previously been used extensively to capture many types of biological relationships such as genetic or protein interactions, application of network methods to structure–function mapping in proteins is relatively new (see Adai et al., 2004; Enright et al., 2002; Finn et al., 2006; Frickey and Lupas, 2004; Medini et al., 2006 for some examples). In our work on functionally diverse enzyme SFs, we have now established the utility of similarity networks for many applications. Examples include choosing targets for experimental analyses (Pieper et al., 2009), showing the relative outlier status of specific families within a large SF (Nguyen et al., 2008), and illustrating the correlation of active site conservation patterns with lineage (Hall et al., 2007). Additionally, as described here, similarity networks currently offer the only practical way to summarize structure–function relationships in large and complex SFs in a single, intuitively accessible representation (see Atkinson and Babbitt, 2009, for example).

Figure 6.3 illustrates some basic principles in the creation and application of thresholded similarity networks such as those provided in this chapter. Although these network figures are static images, the downloadable versions available from the SFLD are fully interactive so that clicking on a node pulls up accession codes to several data resources, species, and other metadata. Such networks can be easily painted with these and other types of information such as the presence of conserved motifs, common genome context, cellular location, or similarity of substrates, making them useful for interpreting structure–function relationships in individual proteins or across entire SFs. Using such orthogonal information to help reveal trends across similarity clusters from these types of functional clues is a major strength of the approach.

Validation of similarity networks using several well-characterized SFs has identified many advantages of network representations, showing that, in general, sequence networks generated from all-by-all pairwise comparisons of a large number of divergent queries correlate well with distances from phylogenetic trees and recapitulate much of the information in them (Atkinson et al., 2009). They also provide new information not available from trees, as they can capture multiple similarity connections between nodes, rather than showing only the two best connections for any one leaf, as is provided in visualizations of trees. Network topologies are generally robust to missing data and agree with known sequence and structural relationships. In addition, structure-based and sequence-based networks largely agree, providing a useful check especially on sequence-based networks that depict very divergent relationships.

A practical advantage of networks is that they can easily handle many more sequences than can be generally evaluated using trees or multiple alignments. However, networks still have size limitations, especially using laptops or desktop

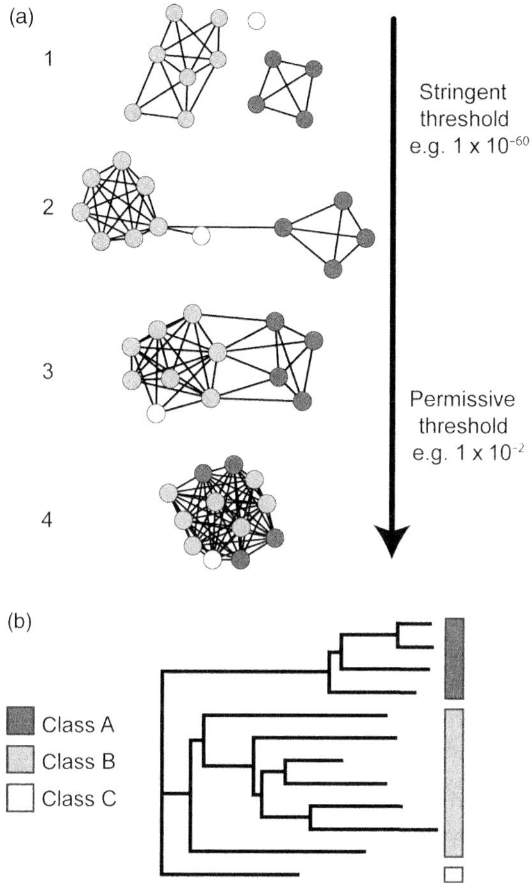

Figure 6.3 Network representation of similarity relationships. (a) Thresholded similarity networks represent sequences as nodes (circles) and all pairwise sequence relationships (alignments) better than a threshold as edges (lines). In this example, thresholds define the minimum significance score at which edges are drawn. For the networks shown in this chapter, network similarity is defined by all-by-all pairwise BLAST comparisons (Altschul et al., 1997), using the BLAST *E*-values as scores. The same network, depicting three simulated protein classes, is shown here at four different threshold cutoffs, labeled 1–4. At the most stringent threshold (top, 1), the sequences are broken up into disconnected groups; within each group the sequences are highly similar and are colored by the same function. As the threshold is relaxed (following the direction of the arrow) and edges associated with less significant relationships are added to the network, groups merge together and eventually become completely interconnected. The relative positioning of disconnected groups has no meaning in this visualization. The networks in this chapter are visualized using the organic layout provided by Cytoscape (Cline et al., 2007); lengths of connecting edges tend to correlate with the relative dissimilarities of each pair of sequences. (b) Simulated dendrogram for a sequence set that might give rise to the network in (a). *Source*: This figure and legend are adapted from Atkinson et al. (2009).

computers. Currently, the practical limit for opening and exploring networks using Cytoscape is approximately 10^6 edges, which typically correlates to a few thousand fairly divergent sequences.

Similarity networks are not without issues, however, and several significant limitations can affect their use. For example, it is important to confirm that the similarity signal captured by BLAST (Altschul et al., 1997) (or even more complex similarity metrics such as hidden Markov models (Eddy, 1996), or some of those used in structure comparison algorithms) is comparable across a target set of highly diverse proteins. For interpreting similarity networks, we also caution that coming to a final conclusion based on networks, especially those generated by pairwise comparisons using algorithms such as BLAST, should usually be accompanied by additional analyses. Unlike phylogenetic trees, BLAST-based or structure-based networks are not based on an explicit evolutionary model and therefore cannot be used as a substitute for a phylogenetic tree. As with other comparisons of very divergent proteins, variations between sequence or structural families associated with domain organization, the presence or absence of multiple domains, or inserts that may differ widely across a divergent set of proteins can also complicate interpretation of relationships. Such issues can be mitigated if care is used in interpreting the results.

6.3 SEQUENCE/STRUCTURE/FUNCTION RELATIONSHIPS IN THE NUCLEOPHILIC ATTACK 6-BLADED β-PROPELLER (N6P) SUPERFAMILY

6.3.1 Classification of Proteins in the 6-Bladed β-Propeller Fold Class and the N6P Superfamily

As discussed in other chapters of this book (see Chapter 3 and 7), CATH (class, architecture, topology, homology) (Cuff et al., 2011) and SCOP (structural classification of proteins) (Andreeva et al., 2008) provide frameworks for identifying similarities in structure across the protein universe. In this section, we use these resources to provide a context for our definition of the N6P SF within the larger scope of the 6-bladed β-propeller fold.

Examination of similarity relationships between a set of 41 nonredundant crystal structures of six-bladed β-propeller fold proteins (Fig. 6.4) identified one major cluster and several smaller clusters of related proteins. Mapping of functional information designating major functional classes (defined as Pfam families (Finn et al., 2010)) onto the network shows that structures of known functional groups cluster together. Additionally, this view allows us to begin to see connections between proteins of different functions. In the major cluster of 18 protein structures, eight families are represented: glucose-sorbosone dehydrogenase (GSDH), senescence marker protein-30/gluconolactonase/luciferin-regenerating enzyme-like region (SGL), arylesterase, strictosidine synthase (SS), major royal jelly protein (MRJP), NHL, PD40, and low-density lipoprotein repeat class B (Ldl_recept_b).

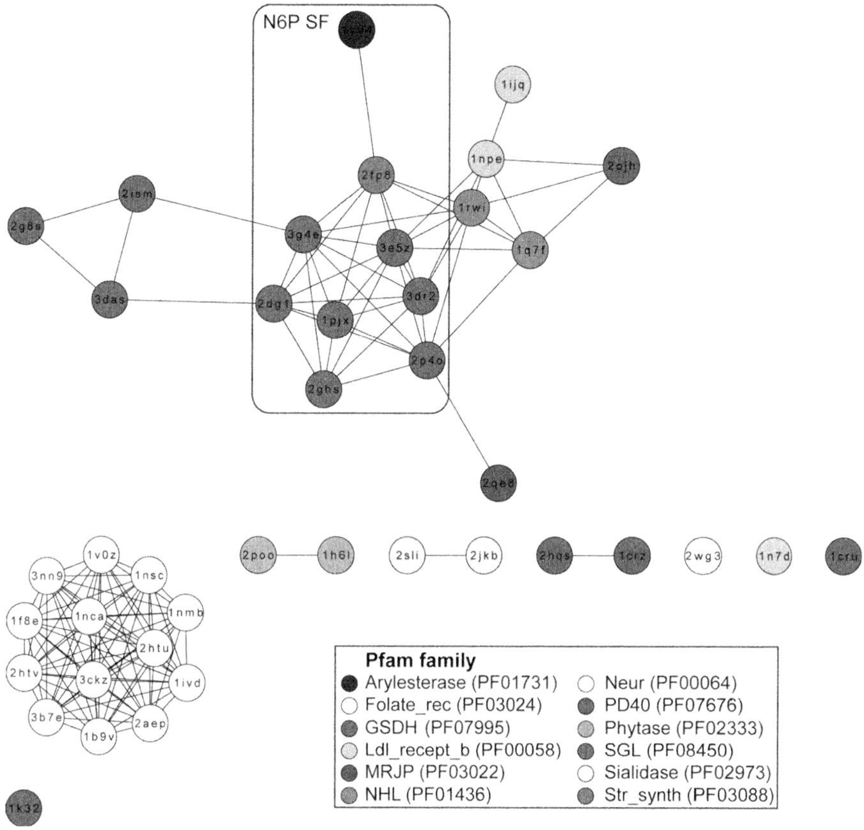

Figure 6.4 Structure-based similarity network of crystal structures sharing the six-bladed β-propeller fold. Forty-one nonredundant structures were compared pairwise using the algorithm TM-align (Zhang and Skolnick, 2005). The structural representatives of the N6P SF are boxed. Structures are colored by Pfam family as shown. Each node represents a structure and is labeled with its Protein Data Bank (PDB) ID. Edges are drawn between nodes only if the alignment score is greater than 0.7, a score considered statistically significant by the authors of TM-align. For this network, the median alignment length for those edges displayed is 233 residues and the median root-mean-square deviation (RMSD) is 2.75 Å. (*See insert for color representation of the figure.*)

Classification of this fold class and of the proteins of the N6P SF are generally similar among the three comprehensive structure and sequence resources (SCOP, CATH, and Pfam) and are generally consistent with those shown in Fig. 6.4. Of the structures represented in this figure, 61% are classified in SCOP v1.75 and 78% in CATH v3.4.0. Among these, several belong to an identical homologous SF in CATH: the TolB, C-terminal domain SF (2.120.10.30). These include proteins from the SGL (1pjx.pdb, 2ghs.pdb, 2p4o.pdb), arylesterase (1v04.pdb), MRJP (2qe8.pdb), NHL (1rwi.pdb, 1q7f.pdb), PD40 (2ojh.pdb), and Ldl_recept_b (1npe.pdb, 1ijq.pdb) Pfam families. CATH also includes proteins not in this central cluster as belonging to the same homologous SF, notably two additional PD40 proteins (1crz.pdb, 2hqs.pdb), an additional Ldl_recept_b (1n7d.pdb), and a GSDH protein (1cru.pdb) that does not connect into our central cluster at this threshold (though three other pdb structures that are not currently classified in CATH do connect). Other proteins in the network are classified into different homologous SFs in CATH: the neuraminidases and sialidases (2.120.10.10) and the phytases (2.120.10.20). SS (str_synth) and the six-bladed β-propeller domain of the folate receptor (Folate_rec) are not yet classified in CATH.

The classification of SFs by SCOP is mostly in agreement with CATH, though there are key differences. Like CATH, SCOP classifies the SGL (2dg1.pdb, 1pjx.pdb, 2ghs.pdb, 2p4o.pdb) and arylesterase (1v04.pdb) structures in the same SF (b.68.6). However, SCOP differs from CATH in that NHL (1rwi.pdb, 1q7f.pdb) and Ldl_recept_b (1npe.pdb, 1ijq.pdb) proteins belong to two distinct SFs (b.68.9 and b.68.5, respectively).

The definition of the N6P SF that is used in the SFLD differs from the SF definition used by SCOP, CATH, and Pfam principally because the SFLD adds criteria associated with conserved catalytic strategies to refine further definitions at the SF level. In some SFs, such strategies can be associated with specific features conserved in members' active sites as well. Our definition of the N6P SF is most similar to the SCOP definition, but includes SS (neither CATH nor SCOP has yet classified the SS structures (2fp8.pdb in the network)) and the subgroup we describe as the strictosidine synthase-like (SSL) proteins. Consistent with our definition, in this chapter we evaluate the boundaries of the N6P SF with respect to its nearest neighbors in sequence, structure, and features of the chemical strategy by which catalysis is enabled. Below, the N6P SF and its constituent subgroups are defined as they are classified in the SFLD.

6.3.2 Definition of the N6P Superfamily

The 2500 members of the N6P SF can be distinguished from the other proteins of the six-bladed β-propeller fold class (~36,000 total members for the families listed in Fig. 6.4 from Pfam 25.0) by sequence and structure similarity and by similarity in the fundamental chemical strategy by which catalysis is enabled. A small number of proteins within the SF, that is, members of the SFLD-defined SS family, differ in additional features that distinguish them from the rest of the SF as well. These are described in detail below.

Three criteria distinguish the N6P SF from the rest of the fold class. First, proteins in the three subgroups that comprise the N6P SF share higher overall sequence similarity with each other than they do with any other protein sharing the same fold. Second, despite the quite different overall reactions represented by the few members of the SF that have been biochemically characterized, all members of the N6P SF can be united by a common catalytic step, that is, nucleophilic attack on an electrophilic substrate (Fig. 6.5). Finally, with the exception of the small number of enzymes that include those biochemically characterized (or predicted) to catalyze the SS reaction, nearly all members of the SF appear to be

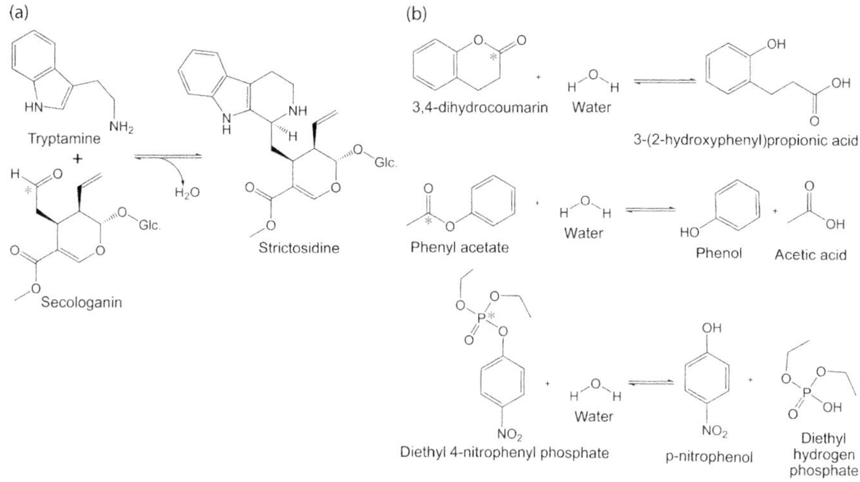

Figure 6.5 Examples of chemical reactions catalyzed by the N6P SF. Asterisk indicates electrophilic atom that is attacked. (a) SS reaction. (b) Examples of lactonase, esterase, and phosphotriesterase reactions catalyzed by members of the senescence marker protein-30/gluconolactonase/luciferin-regenerating enzyme-like region (SGL) and arylesterase-like subgroups. *Source*: Figure and legend adapted from Hicks et al. (2011).

metal dependent, sharing in their active site architectures four highly conserved metal-binding residues that coordinate to a divalent metal. These conserved metal-binding ligands can be identified in multiple sequence alignments representing the huge majority of the 2500 members of the N6P SF (not shown). Although other members of the fold class are metal dependent, the particular metal-dependent binding modality found in the N6P SF distinguishes it from all of these other proteins.

The members of the N6P SF are represented by three highly diverse subgroups, designated by the different colors within the rounded rectangle shown in Fig. 6.4. Each of these subgroups was previously identified as part of the same SF by sequence similarity and by one of the known reactions its members catalyze (Kobayashi et al., 1998), some of which are depicted in Fig. 6.5. A more detailed description of structure–function relationships among these subgroups is given below.

6.4 WHAT WE KNOW AND DO NOT KNOW ABOUT THE ENZYMES OF THE N6P SUPERFAMILY

6.4.1 The Senescence Marker Protein-30/Gluconolactonase/ Luciferin-Regenerating Enzyme-Like (SGL) Subgroup

The SGL subgroup includes about 1500 members (Fig. 6.6), several of which are known to catalyze lactonase, esterase, and organophosphatase activities

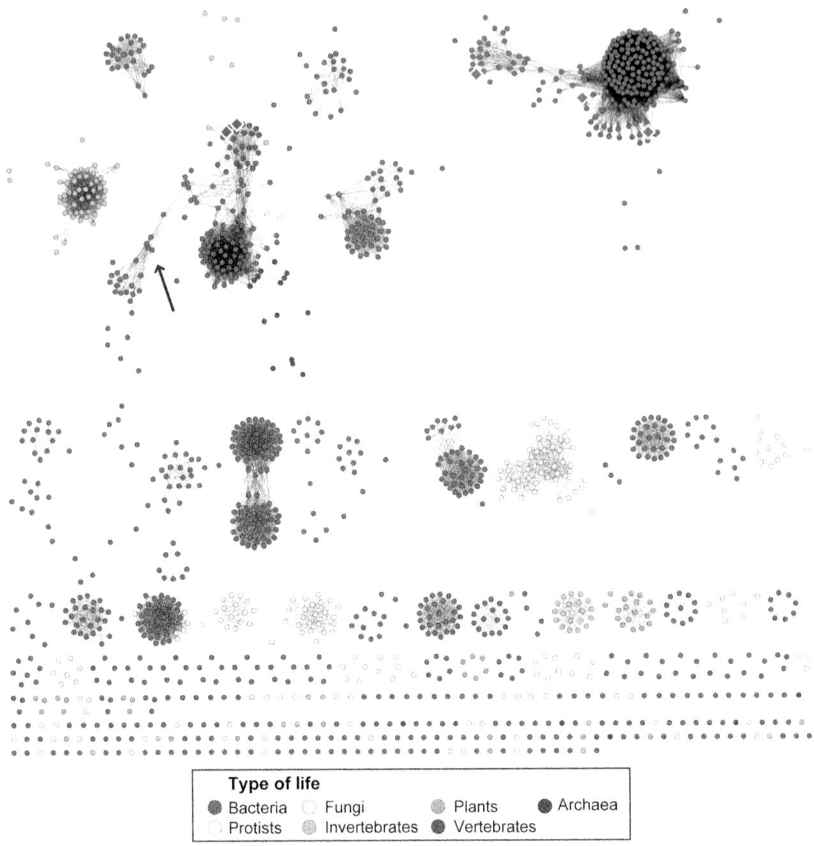

Figure 6.6 Sequence similarity network for the SGL subgroup downloaded from the current SFLD. The 1449 proteins of the SGL subgroup of the N6P SF are shown. Sequence similarity was computed from pairwise BLAST comparisons of all of the members of the subgroup using the SFLD as a background model. Nodes are colored by the type of life as designated in the key. Small circular nodes represent biochemically uncharacterized proteins; large diamond-shaped nodes with white outlines indicate biochemically characterized proteins. Edges are drawn if the pairwise similarity score between two nodes is less than 1×10^{-60}. An arrow indicates the *Burholderia sp. H160* discussed in the text. (*See insert for color representation of the figure.*)

(Fig. 6.5b). Senescence marker protein-30 (SMP-30), also known as *regucalcin*, functions as a lactonase in the biosynthesis of L-ascorbic acid in nonprimate mammals (Kondo et al., 2006) and has been shown to hydrolyze toxic organophosphates in mouse liver (Billecke et al., 1999; Kondo et al., 2004). Luciferin-regenerating enzyme (LRE), as the name suggests, catalyzes the regeneration of luciferin in fireflies through hydrolysis of oxyluciferin to 2-cyano-6-hydroxybenzothiazole, which then is condensed with D-cysteine (Gomi and Kajiyama, 2001). Additional biochemically characterized proteins from this subgroup include drug-responsive protein 35 (Drp35), which functions in the resistance to antibiotics by *Staphylococcus aureus* (Murakami et al., 1999; Tanaka et al., 2007), and diisopropylfluorophosphatase (DFPase) (Scharff et al., 2001), whose native activity is unknown but which can break down organophosphates, including nerve toxins such as soman (Hoskin and Roush, 1982).

A conserved set of metal-coordinating ligands is present in the active site of all structurally characterized proteins of the SGL subgroup. Detailed mechanistic studies done on the DFPase enzyme have implicated these residues as critical for its organophosphatase activity (Blum and Chen, 2010; Blum et al., 2006; Blum et al., 2008; Melzer et al., 2009). In the proposed DFPase mechanism, an aspartic acid coordinated to a divalent metal performs a direct nucleophilic attack on the phosphorous atom of the phosphoryl group of diisopropyl fluorophosphate (DFP), forming a phosphoenzyme intermediate. Water then completes the hydrolytic reaction through regeneration of the aspartic acid, as demonstrated by $H_2{}^{18}O$ incorporation experiments (Blum et al., 2006).

Mutation studies of Drp35 suggest that the structurally equivalent metal-coordinating ligands are necessary to catalyze the lactonase reaction, although the identity of one of the ligands differs from that of DFPase (Asn120 in DFPase) and is in a position structurally equivalent to Asp138 in Drp35. In this case, it is proposed that Asp138 activates a water molecule, leading to a nucleophilic attack on the carbon atom of the carbonyl group of the lactone. The oxygen atom next to the scissile bond is proposed to be protonated by the other metal-coordinated aspartate (Tanaka et al., 2007).

For other enzymes in this subgroup, the details of their reaction mechanisms are unclear. Though the metal-coordinating ligands likely play a role, additional residues may act as specificity determinants. Furthermore, the vast majority of proteins in the SGL subgroup are biochemically uncharacterized. Gene context analysis of SGL members from bacteria suggests that several proteins likely play roles in sugar metabolism and transport (Hicks M. et al., Unpublished), though their reaction specificities are also currently unknown.

6.4.2 The Arylesterase-Like Subgroup

The arylesterase-like subgroup has one defined family in the SFLD: the paraoxonases (which include PON1, PON2, and PON3), with the reaction specificities of the rest of the subgroup classified as unknowns (Fig. 6.7). Sequences from this family share about 60% identity and conserve a set of seven residues thought to be important in catalysis. Five coordinate to the catalytic calcium ion (four of which are conserved at the SF level) and the other two, a His–His dyad, appear to be critical for their lactonase and esterase activities. These proteins differ both in their expression patterns and the degree to which they catalyze their respective reactions. In humans, PON1 expression is found in the liver, PON3 is found in both the liver and kidney, and PON2 is found in most tissues, including the heart, liver, kidney, lung, small intestine, spleen, stomach, placenta, and testis (for a review, see Ng et al., 2005). PON1 catalyzes lactonase, esterase, and

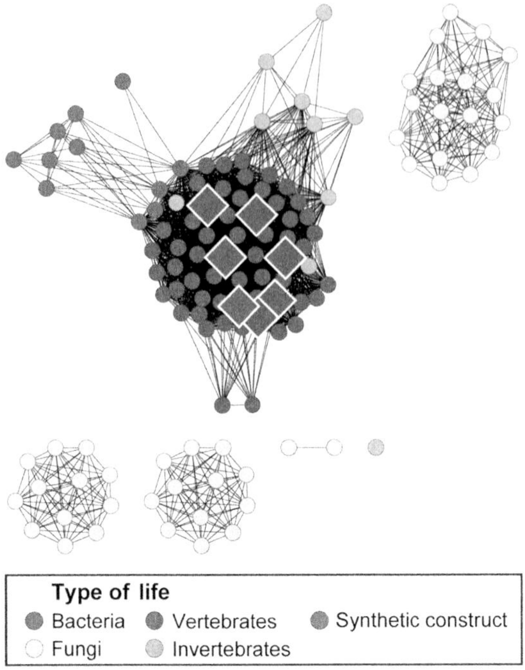

Figure 6.7 Sequence similarity network for the arylesterase-like subgroup downloaded from the SFLD. The 123 proteins of the arylesterase-like subgroup of the N6P SF are shown. Sequence similarity was computed from pairwise BLAST comparisons of all of the members of the subgroup using the SFLD as a background model. Small circular nodes represent biochemically uncharacterized proteins; large diamond-shaped nodes with white outlines indicate biochemically characterized proteins. Edges are drawn if the pairwise similarity score between two nodes is less than 1×10^{-30}. (*See insert for color representation of the figure.*)

organophosphatase reactions, though lactonase appears to be its native activity (Billecke et al., 2000; Draganov et al., 2005; Khersonsky and Tawfik, 2005). PON2 and PON3 also appear to be lactonases, yet differ from PON1 in that they have very limited arylesterase activity and almost no organophosphatase activity (Draganov, 2010). Additionally, high density lipoprotein-associated PON1, the best studied member, exerts its antiatherogenic properties through the inhibition of cell-mediated oxidation of low-density lipoproteins and stimulation of cholesterol efflux, likely via its lactonase activity (Rosenblat et al., 2006). Furthermore, paraoxonases play a role in acylhomoserine lactone-dependent quorum-sensing systems, helping to prevent microbial infection through their lactonase activity (Draganov et al., 2005). For example, in *Pseudomonas aeruginosa*, PON1 in serum has been shown to prevent the formation of bacterial biofilm *in vitro*. Interestingly, PON1-deficient mice show enhanced resistance to *P. aeruginosa* infection, likely due to increased expression of PON2 and PON3 (Ozer et al., 2005).

Structural and mutational analyses indicate that the lactonase and esterase reactions catalyzed by the PON enzymes are likely due to a proton-shuttle mechanism involving the activation of a metal-coordinated water by a His–His dyad, leading to a nucleophilic attack on the carbon atom of a carbonyl group (Harel et al., 2004). It is believed that the organophosphatase reaction catalyzed by PON1 may be performed in a manner analogous to DFPase of the SGL subgroup, which involves a direct nucleophilic attack by a metal-coordinated aspartic acid (Blum et al., 2006).

As the vast majority of proteins in the arylesterase-like subgroup remain uncharacterized, both biochemically and structurally, other functions likely remain to be discovered.

6.4.3 The Strictosidine Synthase-like (SSL) Subgroup

The SSL subgroup represents a group of nearly 400 proteins and is named for the small number of proteins within the subgroup that have been shown to catalyze the SS reaction (Fig. 6.8). Strictosidine is the precursor to the monoterpenoid indole alkaloid biosynthesis pathway, which produces about 2000 known compounds, several of which are used in the treatment of cancer, malaria, hypertension, schizophrenia, and arrhythmic heart disorders (Ma et al., 2006). The huge majority of the rest of the subgroup, comprising approximately 400 proteins, are uncharacterized.

6.5 FUNCTIONAL PREDICTIONS AND PREDICTION OF MISANNOTATION OF SSL SUBGROUP ENZYMES

SS catalyzes a Pictet–Spengler reaction, resulting in the condensation of tryptamine and secologanin to form the strictosidine product (Maresh et al., 2008). This bisubstrate condensation reaction is quite different from the other known hydrolytic reactions catalyzed by the SGL and arylesterase-like subgroups of the SF (Fig. 6.5). And although the condensation reaction catalyzed by SS requires a nucleophilic attack on an electrophilic substrate, which is the fundamental chemistry common to the N6P SF, it differs from most of the other SF members in that it is metal independent and lacks all four metal-coordinating ligands common to both the SGL and arylesterase-like subgroups. Thus, the mechanism by which SS catalyzes this fundamental step is quite distinct from the reactions of the SGL and arylesterase-like subgroups. Instead of metal-assisted catalysis, SS uses a functionally required glutamate, Glu309, which activates tryptamine by abstracting a proton, allowing for a nucleophilic attack on a carbon atom of the aldehyde of a second substrate, secologanin, forming an iminium species. The indole then attacks the iminium moiety to form the strictosidine product (Maresh et al., 2008).

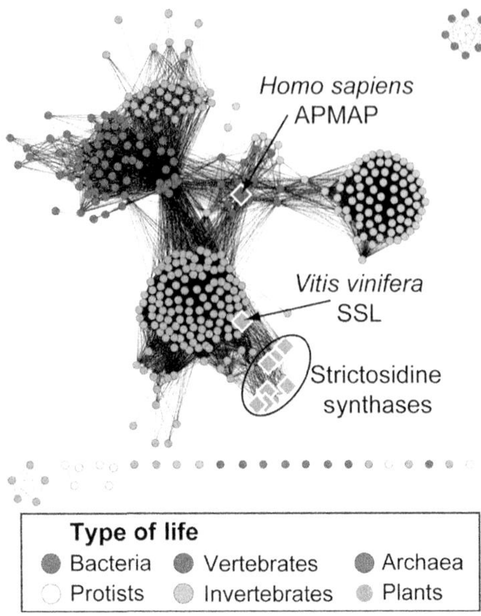

Figure 6.8 Sequence similarity network for the strictosidine synthase-like (SSL) subgroup. The 393 proteins of the SSL subgroup of the N6P SF are shown. Sequence similarity was computed from pairwise BLAST comparisons of all of the members of the subgroup using the SFLD as a background model. Small circular nodes represent biochemically uncharacterized proteins; large diamond-shaped nodes with white outlines indicate biochemically characterized proteins. Edges are drawn if the pairwise similarity score between two nodes is less than 1×10^{-50}. Strictosidine synthases are marked by the oval. Arrows indicate the *Vitis vinifera* SSL and *Homo sapiens* APMAP proteins discussed in the text. (*See insert for color representation of the figure.*)

Consistent with these functional and mechanistic differences, both a multiple alignment of the SSL subgroup proteins (not shown) and the sequence similarity network provided in Fig. 6.9 indicate that most SSL proteins appear to have the same metal-coordinating ligands common to the SGL and arylesterase-like subgroups. Taken together with the known mechanism of SS, these observations suggest three important conclusions. First, despite their putative annotations as SSs, these uncharacterized SSL proteins are unlikely to catalyze the SS reaction, as SS proceeds through a metal-independent mechanism. Moreover, steric clashes with metal ligands would likely make productive binding of tryptamine and secologanin impossible in these SSL subgroup proteins. Second, SSL enzymes are more likely to catalyze hydrolytic reactions similar to those of the other two subgroups. Support for this idea is provided by initial experimental studies of a human protein from the SSL subgroup, namely adipocyte plasma membrane-associated protein (APMAP). Those results suggest that APMAP exhibits at least a low level of esterase activity using both beta-naphthyl acetate and phenyl acetate as substrates (Ilhan et al., 2008). Like most other SSL proteins, APMAP possesses

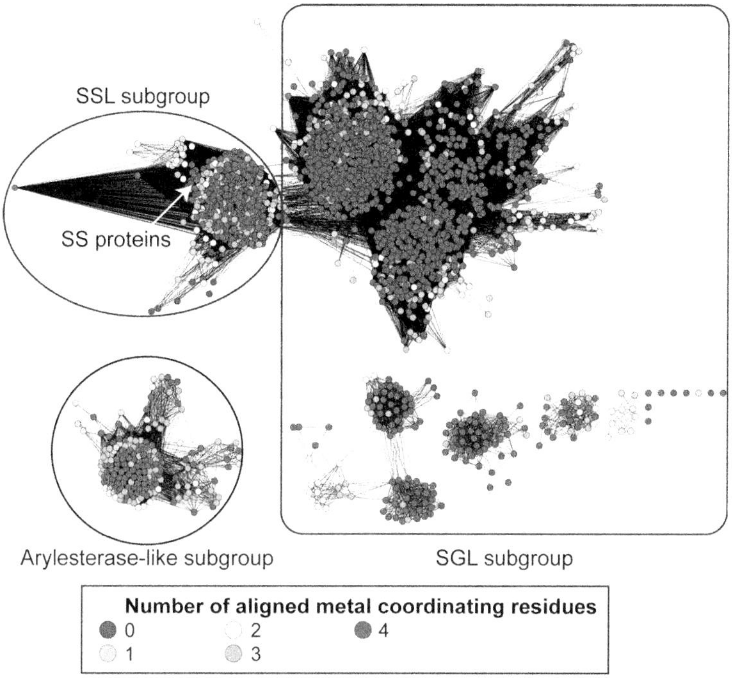

Figure 6.9 Sequence similarity network showing conservation of metal-coordinating active site residues for 2567 proteins of the N6P SF. The number of metal-coordinating active site residues are predicted using the multiple sequence alignment described in Fig. 6.12. Edges are drawn if the pairwise similarity score has an *E*-value greater than 1×10^{-10} using UniRef100 as the background model. At this threshold, the median alignment length is 238 residues and the median identity is 30.0%. Subgroups of the N6P SF are circled or boxed and labeled. The SS proteins are indicated by a white arrow. (The outlier to the far left of the SSL subgroup is an artifact of the organic layout.) (*See insert for color representation of the figure.*)

all four metal-coordinating ligands (Fig. 6.9). Though no additional mutational or mechanistic data for this enzyme is presently available, we predict that these residues play an essential role in catalyzing these esterase reactions, on the basis of the analogy to the reaction mechanisms of characterized SGL subgroup members. Third, as most of the proteins in the SSL subgroup have been annotated as "SS" or "SSL" in public sequence databases such as GenBank (Benson et al., 2010), we can conclude that the annotation of nearly 400 proteins in GenBank as "SS" or "SSL" is misleading to the point that they could be considered to be misannotated.

While these observations confirm the classification of the arylesterase-like, SGL, and the huge majority of SSL subgroup proteins as members of a single SF, they also raise a question whether, despite their overall sequence and structural similarity to the rest of the proteins of the N6P SF, the metal-independent SSs should be included in the SF as well. We argue that the preponderance of evidence supports the inclusion of the SSs in the N6P SF, as discussed below.

6.6 DO THE STRICTOSIDINE SYNTHASES REALLY BELONG TO THE N6P SUPERFAMILY?

As is clear from our analysis of the SSL subgroup, complex variations in sequence among seemingly related proteins can complicate function annotation transfer from biochemically characterized to uncharacterized proteins, resulting in misannotation of function for many members of this subgroup. Moreover, defining what to include in a functionally diverse SF is challenging, as exemplified especially for the SS enzymes.

With respect to the first criterion defined by the SFLD for inclusion of proteins in the N6P SF (see Section 6.3.2 "Definition of the N6P Superfamily"), the SS proteins can be classified as members of the N6P SF on the basis of their overall similarities in sequence and structure to other SF members. For the second similarity in the fundamental catalytic strategy by which catalysis is enabled, the SSs also fit as members of the N6P SF. As shown in Fig. 6.5, like the reactions of other functionally characterized members of the SF, the SS reaction involves a nucleophilic attack on an electrophilic substrate, albeit using a different mechanism than the other known SF members. At first pass, the SSs appear to fail the last required criterion for inclusion in the SF, however: that is, similarity in active site architecture or functionally important residues. Figure 6.10, which compares the active site of SS with those of metal-dependent enzymes from the SGL and arylesterase-like subgroups, illustrates how very different the active sites of the SSs are relative to the rest of the SF. It is these differences in the active sites, taken together with the dissimilarity between the condensation reaction catalyzed by SS and the hydrolytic reactions of the rest of the SF, that make it difficult to rationalize the inclusion of the SSs as members of the N6P SF.

Further comparison of the SSs with other N6P SF members moderates this view, however, leading us to classify the SSs as members of the N6P SF after all. First, a phylogenetic analysis of the SSL subgroup (not shown) suggests that SS likely arose from a metal-dependent ancestor (Hicks et al., 2011). Second, as shown in Fig. 6.11a, comparison of a liganded structure from the SGL subgroup (DFPase) with liganded SS proteins shows that both enzymes initiate nucleophilic attack from the same position in the active site. These similarities in active sites allow us to speculate that SS evolved to bind tryptamine in the region of the active site occupied by two of the four metal binding-ligands present in other SF members such as DFPase and leads to a hypothesis for a mechanistic role of tryptamine that is analogous to the roles of the metals in DFPase. The mechanistic diagrams shown on the right of Fig. 6.11 describe this mechanistic hypothesis. In DFPase (Fig. 6.11b), the reaction is initiated by a metal-assisted nucleophilic attack of the oxygen of Asp229 of

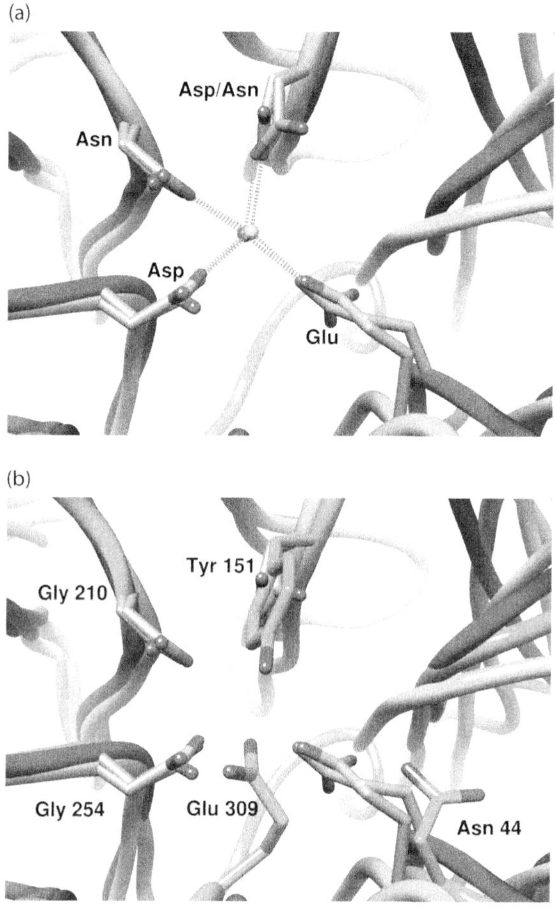

Figure 6.10 Active site superposition of (a) SGL (Drp35, pdb_id: 2dg1; red) and arylesterase-like (PON1, pdb_id: 1v04; cyan) subgroup proteins. Conserved metal-coordinating residues are labeled and colored by element; gray: carbon, blue: nitrogen, red: oxygen. (b) The same superposition as shown in (a), but with the metals removed for clarity and SS (pdb_id: 2fpb; green) added. Four SS residues that superimpose with alpha carbon positions for metal-coordinating residues in Drp35 and PON1 are labeled black. The glutamate (Glu309) required for SS activity in 2fpb is also labeled black. These five SS residues are colored by element; green: carbon, blue: nitrogen, red: oxygen. *Source*: Figure and legend are adapted from Hicks et al. (2011). (*See insert for color representation of the figure.*)

DFPase on the phosphorus of the substrate, DFP. We speculate that the analogy in SS might involve substrate-assisted catalysis in which the amine group of tryptamine is deprotonated by Glu309. This step is then followed by the subsequent attack of the amine on the aldehyde moiety of secologanin (Fig. 6.11c). Mutagenesis of Glu309 has previously established that it is required for SS activity.

6.7 RE-EXAMINATION OF THE BOUNDARIES OF THE N6P SF IN THE CONTEXT OF OTHER β-PROPELLER FOLD PROTEINS

The inclusion of SS in the N6P SF raises new questions for understanding the relationship of this SF to other β-propeller proteins. In the context of the N6P SF, we were able to rationalize the loss of metal-coordinating residues in the mechanism and active site of SSs, allowing us to classify these proteins as members

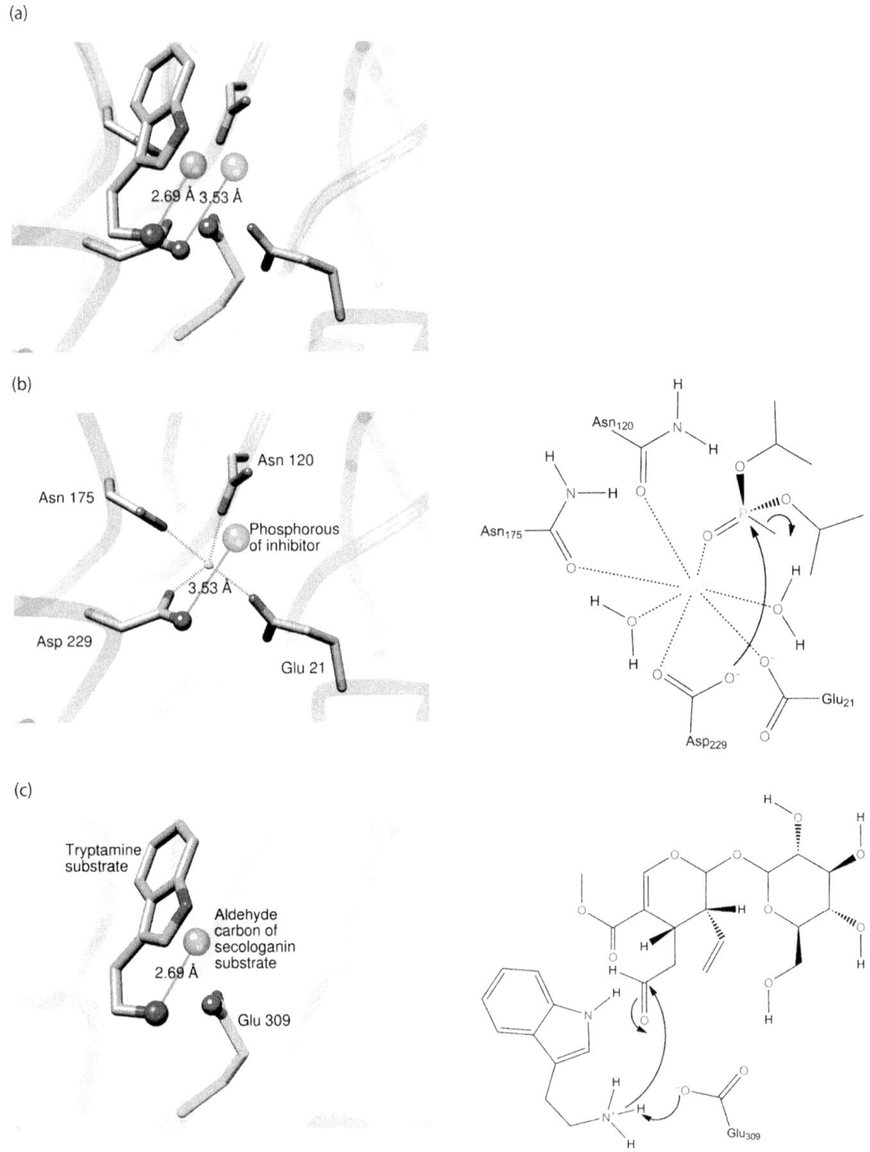

of the SF. But this interpretation is complicated in the context of the larger fold class, which shows that other potentially homologous members of the fold class also lack conservation of metal-binding ligands typical of the N6P SF. Thus, this larger view forces us to re-evaluate our initial criteria for excluding from the SF other proteins in the fold class on the basis of their lack of metal-binding ligands (Fig. 6.12). We also note that different databases classify relationships among especially highly divergent groups of proteins differently. For example, we noted earlier that CATH classifies protein domains from the SGL subgroup of the N6P SF together with members of the Pfam families MRJP, NHL, PD40, and Ldl_recept_b, and a GSDH protein in a single structural SF they designate as the TolB C-terminal domain SF. This differs from the N6P SF definition currently used by the SFLD. We argue, however, that neither definition should be considered as more "correct" than the other. Rather, each merely reflects the different criteria each resource has chosen to use in the classification of sequence, structural, and functional relationships.

From the perspective of the SFLD classification, these relationships become more complicated than classifications based primarily on sequence and structural similarity, as the SFLD attempts to address how functional features are best included in the formal classification of functionally diverse enzyme SFs. Here, we examine further whether other proteins sharing the six-bladed β-propeller fold that also catalyze hydrolytic reactions should be considered as members in the SF. These include phytases, which catalyze the hydrolysis of phytic acid to phosphate, and neuraminidases, which hydrolyze glycosidic linkages of neuraminic acids,

Figure 6.11 A related catalytic strategy unites the SS enzymes with the rest of the N6P SF. Left panels: active site representations; right panels: diagrams depicting steps in the catalytic mechanisms as described in the text. (a) Superposition of the active site of DFPase (pdb_id: 2gvv; SGL subgroup) with dicyclopentyl phosphoramidate inhibitor bound (only the phosphorous of the inhibitor is depicted) (as seen in (b)) and of SS (green backbone) with tryptamine bound (pdb_id: 2fpb) and with secologanin bound (pdb id: 2fpc; only the aldehyde carbon is depicted) (as seen in (c)). Atoms and ligands are depicted as in (b) and (c), with the metal from (b) removed for clarity. (b) The left panel shows the active site of DFPase as in (a) except that the coordination of the metal to the metal binding ligands is shown. The nucleophilic oxygen of the catalytic Asp229, and the electrophilic phosphorous atom of the inhibitor are depicted as ball and stick. Two water molecules and the phosphoryl oxygen coordinated to the metal have been removed for clarity. The right panel depicts the nucleophilic attack of the oxygen of Asp229 on the phosphorous of the substrate DFP. (c) The left panel shows the superposition of SS with tryptamine bound and with secologanin as in (a). The acidic oxygen atom of the catalytic Glu309, the reactive nitrogen atom of tryptamine, and the electrophilic carbon of secologanin are shown as ball and stick. The right panel shows the deprotonation of the amine group of tryptamine by Glu309 and the subsequent nucleophilic attack of the amine on the aldehyde of secologanin. *Source*: Figure and legend are adapted from Hicks et al. (2011).

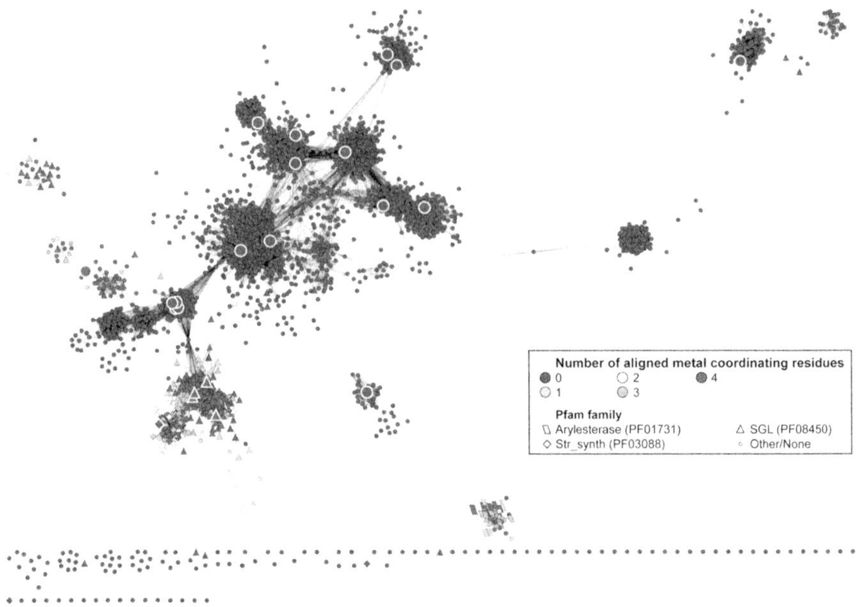

Figure 6.12 Sequence similarity network of those six-bladed β-propeller proteins from the largest cluster of the structure-similarity network in Fig. 6.4. Each node (designated as a "representative node") signifies a group of similar proteins, ranging from 1 to 200 proteins, where each protein shares at least 40% identity with at least one other protein in the group, as calculated using CD-HIT (Li and Godzik, 2006). Four thousand four hundred and fifty-four representative nodes are shown (relative node sizes not designated), representing a total of 17,281 sequences. Sequence similarity was computed from all-by-all pairwise BLAST comparisons among the sequences in and between each representative node. Edges are drawn between representative nodes only if the *E*-value of the best alignment score of a protein represented by that node is at least 1×10^{-7}, using as a background model UniRef100 (Suzek et al., 2007). At this threshold, the median alignment length is 211 residues and the median identity is 29.8%. The number of conserved metal-coordinating residues in the active sites of the majority of proteins represented by a representative node was determined from a multiple sequence alignment of the N6P SF generated using the PROMALS program (Pei et al., 2008) (alignment not shown). Large representative nodes with a light colored outline indicate that the representative node contains at least one structure. The shape of each representative node designates its Pfam assignment as determined by its best hit to Pfam models for members of the N6P SF only if at least one protein of the representative node hit can be assigned to that model. If the proteins corresponding to a representative node match to multiple Pfam models, then the majority dictates the shape. (Note that the group of red colored representative nodes isolated from the majority of the SGL subgroup in the Figure is still considered members of the subgroup. Their apparent separation results from the inadequacy of this 2D depiction to provide an accurate visual representation of high dimensional connections.) (*See insert for color representation of the figure.*)

neither of which connects to the main cluster of proteins in the structure-based similarity network (Fig. 6.4).

To evaluate whether phytases are good candidates for inclusion in the N6P SF, a superposition of phytase with DFPase was examined (not shown). Although this superposition revealed a couple of calcium-coordinating ligands in the phytase active site that are somewhat similar in position to the metal-binding ligands of DFPase, these side chains align poorly relative to alignments between structures classified as members of the SF. Overall, this alignment suggests a different binding modality relative to the N6P SF and suggests a distinct structure–function mapping in phytases. Superposition of this phytase structure with that of SS is not any better (also not shown), further arguing against the inclusion of the phytases as members of the N6P SF. Structural alignments between neuraminidases and N6P SF members are also poor and suggest that these proteins are too dissimilar to include them in the SF. Outside of the families that catalyze hydrolytic reactions, including the NHL, PD40, GSDH, and MRJP Pfam families, a further examination of those structures most similar to the N6P SF in Fig. 6.4 suggests that these structural relationships are too weak to justify inclusion in the N6P SF (not shown).

Interestingly, we find strong conservation patterns between SGL subgroup proteins and seven-bladed β-propeller proteins (virginiamycin B lyases), which may indicate that our initial search for SF members, limited to using the canonically defined 6-bladed β-propeller fold, was insufficient. In β-propeller proteins, blade count may not be highly relevant in defining characteristics of an SF, as sequence similarity between proteins of different blade counts can be more significant than that within proteins sharing the same blade count (Chaudhuri et al., 2008). For example, sequences of the sialidases, which share the six-bladed β-propeller fold with the N6P SF, are more distant from the SGL subgroup proteins than the SGL subgroup is to the seven-bladed virginiamycin B lyases. However, given the current lack of functional information for the SGL subgroup, a strong link between the N6P SF and the virginiamycin B lyases cannot be established.

6.8 USING THE SUPERFAMILY CONTEXT TO SELECT PROTEIN TARGETS FOR EXPERIMENTAL CHARACTERIZATION

Sequence similarity networks that map characterized proteins together with unknowns allow identification of clusters of unknowns that have not yet been characterized, enabling a strategy for choosing proteins for characterization that could be especially informative for functional inference (Pieper et al., 2009). For example, sequence similarity networks can help identify "functional boundaries" within a given SF and provide an estimation of how many functions remain to be discovered. Addressing this problem is a significant challenge for functional inference, as clustering by sequence and structure alone frequently does not track with functional properties (see Glasner et al., 2006 and Schnoes et al., 2009 for some examples). Additionally, knowledge gained from the SF perspective can suggest fundamental catalytic capabilities and constraints that, along with gene context and other orthogonal information, can be used to improve predictions of the functions of unknowns.

An example in the N6P SF is provided by the SSL subgroup protein from *Vitis vinifera*. The sequence similarity network for this subgroup (Fig. 6.8) shows that this uncharacterized protein shares a high degree of overall sequence similarity with known SSs. As reported previously (Hicks et al., 2011), a structure-guided sequence alignment representing structures from all three N6P subgroups, several other sequences that sample the diversity of the SSL subgroup, and the *V. vinifera* protein revealed active site properties in the latter protein that appeared "transitional" between the true SSs and majority of the SSL subgroup proteins that conserve all four of the canonical metal-binding ligands. The *V. vinifera* protein conserves three of the four metal-coordinating ligands, with the missing ligand substituted by a glycine, a residue it shares in common with all known SSs. As predicted by the presence of these metal-binding ligands especially, subsequent biochemical characterization of this protein demonstrated that it does not catalyze the SS reaction at a measurable level but instead exhibits a low level of esterase activity using *p*-nitrophenyl acetate as a substrate. This result suggests properties of a functional boundary between SSs and SSL proteins and provides insight about possible paths by which SSs may have evolved within the background of the hydrolytic and metal-dependent SSL proteins.

Using a more stringent *E*-value cutoff than the one used for the SSL subgroup network presented in Fig. 6.8, the similarity network in Fig. 6.13 shows multiple discrete clusters, one of which cleanly separates all of the SSs into a single cluster. Although insufficient information currently exists to infer how many of the other clusters shown in this network are isofunctional, we hypothesize from this network view that at least several different functional families may remain to be identified in the subgroup. Further comparison of sequence profiles representing these clusters may be useful for hypothesizing additional functional boundaries within the subgroup and for choosing sequences for structural characterization that would likely be useful in answering these questions.

Like the SSL subgroup, the SGL subgroup, which comprises the vast majority of the N6P SF, is also poorly defined in terms of functional families and identification of the reactions they catalyze. However, knowing some types of reactions the characterized members of this subgroup catalyze (lactonase, esterase, organophosphatase) provides hints about specific reactions that may be represented. When evaluated together with orthogonal information such as well-defined gene context, new targets for biochemical and structural characterization can be identified. For example, a potentially attractive target for characterization is the predicted SGL protein encoded in the genome of *Burkholderia sp. H160* (indicated by the arrow in Fig. 6.6), for which the gene context suggests a role in a D-xylose uptake and metabolism pathway (Fig. 6.14). Putting this together with its likely function as a hydrolase, we speculate that this protein may catalyze the breakdown of xylono-1,4 lactone to produce D-xylonate.

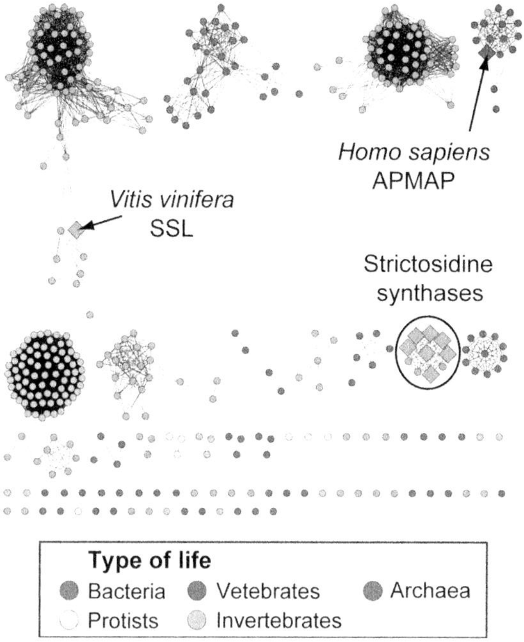

Figure 6.13 Sequence similarity network of 393 proteins of the strictosidine synthase-like (SSL) subgroup downloaded from the SFLD, shown as in Fig 6.8, but edges are drawn if the pairwise similarity score between two nodes is less than 1×10^{-80} (more stringent cutoff). Sequence similarity was computed from pairwise BLAST comparisons of all of the members of the subgroup using the SFLD as a background model. Small circular nodes represent biochemically uncharacterized proteins; large diamond-shaped nodes indicate biochemically characterized proteins. Biochemically characterized SS is marked by the oval. (*See insert for color representation of the figure.*)

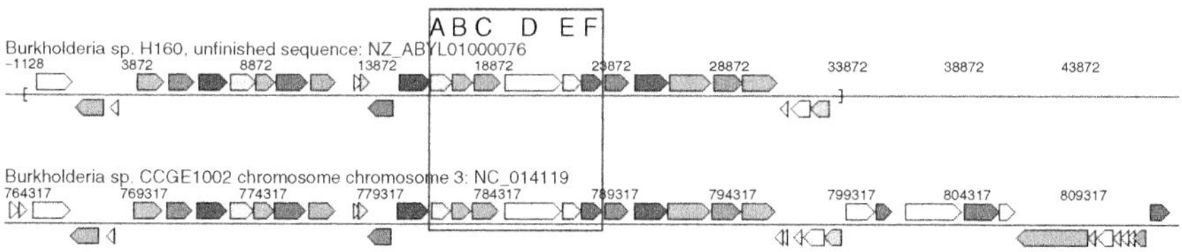

Figure 6.14 Gene context for an uncharacterized SGL subgroup member from *Burkholderia sp. H160* as seen on the Integrated Microbial Genomes (IMGs) system (http://img.jgi.doe.gov/cgi-bin/w/main.cgi) (Markowitz et al., 2010). Genes are labeled by predicted function: (labeled F)—Xylono-1,4-lactonase (SGL member). To the left, (labeled E)—D-xylose dehydrogenase (labeled D)—glycoside hydrolase family 3 domain protein (labeled C)—ABC transporter related (labeled B)—ABC transporter membrane protein 2, carbohydrate uptake transporter-1 family (labeled A)—ABC transporter membrane protein 1, carbohydrate uptake transporter-1 family.

6.9 CONCLUSION

This chapter described the links between structural and functional properties of the proteins of the N6P SF and placed their relationships in the larger context of the six-bladed β-propeller fold proteins. Further dissection of differences among the proteins of the three subgroups of the SF, namely the arylesterase-like, SGL, and SSL subgroups, revealed clues about the functions of uncharacterized proteins in each, allowing the classification of all of the SF members in the SFLD at least at the subgroup level. The limits described here for the prediction and annotation of reaction specificity families also addressed the difficulty of predicting functional boundaries and illustrated some important challenges for discriminating functions based on sequence or structural similarities, especially for functionally diverse enzyme SFs. These issues complicate our understanding of sequence, structure, and function relationships in such systems and illustrate some of the significant challenges for achieving high-quality annotations in the absence of experimental validation.

ACCESS TO DATA FROM THIS WORK

The N6P SF has been added to the SFLD and curated into subgroups and families where possible, based on available high-quality experimental information. SFs are automatically updated on a regular basis, and new sequences and structures are classified according to the SFLD schema using automated protocols to the extent possible, which are then further checked by human curators. These sequence and structure data are freely available at http://sfld.rbvi.ucsf.edu, along with full-length alignments of representative sequences identifying key conserved amino acids. Interactive versions of SF, subgroup, and family networks are also available for download and can be viewed using the freely available Cytoscape program which can be downloaded from http://www.cytoscape.org/.

ACKNOWLEDGMENTS

This work was supported by NIH R01 GM60595 to PCB. AEB was supported by the ARCS Foundation and the PhRMA Foundation Predoctoral Informatics fellowship. Molecular graphics images were produced using the UCSF Chimera package from the Resource for Biocomputing, Visualization, and Informatics at the University of California, San Francisco (supported by NIH P41 RR001081). The authors wish to thank Jimin Pei and Nick Grishin for their help in generating the PROMALS3D alignment.

REFERENCES

Adai, A.T., Date, S.V., Wieland, S., and Marcotte, E.M. (2004) LGL: creating a map of protein function with an algorithm for visualizing very large biological networks. J Mol Biol, 340, 179–190.

Almonacid, D.E. and Babbitt, P.C. (2011) Toward mechanistic classification of enzyme functions. Curr Opin Chem Biol, 15, 435–442.

Altschul, S.F., Madden, T.L., Schaffer, A.A., Zhang, J., Zhang, Z., Miller, W., and Lipman, D.J. (1997) Gapped BLAST and PSI-BLAST: a new generation of protein database search programs. Nucleic Acids Res, 25, 3389–3402.

Andreeva, A., Howorth, D., Chandonia, J.M., Brenner, S.E., Hubbard, T.J., Chothia, C., and Murzin, A.G. (2008) Data growth and its impact on the SCOP database: new developments. Nucleic Acids Res, 36, D419–D425.

Atkinson, H.J. and Babbitt, P.C. (2009) An atlas of the thioredoxin fold class reveals the complexity of function-enabling adaptations. PLoS Comput Biol, 5, e1000541.

Atkinson, H.J., Morris, J.H., Ferrin, T.E., and Babbitt, P.C. (2009) Using sequence similarity networks for visualization of relationships across diverse protein superfamilies. PLoS ONE, 4, e4345.

Babbitt, P.C. and Gerlt, J.A. (1997) Understanding enzyme superfamilies chemistry as the fundamental determinant in the evolution of new catalytic activities. J Biol Chem, 272, 30591–30594.

Babbitt, P.C., Hasson, M.S., Wedekind, J.E., Palmer, D.R., Barrett, W.C., Reed, G.H., Rayment, I., Ringe, D., Kenyon, G.L., and Gerlt, J.A. (1996) The enolase superfamily: a general strategy for enzyme-catalyzed abstraction of the alpha-protons of carboxylic acids. Biochemistry, 35, 16489–16501.

Benson, D.A., Karsch-Mizrachi, I., Lipman, D.J., Ostell, J., and Sayers, E.W. (2010) GenBank. Nucleic Acids Res, 38, D46–D51.

Billecke, S., Draganov, D., Counsell, R., Stetson, P., Watson, C., Hsu, C., and Du, B.N.L. (2000) Human serum paraoxonase (pon1) isozymes Q and R hydrolyze lactones and cyclic carbonate esters. Drug Metabolism and Disposition, 28, 1335–1342.

Billecke, S.S., Primo-Parmo, S.L., Dunlop, C.S., Doorn, J.A., La Du, B.N., and Broomfield, C.A. (1999) Characterization of a soluble mouse liver enzyme capable of hydrolyzing diisopropyl phosphorofluoridate. Chem Biol Interact, 119–120, 251–256.

Blum, M.-M. and Chen, J.C.H. (2010) Structural characterization of the catalytic calcium-binding site in diisopropyl fluorophosphatase (DFPase)–Comparison with related [beta]-propeller enzymes. Chem Biol Interact, 187, 373–379.

Blum, M.-M., Löhr, F., Richardt, A., Rüterjans, H., and Chen, J.C.H. (2006) Binding of a designed substrate analogue to diisopropyl fluorophosphatase: implications for the phosphotriesterase mechanism. J Am Chem Soc, 128, 12750–12757.

Blum, M.M., Timperley, C.M., Williams, G.R., Thiermann, H., and Worek, F. (2008) Inhibitory potency against human acetylcholinesterase and enzymatic hydrolysis of fluorogenic nerve agent mimics by human paraoxonase 1 and squid diisopropyl fluorophosphatase. Biochemistry, 47, 5216–5224.

Chaudhuri, I., Söding, J., and Lupas, A.N. (2008) Evolution of the β-propeller fold. Proteins Struct Funct Bioinformatics, 71, 795–803.

Cline, M.S., Smoot, M., Cerami, E., Kuchinsky, A., Landys, N., Workman, C., Christmas, R., Avila-Campilo, I., Creech, M., Gross, B. et al. (2007) Integration of biological networks and gene expression data using Cytoscape. Nat Protoc, 2, 2366–2382.

Cuff, A.L., Sillitoe, I., Lewis, T., Clegg, A.B., Rentzsch, R., Furnham, N., Pellegrini-Calace, M., Jones, D., Thornton, J., and Orengo, C.A. (2011) Extending CATH: increasing coverage of the protein structure universe and linking structure with function. Nucleic Acids Res, 39, D420–D426.

Draganov, D.I. (2010) Lactonases with organophosphatase activity: structural and evolutionary perspectives. Chem Biol Interact, 187, 370–372.

Draganov, D.I., Teiber, J.F., Speelman, A., Osawa, Y., Sunahara, R., and La Du, B.N. (2005) Human paraoxonases (PON1, PON2, and PON3) are lactonases with overlapping and distinct substrate specificities. J Lipid Res, 46, 1239–1247.

Eddy, S.R. (1996) Hidden Markov models. Current Opin Struct Biol, 6, 361–365.

Enright, A.J., Van Dongen, S., and Ouzounis, C.A. (2002) An efficient algorithm for large-scale detection of protein families. Nucleic Acids Res, 30, 1575–1584.

Finn, R.D., Mistry, J., Schuster-Bockler, B., Griffiths-Jones, S., Hollich, V., Lassmann, T., Moxon, S., Marshall, M., Khanna, A., Durbin, R. et al. (2006) Pfam: clans, web tools and services. Nucleic Acids Res, 34, D247–D251.

Finn, R.D., Mistry, J., Tate, J., Coggill, P., Heger, A., Pollington, J.E., Gavin, O.L., Gunasekaran, P., Ceric, G., Forslund, K. et al. (2010) The Pfam protein families database. Nucleic Acids Res, 38, D211–D222.

Frickey, T. and Lupas, A. (2004) CLANS: a Java application for visualizing protein families based on pairwise similarity. Bioinformatics, 20, 3702–3704.

Frishman, D. (2007) Protein annotation at genomic scale: the current status. Chem Rev, 107, 3448–3466.

Gerlt, J.A. and Babbitt, P.C. (2001) Divergent evolution of enzymatic function: Mechanistically diverse superfamilies and functionally distinct suprafamilies. Annu Rev Biochem, 70, 209–246.

Gerlt, J.A., Babbitt, P.C., Jacobson, M.P., and Almo, S.C. (2012) Divergent evolution in the enolase superfamily: strategies for assigning functions. J Biol Chem, 287, 29–34.

Gerlt, J.A., Babbitt, P.C., and Rayment, I. (2005) Divergent evolution in the enolase superfamily: the interplay of mechanism and specificity. Arch Biochem Biophys, 433, 59–70.

Glasner, M.E., Fayazmanesh, N., Chiang, R.A., Sakai, A., Jacobson, M.P., Gerlt, J.A., and Babbitt, P.C. (2006) Evolution of structure and function in the o-succinylbenzoate synthase/N-acylamino acid racemase family of the enolase superfamily. J Mol Biol, 360, 228–250.

Glasner, M.E., Gerlt, J.A., and Babbitt, P.C. (2007) Mechanisms of protein evolution and their application to protein engineering. Adv Enzymol Relat Areas Mol Biol, 75, 193–239 xii–xiii.

Gomi, K. and Kajiyama, N. (2001) Oxyluciferin, a luminescence product of firefly luciferase, is enzymatically regenerated into luciferin. J Biol Chem, 276, 36508–36513.

Hall, R.S., Brown, S., Fedorov, A.A., Fedorov, E.V., Xu, C., Babbitt, P.C., Almo, S.C., and Raushel, F.M. (2007) Structural diversity within the mononuclear and binuclear active sites of N-acetyl-D-glucosamine-6-phosphate deacetylase. Biochemistry, 46, 7953–7962.

Harel, M., Aharoni, A., Gaidukov, L., Brumshtein, B., Khersonsky, O., Meged, R., Dvir, H., Ravelli, R.B., McCarthy, A., Toker, L. et al. (2004) Structure and evolution of the serum paraoxonase family of detoxifying and anti-atherosclerotic enzymes. Nat Struct Mol Biol, 11, 412–419.

Hicks, M.A., Barber, A.E., Giddings, L.-A., Caldwell, J., O'Connor, S.E., and Babbitt, P.C. (2011) The evolution of function in strictosidine synthase-like proteins. Proteins Struct Funct Bioinformatics, 79, 3082–3098.

Hoskin, F.C.G. and Roush, A.H. (1982) Hydrolysis of nerve gas by squid-type diisopropyl phosphorofluoridate hydrolyzing enzyme on agarose resin. Science, 215, 1255–1257.

Ilhan, A., Gartner, W., Nabokikh, A., Daneva, T., Majdic, O., Cohen, G., Böhmig, G.A., Base, W., Hörl, W.H., and Wagner, L. (2008) Localization and characterization of the novel protein encoded by C20orf3. Biochem J, 414, 485–495.

International Union of Biochemistry and Molecular Biology: Nomenclature Committee, and Webb. (1992). Enzyme nomenclature 1992: recommendations of the Nomenclature Committee of the International Union of Biochemistry and Molecular Biology on the nomenclature and classification of enzymes. (San Diego: Academic Press).

Khersonsky, O. and Tawfik, D.S. (2005) Structure-reactivity studies of serum paraoxonase PON1 suggest that its native activity is lactonase. Biochemistry, 44, 6371–6382.

Kobayashi, M., Shinohara, M., Sakoh, C., Kataoka, M., and Shimizu, S. (1998) Lactone-ring-cleaving enzyme: Genetic analysis, novel RNA editing, and evolutionary implications. Proc Natl Acad Sci USA, 95, 12787–12792.

Kondo, Y., Inai, Y., Sato, Y., Handa, S., Kubo, S., Shimokado, K., Goto, S., Nishikimi, M., Maruyama, N., and Ishigami, A. (2006) Senescence marker protein 30 functions as gluconolactonase in L-ascorbic acid biosynthesis, and its knockout mice are prone to scurvy. Proc Natl Acad Sci USA, 103, 5723–5728.

Kondo, Y., Ishigami, A., Kubo, S., Handa, S., Gomi, K., Hirokawa, K., Kajiyama, N., Chiba, T., Shimokado, K., and Maruyama, N. (2004) Senescence marker protein-30 is a unique enzyme that hydrolyzes diisopropyl phosphorofluoridate in the liver. FEBS Lett, 570, 57–62.

Li, W. and Godzik, A. (2006) Cd-hit: a fast program for clustering and comparing large sets of protein or nucleotide sequences. Bioinformatics, 22, 1658–1659.

Loewenstein, Y., Raimondo, D., Redfern, O.C., Watson, J., Frishman, D., Linial, M., Orengo, C., Thornton, J., and Tramontano, A. (2009) Protein function annotation by homology-based inference. Genome Biol, 10, 207.

Ma, X., Panjikar, S., Koepke, J., Loris, E., and Stockigt, J. (2006) The structure of Rauvolfia serpentina strictosidine synthase is a novel six-bladed beta-propeller fold in plant proteins. Plant Cell, 18, 907–920.

Maresh, J.J., Giddings, L.A., Friedrich, A., Loris, E.A., Panjikar, S., Trout, B.L., Stockigt, J., Peters, B., and O'Connor, S.E. (2008) Strictosidine synthase: mechanism of a Pictet-Spengler catalyzing enzyme. J Am Chem Soc, 130, 710–723.

Markowitz, V.M., Chen, I.M.A., Palaniappan, K., Chu, K., Szeto, E., Grechkin, Y., Ratner, A., Anderson, I., Lykidis, A., Mavromatis, K. et al. (2010) The integrated microbial genomes system: an expanding comparative analysis resource. Nucleic Acids Res, 38, D382–D390.

Martin, A.C. (2004) PDBSprotEC: a web-accessible database linking PDB chains to EC numbers via SwissProt. Bioinformatics, 20, 986–988.

Medini, D., Covacci, A., and Donati, C. (2006) Protein homology network families reveal step-wise diversification of Type III and Type IV secretion systems. PLoS Comput Biol, 2, e173.

Melzer, M., Chen, J.C.H., Heidenreich, A., Gäb, J.r., Koller, M., Kehe, K., and Blum, M.-M. (2009) Reversed enantioselectivity of diisopropyl fluorophosphatase against organophosphorus nerve agents by rational design. J Am Chem Soc, 131, 17226–17232.

Murakami, H., Matsumaru, H., Kanamori, M., Hayashi, H., and Ohta, T. (1999) Cell wall-affecting antibiotics induce expression of a novel gene, drp35, in Staphylococcus aureus. Biochem Biophys Res Comm, 264, 348–351.

Ng, C.J., Shih, D.M., Hama, S.Y., Villa, N., Navab, M., and Reddy, S.T. (2005) The paraoxonase gene family and atherosclerosis. Free Radical Biol Med, 38, 153–163.

Nguyen, T.T., Brown, S., Fedorov, A.A., Fedorov, E.V., Babbitt, P.C., Almo, S.C., and Raushel, F.M. (2008) At the periphery of the amidohydrolase superfamily: Bh0493 from Bacillus halodurans catalyzes the isomerization of D-galacturonate to D-tagaturonate. Biochemistry, 47, 1194–1206.

Ozer, E.A., Pezzulo, A., Shih, D.M., Chun, C., Furlong, C., Lusis, A.J., Greenberg, E.P., and Zabner, J. (2005) Human and murine paraoxonase 1 are host modulators of Pseudomonas aeruginosa quorum-sensing. FEMS Microbiol Lett, 253, 29–37.

Pegg, S.C., Brown, S., Ojha, S., Huang, C.C., Ferrin, T.E., and Babbitt, P.C. (2005) Representing structure-function relationships in mechanistically diverse enzyme superfamilies. Pac Symp Biocomput, 358–369.

Pegg, S.C., Brown, S.D., Ojha, S., Seffernick, J., Meng, E.C., Morris, J.H., Chang, P.J., Huang, C.C., Ferrin, T.E., and Babbitt, P.C. (2006) Leveraging enzyme structure-function relationships for functional inference and experimental design: the structure-function linkage database. Biochemistry, 45, 2545–2555.

Pei, J., Kim, B.-H., and Grishin, N.V. (2008) PROMALS3D: a tool for multiple protein sequence and structure alignments. Nucleic Acids Res, 36, 2295–2300.

Pieper, U., Chiang, R., Seffernick, J.J., Brown, S.D., Glasner, M.E., Kelly, L., Eswar, N., Sauder, J.M., Bonanno, J.B., Swaminathan, S. et al. (2009) Target selection and annotation for the structural genomics of the amidohydrolase and enolase superfamilies. J Struct Funct Genomics, 10, 107–125.

Rentzsch, R. and Orengo, C.A. (2009) Protein function prediction - the power of multiplicity. Trends Biotechnol, 27(4), 210–219.

Rosenblat, M., Gaidukov, L., Khersonsky, O., Vaya, J., Oren, R., Tawfik, D.S., and Aviram, M. (2006) The catalytic histidine dyad of high density lipoprotein-associated serum paraoxonase-1 (PON1) is essential for PON1-mediated Inhibition of low density lipoprotein oxidation and stimulation of macrophage cholesterol efflux. J Biol Chem, 281, 7657–7665.

Scharff, E.I., Koepke, J., Fritzsch, G., Lucke, C., and Ruterjans, H. (2001) Crystal structure of diisopropylfluorophosphatase from Loligo vulgaris. Structure, 9, 493–502.

Schnoes, A.M., Brown, S.D., Dodevski, I., and Babbitt, P.C. (2009) Annotation error in public databases: misannotation of molecular function in enzyme superfamilies. PLoS Comput Biol, 5, e1000605.

Suzek, B.E., Huang, H., McGarvey, P., Mazumder, R., and Wu, C.H. (2007) UniRef: comprehensive and non-redundant UniProt reference clusters. Bioinformatics, 23, 1282–1288.

Tanaka, Y., Morikawa, K., Ohki, Y., Yao, M., Tsumoto, K., Watanabe, N., Ohta, T., and Tanaka, I. (2007) Structural and mutational analyses of Drp35 from Staphylococcus aureus: a possible mechanism for its lactonase activity. J Biol Chem, 282, 5770–5780.

Zhang, Y. and Skolnick, J. (2005) TM-align: a protein structure alignment algorithm based on the TM-score. Nucleic Acids Res, 33, 2302–2309.

7

FUNCTIONAL DIVERSITY OF THE HUP DOMAIN SUPERFAMILY

Benoit H. Dessailly
National Institute of Biomedical Innovation, Osaka, Japan

Christine Orengo
Institute of Structural and Molecular Biology, University College London, London, UK

CHAPTER SUMMARY

A small number of protein domain superfamilies account for a large number of protein domains in genomes, with many different functions and structures. The HUP domain superfamily is one such superfamily. In this chapter, we review the functional diversity of HUP domains, and link it to structural differences between domains.

7.1 INTRODUCTION

Proteins are made up of independent evolutionary units called *domains*. Domains within proteins often adopt compact globular structures, and sometimes perform specific functions. A large number of proteins consist of several such domains. The recognition of domains within proteins often allows the detection of more remote evolutionary relationships than when comparing the proteins

Protein Families: Relating Protein Sequence, Structure, and Function, First Edition.
Edited by Christine Orengo and Alex Bateman.

themselves. Because of the usefulness of domains in detecting remote evolutionary relationships, resources have been developed to organize domains by similarities. Some of these resources, such as Pfam, are based on the recognition of remote sequence similarities between domains to group them into domain families (Finn et al., 2010) (see also Chapter 2), whereas others (SCOP, CATH, see Chapter 3) provide hierarchical classifications that group homologous domains of known 3-D structure into superfamilies using a combination of structural, functional, and sequence information. Those structure-based resources also group domains in higher level assemblies on the basis of structural similarities only (Andreeva et al., 2008; Cuff et al., 2011).

Both SCOP and CATH classifications of domains have been extended to include sequences with unknown 3-D structures by deriving hidden markov model (HMM) profiles from superfamilies of domains of known 3-D structure and using these profiles to detect more superfamily members in sequence databases (de Lima Morais et al., 2011; Lees et al., 2010) (see also Chapter 4). The resource that assigns CATH superfamily domains to sequences is called *Gene3D* (Lees et al., 2010) and contains about 16 million domains, in sequences from Uniprot (The Uniprot Consortium, 2012), RefSeq (Pruitt et al., 2007), and Ensembl (Flicek et al., 2012).

Analysis of domain superfamilies in Gene3D indicates that a small number of superfamilies account for a very large number of sequences. These large superfamilies also tend to be very diverse in structure and function. The reasons for the extreme diversity and success of such superfamilies during evolution have been explored in a number of reviews (Goldstein, 2008; Redfern et al., 2008). Among other explanations, it has been proposed that such superfamilies might represent ideal scaffolds for developing new functions via the easy addition of structural elements to the structural core (Reeves et al., 2006).

The HUP superfamily is an example of such a large and diverse superfamily, in which large embellishments to the structural core seem to play an important role in function diversification (Dessailly et al., 2010). The HUP superfamily is thought to be very ancient, and may already have been highly diversified before the emergence of the Last Universal Common Ancestor (Aravind et al., 2002). The structural core of HUP domains is a Rossmann fold, which consists of a central parallel β-sheet surrounded by α-helices on both sides. This domain superfamily was designated "HUP" after different members of the superfamily, named "High signature" proteins, "Universal stress protein A", and "PP-ATPase" (Aravind et al., 2002). HUP domains can have very different functions, including, for example, aminoacyl-tRNA synthetases that attach amino acids to tRNAs in the context of protein translation, and electron transfer flavoproteins (ETFs) that transfer electrons between proteins in several biological processes.

In this chapter, we summarize the functional diversity encountered in the HUP superfamily, and link it to structural variation among HUP domains.

7.2 FUNCTION DIVERSITY IN THE HUP SUPERFAMILY

7.2.1 Functional Groups

In a previous work, we manually analyzed the diversity of functions encountered among 85 nonredundant domains of known structure in the HUP superfamily (Dessailly et al., 2010). These domains were retrieved from the CATH database (Cuff et al., 2011), version 3.2, using a protocol published in that work. Domains were manually classified according to their function. In this process, functions were defined on the basis of Enzyme Commission (EC) numbers, gene ontology (GO) terms, UniProt descriptions, and further functional information collected from published articles describing these domains.

Domains with the same function were put together into "specificity groups," and specificity groups that shared similar functional mechanisms in spite of different ligand specificities were further grouped together into "functional groups."

Here is a concrete example. Aminoacyl-tRNA synthetases catalyze the attachment of amino acids to their cognate tRNA to form aminoacyl-tRNA molecules, which are then used in protein translation. Each amino acid, for example, tyrosine, is processed by a specific aminoacyl-tRNA synthetase, for example, tyrosyl-tRNA synthetase. Aminoacyl-tRNA synthetases are multidomain proteins and are subdivided into classes I and II on the basis of the fold of their catalytic domain. The catalytic domain of Class I aminoacyl-tRNA synthetases belongs to the HUP superfamily. According to our function classification scheme, Class I aminoacyl-tRNA synthetase catalytic domains are grouped into different specificity groups according to the specific amino acid on which they act. For example, tyrosyl-tRNA synthetases form one specificity group, whereas tryptophanyl-tRNA synthetases constitute another. However, all catalytic domains of Class I aminoacyl-tRNA synthetase are grouped together into the same functional group, because they all catalyze the same general reaction, that is, the attachment of an amino acid to its cognate tRNA. On the other hand, members of another functional group will have a completely unrelated function. For example, HUP domains of Electron Transfer Flavoproteins (ETFs) are nonenzymatic, and are notably involved in the transfer of electrons between fatty acid catabolism pathway components.

In general, two catalytic domains that share the same EC number would be put together in the same specificity group. And, if those domains share the same EC number down to the third digit with members of another specificity group, those two specificity groups would be put together into the same functional group.

By grouping domains in that way, we obtained a classification comprising nine functional groups, and 35 specificity groups within them. The different functional groups in this superfamily consist of proteins with very different functions, which are described below. Table 7.1 also summarizes the functional groups, together with a short description of their function, a representative with a typical 3-D structure for the group, and a listing of individual functions found within the group.

TABLE 7.1 Description of Functional Groups

Domain ID	Specificity	EC Number	Multiple Domain Architecture (CATH Superfamily Codes)
FSG 1. Electron transfer flavoproteins (ETFs)			
2a1uA01 1efpA02 1o97D01	ETFα	Nonenzyme	3.40.50.620, 3.40.50.1220
1efvB00 1efpB00 1o97C00	ETFβ	Nonenzyme	3.40.50.620
FSG 2. Aminoacyl-tRNA synthetases (AATRS) [E]			
2cyaA01 2cybA01 1zh0A01 2cycA01	Tyr-tRNA synthetase	6.1.1.1	3.40.50.620, 1.10.240.10
2pidA01	Tyr-tRNA synthetase	6.1.1.1	3.40.50.620, 1.10.240.10, NA.NA.NA.NA
1n3lA01	Tyr-tRNA synthetase	6.1.1.1	3.40.50.620, 1.10.240.10, 2.40.50.140
1h3fB01 1jilA01 1wq4A01 2ts1A01	Tyr-tRNA synthetase	6.1.1.1	3.40.50.620, 1.10.240.10, 3.10.290.10
1y42X01	Tyr-tRNA synthetase	6.1.1.1	3.40.50.620, 1.10.240.10, 3.10.290.10, NA.NA.NA.NA
1i6kA01 2ip1A01 2yy5A01	Trp-tRNA synthetase	6.1.1.2	3.40.50.620, 1.10.240.10
1r6uB01	Trp-tRNA synthetase	6.1.1.2	1.10.287.10, 3.40.50.620, 1.10.240.10
1wkbA01	Leu-tRNA synthetase	6.1.1.4	3.40.50.620, 3.90.740.10, 1.10.730.10, NA.NA.NA.NA
1obhA01	Leu-tRNA synthetase	6.1.1.4	3.40.50.620, 3.90.740.10, 2.30.210.10, 1.10.730.10
1ileA01	Ile-tRNA synthetase	6.1.1.5	3.40.50.620, 3.90.740.10, 1.10.730.10
1irxA01	Lys-tRNA synthetase	6.1.1.6	3.40.50.620, NA.NA.NA.NA, NA.NA.NA.NA, NA.NA.NA.NA, 1.10.10.350
1gaxA01	Val-tRNA synthetase	6.1.1.9	1.10.730.10, 3.40.50.620, 3.90.740.10, 3.30.1170.10, 1.10.287.380
2csxA01	Met-tRNA synthetase	6.1.1.10	3.40.50.620, 2.170.220.10, 1.10.730.10
2d5bA01	Met-tRNA synthetase	6.1.1.10	3.40.50.620, 2.170.220.10, 1.10.730.10, 2.40.50.140

TABLE 7.1 (*Continued*)

Domain ID	Specificity	EC Number	Multiple Domain Architecture (CATH Superfamily Codes)
1pg2A01	Met-tRNA synthetase	6.1.1.10	3.40.50.620, 2.20.28.20, 1.10.730.10
1li5A01	Cys-tRNA synthetase	6.1.1.16	3.40.50.620, NA.NA.NA.NA
1j09A01	Glu-tRNA synthetase	6.1.1.17	3.40.50.620, 3.90.800.10, 1.10.1160.10, 1.10.8.70, 1.10.10.350
2o5rA01	Glu-tRNA synthetase	6.1.1.17	3.40.50.620, 3.90.800.10, 1.10.1160.10, 1.10.8.70, 1.10.10.350
1qtqA05	Gln-tRNA synthetase	6.1.1.18	3.40.50.620, 3.90.800.10, 1.10.1160.10, 2.40.240.10
1f7uA01	Arg-tRNA synthetase	6.1.1.19	3.30.1360.70, 3.40.50.620, 1.10.730.10
1iq0A01	Arg-tRNA synthetase	6.1.1.19	3.30.1360.70, 3.40.50.620, 1.10.730.10
FSG 3. ATP-pyrophosphatases (ATP-PPASE) [E]			
1kqpA00 1wxiA00	NAD synthetase	6.3.1.5	3.40.50.620
1m1zA02	beta-lactam-synthase	6.3.3.4	3.60.20.10, 3.40.50.620
1k92A01 1korB01 1vl2A01 1kh1C01	Argininosuccinate synthase	6.3.4.5	3.40.50.620, 3.90.1260.10, 1.20.5.470
2dplA01	GMP synthase	6.3.5.2	3.40.50.620, 3.30.300.10
1gpmA02 2ywbA02	GMP synthase	6.3.5.2	3.40.50.880, 3.40.50.620, 3.30.300.10
1ct9B02	Asparagine synthetase B	6.3.5.4	3.60.20.10, 3.40.50.620
1q15D02	CarA	NA	3.60.20.10, 3.40.50.620
FSG 4. Cryptochromes/DNA photolyases (C-DNAP)			
1tezA01	Deoxyribodipyrimidine photo-lyase [8-HDF]	4.1.99.3	3.40.50.620, 1.25.40.80, 1.10.579.10
2j07A01	Deoxyribodipyrimidine photo-lyase [FMN]	4.1.99.3	3.40.50.620, 1.25.40.80, 1.10.579.10
1dnpA01	Deoxyribodipyrimidine photo-lyase [MTHF]	4.1.99.3	3.40.50.620, 1.25.40.80, 1.10.579.10
1np7B01	Cryptochrome DASH	Non-enzyme	3.40.50.620, 1.25.40.80, 1.10.579.10
FSG 5. Nucleotidyltransferases (NT) [E]			
1ej2A00 1f9aA00	Nicotinamide-nucleotide adenylyltransferase	2.7.7.1	3.40.50.620

(*Continued*)

TABLE 7.1 ***(Continued)***

Domain ID	Specificity	EC Number	Multiple Domain Architecture (CATH Superfamily Codes)
1kqnB00	Nicotinamide-nucleotide adenylyltransferase 1	2.7.7.1	3.40.50.620
1nupB00	Nicotinamide-nucleotide adenylyltransferase 3	2.7.7.1	3.40.50.620
1lw7A01	Nicotinamide-nucleotide adenylyltransferase [bi]	2.7.7.1	3.40.50.620, 3.40.50.300
1mrzA01	Riboflavin kinase/FMN adenylyltransferase	2.7.7.2	3.40.50.620, 2.40.30.30
1qjcA00 1tfuA00 1o6bA00 1vlhB00 1od6A00	Phosphopantetheine adenylyltransferase	2.7.7.3	3.40.50.620
1v47B02	Sulfate adenylyltransferase	2.7.7.4	3.10.400.10, 3.40.50.620
1jhdA01	Sulfate adenylyltransferase	2.7.7.4	3.10.400.10, 3.40.50.620
1r6xA02	Sulfate adenylyltransferase	2.7.7.4	3.10.400.10, 3.40.50.620, 3.40.50.300
1yumA00 1k4kB00	Nicotinate-nucleotide adenylyltransferase	2.7.7.18	3.40.50.620
1kamA00	Nicotinate-nucleotide adenylyltransferase	2.7.7.18	3.40.50.620
1cozA00	Glycerol-3-phosphate cytidylyltransferase	2.7.7.39	3.40.50.620
FSG 6. Pantothenate synthetases (PS) [E]			
1mopA01 1ihoA01	Pantothenate synthetase	6.3.2.1	3.40.50.620, 3.30.1300.10
1v8fA01 2ejcA01	Pantothenate synthetase	6.3.2.1	3.40.50.620, 3.30.1300.10
FSG 7. tRNA specific 2-thiouridylases (TRMU) [E]			
2derB01 2hmaA01	tRNA-specific 2-thiouridylase	2.1.1.61	3.40.50.620, 2.30.30.280, 2.40.30.10
FSG 8. Phosphoadenylyl-sulfate reductases (PAPSR) [E]			
1surA00	Phosphoadenylyl-sulfate reductase	1.8.4.8	3.40.50.620
FSG 9. Universal stress proteins A (USPA)			
1mjhB00	MJ0577	NA	3.40.50.620

TABLE 7.1 *(Continued)*

Domain ID	Specificity	EC Number	Multiple Domain Architecture (CATH Superfamily Codes)
1jmvA00	Universal Stress Protein A	Non-enzyme	3.40.50.620
Poorly characterized proteins			
1tq8A00	Uncharacterized protein	NA	3.40.50.620
1q77A00	Uncharacterized protein	NA	3.40.50.620
1ru8A01	Uncharacterized protein	NA	3.40.50.620, 3.90.1490.10
2dumC00	Uncharacterized protein	NA	3.40.50.620
2pg3A00	Uncharacterized protein	NA	3.40.50.620
2ielA00	Uncharacterized protein	NA	3.40.50.620
2pfsA01	Uncharacterized protein	NA	3.40.50.620

This table summarizes the classification of HUP domains in functional groups and function specificity groups as previously presented (Dessailly et al., 2010). Domains are grouped together under headings that describe their functional groups. The first column provides the CATH Domain IDs, the second column describes the function of the protein that contains those domains, and the third column gives the ec number of those proteins ("NA" if unknown, "nonenzyme" for nonenzymes). The fourth column provides a simple description of the multidomain architecture of the protein, with a comma-separated string consisting of the CATH codes of the superfamilies to which domains in the protein belong and "NA.NA.NA.NA." in the multidomain architecture string indicates the likely presence of yet uncharacterised domains in the protein.

7.3 DESCRIPTION OF FUNCTIONAL GROUPS

7.3.1 Electron-Transfer Flavoproteins

ETFs are small electron-transfer proteins that are found in all domains of life (Toogood et al., 2007). All ETFs share an identical molecular activity of electron transfer but can be involved in a variety of cellular processes. For example, these proteins can be active in the transfer of electrons to nitrogen-fixing enzymes in some prokaryotes (Scott and Ludwig, 2004). In mammals, ETFs act as electron acceptors for several primary flavoprotein dehydrogenases, in the context of several processes including fatty-acid metabolism or catabolism of amino acids and other compounds. ETFs always consist of a heterodimer in which each subunit, designated α or β, contains a HUP domain. The α subunit also contains an extra C-terminal domain with a different fold. ETFs bind adenosine monophosphate (AMP) as well as a flavin adenine dinucleotide (FAD) cofactor.

7.3.2 Aminoacyl-tRNA Synthetases

Aminoacyl-tRNA synthetases catalyze the attachment of amino acids to the acceptor end of their cognate tRNAs. This is completed in a highly specific two-step reaction. First, an ATP molecule is hydrolyzed in its $\alpha-\beta$ bond and is used with an amino acid to form an aminoacyl–adenylate intermediate. After that, the aminoacyl moiety is transferred to the tRNA (O'Donoghue and Luthey-Schulten, 2003). These enzymes are found in all living organisms and subdivide into two main classes (termed class I and class II), depending on their cognate amino acids (Woese et al., 2000). As mentioned earlier, both classes are characterized by a different catalytic domain. Class I aminoacyl-tRNA synthetases have a HUP catalytic domain, and are responsible for binding amino acids C, E, I, K, L, M, Q, R, Y, V, and W to their cognate tRNAs. Because of their central function in protein translation, aminoacyl-tRNA synthetases have been suggested to be very ancient enzymes that might have arisen as far back as the RNA world (Ribas de Pouplana and Schimmel, 2000). The typical reaction catalyzed by members of this group is

$$\text{ATP} + \text{L-AA} + \text{tRNA(AA)} = \text{AMP} + \text{PPi} + \text{L-AA-tRNA(AA)}.$$

7.3.3 N-Type ATP-pyrophosphatases

N-type ATP-pyrophosphatases are a set of metabolic enzymes that catalyze strikingly similar reactions, although acting on different substrates. The enzymes in this group include argininosuccinate synthetase and nicotinamide adenine dinucleotide (NAD) synthetase (see Table 7.1 for complete list of individual enzymes). These enzymes commonly activate a substrate with ATP to form an adenylated intermediate, which is then de-adenylated by the attack of a nucleophilic nitrogen from a second substrate. Enzymes in this functional group also share a common fingerprint sequence in the so-called ATP-binding PP loop (Bork and Koonin, 1994), which is located in the structural core between core β-strand 1 and core α-helix 1 (Aravind et al., 2002). As for aminoacyl-tRNA synthetases, the first step of the catalytic reaction involves an initial hydrolysis of the $\alpha-\beta$ bond of ATP. An example reaction for enzymes of this functional group is that catalyzed by NAD^+ synthetase:

$$\text{ATP} + \text{deamido-NAD}^+ + \text{NH}_3 = \text{AMP} + \text{PPi} + \text{NAD}^+.$$

7.3.4 Cryptochromes and DNA Photolyases

This functional group groups together photoreceptor proteins with very similar structures. DNA photolyases are light-activated enzymes that repair UV-induced lesions in DNA, whereas cryptochromes act as photoreceptors in different contexts depending on their taxonomic range (Lin and Todo, 2005). Cryptochromes seem to be involved in the regulation of the circadian clock by light in many eukaryotes. The molecular mechanism of cryptochromes is generally not very well understood, but it appears that some bind DNA whereas other do not seem to be able to bind DNA directly (e.g., plant cryptochromes) (Lin and Todo, 2005). All members of this functional group bind FAD and another ligand called a *chromophore*; the chromophore absorbs light, and the excitation energy is then transferred from it to the FAD. DNA photolyases consist of an HUP domain and another α-helical domain at the C-terminus. Eukaryotic cryptochromes usually have an extra C-terminal domain which modulates their function, as compared with DNA photolyases. The HUP domain of cryptochromes and DNA photolyases is located at the N-terminus of the protein. The chromophore binds in the HUP domain, whereas the FAD binds in another α-helical domain of the same protein (Deisenhofer, 2000). It is worth noting that cryptochromes do not seem to possess catalytic activity, and that the DNA repair enzymatic activity of DNA photolyases is not performed by the HUP domain.

7.3.5 Nucleotidyltransferases

These enzymes catalyze a wide variety of reactions that involve the cleaving of the $\alpha-\beta$ phosphodiester bond of nucleotide triphosphate (NTP), followed by the transfer of the nucleotidyl moiety (nucleoside monophosphate, NMP) to another substrate and the release of pyrophosphate. Examples of such enzymes include nicotinate mononucleotide adenylyltransferase, sulfate adenylyltransferase, and CTP:glycerol-3-phosphate cytidylyltransferase. ATP is the bound nucleotide triphosphate for most known members of this functional group.

7.3.6 Pantothenate Synthetase

Pantothenate synthetase is the last enzyme in the biosynthesis pathway of pantothenate (vitamin B5), which is an essential precursor of coenzyme A (von Delft et al., 2001). This enzyme is found in prokaryotes as well as plants and fungi but not in animals, making it a potentially interesting drug target. Pantothenate synthetase constitutes a functional group by itself. It is functionally related to nucleotidyltransferases, in particular phosphopantetheine adenylyltransferase, which is also part of the coenzyme A biosynthesis pathway (Izard, 2002) but differs from them in that only the reaction intermediate and not the product is adenylated in pantothenate synthetase. The formation of an adenylated intermediate in its catalytic mechanism is also reminiscent of aminoacyl-tRNA synthetases. The enzyme forms a homodimer in which each subunit contains two domains, with the HUP domain at the N-terminus. The additional C-terminal domain is inserted after core β-strand 5 of the HUP domain, and acts as a lid over the main active site in the HUP domain (von Delft et al., 2001).

7.3.7 tRNA 2-Thiouridylases

Transfer-RNA 2-thiouridylases catalyze the conversion of tRNA anticodon uridine nucleotides to modified 2-thiouridines, via the formation of an adenylated intermediate (Numata et al., 2006). Together with subsequent modifications of the same nucleotide, this step is necessary for correct anticodon–codon recognition during translation (Roovers et al., 2008).

7.3.8 Phosphoadenylyl Sulfate Reductase

This functional group consists of a single enzyme that catalyzes the reduction of phosphoadenylyl sulfate to sulfite, using thioredoxin as an electron donor (Savage et al., 1997). This reaction is part of the general pathway that supplies many organisms with reduced sulfur for the biosynthesis of all sulfur-containing metabolites, by converting inorganic sulfate to sulfide. This pathway also contains ATP-sulfurylase, another HUP domain protein from the functional group of nucleotidyltransferases.

7.3.9 Universal Stress Protein A

Universal stress proteins are proteins whose expression is stimulated under various stress conditions, but their molecular function is not very well characterized (Nachin et al., 2008; Sousa and McKay, 2001). They are found in all three domains of life. Several structures have been solved for members of this superfamily, mostly in the context of PSI-2 structural genomics project. On the basis of existing structures in the PDB, it has been suggested that the universal stress protein A functional group subdivides into two subfamilies, one of which binds ATP (see PDB entry 1mjh) and the other does not (see PDB entry 1jmv) (Nachin et al., 2008; Sousa and McKay, 2001; Zarembinski et al., 1998).

7.4 FUNCTIONAL DIVERSITY IN Gene3D HUP DOMAINS

The above functional classification is based only on a small set of nonredundant HUP domains of known structure from the CATH database (Cuff et al., 2011). To get an idea of the functional diversity and the number of members in each function category, it is useful to consider all domains in Gene3D (Lees et al., 2010). This cannot be done manually because of the large number of domains contained in Gene3D. For this analysis, we used Gene3D version 9.2, which contains 76,860 HUP superfamily domains.

A large-scale functional classification of HUP domains can be performed using a novel method that automatically clusters proteins with the same function together in functional families. The GeMMA algorithm derives profiles from groups of protein domains that share similar functions as defined by their sets of high quality GO terms. This is done using the COMPASS method developed in the Grishin group (Sadreyev and Grishin, 2003). Profiles are then compared and merged in an iterative manner (Lee et al., 2010). This new method has been applied to all known sequences of the HUP superfamily, in which we have identified 73 functional families, which together account for 71,302 sequences in the superfamily. Figure 7.1 shows the distributions of domains in these different functional families, and Table 7.2 provides a brief description of the different

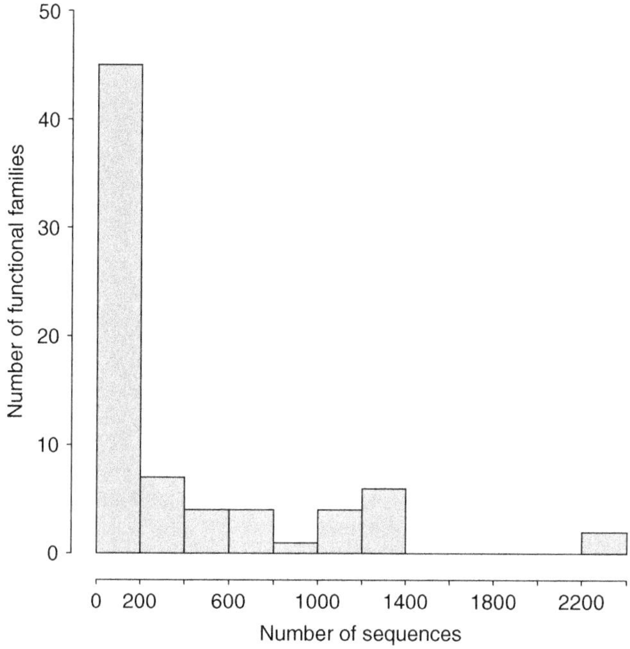

Figure 7.1 Distribution of HUP domains in functional families. This figure shows the distribution of HUP domains from the Gene3D database among the functional families identified by GeMMA.

functional families identified. Most (30/35) functions present in our manual analysis of functional diversity are represented by at least one functional family as identified automatically by GeMMA. The missing ones have probably been merged into functional families with other domains of similar functions.

Furthermore, running GeMMA on Gene3D members of this superfamily readily identifies three new functional families whose members have functions that are different from those of existing functional groups, and may therefore constitute novel functional groups (see Table 7.2). An additional functional family represents 7-cyano-7-deazaguanine synthase, which shares similarities with functions of members of the functional group of N-type ATP-pyrophosphatases. However, the reaction mechanism of 7-cyano-7-deazaguanine is not completely understood and may include steps that would justify putting it in its own separate functional group.

Another seven functional families represent new additional functions of pre-existing functional groups. Again, these are listed, together with the functional group to which they belong, in Table 7.2. Some of these functions are subdivided among more than one functional family (see below).

TABLE 7.2 Mapping of functional families automatically detected by GeMMA to manual classification of functions in the HUP superfamily

Function specificity	GeMMA functional family name	#Sequences
FSG 1. Electron transfer flavoproteins (ETFs)		
ETF α	Electron transfer flavoprotein subunit α	458
ETF β	Not found	
FSG 2. Aminoacyl-tRNA synthetases (AATRS)		
6.1.1.1	Tyrosyl-tRNA synthetase, mitochondrial	1216
6.1.1.2	Tryptophanyl tRNA synthetase 2, mitochondrial	763
6.1.1.4	Leucyl-tRNA synthetase	4
6.1.1.4	Leucyl-tRNA synthetase	142
6.1.1.4	Leucyl-tRNA synthetase, cytoplasmic	127
6.1.1.4	Leucyl-tRNA synthetase, mitochondrial	1235
6.1.1.5	Isoleucyl-tRNA synthetase	1
6.1.1.5	Isoleucyl-tRNA synthetase	7
6.1.1.5	Isoleucyl-tRNA synthetase	7
6.1.1.5	Isoleucyl-tRNA synthetase	15
6.1.1.5	Isoleucyl-tRNA synthetase, putative	1
6.1.1.5	Isoleucyl-tRNA synthetase, mitochondrial	12
6.1.1.5	Isoleucyl-tRNA synthetase, mitochondrial	1374
6.1.1.6	Lysyl-tRNA synthetase 1	93
6.1.1.9	Valyl-tRNA synthetase	2
6.1.1.9	Valyl-tRNA synthetase	4
6.1.1.9	Valyl-tRNA synthetase	6
6.1.1.9	Valyl-tRNA synthetase	95
6.1.1.9	Valyl-tRNA synthetase, mitochondrial	1057
6.1.1.10	Methionine-tRNA synthetase 2, mitochondrial	1188
6.1.1.16	Cysteinyl-tRNA synthetase, mitochondrial	2296
6.1.1.16	Cysteinyl-tRNA synthetase	285
6.1.1.17	Glutamyl-tRNA synthetase	1
6.1.1.17	Glutamyl-tRNA synthetase	3
6.1.1.17	Glutamyl-tRNA synthetase, mitochondrial	2356
6.1.1.18	Probable glutaminyl-tRNA synthetase	4
6.1.1.19	Arginyl-tRNA synthetase, mitochondrial	1327
FSG 3. ATP-pyrophosphatases (ATP-PPASE)		
6.3.1.5	Glutamine-dependent NAD(+) synthetase	502
6.3.3.4	Carboxyethyl-arginine β-lactam-synthase	11
6.3.4.5	Argininosuccinate synthase, chloroplastic	1110
6.3.5.2	GMP synthase [glutamine-hydrolyzing] subunit B	1193
6.3.5.4	Asparagine synthetase B [glutamine-hydrolyzing]	204
6.3.5.4	Asparagine synthetase domain-containing protein 1	34
6.3.4.19 (novel)	tRNA(Ile)-lysidine synthase	2
6.3.4.19 (novel)	tRNA(Ile)-lysidine synthase	3
CarA	Not found	

TABLE 7.2 (*Continued*)

Function specificity	GeMMA functional family name	#Sequences
FSG 4. Cryptochromes/DNA photolyases (C-DNAP)		
4.1.99.3	Deoxyribodipyrimidine photo-lyase, mitochondrial	382
Cryptochrome DASH	Not found	
FSG 5. Nucleotidyltransferases (NT)		
2.7.7.1	Nicotinamide mononucleotide adenylyltransferase 1	665
2.7.7.1	Nicotinamide-nucleotide adenylyltransferase 2	68
2.7.7.2	Riboflavin biosynthesis protein ribF	57
2.7.7.3	Phosphopantetheine adenylyltransferase	55
2.7.7.3	Phosphopantetheine adenylyltransferase	751
2.7.7.4	Sulfate adenylyltransferase subunit 2 2	314
2.7.7.4	3′-Phosphoadenosine 5′-phosphosulfate synthase 2	423
2.7.7.18	Not found	
2.7.7.39	Glycerol-3-phosphate cytidylyltransferase	14
2.7.7.14 (novel)	Probable ethanolamine-phosphate cytidylyltransferase	1
2.7.7.14 (novel)	Probable ethanolamine-phosphate cytidylyltransferase domain 2	2
2.7.7.14 (novel)	Probable ethanolamine-phosphate cytidylyltransferase domain 1	2
2.7.7.14 (novel)	Ethanolamine-phosphate cytidylyltransferase	2
2.7.7.14 (novel)	Ethanolamine-phosphate cytidylyltransferase domain 2	61
2.7.7.14 (novel)	Ethanolamine-phosphate cytidylyltransferase domain 1	88
2.7.7.−(novel)	CTP:phosphorylcholine cytidylyltransferase	166
2.7.7.−(novel)	Bifunctional protein hldE	2
2.7.7.−(novel)	Bifunctional protein hldE	296
2.7.7.2 (novel, previously only in a bifunctional enzyme)	FAD synthase	50
FSG 6. Pantothenate synthetases (PS)		
6.3.2.1	Not found	
6.3.2.1 + 2.7.4.14 (novel)	Bifunctional pantoate ligase/cytidylate kinase	883
FSG 7. tRNA specific 2-thiouridylases (TRMU)		
2.1.1.61	tRNA-specific 2-thiouridylase mnmA	4
2.1.1.61	Mitochondrial tRNA-specific 2-thiouridylase 1	1326
FSG 8. Phosphoadenylyl-sulfate reductases (PAPSR)		
1.8.4.8	Phosphoadenosine phosphosulfate reductase	2

(*Continued*)

TABLE 7.2 (*Continued*)

Function specificity	GeMMA functional family name	#Sequences
N.A. (novel)	5′-Adenylylsulfate reductase 2, chloroplastic	254
FSG 9. Universal stress proteins A (USPA)		
Universal stress protein A	Universal stress protein Rv2624c/MT2699	9
Universal stress protein A	Universal stress protein A family protein C25B2.10	343
FSG. 10 (novel)		
2.8.1.4 (novel)	tRNA sulfurtransferase	503
FSG. 11 (novel)		
6.3.4.−(novel)	7-Cyano-7-deazaguanine synthase 1	682
FSG. 12 (novel)		
N.A. (novel)	tRNA 2-thiocytidine biosynthesis protein TtcA 2	1205
FSG. 13 (novel)		
6.2.1.22 (novel)	[Citrate [pro-3S]-lyase] ligase	46
Poorly characterized		
N.A.	Uncharacterized 37.8 kDa protein in gpa 5′ region	1
N.A.	Microcin J25-processing protein mcjC	1
N.A.	Uncharacterized protein C57A10.07	128
N.A.	Meiotically up-regulated gene 71 protein	7
N.A.	UPF0647 protein C694.03	12
N.A.	Sensor protein kdpD	15
N.A.	NADPH-dependent 7-cyano-7-deazaguanine reductase (Misannotation)	12

Each line in the table corresponds to a match (or mismatch) between a functional family identified by the automated method GeMMA and a specific function from our manual classification of HUP domains (see Table 7.1). Functional families are grouped according to the functional group to which they match. Novel functional groups, as compared with those listed in Table 7.1, are annotated as such. For each line in the table, Column 1 gives a short description of the functional family. This description is usually the EC number for enzymes or a short description taken from our manual classification of functions (see Table 7.1) when there is a match. When there is no match and the EC number is unknown, "N.A." (not assigned) is used. Column 2 gives the function family name automatically assigned by GeMMA. Column 3 gives the number of sequences in the GeMMA family. Novel GeMMA families are annotated as such. Some functions from our manual functional classification are not recovered by GeMMA, and those are indicated as "Not found" in Column 2.

This automated protocol of functional family identification has been developed so as to maximize purity. Because of that, functional families are sometimes so specific that domains with a given specificity (e.g., Valyl-tRNA synthetases) are subdivided into different functional families. This overdivision can sometimes be based on phylogenetic distribution, with bacterial and mitochondrial Valyl-tRNA synthetases tending to cluster in one functional family whereas eukaryotic nuclear

Valyl-tRNA synthetases tend to group into another. A total of 11 functions are thus subdivided into several functional families (see Table 7.2).

7.5 FUNCTION DIVERSITY AND EVOLUTION

As explained previously, different functional groups represent major functional shifts of HUP domains. In spite of this, some aspects of functional mechanisms are conserved between several different functional groups. These shared functional properties are possibly giving us insights into functional features of ancestral HUP domains. First of all, the HUP domain is catalytic in all functional groups but three, that is, ETFs, cryptochromes/DNA photolyases, and universal stress proteins. This suggests that the ancestral HUP domain may have been catalytically active.

Furthermore, HUP domains from all enzymatic functional groups seem to bind a nucleotide-triphosphate, usually ATP. The binding site for this nucleotide-triphosphate is always located at the C-terminal tip of the core β-sheet. moreover, aminoacyl-tRNA synthetases, N-type ATP-pyrophosphatases, nucleotidyltransferases, pantothenate synthetases, and tRNA 2-thiouridylases all share a common mechanistic step in that they hydrolyze the $\alpha-\beta$ phosphodiester bond of nucleotide-triphosphates, and transfer the nucleotidyl moiety onto a second substrate molecule to form an adenylated intermediate. Together, these enzymatic groups cover the vast majority of domains in the superfamily, suggesting that the ancestral HUP domain might have been able to catalyze a reaction with a similar mechanistic step.

In addition, members of several functional groups are able to bind and process tRNAs, even though they often catalyze very different reactions. These functional groups include aminoacyl-tRNA synthetases and tRNA 2-thiouridylases, as well as three novel functional groups detected by GeMMA (see Table7. 2), that is, tRNA(Ile)-lysidine synthases, tRNA sulfurtransferases, and tRNA 2-thiocytidine biosynthesis protein. An additional novel functional group mentioned in Table 7.2, that is, 7-cyano-7-deazaguanine, is involved in the processing of precursor molecules for the modification of tRNAs, although it does not seem to bind to tRNA itself. The presence of so many functions related to the maturing of tRNAs suggests that all these functional groups might have evolved from a common precursor involved in a similar process. According to the retrograde model of pathway evolution ((Rison and Thornton, 2002) and references therein), pathways can evolve by diversification of enzymes within a biological process to catalyze novel steps in that process. A famous example of retrograde pathway evolution is the histidine biosynthesis pathway, in which several steps are catalyzed by apparently related triosephosphate isomerase (TIM) barrel enzymes. A similar process might have taken place here, where many different tRNA-processing HUP domain enzymes would have diversified from a single ancestral one.

Interestingly, the HUP superfamily contains other examples of enzymes from different functional groups catalyzing reactions within the same pathway. These include NAD synthetase (N-type ATP-pyrophosphatases functional group) and nicotinate/nicotinamide-nucleotide adenylyltransferases (nucleotidyltransferases functional group), which catalyze subsequent steps of the NAD biosynthesis pathway. It is also the case of phosphopantetheine adenylyltransferase (nucleotidyltransferases functional group) and pantothenate synthetase (pantothenate synthetase functional group), which catalyze different steps of both the pantothenate and coenzyme A biosynthesis pathways. It is worth noting that both the coenzyme A and the NAD biosynthesis pathways are cofactor biosynthesis pathways, and that cofactors are thought to have had an important role in ancient evolutionary times (Ji et al., 2007). The involvement of HUP domains in both cofactor biosynthesis and tRNA maturing, as well as the widespread ability of HUP domains to bind ATP, seems to point at a very ancient origin for this domain superfamily.

As mentioned above, most HUP domains, across several different FSGs, bind to nucleotide-triphosphates, ATP in most cases. However, the other ligands are very diverse, ranging from rather large molecules such as tRNAs or NAD^+ to very small molecules such as sulfate or ammonium. In many proteins, HUP domains are combined with other domains, and these extra domains often play an important role in ligand binding (see section on multidomain architectures (MDAs) below). For example, extra domains are important in mediating the interaction of aminoacyl-tRNA synthetases with tRNAs.

7.6 MULTIDOMAIN ARCHITECTURES IN THE DIFFERENT FSGs

HUP domains are found in proteins with a wide variety of MDAs. This is in line with previous observations, which suggest that domains from superfamilies with many members are generally also combined in proteins with domains from many different superfamilies (Bornberg-Bauer et al., 2005). In addition, on the basis of structural data, we observe that most functional groups are also characterized by different MDAs. In aminoacyl-tRNA synthetases and N-type ATP-pyrophosphatases, HUP domains are found in combination with several different types of other domains. In nucleotidyl-transferases, on the other hand, most proteins consist of the HUP domain only. The other functional groups all have few members and their domain organization is homogeneous: ETFs and cryptochromes/DNA photolyases contain multiple domains, which are consistent across the functional family, whereas phosphoadenosine phosphosulfate reductases and universal stress proteins always consist of a single domain. MDAs for all proteins in the dataset are shown in Table 7.1. Extra domains can affect the function of proteins with HUP domains in a number of ways. For example, they can contribute residues to the main active site from the HUP domain and thereby modulate the catalytic activity of the HUP domain containing protein. They can also participate in binding ligands as is observed in aminoacyl-tRNA synthetases, where other domains are involved in binding the tRNA (see above).

Figure 7.2 shows the insertion points of extra domains relative to the HUP domain core. In most cases, the extra domains are located at the N- or C-terminus

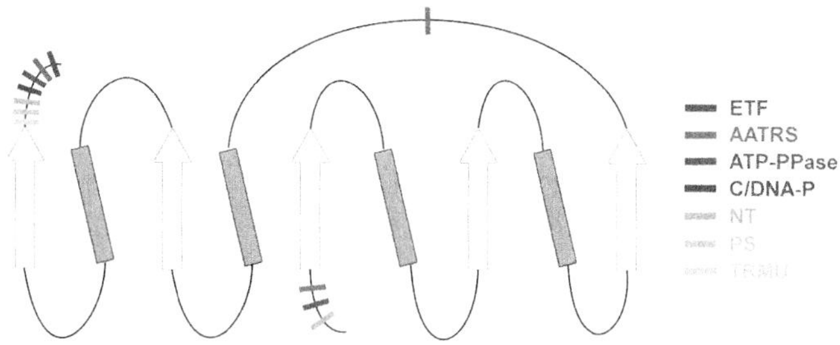

Figure 7.2 Insertion points of extra domains relative to the HUP domain core. The HUP domain core is represented using yellow arrows for the core β-strands and pink rectangles for α-helices. Insertion points for extra domains are represented as colored short lines crossing the loops where the domains are inserted. The lines are colored according to the functional group in which the insertion point is used. The color used for each functional group is given at the right of the figure. For some functional groups, no members in our dataset have extra domains, and these functional groups therefore do not appear in the figure. (*See insert for color representation of the figure.*)

of HUP domains. One exception is the editing domain of Ile/Leu/Val-tRNA synthetases, which is inserted between core β-strand 3 and core α-helix 3 of the HUP domain. This extra domain is not found in other contexts, so it is possible that it might have evolved gradually from structural embellishments in that region of the HUP domain. Previous studies have suggested that domains inserted within the boundaries of other domains might result from gradual embellishments of the parent domains (Aroul-Selvam et al., 2004; Jiang and Blouin, 2007).

7.7 STRUCTURAL DIVERSITY OF HUP DOMAINS

In this section, we first describe the common structural core of all domains in the HUP superfamily. Then we briefly present the structural diversity across the whole superfamily. Finally, we explore structural diversity within each of the functional groups defined above. All structural comparisons were performed using SSAP (Orengo and Taylor, 1996) for pairwise structure alignments, and CORA (Orengo, 1999) for multiple structure alignments.

7.7.1 Structural Core

The structural core consists of all secondary structure elements that are present in at least 80% of the members of this superfamily. The structural core of the HUP domain superfamily consists of a Rossmann fold with five central β-strands arranged in a parallel sheet, surrounded by α-helices on both sides of the sheet (Aravind et al., 2002). The active site of catalytic HUP domains is always located in a cleft at the C-terminal tips of the central β-strands. Figure 7.3 shows the HUP domain core with the location of its active site. In this work, the secondary

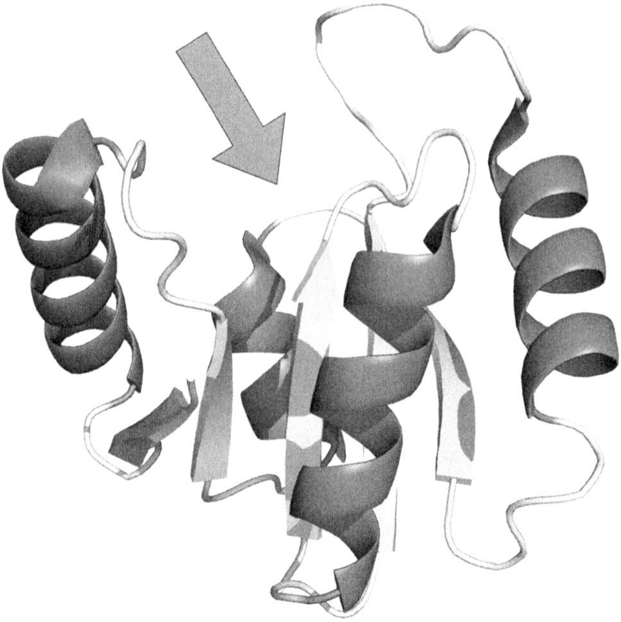

Figure 7.3 HUP domain structural core. Three-dimensional structure of the HUP domain structural core. Core β-strands are colored light grey and core α-helices are colored dark grey. An arrow indicates the typical location of the main active site for most members of the superfamily.

structure elements that constitute the HUP domain core are sometimes labeled cB1 to cB5 for β-strands and cH1 to cH4 for α-helices. In these labels, "c" stands for "core", "B" for "β-strand", "H" for "α-helix", and the numbers refer to the order of the respective strand or helix with respect to the N-terminus. In the reference orientation used in Fig. 7.3, the β-strands follow the left-to-right order cB5–cB4–cB1–cB2–cB3. Core β-strand 1 is therefore the most N-terminal core β-strand, and it is also the central strand of the core β-sheet. It is followed by a long loop that connects it to core α-helix 1 and is generally very important for the function of HUP domains since it is involved in binding nucleotides.

7.7.2 Structural Diversity Across the Superfamily

In spite of this common structural core, pairwise structural comparisons of HUP domains suggest that they can be very different from one another. An example of this is given in Fig. 7.4, where two domains from different functional groups are shown in the same orientation, with the structural differences between them highlighted. Structural diversity in the superfamily is further illustrated in Fig. 7.5, which gives a histogram of normalized root-mean-square deviation (RMSD) values between pairs of HUP domains, with many pairs of HUP domains having normalized RMSD values larger than 10 Å. The normalized RMSD score used is the RMSD normalized by the number of aligned residues between the two proteins being compared, and multiplied by the length of the largest protein ((Dessailly et al., 2010) and references therein). Manual verification of some of these cases suggests that those very large RMSD values are generally due

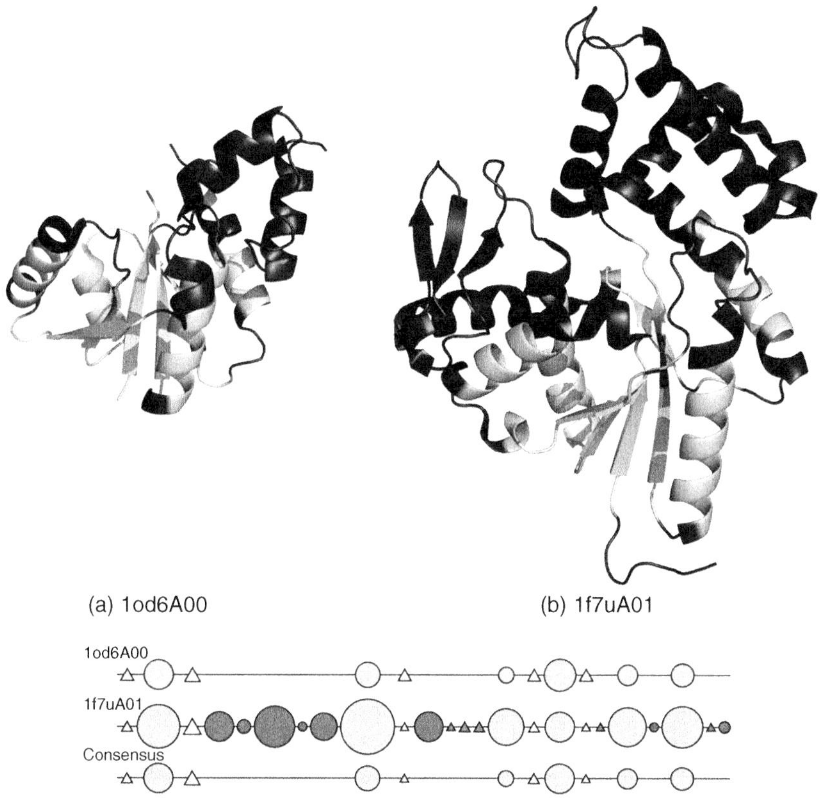

Figure 7.4 Structural diversity between two HUP domains. This figure illustrates the extent of structural differences between domains from the same superfamily. The top panel gives three-dimensional representations of the HUP domains, oriented in the same way, with the structural core colored in gray and the structural differences highlighted in black. The bottom panel emphasizes differences and similarities between secondary structures of these two domains. This diagram was generated using the program 2DSEC, introduced in a previous work (Reeves et al., 2006). Each line in the diagram represents a domain, and the corresponding domain ID is indicated. Circles represent α-helices and triangles represent β-strands. Light gray β-strands and α-helices are those that are conserved across a majority of domains in the group. Domain 1od6A00 (a) is from phosphopantetheine adenylyltransferase, and domain 1f7uA01 (b) is from arginyl-tRNA synthetase.

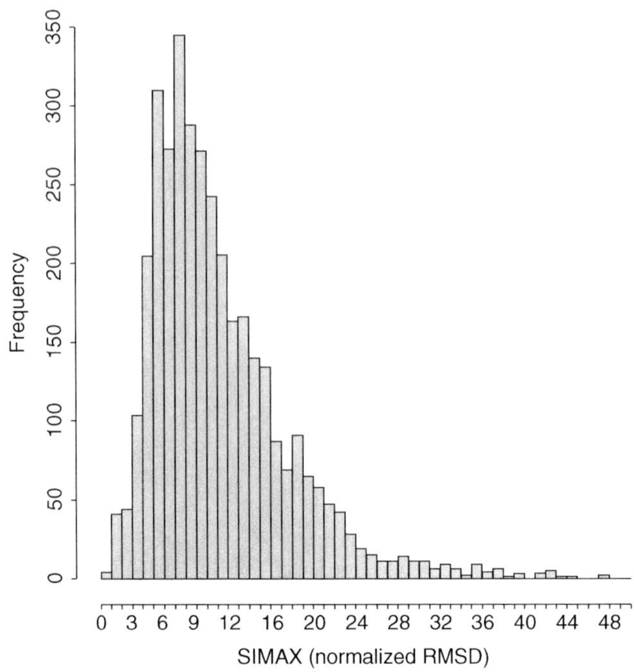

Figure 7.5 Structural similarities between HUP domains. Histogram of normalized RMSD values between pairs of domains in the HUP superfamily. All possible pairwise comparisons were made for the 85 domains in our dataset, resulting in a total of 3570 pairs.

to misalignments of the HUP domains. But the fact that structure superposition programs fail at superposing the HUP domains reflects the large structural differences between them. Such structural diversity is notably explained by the diversity of structural embellishments that are usually inserted in the connections between the elements of secondary structures of the HUP domain core. These embellishments are often important for the function of HUP domains. Embellishments and their involvement in function are described in the next section. As we will see, members of the different functional groups described in Table 7.1 tend to be characterized by specific embellishments, even though there can be important variation even within functional groups.

7.7.3 Structural Diversity Within Functional Groups

To analyze the structural diversity of HUP domains within functional groups, all HUP domains can be grouped into structural clusters on the basis of their structural similarity.

Structures can be clustered on the basis of structural similarity measured by normalized RMSD following structural superpositions made using SSAP (Orengo

TABLE 7.3 Comparison of Structural Clusters with Functional Groups

Functional Groups	SSAP SC5
ETF	1
AATRS	8
ATPPPASE	4
CDNAP	1
NT	5
PS	1
TRMU	1
PAPSR	1
USPA	2

This table is adapted from a previous publication (Dessailly et al., 2010).
This table shows the number of structural clusters ("SC") for each functional group in our dataset. Structural clusters were defined using SSAP and a clustering cutoff of normalized RMSD equal to 5 Å. For Example, HUP domains in the USPA functional group belong to two different structural clusters. Functional groups are referred to using the acronyms introduced in the functional group headers of Table 7.1.

and Taylor, 1996). Analyses of homologous domains in the CATH database suggest that a normalized RMSD value of 5 Å is an appropriate threshold to define structural similarity (Cuff et al., 2009). Two domains with a normalized RMSD above 5 Å can be deemed to carry significant structural differences. By performing multilinkage clustering, with a normalized RMSD cutoff of 5 Å, it is possible to identify a total of 17 structural clusters in the HUP superfamily (CATH version 3.2) (Dessailly et al., 2010).

Comparing structural clusters with functional groups in terms of domain content allows us to analyze quantitatively structural diversity within Functional Groups. A comparison of these structural clusters with functional groups is presented in Table 7.3. Domains within small and less functionally diverse functional groups such as cryptochromes/DNA photolyases and pantothenate synthetases are structurally very similar and are generally found within the same structural cluster. In contrast, structural diversity can be important within larger functional groups. For example, aminoacyl-tRNA synthetases are found in eight different structural clusters. All large and diverse functional groups have at least four different structural clusters within them.

In addition, domains from different functional groups can sometimes be found in the same structural cluster, further reflecting the complexity of the relation between structure and function in this superfamily.

The discrepancy in terms of structural diversity within functional groups is reflected in Fig. 7.6 with 2DSEC plots (Reeves et al., 2006) for one structurally homogeneous functional group, that is, cryptochrome/DNA-photolyase, and one structurally diverse functional group, that is, ATP-pyrophosphatases. 2DSEC plots allow the visualization of a multiple structure alignment in the form of a bidimensional secondary structure alignment (Reeves et al., 2006).

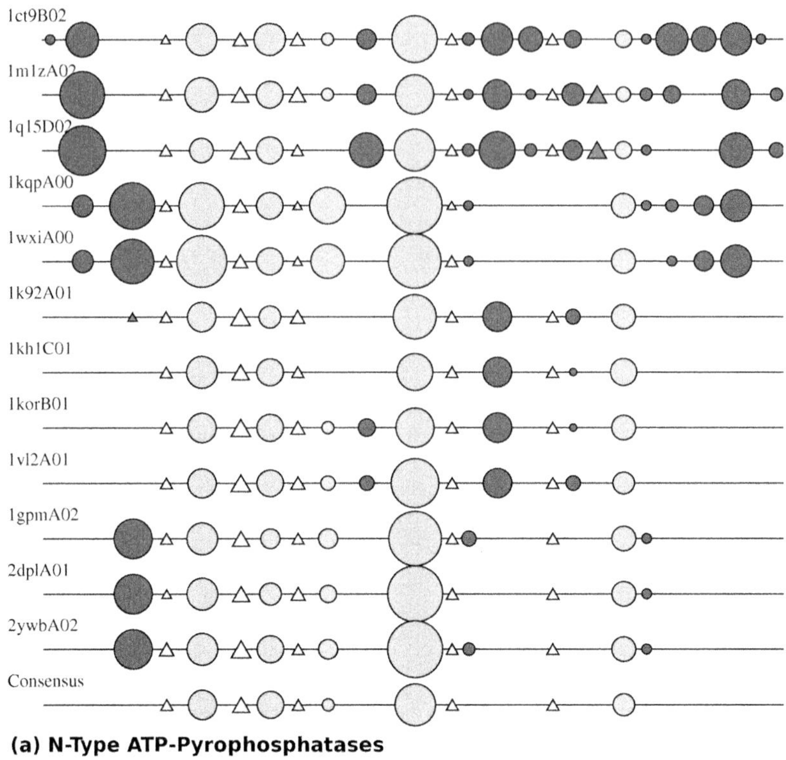

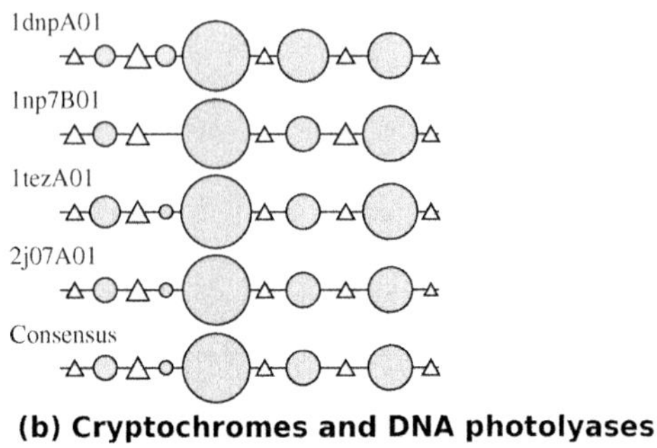

Figure 7.6 Different patterns of structural diversity in different functional groups. 2DSEC diagrams representing secondary structure diversity among members of functional groups. N-type ATP-pyrophosphatases are chosen as a structurally diverse functional group (a), and cryptochromes and DNA photolyases are chosen as an example of a structurally homogeneous functional group (b). See legend of Fig. 7.4 for more information about 2DSEC.

7.8 STRUCTURAL EMBELLISHMENTS IN THE HUP SUPERFAMILY AND THEIR ROLE IN DETERMINING THE FUNCTION DIVERSITY OF HUP DOMAINS

As briefly mentioned in the previous sections, HUP domains within a functional group can be significantly diverse structurally, in spite of functional similarities. In general, structural diversity in the HUP superfamily is due to the presence of embellishments to the common structural core. These structural embellishments are polypeptide segments of at least three residues that are not shared among all members of the superfamily, that is, that are not part of the structural core. In the case of the HUP superfamily, these embellishments generally adopt regular secondary structures. As illustrated in Fig. 7.7, these structural embellishments can be extensive and account for large structural differences between members of the superfamily.

Embellishments in the different functional groups tend to occur in specific insertion points relative to the core. For example, domains in the functional group of aminoacyl-tRNA synthetases always harbor structur al additions inserted between core β-strand 3 and core α-helix 3.

Furthermore, it appears that embellishments tend to be preferentially inserted in a specific subset of topological locations of the domain core, with some locations almost never occupied by insertions. Embellishments inserted at different topological locations often cluster together in the 3-D structure. As noted in our previous work, embellishments are thus often inserted in such locations that allow them to influence the shape of the molecular surface in the vicinity of the main active site, towards the C-terminal tips of the central core β-strands. In contrast, very few structural embellishments are inserted at the opposite end of the central core β-strands, and when they are, they are generally long, loose linkers that connect the HUP domain to another domain in the same protein chain (Dessailly et al., 2010).

Structural embellishments vary significantly in shape, size, and secondary structure contents. An example of a very large embellishment found among the HUP domain structures is the 66-residue-long insertion after core β-strand 2 in yeast arginyl-tRNA synthetase. The smallest bound for the size of embellishments is arbitrary since we define structural embellishments as "sizeable" insertions relative to the core, which contain at least three residues. From the data observed in the HUP superfamily, there do not seem to be any preferences in terms of shape and secondary structure contents for structural embellishments. Figure 7.8 shows the insertion points of structural embellishments relative to the core, together with representative embellishments taken from different superfamily members.

7.8.1 Functional Role of Structural Embellishments

In this section, we explore the functional importance of structural embellishments of HUP domains. In most cases, the function of proteins is ultimately a matter of interactions between molecules. Structural embellishments in the HUP

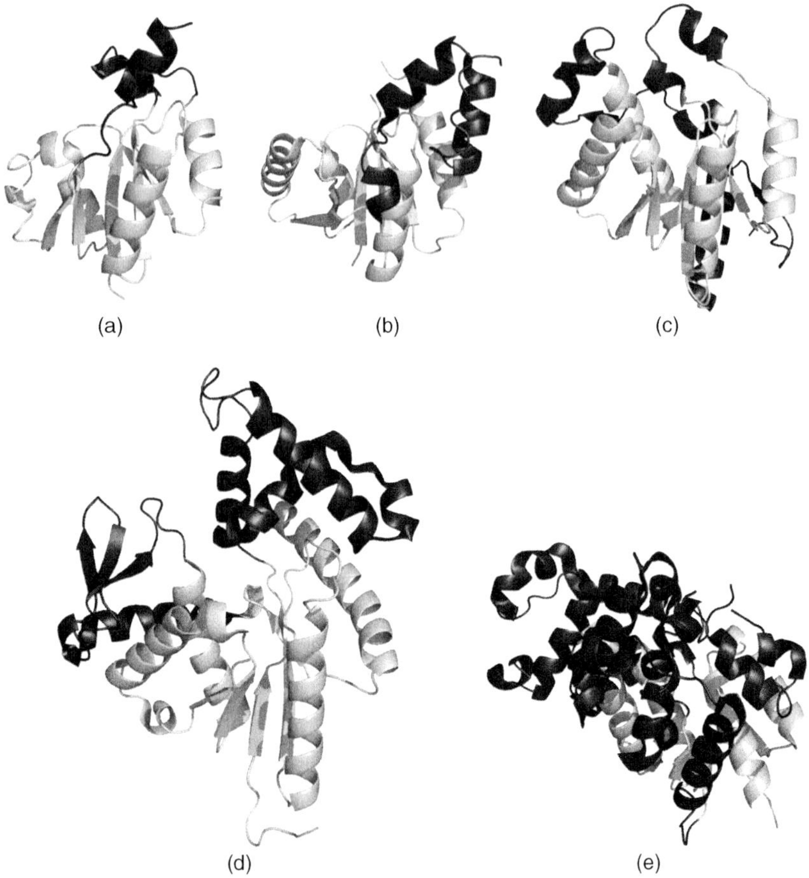

Figure 7.7 Structural core and embellishments in HUP domains. This figure shows five HUP domains from diverse functional groups, with the core colored grey and the embellishments colored black. All domains are presented in the same orientation. The following domains were used: (a) glycerol 3-phosphate cytidylyltransferase (1cozA00—functional group of nucleotidyltransferases), (b) phosphopantetheine adenylyltransferase (1od6A00—functional group of nucleotidyltransferases), (c) pantothenate synthetase (1mopA01—functional group of pantothenate synthetase), (d) arginyl-tRNA synthetase (1f7uA01—functional group of aminoacyl-tRNA Synthetases), (e) Asparagine synthetase B (1ct9B02—functional group of N-type ATP-pyrophosphatases).

superfamily are very often involved in mediating such interactions. These data are summarized in Table 7.4. Below, we consider each type of functional site separately. Functional site information is available from a number of different sources. Catalytic site residues can be obtained from the Catalytic Site Atlas (Porter et al., 2004). Information on binding sites, including binding sites for other domains in the same protein, other proteins, and other molecules, can be

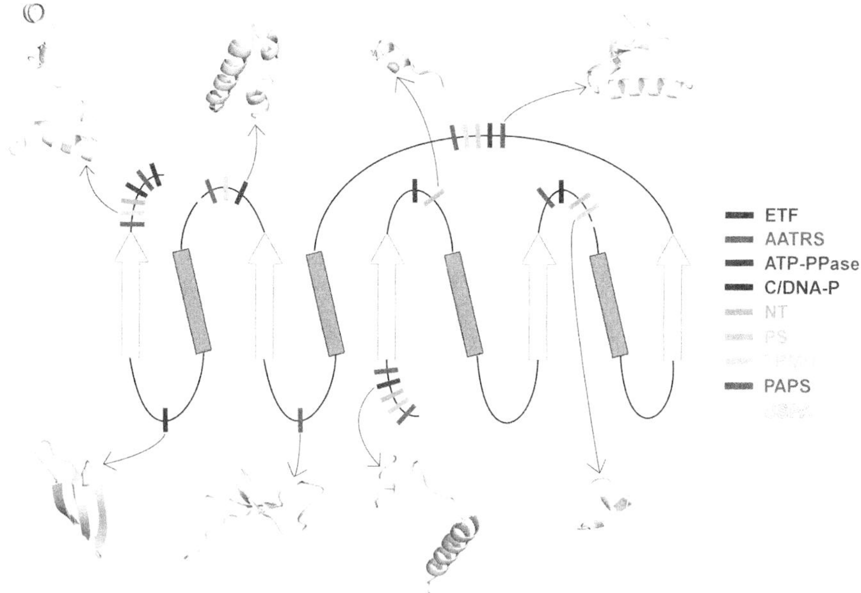

Figure 7.8 Examples of embellishments and their topological insertion points relative to the HUP superfamily domain core. The HUP domain core is schematically represented with yellow arrows for β-strands and pink rectangles for α-helices as in Fig. 7.2. Insertion points of embellishments are indicated by short colored lines crossing the core loops. These insertion points are colored according to the functional group where an embellishment is found at that insertion point. The colors assigned to each functional group are given on the right of the figure. For some insertion points, the 3-D structure of an embellishment found at that point is shown. Three-dimensional structural images were generated using PyMol. (*See insert for color representation of the figure.*)

extracted from PDB entries. Extra information on functional residues is also available from the published literature describing the HUP domains.

7.8.2 Catalytic Sites/Ligand-Binding Sites

As mentioned previously, a large number of structural embellishments are located in the vicinity of the active site. This is notably due to the preferred topological insertion points of embellishments relative to the structural core, as illustrated in Fig. 7.8. Those embellishments seem to play an important role in mediating the functional specificity of different HUP domains by modulating the shape of the active site. Typical examples of structural embellishments influencing the shape of the active site include the C-terminal embellishments of the glycerol-3-phosphate cytidylyltransferase HUP domain and asparagine synthetase B HUP domains, as illustrated in Fig. 7.9, where the effect of these embellishments on the shape of the active site cavity is indicated.

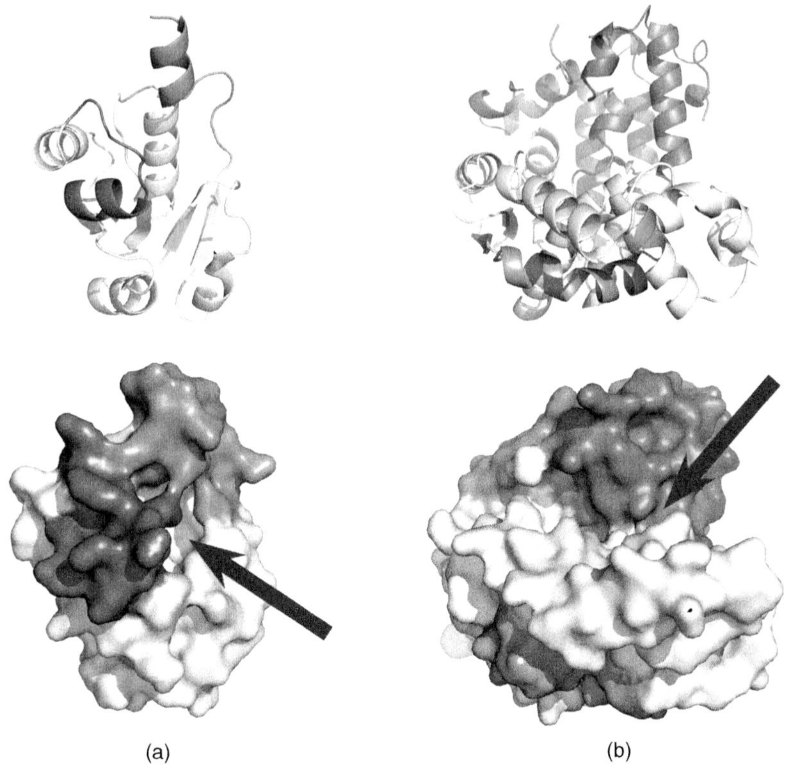

Figure 7.9 Role of structural embellishments in shaping catalytic sites. Illustration of the role of structural embellishments in modifying the shape of the active site cavity, for HUP domains from (a) glycerol-3-phosphate cytidylyltransferase (CATH domain ID 1cozA00), and (b) asparagine synthetase B (CATH domain ID 1ct9B02). In each subfigure, the top panel shows a cartoon representation of the domain, and the bottom panel shows a molecular surface representation. Embellishment residues are colored, whilst other residues are in white. This figure was generated using PyMol. The location of the active site is indicated by an arrow. (*See insert for color representation of the figure.*)

7.8.3 Protein–Protein Interfaces

Very little structural data is available on the interactions between HUP domains and other proteins. Therefore, very few examples can be used in this superfamily to illustrate the involvement of structural embellishments in protein–protein interactions. However, it is interesting to note that the few available examples of known protein–protein interfaces involving HUP domains do indeed involve structural embellishments. That is, for example, the case of ETFs where β-sheet embellishments on the side of the core β-sheet mediate the interaction between the two distinct subunits of the ETF complex (see Fig. 7.10).

More structural data is available on interactions between HUP domains and other identical subunits within homomultimer complexes. In those cases, structural embellishments appear to be often involved in the interactions. For example, a region called the *connective polypeptide* in aminoacyl-tRNA synthetases, which is an insert between core β-strand 3 and core α-helix 3, is crucial for mediating interactions between identical subunits in those enzymes. Embellishments in the same region are also important for such inter-subunit interactions in N-type ATP-pyrophosphatases (asparagine synthetase B and NAD synthetase), pantothenate synthetases, and phosphoadenylyl sulfate reductases. N-terminal embellishments are also important for subunit interactions in some

Figure 7.10 Role of the structural embellishment in protein–protein interaction. This figure represents the ETF complex (PDB ID 1o97), consisting of subunits α (on the left, in white) and β (on the right, in grey). Subunit α consists of a HUP domain (bottom, shown as cartoon), and another domain (top, shown as C-α trace). Subunit β consists of a single HUP domain. Embellishments in both subunits are colored. The embellishment inserted between core β-strand 1 and core α-helix 1 (only present in subunit β) is colored green. The embellishment inserted between core α-helix 4 and core β-strand 5 is colored light blue for subunit α and dark blue for subunit β. The embellishment inserted after core β-strand 5 is colored light red for subunit α and dark red for subunit β. As can be seen in the figure, the embellishments colored in shades of red and blue are important for mediating the interaction between the two subunits. (*See insert for color representation of the figure.*)

TABLE 7.4 Major functional roles and insertion points of structural embellishments in the HUP superfamily

	Before cB1	cB1–cH1	cH1–cB2	cB2–cH2	cH2–cB3	cB3–cH3	cH3–cB4	cB4–cH4	cH4–cB5	After cB5
LIGAND	AAtRS			C/DNAP PS		AAtRS TRMU		PAPS		ATP-PPase NT TRMU PAPS
DOMAIN	ATP-PPase			C/DNAP		AAtRS ATP-PPase TRMU		ATP-PPase TRMU		ATP-PPase
SUBUNIT	NT					AAtRS ATP-PPase PS PAPS			ETF	ETF ATP-PPase
PROTEIN		ETF								ETF PAPS

This table is adapted from a previous publication (Dessailly et al., 2010).
Lines in the table refer to the major functions that embellishments can be involved in. Columns indicate the insertion points, relative to the HUP domain core, where the embellishments are located. Each box in the table lists the different functional groups that have an embellishment at that insertion point and with that function. Function "Ligand" refers to ligand binding or involvement in the catalytic reaction, "Domain" refers to domain–domain interfaces, "Subunit" refers to interfaces between subunits within a homo-multimeric complex, and "Protein" refers to interfaces between different proteins. Core secondary structures in the column headers are those introduced in Section 7.7.1, where cB1 stands for the first core β-strand, cH1 stands for the first core α-helix, etc. Functional groups are referred to using the acronyms introduced in the functional group headers of Table 7.1.

nucleotidyltransferases (sulfate adenylyltransferases), as well as C-terminal embellishments in some N-type ATP-pyrophosphatases (GMP synthetases).

7.8.4 Domain–Domain Interfaces

Interfaces between different domains of the same polypeptide chain are often mediated by structural embellishments on the HUP domain as well. As mentioned previously, proteins containing HUP domains are very diverse in terms of multi-domain architectures (MDAs). In particular, HUP domains of N-type ATP-pyrophosphatases and aminoacyl-tRNA synthetases occur in a large number of different multidomain contexts (see Table 7.1). Structural embellishments that are involved in mediating interactions with other domains of the protein can be inserted at several different points of the HUP domain (see Table 7.4).

7.9 CONCLUSION

Systematic analyses of functional and structural diversity in the HUP superfamily suggest that HUP domains with different functions tend to adopt different structures as well. Structural embellishments to the superfamily core are mostly responsible for these structural differences. Different functional groups are characterized by specific structural embellishments. These embellishments are often involved in some aspect of function, such as binding to different partner molecules. Structural insertions may therefore play an important role during evolution in shaping the functional diversity of protein domain superfamilies. Some domain structures may be more tolerant than others to sizeable insertions, thus explaining in part why some domain superfamilies seem to have diversified much more than others during evolution (Goldstein, 2008).

REFERENCES

Andreeva, A., Howorth, D., Chandonia, J.M., Brenner, S.E., Hubbard, T.J., Chothia, C., and Murzin, A.G. (2008) Data growth and its impact on the SCOP database: new developments. Nucleic Acids Res, 36, D419–D425.

Aravind, L., Anantharaman, V., and Koonin, E.V. (2002) Monophyly of class I aminoacyl tRNA synthetase, USPA, ETFP, photolyase, and PP-ATPase nucleotide-binding domains: implications for protein evolution in the RNA. Proteins, 48, 1–14.

Aroul-Selvam, R., Hubbard, T., and Sasidharan, R. (2004) Domain insertions in protein structures. J Mol Biol, 338, 633–641.

Bork, P. and Koonin, E.V. (1994) A P-loop-like motif in a widespread ATP pyrophosphatase domain: implications for the evolution of sequence motifs and enzyme activity. Proteins, 20, 347–355.

Bornberg-Bauer, E., Beaussart, F., Kummerfeld, S.K., Teichmann, S.A., and Weiner, J. 3rd (2005) The evolution of domain arrangements in proteins and interaction networks. Cell Mol Life Sci, 62, 435–445.

Consortium, T.U (2012) Reorganizing the protein space at the Universal Protein Resource (UniProt). Nucleic Acids Res, 40, D71–D75.

Cuff, A., Redfern, O.C., Greene, L., Sillitoe, I., Lewis, T., Dibley, M., Reid, A., Pearl, F., Dallman, T., Todd, A. et al. (2009) The CATH hierarchy revisited-structural divergence in domain superfamilies and the continuity of fold space. Structure, 17, 1051–1062.

Cuff, A.L., Sillitoe, I., Lewis, T., Clegg, A.B., Rentzsch, R., Furnham, N., Pellegrini-Calace, M., Jones, D., Thornton, J., and Orengo, C.A. (2011) Extending CATH: increasing coverage of the protein structure universe and linking structure with function. Nucleic Acids Res, 39, D420–D426.

de Lima Morais, D.A., Fang, H., Rackham, O.J., Wilson, D., Pethica, R., Chothia, C., and Gough, J. (2011) SUPERFAMILY 1.75 including a domain-centric gene ontology method. Nucleic Acids Res, 39, D427–D434.

Deisenhofer, J. (2000) DNA photolyases and cryptochromes. Mutat Res, 460, 143–149.

Dessailly, B.H., Redfern, O.C., Cuff, A.L., and Orengo, C.A. (2010) Detailed analysis of function divergence in a large and diverse domain superfamily: toward a refined protocol of function classification. Structure, 18, 1522–1535.

Finn, R.D., Mistry, J., Tate, J., Coggill, P., Heger, A., Pollington, J.E., Gavin, O.L., Gunasekaran, P., Ceric, G., Forslund, K. et al. (2010) The Pfam protein families database. Nucleic Acids Res, 38, D211–D222.

Flicek, P., Amode, M.R., Barrell, D., Beal, K., Brent, S., Carvalho-Silva, D., Clapham, P., Coates, G., Fairley, S., Fitzgerald, S. et al. (2012) Ensembl 2012. Nucleic Acids Res, 40, D84–D90.

Goldstein, R.A. (2008) The structure of protein evolution and the evolution of protein structure. Curr Opin Struct Biol, 18, 170–177.

Izard, T. (2002) The crystal structures of phosphopantetheine adenylyltransferase with bound substrates reveal the enzyme's catalytic mechanism. J Mol Biol, 315, 487–495.

Ji, H.F., Kong, D.X., Shen, L., Chen, L.L., Ma, B.G., and Zhang, H.Y. (2007) Distribution patterns of small-molecule ligands in the protein universe and implications for origin of life and drug discovery. Genome Biol, 8, R176.

Jiang, H. and Blouin, C. (2007) Insertions and the emergence of novel protein structure: a structure-based phylogenetic study of insertions. BMC Bioinformatics, 8, 444.

Lee, D.A., Rentzsch, R., and Orengo, C. (2010) GeMMA: functional subfamily classification within superfamilies of predicted protein structural domains. Nucleic Acids Res, 38, 720–737.

Lees, J., Yeats, C., Redfern, O., Clegg, A., and Orengo, C. (2010) Gene3D: merging structure and function for a Thousand genomes. Nucleic Acids Res, 38, D296–D300.

Lin, C. and Todo, T. (2005) The cryptochromes. Genome Biol, 6, 220.

Nachin, L., Brive, L., Persson, K.C., Svensson, P., and Nystrom, T. (2008) Heterodimer formation within universal stress protein classes revealed by an in silico and experimental approach. J Mol Biol, 380, 340–350.

Numata, T., Ikeuchi, Y., Fukai, S., Suzuki, T., and Nureki, O. (2006) Snapshots of tRNA sulphuration via an adenylated intermediate. Nature, 442, 419–424.

O'Donoghue, P. and Luthey-Schulten, Z. (2003) On the evolution of structure in aminoacyl-tRNA synthetases. Microbiol Mol Biol Rev, 67, 550–573.

Orengo, C.A. (1999) CORA–topological fingerprints for protein structural families. Protein Sci, 8, 699–715.

Orengo, C.A. and Taylor, W.R. (1996) SSAP: sequential structure alignment program for protein structure comparison. Methods Enzymol, 266, 617–635.

Porter, C.T., Bartlett, G.J., and Thornton, J.M. (2004) The Catalytic Site Atlas: a resource of catalytic sites and residues identified in enzymes using structural data. Nucleic Acids Res, 32, D129–D133.

Pruitt, K.D., Tatusova, T., and Maglott, D.R. (2007) NCBI reference sequences (RefSeq): a curated non-redundant sequence database of genomes, transcripts and proteins. Nucleic Acids Res, 35, D61–D65.

Redfern, O.C., Dessailly, B., and Orengo, C.A. (2008) Exploring the structure and function paradigm. Curr Opin Struct Biol, 18, 394–402.

Reeves, G.A., Dallman, T.J., Redfern, O.C., Akpor, A., and Orengo, C.A. (2006) Structural diversity of domain superfamilies in the CATH database. J Mol Biol, 360, 725–741.

Ribas de Pouplana, L. and Schimmel, P. (2000) A view into the origin of life: aminoacyl-tRNA synthetases. Cell Mol Life Sci, 57, 865–870.

Rison, S.C. and Thornton, J.M. (2002) Pathway evolution, structurally speaking. Curr Opin Struct Biol, 12, 374–382.

Roovers, M., Oudjama, Y., Kaminska, K.H., Purta, E., Caillet, J., Droogmans, L., and Bujnicki, J.M. (2008) Sequence-structure-function analysis of the bifunctional enzyme MnmC that catalyses the last two steps in the biosynthesis of hypermodified nucleoside mnm5s2U in tRNA. Proteins, 71, 2076–2085.

Sadreyev, R. and Grishin, N. (2003) COMPASS: a tool for comparison of multiple protein alignments with assessment of statistical significance. J Mol Biol, 326, 317–336.

Savage, H., Montoya, G., Svensson, C., Schwenn, J.D., and Sinning, I. (1997) Crystal structure of phosphoadenylyl sulphate (PAPS) reductase: a new family of adenine nucleotide alpha hydrolases. Structure, 5, 895–906.

Scott, J.D. and Ludwig, R.A. (2004) Azorhizobium caulinodans electron-transferring flavoprotein N electrochemically couples pyruvate dehydrogenase complex activity to N2 fixation. Microbiology, 150, 117–126.

Sousa, M.C. and McKay, D.B. (2001) Structure of the universal stress protein of Haemophilus influenzae. Structure, 9, 1135–1141.

Toogood, H.S., Leys, D., and Scrutton, N.S. (2007) Dynamics driving function: new insights from electron transferring flavoproteins and partner complexes. FEBS J, 274, 5481–5504.

von Delft, F., Lewendon, A., Dhanaraj, V., Blundell, T.L., Abell, C., and Smith, A.G. (2001) The crystal structure of E. coli pantothenate synthetase confirms it as a member of the cytidylyltransferase superfamily. Structure, 9, 439–450.

Woese, C.R., Olsen, G.J., Ibba, M., and Soll, D. (2000) Aminoacyl-tRNA synthetases, the genetic code, and the evolutionary process. Microbiol Mol Biol Rev, 64, 202–236.

Zarembinski, T.I., Hung, L.W., Mueller-Dieckmann, H.J., Kim, K.K., Yokota, H., Kim, R., and Kim, S.H. (1998) Structure-based assignment of the biochemical function of a hypothetical protein: a test case of structural genomics. Proc Natl Acad Sci USA, 95, 15189–15193.

8

THE NAD BINDING DOMAIN AND THE SHORT-CHAIN DEHYDROGENASE/REDUCTASE (SDR) SUPERFAMILY

NICHOLAS FURNHAM, GEMMA L. HOLLIDAY, AND JANET M. THORNTON

European Bioinformatics Institute, Wellcome Trust Genome Campus, Cambridge, UK

CHAPTER SUMMARY

Structurally defined protein domain superfamilies can be functionally very diverse. To be able to establish the routes by which functions, and in particular enzyme functions, emerge through evolution requires exploring the relationships between protein sequence, structures, and their chemistries. Publicly available resources, such as FunTree, catalog many of these domain superfamilies by integrating sequence and structural relationships through phylogenetic analysis and similarities of reaction chemistries and enzyme mechanisms to be able to explore the evolution of enzyme function. An example of such a functionally diverse superfamily is the nicotinamide adenine dinucleotide (NAD) binding domain, one of the largest domain superfamilies and participates in all the major classes of enzyme function. This chapter describes the NAD binding domain superfamily using information held in FunTree to analyze the range of functions undertaken by the superfamily and what might be driving the ability of the domain to participate in such a wide range of functions. In addition, we focus on a large subclassification of the superfamily, the

Protein Families: Relating Protein Sequence, Structure, and Function, First Edition.
Edited by Christine Orengo and Alex Bateman.

short-chain dehydrogenase/reductase (SDR), which exemplifies the changes in function of the entire superfamily, finally describing the range of SDRs found in the human genome.

8.1 THE NAD BINDING DOMAIN

The NAD Rossmann-like binding domain superfamily is an ancient protein superfamily that is found in all kingdoms of life. As the name suggests, it binds NAD or its phosphorylated form NADP (nicotinamide adenine dinucleotide phosphate), which, acting as a cofactor or substrate, has been found to play a crucial role in the growth and reproduction of living organisms (Felczak and Pankiewicz, 2011).

Three-dimensional atomic structures of the domain reveal a Rossmann-like fold, first defined in 1974 (Rossmann et al., 1974), consisting of a three-layered

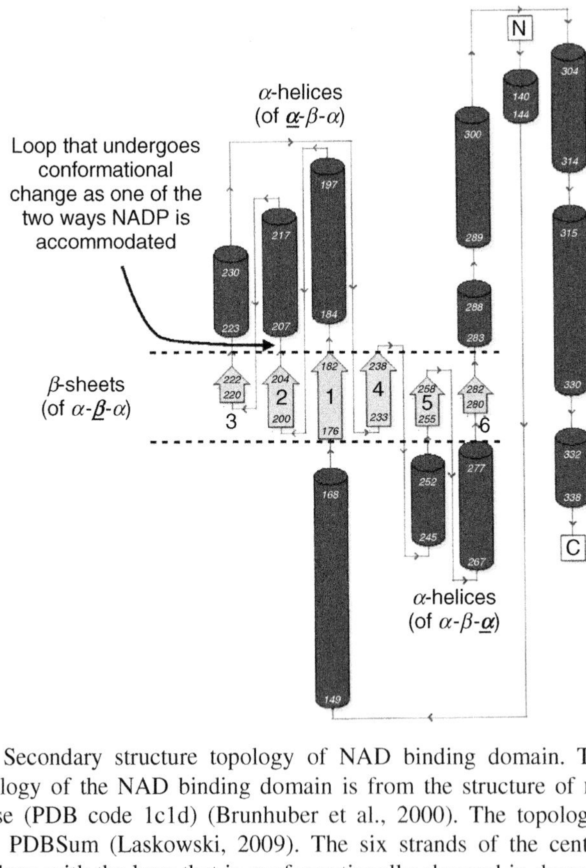

Figure 8.1 Secondary structure topology of NAD binding domain. The characteristic $\alpha\beta\alpha$ topology of the NAD binding domain is from the structure of L-phenylalanine dehydrogenase (PDB code 1c1d) (Brunhuber et al., 2000). The topology diagram was generated by PDBSum (Laskowski, 2009). The six strands of the central β-sheet are highlighted along with the loop that is conformationally changed in domain relatives that accommodate NADP.

alpha/beta/alpha domain, where the six beta strands are parallel. The strands are in the order of 6–5–4–1–2–3, with a long loop between strands 3 and 4, which creates a natural cavity that binds the adenine ring of the NAD (Lesk, 1995) (see Fig. 8.1). There is an extensive network of hydrogen bonds and van der Waals interactions that give rise to a consensus sequence associated with NAD binding containing a three-glycine pattern GXGXXG, with the first two glycine residues involved in binding the NAD, while the third is involved in the protein packing (Wierenga et al., 1985; Wierenga et al., 1986). As well as binding NAD, NADP can also be accommodated by either a conformational change in the loop that connects the second strand to the second helix or by a mutation from a conserved aspartate to an asparagine, resulting in extra hydrogen bonds that push the entire substrate away from the domain (Lesk, 1995).

Many different enzymes contain the NAD binding domain, with the InterPro (Hunter et al., 2009) resource (InterPro ID: IPR016040) reporting 12 enzyme superfamilies including several dehydrogenases such as the formate/glycerate dehydrogenases (Lamzin et al., 1994), lactate dehydrogenase (Read et al., 2001), and tyrosine-dependent oxidoreductases (also known as *short-chain dehydrogenases*) (Jornvall et al., 1995), as well as the domain being part of the transcriptional repressor Rex (Sickmier et al., 2005) and forming the CoA-binding N-terminal domain of the alpha chain of succinyl-CoA synthetase (Fraser et al., 1999).

The CATH (Orengo et al., 1997) classification of domain structures (see Chapter 3) catalogs 3589 NAD binding domains from experimentally resolved three-dimensional structures (CATH-ID: 3.40.50.720). These structures are used to generate hidden Markov models (HMMs) to search sequence databases in CATH-Gene3D (Yeats et al., 2008), which reports 224,584 sequences of the domain. This makes the NAD binding domain superfamily one of the largest superfamilies in CATH (Cuff et al., 2011).

8.2 NAD BINDING DOMAIN IN MULTIDOMAIN ARCHITECTURES

Proteins can be made up of one or more domains, with the order in which the domains are arranged defining its multidomain architecture (MDA). One of the notable features of the NAD binding domain is its association with many different MDAs. In evolution, this occurs through a process of domain duplication and combination, and has been shown to be a mechanism to increase the functional repertoire of a domain superfamily (Bashton and Chothia, 2007). The complexity in MDAs that this process generates can be realized by the analysis of MDAs of sequences from UniProtKB (Consortium, 2009) with the NAD binding domain identified by CATH/CATH-Gene3D using ArchSchema (Tamuri and Laskowski, 2010). This identifies 2554 unique domain architectures, with a maximum of 113 domains in a single MDA, though the distribution is skewed toward architectures with less than 5 domains, which account for 60% of MDAs. Most sequences have only a few domains, with approximately 50% comprising just two domains. If

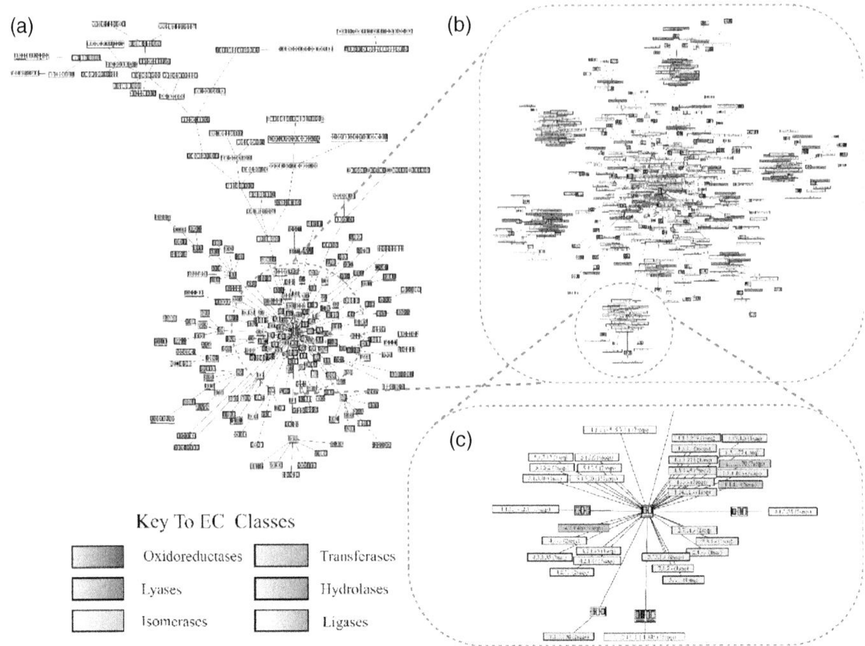

Figure 8.2 MDAs that include the NAD binding domain. (a) An ArchSchema graph showing the multidomain architectures (MDAs) associated with the NAD Binding domain superfamily. Each node in the graph represents an MDA, with each domain represented by a colored box. The NAD binding domain is colored green, with other domains given a unique color. The "edges" joining the nodes show the relationships between the architectures, based on the similarity of their domain compositions. Only MDAs from sequences in the reviewed section of the UniProtKB are shown. (b) The central section of the ArchSchema graph centered on the single-domain enzymes is shown with the Enzyme Commission (EC) codes annotated by UniProtKB. Note that, when the annotations are added, the graph is reoptimized and thus the positions of nodes will have changed. (c) A few nodes of the graph are highlighted with EC annotations. Each EC class is given a color, with the intensity of the color indicating the proportion of sequences in that MDA that share the function. Some sequences have multiple functions. These are shown as white and the ECs are listed together. A red bar under the domains indicates where experimentally determined three-dimensional atomic structures exist. (*See insert for color representation of the figure.*)

only those sequences found in the reviewed section of UniProtKB are analyzed, 237 MDAs are found with a maximum of 26 domains in a single MDA.

A directed graph representation of the different MDAs found in the reviewed section of UniProtKB is shown in Fig. 8.2a. It is possible to overlay onto the graph the enzyme functions, as defined by the Enzyme Commission (EC) code (International Union of Biochemistry and Molecular Biology. Nomenclature and Webb, 1992), as shown in Fig. 8.2b and c. Only the reviewed section of UniProtKB is used because of issues regarding the accuracy of the functional annotations in public repositories derived from automated methods (Furnham et al., 2009; Schnoes et al., 2009). It can be seen that the oxidoreductase class of enzymes dominates the superfamily, but all classes of enzymes are found. Some domain combinations, for example, the NAD binding domain with the uridine diphosphogalactose (UDP)-galactose 4-epimerase domain, have 19 unique fully described functions from four of the six EC classes and incomplete classifications from a further class. One of the MDAs is annotated as bifunctional, possessing both oxidoreductase and isomerase activity. A total of 275 unique, fully described enzyme functions are found in this superfamily, with a further 42 partially described functions ascribed. It is possible that not all of these functions are associated uniquely with the presence of the NAD binding domain in the MDA. Some functions could be carried out by another domain or group of domains within the MDA but without the presence of the NAD binding domain. These could be considered as "confusion" domains. It is possible, though a combinatorial search, to identify whether any of the domains or groups of domains (in the same linear order) possess any of the functions associated with the MDA. Analysis of the superfamily shows that most functions are uniquely associated in the context of the NAD binding domain, but 10% of functions are found without the presence of the domain (see Fig. 8.3). Any subsequent analysis of enzyme functions in this superfamily needs to account for functions possessed by these "confusion" domains to remove any ambiguity about their contribution to an enzyme's catalytic function.

8.3 TRANSITIONS IN FUNCTION

In evolution, there are many different routes being taken to obtain different reactions, mechanisms, and specificities within a superfamily (Bartlett et al., 2003). Such routes involve gene duplication followed by subfunctionalization. The basis

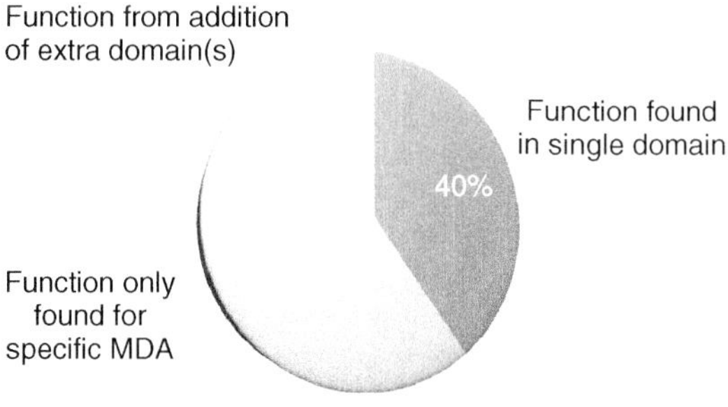

Figure 8.3 Contribution of functions from domains in multidomain architectures and single-domain enzymes. The pie chart shows the contribution to the functional repertoire of the NAD binding domain superfamily from functions associated with the single domain (dark gray), uniquely when it is in association with one or more domain in a multidomain architecture (mid gray) and when the function is found with and without the presence of the domain.

of such subfunctionalization can be twofold: first, by alteration of the enzyme structure, by mutations, local insertions, and/or deletions within a domain, or second, by changes in MDA.

Within a superfamily, it is possible to analyze the exchanges that occur in function by looking at the range of EC classes to which the superfamily members belong. For example, if some superfamily members belong to EC 1 (oxidoreductases) while others belong to EC 2 (transferases), it is reasonable to assume that a transition between one class and the other has occurred since the protein diverged from some common ancestor. As well as looking for exchanges between classes (overall chemical function), it is also possible to identify exchanges within a class (type of chemistry or substrate specificity). A matrix of all the possible exchanges observed within the NAD binding domain superfamily is shown in Fig. 8.4.

Exchanges occur between all classes, demonstrating the huge potential range of functional diversity in this superfamily. Most changes occur within the oxidoreductase class, with just a few exchanges occurring within the other classes. If the in-class exchanges are grouped by each level of the EC classification, it can be seen that most exchanges occur at the fourth level. As this level of the classification is a serial number describing the substrate specificity of the enzyme reaction, with the first three levels describing the chemistry, it indicates that changes in chemistry are much less likely than changes to the specificity.

Of course, not all changes in function would have occurred during the course of evolution, as a change from one function to another may not have occurred directly but via an intermediary function. The evolutionary route taken within a superfamily to exchange the enzyme function from one reaction to another can

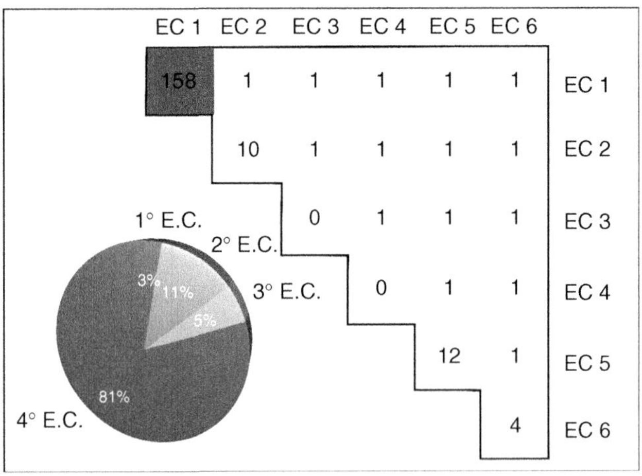

Figure 8.4 Changes in function for the NAD binding domain superfamily. A matrix of all possible changes between EC class (off the diagonal) and within class (along the diagonal). The inset pie chart shows the proportion of changes that occur for each level of the EC classification.

be determined by combining functional information with phylogenetic analysis. This is difficult because the relationships between the diverse members of this ancient superfamily are hard to determine. A solution to these difficulties is to subdivide the superfamily into structurally similar groups (Cuff et al., 2011). These are groups of structures of domain relatives that can be well aligned and superimposed in three dimensions. Sequence relatives can be added to each structural grouping and then all sequences can be multiply aligned and used to derive the phylogenetic tree for that structurally similar group. The tree can then be systematically traversed from leaf to root to identify changes in function at tree branch points.

In addition to limiting the functional changes to those observed occurring directly between one and another, multiple occurrences of functional change can be captured. However, because the superfamily is split between several trees, some ancient changes are missed. This type of analysis has been performed across 276 superfamilies (see Furnham et al. (2012) for methodology) and has been applied here on the NAD binding domain superfamily (see Fig. 8.5). This shows that some in-class changes, especially in the oxidoreductases, have occurred more than once in evolution, while other changes such as those occurring within the transferase class have occurred so far back in evolution that they cannot be captured by the grouping of structurally similar domains.

EC 1	EC 2	EC 3	EC 4	EC 5	EC 6	
191	1	1	2	4	2	EC 1
	0	0	1	0	1	EC 2
		0	0	0	1	EC 3
			1	0	2	EC 4
				2	1	EC 5
					9	EC 6

Figure 8.5 Changes in function for the NAD binding domain superfamily using the phylogenetic trees of structurally similar groups. A matrix of all changes between EC class (off the diagonal) and within class (along the diagonal) as defined by the changes in function within structurally similar groups of the NAD domain. By traversing the tree from branch to root, it is possible to identify the route by which change has occurred in evolution.

As we have seen, the NAD binding domain is observed as part of a biological unit that performs functions in all six of the EC classes, although the domain is most often active as an oxidoreductase. This is not surprising, given the fact that a large proportion of oxidoreductases actually require NAD or NADP for their function. The primary role of NAD(P) in enzyme reactions is that of a hydride (H^-) acceptor/donor, which is a good source of reducing equivalents (hydride ions) and hence its overoccurrence in the oxidoreductase class. Of course, although enzymes can be classified by their overall reaction, this is achieved by a combination of different steps that carry out the transformation, known as the *mechanism*. MACiE (Holliday et al., 2007) is a database of such mechanisms, and contains 16 entries in the current version (3.0) of MACiE, which include the NAD(P) binding domain. In 12 of these entries, the NAD(P) binding domain is annotated as being catalytic and they cover EC classes 1, 3, 4, and 5.

Focusing on these 12 enzymes, the catalytic residues are entirely charged or neutral (hydrophilic) in nature, with no instances currently annotated of hydrophobic residues being catalytically important. However, despite the common domain, there is very little similarity between the catalytic machinery (those residues that are catalytic) for the entries in MACiE (see Table 8.1). Here, we can see that the complement of catalytic residues (the presence of absence of the individual residue types) range from totally dissimilar to identical, with no single residue in common between all entries. From Table 9.1, it can be seen that, although the complement of catalytic residues might be very similar (see the cells highlighted in gray), when the catalytic residues are superimposed (the bottom half of the table), the catalytic machinery is markedly different. The most obvious case of this is M0237 (EC 1.5.1.3) and M0021 (EC 1.1.1.38). In these two cases, there are three residues in common (Arg, Asp, and Tyr), and, whereas the Asp and Tyr in M0237 are involved in a dyad, in M0021 the Asp is in a dyad with a Lys, not the Tyr.

In general, the catalytic machinery of enzymes utilising this domain lack significant similarity between each other (with the majority being less than 50% similar to each other). However, there is one interesting case in which the catalytic residues are identical (both in terms of their composition and superimposition) to each other; further, these enzymes differ at the class level of the EC nomenclature (M0255 and M0188, EC 1.1.1.1 and EC 5.1.3.2). In Fig. 8.6, the mechanisms of these two enzymes are shown side by side, and it can be seen that they share a very similar first step (the transfer of a hydride from the substrate to the NAD) but whereas the products of the first step in M0255 dissociate immediately from the active site, the products of this first step in M0188 remain, with the UDP substrate rotating so that the hydride is added onto the substrate in a different orientation, resulting in a stereoisomerization, thus changing the enzyme's class from an oxidoreductase to an isomerase.

However, despite these seeming differences, some residues that are "more" common perform similar functions:

- Tyrosine (66%)—primarily involved in proton shuttling

TABLE 8.1 The complement of amino acid residues (top half the table), calculated by taking the Tanimoto similarity between the counts of different residue types present as catalytic residues in MACiE, and the 3D superimposition of the catalytic residues (bottom half), calculated using IsoCleft

	Amino Acid Residue Complement															
MACiE Id	M0256	M0255	M0092	M0021	M0227	M0108	M0110	M0237	M0114	M0116	M0142	M0090	M0226	M0228	M0188	M0317
M0256		0.25	0.00	0.00	0.43	0.17	0.25	0.00	0.25	0.00	0.20	0.27	0.40	0.00	0.25	0.25
M0255	0.00		0.13	0.33	0.57	0.14	0.2	0.20	0.20	0.13	0.00	0.15	0.60	0.33	1.00	0.20
M0092	0.00	0.00		0.38	0.17	0.10	0.13	0.13	0.13	0.20	0.25	0.38	0.22	0.57	0.13	0.29
M0021	0.08	0.00	0.00		0.18	0.25	0.14	0.60	0.14	0.57	0.13	0.31	0.25	0.43	0.33	0.33
M0227	0.00	0.00	0.00	0.11		0.18	0.22	0.10	0.38	0.08	0.09	0.43	0.63	0.18	0.57	0.22
M0108	0.13	0.10	0.15	0.11	0.00		0.14	0.33	0.33	0.22	0.13	0.13	0.25	0.25	0.14	0.14
M0110	0.11	0.00	0.22	0.12	0.00	0.141		0.00	0.00	0.00	0.00	0.15	0.14	0.00	0.20	0.20
M0237	0.13	0.43	0.18	0.08	0.00	0.00	0.00		0.20	0.80	0.17	0.15	0.14	0.33	0.20	0.00
M0114	0.13	0.26	0.15	0.08	0.16	0.22	0.00	0.00		0.13	0.17	0.25	0.33	0.14	0.20	0.20
M0116	0.00	0.3	0.00	0.17	0.21	0.00	0.00	0.00	0.00		0.25	0.29	0.10	0.38	0.13	0.00
M0142	0.00	0.21	0.00	0.11	0.00	0.00	0.00	0.22	0.00	0.00		0.33	0.29	0.29	0.00	0.17
M0090	0.00	0.13	0.11	0.10	0.14	0.00	0.00	0.00	0.11	0.08	0.16		0.31	0.21	0.15	0.36
M0226	0.00	0.49	0.00	0.11	0.25	0.21	0.16	0.30	0.00	0.00	0.16	0.00		0.43	0.60	0.33
M0228	0.00	0.59	0.16	0.00	0.53	0.18	0.00	0.32	0.00	0.22	0.00	0.00	0.38		0.33	0.14
M0188	0.00	1.00	0.00	0.00	0.00	0.00	0.34	0.43	0.00	0.00	0.18	0.14	0.49	0.55		0.20
M0317	0.10	0.16	0.00	0.22	0.30	0.12	0.22	0.00	0.00	0.00	0.00	0.16	0.00	0.15	0.18	
	Superimposition of the catalytic residues															

The two entries that have identical catalytic machineries are highlighted in light gray; the ones that have similar complements but not similar superimpositions are highlighted in gray

M0255 (EC 1.1.1.1) Alcohol Dehydrogenase

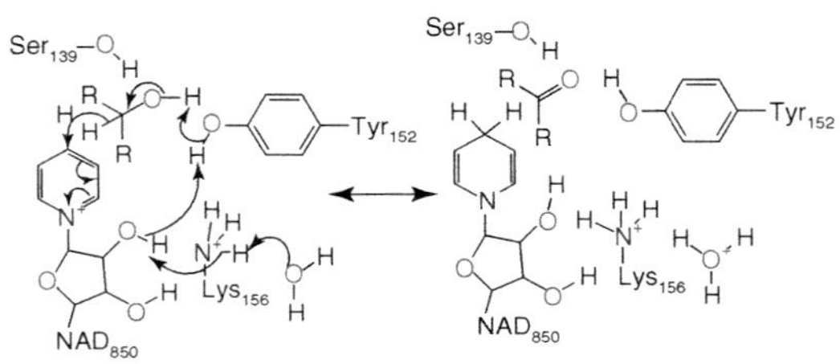

Step 1

Bulk water initiates the deprotonation of Tyr152 via Lys156, the ribose ring of NAD850. Tyr152 deprotonates the alcohol group, initiating hydride transfer to NAD850.

M0188 (EC 5.1.3.2) UDP-glucode 4-epimerase

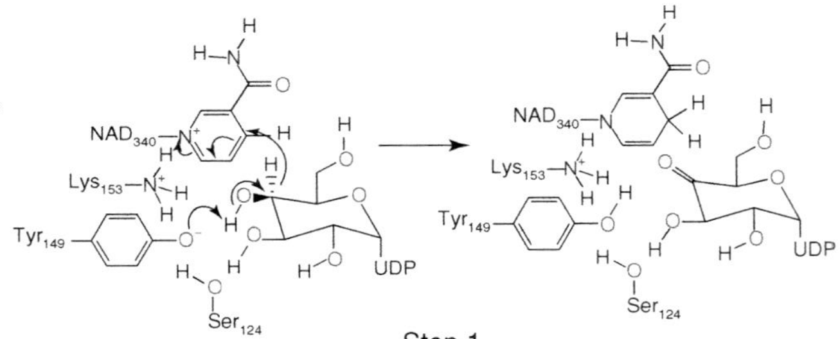

Step 1

Tyr149 deprotonates the 4-hydroxide of the substrate, resulting in an elimination of the hydride from the C4 position. The hydride then attacks the NAD cofactor in a nucleophilic addition reaction.

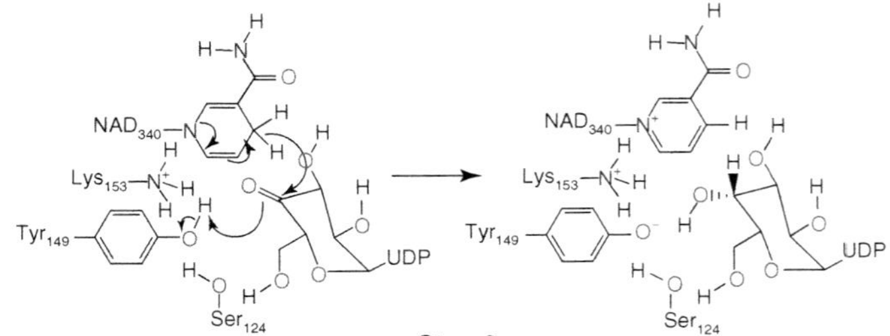

Step 2

The intermediate undergoes a conformational change before the NAD initiates an elimination of the hydride back to the intermediate, which in turn deprotonates Tyr149.

Figure 8.6 The reaction steps for MACiE entries M0255 (alcohol dehydrogenase, EC 1.1.1.1) and M0188 (UDP-glucose 4-epimerase, EC 5.3.1.2), which, despite having identical catalytic machinery, perform distinct functions via different mechanisms.

- Lysine (66%)—primarily involved in stabilization and activation
- Aspartic acid (58%)—primarily involved in proton shuttling.

Despite these differences, the mechanisms by which the reaction is carried out share several common motifs. There are certain bond changes that are ubiquitous, in particular the C–C and C–N bond order changes, which allow the active portion of NAD(P) to accept/donate a hydride and the C–H bond formation/cleavage. However, once the NAD(P) has played its role in the reaction, there is a great deal of potential for variability in the rest of the mechanisms, which leads to a lack of similarity in the overall reaction chemistry. This variability is often the result of the second catalytic domain within the enzyme.

As might be expected, given the variation in EC class found for this CATH domain, there is little similarity in the overall reactions in which the domain is involved. What similarity there is comes from the reduction of a double bond with the addition of a hydride ion onto the primary substrate. However, this is only the case in the oxidoreductases (EC 1), with the other classes utilizing the NAD(P) as a catalyst such that the NAD(P) is used and regenerated within a single turnover of the enzyme.

8.4 CHARACTERIZATION AND OVERVIEW OF THE SDR SUPERFAMILY

One of the major subclassifications within the NAD binding domain superfamily is the short-chain dehydrogenase/reductases (Kavanagh et al., 2008b). They are, like their parent superfamily, found in all kingdoms of life and consist of mostly just the Rossmann fold of the cofactor-binding domain with short extensions and decorations (Jornvall et al., 2010) and have enough relatives to have been considered a superfamily in their own right. This superfamily is distinguished by a highly conserved motif of specific residues in the active site consisting of a Ser, Tyr, and Lys triad (Oppermann et al., 1997). As of 2010, there are more than 47,000 sequences identified as SDRs, which can be subclassified by HMMs into 314 separate SDR subfamilies covering approximately 70% of SDR sequences, with the rest belonging to families with too few members to establish a model (Kallberg et al., 2010). These families generally can be grouped into two major types denoted "Classical" and "Extended," distinguishable by the size and sequence conservation patterns in the NAD binding site, with four minor types denoted "Intermediate," "Divergent," "Complex," and "Atypical" (Persson et al., 2009). This has been encapsulated into a nomenclature scheme comprising the type, family, and gene into the SDR naming convention.

The reactions undertaken by the SDRs often proceed through an ordered mechanism, with the NAD binding first and leaving last and binding in an extended conformation that allows for the transfer of the "4-pro-S" hydride (Kavanagh et al., 2008a). The most common type of SDR, namely the "classical" SDR, has a one domain architecture and mostly undertakes NAD-dependent oxidoreduction

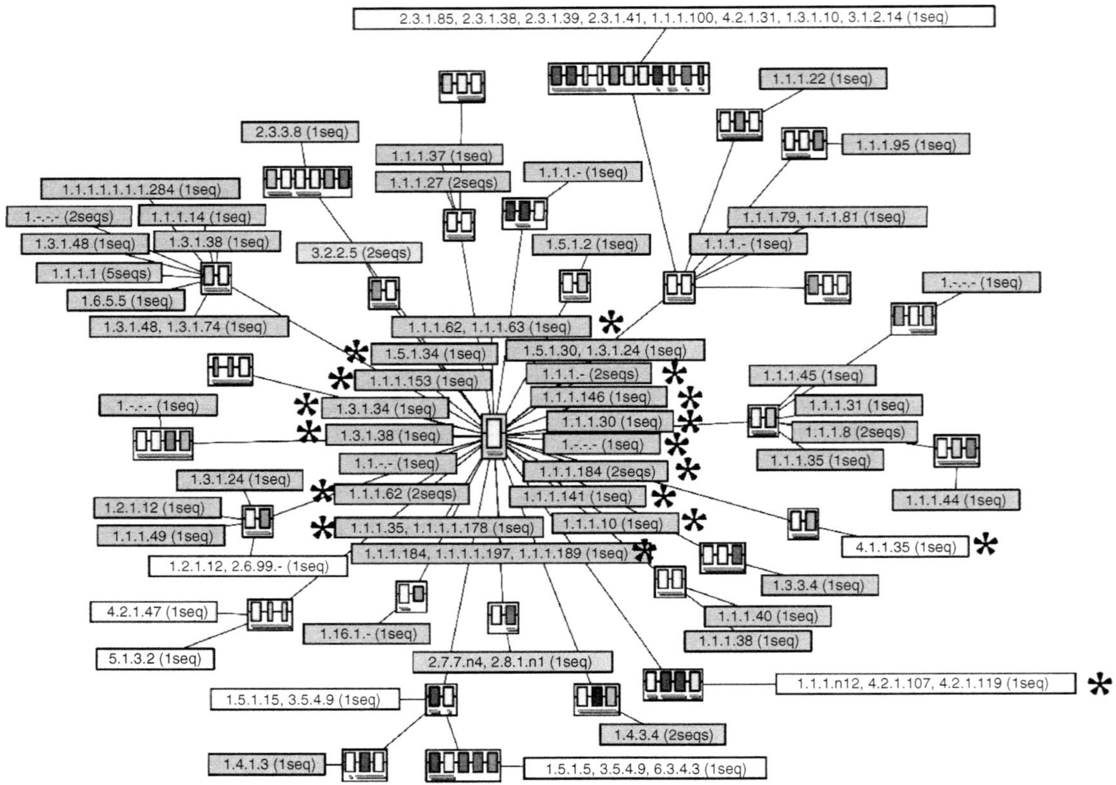

Figure 8.7 The NAD Binding Domain In *Homo sapiens*. The various MDAs and their associated functions for human sequences in the reviewed section of UniProtKB which contain the NAD binding domain and have an experimentally determined atomic structure. Each domain is assigned a unique tone on a grayscale. Those that have been identified as belonging to the short-chain dehydrogenase/reeducate superfamily (SDRs) are highlighted by an asterisk. For each MDA, node functional annotation is provided as EC numbers, with EC numbers corresponding to functions that are not oxidoreductases highlighted with a thick border and those thst are multifunctional with a white background. (*See insert for color representation of the figure*)

of hydroxy and keto groups found within a highly diverse set of metabolites including steroids, alcohols, polycols, growth factors, xenobiotics, and secondary metabolites (Kavanagh et al., 2008a). The other major class, namely the "extended" type, has an extension of the core NAD binding domain, which can be large enough to be classified as a second domain and is involved in substrate binding (Labesse et al., 1994). This class of SDR brings a greater level of functional diversity with the addition of two further enzymatic classes: lyases (Bar-Peled et al., 2001) and isomerases (Mason et al., 1998) or epimerases (Thoden et al., 2001).

8.5 THE SDRS IN HUMANS

The NAD binding domain has been identified in a number of proteins in the human genome, often associated with one or more other domains. They participate in a wide range of enzyme functions and many have had their three-dimensional atomic structures resolved (see Fig. 8.8). In total, 73 SDR superfamily members have been identified in the human genome, of which 54 are of the classical type and 11 are of the extended type (Bray et al., 2009). In total, 23 have been structurally determined and are highlighted in Fig. 8.7. Of these, 19 have a confident functional annotation in the reviewed section of UniProtKB, and their structures, superimposed on each other, are shown in

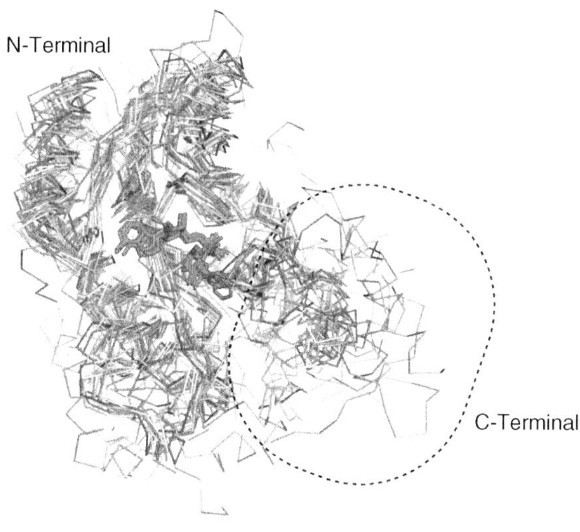

Figure 8.8 The superimposed structures of human SDRs. The 19 experimentally determined atomic structures of human SDRs, as highlighted in Fig. 8.7, are shown as a ribbon and superimposed against each other. The variability of the C-terminal area which is involved in substrate recognition and binding is highlighted by a dashed circle. The NAD(H), found bound in many of the structures, is shown as a stick presentation and highlighted in black.

Fig. 8.8. Most have a small extension to the NAD domain at the C-terminal end, which seems to be implicated in substrate recognition and binding. For two enzymes, the structures show that the C-terminal extension is large enough to constitute a domain. This addition of a domain is associated with a change in enzymatic class from oxidoreductase, which the rest of the enzymes undertake, to lyase. In fact, one of the structures is composed of the NAD domain and three other domains and is a multifunctional enzyme undertaking three reactions, one of which is an oxidoreductase (EC 1.1.1.n12; note the n12 is a UniProtKB substrate identifier and not part of the formal EC) and two are lyase reactions (EC 4.2.1.107 and 4.2.1.119).

8.6 CONCLUSIONS

The NAD binding domain superfamily is one the of the largest domain superfamilies and participates in a wide range of enzyme functions. Most of these are oxidoreductases, but all six functional classes are represented. It seems that this extraordinary ability to participate in such a wide range of reactions comes from the domain's ability to interact with a wide range of substrates through the plasticity of its C-terminal sequence and its ability to combine with other domains. In this context, it is possible to consider the NAD binding domain to be providing the reductive abilities of the NAD in a variety of different contexts. Though this superfamily has been studied for many years, there is a continual expansion, adding new members with new functions, many of which participate in important and fundamental processes in all kingdoms of life.

REFERENCES

Bar-Peled, M., Griffith, C.L., and Doering, T.L. (2001) Functional cloning and characterization of a UDP- glucuronic acid decarboxylase: the pathogenic fungus Cryptococcus neoformans elucidates UDP-xylose synthesis. Proc Natl Acad Sci USA, 98, 12003–12008.

Bartlett, G.J., Borkakoti, N., and Thornton, J.M. (2003) Catalysing new reactions during evolution: economy of residues and mechanism. J Mol Biol, 331, 829–860.

Bashton, M. and Chothia, C. (2007) The generation of new protein functions by the combination of domains. Structure, 15, 85–99.

Bray, J.E., Marsden, B.D., and Oppermann, U. (2009) The human short-chain dehydrogenase/reductase (SDR) superfamily: a bioinformatics summary. Chem Biol Interact, 178, 99–109.

Brunhuber, N.M., Thoden, J.B., Blanchard, J.S., and Vanhooke, J.L. (2000) Rhodococcus L-phenylalanine dehydrogenase: kinetics, mechanism, and structural basis for catalytic specificity. Biochemistry, 39, 9174–9187.

Consortium, U (2009) The Universal Protein Resource (UniProt) 2009. Nucleic Acids Res, 37, D169–D174.

Cuff, A.L., Sillitoe, I., Lewis, T., Clegg, A.B., Rentzsch, R., Furnham, N., Pellegrini-Calace, M., Jones, D., Thornton, J., and Orengo, C.A. (2011) Extending CATH: increasing coverage of the protein structure universe and linking structure with function. Nucleic Acids Res, 39, D420–D426.

Felczak, K. and Pankiewicz, K.W. (2011) Rehab of NAD(P)-dependent enzymes with NAD(P)-based inhibitors. Curr Med Chem, 18, 1891–1908.

Fraser, M.E., James, M.N., Bridger, W.A., and Wolodko, W.T. (1999) A detailed structural description of Escherichia coli succinyl-CoA synthetase. J Mol Biol, 285, 1633–1653.

Furnham, N., Garavelli, J.S., Apweiler, R., and Thornton, J.M. (2009) Missing in action: enzyme functional annotations in biological databases. Nat Chem Biol, 5, 521–525.

Furnham, N., Sillitoe, I., Holliday, G.L., Cuff, A.L., Rahman, S.A., Laskowski, R.A., Orengo, C.A., and Thornton, J.M. (2012) FunTree: a resource for exploring the functional evolution of structurally defined enzyme superfamilies. Nucleic Acids Res, 40, D776–D782.

Holliday, G.L., Almonacid, D.E., Bartlett, G.J., O'Boyle, N.M., Torrance, J.W., Murray-Rust, P., Mitchell, J.B., and Thornton, J.M. (2007) MACiE (mechanism, annotation and classification in enzymes): novel tools for searching catalytic mechanisms. Nucleic Acids Res, 35, D515–D520.

Hunter, S., Apweiler, R., Attwood, T.K., Bairoch, A., Bateman, A., Binns, D., Bork, P., Das, U., Daugherty, L., Duquenne, L. et al. (2009) InterPro: the integrative protein signature database. Nucleic Acids Res, 37, D211–D215.

International Union of Biochemistry and Molecular Biology. Nomenclature, C., and Webb, E.C. Enzyme nomenclature 1992: recommendations of the Nomenclature Committee of the International Union of Biochemistry and Molecular Biology on the nomenclature and classification of enzymes / prepared for NC-IUBMB by Edwin C. Webb 1992; (San Diego Published for the International Union of Biochemistry and Molecular Biology by Academic Press).

Jornvall, H., Hedlund, J., Bergman, T., Oppermann, U., and Persson, B. (2010) Superfamilies SDR and MDR: from early ancestry to present forms. Emergence of three lines, a Zn-metalloenzyme, and distinct variabilities. Biochem Biophys Res Commun, 396, 125–130.

Jornvall, H., Persson, B., Krook, M., Atrian, S., Gonzalez-Duarte, R., Jeffery, J., and Ghosh, D. (1995) Short-chain dehydrogenases/reductases (SDR). Biochemistry, 34, 6003–6013.

Kallberg, Y., Oppermann, U., and Persson, B. (2010) Classification of the short-chain dehydrogenase/reductase superfamily using hidden Markov models. Febs J, 277, 2375–2386.

Kavanagh, K. L., Rnvall, H, Persson, B, Oppermann U (2008a). The SDR superfamily: functional and structural diversity within a family of metabolic and regulatory enzymes, Cell Mol Life Sci, Vol 65 (Heidelberg, Allemagne: Springer).

Kavanagh, K.L., Jornvall, H., Persson, B., and Oppermann, U. (2008b) Medium- and short-chain dehydrogenase/reductase gene and protein families : the SDR superfamily: functional and structural diversity within a family of metabolic and regulatory enzymes. Cell Mol Life Sci, 65, 3895–3906.

Labesse, G., Vidal-Cros, A., Chomilier, J., Gaudry, M., and Mornon, J.P. (1994) Structural comparisons lead to the definition of a new superfamily of NAD(P)(H)-accepting oxidoreductases: the single-domain reductases/epimerases/dehydrogenases (the 'RED' family). Biochem J, 304(Pt 1), 95–99.

Lamzin, V.S., Dauter, Z., Popov, V.O., Harutyunyan, E.H., and Wilson, K.S. (1994) High resolution structures of holo and apo formate dehydrogenase. J Mol Biol, 236, 759–785.

Laskowski, R.A. (2009) PDBsum new things. Nucleic Acids Res, 37, D355–D359.

Lesk, A.M. (1995) NAD-binding domains of dehydrogenases. Curr Opin Struct Biol, 5, 775–783.

Mason, J.I., Naville, D., Evans, B.W., and Thomas, J.L. (1998) Functional activity of 3beta-hydroxysteroid dehydrogenase/isomerase. Endocr Res, 24, 549–557.

Oppermann, U.C., Persson, B., Filling, C., and Jornvall, H. (1997) Structure-function relationships of SDR hydroxysteroid dehydrogenases. Adv Exp Med Biol, 414, 403–415.

Orengo, C.A., Michie, A.D., Jones, S., Jones, D.T., Swindells, M.B., and Thornton, J.M. (1997) CATH–a hierarchic classification of protein domain structures. Structure, 5, 1093–1108.

Persson, B., Kallberg, Y., Bray, J.E., Bruford, E., Dellaporta, S.L., Favia, A.D., Duarte, R.G., Jornvall, H., Kavanagh, K.L., Kedishvili, N. et al. (2009) The SDR (short-chain dehydrogenase/reductase and related enzymes) nomenclature initiative. Chem Biol Interact, 178, 94–98.

Read, J.A., Winter, V.J., Eszes, C.M., Sessions, R.B., and Brady, R.L. (2001) Structural basis for altered activity of M- and H-isozyme forms of human lactate dehydrogenase. Proteins, 43, 175–185.

Rossmann, M.G., Moras, D., and Olsen, K.W. (1974) Chemical and biological evolution of nucleotide-binding protein. Nature, 250, 194–199.

Schnoes, A.M., Brown, S.D., Dodevski, I., and Babbitt, P.C. (2009) Annotation error in public databases: misannotation of molecular function in enzyme superfamilies. PLoS Comput Biol, 5, e1000605.

Sickmier, E.A., Brekasis, D., Paranawithana, S., Bonanno, J.B., Paget, M.S., Burley, S.K., and Kielkopf, C.L. (2005) X-ray structure of a Rex-family repressor/NADH complex insights into the mechanism of redox sensing. Structure, 13, 43–54.

Tamuri, A.U., and Laskowski, R.A. (2010). ArchSchema: a tool for interactive graphing of related Pfam domain architectures. Bioinformatics Oxford, England 26, 1260–1261.

Thoden, J.B., Wohlers, T.M., Fridovich-Keil, J.L., and Holden, H.M. (2001) Molecular basis for severe epimerase deficiency galactosemia. X-ray structure of the human V94m-substituted UDP-galactose 4-epimerase. J Biol Chem, 276, 20617–20623.

Wierenga, R.K., De Maeyer, M.C.H., and Hol, W.G.J. (1985) Interaction of pyrophosphate moieties with .alpha.-helixes in dinucleotide-binding proteins. Biochemistry, 24, 1346–1357.

Wierenga, R.K., Terpstra, P., and Hol, W.G. (1986) Prediction of the occurrence of the ADP-binding beta alpha beta-fold in proteins, using an amino acid sequence fingerprint. J Mol Biol, 187, 101–107.

Yeats, C., Lees, J., Reid, A., Kellam, P., Martin, N., Liu, X., and Orengo, C. (2008) Gene3D: comprehensive structural and functional annotation of genomes. Nucleic Acids Res, 36, D414–D418.

9

THE GLOBIN FAMILY

ARTHUR M. LESK

Department of Biochemistry and Molecular Biology, Huck Institute for Genomics, Proteomics and Bioinformatics, The Pennsylvania State University, University Park, PA, USA

JULIETTE T.J. LECOMTE

Department of Biophysics, Johns Hopkins University, Baltimore, MD, USA

CHAPTER SUMMARY

Globins are probably most widely known for their functions in oxygen transport and storage. The early determination of the crystal structures of globins, and their amino acid sequences, made possible pioneering investigations of protein evolution—at the level of sequence and of structure, the mechanism of allosteric changes, and the implication of mutations in disease. Subsequently, our knowledge of the range of the globin family has been greatly extended. Many homologs from a wide range of species are now known, with a wide range of functions. This chapter surveys what has been learnt about this family, and what topics continue to be active in current research. It describes the basic globin structure and its variations; the taxonomic distribution of different types of globins; and the variety of known functions, focusing on the mechanism of the allosteric change in mammalian tetrameric hemoglobins and on the effects of mutations with clinical consequences.

Protein Families: Relating Protein Sequence, Structure, and Function, First Edition.
Edited by Christine Orengo and Alex Bateman.

9.1 INTRODUCTION

The globins are an ancient family of proteins, appearing in archaea, bacteria, and eukarya. Globins have many functions, of which the best known are O_2 transport and receipt in vertebrates by hemoglobin and myoglobin, and the reverse transport of CO_2. The secondary structure of globins is α-helical. The tertiary structure includes a common geometric pattern of assembly of helices, known as the "globin fold." Observed quaternary structures include monomers, dimers, tetramers, and higher aggregates.

Globins have played prominent roles in the history of molecular biology, arguably well in excess of their biological significance. Hemoglobin was the first protein crystallized. Early studies of the crystallography of hemoglobins, before the discovery of X-ray diffraction, provided evidence that proteins adopt native structures and that they evolve. Sperm whale (*Physeter macrocephalus, a.k.a. Physeter catodon*) myoglobin was the first protein solved by X-ray crystallography, in 1959, followed shortly by horse hemoglobin (see retrospective review by Brunori, 2010). The similarity in folding pattern between myoglobin and the hemoglobin monomers provided direct evidence for similarity and divergence of proteins at the level of detailed atomic structure.

The hemoglobin structure also revealed that the four heme groups were so far apart in the tetramer that direct interaction could not explain the cooperativity of O_2 binding. This puzzle stimulated the development of the Monod–Wyman–Changeux model of allosteric change.

The first human disease assignable to a specific mutation in a protein was sickle-cell anemia, which is the effect of a Gln→Val mutation in the β chain of hemoglobin.

Early amino acid sequence determinations were rich in globins. Globins were often the first, and sometimes the only, protein sequence known from a species. It is noteworthy that the crystal structures of myoglobin and hemoglobin were solved before the amino acid sequences were determined.

Globins have continued to contribute to many studies illuminating protein structure, folding, function, and evolution. They are responsible for several genetic diseases in addition to sickle-cell anemia. Neonatal and even prenatal screening for hemoglobin mutants is common.

9.2 EARLY HISTORY OF GLOBIN RESEARCH

Globins have a distinguished history in biochemical science, going back as far as Lavoisier (Holmes, 1995). Nineteenth century globin studies included the identification of hemoglobin as an example of the class of chemicals that Berzelius named proteins, and the recognition of its role in respiration:

> Experiments have proven that in vertebrates the most important interaction between the air and the blood concerns the hämatin [= haem]. Blood serum ... transforms ... a

> portion of oxygen gas to carbonic acid gas ... when the serum is mixed with hämatin, whose brown colour is thereby transformed into a very vivid red. It is believed that the oxidation state of the iron is thereby also changed ... It is difficult to decide ... whether these colour changes are accompanied by a chemical change, or consist only in a change in the physical constitution of the hämatin.
>
> Berzelius (1840), pp. 123–124.

This quotation from Berzelius reflects the older idea that blood in general, and perhaps hemoglobin in particular, might be the *site* of the transformation of O_2 to CO_2. By 1866, Hoppe-Seyler was able correctly to assign the role of hemoglobin as a *carrier* of oxygen in the blood.

In the same year as Berzelius's publication, Hünefeld observed the first protein crystals of earthworm hemoglobin, followed shortly by a more systematic preparation of hemoglobin crystals by O. Funke from different species.

Hemoglobin crystals were to make several important contributions to our understanding of the biology of proteins.

Just over a century ago, Reichert and Brown (1909) published a study of crystals of hemoglobin isolated from different species of fishes. Hemoglobin crystallography, three years before the discovery of X-ray diffraction, was limited to measuring the angles between the crystal faces. Steno's law (1669) states that the interfacial angles of all crystals of a substance are the same, independent of the size and macroscopic shape of the crystal. Therefore, these angles characterize the substance. Reichert and Brown found that the patterns of divergence of these angles correlated with the evolutionary tree of the species of fishes. They even found differences between crystals of deoxy- and oxyhemoglobin.

We can now interpret and appreciate these observations. The formation of crystals implies that the molecules can take up a definite structure that is able to pack into regular arrays. The differences in interfacial angles imply that crystals of hemoglobins from different fishes have different structures. The correlation of the divergence patterns of the crystals and the species implies that evolution is shaping molecules as well as body forms, in parallel processes. The differences between crystals of deoxy- and oxyhemoglobin imply that the protein undergoes a conformational change upon binding oxygen.

Fifty years later, Perutz announced the solution of the X-ray crystal structure of hemoglobin.

How did Perutz come to take up the problem of hemoglobin? Felix Haurowitz, a professor in Prague before World War II, was married to Perutz's cousin Gina. During a family visit while Perutz was en route from Vienna to England to begin graduate studies at Cambridge, Haurowitz called Perutz's attention to the change in crystal form of hemoglobin on the transition between the oxy and deoxy forms. Perutz took this on as his Ph.D. thesis topic, beginning a lifetime dedication to this molecule.

Cambridge awarded Perutz the Ph.D. in 1939. For his thesis Perutz prepared and mounted hemoglobin crystals, determined their unit cell dimensions and degrees of hydration, and studied the effects of denaturation. The thesis also included a Patterson analysis of a hemoglobin derivative studied by Denis Riley. Fortunately, in those days it was not necessary to solve the structure to obtain the degree.

In the March 19, 1938, issue of *Nature*, Perutz, J.D. Bernal, and I. Fankuchen published a paper about X-ray diffraction from hemoglobin crystals. This was exactly 1 week after the Anschluss of Austria and Germany, as a result of which Perutz—Austrian in origin—became an enemy alien in England, and was subsequently interned. The development of the method of isomorphous replacement to address the phase problem, and the structures, first of myoglobin and then hemoglobin, had to await the end of the war.

9.3 GLOBIN STRUCTURES

The structures of globins from many organisms have been determined. The early globin structures were all what are now called full-length globins, that is, with monomers having approximately 150 residues. In 1989, a family of smaller globins (somewhat unfortunately called truncated) was discovered. The first identified member, from *Paramecium caudatum*, is 116 residues long (Iwaasa et al., 1989). The globin domain of truncated hemoglobins is shorter than that of the full-length proteins. Both full-length and truncated hemoglobin domains, however, vary considerably in the number of amino acids.

9.3.1 Structures of the Monomers

9.3.1.1 Full-Length Globins The "globin fold" consists of eight helices assembled into a common tertiary structure, and enclosing a heme group. The helices are conventionally designated A–H and form a "3/3 sandwich"; the A-, E-, and F-helices stack over the B-, G-, and H-helices. Figure 9.1 shows the structure of sperm whale myoglobin. Figure 9.2 contains a superposition of chain traces connecting successive Cα atoms, showing the structural variability among three full-length globins: sperm whale (black), insect (*Chironomus*) (light gray) and plant (yellow lupin) (dark gray).

All helices in globins are α-helices except for C, which is a 3_{10} helix. In some globins, the D-helix has unwound or is largely deleted, for instance, in the α chain of mammalian hemoglobins. In mammalian globins, the F-helix is broken into F$'$ and F (see Fig. 9.1).

Globin monomers are stabilized by packings of residues at interfaces between helices, forming the hydrophobic core of the structure. The helix interfaces are well packed. Through the hydrophobic effect and van der Waals interactions, such packings stabilize tertiary structures, and partially control the relative geometries of packed helices.

The ridges-into-grooves model accounts for the nature of side-chain packing at helix interfaces in globular proteins (Chothia et al., 1977, 1981):

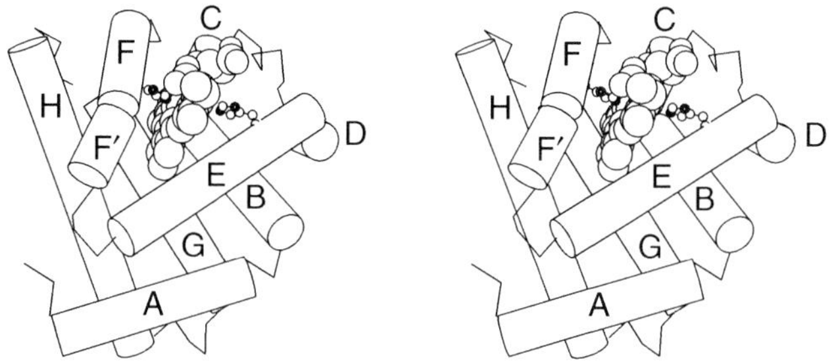

Figure 9.1 The structure of sperm whale myoglobin. Letters indicate helices, in order along the chain. As in other mammalian globins, the F-helix is broken, into F′ and F segments.

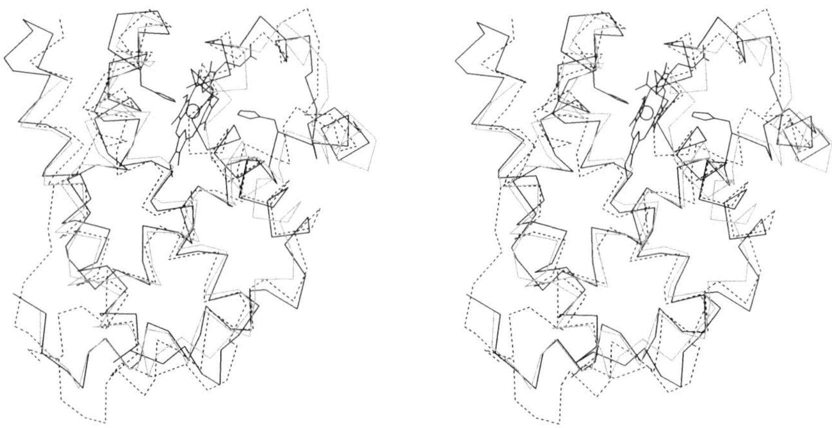

Figure 9.2 Superposition of three full-length globins. From sperm whale (black), insect (*Chironomus*) (gray), and plant (broken lines).

- Side chains on the surfaces of helices form ridges protruding from the helix surface.
- These ridges may form from residues separated by four positions in the sequence (most common), three positions, or one position.
- Ridges on the surface of one helix pack into grooves of the other.
- Different combinations of ridge–groove structures produce different interaxial angles in packed helices. The most common packing—$i \pm 4$ ridges from both helices—corresponds to an interaxial angle of about -40°.

In full-length globins, five helix packings regularly occur, with extensive interfacial contact: A/H, B/E, B/G, F/H, and G/H. Other helix packings occur in some but not all. The structures of the helix interfaces—the ridge and groove patterns—are conserved, conferring sequence and structural specificity.

Some ridges contain "notches" created by a residue smaller than its neighbors. Bringing two such notches together can form exceptional "crossed-ridge" structures, in which the globins are unusually rich.

Comparative analyses of structures from many different organisms suggested the following picture of sequence–structure relationships in full-length globins (Lesk and Chothia, 1980):

- The principal determinants of the three-dimensional structure of the full-length globins lie in approximately 60 residues that are involved in the packing of helices and in the interactions between the helices and the heme group.
- Although mutations of the buried residues keep the side chains nonpolar, the side chains vary in *size*.
- In response to mutations at interfaces, the assembly of helices makes adjustments. Shifts in the relative position and orientation of homologous pairs of packed helices may be as much as 7 Å and 30° (among the structures considered in the 1980 work).
- Despite the large changes in the relative positions and orientations of the helices, a subtle feature of the interface structure is preserved. Although changes in the volume of residues at the helix interfaces cause shifts and rotations of the helices packed, there is substantial conservation of the *reticulation* of the residues; that is, homologous residues tend to make homologous contacts. The pattern of residue–residue contacts at interfaces tends to remain even if the residues that make the contact mutate. (This observation, in globins and other families, is at the basis of the well-known DALI (distance-matrix alignment) approach to structural alignment by matching interresidue contact maps (Holm and Sander, 1993).)

How do the globins reconcile the large changes in individual helix–helix contacts to preserve function? Despite the large changes in the relative positions and orientations of the helices, the structures of the heme pockets are very similar. The shifts in the helix packings produced by mutations are coupled to maintain the relative geometry of the residues that form the heme pocket. It is not uncommon among homologous proteins for the active site to be more tightly conserved than other parts of the structure.

These conclusions, first proposed over 30 years ago, were challenged when the structures of a class of shortened globin structures appeared.

9.3.1.2 Truncated Globins Truncated globins are short proteins occurring in prokaryotes, protozoans, fungi, and Archaeplastida. They maintain a recognizable globin fold despite containing only approximately 130 residues (in some cases even fewer), which is substantially smaller than the approximately 150 residues of typical full-length globins. They have been implicated in diverse functions, including detoxification of NO and photosynthesis.

Which residues of full-length globins, and which structural elements, are cast out in these shorter structures? Truncated globins retain most but not all of the helices of the standard globin fold (Nardini et al., 2007). They show a severe shortening or complete deletion of the A-helix, no D-helix, a variable CE region, and an extended EF loop. The length of the F-helix, which contains the iron-linked histidine, varies. The helical topology is referred to as "2/2"; the "sandwich" is composed of the E- and B-helices over the G- and H-helices. Of the 59 sites involved in conserved helix-to-helix or helix-to-heme contacts in full-length globins, 41 appear, with conserved contacts, in truncated globins. The helix–helix interfaces have ridge–groove packing patterns similar to those of the full-length globins, with the exception of the B/E contact, which has an unusual crossed-ridge structure in full-length globins but is normal in truncated globins.

Despite the large differences, it is possible to align the sequences and superpose the structures of the truncated and full-length globins (Figs. 9.3 and 9.4).

9.3.1.3 Globin Dynamics and Hexacoordinate Globins All globins, be they full-length or truncated, anchor the *b* heme (= iron protoporphyrin IX) by coordination of the iron with the strictly conserved "proximal" histidine. Oxygen (O_2), the functional ligand, attaches to the ferrous iron on the opposite or "distal" side. On elucidation of the vertebrate hemoglobin fold, it was a surprise to discover that the heme group was deeply sheltered in the protein matrix. Perutz and Mathews (1966) noted the bulk of the "amino acid side chains which block the entrance to the pocket of the α-chain" and further commented: "It is hard to see how the ligand can get past them." Blocking the entrance to the heme pocket protected the ferrous iron from detrimental oxidation, but also impeded oxygen access. Thus became apparent the necessity of protein dynamics for proper

```
                                    AAAAAAAAAAAAAAAAA   BBBBBBBBBBBBBBB     CCCCCCC
      Sperm whale myoglobin         MVLSEGEWQLVLHVWAKVEA--DVAGHGQDILIRLFK---SHPETLEKFD
Paramecium caudatum trHbN                      MSLFEQLGGQAAVQAVTAQFYANIQADA----TVATFFN
Mycobacterium tuberculosis trHbO             MPKSFYDAVGGAKTFDAIVSRFYAQVAEDE----VLRRVYP
Campylobacter jejuni trHbP                      MKFETINQ-ESIAKLMEIFYEKVRKDK----DLGPIFN

                DDDDDD EEEEEEEEEEEEEEEEEEE                 FFFFFFFF
Sw Mb     RFKHLKTEAEMKASEDLKKHGVTVLTALGAILKKKGHHEAE----LKPLAQSHATKHK------I
Pc trHbN  -------------GIDMPNQTNKTAAFLCAALGGP--NAWT---G-RNLKEVHA--NMGV-----
Mt trHbO  -------------EDDLAGAEERLRMFLEQYWGGP--RTYSEQRGHPRLRMRHA--PFRIS----
Cj trHbP  ---NAIG----TSDEEWKEHKAKIGNFWAGMLLGE--GDYN---GQP--LKKHL--DLPPFP---

          GGGGGGGGGGGGGGGGGGG          HHHHHHHHHHHHHHHHHHHHHHHHHH
Sw Mb     PIKYLEFISEAIIHVLHSRHPG----DFGADAQ-GAMNKALELFRKDIAAKYKELGYQG
Pc trHbN  SNAQFTTVIGHLRSALTGA-------GVAAALVEQTVAVAETVRGDVVTV
Mt trHbO  -LIERDAWLRCMHTAVASIDSETLDDEHRRELLDYLEMAAHSLVNSPF
Cj trHbP  -QEFFEIWLKLFEESLNIVY-NEEMKNVILQRAQMIASHFQNMLYKYGGH
```

Figure 9.3 Alignment of the sequences of sperm whale myoglobin, and truncated globins from *Paramecium caudatum*, *Mycobacterium tuberculosis, and Campylobacter jejuni.* Residues at position B10, E7, and F8 (proximal histidine) are in gray. Letters on the top line indicate the extents of the helices in the sperm whale myoglobin structure.

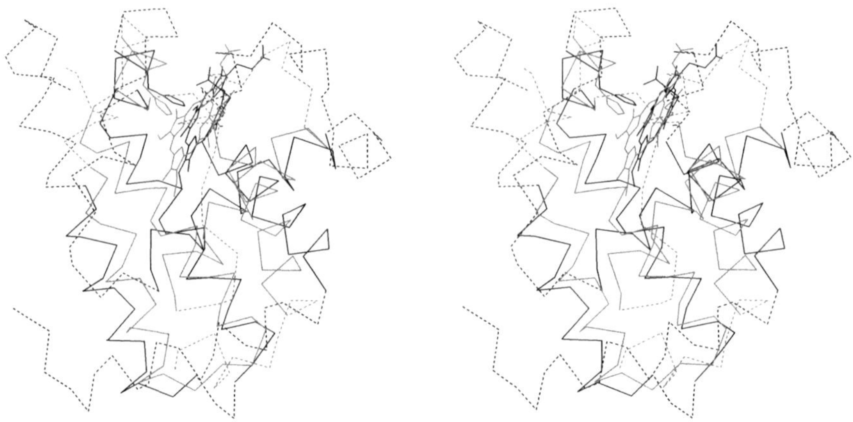

Figure 9.4 Superposition of the structures of sperm whale myoglobin [1mbo] and truncated globin from *Paramecium caudatum* [1dlw]. Solid lines indicate the parts of the structure common to full-length and truncated globins. Broken lines indicate the parts of the structure absent from truncated globins or changed in conformation.

protein function. Obstruction of the heme is partially alleviated in several 2/2 globins where tunnels and cavities connect the exterior solvent to the distal heme pocket (Nardini et al., 2007).

How a ligand in solution accesses the iron is particularly problematic and dependent on side-chain motions in a growing number of 3/3 and 2/2 globins, the geometry of which allows a protein residue, most frequently a histidine, to serve as a sixth ligand to the iron. These "hexacoordinate" globins require the displacement of the distal ligand to bind oxygen. Neutral histidine has a high affinity for iron, and the strength of the interaction modulated by the protein ranges from weak, as in *Oryza sativa* nonsymbiotic hemoglobin 1, to strong, as in human neuroglobin (Kakar et al., 2010). A full understanding of the globin reactivity must take into account not only the energetics of exogenous and endogenous ligand binding but also the rates of association and dissociation as dictated in part by the plasticity of the protein molecule. (Typical diatomic ligands such as O_2 and CN^- are exogenous ligands; endogenous ligands are side chains contributed by the protein itself.)

9.3.1.4 Phycocyanins and the Globins Phycocyanins are the major constituents of the phycobilisome, the supramolecular light-harvesting complexes of cyanobacteria and red algae. They bind a chromophore, which is an open-chain tetrapyrrole structure similar to a bile pigment (the heme group is a tetrapyrrole structure *closed* into a macrocycle).

When the first phycocyanin structure was solved, it showed an entirely unsuspected similarity in folding pattern to the globins, despite the difference in prosthetic group and function, and the absence of any apparent similarity between globin and phycocyanin sequences (Fig. 9.5). (The phycocyanin

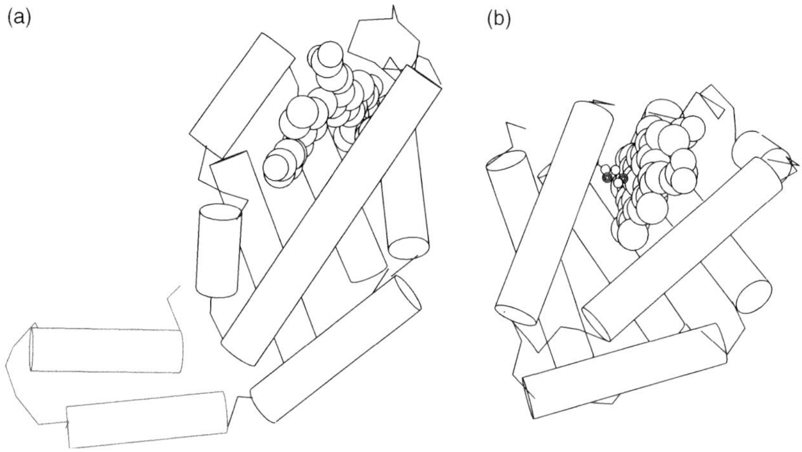

Figure 9.5 Structures of (a) phycocyanin from *Mastigocladus laminosus* and (b) myoglobin from Aplysia limacina (sea hare). The two N-terminal extension helices in phycocyanin, which have no counterparts in globin structures, are in gray. The ligand in phycocyanin is an open-chain tetrapyrrole, the chromophore.

structure contains two N-terminal helices in addition to the set displaying the same structural relationships as the helices in full-length globins. These are involved in oligomerization to form the phycobilisome, a feature that phycocyanins do not share with the globins.)

What is the origin of the similarity of the folds? Do globins and phycocyanins share a common ancestor, or did their very different functions somehow require the independent evolution of the same folding pattern? It is difficult to answer these questions because (i) the sequences are so different that it is not easy to align them with any confidence and (ii) some of the structural differences between globins and phycocyanins are much larger than those between even distantly related full-length globin structures.

To try to distinguish between true evolutionary relationship or convergence, we can only observe the similarities of the structures and try to interpret them. The basic dilemma is the distinction between points of similarity—which may be consistent with either homology or convergence—and criteria for true relationship. W.E. Le Gros Clark, the paleontologist, wrote "While it may be broadly accepted that, as a general proposition, degrees of genetic relationship can be assessed by noting degrees of resemblance in anatomical details, it needs to be emphasized that morphological characters vary considerably in their significance for this assessment." For the comparison of phycocyanin with the globins, the implication is that we must give different weights to different sorts of structural similarities, disregarding features common to many classes of proteins, and emphasizing the unusual ones. We must look for *structural similarities that are not specifically required by structure or function*. That is the principle. The problem with putting it into practice is that it is impossible to be sure whether any feature of a protein structure is not required for structure or function.

The C-helices in globins and phycocyanins are both of the rare 3_{10} structure; the rest are all α-helices. The C-helix does not appear to play a role in the function of monomeric globins. (It is important in the allosteric change in hemoglobin, but that was surely a much later development.) If, as seems likely, a sequence compatible with an α-helix in this region could produce a viable globin or phycocyanin, then the fact that the C-helices in both globins and phycocyanins are of the unusual 3_{10} type is a structural similarity not specifically required by structure or function. (This argument was advanced before the discovery of the truncated globins, and it is supported by the appearance of an α-helix C in the structure of the 2/2 globin from *Campylobacter jejuni* [2ig3].)

What about the pattern of interactions of residues at the helix interfaces? The observation of common types of ridges-into-grooves packing at interfaces between corresponding pairs of helices in globins and phycocyanins is not evidence for evolutionary relationship, as these structures appear in many unrelated proteins (Chothia et al., 1977, 1981). More significant are the interfaces with unusual structures: the B–E and G–H packings. In full-length globins, these interfaces have an unusual "crossed-ridge" structure. It is, therefore, significant that the B–E interface in phycocyanins has a similar crossed-ridge packing. A similar crossed-ridge structure is also observed in the G–H contact in phycocyanins. But can we argue that this type of interface is not required by the structure? In phycocyanins, the interaxial angles of the B/E-helices are all in the region of -50° to -60°, which would be expected from a *normal* "4–4" packing. Therefore, the phycocyanin structure cannot require the special crossed-ridge structure to achieve an unusual interhelix-axis angle. (This argument was also advanced before the discovery of the truncated globins, and it is also supported by the appearance of the normal non-crossed-ridge packing in the B/E interfaces in truncated globins.)

The observation that the structural similarities between globins and phycocyanins extends down to the minute structural details, apparently not required for the fold or for function, suggests that the structures are homologous rather than convergent (Pastore and Lesk, 1990).

Subsequent to that work, the increase in the power of fold-recognition methods such as threading or profile–profile techniques gives significant matches between globins and phycocyanins. Had these methods been available, and applied, before the structure of phycocyanin was solved, it would not have come as such a surprise to see the similarity to the globin fold. Whether these results would have settled the question of homology is another matter; after all, even after the structure was known, the answer was not obvious.

9.3.2 Multimeric States of Globins

Many globins are biologically active either in the form of monomers or as single domains within multidomain proteins. However, some globins associate to form multimeric complexes (see Fig. 1 of Weber and Vinogradov, 2001). The best known, of course, are the tetrameric hemoglobins.

In humans, several preadult tetramers appear during the course of early development. A human embryo (up to 6 weeks after conception) primarily synthesizes two hemoglobin chains: ζ and ε. These form a $\zeta_2\varepsilon_2$ tetramer. From 6 weeks after conception until about 8 weeks after birth, the predominant species shifts to the fetal hemoglobin $\alpha_2\gamma_2$. This is succeeded by adult hemoglobin, $\alpha_2\beta_2$.

Several globins appear to form homodimers. These include globins from several molluscs, echinoderms, insects, and lampreys (Royer et al., 2005). In many cases, the intersubunit interfaces differ from those of vertebrate tetramers.

Earthworms produce extracellular globins in the form of high-molecular-weight ($\sim 3 \times 10^6$) polymers. In addition, large aggregates of deoxyhemoglobin are associated with sickle-cell anemia.

9.4 SPECIES DISTRIBUTION IN GLOBINS

The past two decades have seen a phenomenal increase in the number of hemoglobin sequences deposited in databases. The additions belong to the three kingdoms of life and reveal the broad distribution of the superfamily. In fact, the absence of a globin gene is the exception rather than the rule. Globins appear to be underrepresented or absent only from pathogenic unicellular eukaryotes with reduced genomes, and from a few bacterial phyla, for example, Spirochaetes and Bacteroidales/Chlorobi. In 2005, Serge Vinogradov and coworkers published a phylogenetic analysis that discerned three separate globin lineages (Vinogradov et al., 2005): flavohemoglobin, globin-coupled sensors (GCSs), and 2/2 hemoglobins. The classification holds to this day.

9.4.1 The flavohemoglobin family

The flavohemoglobin family comprises chimeric proteins (FHbs) containing an N-terminal 3/3 globin domain and a C-terminal flavoprotein, other chimeric combinations of the globin and reductase domains, and single-domain 3/3 globins related to the globin domain of flavohemoglobin (SDFgbs) (Bonamore and Boffi, 2008). FHbs are widely distributed in bacteria and nonmetazoan eukaryotes (Table 9.1). A few nonmetazoan eukaryotes harbor SDFgbs, but these proteins have been found mostly in bacteria and metazoans. Vertebrate globins are SDFgbs (except for androglobin, see Section 9.4.4).

In the pre-molecular biology days, studies of the globin superfamily were conditioned by the availability of material from natural sources. It is therefore no surprise that many laboratories focused on the characterization of plentiful human and horse hemoglobins and sperm whale myoglobin (sperm whale myoglobin was chosen for the structural work because of the quality of the crystals it produced). Hemoglobin and myoglobin, however, are not the only vertebrate SDFgbs. In 2000–2001, genomic and proteomic analyses revealed two additional

TABLE 9.1 Distribution of Globins

	Flavohemoglobins			Sensor Globins			2/2 Globins			
	FHb	SDFgb	Chim	GCS	SDSgb	Pgb	I	II	III	Chim
Bacteria										
Cyano		•					•			
α-Proteo	•	•		•	•		•	•	•	
Other	•	•		•	•	•	•	•	•	•
Archaea										
Crenarcheota						•				
Euryarchaeota				•		•	•			
Eukaryotes										
Non-metazoan	•	•	•		•		•	•		
Metazoan		•	•							

FHb, flavohemoglobins (2-domains); SDFgb, single-domain globin related to the N-terminal globin domain of FHbs; Chim, other chimeras; GCS, globin-coupled sensor (two-domain); SDSgb, single-domain globin related to the N-terminal globin domain of GCSs; Pgb, protoglobin; I, II, and III, Group I, II, and III, respectively.

members of the lineage: neuroglobin and cytoglobin, both of which are hexacoordinate. Human neuroglobin is present at low levels in the cytoplasm of neurons, including the retina, and endocrine cells. Human cytoglobin, originally called the stellate cell activation-associated protein, exists at low levels in the cytoplasm and nucleus of fibroblasts and similar cells. Neuroglobin branched out approximately 800 million years ago from an ancestor common to the intracellular globin of annelids. Cytoglobin arose from a second branch of vertebrate globins. Figure 9.6 shows an alignment of the sequences of human SDFgbs.

In the past few years, globin E (birds), globin X (fishes and amphibians, hexacoordinate), and globin Y (monotremes and amphibians) have made their appearance in the list of vertebrate homologs. Along with agnathan hemoglobins, they raise to nine the number of distinct vertebrate globin types. Globin X is most closely related to neuroglobin, whereas the others group with hemoglobin, myoglobin, and cytoglobin (Storz et al., 2011).

9.4.2 The Globin-Coupled Sensor Family

The GCS family is considerably less populated than the flavohemoglobin family. It derives its name from chimeric proteins containing an N-terminal 3/3 globin domain (sensing oxygen or other small ligands) and one or more C-terminal domains (transducing the state of the sensor). GCSs are found in bacteria and Euryarchaeota (Freitas et al., 2003). The corresponding single-domain globin occurs in bacteria, Euryarchaeota, and fungi. The third branch of the GCS family comprises protoglobins. These nonchimeric proteins are found in bacteria and Archaea

```
                                    AAAAAAAAAAAAAAAA BBBBBBBBBBBBBBBBCCCCCCC
Haemoglobin γ1 ---------------MGHFTEEDKATITSLWGKV--NVEDAGGETLGRLLVVYPWTQRFFDS-
Haemoglobin γ2 ---------------MGHFTEEDKATITSLWGKV--NVEDAGGETLGRLLVVYPWTQRFFDS-
Haemoglobin ε  ---------------MVHFTAEEKAAVTSLWSKM--NVEEAGGEALGRLLVVYPWTQRFFDS-
Haemoglobin δ  ---------------MVHLTPEEKTAVNALWGKV--NVDAVGGEALGRLLVVYPWTQRFFES-
Haemoglobin β  ---------------MVHLTPEEKSAVTALWGKV--NVDEVGGEALGRLLVVYPWTQRFFES-
Haemoglobin θ  ----------------MALSAEDRALVRALWKKLGSNVGVYTTEALERTFLAFPATKTYFSH-
Haemoglobin α  ----------------MVLSPADKTNVKAAWGKVGAHAGEYGAEALERMFLSFPTTKTYFPH-
Haemoglobin ζ  ----------------MSLTKTERTIIVSMWAKISTQADTIGTETLERLFLSHPQTKTYFPH-
Haemoglobin μ  -----------------MLSAQERAQIAQVWDLIAGHEAQFGAELLLRLFTVYPSTKVYFPH-
Cytoglobin     MEKVPGEMEIERRERSEELSEAERKAVQAMWARLYANCEDVGVAILVRFFVNFPSAKQYFSQ-
Myoglobin      ----------------MGLSDGEWQLVLNVWGKVEADIPGHGQEVLIRLFKGHPETLEKFDK-
Neuroglobin    ------------------MERPEPELIRQSWRAVSRSPLEHGTVLFARLFALEPDLLPLFQYN

                    DDDDDDDEEEEEEEEEEEEEEEEEEEEE          FFFFFFFFF     GGGGGGG
Haemoglobin γ1 FGNLSSASAIMGNPKVKAHGKKVLTSLGDAIKHLDD---LKGTFAQLSELHCDKLHVDPENFKL
Haemoglobin γ2 FGNLSSASAIMGNPKVKAHGKKVLTSLGDAIKHLDD---LKGTFAQLSELHCDKLHVDPENFKL
Haemoglobin ε  FGNLSSPSAILGNPKVKAHGKKVLTSFGDAIKNMDN---LKPAFAKLSELHCDKLHVDPENFKL
Haemoglobin δ  FGDLSSPDAVMGNPKVKAHGKKVLGAFSDGLAHLDN---LKGTFSQLSELHCDKLHVDPENFRL
Haemoglobin β  FGDLSTPDAVMGNPKVKAHGKKVLGAFSDGLAHLDN---LKGTFATLSELHCDKLHVDPENFRL
Haemoglobin θ  LDLSP------GSSQVRAHGQKVADALSLAVERLDD---LPHALSALSHLHACQLRVDPASFQL
Haemoglobin α  FDLSH------GSAQVKGHGKKVADALTNAVAHVDD---MPNALSALSDLHAHKLRVDPVNFKL
Haemoglobin ζ  FDLHP------GSAQLRAHGSKVVAAVGDAVKSIDD---IGGALSKLSELHAYILRVDPVNFKL
Haemoglobin μ  LSACQ------DATQLLSHGQRMLAAVGAAVQHVDN---LRAALSPLADLHALVLRVDPANFPL
Cytoglobin     FKHMEDPLEMERSPQLRKHACRVMGALNTVVENLHDPDKVSSVLALVGKAHALKHKVEPVYFKI
Myoglobin      FKHLKSEDEMKASEDLKKHGATVLTALGGILKKKGH---HEAEIKPLAQSHATKHKIPVKYLEF
Neuroglobin    CRQFSSPEDCLSSPEFLDHIRKVMLVIDAAVTNVEDLSSLEEYLASLGRKHR-AVGVKLSSFST

               GGGGGGGGGGGG     HHHHHHHHHHHHHHHHHHHHHHHHHH
Haemoglobin γ1 LGNVLVTVLAIHFGKEFTPEVQASWQKMVTAVASALSSRYH-----------------------
Haemoglobin γ2 LGNVLVTVLAIHFGKEFTPEVQASWQKMVTGVASALSSRYH-----------------------
Haemoglobin ε  LGNVMVIILATHFGKEFTPEVQAAWQKLVSAVAIALAHKYH-----------------------
Haemoglobin δ  LGNVLVCVLARNFGKEFTPQMQAAYQKVVAGVANALAHKYH-----------------------
Haemoglobin β  LGNVLVCVLAHHFGKEFTPPVQAAYQKVVAGVANALAHKYH-----------------------
Haemoglobin θ  LGHCLLVTLARHYPGDFSPALQASLDKFLSHVISALVSEYR-----------------------
Haemoglobin α  LSHCLLVTLAAHLPAEFTPAVHASLDKFLASVSTVLTSKYR-----------------------
Haemoglobin ζ  LSHCLLVTLAARFPADFTAEAHAAWDKFLSVVSSVLTEKYR-----------------------
Haemoglobin μ  LIQCFHVVLASHLQDEFTVQMQAAWDKFLTGVAVVLTEKYR-----------------------
Cytoglobin     LSGVILEVVAEEFASDFPPETQRAWAKLRGLIYSHVTAAYKEVGWVQQVPNATTPPATLPSSGP
Myoglobin      ISECIIQVLQSKHPGDFGADAQGAMNKALELFRKDMASNYKELGFQG-----------------
Neuroglobin    VGESLLYMLEKCLGPAFTPATRAAWSQLYGAVVQAMSRGWDGE---------------------
```

Figure 9.6 The aligned sequences of human single-domain globins. Conserved residues at position B10 (Leu), E7 (distal histidine), and F8 (proximal histidine) are in gray.

and are likely derived from the single-domain ancestral version of the GCS. The globin domain GCSs are longer than the canonical globin domain by some 40 residues, located in a pre-A-helix (Z-helix), the CE loop, and the FG loop.

9.4.3 The Truncated Globin Family

The truncated or 2/2 globin family has members in prokaryotes, protozoans, fungi, and Archaeplastida (Vuletich and Lecomte, 2008; Hoogewijs et al., 2012a). It is divided into three paralogous groups exhibiting distinctive distribution and amino acid conservation patterns. Group I proteins are the only ones found in Archaea, Cyanobacteria, Ciliates, and fungi; Group II proteins are the principal plant 2/2 globins; Group III proteins appear largely restricted to bacteria. As in the other two families, there exist chimeric versions of the truncated globins. To date, only chimeras of Group II proteins have been identified. The added domain is N-terminal and is annotated as a member of the antibiotic biosynthesis monooxygenase family.

9.4.4 Androglobin: a Protein Containing a Circularly Permuted Globin Domain

The variety of globin genes has recently been extended by the discovery of androglobin (Hoogewijs et al., 2012b). Sequence alignment shows this protein to contain an internal circularly permuted globin structure. The sequence contains a region corresponding to globin helices C–H, followed by a calmodulin-binding IQ motif, followed by globin helices A and B.

9.4.5 A Model of Globin Evolution

As of 2009, two-thirds of published bacterial genomes contained one or more globin genes. Such broad distribution supports the emergence of the globin superfamily 3.5–3.8 billion years ago. A tentative model of evolution is summarized in Fig. 9.7, where horizontal gene transfers are indicated by arrows (Vinogradov et al., 2007).

The flavohemoglobin family has two branches, single-domain and chimeric. As in the other two families, single-domain proteins appeared first, to be transformed into chimeric proteins at a later stage. The GCS family has three branches, with protoglobins appearing last. In the 2/2 family, Group III proteins were a late invention derived from the Group II branch. There is some uncertainty as to which of Group I and Group II is ancestral (Vuletich and Lecomte, 2008); an up-to-date assessment (Vinogradov et al., 2011) favors Group I.

Whether the three families of globins illustrate divergent or convergent evolution is difficult to determine. It seems likely that the full-length domain originated only once, with a fold related to that of the current SDFgbs. The 2/2 domain may have arisen independently, although current models propose that it too was derived from the SDFgb line. Group I and Group II proteins have a 3_{10} C-helix, which argues for common ancestry with SDFgbs.

9.5 GLOBIN FUNCTIONS

Vertebrate hemoglobin (a tetramer) and myoglobin (a monomer) illustrate the most familiar globin function. Oxygen binds reversibly to the ferrous heme iron and is transported and stored for use in respiration. Heme, however, is a highly reactive cofactor. Hemoglobin and myoglobin have achieved the delicate balance

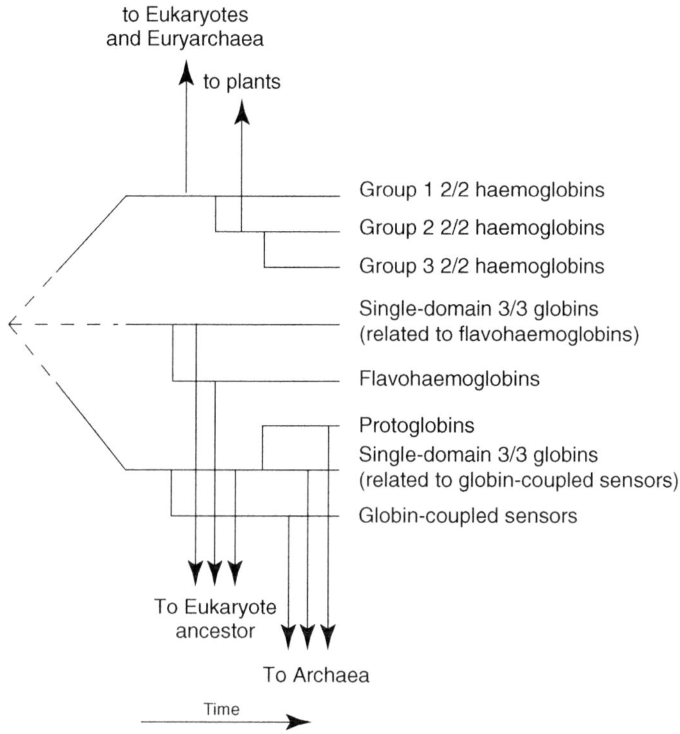

Figure 9.7 Model of globin evolution according to Vinogradov and coworkers (2007). Single-domain 3/3 globins related to the N-terminal domain of flavohemoglobins (indicated by a box) are considered to be the ancestors of all globins. The arrows indicate the occurrence of horizontal gene transfers.

of binding and releasing O_2 at rates compatible with the particular demands of each organism's oxidative metabolism while minimizing the frequency of processes that consume oxygen and produce toxic reactive oxygen species.

While O_2 combines with heme rapidly, generally at near diffusion-limited speed ($k_{on} = 10^8\ M^{-1}\ s^{-1}$) depending on iron accessibility, rates of O_2 dissociation vary over orders of magnitude. To modulate the dissociation rate constant (k_{off}), globins use more or less extensive networks of polar and nonpolar residues. Two positions stand out across all lineages: B10 and E7, which are occupied by a leucine and a histidine, respectively, in practically all vertebrate hemoglobins and myoglobins. His E7 forms in most cases a hydrogen bond with bound O_2 that stabilizes the oxy state. In many nonvertebrate globins, a combination of Tyr B10 and Gln E7 holds the ligand tightly, slows down k_{off}, and thereby increases O_2 affinity (k_{on}/k_{off}).

Because globins can display widely different O_2 affinities, variations on the transport and storage theme abound. Leghemoglobin (LegHb, where leg refers to legume) is a symbiotic hemoglobin expressed at millimolar levels in the nitrogen-fixing nodules caused by infection of plant roots with rhizobia. LegHb delivers O_2 to a terminal oxidase of the symbiont, but Tyr B10 and His E7 impart high affinity. LegHb's role is to sequester O_2: free O_2 is kept at a very low cellular concentration to allow the O_2-sensitive nitrogenase complex to function, undamaged. Conversely, the role of SDFgb of the gram-negative and obligate aerobe *Vitreoscilla* is delivery: this globin has a fast k_{off} despite containing Tyr B10 and Gln E7. *Vitreoscilla* Hb binds to subunit I of cytochrome *bo* oxidase and supports respiration under hypoxic conditions.

Oxygen binding and the ensuing conformational or electronic perturbations are exploited by the GCSs. Aerotactic GCSs such as the bacterial and archaeal HemAT sense the level of O_2 in the environment of the cell and mediate an aerophobic or aerophilic response via a signaling domain interacting with the chemotaxis system. Gene regulatory GCSs are fused to various C-terminal transducer domains (e.g., GAF and STAS) and associate with O_2 to report on its cellular levels.

Is reversible O_2 binding the only basis of function of all vertebrate globins? Myoglobin knock-out mice are minimally inconvenienced by the absence of an O_2 storage protein in their "red" muscles (Garry et al., 2000). This unexpected observation has prompted a closer look at the role of myoglobin. Aside from facilitated O_2 diffusion (Gros et al., 2010), there is good evidence that myoglobin participates in the management of reactive nitrogen species, and it is now well established that both hemoglobin and myoglobin can function as dioxygenases. They combine bound O_2 with NO to produce the relatively innocuous nitrate ion (Gardner, 2005).

What then controls the fate of a bound ligand in a globin? Different sets of nonpolar and hydrogen-bonding residues are observed in the different lineages, presumably regulating ligand k_{on} and k_{off} after their own fashion, but redistributing electronic density for bond-breaking and bond-forming. Further structural features, such as the "doming" of the heme macrocycle, interactions between the propionate substituents and protein side chains, and hydrogen bonding of the proximal histidine imidazole ring, influence the electronic distribution of the heme and its redox potential, further conditioning its reactivity; and also contributing to the mechanism of the allosteric change in hemoglobin.

The ancient origin of the globin superfamily, traced back to the community of organisms that formed the last universal common ancestor, suggests several possibilities for early globin function. These relate to the toxicity of O_2 and byproducts to anaerobic life, the reactivity of nitrogen oxides, and the availability of iron. Reversible O_2 binding is thought to be a late development in the adaptation of the globin fold, exploited to the fullest by multicellular organisms.

One might hope that extant nonvertebrates might retain clues to the original function of globins. Unfortunately, the minute amounts of globin present in the cells of most extant nonvertebrate organisms have made their functional characterization difficult.

Besides low expression levels, three additional problems confound functional studies. First, in organisms that lend themselves to genetic manipulation, phenotypic differences associated with inactivation of globin genes tend to be subtle, often because of functional redundancy. Phenotypes become informative only under specific growth conditions that may not be easy to foresee. Second, a globin may operate indirectly, for example, as a regulator rather than an enzyme. Transcriptomic, proteomic, and other complementary data may be required to identify the step in which the globin participates. Finally, the reactivity of the heme cofactor complicates matters in two ways. A purified protein may display a tantalizing activity in the test tube, and yet have a physiological role unrelated to that activity; and, because structural adjustments as slight as the weakening of a hydrogen bond can alter heme chemistry, apparently innocent amino acid replacements may have major consequences. As a result, it is particularly difficult to reach definitive functional conclusions either from structural information alone or from *in vitro* experiments.

Despite these reservations and bearing in mind the cautionary tale of myoglobin, a few nontraditional globin functions have been demonstrated with reasonable certainty.

Escherichia coli flavohemoglobin (HMP) plays a role in cellular redox homeostasis and in the response to oxidative and nitrosative stresses (Bonamore and Boffi, 2008). Under aerobic conditions, HMP shows a robust NO scavenging and dioxygenase activity. The reaction oxidizes the heme iron to the ferric state, and the flavin-containing reductase domain provides the electrons to the heme group. This protective function, along with detoxification of other reactive oxygen species and reactive nitrogen species, is likely to be widespread among FHbs and other globins, the single-domain proteins requiring reducing equivalents from a separate source.

Peroxidase activity appears to be general, but rather as a detrimental side reaction. In one example of the contrary, a 3/3 globin turned its oxidative capability into a defense mechanism. The intracellular dehaloperoxidase of the marine polychaete *Amphitrite ornata* is used for detoxification purposes as well as oxygen binding.

Sulfide transport, instead of oxygen transport, is observed in certain invertebrate SDFgbs and perhaps certain 2/2 globins as well.

A different kind of adaptation is demonstrated by the extraordinary SDFgb of the nematode *Mermis nigrescens*, concentrated in its oxy state at the tip of the female worm to shadow a photoreceptor, and direct phototaxis (Burr et al., 2000).

Diversity of physiological roles is beautifully illustrated by *Caenorhabditis elegans*. The genome of *C. elegans*, a heme auxotroph, contains 33 globin genes. Most of the gene products have yet to be studied in detail, but recent work provides evidence for oxygen sensing and a role in development (Tilleman et al., 2011). At least one of the 3/3 globins is hexacoordinate, and its properties suggest that it may function as an electron transport protein. This would constitute the first example of a cytochrome-like functionality in a globin fold.

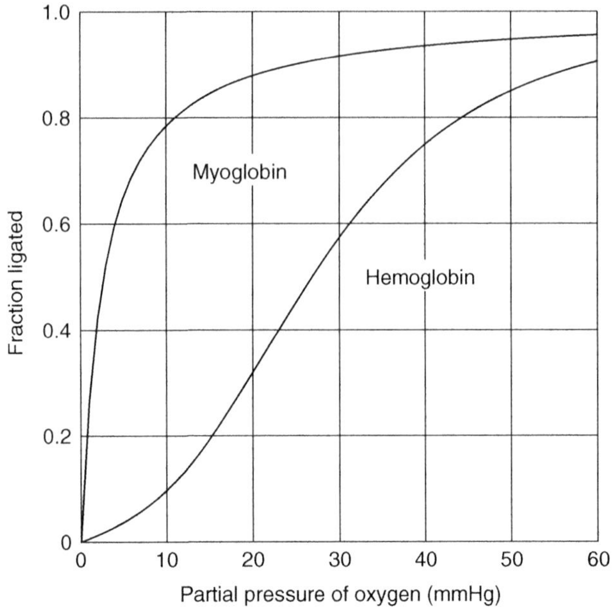

Figure 9.8 Oxygen-dissociation curves for myoglobin and hemoglobin. Myoglobin shows a simple equilibrium, with a binding constant independent of oxygen concentration. Hemoglobin shows positive cooperativity, the binding constant for the first oxygen being several orders of magnitude smaller than the binding constant for the fourth oxygen. The units of mmHg for partial pressure are traditional in the literature on this topic. 760 mmHg = 1 atm = 101,325 Pa.

9.5.1 The Allosteric Change of Hemoglobin

Allosteric proteins show "action at a distance": ligand binding at one site affects activity at another. An impulse at the first site must cause a conformational change affecting the second.

Some allosteric changes produce cooperativity. Hemoglobin is a classic case: binding of oxygen at any site enhances the affinity of other sites. Figure 9.8 compares noncooperative and cooperative binding curves. Hemoglobin shows positive cooperativity: binding of oxygen increases the affinity for additional oxygen. Other proteins show negative cooperativity: binding of the first ligands reduces the affinity for additional ligand.

The Hill plot is a method for analyzing binding data that reveal cooperativity.

Suppose that a protein with n binding sites can exist in only two states: no ligand at all, or all sites occupied. The overall dissociation equilibrium is

$$\mathrm{PL}_n \rightleftharpoons \mathrm{P} + n\mathrm{L}, \qquad K_D = \frac{[\mathrm{P}][\mathrm{L}]^n}{[\mathrm{PL}_n]}$$

Let $\overline{v}$ be the fractional saturation:

$$\overline{v} = \frac{n[PL_n]}{n([P] + [PL_n])} = \frac{\text{Number of sites occupied}}{\text{Total number of sites}}$$

This is equivalent to the Hill equation

$$\overline{v} = \frac{[L]^n}{K_D + [L]^n}$$

A form of the Hill equation suitable for plotting data is

$$\log\left(\frac{\overline{v}}{1 - \overline{v}}\right) = n \log[L] - \log K_D$$

In ideal cases, the graph of $\log(\overline{v}/(1 - \overline{v}))$ against $\log[L]$ is a straight line with slope n and intercept on the abscissa $(1/n) \log K_D$. A deviation from linearity in a Hill plot reveals the degree of cooperativity in binding.

Figure 9.9 shows a Hill plot of myoglobin and hemoglobin. Corresponding, to the free ligand concentration [L] is the partial pressure of O_2, pO_2. Myoglobin is a monomeric globin with one O_2 binding site. Correspondingly the Hill plot shows a slope of 1, with no deviation from linearity. In contrast, the Hill plot for hemoglobin describes a curve at intermediate pO_2 values. Hemoglobin has four binding sites. If the binding were perfectly cooperative, a straight line would be obtained with a slope of 4 throughout. Instead, the maximum slope in the transition region is approximately 3.0. At low and high values of pO_2, the slope of the hemoglobin Hill plot approaches 1.0. This is because at very low pO_2 only weak affinity sites are available for binding and at very high pO_2 only high affinity sites remain. These binding events are not cooperative.

The hemoglobin curve is mathematically complex, but the two asymptotes of the Hill plot convey useful information. Their intercept with the vertical line that corresponds to $pO_2 = 0$ is the log of the microscopic dissociation constant of the weak site (lower asymptote, κ_1) or the strong site (higher asymptote, κ_4). The intercept of the asymptotes with the horizontal line at $\log(\overline{v}/(1 - \overline{v})) = 0$ returns the negative log of the corresponding microscopic dissociation constant. The microscopic dissociation constant for the binding of the first oxygen to hemoglobin is $k_1 = [\text{Hb}]pO_2/[\text{Hb}(O_2)_1]$. That for the binding of the fourth oxygen to hemoglobin is $k_4 = [\text{Hb}(O_2)_3]pO_2/[\text{Hb}(O_2)_4]$. For hemoglobin (under defined experimental conditions), extrapolation of the asymptote of the deoxy state gives $\kappa_1 = 0.0218$ mm Hg, and extrapolation of the asymptote of the oxy state gives $\kappa_4 = 3.45$ mm Hg (traditional units). The ratio is 158.

In contrast to a ligand that induces a cooperative effect on activity toward the same ligand, an effector is a ligand that alters the activity of a protein toward a different ligand. 2,3-Bisphosphoglycerate (BPG, also known as diphosphoglycerate, DPG) is an effector for hemoglobin. It binds preferentially to the deoxy form, thereby preferentially stabilizing that form. This has the effect

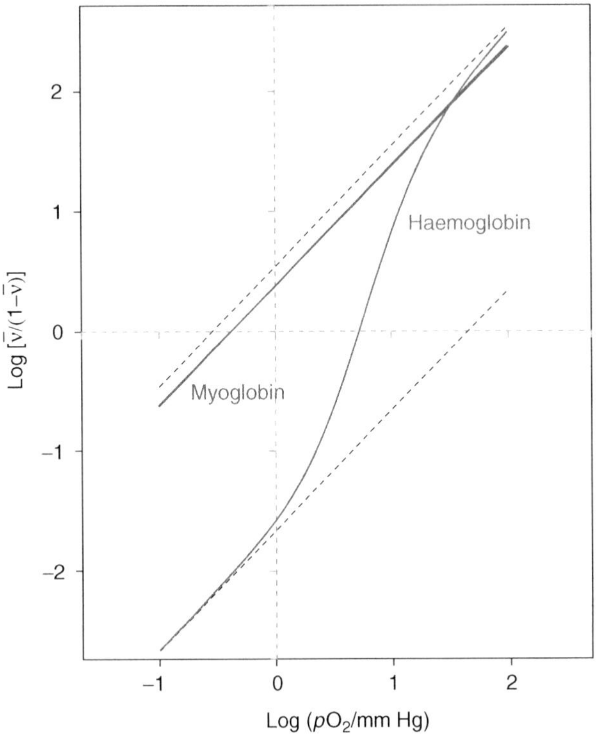

Figure 9.9 A Hill plot of the binding of O_2 to myoglobin and hemoglobin. The partial pressure of O_2, on the abscissa, is analogous to the concentration of a ligand. A Hill plot indicates the extent of cooperativity in binding of multiple ligands to proteins. Myoglobin is a monomer, showing no cooperative effects in binding O_2. The Hill plot is linear with a slope equal to 1. Hemoglobin deviates from linearity. Over the intermediate range of pO_2, it shows a slope of ~3.0. This implies that the cooperativity among the four subunits in binding O_2 is substantial but not perfect. The fact that the slope is greater than 1 means hemoglobin shows positive cooperativity; that is, the binding of some O_2 *increases* the affinity for additional O_2 binding. A slope of less than 1 would imply negative cooperativity: the binding of some ligand would *decrease* the affinity for additional ligand. It is interesting to note that the deoxyhemoglobin has an O_2 affinity much lower than that of myoglobin. Oxyhemoglobin has comparable affinity. The hemoglobin curve was constructed using the microscopic binding constants determined by Imai (1982). The myoglobin curve used a value consistent with those reported by Wittenberg and Wittenberg (2003).

of decreasing the oxygen affinity and enhancing O_2 delivery. People and animals living at high altitude have higher concentrations of BPG than their sea-level relatives. It is BPG binding, rather than the intrinsic difference in oxygen affinity between fetal and maternal hemoglobin, that ensures the transfer of oxygen from mother to fetus. Protons have the same effect as BPG. They decrease the oxygen affinity when their concentration increases. This "Bohr effect" enhances the release of O_2 when the pH of the blood becomes more acidic.

A general model for cooperativity, explaining the influence extending over long distances between binding sites, as in hemoglobin, is that the subunits of a protein are in equilibrium between two forms, with different binding constants, and that ligands shift the equilibrium between them. One form, called *tense* (T), is structurally constrained in a way that inhibits binding. The other, *relaxed* (R) form is unconstrained. This model (Monod et al., 1965) rationalizes the properties of hemoglobin:

- At low partial pressures of O_2, all the subunits of hemoglobin are in the T form. The binding constant for oxygen is low, because binding to the T state is inhibited.
- At high partial pressures of O_2, all the subunits of hemoglobin are in the R form. The binding constant for oxygen is high, comparable to that of simple globins such as myoglobin, or dissociated α- or β-chains of hemoglobin, because binding of O_2 to the R state is unconstrained.
- Binding of between two and three oxygen molecules per tetramer shifts the subunits *concertedly* from all T state to all R state. All subunits, ligated or not, change conformation together.

Writing the equilibrium between T and R states in the free and fully ligated forms

$$\mathrm{T} \rightleftharpoons \mathrm{R}$$

$$\mathrm{TL}_n \rightleftharpoons \mathrm{RL}_n$$

a ligand-induced switch between T and R states is consistent with the free-energy relationships shown in Fig. 9.10 implying that in the absence of ligand the T state is more stable than the R, but in the ligated state RL_n is more stable than TL_n. Although not all allosteric proteins fit this picture, and we are discovering that states other than T and R can be populated by the hemoglobin tetramer, the terminology of T and R states continues to be used.

The interpretation of this scheme in structural terms was one of the early triumphs of protein crystallography. It involves coupled changes in tertiary and quaternary structure.

9.5.2 The Mechanism of the Allosteric Change of Hemoglobin

In the erythrocyte, hemoglobin is an equilibrium mixture of deoxy and oxy forms; the concentration of partially ligated forms is tiny. Binding of oxygen induces structural changes that alter the relative free energy of the two forms, shifting the equilibrium toward the high affinity form. Starting from the deoxy state, partial ligation (between two and three oxygens) is enough to shift the equilibrium

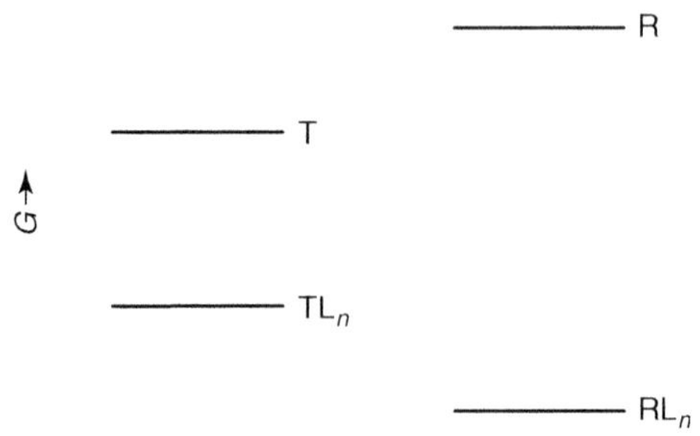

Figure 9.10 Free-energy relationships among fully liganded and unliganded tense (T) and relaxed (R) states of hemoglobin. L = Ligand; for the hemoglobin–O_2 system, n = 4. Note that the stabilization of liganded T (not normally observed) with respect to unliganded T is smaller than the stabilization of liganded R with respect to unliganded R (not normally observed), reflecting the higher oxygen affinity of the R state.

to the oxy state, which will then pick up the remaining oxygens with greater affinity. Conversely, starting with the fully ligated oxy structure, partial loss of oxygen will shift the equilibrium in the opposite direction, stimulating release of the remaining oxygen. Other molecules that modify oxygen affinity, such as the natural allosteric effector BPG, operate in part by shifting this equilibrium, by preferentially stabilizing one of the two forms.

The oxygen affinity of the oxy form of hemoglobin is similar in magnitude to that of isolated α and β subunits and to that of myoglobin. The oxygen affinity of the deoxy form is much less: the ratio of the binding constants for the first and fourth oxygens is 1 : 150–300, depending on conditions. Therefore, it is the deoxy form which is special, that has had its oxygen affinity "artificially" reduced. In the terminology of Monod et al. (1965), the reduced oxygen affinity of the deoxy form of hemoglobin arises from structural constraints that hold the structure in a "tense" (T), internally inhibited state; the oxy form is in a "relaxed" (R) state, as free to bind oxygen as the isolated monomer.

The binding of oxygen is accompanied by a change in the state of the iron. In the deoxy state, the iron has five ligands—the four pyrrole nitrogens of the heme group and the proximal histidine—and is in a high spin Fe (II) state with an ionic radius of 2.06 Å. In the oxy state, the iron has six ligands—oxygen being the sixth—and is in a low spin Fe (II) state with an ionic radius of 1.94 Å. These radii are important because the distance from the pyrrole nitrogens to the center of the heme is 2.03 Å; this implies that the iron will fit in the plane of the pyrrole nitrogens in the oxy state but not in the deoxy state. The difference in color between arterial and venous blood arises because oxygen binding affects the electronic state of the iron. This, by the way, is the answer to the question raised at the end of the quote from Berzelius at the beginning of this chapter.

Comparison of the oxy and deoxy structures has defined the changes in tertiary structures of individual subunits, and in the quaternary structure, or the relative geometry of the subunits and the interactions at their interfaces (Baldwin and

Chothia, 1979). Here is a simple description of how these changes are coupled. The quaternary structure is determined by the way the subunits fit together. This fit depends on the shapes of their surfaces. Oxygen binding alters the energies of alternative tertiary structures of the ligated subunit. The tertiary structural changes alter the shapes of the surfaces of the subunits, changing the way they fit together. This alters the quaternary structure, inducing tertiary structural changes in other subunits.

The tertiary structural changes in α and β subunits are similar but not identical.

At the heme group itself, in the deoxy form the iron atom is out of the plane of the four pyrrole nitrogens of the heme group. There are two reasons for this: the larger radius of the iron in its high spin state, and steric repulsions between the Nε of the proximal histidine and the pyrrole nitrogens. The heme group is "domed;" that is, the iron-bound nitrogens of the pyrrole rings are out of the plane of the carbon atoms of the porphyrin ring of the heme, by 0.16 Å in the α subunits and 0.10 Å in the β subunits.

Forming the link between iron and oxygen would, in the absence of tertiary structural change, create strain in the structure. Without constraint, the heme would become planar, and the iron would move into this plane. These changes are resisted by the steric interactions between the proximal histidine and the heme group, and the packing of the FG corner against the heme group (The FG corner is the region between the C-terminus of the F-helix and the N-terminus of the G-helix). In the β subunit, an additional barrier to the binding of oxygen without tertiary structural change is the position of Val E11 in the region of space to be occupied by the oxygen itself. (Val E11 corresponds to the eleventh residue of the E-helix in the reference structure, sperm whale myoglobin.)

In the oxy form, these impediments are relieved by changes in tertiary structure. There is in both α and β subunits a shift of the F-helix across the heme plane by about 1 Å, and a rotation relative to the heme plane. The effect is to permit a reorientation of the proximal histidine so that the iron atom can enter the heme plane. Associated with this shift in the F-helix, there are conformational changes in the FG corner.

These tertiary structural changes are coupled to changes in quaternary structure.

The hemoglobin tetramer can be thought of as a pair of dimers: $\alpha_1\beta_1$ and $\alpha_2\beta_2$. In the allosteric change, the $\alpha_1\beta_1$ and $\alpha_2\beta_2$ interfaces retain their structure, as does a portion of the molecule adjoining these interfaces, including the B, C, G, and H regions of both subunits and the D-helix of the β subunit. The overall allosteric change involves a rotation of 15° of the $\alpha_1\beta_1$ dimer with respect to the $\alpha_2\beta_2$, around an axis approximately perpendicular to their interface. (The motion is like that of a pair of shears in which α_1 and α_2 are the blades and β_1 and β_2 are the handles.)

To understand the quaternary structural change, we must describe how the interface between the $\alpha_1\beta_1$ and $\alpha_2\beta_2$ dimers changes. The most important intersubunit contacts are between the α_1–β_2 and α_2–β_1 subunits. (In the open-shears image, the important variable contacts are between each blade and the opposite handle. The contacts between each blade and its own handle are—in hemoglobin as well as in shears—rigid.)

The interacting regions are the α_1 FG corner–β_2 C-helix and β_2 FG corner–α_1 C-helix. These are not identical. (However, these interactions are the same, by symmetry, as those of the α_2 FG corner–β_1 C-helix and β_1 FG corner–α_2 C-helix, respectively.)

The α_1 FG–β_2 C interaction is very similar in oxy and deoxy structures. Residues Arg 92 FG4, Asp 94 G1, and Pro 95 G2 of the α_1 subunit are in contact with Tyr 37 C3 and Arg 40 C6 in the β_2 subunit. In comparing oxy and deoxy structures, there are small conformational changes in these residues but the pattern of interactions is retained.

The other region of contact, β_2 FG corner–α_1 C-helix, differs substantially between oxy and deoxy structures (Fig. 9.10b). In the deoxy structure, His β_2 97 (FG4) packs between Thr α_1 41 (C6) and Pro α_1 (44) (CD2), and there is a hydrogen bond between the side chains of Asp β_2 99 (G1) and Tyr α_1 42 (C7). In the oxy structure, His β_2 97 packs between Thr α_1 38 (C3) and Thr α_1 41 (C6). (Because the C-helix is a 3_{10} helix, this corresponds to a jump of one turn relative to the His β_2 97 against which it packs. The shift in β_2 FG corner relative to α_1 C is approximately 6 Å.) The Asp–Tyr hydrogen bond is not made in the oxy structure.

This explains the two discrete quaternary states. The contact between the β_2 FG corner and the α_1 C-helix has two possible states, depending on the subunit shape presented by the tertiary structural state. The other contact, α_1 FG–β_2 C changes only slightly.

The requirement for the quaternary structural change arises from the tertiary structural changes; in particular, from the shifts that bring the FG corners of the α_1 and β_1 subunits in the $\alpha_1\beta_1$ dimer 2.5 Å closer together in the oxy structure relative to the deoxy structure (Fig. 9.11). One tertiary structural state of the $\alpha_1\beta_1$ and $\alpha_2\beta_2$ dimers can form a tetramer with one state of packing at the interface; the other tertiary structural state is compatible with the alternative packing.

Summarizing, starting from the deoxy structure, ligation of oxygen requires relief of strain around the heme group by shifting the F-helix and the FG corner. To accommodate these changes, there must be a set of tertiary structural changes that change the overall shape of the $\alpha_1\beta_1$ and $\alpha_2\beta_2$ dimers, notably the shifting of the relative positions of the FG corners. This destabilizes the deoxy quaternary structure because the dimers, having changed their shape, no longer fit together properly. Adopting the alternative quaternary structure requires the tertiary structural changes to take place even in subunits not yet liganded. As a result of the quaternary structural change, these unliganded subunits have been brought to a state of enhanced oxygen affinity.

The work of Baldwin and Chothia, based on crystal structures of the R and T states of hemoglobin, gave a description of the structural differences between the *endpoints* of the allosteric transition in hemoglobin. The detailed mechanism of allostery and cooperativity in hemoglobin requires experiments or computations that access the pathway between the end states. Investigation of hemoglobins

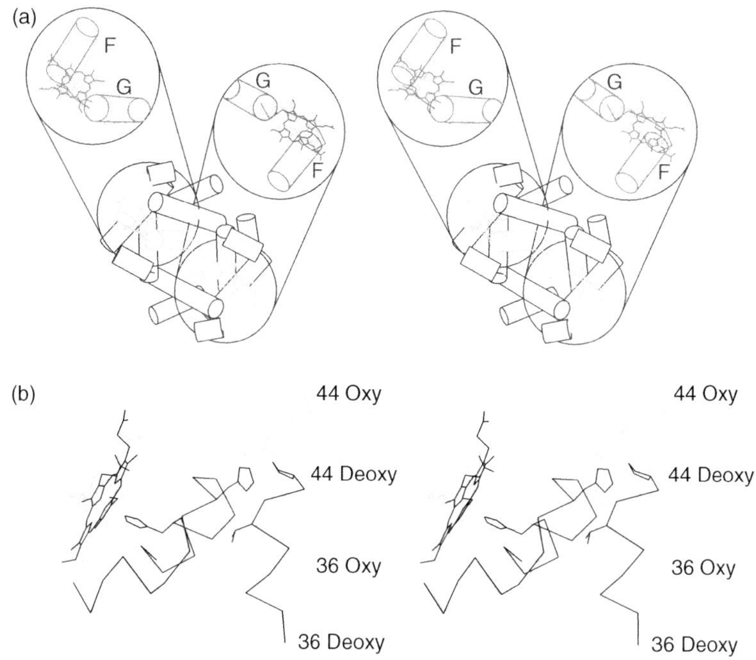

Figure 9.11 Some important structural differences between oxy- and deoxyhemoglobin [1hho,2hhb]. (a) Main picture: the $\alpha 1\beta 1$ dimer of human hemoglobin. The subsidiary pictures (circles) show F-helix, FG corner, G-helix, and heme group of the oxy form in light gray and deoxy in dark gray. The oxy and deoxy $\alpha_1\beta_1$ dimers have been superposed on their interface; in this frame of reference, there is a small shift in the heme groups, and a shift and conformational change in the FG corners. (b) Alternative packings of α_1 and β_1 subunits in oxyhemoglobin (gray) and deoxyhemoglobin (black). The oxy and deoxy structures have been superposed on the F- and G-helices of the α_1 monomer. Although for purposes of this illustration we have regarded the α_1 subunit as fixed and the β_2 subunit as mobile, only the relative motion is significant.

from various sources continues to be an active field of research (Bellelli and Brunori, 2011; Peracchi and Mozzarelli, 2011).

Powerful experimental and computational tools now make possible an exploration of the R and T ensembles with respect not only to enthalpic properties, as deducible from X-ray structures, but also to entropic contributions, as extracted from measurement of protein dynamics. Hemoglobin remains an ideal testing ground for refined thermodynamic and kinetic models of ligand binding.

In a recent important paper appearing after completion of this manuscript, Fischer et al. (2011) computed a pathway for the allosteric change in adult human hemoglobin ($\alpha_2\beta_2$). Their results indicate that the structural change has two components, with different time constants, and they give a detailed characterization of the intermediates along the pathway.

9.6 HEMOGLOBINOPATHIES—"MOLECULAR DISEASES" CAUSED BY ABNORMAL HEMOGLOBINS

Proteins with altered sequences can have altered structures and altered functions. Mutations associated with disease are collected by the organization, *Online Mendelian Inheritance in Man (OMIM)*™ (http://www/ncbi.nlm.nih.gov/omim).

9.6.1 Sickle-Cell Anemia

The first disease recognized to arise from a protein of altered structure was sickle-cell anemia. L. Pauling and coworkers showed in 1949 that hemoglobin isolated from patients of sickle-cell anemia differed in electric charge from normal hemoglobin (Pauling et al., 1949). This was the first evidence that genes precisely control the structures of proteins. Recognize that Pauling's work preceded the first determination of the amino acid sequence of a protein.

The cause of sickle-cell anemia is a single amino acid substitution in the β chain of hemoglobin. A charged residue, $\beta6$ glutamic acid, changes to a neutral one, valine. As a result, a patch on the surface of the molecule becomes electrically neutral and "sticky." The hemoglobin molecules can clump together. In the unligated or deoxy state (in the deoxy form, typical of venous blood, hemoglobin is not binding oxygen), the mutant hemoglobin forms polymers within the erythrocyte. To flow through small capillaries, erythrocytes must be deformable, as their typical size, 7.8 μm (in humans), is larger than the diameter of small capillaries. The formation of the polymers has a rigidifying effect on the erythrocytes, impeding their flow and blocking capillaries. In the traffic jam building up behind a plugged capillary, arriving red cells release their oxygen to surrounding hypoxic tissues, become deoxygenated, and thereby aggravate the problem.

9.6.2 Thalassaemias

Thalassaemias are genetic diseases associated with defective or deleted hemoglobin genes (Weatherall et al., 2010).

Most Caucasians have four genes for the α chain of normal adult hemoglobin, two alleles of each of the two tandem genes α_1 and α_2. Therefore, α-thalassaemias can present clinically in different degrees of severity, depending on how many genes express normal α chains. Only deletions leaving fewer than two active genes present as symptomatic under normal conditions. Observed genetic defects include deletions of both genes (a process made more likely by the tandem gene arrangement and sequence repetition, which make crossing over more likely) and loss of chain termination leading to transcriptional "read through," creating extended polypeptide chains that are unstable.

β-Thalassaemias are usually point mutations. These may be

- missense mutations (amino acid substitutions);
- nonsense mutations (changes from a triplet coding for an amino acid to a stop codon) leading to premature termination and a truncated protein;

- mutations in splice sites;
- mutations in regulatory regions;
- certain deletions including the normal termination codon and the intergenic region between δ and β genes that create $\delta-\beta$ fusion proteins.

9.6.3 Other Human Hemoglobin Mutants

Many other mutants are known. For adult hemoglobin, OMIM currently lists 294 mutants of the α chain and 538 of the β chain. Most but not all of these are single-nucleotide polymorphisms. In most cases, the number of affected people is very small, compared to the prevalence of sickle-cell anemia and thalassemias. Many of these mutants have minimal or no observed clinical consequences.

9.7 CONCLUSIONS

Globins have been at the forefront of many important developments in molecular biology, from the first identification of proteins as a class of biochemicals, through structure and sequence determinations, to the mechanism of function in general and allosteric regulation in particular, to molecular disease and clinical applications, to appreciation of protein evolution and the diversity of protein families, and to the control of expression. Now—largely as the result of methods and insights developed originally with the globins—they are but one protein family among many others. But just as the discovery that the Earth is not the center of the heavens enriched rather than impoverished astronomy, we are confident that the chemistry and biology of globins have not exhausted the messages that we remain poised to receive.

REFERENCES

Baldwin, J. and Chothia, C. (1979) Haemoglobin: the structural changes related to ligand binding and its allosteric mechanism. J Mol Biol, 129, 175–220.

Bellelli, A. and Brunori, M. (2011) Hemoglobin allostery: variations on the theme. Biochim Biophys Acta, 1807, 1262–1272.

Berzelius, J.J. (1840) Lehrbuch der Chemie, 3rd. edn, Arnold, Dresden.

Bonamore, A. and Boffi, A. (2008) Flavohemoglobin: structure and reactivity. IUBMB Life, 60, 19–28.

Brunori, M. (2010) 1960 *Annus mirabilis*: the birth of structural biology. Rend Fis Acc Lincei, 21, 335–342.

Burr, A.H., Hunt, P., Wagar, D.R., Dewilde, S., Blaxter, M.L., Vanfleteren, J.R., and Moens, L. (2000) A hemoglobin with an optical function. J Biol Chem, 275, 4810–4815.

Chothia, C., Levitt, M., and Richardson, D. (1977) Structure of proteins: packing of α-helices and pleated sheets. Proc Natl Acad Sci USA, 74, 4130–4134.

Chothia, C., Levitt, M., and Richardson, D. (1981) Helix to helix packing in proteins. J Mol Biol, 145, 215–250.

Fischer, S., Olsen, K.W., Nam, K., and Karplus, M. (2011) The unsuspected pathway of the allosteric transition in hemoglobin. Proc Natl Acad Sci USA, 108, 5608–5613.

Freitas, T.A., Hou, S., and Alam, M. (2003) The diversity of globin-coupled sensors. FEBS Lett, 552, 99–104.

Gardner, P.R. (2005) Nitric oxide dioxygenase function and mechanism of flavohemoglobin, hemoglobin, myoglobin and their associated reductases. J Inorg Biochem, 99, 247–266.

Garry, D.J., Meeson, A., Yan, Z., and Williams, R.S. (2000) Life without myoglobin. Cell Mol Life Sci, 57, 896–898.

Gros, G., Wittenberg, B.A., and Jue, T. (2010) Myoglobin's old and new clothes: from molecular structure to function in living cells. J Exp Biol, 213, 2713–2725.

Holm, L. and Sander, C. (1993) Protein structure comparison by alignment of distance matrices. J Mol Biol, 233, 123–138.

Holmes, F.L. (1995) Crystals and carriers: the chemical and physiological identification of hemoglobin, in *No Truth Except in the Details*: *Essays in Honor of Martin J. Klein* (eds A.J. Kox and D.M. Siegel), Kluwer Academic Publishers, Dordrecht, pp. 191–243.

Hoogewijs, D., Dewilde, S., Vierstraete, A., Moens, L., and Vinogradov, S.N. (2012a) A phylogenetic analysis of the globins in fungi. PLoS One, 7, e31856.

Hoogewijs, D., Ebner, B., Germani, F., Hoffmann, F.G., Fabrizius, A., Moens, L., Burmester, T., Dewilde, S., Storz, J.F., Vinogradov, S.N., and Hankeln, T. (2012b) Androglobin: a chimeric globin in metazoans that is preferentially expressed in mammalian testes. Mol Biol Evol, 29, 1105–1114.

Imai, K. (1982) Allosteric Effects in Haemoglobin, Cambridge University Press, Cambridge (MA), pp. 129–137.

Iwaasa, H., Takagi, T., and Shikama, K. (1989) Protozoan myoglobin from *Paramecium caudatum*. Its unusual amino acid sequence. J Mol Biol, 208, 355–358.

Kakar, S., Hoffman, F.G., Storz, J.F., Fabian, M., and Hargrove, M.S. (2010) Structure and reactivity of hexacoordinate hemoglobins. Biophys Chem, 152, 1–14.

Lesk, A.M. and Chothia, C. (1980) How different amino acid sequences determine similar protein structures: The structure and evolutionary dynamics of the globins. J Mol Biol, 136, 225–270.

Monod, J., Wyman, J., and Changeux, J.-P. (1965) On the nature of allosteric transitions: a plausible model. J Mol Biol, 12, 88–118.

Nardini, M., Pesce, A., Milani, M., and Bolognesi, M. (2007) Protein fold and structure in the truncated (2/2) globin family. Gene, 398, 2–11.

Pastore, A. and Lesk, A.M. (1990) Comparison of the structures of globins and phycocyanins: evidence for evolutionary relationship. Proteins, 8, 133–155.

Pauling, L., Itano, H.A., Singer, S.J., and Wells, I.C. (1949) Sickle cell anemia, a molecular disease. Science, 110, 543–548.

Peracchi, A. and Mozzarelli, A. (2011) Exploring and exploiting allostery: models, evolution, and drug targeting. Biochim Biophys Acta, 1814, 922–933.

Perutz, M.F. and Mathews, F.S. (1966) An X-ray study of azide methaemoglobin. J Mol Biol, 21, 199–202.

Reichert, E.T. and Brown, A.P. (1909) The Differentiation and Specificity of Corresponding Proteins and other Vital Substances in Relation to Biological Classification and Organic Evolution: The Crystallography of Hemoglobins, Carnegie Institute, Washington.

Royer, W.E. Jr., Zhu, H., Gorr, T.A., Flores, J.F., and Knapp, J.E. (2005) Allosteric hemoglobin assembly: diversity and similarity. J Biol Chem, 280, 27477–27480.

Storz, J.F., Opazo, J.C., and Hoffmann, F.G. (2011) Phylogenetic diversification of the globin gene superfamily in chordates. IUBMB Life, 63, 313–322.

Tilleman, L., Germani, F., De Henau, S., Geuens, E., Hoogewijs, D., Braeckman, B.P., Vanfleteren, J.R., Moens, L., and Dewilde, S. (2011) Globins in *Caenorhabditis elegans*. IUBMB Life, 63, 166–174.

Vinogradov, S.N., Hoogewijs, D., Bailly, X., Arredondo-Peter, R., Gough, J., Guertin, M., Dewilde, S., Moens, L., and Vanfleteren, J.R. (2005) Three globin lineages belonging to two structural classes in genomes from the three kingdoms of life. Proc Natl Acad Sci USA, 102, 11385–11389.

Vinogradov, S.N., Hoogewijs, D., Bailly, X., Mizuguchi, K., Dewilde, S., Moens, L., and Vanfleteren, J.R. (2007) A model of globin evolution. Gene, 398, 132–142.

Vinogradov, S.N., Fernandez, I., Hoogewijs, D., and Arredondo-Peter, R. (2011) Phylogenetic relationships of 3/3 and 2/2 hemoglobins in Archaeplastida genomes to bacterial and other eukaryote hemoglobins. Mol Plant, 4, 42–58.

Vuletich, D.A. and Lecomte, J.T.J. (2008) The phylogeny and structural properties of 2/2 haemoglobins, in Protein Reviews, Dioxygen binding and sensing proteins, Vol. 9 (eds M. Bolognesi, G. Di Prisco, and C. Verde), Springer-Verlag Italia, Italy, pp. 31–43.

Weatherall, D.J., Williams, T.N., Allen, S.J., and O'Donnell, A. (2010) The population genetics and dynamics of the thalassemias. Hematol Oncol Clin North Am, 24, 1021–1031.

Weber, R.E. and Vinogradov, S.N. (2001) Nonvertebrate hemoglobins: functions and molecular adaptations. Physiol Rev, 81, 569–628.

Wittenberg, J.B. and Wittenberg, B.A. (2003) Myoglobin function reassessed. J Exp Biol, 206, 2011–2020.

SECTION III

REVIEW OF PROTEIN FAMILIES IN IMPORTANT BIOLOGICAL SYSTEMS

10

FUNCTIONAL ADAPTATION AND PLASTICITY IN CYTOSKELETAL PROTEIN DOMAINS: LESSONS FROM THE ERYTHROCYTE MODEL

ANTHONY J. BAINES
School of Biosciences, University of Kent, Canterbury, UK

CHAPTER SUMMARY

The evolution of the cytoskeleton in eukaryotic cells has frequently been accompanied by sequential duplication and adaptation of pre-existing folding modules, many of which arose in prokaryotes. In this chapter, I shall discuss the nature and adaptation of some common protein domains found in cytoskeletal proteins, taking as a theme, the cytoskeleton of one of the most thoroughly understood biological systems, namely the mammalian red blood cell. Mammalian red blood cells have adapted to survive the rigours of circulation in the absence of major biosynthetic repair mechanisms through the presence of a plasma membrane-associated cytoskeleton. This is organized around short actin filaments, which are linked to the membrane via adaptor and cross-linking proteins that include spectrin, ankyrin and protein 4.1. Each of these has a modular structure comprising domains that are widely used in eukaryotic systems: among these are spectrin triple helical bundles, calponin homology domains, ankyrin repeats, and four point one, ezrin, radixin, moesin

Protein Families: Relating Protein Sequence, Structure, and Function, First Edition.
Edited by Christine Orengo and Alex Bateman.

(FERM) domains. Lessons learnt from their structural and functional adaptations have informed understanding of the cytoskeleton and its evolution through the adaptation of pre-existing folding structures to new functions.

10.1 INTRODUCTION

The eukaryotic cytoskeleton contains several large filament systems—microtubules, actin filaments, intermediate filaments, and myosin filaments. Connecting these to other cellular structures (including organelle and plasma membranes, and chromosomes) are a diverse range of linker and motor proteins that make up the cytoskeletal infrastructure of the cell. The cytoskeleton is essential for the stability of animal membranes against the forces generated by movement, for transport around the cell, for organising signaling and cell adhesion systems, and for organizing intracellular compartments.

Many of the wide range of linker and motor proteins that connect to the major filament systems have evolved by the duplication and adaptation of protein domains that arose in early eukaryotic lineages, or that were inherited from prokaryotic ancestors. Thus, the cytoskeleton has been an important source for the discovery of protein domains, and also for the discovery of the evolutionary principles that allow old folds to adapt to new functions.

One of the most thoroughly understood cytoskeletal systems is that of the erythrocyte plasma membrane (for recent reviews see Baines, 2010b; Mohandas and Gallagher, 2008). As it became possible to isolate high quality, structurally intact erythrocyte membranes (ghosts) from human blood samples (Dodge et al., 1963), in the past 50 years, erythrocytes have been the focus of intense biochemical investigation, both because, human hereditary anemia can arise from defects in the membrane, and also because erythrocytes are a unique source of pure human membranes that can be isolated in "biochemical quantities."

Mammalian red blood cells have evolved as enucleated remnants from an asymmetric cell division that gives rise to a plasma membrane that acts as a "carrier bag" for proteins and enzymes of gas transport. These have a lifetime in circulation of about 120 days, and must survive in the absence of major biosynthetic repair mechanisms. Even though, at the level of resolution of the light microscope, there are no substantial structures within the cell to stop it breaking up under the shearing forces imposed by circulation, the delicate lipid bilayer has extraordinary resilience.

A key observation in understanding the basis for red cell survival is that, while erythrocyte ghosts can be isolated intact in hypotonic solutions (e.g., 5 mM sodium phosphate, pH 7.4), extraction of ghosts with very low salt solutions (essentially distilled water with just enough buffer to maintain a slightly alkaline pH) leads to rapid endovesiculation and loss of membrane integrity (Steck and Kant, 1974). Several peripheral membrane proteins become soluble under these conditions, among them being actin (Tilney and Detmers, 1975) and a number of actin-binding proteins including spectrin, protein 4.1R (Tyler et al., 1979; Ungewickell et al., 1979) and adducin (Gardner and Bennett, 1987).

Figure 10.1 summarizes in schematic form current view of the red cell cytoskeleton.

Erythrocyte membranes contain short filaments of F-actin (~12–16 monomers per filament) (Pinder and Gratzer, 1983) joined to the membrane by a number of adapter proteins. The filaments are cross-linked (Cohen et al., 1980) by a long fibrous protein named *spectrin* (Marchesi and Steers, 1968) (the name comes from the observation that it is abundant in ghosts—another name for a ghost is a spectre, hence spectrin).

In vivo, spectrin is primarily tetrameric, and composed of elongated dimers that associate head-to-head. Each dimer contains an α-spectrin and a β-spectrin chain, arranged side-by-side and antiparallel. Each tetramer is therefore an $(\alpha\beta)_2$ heterotetramer. As we shall see later, erythrocyte spectrin tetramers can dissociate *in vivo*, and it appears that the dimer–tetramer interconversion in circulating red cells is dynamic. Smaller amounts of higher order oligomers ($(\alpha\beta)_3$, $(\alpha\beta)_4$, and so on) can form (Morrow et al., 1981; Shahbakhti and Gratzer, 1986).

Approximately 5–7 spectrin molecules bind to each short F-actin filament, so electron microscopy reveals a roughly hexagonal array on the cytoplasmic face of the erythrocyte plasma membrane of spectrin molecules cross-linked at junction points that contain actin filaments (Byers and Branton, 1985; Liu et al., 1987; Ursitti and Fowler, 1994), (Fig. 10.1a).

Both spectrin and actin are peripheral membrane proteins, and require adapters to link them to the membrane. A major mode of membrane linkage for spectrin is the protein ankyrin that links to a macro-complex of transmembrane proteins that forms a metabolon dedicated to carbon dioxide transport (Bennett and Stenbuck, 1979; Bruce et al., 2003), (Fig. 10.1b).

The interaction of spectrin with F-actin is weak in the absence of the regulatory adapters, proteins 4.1R, and adducin. Each of these promotes the interaction of spectrin with actin (Gardner and Bennett, 1987; Ungewickell et al., 1979), and this activity is regulated by phosphorylation (Chao and Tao, 1991; Gardner and Bennett, 1987; Gauthier et al., 2011). Proteins 4.1R and adducin also bind directly to a number of transmembrane proteins, and thus form a mechanism for linking F-actin to the membrane (Anong et al., 2009; Khan et al., 2008; Salomao et al., 2008). The adducin–spectrin pair also acts as a plus-end capping complex on actin filaments (Kuhlman et al., 1996), and the minus end of the actin filament is capped by tropomodulin (Fowler, 1990). Tropomyosin lies along the side of the filament (the length of the filament can accommodate a single tropomyosin) (Fowler and Bennett, 1984) (Fig. 10.1c).

Spectrin was one of the earliest nonmuscle actin-binding proteins to be identified (Marchesi and Steers, 1968). It was also one of the earliest nonmotor cytoskeletal proteins to be subjected to limited proteolysis to define its domain structure (e.g., Speicher et al., 1980). Limited proteolysis has been a powerful tool for the dissection of folding elements within the overall protein structure (Neurath, 1980). Each of spectrin (Karinch et al., 1990; Speicher and Marchesi, 1982;

(a)

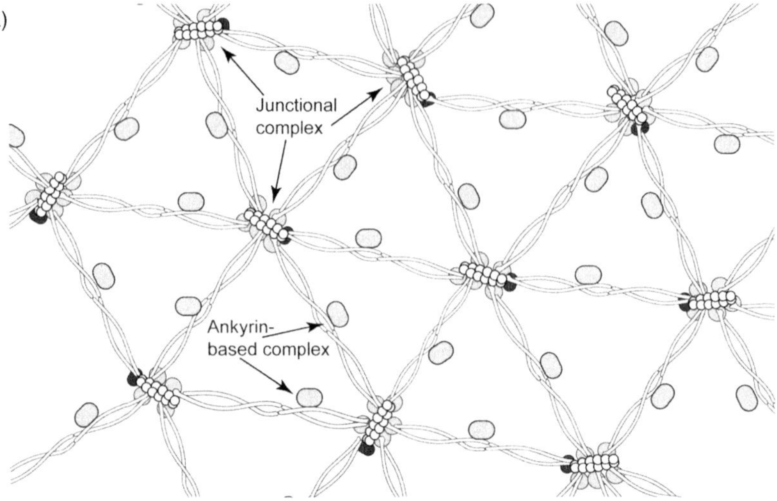

Figure 10.1 The red cell cytoskeleton. (a) A schematic representation of the roughly hexagonal array of protein filaments arranged on the cytoplasmic face of the membrane. The long filaments are spectrin tetramers (*in vivo*, there is a smaller proportion of higher order oligomers) cross-linked at junction points that contain F-actin. These junctional complexes form a link between spectrin and the membrane. Spectrin is also linked to the plasma membrane via the ankyrin-based complex. (b) The ankyrin-based complex. Ankyrin coordinates a complex of membrane proteins that include the anion exchanger (AE1 or band 3), the rhesus proteins Rh and RhAG, as well as LW, CD47 and glycophorins A and B. Protein 4.2 is also present as a link between membrane proteins and the cytoskeleton. Note also that AE1 binds numerous proteins on cytoplasmic face including carbonic anhydrase II, several glycolytic enzymes (aldolase, phosphofructokinase, glyceraldehyde-3-phosphate dehydrogenase) and deoxyhaemoglobin (d-Hb). As noted in the text, the complex of Rh/RhAG, AE1 and carbonic anhydrase has been hypothesised to form a metabolon for carbon dioxide transport (Bruce et al., 2003). Recently, competitive interactions between deoxyhaemoglobin and glycolytic enzymes have been suggested to control glycolytic flux in relation to oxygen tension (Lewis et al., 2009): this allows regulation of the production of 2,3-bisphosphoglycerate, which controls the interaction of haemoglobin with oxygen. For simplicity, spectrin is not shown in this figure, but it can bind ankyrin via the β-chain, and 4.2 via the α-chain. (c) The junctional complex. A short filament of F-actin (approximately 12–16 monomers) is approximately the same length as a single tropomyosin. The minus end is capped by tropomodulin. A complex of spectrin (not shown) and adducin caps the plus end. Adducin and protein 4.9 (dematin) can interact with the glucose transporter Glut1. Adducin also interacts with the anion exchanger. Several molecules of protein 4.1 are also present in the complex and they can link to proteins such as Duffy, Rh, glycophorin C (XK/kell complex, not sown), and also the anion exchanger. For simplicity, spectrin is not shown in this figure, but at junction points approximately 5–7 spectrin molecules interact with each actin filament (see (a)).

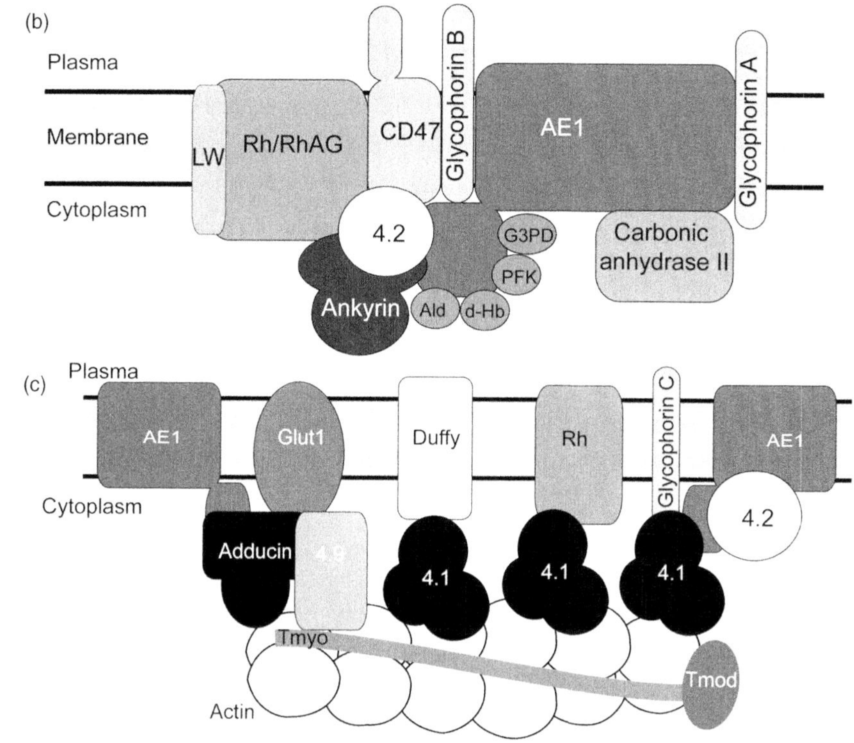

Figure 10.1 (*Continued*)

Speicher et al., 1982), ankyrin (Bennett, 1978; Hall and Bennett, 1987; Weaver and Marchesi, 1984; Weaver et al., 1984), protein 4.1R (Correas et al., 1986; Hall, 1993; Leto and Marchesi, 1984), and adducin (Joshi and Bennett, 1990) has been subjected to extensive ranges of experiments using, principally, trypsin or chymotrypsin to dissect out protease-resistant folded structures. These experiments have led to the definition of numerous functional and regulatory domains that are now recognised as widely used throughout eukaryotic cytoskeletons.

Among the domains discovered in the erythrocyte cytoskeleton were the spectrin-type triple helical repeat domain, the FERM (supra-)domain (in 4.1R), ankyrin repeats and the calponin homology domain.

These proteins are not just expressed in red blood cells. As we shall see later, they evolved in the simplest animals, and are expressed in every animal tissue. The β-spectrin found in erythrocytes represents a type of spectrin polypeptide found in all animals. A second type of β is encoded in all animals in a separate gene: this is a high molecular mass form, known as β_{Heavy}*-spectrin*. Spectrin tetramers can form from α chains interacting with either a conventional β-spectrin, or with β_{Heavy}.

Given the importance of the erythrocyte paradigm in the history of the discovery of protein domains, this chapter will focus on the lessons that have been drawn from their analyses. Key themes that will be discussed include functional flexibility in the adaptation of a single fold, and the significance of folded domains as targets for multiple converging modes of regulation.

10.2 THE SPECTRIN SUPERFAMILY: COMMON FOLDING UNITS ADAPTED TO VARYING FUNCTIONS

Sequence analysis arising from the cDNA cloning of multiple cytoskeletal proteins revealed a superfamily of actin-binding proteins conserved among all animals, whose prototype was spectrin (Baron et al., 1987; Byers et al., 1989; Karinch et al., 1990; Levine et al., 1992; Winder et al., 1995). Members of the spectrin superfamily include α-actinin, dystrophin, utrophin, and a group of giant proteins collectively known as *spectraplakins*, ACF7/MACF1, dystonin/BPAG1 (Bennett and Baines, 2001; Roper et al., 2002).

Structures of example members of the spectrin superfamily are summarized in Figure. 10.2 and Table 10.1.

Although spectrins evolved as single genes (invertebrates have one α-spectrin, one β-spectrin, and one β_{Heavy}-spectrin), with the advent of vertebrates, two whole-genome genome duplications (Blomme et al., 2006; Dehal and Boore, 2005) resulted in four β-spectrin genes in all vertebrates. The same applies to α-actinins. However, retention of all the paralogues in the other genes was evidently not advantageous, so only one α-spectrin, one β_{Heavy}-spectrin, and pairs of dytrophin/utrophin and spectraplakins remain. One further gene duplicaton occurred with the advent of mammals: the α-spectrin gene was duplicated giving rise to αI-spectrin (the form abundant in erythrocytes) and αII-spectrin (the major tissue form) (Baines, 2009, 2010a; Salomao et al., 2006).

The defining features of spectrin superfamily proteins are an actin-binding domain comprising a pair of calponin homology domains (CH1 and CH2), as well as numerous triple helical repeats of about 106 amino acids known as spectrin repeats. In addition, 2–4 EF hands that form a calmodulin-like domain are present in canonical members of the superfamily. Note that in the prototype, spectrin, not all these domains are present in a single polypeptide, but the quaternary structure of the whole protein includes them all (Baines, 2010b; Bennett and Baines, 2001).

Additional domains are characteristic of each group: α-spectrins contain an SH3 domain; β- and β_{Heavy} spectrins contain pleckstrin homology domains; β_{Heavy}-spectrins also contain an SH3-like domain; dystrophin/utrophin contains WW and ZZ domains; spectraplakins contain GAS2 domains. Only α-actinin contains the minimal complement of CH, spectrin repeats and calmodulin-like domain.

Further proteins related to the family include nesprins and plectin (Allen and Shah, 1999; Warren et al., 2005), but as these lack some of the canonical features of the spectrin superfamily they are not discussed here in any detail.

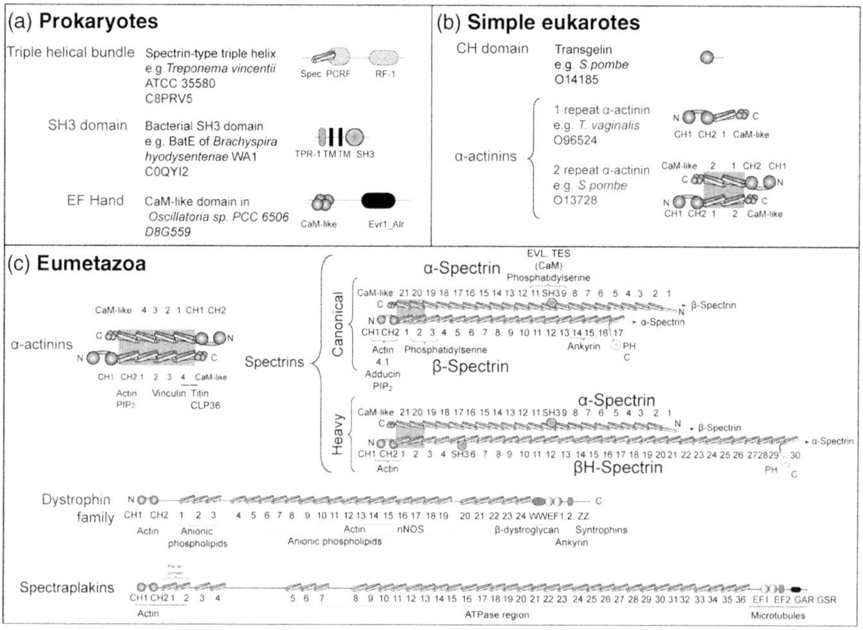

Figure 10.2 The spectrin superfamily, its domains and hints at an evolutionary lineage. (a) Domains characteristic of the spectrin superfamily are found in prokaryotes. Although their origin—whether they evolved in prokaryotes or were acquired by horizontal gene transfer—is not always clear, spectrin triple helices, SH3 and calmodulin-like domains are found in prokaryotes. (b) Calponin homology (CH) domains appear to have evolved with the eukaryotes. They can be found in single copies (for example, in transgelin), or in the simplest of the spectrin superfamily proteins, α-actin. α-actinin contains 1–4 spectrin triple helices, a pair of CH domains, and a calmodulin-like domain. (c) The metazoan spectrin superfamily. Metazoa have α-actinins with four spectrin triple helices, as well as pairs of CH domains and calmodulin-like domains. In spectrin, α chains can be paired with either conventional β chains (distinguished by 16 full triple helices) or β heavy-spectrin chains (30 triple helices) (drawn after Baines, 2009). Further members of the family include dystrophin (and its vertebrate paralogue utrophin), and the spectraplakins (drawn after respectively Mirza et al., 2010; Wu et al., 2008). These all contain the characteristic CH domain pairs, multiple triple helices and EF hands. In addition, they contain further domains characteristic to each protein.

10.3 ORIGINS OF DOMAINS IN THE SPECTRIN LINEAGE

The characteristic domains that define spectrins are calponin homology domains, spectrin triple helices, SH3 domains, and four tandem EF-hand domains (calmodulin-like domain) (Bennett and Baines, 2001) (Fig. 10.2). All of these have origins in simpler organisms than metazoa, so presumably they have been combined during evolution to form proteins with new functions.

TABLE 10.1 Human Members of the Spectrin Superfamily, Ankyrin and Protein 4.1

Spectrin Polypeptide	Gene	Uniprot Accession	Comments
αI-spectrin	SPTA1	P02549	Can form tetramer with any β-spectrin. Erythrocyte spectrin is αI-β I
αII-spectrin	SPTAN1	Q13813	Most abundant form in cells other than red blood cells. Can form tetramer with any β-spectrin
β I-spectrin	SPTB	P11277	Erythrocyte spectrin is αI-β I
β II-spectrin	SPTBN1	Q01082	Most abundant form in cells other than red blood cells. Can form tetramer with any α-spectrin
β III-spectrin	SPTBN2	O15020	Can form tetramer with any α-spectrin
β IV-spectrin	SPTBN4	Q9H254	Can form tetramer with any α-spectrin
β V-spectrin	SPTBN5	Q9NRC6	β_{Heavy}-spectrin polypeptide. Most frequently forms tetramers with αII-Spectrin
α-actinin 1	ACTN1	P12814	
α-actinin 2	ACTN2	P35609	
α-actinin 3	ACTN3	Q08043	
α-actinin 4	ACTN4	O43707	
Dystrophin	DMD	P11532	
Utrophin	UTRN	P46939	
ACF7/MACF1	MACF1	E9PJT0	Spectraplakin
Dystonin/BPAG1	DST	Q03001	Spectraplakin
Ankyrin polypeptide	Gene	Uniprot accession	Comments
Ankyrin-R	ANK1	P16157	Erythrocyte ankyrin
Ankyrin-B	ANK2	Q01484	
Ankyrin-G	ANK3	Q12955	
Protein 4.1 polypeptide	Gene	Uniprot accession	Comments
4.1R	EPB41	P11171	Erythrocyte 4.1
4.1N	EPB41L1	Q9H4G0	
4.1G	EPB41L2	O43491	
4.1B	EPB41L3	Q9Y2J2	

An attractive hypothesis for the origin of eukaryotes is that simple eukaryotes are formed by the fusion of a eubacterium and an archean (Margulis et al., 2000). In this hypothesis, the last eukaryotic common ancestors derived from organisms are considered to be similar to the present day spirochetes and thermoplasmas. Genetic evidence indicates that many of the proteins of the motility apparatus derive from the spirochete ancestor (Margulis et al., 2006).

The genomes of a number of spirochetes are now available, and it is possible to examine predicted proteins available in Uniprot (Bairoch et al., 2005) for the presence of spectrin-related domains using Pfam (Bateman et al., 2002) hidden Markov models (HMMs). (See Chapter 2 and Chapter 4).

EF hands are widely represented in both eubacteria and archea. Indeed, small proteins with four tandem EF-hand domains are found in multiple bacterial lineages, and these may be the ancestor of a domain common to the spectrin superfamily. Moreover, there are examples of large bacterial proteins that contain a calmodulin-like domain. For example, Uniprot:D8G559 from the cyanobacterium *Oscillatoria* sp. PCC 6506 contains an N-terminal calmodulin-like domain that represents about 1/8 of the entire protein. However, Ponting et al. (1999) suggested that EF hands may have been acquired by horizontal gene transfer, with some bacterial lineages being particularly acquisitive of such domains.

Similarly, although spectrin-type triple helical bundles are also represented in bacteria and archea, the distribution is overwhelmingly in eukaryotes (Pfam annotates their presence in only four bacterial species, compared to over 200 eukaryotic species). Spectrin repeats may possibly be a eukarytic invention, acquired by some bacteria via horizontal transfer.

The only exception to a potential bacterial origin for the defining domains of the spectrin superfamily is the calponin homology domain. Pfam annotates no prokaryotic proteins with the CH domain HMM (PF00307). However, CH domains seem to be common to all eukaryotes as they are annotated in all branches. A simple hypothesis might be that CH domains are a eukaryotic invention, although the four helical orthogonal bundles could, conceptually, be an adaptation of an earlier structure.

In any case, the appearance of the CH domain seems to be accompanied by the presence of several actin-binding proteins characterised by pairs of CH domains in all eukaryotic genomes available. α-actinin family members are present throughout the fungi and metazoa (Virel and Backman, 2004). These proteins contain a pair of CH domains in the N-terminal region, one, two, or four spectrin-type triple helical repeats and a calmodulin-like domain near the C-terminus.

The triple helical domains of α-actinin seem to have arisen by sequential duplication of a single domain within an archetypal α-actinin gene (Virel and Backman, 2004, 2007). The simplest α-actinins, such as those from *Trichomonas vaginalis* contain just one helical region, although it is by no means certain that this folds into a triple helix (Virel and Backman, 2007). Throughout the fungi, α-actinins generally have two triple helical domains (for example, in the *Schizosaccharomyces pombe* Ain1 protein (Wu et al., 2001)). A subsequent intragenic duplication seems to have given rise to the four helical rod domain of choanoflagellate and metazoan α-actinins.

Choanoflagellates such as *Monosiga brevicollis* represent a lineage of colonial protists from which it is thought that the metazoa are descended (King, 2005). The genome of *M. brevicollis* encodes larger and more complex structures containing spectrin-type triple helical bundles (Baines, 2009). In addition to proteins that are similar to animal α-actinins (Uniprot:A9V1P2_MONBE,

A9V6X3_MONBE), there are genes encoding an α-spectrin (A9UST9_MONBE), a β-spectrin (A9UR27_MONBE), and a β-heavy-spectrin (A9UST0_MONBE). As the spectrins are ubiquitous in metazoa, and essential for normal metazoan physiology, it might be conjectured that the evolution of spectrin was an essential prerequisite—although, clearly, not sufficient—for the subsequent appearance of metazoa (Baines, 2009, 2010a).

On the other hand, sequences representing ankyrin, protein 4.1 or adducin only appear in their fully developed forms in true metazoa (Baines, 2010a). Thus, it seems likely that spectrin proteins appeared originally as actin-binding or cross-linking proteins, and only achieved their potential in tissue formation with the appearance of binding partners as tissues evolved.

Of the other branches of the spectrin superfamily, the lineages leading to current vertebrate dystrophin and the spectraplakin branches seem, similarly, to have arisen with the evolution of tissues. The genomes of Cnidaria, some of the simplest animals with tissues, contain representatives of these branches.

In summary, the spectrin superfamily seems to have arisen from combinations of domains; some of these may have appeared very early in evolution, others appeared only with the eukaryotes. With the advent of the eukaryotes, actin cross-linking proteins, such as α-actinin, appeared. These were further adapted, presumably by intragenic duplications, to form spectrins as the precursors of animals appeared. Subsequently, with the appearance of tissues, the large actin-binding proteins that include dystrophin and the spectraplakins evolved.

10.4 SPECTRIN TRIPLE HELICES—A COMMON FOLD BUT WITH MANY FUNCTIONAL ADAPTATIONS

The triple helical bundle of about 106 amino acids that defines a spectrin-type triple helix was first identified in spectrin in the 1980s. Direct protein sequencing by Speicher and Marchesi (1984) revealed repeats of this size that were predicted to form triple helices. Subsequent nuclear magentic resonance (NMR) and X-ray analysis of recombinant fragments of spectrin fulfilled this prediction (Pascual et al., 1996; Yan et al., 1993). The three helices (named A, B, and C in the direction from N to C) are linked by turns, and so fold back on each other. (Scop: Class: All alpha proteins [46456]; Fold: Spectrin repeat-like [46965]; Superfamily: Spectrin repeat [46966]. 15 other superfamilies in the Scop (Murzin et al., 1995) database are also classified with this fold).

At the time of writing, the only published full structure of a spectrin superfamily protein is that of chicken α-actinin (Liu et al., 2004). Figure 10.3a shows the structure revealed by electron microscopic analysis. Four tandem triple helical bundles are arranged end to end with an actin-binding domain at the N-terminus comprising two CH domains, and four EF hands in a calmodulin-like domain in the C-terminal region.

When large numbers of triple helical sequences became available, it was apparent that very few residues were conserved throughout. Indeed, in the large

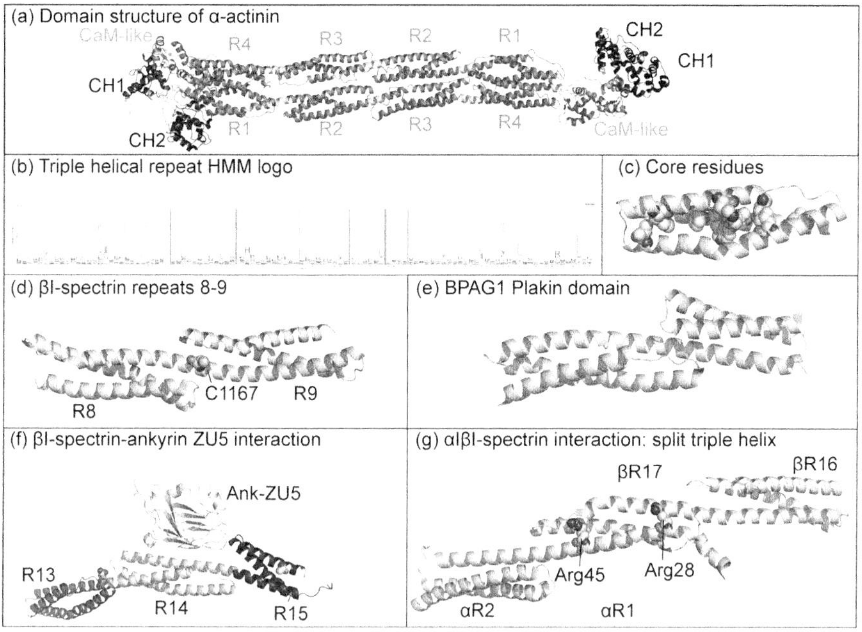

Figure 10.3 Triple helical repeats. (a) The structure of chicken α-actinin. This protein contains four triple helices linked end to end via short linkers: this structure forms an antiparallel dimer that makes up the central rod domain (PDB: 1SJJ). (b). Multiple sequence alignment of many triple helices reveal that very few residues are fully conserved. The most highly conserved are typically hydrophobic, as indicated in this sequence logo. (c) Residues highly conserved as displayed in the sequence logo (b) are shown on a molecular graphic of a triple helix (PDB: 1U4Q) and these form the core of the triple helix. (d) In β I spectrin, a cysteine residue (C1167) sits in the linker region between repeats 8 and 9. In erythrocyte membranes subjected to mechanical shearing forces, the cysteine becomes available to modification by chemical reagents, indicating and folding of this structure under tension. (PDB: 1S35) (e) The plakin domain. Plakin domains are found in the spectraplakins and in proteins such as Plectin. These have the same overall fold as spectrin triple helices, but with very low sequence identity to canonical spectrin helices. (PDB: 2IAK) (f) Spectrin triple helices as protein binding sites. In the β-spectrins, repeats 14 and 15 bind ankyrin through its ZU5 domain. This interacts with both repeats 14 and 15, a tyrosine that is highly conserved throughout β spectrins on the turn between helices B and C of repeat 15 is essential for this. (PDB:3KBT) (g) A split triple helix forms the interaction site between α and β spectrins in the formation of tetramers. A single helix from α-spectrin (αR1) interacts with two helices from β-spectrin (βR17) two we constitute a triple helix. Critical interactions include those formed by charged residues (such as Arg45 and Arg28): mutations in these are found in certain hemolytic anemias (PDB: 3LBX). (*See insert for color representation of the figure.*)

compilations of triple helical sequences given in Pfam (Bateman et al., 2002) and SMART (Letunic et al., 2002), no single residue is present throughout. The most strikingly conserved residues are indicated in Fig. 10.3b as a sequence logo: note that two tryptophan residues, one in helix A and one in helix C, are the most frequently conserved residues. These, and the other most closely conserved residues are very hydrophobic, and they form the core of the folding structure (Fig. 10.3c). In this sense, the helices in these triple helical bundles have an amphipathic character. Pairs of repeats are joined by the fusion of helices C and A via a short linker (Fig. 10.3d).

In the spectraplakins, limited proteolysis can yield fragments that represent pairs of highly stable spectrin repeats. These were not originally recognised as such; only when crystallography revealed their structure did this become clear (Fig. 10.3e) (Jefferson et al., 2007). These pairs of repeats, known as plackin domains (Jefferson et al., 2007), again reveal the lack of sequence conservation that underlies this fold.

Spectrin triple helices provide one of the most striking examples of functional adaptation of a common fold. Examples of this simple fold functioning as a spacer now exist, as a binding platform for protein and lipid ligands, as a spring and as enzymes.

10.5 FUNCTION AS A SPACER MODULE

As α-actinin is an actin cross-linking protein, the four triple helices form a spacer between actin filaments approximately 24 nm in length, forming a rod that twists 90° along its length (Fig. 10.3a). In other molecules of the spectrin superfamily, the extended chain of triple helices means that the spacing between actin-binding domains is much greater—approximately 200 nm in "conventional" spectrins and 265 nm in "heavy" spectrins (Glenney et al., 1982; Shotton et al., 1979).

The longer chains are much more flexible than the shorter α-actinin chains: although molecular dynamics reveals some flexibility in the α-actinin chain (Golji et al., 2009), the antiparallel dimer is usually considered to be a rod (Ylanne et al., 2001).

10.6 FUNCTION AS A PROTEIN BINDING MODULE

α-actinin is an antiparallel dimer, arranged such that repeat 1 of one chain interacts with repeat 4 of the opposite chain; correspondingly repeat 2 of one chain interacts with repeat 3 of the other. Importantly, no other interactions are possible (i.e., repeat 1 will not interact with repeats 2 or 3 and so on) (Ylanne et al., 2001).

The specificity of the interactions is mirrored in the low sequence identity between the repeats. Within any one protein, individual repeats are only 25–35% identical to each other (Baines, 2003). Any particular repeat in any of the proteins is much more closely related to equivalent repeats from different (and even distant) species than it is to other repeats in the same protein.

Repeats 1 and 2 of α-actinin are adjacent to the actin-binding domain, and these are homologous to the equivalent repeats in β- and β_{Heavy}-spectrin (Baines, 2003). In α-spectrin, repeats 20 and 21 are homologous to α-actinin repeats 3 and 4, thus spectrin dimers can form by interaction of these repeats, presumably in an analogous way to the formation of α-actinin dimers (An et al., 2011; Baines, 2003; Li et al., 2008; Ursitti et al., 1996).

However, most other repeats in spectrin molecules do not interact with each other, so even though they probably all descend from an α-actinin-like ancestor, they have been adapted in evolution not to interact with each other. Presumably, an advantage of this adaptation is to allow additional flexibility in spectrin chains. Repeat 1 is very specific to spectrin and α-actinin—there is no equivalent repeat in the spectraplakins or dystrophin branches of the spectrin superfamily, so it appears that the dimerisation mechanism of the earliest members of the spectrin superfamily was not advantageous and was lost in the subsequent evolution of spectraplakin/dystrophin branches (Baines, 2003).

An additional point about the low sequence identity between repeating units is that it represents a potential mechanism to prevent misfolding. High sequence identity between adjacent repeats might allow, for instance, helix A of one repeat to interact with helices B and C of the neighboring repeat, resulting in misfolding. Clarke and colleagues have tested this hypothesis using tandem immunoglobulin domains from the I band of titin (Borgia et al., 2011). They found that low sequence identity between neighboring domains did indeed reduce incidence of misfolding. The general principle here seems highly likely to apply to spectrin repeats.

Spectrin-type triple helices have been adapted to bind to many other protein ligands. Numerous ligands are known for α-actinin, among others vinculin (Bois et al., 2005), endothelial nitric oxide synthase (Hiroi et al., 2008), β1-integrin (Kelly and Taylor, 2005) and the actinin-associated lim protein (Klaavuniemi et al., 2009). It seems highly likely that each interaction is characterised by different structural features, indicating a separate evolutionary path to each interaction, and there is no single interface or mode of interaction.

In the case of vinculin, an α helix from repeat 4 of α-actinin interacts with a bundle of helices in the vinculin head domain (PDB:1YDI). The α-actinin helix fits into a groove in the vinculin helical bundle and forms extensive interactions with them.

By way of contrast, β-spectrin interacts with a globular domain of ankyrin (the ZU5 domain) through a limited number of amino acid side chains on the surface of repeats 14 and 15 (Ipsaro and Mondragon, 2010). Mutagenesis has identified key residues here as including a conserved tyrosine that sits on the turn between helices B and C of repeat 15 (Davis et al., 2008; Stabach et al., 2009) (Fig. 10.3f).

Perhaps, the most extreme example of an adaptation of a triple helix comes from the structure of the interaction site between α and β spectrins where they form a tetramer (Ipsaro et al., 2010). Here, two helices that form a hairpin from β spectrin (partial repeat 16) interact with a single helix from α spectrin (partial repeat 1) to recapitulate a full triple helix (Fig. 10.3g).

The evolution of nonnucleated mammalian erythrocytes was accompanied by duplication of the α-spectrin gene. One of the resulting pair, αI-spectrin was adapted for function in these cells (Salomao et al., 2006). Mammalian erythrocyte spectrin comprises only αI- and β I-spectrin polypeptides. The interaction of this polypeptide with β I-spectrin at the site where tetramers form is much weaker than that of the archetypal α-spectrin (αII-spectrin in mammals) for the abundant β II-spectrin of tissues: K_D for αI-β I interaction is about 800 nM, for αII-β II it is about 10 nM (Bignone and Baines, 2003). The low affinity for tetramer formation by erythrocyte αI-β I suggests that under physiological shear, erythrocyte spectrin tetramers may dissociate. αI-spectrin represents a neofunctionalized spectrin adapted to the rapid make and break of tetramers, so that the membrane accommodates the enormous distortions imposed on it during the passage of the cell through the microvasculature by means of local dissociation of spectrin tetramers to dimers. This is considered to be essential for allowing mammalian red blood cells to survive in circulation (An et al., 2002; Salomao et al., 2006).

In summary, the conclusion to be drawn from the various analyses of interaction between spectrin-type triple helices and protein ligands is that they are an extremely adaptable platform for protein assemblies (see also Djinovic-Carugo et al., 2002).

10.7 LIPID INTERACTIONS

Certain spectrin triple helices have been adapted to bind phospholipids. Repeats 8 and 9–10 from αI-spectrin and repeats 2, 3, 4, 12, 13, and 14 of βI-spectrin bind to phosphatidylserine (PS) and phosphatidylethanolamine (PE) (An et al., 2004; Grzybek et al., 2006). These lipids are characteristic of the cytoplasmic leaflet of the erythrocyte membrane. The specificity of these interactions is indicated by the fact that these repeats do not bind phosphatidylcholine. Physiologically, it seems that the interaction of spectrin with these lipids contributes to the strength of the membrane interaction so that triple helical interactions with both proteins and lipids are required for full membrane interaction (Manno et al., 2002).

A further point is that there is no reason to suppose that the interactions of the triple helical fold are restricted to proteins and lipids. Spectrin forms tight interactions with hemin (Jarolim et al., 1990; Liu et al., 1985), for instance, although the modes of interaction have not been mapped.

10.8 ACCOMODATING DEFORMATION

Erythrocytes must withstand the shearing forces encountered during circulation and be extensively deformable to squeeze through very narrow blood vessels. It is thought that the flexibility, and indeed elasticity, of spectrin polypeptides is essential for this (Hansen et al., 1997).

Analyses of the folding properties of individual triple helices from spectrin revealed that individual repeats denature at different temperatures (An et al., 2006a; An et al., 2006c). At least two repeats in human erythrocyte spectrin are at the very least on the edge of being unfolded at physiological temperature. Atomic force microscopy reveals that under tension, in the range encountered physiologically, particular spectrin repeats unfold reversibly (Law et al., 2003a; Law et al., 2004; Law et al., 2003b). Two repeats, βI-spectrin repeats 8 and 9, although they resist pulling at low temperatures, fail to do so at 37° (An et al., 2006a).

Figure 10.3d shows the structure of βI-spectrin repeats 8 and 9 (Kusunoki et al., 2004). Note how a single helix connects the two, and forms helix C of repeat 8 and helix A of repeat 9. The part of this helix that forms a link between the two triple helical units contains a single cysteine residue (C1167). This residue is inaccessible to chemical modification in unsheared erythrocytes ghosts. However, when subjected to physiological shearing forces, this residue becomes available to chemical modification indicating stretching (or partial unfolding) of the triple helices in response to mechanical tension (Johnson et al., 2007b; Krieger et al., 2011).

A further corollary of this is that there are now several cases of hereditary anemia that are characterised by alterations in the folding of spectrin triple helices (Johnson et al., 2007a). In particular, weakening of the linker regions between them can lead to thermal denaturation at lower temperatures *in vitro*. This weakening of the folding structure correlates with weakened membrane structure.

Electron microscopy of stretched erythrocytes cytoskeletons reveals elongated tetramers about 200 nm in length (Byers and Branton, 1985; Liu et al., 1987). However, the density of spectrin molecules in living cells means that they can be little more than 60–80 nm in length (Byers and Branton, 1985).

Lateral association between repeats in the formation of dimers allows spectrin triple helices to resist higher shear forces than monomers. This is probably a functional distinction between proteins such as α-actinin and spectrin, which are adapted to accommodate mechanical forces, and monomeric proteins such as dystrophin (Law et al., 2004).

In the absence of spectrin, membranes and their connections to other cells in simple animals are unable to resist the forces engendered by muscle contraction (Hammarlund et al., 2000; Hammarlund et al., 2007; Moorthy et al., 2000). Perhaps a fundamental requirement for spectrin in the evolution of tissues was in resistance to the forces encountered as cells adapted to communal life.

The stability of individual dystrophin triple helices is reminiscent of that of spectrin. The thermal and solvent denaturation patterns of individual triple helices vary widely across the molecule. Interestingly, in both spectrin and dystophin, there is some correlation between the most stable repeats and points of critical interactions. In dystrophin, repeats 16 : 17 bind nNOS and are extremely stable (T_m 71° and 68°, respectively) (Mirza et al., 2010); in α I-spectrin, repeat 16 is required for stabilizing the formation of tetramers, and is the most stable of the repeats (Tm 72°) (An et al., 2006a).

Ervasti (2007) has suggested that the most unstable regions of dystrophin may unfold reversibly under mechanical tension at the muscle plasma membrane, and has likened this aspect of dystrophin to being a "molecular shock absorber."

10.9 TRIPLE HELICAL REPEATS AS THE BASIS FOR ENZYME STRUCTURES

Goodman and coworkers reported that erythrocyte αI-spectrin is an E2/E3 ubiquitin-conjugating enzyme (Hsu and Goodman, 2005; Kakhniashvili et al., 2001) and that αI-spectrin repeats 20 and 21, ankyrin and the anion exchanger were substrates for this (Chang et al., 2004a; Chang et al., 2004b; Ghatpande and Goodman, 2004). Mutagenesis and recombinant protein analysis revealed that the activity was contained in repeat 20.

Wu et al. (2008) reported that a large central region of the spectraplakins MACF1/ACF7 has an ATPase activity, and that this activity is stimulated by F-actin.

An exciting and hitherto under-explored prospect is the potential for enzymic adaptations of spectrin-type triple helices.

10.10 CALPONIN HOMOLOGY DOMAINS

10.10.1 Discovery

When the sequences of β-spectrin (Byers et al., 1992; Hu et al., 1992; Moorthy et al., 2000; Stankewich et al., 1998; Tse et al., 2001; Winkelmann et al., 1990), β heavy-spectrin (Dubreuil et al., 1990; McKeown et al., 1998; Stabach and Morrow, 2000), α-actinin (Baron et al., 1987), utrophin (Guo et al., 1996), and dystrophin (Koenig et al., 1988; Neuman et al., 2001) were determined from cDNA analysis, it became apparent that close to their N-termini, there were sequence domains shared between them that were distinct from the triple helices (Baron et al., 1987). As these were very close to where each protein binds F-actin, these were assumed to be actin-binding sites.

Proteolytic analysis of α-actinin revealed that this N-terminal region contained an actin-binding activity in a 27 or 36 kDa fragment, depending on the precise protease used (Imamura et al., 1988; Mimura and Asano, 1986). When proteolytic digests of spectrin were analyzed for actin-binding fragments, a smaller fragment representing residues 47–186 of βI-spectrin was detected (Karinch et al., 1990). This localised the actin-binding site in both proteins close to the N-terminus. The extent of sequence similarity indicated that this was indeed a common domain in the spectrin superfamily. It was notable that the spectrin fragment was about half the size of the α-actinin fragment, and it appeared on the basis of sequence analysis that the spectrin fragment represented the first of two homologous domains approximately 100 amino acids in length present in the actin-binding region of the spectrin superfamily.

Small and coworkers (Vancompernolle et al., 1990) isolated a proteolytic fragment of an otherwise unrelated actin-binding protein calponin, and noticed a weak sequence similarity to the spectrin superfamily actin-binding domain. However, calponin does not bind actin through this region (Galkin et al., 2006; Gimona and Mital, 1998).

As more and more sequences became available over the subsequent years, it became clear that there was a "genetically mobile module" embodied in this common region of sequence. A review of the sequences by Castresana and Saraste (1995) identified the wide spread of this domain, and they named it a calponin homology domain. Clustering analysis revealed multiple classes of CH domains (Gimona et al., 2002). It is now clear that this domain is present in a large number of eukaryotic proteins, not just actin-binding proteins but also certain microtubule-interactive proteins, components of small G protein signaling pathways, and certain enzymes (Sjöblom et al., 2008).

Calponin homology domains can exist in 1–4 copies within any single polypeptide. In the spectrin superfamily, there is a tandem pair classified as CH1 and CH2. These are now considered to represent archetypes of components of actin-binding domains (Gimona et al., 2002; Sjöblom et al., 2008). However, calponin homology domains have been adapted to such an extent in evolution that no single function unifies protein modules with this common fold.

10.10.2 Prevalence

Numerous CH domain structures are now available through the Protein Data Bank. These all conform to a fold comprising a four helical orthogonal bundle (Cath:1.10.418.10). The structure is recognized in Scop (Murzin et al., 1995) as part of the calponin-homology domain superfamily [Scop:47576] that also contains the X-Prolyl dipeptidyl aminopeptidase PepX, N-terminal domain [Scop:81761], and Hook domain [Scop:116907]

No clear representatives of this domain are annotated in prokaryotes, but all simple eukaryotes contain several annotated domains. In fission yeast *S. pombe*, for example, five representatives are annotated that cover many of the functional classes represented very widely by eukaryotes including metazoa.

S. pombe has the α-actinin-like protein ain1 that contains an N-terminal pair of CH1 and CH2 domains. A small protein named trangelin (stg1) contains a single CH domain, which can bind actin and is required for cytokinesis and morphogenesis (Nakano et al., 2005). A Ras GTPase-activating protein rng2 contains a single CH domain and is required for cytokinesis (Eng et al., 1998). It is a component of the actomyosin ring and the spindle pole body. Fimbrin is another actin-binding protein, but distinctive in that it has four tandem CH domains (designated CH1, CH2, CH1.1, and CH2.2). It also has a pair of EF hands, although they are at the opposite end of the polypeptide in comparison with the spectrin superfamily. A further uncharacterised open reading frame (Uniprot:Q10201) encodes a 112 kDa protein with a single CH domain.

In mammals, additional classes include CH domains found in VAV proteins, the CAMSAP family and the microtubule end-binding (EB) proteins (Baines et al., 2009b; Castresana and Saraste, 1995; Hayashi and Ikura, 2003; Sjöblom et al., 2008; Yabana and Shibuya, 2002). These do not bind actin, although they appear to be interactive sites for other proteins. Of particular, interest in the EB proteins bind to microtubules and form part of a complex that tracks the plus ends of microtubules (Sjöblom et al., 2008). Vav3 appears to have an autoinhibitory CH domain; autoinhibition is overcome by binding to the pleckstrin homology domain in the adaptor protein APS (Yabana and Shibuya, 2002).

10.11 FUNCTIONAL INTERACTIONS AND REGULATION

The nature of the interaction of the CH1–CH2 pair with F-actin has been studied intensively for utrophin, dystrophin, and α-actinin (e.g., Galkin et al., 2010; Galkin et al., 2002; Keep et al., 1999; Moores et al., 2000; Norwood et al., 2000). The pairs of CH domains seem to adopt "open" (e.g., PDB : 1QAG, PDB : 1DXX) and "closed" (e.g., PDB : 1TJT, PDB : 2EYN) conformations. Earlier interpretations suggested that there were two sites in CH1 that bound F-actin, and one in CH2 that provided specificity between populations of actin filaments (among others Moores et al., 2000; Moores and Kendrick-Jones, 2000). However, a recent interpretation of the high resolution electron microscopic reconstruction of α-actin CH1–CH2 pair interaction with F-actin has thrown novel light on the mechanism. Now it seems that CH2 may not bind actin at all, but rather serves as an autoinhibitory domain that binds CH1 (closed conformation) and prevents CH1 binding unless the structure of the pair is opened (Galkin et al., 2010). The binding of α-actinin CH1 to actin is much weaker than the whole protein (Way et al., 1992), so Galkin et al. (2010) suggest that CH2 also serves a positive regulatory role, once in the open conformation. What might regulate the conformation? The CH2 domain of α-actinin binds phosphatidylinositol-4, 5-bisphosphate (PIP_2) (Fraley et al., 2003; Full et al., 2007). This inhibits actin binding: its precise effect on the structure of the pair remains to be seen.

In spectrin, the CH1 domain was first identified as an actin-binding fragment, but its interaction with actin is actually rather weak (K_D approximately 10^{-4} M) (Karinch et al., 1990). Strong physiological complexes with actin require the presence of protein 4.1R, the spectrin-actin-binding (SAB) domain of which forms a ternary complex with spectrin and F-actin (K_D 10^{-12} M) (Ohanian et al., 1984). The SAB binds to both the CH1 and CH2 domains of β-spectrin (An et al., 2005a). One hypothesis—as an extension of Galkin et al. (2010)—might be that the SAB opens the conformation of the spectrin CH1–CH2 pair and stabilizes an active conformation. (α-actinin has no requirement for protein 4.1R in the formation of strong complexes.) The CH1–CH2 pair in spectrin also binds

PIP_2. As with α-actinin, this inhibits the ability of spectrin to cross-link and gel solutions of actin filaments (An et al., 2005a).

It also appears that other classes of calponin homology domain function in tandem pairs. In the microtubule-binding protein EB1, the protein is a dimer with closely adjacent domains (CHe class) (Sjöblom et al., 2008).

The Ncd80 complex, a core component of the kinetochore that links chromosomes to the microtubules of mitotic spindles, contains two polypeptides, Nuf2 and Ncd80: each of these contributes a CH domain to the microtubule-binding structure (Sjöblom et al., 2008).

Beyond the cytoskeleton, many further interactions of CH domains have been annotated. For example, the kinase ERK interacts with the CH domains of both calponin and α-actinin (Leinweber et al., 1999). The type 5 CH domain of α-parvin interacts with paxillin LD1 (Lorenz et al., 2008; Wang et al., 2008).

In summary, as with spectrin repeats, a simple fold common to all eukaryotes has been adapted to numerous different functions during the course of eukaryotic evolution.

10.12 THE CALMODULIN-LIKE DOMAIN

The C-terminal regions of both α-spectrins and α-actinin contain a domain that can be aligned with the whole of calmodulin (Tang et al., 2001; Trave et al., 1995). This region contains four EF hands, small helix-loop-helix structures best known as Ca^{2+}-binding sites (for review, see Gifford et al., 2007). They generally occur in pairs: spectrin and α-actinin have two such pairs (See Fig. 10.3). The Scop (Murzin et al., 1995) classification is family: calmodulin-like [47502].

However, not all the EF hands bind Ca^{2+}. αI-spectrin only has a single EF hand capable of specifically binding Ca^{2+}, and even so, the binding is so weak as to be physiologically questionable (Lundberg et al., 1995; Travé et al., 1995). In αII-spectrin, one of the pairs is fully active, and binding of concentrations of Ca^{2+} likely to be encountered physiologically in synapses induces conformational changes (Lundberg et al., 1997).

Vertebrates have four α-actinin genes (Table 10.1). Two of these (the nonmuscle isoforms 1 and 4) have calmodulin-like domains that bind calcium through their EF hands, and this reduces their affinity for F-actin, which may regulate their ability to link stress fibers to focal adhesions (Burridge and Feramisco, 1981). By contrast, the EF hands of the two skeletal muscle isoforms (2 and 3) have evolved so as not to bind calcium at physiological concentrations.

The calmodulin-like domains are required for the function of each protein, and they are nonredundant. The calmodulin-like domain of fruit fly α-actinin is required for viability (Dubreuil and Wang, 2000). A hybrid α-actinin containing the calmodulin-like domain of fly α-spectrin cannot rescue the null mutant, unlike the native protein (Dubreuil and Wang, 2000).

Mutation of the calmodulin-like domain of mammalian erythrocyte αI-spectrin or its truncation can give rise to hemolytic anemia (Robledo et al., 2010). The basis for this seems to be twofold. First, the calmodulin-like domain is required for optimal spectrin-actin interaction (Korsgren and Lux, 2010). Second, this domain binds to protein 4.2 (Korsgren et al., 2010), a linker between spectrin and transmembrane proteins, including the erythrocyte anion exchanger and CD47 (Satchwell et al., 2009) (See Fig. 10.1b and c). Human patients lacking protein 4.2 have hereditary spherocytosis (An and Mohandas, 2008).

The α-actinin calmodulin-like domain is required for targeting α-actinin to the skeletal muscle Z-disc (Young and Gautel, 2000). This domain can interact with the opposite chain in the α-actinin dimer, where it binds to an α helix that links between the CH2 domain and the first spectrin-type triple helical repeat. This interaction is blocked by binding of PIP_2 to the CH domains: PIP_2 is present in the Z-disc, thus interaction of α-actinin with the Z-disc releases this intramolecular interaction. A sequence within the Z-disc of the giant protein titin has sequence similarity to the part of α-actinin that this domain binds to, and can bind it with nM affinity. Thus in the Z-disc, a rearrangement induced by PIP_2 causes a change of interaction of the calmodulin-like domain from interacting with the opposite strand to interaction with titin (Young and Gautel, 2000).

In the dystrophin and spectraplakin branches, there seems only to be one pair of EF hands. The C-terminal region of dystrophin and utrophin contains several adjacent folded regions: in addition to the EF hands, there is a WW domain (named after its two tryptophan residues) and a ZZ domain that binds zinc ions. The EF-hand region stabilises the WW domain that binds dystroglycan (Huang et al., 2000; Ilsley et al., 2002; Rentschler et al., 1999).

These examples indicate the range of physiological functions of the EF hands in the spectrin superfamily. They encompass stabilisation of other folding structures, intracellular targeting, and giving resilience to a plasma membrane.

10.13 ANKYRIN

10.13.1 Overview

Ankyrin is an adapter between numerous transmembrane proteins and the cytoplasmic cytoskeleton, principally spectrin (Baines, 2010b; Bennett and Baines, 2001). Ankyrin was discovered as a peripheral membrane protein on the cytoplasmic face of the erythrocyte membrane that can act as a "receptor" for spectrin (Bennett and Branton, 1977; Bennett and Stenbuck, 1980). As a point of historical interest in the discovery of domains, ankyrin was discovered initially through the use of controlled, limited proteolysis of erythrocyte membranes to liberate an active fragment that could compete with the native inside out membrane vesicles for binding to spectrin (Bennett, 1978).

The structure of ankyrin is summarized in Figure 10.3.

Ankyrin is exquisitely sensitive to endogenous erythrocyte proteases, and so it was impossible to purify originally as a whole protein. However, the spectrin-binding region of this protein is sufficiently tightly folded to resist limited treatment with chymotrypsin; chymotrypsin treatment of erythrocyte membranes yielded a 72 kDa fragment that contained the spectrin-binding activity (Bennett, 1978). Identification of this active fragment allowed purification of the whole protein (a series of differentially spliced polypeptides ranging between ~210 and 180 kDa) (Bennett and Stenbuck, 1980). The 72 kDa spectrin-binding fragment was later revealed to contain discrete folding structures within it, critically a sequence domain now recognised to be common to a number of membrane-linked structural proteins, including the tight junction protein Z0-1 and Unc5-like netrin receptors (Ipsaro et al., 2008; Mohler et al., 2004c; Weaver and Marchesi, 1984; Weaver et al., 1984). This domain is now known as a *Zu5 domain*, and is another example of a genetically mobile protein-interactive module (Fig. 10.4a and c).

Subsequent proteolysis of native erythrocyte ankyrin revealed that its membrane-binding activity was in the N-terminal half of the protein (Michaely and Bennett, 1995). This yielded a number of membrane-binding fragments (D1–D4), which show specificity for various membrane proteins (Michaely and Bennett, 1995). Because there are multiple regions within ankyrin that can bind different transmembrane proteins, it has the potential to cross-link different membrane proteins into large complexes. 24 small repeating units of 33 amino acids make up the whole of the membrane-binding region; six were present in each of D1–D4 (Michaely et al., 2002). These repeating units are now referred to as *ankyrin repeats* (Fig. 10.4b).

Further sequence comparison revealed that ankyrin contained a death domain, again another module associated with protein binding activities (Del Rio et al., 2004; Mohler et al., 2002) (Fig. 10.4d).

Three ankyrin genes are common to vertebrates (ANK1, ANK2, and ANK3 encoding ankyrin-R, ankyrin-B, and ankyrin-G poypeptides, respectively: see Table 10.1), but invertebrates contain only a single ankyrin gene. An exception to this is the arthropod lineage in which duplication has led to the evolution of a second gene (Hopitzan et al., 2006; Hortsch et al., 2002).

Ankyrins have a number of other differentially spliced modules, some of which are specific to the various genes. Among these, is a domain designated obscurin/titin-binding-related domain OTBD that can bind to cytoskeletal proteins including obscurin and filamin (Bagnato et al., 2003; Busby et al., 2010; Hopitzan et al., 2006; Maiweilidan et al., 2011). Ankyrin also exemplifies the extremes of differential splicing of cytoskeletal proteins. Canonical ankyrins are approximately 210 kDa: some differentially spliced inserts generate giant ankyrins more than 400 kDa (Chan et al., 1993); small ankyrins lacking the membrane-binding domain are approximately 100 kDa (Gagelin et al., 2002; Hopitzan et al., 2005; Peters et al., 1995); in muscle there are some tiny forms as small as 15,000 that lack the membrane-binding domain (Bagnato et al., 2003), and substitute a transmembrane domain: these retained the OTBD domain, so cytoskeletal proteins can be linked to membranes via mechanisms entirely different from those in cannonical ankyrins.

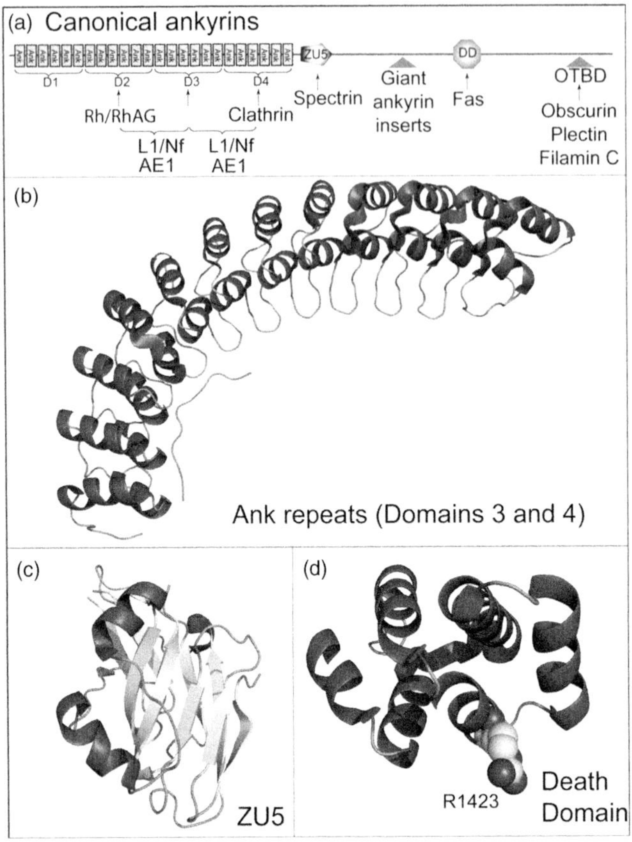

Figure 10.4 Ankyrin: Ankyrins form major linkages between spectrin and plasma membranes, and coordinate formation of multiprotein complexes both in erythrocytes and other tissues. (a) Domain structure of canonical ankyrins. The membrane-binding region is made up of sequential ank repeats: groups of six of these form separate functional domains that bind and cross-link multiple membrane proteins as indicated. A ZU5 domain adjacent to the membrane-binding region finds spectrin and a death (Dd) domain can bind fas. Many splice variants of ankyrins are known: some small variants lack the membrane-binding region; in certain ankyrins giant isoforms can be formed by insertions between ZU5 and DD; in the C- terminal region of muscle ankyrins a domain designated OTBD can bind obscurin, plectin and filamin C. (b) Ank repeats are helix-loop-helix units that together form long curved structures. (PDB: 1N11) (c) The ZU5 domain (PDB: 3F59). (d) The death domain. In ankyrin-R, Arg1423 is required for the interactive properties of this domain (PDB: 2YVI).

10.14 ANKYRIN REPEATS

10.14.1 Structure and Prevalence

Erythrocyte ankyrin-R (ANK1) contains 24 repeats of 33 amino acids (Lambert et al., 1990; Lux et al., 1990). X-ray crystallography revealed that they form a helix-loop-helix structure, with a β-hairpin/loop region projecting from the helices at approximately 90° angle (Michaely et al., 2002) (Scop:Superfamily: Ankyrin repeat [48403]). Long multirepeat constructs of ankyrin repeats form curving structures resembling spirals (Fig. 10.3b).

Genomic analysis, and further structural analysis reveal that ankyrin repeats are ubiquitous in the kingdoms of life (Li et al., 2006). This simple folding structure may have evolved early as examples are known in both bacteria and archea (indeed, 10% of all ankyrin repeats annotated in the SMART database are bacterial); however, it has been suggested that at least some prokaryotic and viral examples were acquired by a horizontal gene transfer from eukaryotes (Al-Khodor et al., 2010).

Ank repeats are widely scattered throughout eukaryotic genomes, and are found not just in cytoskeletal proteins, but in many types of signaling proteins and enzymes. In the context of this review on cytoskeletal proteins, it would not be suitable to review these in detail here: readers are referred to recent reviews by (Ferreiro and Komives, 2010; Mosavi et al., 2004; Tee and Peppelenbosch, 2010) that cover aspects of ank repeat function in signaling and muscle physiology.

10.14.2 Binding Functions

Ankyrin repeats can bind an immense range of different proteins. Mammalian ankyrins can interact, for example, with the erythrocyte Cl^-/HCO_3^- anion exchanger, sodium-potassium ATPase, rhesus complex, voltage gated sodium channels, clathrin heavy chains, and cell adhesion molecules of the L1 family (e.g. Bennett, 2000; Bennett and Chen, 2001; Davis and Bennett, 1990a, 1994; Davis and Bennett, 1990b; Devarajan et al., 1994; Garver et al., 1997; Jenkins et al., 2001; Lopez et al., 2005; Michaely and Bennett, 1995; Mohler et al., 2005; Mohler et al., 2004a; Mohler et al., 2004b; Nicolas et al., 2003; Pan et al., 2006).

As with spectrin repeats, there is no single binding surface on ankyrin repeats. In the case of ankyrin-R, ion transporters such as the erythrocyte anion exchanger associate with a large central cavity of the repeating spiral; clathrin and cell adhesion molecules associate with its outer region (Michaely et al., 2002).

The observation that ankyrin has the potential to bind more than one transmembrane protein through its repeat domains raises the possibility that it might be able to cross-link complexes. Bruce and coworkers (2003) were able to isolate a macromolecular complex from erythrocytes, containing both the anion exchanger and the rhesus proteins. They suggested that ankyrin coordinates the formation of a metabolon dedicated to carbon dioxide transport. The anion exchanger is responsible for the exchange of the products (HCO_3^-) of the carbonic anhydrase reaction across the membrane. Carbon dioxide, the substrate for carbonic anhydrase, may possibly enter the cell at least in part via the rhesus complex (e.g., Boron, 2010). This trimeric transmembrane complex has been suggested to mediate carbon dioxide transport on the basis that Rh_{null} individuals have reduced carbon dioxide permeation into the cell (Endeward et al., 2006, 2008). However, this suggestion remains controversial (Ripoche et al., 2006).

A different function for cross-linking and binding proteins is revealed in the heart. Here, ankyrin-B binds to two of the components of the system that gives fast clearance of calcium ions from the cytoplasm at the end of contraction (Mohler et al., 2005). A sodium gradient generated by the sodium–potassium pump provides energy for the fast exchange of calcium from the cytoplasm via the sodium–calcium exchanger NCX1. These two proteins bind to different domains of the ankyrin repeats, thus a single ankyrin can coordinate a complex of these proteins. Importantly, ankyrin is required to anchor these proteins at the cell surface: cardiomyopathies can arise from defects in ankyrin-B (Mohler and Bennett, 2005).

10.14.3 Making Use of Ankyrin-Repeat Binding Activities: Designed Ankyrin-Repeat Proteins (Darpins)

The wide range of binding activities of ank repeats have prompted a number of researchers to investigate the possibility that these small simple folding structures could be designed to act as modulators of cellular functions, in rather analogous ways to antibodies (Stumpp and Amstutz, 2007). Indeed, a common analogy for ankyrin repeats is that they are the cytoplasmic equivalent of antibodies, in the sense that they can bind to any number of targets with great specificity. They contain no disulphide bonds, and it is possible to express ank repeats with high efficiency in bacterial cytoplasm.

Because of their small size and simple folding, it has been possible to generate phage libraries expressing four repeat modules (a minimum binding unit): these are known as *designed ankyrin-repeat proteins (darpins)*. These libraries can be used to select ankyrin-repeat proteins for particular purposes. Early examples of this included development of protease inhibitors (Kawe et al., 2006; Schweizer et al., 2007). Darpins with very high affinity binding to the growth factor receptor Her2 have been developed. Her2 is highly expressed on certain tumors, and thus Her2-specific darpins should be able to target such tumors. Their pharmacokinetics can be very advantageous (Zahnd et al., 2010) compared to, for example, single chain antibody variable fragment constructs. A recent development has been to use Her2-specific darpins to target adenovirus to tumors (Dreier et al., 2011). The use of darpins for tumor targeting is a fascinating prospect for the future (Munch et al., 2011).

10.14.4 Ankyrin Repeats as Nanosprings

Ankyrin repeats are present in cytoplasmic domains of transient receptor potential channels in stacks of 17 to 29 repeats. One hypothesis is that these act a spring that gates mechano-receptors in systems such as hair cells (Corey et al., 2004) or *Drosophila* bristles (Howard and Bechstedt, 2004).

The structure of ankyrin repeats with their antiparallel α helices arranged in a spiral gave rise to the idea that these might reversibly unfold under mechanical tension, giving spring-like behavior. Single molecule atomic force microscopy revealed this to be the case (Lee et al., 2006). The spring constant for 24 ankyrin-B repeats was approximately 1.9 pN/nm. Ankyrin-B repeats were found to extend reversibly in multiple stretch-relaxation cycles with remarkable resilience.

Since ankyrin was first found in the erythrocyte, an obvious question is to what extent ankyrin repeats unfold when erythrocytes are subjected to shearing forces in circulation. As ankyrin cross-links proteins in a large macromolecular complex, is this structure altered over the course of circulation? Furthermore, does the springiness of ankyrin help to accommodate deformation of red cell membranes? These questions will be challenging to answer.

In the case of transient receptor potential channels, the ankyrin repeats mediate self-association of channels into tetramers (Lepage et al., 2009). There is also evidence for ATP binding to TRPV1 ankyrin-repeat domain and this sensitizes the channel (Phelps et al., 2010).

Ankyrin repeats are not the only examples of tightly packed α helical repeats predicted to form spiral or solenoid structures. Ank, ARM, and HEAT multiple repeats all form spiral-shaped protein domains, and these all show stretch ratios of the order of 10–15 (Kim et al., 2010). In each case, atomic force microscopy identifies unfolding of individual domains as multirepeat proteins are stretched. Multiple repeats of each of these domains are found in over 800 human proteins, so these domains represent a potential widespread mechanism for the formation of mechano-sensitive structures, or perhaps form the basis of springy nanostructures.

10.15 PROTEIN 4.1R: THE FERM (SUPRA-)DOMAIN AS A SITE FOR CONVERGING FORMS OF REGULATION

Protein 4.1R is essential for the formation of the erythrocyte membrane junctional complex (Baines, 2010b; Mohandas and Gallagher, 2008). Here, spectrin molecules are cross-linked to short actin filaments, a reaction promoted by protein 4.1R (Fig. 10.1a and c). 4.1R can also bind a wide variety of membrane proteins (Fig. 10.1c; Baines, 2010b). This means that the approximately half-dozen 4.1 proteins in each junctional complex can link a range of membrane proteins together.

These interactions are essential for membrane protein retention at the cell surface during erythrocyte development: in the absence of 4.1 they are not effectively retained at the cell surface. In this sense, 4.1 and ankyrin each are required to give "permission" for transmembrane proteins to remain at the cell surface. One hypothesis is that this was important in animal evolution to prevent the unwanted activation of signaling pathways or cell adhesions: in this hypothesis, unless membrane proteins are given permission to remain at the cell surface by binding to adapters such as 4.1 or ankyrin, they are internalised (Baines, 2010b; Baines et al., 2009a).

To analyze functional domains of protein 4.1R, Marchesi and coworkers subjected it to controlled, limited proteolysis (Leto and Marchesi, 1984). They found a small 8 kDa domain that bound spectrin and actin (Correas et al., 1986) (now known as the spectrin-actin-binding (SAB) domain), as well as 30 and 18 kDa fragments whose function was not clear at the time. The 18 kDa region is a C-terminal domain (CTD) that binds numerous membrane proteins in complex tissues, although no protein ligands are known in erythrocytes (Scott et al., 2001). The structure of 4.1R is summarized in Figure. 10.5.

Sequence analysis of protein 4.1R as well as that of proteins such as ezrin, radixin, and moesin revealed that the 30 kDa fragment represented a shared structure (Chishti et al., 1998). This became known as the FERM domain (Chishti et al., 1998). The FERM domain is now recognised as a domain abundant in eukaryotes, principally in metazoa. It represents a "genetically mobile" module that has been adapted to numerous proteins, including talin, a cytoskeletal protein at regions of cell–substrate contacts, unconventional myosins X, VIIa

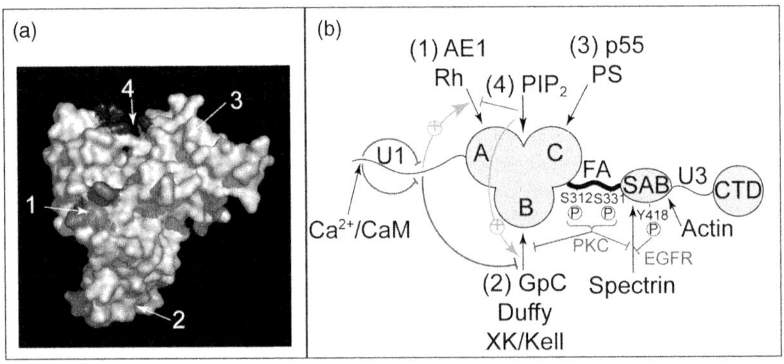

Figure 10.5 4.1R and the regulation of its FERM domain. (a) The 4.1R FERM domain. Four ligand-binding regions are indicated: (1) a region in lobe A binds the anion exchanger AE1 (Band 3) and Rh; (2) in lobe B, a region binds glycophorin C (GpC), the XK/Kell complex, and Duffy; (3) In lobe C, there are binding sites for p55 and phosphatidylserine (PS); (4) PIP_2 binds between lobes A and C. (b) A schematic view of the domains in 4.1R: U1, and N-terminal headpiece; the FERM domain is shown as a three-lobed (grey-shaded) cloverleaf; FA is the FERM adjacent region; SAB is the spectrin actin-binding domain; U3 is a linker between the SAB and the C-terminal domain (CTD). The headpiece promotes interaction at binding site (1), but inhibits site (2). Ca^{2+}/calmodulin binds in the headpiece and abrogates its regulation of the FERM domain. Phosphorylation of ser312 and ser331 by protein kinase C in the FA domain weakens interaction at site (2), as well as interaction with spectrin at the SAB. The SAB is also phosphorylated by EGF receptor, and this also reduces interaction with spectrin. PIP_2, binding at site (4) reduces interaction with ligands at site (1), but promotes interactions at site (2). (*See insert for color representation of the figure.*)

and XV, focal-adhesion kinases, Janus tyrosine kinases, several protein-tyrosine phosphatases (Campbell and Ginsberg, 2004; Frame et al., 2010; Lim et al., 2008; Malinin et al., 2010; Tepass, 2009; Yu et al., 2011). Ensembl (Hubbard et al., 2002) annotates 49 different genes with FERM domains in the human genome.

10.16 STRUCTURE AND REGULATION OF THE 4.1R FERM DOMAIN

Numerous FERM structures are now available in the protein data bank: that of 4.1R acts as a useful prototype (PDB:1GG3) (Han et al., 2000) (Fig. 10.5a).

The domain has three lobes, each of which appears to be an independent folding structure (and indeed each folds independently when expressed in bacteria (Salomao et al., 2008)), although there are extensive contacts between three. The N-terminal lobe (lobe A) has a fold that is a representative of the ubiquitin superfamily (SCOP:54236). The central lobe (lobe B) has an α-helical fold such as acyl-CoA binding protein (Scop Superfamily: Second domain of FERM [47031]). The C-terminal lobe (lobe C) has a fold such as a pleckstrin homology domain (Han et al., 2000) (Scop Superfamily: PH domain-like [50729]). This overall folding structure is conserved throughout known FERM domains, even though the level of sequence conservation is extremely low (Tepass, 2009).

The three lobes of the FERM domain represent independently folded structures: each can be expressed as a functionally active recombinant protein on its own (Salomao et al., 2008). Each lobe therefore has the characteristics of a domain in its own right. However, the contacts between the lobes are so extensive that they form a coherent structure, and, as will be noted below in more detail, two of the lobes together form a single binding site for a phospholipid (An et al., 2006b). In this sense, the FERM "domain" is really a supra-domain in the sense defined by Vogel et al. (2004): a three-domain combination embodying particular functional and spatial relationships. However, the FERM structure is universally described as a "domain" in the literature, so I shall use that term for the purposes of this chapter.

Analysis of the activities of the individual lobes of 4.1R FERM domain has revealed the different lobes interact with different ligands (An et al., 2001; Han et al., 2000; Salomao et al., 2008; Seo et al., 2009). Lobe A binds the erythrocyte anion exchanger and the rhesus complex proteins Rh; lobe B finds the transporter XK, the chemokine receptor Duffy and glycophorin C; the cytoskeletal adapter protein p55 binds across lobes A and C; lobe C also binds PS (Fig. 10.5a).

However, the presence of multiple binding sites does not signify that 4.1R FERM domain can necessarily simultaneously cross-link several proteins. The FERM domain is the site of converging forms of regulation that differentially control whether individual lobes are active.

A further binding interaction is with PIP_2 (An et al., 2006b). The binding site for this phospholipid is provided both by lobes A and C. Although the nature of the interaction appears involve basic residues in these lobes interacting with the phosphate groups of PIP_2, the interaction is clearly with the whole lipid, and not just with the highly charged head group.

PIP_2 turns over rapidly in red cell membranes, but in healthy cells, is present at substantial excess over the number of 4.1R molecules. In the presence of PIP_2, 4.1R binds glycophorin C strongly, but the anion exchanger weakly. By contrast, in metabolically depleted cells that contain little PIP_2, glycophorin C binds weakly to the FERM domain, but the anion exchanger binds strongly (An et al., 2006b).

4.1R is subject to differential mRNA splicing: during erythrocyte development, an N-terminal headpiece (also known as *U1*) is expressed in erythrocyte precursors, but is not present in the major mature erythrocyte form (Conboy, 1999). The headpiece modulates FERM activity by promoting binding to the anion exchanger and inhibiting interaction with glycophorin C. The headpiece contains a Ca^{2+}/calmodulin-binding site (Leclerc and Vetter, 1998; Nunomura et al., 2009); in the presence of Ca^{2+}/calmodulin, neither band 3 nor glycophorin C bind to 4.1R containing the headpiece.

4.1R is extensively phosphorylated (see Fig. 10.5b). The effects of phosphorylation catalyzed by protein kinases C and A have been documented most closely (e.g., Chao and Tao, 1991; Gauthier et al., 2011; Horne et al., 1985; Pinder et al., 1995; Subrahmanyam et al., 1991). These phosphorylate residues in a region directly adjacent to lobe C of the FERM domain in the FERM adjacent (FA) domain (Baines, 2006). Ser-312 and ser-331 in the FA domain are substrates for protein kinase C: phosphorylation here weakens the interaction of lobe A with glycophorin C, Duffy, and XK (Gauthier et al., 2011), and this is associated with weakening membrane mechanical stability (Manno et al., 2005).

10.16.1 Parallels with the ERM Proteins

The FERM domains of ezrin, radixin and moesin (ERM proteins) bind PIP_2: this is essential for their activation at the plasma membrane (e.g., Blin et al., 2008; Fievet et al., 2004; Hamada et al., 2000; Hao et al., 2009; Hirao et al., 1996; Janke et al., 2008; Niggli et al., 1995). The C-terminal domains of these proteins contain a 34 amino acid actin-binding site, but interaction of the C-terminal domain with the FERM domain occludes it. PIP_2 binding between lobes A and C unmasks ERM protein activities by releasing the C-terminal domain. Once released, the C-terminal domain becomes phosphorylated at a conserved threonine. Interestingly, binding of ezrin to PIP_2-containing giant unilamellar vesicles indicates that the FERM domain can block access to neighboring PIP_2 molecules, and thus lower the available concentration (Blin et al., 2008).

The parallels between the functions and regulation of the FERM domains of 4.1R and ERMs reveal a theme that FERM domain activity is regulated by inhibitory interactions of neighboring regions of polypeptide. This is relieved by PIP_2 binding between lobes A and C, which, in addition, activates certain membrane-binding activity. Phosphorylation outside the FERM domain further contributes to regulation of FERM activity. A pattern of complex forms of regulation converging on the FERM domain is readily apparent.

10.17 SUMMARY AND CONCLUSION

In this chapter, I have chosen a few examples of protein domains that were discovered, at least in part, by their function in the mammalian erythrocyte cytoskeleton. They exemplify a number of emerging themes that apply not just to cytoskeletal proteins but also to many other signaling, enzymatic, or regulatory proteins.

With spectrin repeats and ankyrin repeats, we see binding activities that are highly adapted to individual proteins. The binding activities concerned are not restricted to particular surfaces in the way that classic binding activities (such as SH3 domain–proline-rich peptide interactions) exemplify. Rather, the adaptability of these simple folds makes it difficult to predict the nature and existence of possible binding activities. Furthermore, both these two types of repeats display a "springyness" revealed as reversible unfolding under tension. This reversible unfolding seems to be integral to the dynamics of the erythrocyte membrane, and raises questions for the nature of sensing stretch in biological membranes. An interesting observation in this respect is that Discher and others have discerned a role for the tension on plasma membranes in directing differentiation of stem cells (Discher et al., 2005; Discher et al., 2009). An open question is to what extent the unfolding of domains with springy characteristics contributes to such cellular responses. Equally, it may be the case that reversible unfolding allows formation of complexes under condition of low mechanical tension that are then disrupted under increased loading.

Over the past decade, it has also become clear that numerous domains show interactions with phospholipids, and such interactions serve to modulate protein activity. Many folding structures (including those beyond the scope of this review, such as Vps27p FYVE, p47phox-PX, epsin1-ENTH, and CALM-ANTH domains) are now recognised as interacting with lipid bilayers (see Chapter 3 in Luckey, 2008). In the case of the FERM domain, as we have seen, PIP_2 can serve to mask or reveal protein binding sites (An et al., 2006b; Edwards and Keep, 2001; Hamada et al., 2000). PIP_2 also can modulate the activity of pairs of CH domains in relation to binding actin or protein 4.1 (An et al., 2005a; Sjöblom et al., 2008). PS, another phospholipid characteristic of the cytoplasmic leaflet of plasma membranes, interacts with the 4.1R FERM domain (An et al., 2001; Cohen et al., 1988) and with certain spectrin triple helices (An et al., 2005b; An et al., 2004). In the former case, this interaction directs membrane trafficking of 4.1R, and in the latter, is required for stable membrane interaction of spectrin. Some isoforms of spectrin also contain pleckstrin homology domains, which, too, bind PIP_2 (Fukami et al., 1996; Zhang et al., 1995).

Phosphoproteomics has revealed a broad general theme that the most tightly folded domains tend not to be subject to extensive phosphorylation (Collins et al.,

2008). Rather, neighboring less tightly folded (or disordered) regions act to regulate their neighbors. This is exemplified in the FERM domain of 4.1R and the ERM proteins where phosphorylation of neighboring regions controls activity. Nevertheless, it is clear, particularly in pathological conditions that phosphorylation patterns can change, and recent experiments with sickle cell disease have revealed patterns of phosphorylation within the 4.1R FERM domain that are not normally seen (Siciliano et al., 2010). Looking to the future, it will be important to establish the prevalence of anomalous phosphorylation of tightly folded domains in disease processes.

In conclusion, the examples discussed here have shown extraordinary adaptability in the functional evolution of simple folds that arose in unicellular organisms, and have been adapted to the requirements of differentiated tissues in animals. The state of our knowledge of these domains has advanced enormously in the past decade. At the turn of the millennium, the widespread assumption was that homologous sequences had generally similar functions, and that the identification of protein domains as genetically mobile modules would allow inference of function. While this remains true for some domains (e.g., FERM), lessons learned since that point—exemplified by spectrin repeats that can be adapted to be a spacer, to bind numerous protein, lipid, or other ligands (and on no particular surface), to be a spring and possibly even an enzyme—indicate that the concept is now an oversimplification, and serve to remind us that biology will always surprise us. In relation to some of the domains discussed here, structural homology identifies little more than descent from a common ancestor, rather than commonality of biological function.

REFERENCES

Al-Khodor, S., Price, C.T., Kalia, A., and Abu Kwaik, Y. (2010) Functional diversity of ankyrin repeats in microbial proteins. Trends Microbiol, 18, 132–139.

Allen, P.G. and Shah, J.V. (1999) Brains and brawn: plectin as regulator and reinforcer of the cytoskeleton. Bioessays, 21, 451–454.

An, X., Debnath, G., Guo, X., Liu, S., Lux, S.E., Baines, A., Gratzer, W., and Mohandas, N. (2005a) Identification and functional characterization of protein 4.1R and actin-binding sites in erythrocyte beta spectrin: regulation of the interactions by phosphatidylinositol-4,5-bisphosphate. Biochemistry, 44, 10681–10688.

An, X., Guo, X., Gratzer, W., and Mohandas, N. (2005b) Phospholipid binding by proteins of the spectrin family: a comparative study. Biochem Biophys Res Commun, 327, 794–800.

An, X., Guo, X., Sum, H., Morrow, J., Gratzer, W., and Mohandas, N. (2004) Phosphatidylserine binding sites in erythroid spectrin: location and implications for membrane stability. Biochemistry, 43, 310–315.

An, X., Guo, X., Yang, Y., Gratzer, W.B., Baines, A.J., and Mohandas, N. (2011) Intersubunit interactions in erythroid and non-erythroid spectrins. Biochim Biophys Acta - Proteins Proteomics, 1814, 420–427.

An, X., Guo, X., Zhang, X., Baines, A.J., Debnath, G., Moyo, D., Salomao, M., Bhasin, N., Johnson, C., Discher, D. et al. (2006a) Conformational stabilities of the structural repeats of erythroid spectrin and their functional implications. J Biol Chem, 281, 10527–10532.

An, X., Lecomte, M.C., Chasis, J.A., Mohandas, N., and Gratzer, W. (2002) Shear-response of the spectrin dimer-tetramer equilibrium in the red blood cell membrane. J Biol Chem, 277, 31796–31800.

An, X. and Mohandas, N. (2008) Disorders of red cell membrane. Br J Haematol, 141, 367–375.

An, X., Zhang, X., Debnath, G., Baines, A.J., and Mohandas, N. (2006b) Phosphatidylinositol-4,5-biphosphate (PIP2) differentially regulates the interaction of human erythrocyte protein 4.1 (4.1R) with membrane proteins. Biochemistry, 45, 5725–5732.

An, X., Zhang, X., Salomao, M., Guo, X., Yang, Y., Wu, Y., Gratzer, W., Baines, A.J., and Mohandas, N. (2006c) Thermal stabilities of brain spectrin and the constituent repeats of subunits. Biochemistry, 45, 13670–13676.

An, X.L., Takakuwa, Y., Manno, S., Han, B.G., Gascard, P., and Mohandas, N. (2001) Structural and functional characterization of protein 4.1R- phosphatidylserine interaction: potential role in 4.1R sorting within cells. J Biol Chem, 276, 35778–35785.

Anong, W.A., Franco, T., Chu, H., Weis, T.L., Devlin, E.E., Bodine, D.M., An, X., Mohandas, N., and Low, P.S. (2009) Adducin forms a bridge between the erythrocyte membrane and its cytoskeleton and regulates membrane cohesion. Blood, 114, 1904–1912.

Bagnato, P., Barone, V., Giacomello, E., Rossi, D., and Sorrentino, V. (2003) Binding of an ankyrin-1 isoform to obscurin suggests a molecular link between the sarcoplasmic reticulum and myofibrils in striated muscles. J Cell Biol, 160, 245–253.

Baines, A.J. (2003) Comprehensive analysis of all triple helical repeats in beta-spectrins reveals patterns of selective evolutionary conservation. Cell Mol Biol Lett, 8, 195–214.

Baines, A.J. (2006) A FERM-adjacent (FA) region defines a subset of the 4.1 superfamily and is a potential regulator of FERM domain function. BMC Genomics, 7, 85.

Baines, A.J. (2009) Evolution of spectrin function in cytoskeletal and membrane networks. Biochem Soc Trans, 37, 796–803.

Baines, A.J. (2010a) Evolution of the spectrin-based membrane skeleton. Transfus Clin Biol, 17, 95–103.

Baines, A.J. (2010b) The spectrin-ankyrin-4.1-adducin membrane skeleton: adapting eukaryotic cells to the demands of animal life. Protoplasma, 244, 99–131.

Baines, A.J., Bennett, P.M., Carter, E.W., and Terracciano, C. (2009a) Protein 4.1 and the control of ion channels. Blood Cells Mol Dis, 42, 211–215.

Baines, A.J., Bignone, P.A., King, M.D., Maggs, A.M., Bennett, P.M., Pinder, J.C., and Phillips, G.W. (2009b) The CKK domain (DUF1781) binds microtubules and defines the CAMSAP/ssp4 family of animal proteins. Mol Biol Evol, 26, 2005–2014.

Bairoch, A., Apweiler, R., Wu, C.H., Barker, W.C., Boeckmann, B., Ferro, S., Gasteiger, E., Huang, H., Lopez, R., Magrane, M. et al. (2005) The Universal Protein Resource (UniProt). Nucleic Acids Res, 33, D154–D159.

Baron, M.D., Davison, M.D., Jones, P., and Critchley, D.R. (1987) The sequence of chick alpha-actinin reveals homologies to spectrin and calmodulin. J Biol Chem, 262, 17623–17629.

Bateman, A., Birney, E., Cerruti, L., Durbin, R., Etwiller, L., Eddy, S.R., Griffiths-Jones, S., Howe, K.L., Marshall, M., and Sonnhammer, E.L. (2002) The Pfam protein families database. Nucleic Acids Res, 30, 276–280.

Bennett, P.B. (2000) Anchors Aweigh!: ion channels, cytoskeletal proteins, and cellular excitability. Circ Res, 86, 367–368.

Bennett, V. (1978) Purification of an active proteolytic fragment of the membrane attachment site for human erythrocyte spectrin. J Biol Chem, 253, 2292–2299.

Bennett, V. and Baines, A.J. (2001) Spectrin and ankyrin-based pathways: metazoan inventions for integrating cells into tissues. Physiol Rev, 81, 1353–1392.

Bennett, V. and Branton, D. (1977) Selective association of spectrin with the cytoplasmic surface of human erythrocyte plasma membranes. Quantitative determination with purified (32P)spectrin. J Biol Chem, 252, 2753–2763.

Bennett, V. and Chen, L. (2001) Ankyrins and cellular targeting of diverse membrane proteins to physiological sites. Curr Opin Cell Biol, 13, 61–67.

Bennett, V. and Stenbuck, P.J. (1979) The membrane attachment protein for spectrin is associated with band 3 in human erythrocyte membranes. Nature, 280, 468–473.

Bennett, V. and Stenbuck, P.J. (1980) Human erythrocyte ankyrin. Purification and properties. J Biol Chem, 255, 2540–2548.

Bignone, P.A. and Baines, A.J. (2003) Spectrin alphaII and betaII isoforms interact with high affinity at the tetramerization site. Biochem J, 374, 613–624.

Blin, G., Margeat, E., Carvalho, K., Royer, C.A., Roy, C., and Picart, C. (2008) Quantitative analysis of the binding of ezrin to large unilamellar vesicles containing phosphatidylinositol 4,5 bisphosphate. Biophys J, 94, 1021–1033.

Blomme, T., Vandepoele, K., De Bodt, S., Simillion, C., Maere, S., and Van de Peer, Y. (2006) The gain and loss of genes during 600 million years of vertebrate evolution. Genome Biol, 7, R43.

Bois, P.R.J., Borgon, R.A., Vonrhein, C., and Izard, T. (2005) Structural Dynamics of {alpha}-Actinin-Vinculin Interactions. Mol Cell Biol, 25, 6112–6122.

Borgia, M.B., Borgia, A., Best, R.B., Steward, A., Nettels, D., Wunderlich, B., Schuler, B., and Clarke, J. (2011) Single-molecule fluorescence reveals sequence-specific misfolding in multidomain proteins. Nature, 474(7353), 662–665.

Boron, W.F. (2010) Sharpey-Schafer lecture: gas channels. Exp Physiol, 95, 1107–1130.

Bruce, L.J., Beckmann, R., Ribeiro, M.L., Peters, L.L., Chasis, J.A., Delaunay, J., Mohandas, N., Anstee, D.J., and Tanner, M.J. (2003) A band 3-based macrocomplex of integral and peripheral proteins in the RBC membrane. Blood, 101, 4180–4188.

Burridge, K. and Feramisco, J.R. (1981) Non-muscle alpha actinins are calcium-sensitive actin-binding proteins. Nature, 294, 565–567.

Busby, B., Willis, C.D., Ackermann, M.A., Kontrogianni-Konstantopoulos, A., and Bloch, R.J. (2010) Characterization and comparison of two binding sites on obscurin for small ankyrin 1. Biochemistry, 49, 9948–9956.

Byers, T.J., Brandin, E., Lue, R.A., Winograd, E., and Branton, D. (1992) The complete sequence of *Drosophila* beta-spectrin reveals supra-motifs comprising eight 106-residue segments. Proc Natl Acad Sci USA, 89, 6187–6191.

Byers, T.J. and Branton, D. (1985) Visualization of the protein associations in the erythrocyte membrane skeleton. Proc Natl Acad Sci USA, 82, 6153–6157.

Byers, T.J., Husain-Chishti, A., Dubreuil, R.R., Branton, D., and Goldstein, L.S. (1989) Sequence similarity of the amino-terminal domain of *Drosophila* beta spectrin to alpha actinin and dystrophin. J Cell Biol, 109, 1633–1641.

Campbell, I.D. and Ginsberg, M.H. (2004) The talin-tail interaction places integrin activation on FERM ground. Trends Biochem Sci, 29, 429–435.

Castresana, J. and Saraste, M. (1995) Does Vav bind to F-actin through a CH domain? FEBS Lett, 374, 149–151.

Chan, W., Kordeli, E., and Bennett, V. (1993) 440-kD ankyrin-B: structure of the major developmentally regulated domain and selective localization in unmyelinated axons. J Cell Biol, 123, 1463–1473.

Chang, T.L., Cubillos, F.F., Kakhniashvili, D.G., and Goodman, S.R. (2004a) Ankyrin is a target of spectrin's E2/E3 ubiquitin-conjugating/ligating activity. Cell Mol Biol (Noisy-le-Grand), 50, 59–66.

Chang, T.L., Cubillosi, F.F., Kakhniashvili, D.G., and Goodman, S.R. (2004b) Band 3 is a target protein of spectrin's E2/E3 activity: implication for sickle cell disease and normal red blood cell aging. Cell Mol Biol (Noisy-le-Grand), 50, 171–177.

Chao, T.S. and Tao, M. (1991) Modulation of protein 4.1 binding to inside-out membrane vesicles by phosphorylation. Biochemistry, 30, 10529–10535.

Chishti, A.H., Kim, A.C., Marfatia, S.M., Lutchman, M., Hanspal, M., Jindal, H., Liu, S.C., Low, P.S., Rouleau, G.A., Mohandas, N. et al. (1998) The FERM domain: a unique module involved in the linkage of cytoplasmic proteins to the membrane. Trends Biochem Sci, 23, 281–282.

Cohen, A.M., Liu, S.C., Lawler, J., Derick, L., and Palek, J. (1988) Identification of the protein 4.1 binding site to phosphatidylserine vesicles. Biochemistry, 27, 614–619.

Cohen, C.M., Tyler, J.M., and Branton, D. (1980) Spectrin-actin associations studied by electron microscopy of shadowed preparations. Cell, 21, 875–883.

Collins, M.O., Yu, L., Campuzano, I., Grant, S.G., and Choudhary, J.S. (2008) Phosphoproteomic analysis of the mouse brain cytosol reveals a predominance of protein phosphorylation in regions of intrinsic sequence disorder. Mol Cell Proteomics, 7, 1331–1348.

Conboy, J. (1999) The role of alternative pre-mRNA splicing in regulating the structure and function of skeletal protein 4.1. Proc Soc Exp Biol Med, 220, 73–78.

Corey, D.P., Garcia-Anoveros, J., Holt, J.R., Kwan, K.Y., Lin, S.-Y., Vollrath, M.A., Amalfitano, A., Cheung, E.L.M., Derfler, B.H., Duggan, A. et al. (2004) TRPA1 is a candidate for the mechanosensitive transduction channel of vertebrate hair cells. Nature, 432, 723–730.

Correas, I., Leto, T.L., Speicher, D.W., and Marchesi, V.T. (1986) Identification of the functional site of erythrocyte protein 4.1 involved in spectrin-actin associations. J Biol Chem, 261, 3310–3315.

Davis, J.Q. and Bennett, V. (1990a) The anion exchanger and Na+K(+)-ATPase interact with distinct sites on ankyrin in in vitro assays. J Biol Chem, 265, 17252–17256.

Davis, J.Q. and Bennett, V. (1994) Ankyrin binding activity shared by the neurofascin/L1/NrCAM family of nervous system cell adhesion molecules. J Biol Chem, 269, 27163–27166.

Davis, L., Abdi, K., Machius, M., Brautigam, C., Tomchick, D.R., Bennett, V., and Michaely, P. (2008) Localization and structure of the ankyrin-binding site on beta 2-spectrin. J Biol Chem, 283, 26528–26537.

Davis, L.H. and Bennett, V. (1990b) Mapping the binding sites of human erythrocyte ankyrin for the anion exchanger and spectrin. J Biol Chem, 265, 10589–10596.

Dehal, P. and Boore, J.L. (2005) Two rounds of whole genome duplication in the ancestral vertebrate. PLoS Biol, 3, e314.

Del Rio, M., Imam, A., DeLeon, M., Gomez, G., Mishra, J., Ma, Q., Parikh, S., and Devarajan, P. (2004) The death domain of kidney ankyrin interacts with Fas and promotes Fas-mediated cell death in renal epithelia. J Am Soc Nephrol, 15, 41–51.

Devarajan, P., Scaramuzzino, D.A., and Morrow, J.S. (1994) Ankyrin binds to two distinct cytoplasmic domains of Na,K-ATPase alpha subunit. Proc Natl Acad Sci USA, 91, 2965–2969.

Discher, D.E., Janmey, P., and Wang, Y.L. (2005) Tissue cells feel and respond to the stiffness of their substrate. Science, 310, 1139–1143.

Discher, D.E., Mooney, D.J., and Zandstra, P.W. (2009) Growth factors, matrices, and forces combine and control stem cells. Science, 324, 1673–1677.

Djinovic-Carugo, K., Gautel, M., Ylanne, J., and Young, P. (2002) The spectrin repeat: a structural platform for cytoskeletal protein assemblies. FEBS Lett, 513, 119–123.

Dodge, J.T., Mitchell, C., and Hanahan, D.J. (1963) The preparation and chemical characteristics of hemoglobin-free ghosts of human erythrocytes. Arch Biochem Biophys, 100, 119–130.

Dreier, B., Mikheeva, G., Belousova, N., Parizek, P., Boczek, E., Jelesarov, I., Forrer, P., Pluckthun, A., and Krasnykh, V. (2011) Her2-specific multivalent adapters confer designed tropism to adenovirus for gene targeting. J Mol Biol, 405, 410–426.

Dubreuil, R.R., Byers, T.J., Stewart, C.T., and Kiehart, D.P. (1990) A beta-spectrin isoform from *Drosophila* (beta H) is similar in size to vertebrate dystrophin. J Cell Biol, 111, 1849–1858.

Dubreuil, R.R. and Wang, P. (2000) Genetic analysis of the requirements for α-actinin function. J Muscle Res Cell Motil, 21, 705–713.

Edwards, S.D. and Keep, N.H. (2001) The 2.7 A crystal structure of the activated FERM domain of moesin: an analysis of structural changes on activation. Biochemistry, 40, 7061–7068.

Endeward, V., Cartron, J.P., Ripoche, P., and Gros, G. (2006) Red cell membrane CO_2 permeability in normal human blood and in blood deficient in various blood groups, and effect of DIDS. Transfus Clin Biol, 13, 123–127.

Endeward, V., Cartron, J.P., Ripoche, P., and Gros, G. (2008) RhAG protein of the Rhesus complex is a CO_2 channel in the human red cell membrane. FASEB J, 22, 64–73.

Eng, K., Naqvi, N.I., Wong, K.C.Y., and Balasubramanian, M.K. (1998) Rng2p, a protein required for cytokinesis in fission yeast, is a component of the actomyosin ring and the spindle pole body. Curr Biol, 8, 611–621.

Ervasti, J.M. (2007) Dystrophin, its interactions with other proteins, and implications for muscular dystrophy. Biochim Biophys Acta - Mol Basis Dis, 1772, 108–117.

Ferreiro, D.U. and Komives, E.A. (2010) Molecular mechanisms of system control of NF-kappaB signaling by IkappaBalpha. Biochemistry, 49, 1560–1567.

Fievet, B.T., Gautreau, A., Roy, C., Del Maestro, L., Mangeat, P., Louvard, D., and Arpin, M. (2004) Phosphoinositide binding and phosphorylation act sequentially in the activation mechanism of ezrin. J Cell Biol, 164, 653–659.

Fowler, V.M. (1990) Tropomodulin: a cytoskeletal protein that binds to the end of erythrocyte tropomyosin and inhibits tropomyosin binding to actin. J Cell Biol, 111, 471–481.

Fowler, V.M. and Bennett, V. (1984) Erythrocyte membrane tropomyosin. Purification and properties. J Biol Chem, 259, 5978–5989.

Fraley, T.S., Tran, T.C., Corgan, A.M., Nash, C.A., Hao, J., Critchley, D.R., and Greenwood, J.A. (2003) Phosphoinositide binding inhibits α-actinin bundling activity. J Biol Chem, 278, 24039–24045.

Frame, M.C., Patel, H., Serrels, B., Lietha, D., and Eck, M.J. (2010) The FERM domain: organizing the structure and function of FAK. Nat Rev Mol Cell Biol, 11, 802–814.

Fukami, K., Sawada, N., Endo, T., and Takenawa, T. (1996) Identification of a phosphatidylinositol 4,5-bisphosphate-binding site in chicken skeletal muscle alpha-actinin. J Biol Chem, 271, 2646–2650.

Full, S.J., Deinzer, M.L., Ho, P.S., and Greenwood, J.A. (2007) Phosphoinositide Binding Regulates α-actinin CH_2 Domain Structure: Analysis by Hydrogen/Deuterium Exchange Mass Spectrometry, Cold Spring Harbor Laboratory Press, pp. 2597–2604.

Gagelin, C., Constantin, B., Deprette, C., Ludosky, M.A., Recouvreur, M., Cartaud, J., Cognard, C., Raymond, G., and Kordeli, E. (2002) Identification of Ank(G107), a muscle-specific ankyrin-G isoform. J Biol Chem, 277, 12978–12987.

Galkin, V.E., Orlova, A., Fattoum, A., Walsh, M.P., and Egelman, E.H. (2006) The CH-domain of Calponin does not Determine the Modes of Calponin Binding to F-actin. J Mol Biol, 359, 478–485.

Galkin, V.E., Orlova, A., Salmazo, A., Djinovic-Carugo, K., and Egelman, E.H. (2010) Opening of tandem calponin homology domains regulates their affinity for F-actin. Nat Struct Mol Biol, 17, 614–616.

Galkin, V.E., Orlova, A., VanLoock, M.S., Rybakova, I.N., Ervasti, J.M., and Egelman, E.H. (2002) The utrophin actin-binding domain binds F-actin in two different modes: implications for the spectrin superfamily of proteins. J Cell Biol, 157, 243–251.

Gardner, K. and Bennett, V. (1987) Modulation of spectrin-actin assembly by erythrocyte adducin. Nature, 328, 359–362.

Garver, T.D., Ren, Q., Tuvia, S., and Bennett, V. (1997) Tyrosine phosphorylation at a site highly conserved in the L1 family of cell adhesion molecules abolishes ankyrin binding and increases lateral mobility of neurofascin. J Cell Biol, 137, 703–714.

Gauthier, E., Guo, X., Mohandas, N., and An, X. (2011) Phosphorylation-dependent perturbations of the 4.1R-associated multiprotein complex of the erythrocyte membrane. Biochemistry, 50, 4561–4567.

Ghatpande, S.S. and Goodman, S.R. (2004) Ubiquitination of spectrin regulates the erythrocyte spectrin-protein-4.1-actin ternary complex dissociation: implications for the sickle cell membrane skeleton. Cell Mol Biol (Noisy-le-Grand), 50, 67–74.

Gifford, J.L., Walsh, M.P., and Vogel, H.J. (2007) Structures and metal-ion-binding properties of the Ca^{2+}-binding helix-loop-helix EF-hand motifs. Biochem J, 405, 199–221.

Gimona, M., Djinovic-Carugo, K., Kranewitter, W.J., and Winder, S.J. (2002) Functional plasticity of CH domains. FEBS Lett, 513, 98–106.

Gimona, M. and Mital, R. (1998) The single CH domain of calponin is neither sufficient nor necessary for F-actin binding. J Cell Sci, 111, 1813–1821.

Glenney, J.R. Jr., Glenney, P., Osborn, M., and Weber, K. (1982) An F-actin- and calmodulin-binding protein from isolated intestinal brush borders has a morphology related to spectrin. Cell, 28, 843–854.

Golji, J., Collins, R., and Mofrad, M.R.K. (2009) Molecular mechanics of the alpha-actinin rod domain: bending, torsional, and extensional behavior. PLoS Comput Biol, 5, e1000389.

Grzybek, M., Chorzalska, A., Bok, E., Hryniewicz-Jankowska, A., Czogalla, A., Diakowski, W., and Sikorski, A.F. (2006) Spectrin-phospholipid interactions. Existence of multiple kinds of binding sites? Chem Phys Lipids, 141, 133–141.

Guo, W.X., Nichol, M., and Merlie, J.P. (1996) Cloning and expression of full length mouse utrophin: the differential association of utrophin and dystrophin with AChR clusters. FEBS Lett, 398, 259–264.

Hall, A. (1993) Ras-related proteins. [Review]. Curr Opin Cell Biol, 5, 265–268.

Hall, T.G. and Bennett, V. (1987) Regulatory domains of erythrocyte ankyrin. J Biol Chem, 262, 10537–10545.

Hamada, K., Shimizu, T., Matsui, T., Tsukita, S., and Hakoshima, T. (2000) Structural basis of the membrane-targeting and unmasking mechanisms of the radixin FERM domain. EMBO J, 19, 4449–4462.

Hammarlund, M., Davis, W.S., and Jorgensen, E.M. (2000) Mutations in beta-spectrin disrupt axon outgrowth and sarcomere structure. J Cell Biol, 149, 931–942.

Hammarlund, M., Jorgensen, E.M., and Bastiani, M.J. (2007) Axons break in animals lacking beta-spectrin. J Cell Biol, 176, 269–275.

Han, B.G., Nunomura, W., Takakuwa, Y., Mohandas, N., and Jap, B.K. (2000) Protein 4.1R core domain structure and insights into regulation of cytoskeletal organization. Nat Struct Biol, 7, 871–875.

Hansen, J.C., Skalak, R., Chien, S., and Hoger, A. (1997) Spectrin properties and the elasticity of the red blood cell membrane skeleton. Biorheology, 34, 327–348.

Hao, J.J., Liu, Y., Kruhlak, M., Debell, K.E., Rellahan, B.L., and Shaw, S. (2009) Phospholipase C-mediated hydrolysis of PIP2 releases ERM proteins from lymphocyte membrane. J Cell Biol, 184, 451–462.

Hayashi, I. and Ikura, M. (2003) Crystal structure of the amino-terminal microtubule-binding domain of end-binding protein 1 (EB1). J Biol Chem, 278, 36430–36434.

Hirao, M., Sato, N., Kondo, T., Yonemura, S., Monden, M., Sasaki, T., Takai, Y., and Tsukita, S. (1996) Regulation mechanism of ERM (ezrin/radixin/moesin) protein/plasma membrane association: possible involvement of phosphatidylinositol turnover and Rho-dependent signaling pathway. J Cell Biol, 135, 37–51.

Hiroi, Y., Guo, Z., Li, Y., Beggs, A.H., and Liao, J.K. (2008) Dynamic regulation of endothelial NOS mediated by competitive interaction with alpha-actinin-4 and calmodulin. FASEB J, 22, 1450–1457.

Hopitzan, A.A., Baines, A.J., and Kordeli, E. (2006) Molecular evolution of ankyrin: gain of function in vertebrates by acquisition of an obscurin/titin-binding-related domain. Mol Biol Evol, 23, 46–55.

Hopitzan, A.A., Baines, A.J., Ludosky, M.A., Recouvreur, M., and Kordeli, E. (2005) Ankyrin-G in skeletal muscle: tissue-specific alternative splicing contributes to the complexity of the sarcolemmal cytoskeleton. Exp Cell Res, 309, 86–98.

Horne, W.C., Leto, T.L., and Marchesi, V.T. (1985) Differential phosphorylation of multiple sites in protein 4.1 and protein 4.9 by phorbol ester-activated and cyclic amp-dependent protein-kinases. J Biol Chem, 260, 9073–9076.

Hortsch, M., Paisley, K.L., Tian, M.Z., Qian, M., Bouley, M., and Chandler, R. (2002) The axonal localization of large *Drosophila* ankyrin2 protein isoforms is essential for neuronal functionality. Mol Cell Neurosci, 20, 43–55.

Howard, J. and Bechstedt, S. (2004) Hypothesis: a helix of ankyrin repeats of the NOMPC-TRP ion channel is the gating spring of mechanoreceptors. Curr Biol, 14, R224–R226.

Hsu, Y.J. and Goodman, S.R. (2005) Spectrin and ubiquitination: a review. Cell Mol Biol (Noisy-le-Grand), (51 Suppl), OL801–OL807.

Hu, R.J., Watanabe, M., and Bennett, V. (1992) Characterization of human brain cDNA encoding the general isoform of beta-spectrin. J Biol Chem, 267, 18715–18722.

Huang, X., Poy, F., Zhang, R., Joachimiak, A., Sudol, M., and Eck, M.J. (2000) Structure of a WW domain containing fragment of dystrophin in complex with beta-dystroglycan. Nat Struct Biol, 7, 634–638.

Hubbard, T., Barker, D., Birney, E., Cameron, G., Chen, Y., Clark, L., Cox, T., Cuff, J., Curwen, V., Down, T. et al. (2002) The Ensembl genome database project. Nucleic Acids Res, 30, 38–41.

Ilsley, J.L., Sudol, M., and Winder, S.J. (2002) The WW domain: linking cell signalling to the membrane cytoskeleton. Cell Signal, 14, 183–189.

Imamura, M., Endo, T., Kuroda, M., Tanaka, T., and Masaki, T. (1988) Substructure and higher structure of chicken smooth muscle alpha-actinin molecule. J Biol Chem, 263, 7800–7805.

Ipsaro, J.J., Harper, S.L., Messick, T.E., Marmorstein, R., Mondragon, A., and Speicher, D.W. (2010) Crystal structure and functional interpretation of the erythrocyte spectrin tetramerization domain complex. Blood, 115, 4843–4852.

Ipsaro, J.J., Huang, L., Gutierrez, L., and MacDonald, R.I. (2008) Molecular epitopes of the ankyrin-spectrin interaction. Biochemistry, 47, 7452–7464.

Ipsaro, J.J. and Mondragon, A. (2010) Structural basis for spectrin recognition by ankyrin. Blood, 115, 4093–4101.

Janke, M., Herrig, A., Austermann, J., Gerke, V., Steinem, C., and Janshoff, A. (2008) Actin binding of ezrin is activated by specific recognition of PIP2-functionalized lipid bilayers. Biochemistry, 47, 3762–3769.

Jarolim, P., Lahav, M., Liu, S.C., and Palek, J. (1990) Effect of hemoglobin oxidation products on the stability of red cell membrane skeletons and the associations of skeletal proteins: correlation with a release of hemin. Blood, 76, 2125–2131.

Jefferson, J.J., Ciatto, C., Shapiro, L., and Liem, R.K. (2007) Structural analysis of the plakin domain of bullous pemphigoid antigen1 (BPAG1) suggests that plakins are members of the spectrin superfamily. J Mol Biol, 366, 244–257.

Jenkins, S.M., Kizhatil, K., Kramarcy, N.R., Sen, A., Sealock, R., and Bennett, V. (2001) FIGQY phosphorylation defines discrete populations of L1 cell adhesion molecules at sites of cell-cell contact and in migrating neurons. J Cell Sci, 114, 3823–3835.

Johnson, C.P., Gaetani, M., Ortiz, V., Bhasin, N., Harper, S., Gallagher, P.G., Speicher, D.W., and Discher, D.E. (2007a) Pathogenic proline mutation in the linker between spectrin repeats: disease caused by spectrin unfolding. Blood, 109, 3538–3543.

Johnson, C.P., Tang, H.Y., Carag, C., Speicher, D.W., and Discher, D.E. (2007b) Forced unfolding of proteins within cells. Science, 317, 663–666.

Joshi, R. and Bennett, V. (1990) Mapping the domain structure of human erythrocyte adducin. J Biol Chem, 265, 13130–13136.

Kakhniashvili, D.G., Chaudhary, T., Zimmer, W.E., Bencsath, F.A., Jardine, I., and Goodman, S.R. (2001) Erythrocyte spectrin is an E2 ubiquitin conjugating enzyme. Biochemistry, 40, 11630–11642.

Karinch, A.M., Zimmer, W.E., and Goodman, S.R. (1990) The identification and sequence of the actin-binding domain of human red blood cell beta-spectrin. J Biol Chem, 265, 11833–11840.

Kawe, M., Forrer, P., Amstutz, P., and Pluckthun, A. (2006) Isolation of intracellular proteinase inhibitors derived from designed ankyrin repeat proteins by genetic screening. J Biol Chem, 281, 40252–40263.

Keep, N.H., Norwood, F.L., Moores, C.A., Winder, S.J., and Kendrick-Jones, J. (1999) The 2.0 A structure of the second calponin homology domain from the actin-binding region of the dystrophin homologue utrophin. J Mol Biol, 285, 1257–1264.

Kelly, D.F. and Taylor, K.A. (2005) Identification of the beta1-integrin binding site on alpha-actinin by cryoelectron microscopy. J Struct Biol, 149, 290–302.

Khan, A.A., Hanada, T., Mohseni, M., Jeong, J.J., Zeng, L., Gaetani, M., Li, D., Reed, B.C., Speicher, D.W., and Chishti, A.H. (2008) Dematin and adducin provide a novel link between the spectrin cytoskeleton and human erythrocyte membrane by directly interacting with glucose transporter-1. J Biol Chem, 283, 14600–14609.

Kim, M., Abdi, K., Lee, G., Rabbi, M., Lee, W., Yang, M., Schofield, C.J., Bennett, V., and Marszalek, P.E. (2010) Fast and forceful refolding of stretched alpha-helical solenoid proteins. Biophys J, 98, 3086–3092.

King, N. (2005) Choanoflagellates. Curr Biol, 15, R113–R114.

Klaavuniemi, T., Alho, N., Hotulainen, P., Kelloniemi, A., Havukainen, H., Permi, P., Mattila, S., and Ylanne, J. (2009) Characterization of the interaction between Actinin-Associated LIM Protein (ALP) and the rod domain of alpha-actinin. BMC Cell Biol, 10, 22.

Koenig, M., Monaco, A.P., and Kunkel, L.M. (1988) The complete sequence of dystrophin predicts a rod-shaped cytoskeletal protein. Cell, 53, 219–228.

Korsgren, C. and Lux, S.E. (2010) The carboxyterminal EF domain of erythroid alpha-spectrin is necessary for optimal spectrin-actin binding. Blood, 116, 2600–2607.

Korsgren, C., Peters, L.L., and Lux, S.E. (2010) Protein 4.2 binds to the carboxyl-terminal EF-hands of erythroid alpha-spectrin in a calcium- and calmodulin-dependent manner. J Biol Chem, 285, 4757–4770.

Krieger, C.C., An, X., Tang, H.Y., Mohandas, N., Speicher, D.W., and Discher, D.E. (2011) Cysteine shotgun-mass spectrometry (CS-MS) reveals dynamic sequence of protein structure changes within mutant and stressed cells. Proc Natl Acad Sci USA, 108, 8269–8274.

Kuhlman, P.A., Hughes, C.A., Bennett, V., and Fowler, V.M. (1996) A new function for adducin. Calcium/calmodulin-regulated capping of the barbed ends of actin filaments. J Biol Chem, 271, 7986–7991.

Kusunoki, H., MacDonald, R.I., and Mondragon, A. (2004) Structural insights into the stability and flexibility of unusual erythroid spectrin repeats. Structure (Camb), 12, 645–656.

Lambert, S., Yu, H., Prchal, J.T., Lawler, J., Ruff, P., Speicher, D., Cheung, M.C., Kan, Y.W., and Palek, J. (1990) cDNA sequence for human erythrocyte ankyrin. Proc Natl Acad Sci USA, 87, 1730–1734.

Law, R., Carl, P., Harper, S., Dalhaimer, P., Speicher, D.W., and Discher, D.E. (2003a) Cooperativity in forced unfolding of tandem spectrin repeats. Biophys J, 84, 533–544.

Law, R., Harper, S., Speicher, D.W., and Discher, D.E. (2004) Influence of lateral association on forced unfolding of antiparallel spectrin heterodimers. J Biol Chem, 279, 16410–16416 Epub 12004 Feb 16413.

Law, R., Liao, G., Harper, S., Yang, G., Speicher, D.W., and Discher, D.E. (2003b) Pathway shifts and thermal softening in temperature-coupled forced unfolding of spectrin domains. Biophys J, 85, 3286–3293.

Leclerc, E. and Vetter, S. (1998) Characterization of a calcium-dependent calmodulin-binding domain in the 135-kD human protein 4.1 isoform. Eur J Biochem, 258, 567–571.

Lee, G., Abdi, K., Jiang, Y., Michaely, P., Bennett, V., and Marszalek, P.E. (2006) Nanospring behaviour of ankyrin repeats. Nature, 440, 246–249.

Leinweber, B.D., Leavis, P.C., Grabarek, Z., Wang, C.L., and Morgan, K.G. (1999) Extracellular regulated kinase (ERK) interaction with actin and the calponin homology (CH) domain of actin-binding proteins. Biochem J, 344(Pt 1), 117–123.

Lepage, P.K., Lussier, M.P., McDuff, F.-O., Lavigne, P., and Boulay, G. (2009) The self-association of two N-terminal interaction domains plays an important role in the tetramerization of TRPC4. Cell Calcium, 45, 251–259.

Leto, T.L. and Marchesi, V.T. (1984) A Structural Model Of Human-Erythrocyte Protein-4.1. J Biol Chem, 259, 4603–4608.

Letunic, I., Goodstadt, L., Dickens, N.J., Doerks, T., Schultz, J., Mott, R., Ciccarelli, F., Copley, R.R., Ponting, C.P., and Bork, P. (2002) Recent improvements to the SMART domain-based sequence annotation resource. Nucleic Acids Res, 30, 242–244.

Levine, B.A., Moir, A., Patchell, V.B., and Perry, S.V. (1992) Binding-sites involved in the interaction of actin with the n-terminal region of dystrophin. FEBS Lett, 298, 44–48.

Lewis, I.A., Campanella, M.E., Markley, J.L., and Low, P.S. (2009) Role of band 3 in regulating metabolic flux of red blood cells. Proc Natl Acad Sci USA, 106, 18515–18520.

Li, D., Tang, H.Y., and Speicher, D.W. (2008) A structural model of the erythrocyte spectrin heterodimer initiation site determined using homology modeling and chemical cross-linking. J Biol Chem, 283, 1553–1562.

Li, J., Mahajan, A., and Tsai, M.D. (2006) Ankyrin repeat: a unique motif mediating protein-protein interactions. Biochemistry, 45, 15168–15178.

Lim, S.T., Mikolon, D., Stupack, D.G., and Schlaepfer, D.D. (2008) FERM control of FAK function: implications for cancer therapy. Cell Cycle, 7, 2306–2314.

Liu, J., Taylor, D.W., and Taylor, K.A. (2004) A 3-D reconstruction of smooth muscle alpha-actinin by CryoEm reveals two different conformations at the actin-binding region. J Mol Biol, 338, 115–125.

Liu, S.C., Derick, L.H., and Palek, J. (1987) Visualization of the hexagonal lattice in the erythrocyte membrane skeleton. J Cell Biol, 104, 527–536.

Liu, S.C., Zhai, S., Lawler, J., and Palek, J. (1985) Hemin-mediated dissociation of erythrocyte membrane skeletal proteins. J Biol Chem, 260, 12234–12239.

Lopez, C., Metral, S., Eladari, D., Drevensek, S., Gane, P., Chambrey, R., Bennett, V., Cartron, J.P., Le Van Kim, C., and Colin, Y. (2005) The ammonium transporter RhBG: requirement of a tyrosine-based signal and ankyrin-G for basolateral targeting and membrane anchorage in polarized kidney epithelial cells. J Biol Chem, 280, 8221–8228.

Lorenz, S., Vakonakis, I., Lowe, E.D., Campbell, I.D., Noble, M.E., and Hoellerer, M.K. (2008) Structural analysis of the interactions between paxillin LD motifs and alpha-parvin. Structure, 16, 1521–1531.

Luckey, M. (2008) Membrane Structural Biology: With Biochemical and Biophysical Foundations, Cambridge University Press, Cambridge; New York.

Lundberg, S., Bjork, J., Lofvenberg, L., and Backman, L. (1995) Cloning, expression and characterization of two putative calcium-binding sites in human non-erythroid alpha-spectrin. Eur J Biochem, 230, 658–665.

Lundberg, S., Buevich, A.V., Sethson, I., Edlund, U., and Backman, L. (1997) Calcium-binding mechanism of human nonerythroid alpha-spectrin EF-structures. Biochemistry, 36, 7199–7208.

Lux, S.E., John, K.M., and Bennett, V. (1990) Analysis of cDNA for human erythrocyte ankyrin indicates a repeated structure with homology to tissue-differentiation and cell-cycle control proteins. Nature, 344, 36–42.

Maiweilidan, Y., Klauza, I., and Kordeli, E. (2011) Novel interactions of ankyrins-G at the costameres: the muscle-specific Obscurin/Titin-Binding-related Domain (OTBD) binds plectin and filamin C. Exp Cell Res, 317, 724–736.

Malinin, N.L., Plow, E.F., and Byzova, T.V. (2010) Kindlins in FERM adhesion. Blood, 115, 4011–4017.

Manno, S., Takakuwa, Y., and Mohandas, N. (2002) Identification of a functional role for lipid asymmetry in biological membranes: phosphatidylserine-skeletal protein interactions modulate membrane stability. Proc Natl Acad Sci USA, 99, 1943–1948.

Manno, S., Takakuwa, Y., and Mohandas, N. (2005) Modulation of erythrocyte membrane mechanical function by protein 4.1 phosphorylation. J Biol Chem, 280, 7581–7587.

Marchesi, V.T. and Steers, E. Jr. (1968) Selective solubilization of a protein component of the red cell membrane. Science, 159, 203–204.

Margulis, L., Chapman, M., Guerrero, R., and Hall, J. (2006) The last eukaryotic common ancestor (LECA): acquisition of cytoskeletal motility from aerotolerant spirochetes in the Proterozoic Eon. Proc Natl Acad Sci USA, 103, 13080–13085.

Margulis, L., Dolan, M.F., and Guerrero, R. (2000) The chimeric eukaryote: origin of the nucleus from the karyomastigont in amitochondriate protists. Proc Natl Acad Sci USA, 97, 6954–6959.

McKeown, C., Praitis, V., and Austin, J. (1998) sma-1 encodes a betaH-spectrin homolog required for *Caenorhabditis elegans* morphogenesis. Development, 125, 2087–2098.

Michaely, P. and Bennett, V. (1995) Mechanism for binding site diversity on ankyrin. Comparison of binding sites on ankyrin for neurofascin and the Cl-/HCO_3- anion exchanger. J Biol Chem, 270, 31298–31302.

Michaely, P., Tomchick, D.R., Machius, M., and Anderson, R.G. (2002) Crystal structure of a 12 ANK repeat stack from human ankyrinR. EMBO J, 21, 6387–6396.

Mimura, N. and Asano, A. (1986) Isolation and characterization of a conserved actin-binding domain from rat hepatic actinogelin, rat skeletal muscle, and chicken gizzard alpha-actinins. J Biol Chem, 261, 10680–10687.

Mirza, A., Sagathevan, M., Sahni, N., Choi, L., and Menhart, N. (2010) A biophysical map of the dystrophin rod. Biochim Biophys Acta, 1804, 1796–1809.

Mohandas, N. and Gallagher, P.G. (2008) Red cell membrane: past, present, and future. Blood, 112, 3939–3948.

Mohler, P.J. and Bennett, V. (2005) Ankyrin-based cardiac arrhythmias: a new class of channelopathies due to loss of cellular targeting. Curr Opin Cardiol, 20, 189–193.

Mohler, P.J., Davis, J.Q., and Bennett, V. (2005) Ankyrin-B coordinates the Na/K ATPase, Na/Ca exchanger, and InsP3 receptor in a cardiac T-tubule/SR microdomain. PLoS Biol, 3, e423.

Mohler, P.J., Davis, J.Q., Davis, L.H., Hoffman, J.A., Michaely, P., and Bennett, V. (2004a) Inositol 1,4,5-trisphosphate receptor localization and stability in neonatal cardiomyocytes requires interaction with ankyrin-B. J Biol Chem, 279, 12980–12987.

Mohler, P.J., Gramolini, A.O., and Bennett, V. (2002) The ankyrin-B C-terminal domain determines activity of ankyrin-B/G chimeras in rescue of abnormal inositol 1,4,5-trisphosphate and ryanodine receptor distribution in ankyrin-B (/) neonatal cardiomyocytes. J Biol Chem, 277, 10599–10607.

Mohler, P.J., Rivolta, I., Napolitano, C., LeMaillet, G., Lambert, S., Priori, S.G., and Bennett, V. (2004b) Nav1.5 E1053K mutation causing Brugada syndrome blocks binding to ankyrin-G and expression of Nav1.5 on the surface of cardiomyocytes. Proc Natl Acad Sci USA, 101, 17533–17538.

Mohler, P.J., Yoon, W., and Bennett, V. (2004c) Ankyrin-B targets beta2-spectrin to an intracellular compartment in neonatal cardiomyocytes. J Biol Chem, 279, 40185–40193 Epub 42004 Jul 40119.

Moores, C.A., Keep, N.H., and Kendrick-Jones, J. (2000) Structure of the utrophin actin-binding domain bound to F-actin reveals binding by an induced fit mechanism. J Mol Biol, 297, 465–480.

Moores, C.A. and Kendrick-Jones, J. (2000) Biochemical characterisation of the actin-binding properties of utrophin. Cell Motil Cytoskeleton, 46, 116–128.

Moorthy, S., Chen, L., and Bennett, V. (2000) *Caenorhabditis elegans* beta-G spectrin is dispensable for establishment of epithelial polarity, but essential for muscular and neuronal function. J Cell Biol, 149, 915–930.

Morrow, J.S., Haigh, W.B. Jr., and Marchesi, V.T. (1981) Spectrin oligomers: a structural feature of the erythrocyte cytoskeleton. J Supramol Struct Cell Biochem, 17, 275–287.

Mosavi, L.K., Cammett, T.J., Desrosiers, D.C., and Peng, Z.Y. (2004) The ankyrin repeat as molecular architecture for protein recognition. Protein Sci, 13, 1435–1448.

Munch, R.C., Muhlebach, M.D., Schaser, T., Kneissl, S., Jost, C., Pluckthun, A., Cichutek, K., and Buchholz, C.J. (2011) DARPins: an efficient targeting domain for lentiviral vectors. Mol Ther, 19, 686–693.

Murzin, A.G., Brenner, S.E., Hubbard, T., and Chothia, C. (1995) SCOP: a structural classification of proteins database for the investigation of sequences and structures. J Mol Biol, 247, 536–540.

Nakano, K., Bunai, F., and Numata, O. (2005) Stg1 is a novel SM22/transgelin-like actin-modulating protein in fission yeast. FEBS Lett, 579, 6311–6316.

Neuman, S., Kaban, A., Volk, T., Yaffe, D., and Nudel, U. (2001) The dystrophin/utrophin homologues in Drosophila and in sea urchin. Gene, 263, 17–29.

Neurath, H. (1980) Limited proteolysis, protein folding and physiological regulation, in Protein Folding (ed R. Jaenicke), Elsevier/North Holland Biomedical Press, Amsterdam-New York, pp. 501–504.

Nicolas, V., Le Van Kim, C., Gane, P., Birkenmeier, C., Cartron, J.P., Colin, Y., and Mouro-Chanteloup, I. (2003) Rh-RhAG/ankyrin-R, a new interaction site between the membrane bilayer and the red cell skeleton, is impaired by Rh(null)-associated mutation. J Biol Chem, 278, 25526–25533.

Niggli, V., Andreoli, C., Roy, C., and Mangeat, P. (1995) Identification of a phosphatidylinositol-4,5-bisphosphate-binding domain in the N-terminal region of ezrin. FEBS Lett, 376, 172–176.

Norwood, F.L., Sutherland-Smith, A.J., Keep, N.H., and Kendrick-Jones, J. (2000) The structure of the N-terminal actin-binding domain of human dystrophin and how mutations in this domain may cause Duchenne or Becker muscular dystrophy. Struct Fold Des, 8, 481–491.

Nunomura, W., Parra, M., Hebiguchi, M., Sawada, K.I., Mohandas, N., and Takakuwa, Y. (2009) Marked difference in membrane protein binding properties of the two isoforms of protein 4.1R expressed at early and late stages of erythroid differentiation. Biochem J, 417, 141–148.

Ohanian, V., Wolfe, L.C., John, K.M., Pinder, J.C., Lux, S.E., and Gratzer, W.B. (1984) Analysis of the ternary interaction of the red cell membrane skeletal proteins spectrin, actin, and 4.1. Biochemistry, 23, 4416–4420.

Pan, Z., Kao, T., Horvath, Z., Lemos, J., Sul, J.Y., Cranstoun, S.D., Bennett, V., Scherer, S.S., and Cooper, E.C. (2006) A common ankyrin-G-based mechanism retains KCNQ and NaV channels at electrically active domains of the axon. J Neurosci, 26, 2599–2613.

Pascual, J., Pfuhl, M., Rivas, G., Pastore, A., and Saraste, M. (1996) The spectrin repeat folds into a three-helix bundle in solution. FEBS Lett, 383, 201–207.

Peters, L.L., John, K.M., Lu, F.M., Eicher, E.M., Higgins, A., Yialamas, M., Turtzo, L.C., Otsuka, A.J., and Lux, S.E. (1995) Ank3 (epithelial ankyrin), a widely distributed new member of the ankyrin gene family and the major ankyrin in kidney, is expressed in alternatively spliced forms, including forms that lack the repeat domain. J Cell Biol, 130, 313–330.

Phelps, C.B., Wang, R.R., Choo, S.S., and Gaudet, R. (2010) Differential regulation of TRPV1, TRPV3, and TRPV4 sensitivity through a conserved binding site on the ankyrin repeat domain. J Biol Chem, 285, 731–740.

Pinder, J.C., Gardner, B., and Gratzer, W.B. (1995) Interaction of protein 4.1 with the red cell membrane: effects of phosphorylation by protein kinase C. Biochem Biophys Res Commun, 210, 478–482.

Pinder, J.C. and Gratzer, W.B. (1983) Structural and dynamic states of actin in the erythrocyte. J Cell Biol, 96, 768–775.

Ponting, C.P., Aravind, L., Schultz, J., Bork, P., and Koonin, E.V. (1999) Eukaryotic signalling domain homologues in archaea and bacteria. Ancient Ancestry and Horizontal Gene Transfer. J Mol Biol, 289, 729–745.

Rentschler, S., Linn, H., Deininger, K., Bedford, M.T., Espanel, X., and Sudol, M. (1999) The WW domain of dystrophin requires EF-hands region to interact with beta-dystroglycan. Biol Chem, 380, 431–442.

Ripoche, P., Goossens, D., Devuyst, O., Gane, P., Colin, Y., Verkman, A.S., and Cartron, J.P. (2006) Role of RhAG and AQP1 in NH_3 and CO_2 gas transport in red cell ghosts: a stopped-flow analysis. Transfus Clin Biol, 13, 117–122.

Robledo, R.F., Lambert, A.J., Birkenmeier, C.S., Cirlan, M.V., Cirlan, A.F., Campagna, D.R., Lux, S.E., and Peters, L.L. (2010) Analysis of novel sph (spherocytosis) alleles in mice reveals allele-specific loss of band 3 and adducin in alpha-spectrin-deficient red cells. Blood, 115, 1804–1814.

Roper, K., Gregory, S.L., and Brown, N.H. (2002) The 'Spectraplakins': cytoskeletal giants with characteristics of both spectrin and plakin families. J Cell Sci, 115, 4215–4225.

Salomao, M., An, X., Guo, X., Gratzer, W.B., Mohandas, N., and Baines, A.J. (2006) Mammalian alphaI-spectrin is a neofunctionalized polypeptide adapted to small highly deformable erythrocytes. Proc Natl Acad Sci USA, 103, 643–648.

Salomao, M., Zhang, X., Yang, Y., Lee, S., Hartwig, J.H., Chasis, J.A., Mohandas, N., and An, X. (2008) Protein 4.1R-dependent multiprotein complex: new insights into the structural organization of the red blood cell membrane. Proc Natl Acad Sci USA, 105, 8026–8031.

Satchwell, T.J., Shoemark, D.K., Sessions, R.B., and Toye, A.M. (2009) Protein 4.2: a complex linker. Blood Cells Mol Dis, 42, 201–210.

Schweizer, A., Roschitzki-Voser, H., Amstutz, P., Briand, C., Gulotti-Georgieva, M., Prenosil, E., Binz, H.K., Capitani, G., Baici, A., Pluckthun, A. et al. (2007) Inhibition of caspase-2 by a designed ankyrin repeat protein: specificity, structure, and inhibition mechanism. Structure, 15, 625–636.

Scott, C., Phillips, G.W., and Baines, A.J. (2001) Properties of the C-terminal domain of 4.1 proteins. Eur J Biochem, 268, 3709–3717.

Seo, P.S., Jeong, J.J., Zeng, L., Takoudis, C.G., Quinn, B.J., Khan, A.A., Hanada, T., and Chishti, A.H. (2009) Alternatively spliced exon 5 of the FERM domain of protein 4.1R encodes a novel binding site for erythrocyte p55 and is critical for membrane targeting in epithelial cells. Biochim Biophys Acta, 1793, 281–289.

Shahbakhti, F. and Gratzer, W.B. (1986) Analysis of the self-association of human red cell spectrin. Biochemistry, 25, 5969–5975.

Shotton, D.M., Burke, B.E., and Branton, D. (1979) The molecular structure of human erythrocyte spectrin. Biophysical and electron microscopic studies. J Mol Biol, 131, 303–329.

Siciliano, A., Turrini, F., Bertoldi, M., Matte, A., Pantaleo, A., Olivieri, O., and De Franceschi, L. (2010) Deoxygenation affects tyrosine phosphoproteome of red cell membrane from patients with sickle cell disease. Blood Cells Mol Dis, 44, 233–242.

Sjöblom, B., Ylänne, J., and Djinovic-Carugo, K. (2008) Novel structural insights into F-actin-binding and novel functions of calponin homology domains. Curr Opin Struct Biol, 18, 702–708.

Speicher, D.W. and Marchesi, V.T. (1982) Spectrin domains: proteolytic susceptibility as a probe of protein structure. J Cell Biochem, 18, 479–492.

Speicher, D.W. and Marchesi, V.T. (1984) Erythrocyte spectrin is comprised of many homologous triple helical segments. Nature, 311, 177–180.

Speicher, D.W., Morrow, J.S., Knowles, W.J., and Marchesi, V.T. (1980) Identification of proteolytically resistant domains of human erythrocyte spectrin. Proc Natl Acad Sci USA, 77, 5673–5677.

Speicher, D.W., Morrow, J.S., Knowles, W.J., and Marchesi, V.T. (1982) A structural model of human erythrocyte spectrin. Alignment of chemical and functional domains. J Biol Chem, 257, 9093–9101.

Stabach, P.R. and Morrow, J.S. (2000) Identification and characterization of beta V spectrin, a mammalian ortholog of *Drosophila* beta H spectrin. J Biol Chem, 275, 21385–21395.

Stabach, P.R., Simonovic, I., Ranieri, M.A., Aboodi, M.S., Steitz, T.A., Simonovic, M., and Morrow, J.S. (2009) The structure of the ankyrin-binding site of beta-spectrin reveals how tandem spectrin-repeats generate unique ligand-binding properties. Blood, 113, 5377–5384.

Stankewich, M.C., Tse, W.T., Peters, L.L., Ch'ng, Y., John, K.M., Stabach, P.R., Devarajan, P., Morrow, J.S., and Lux, S.E. (1998) A widely expressed betaIII spectrin associated with Golgi and cytoplasmic vesicles. Proc Natl Acad Sci USA, 95, 14158–14163.

Steck, T.L. and Kant, J.A. (1974) Preparation of impermeable ghosts and inside-out vesicles from human erythrocyte membranes. Methods Enzymol, 31, 172–180.

Stumpp, M.T. and Amstutz, P. (2007) DARPins: a true alternative to antibodies. Curr Opin Drug Discov Dev, 10, 153–159.

Subrahmanyam, G., Bertics, P.J., and Anderson, R.A. (1991) Phosphorylation of protein 4.1 on tyrosine-418 modulates its function in vitro. Proc Natl Acad Sci USA, 88, 5222–5226.

Tang, J., Taylor, D.W., and Taylor, K.A. (2001) The three-dimensional structure of alpha-actinin obtained by cryoelectron microscopy suggests a model for $Ca^{(2+)}$-dependent actin binding. J Mol Biol, 310, 845–858.

Tee, J.M. and Peppelenbosch, M.P. (2010) Anchoring skeletal muscle development and disease: the role of ankyrin repeat domain containing proteins in muscle physiology. Crit Rev Biochem Mol Biol, 45, 318–330.

Tepass, U. (2009) FERM proteins in animal morphogenesis. Curr Opin Genet Dev, 19, 357–367.

Tilney, L.G. and Detmers, P. (1975) Actin in erythrocyte ghosts and its association with spectrin. Evidence for a nonfilamentous form of these two molecules in situ. J Cell Biol, 66, 508–520.

Travé, G., Lacombe, P.J., Pfuhl, M., Saraste, M., and Pastore, A. (1995) Molecular mechanism of the calcium-induced conformational change in the spectrin EF-hands. EMBO J, 14, 4922–4931.

Trave, G., Pastore, A., Hyvonen, M., and Saraste, M. (1995) The C-terminal domain of alpha-spectrin is structurally related to calmodulin. Eur J Biochem, 227, 35–42.

Tse, W.T., Tang, J., Jin, O., Korsgren, C., John, K.M., Kung, A.L., Gwynn, B., Peters, L.L., and Lux, S.E. (2001) A new spectrin, beta IV, has a major truncated isoform that associates with promyelocytic leukemia protein nuclear bodies and the nuclear matrix. J Biol Chem, 276, 23974–23985.

Tyler, J.M., Hargreaves, W.R., and Branton, D. (1979) Purification of two spectrin-binding proteins: biochemical and electron microscopic evidence for site-specific reassociation between spectrin and bands 2.1 and 4.1. Proc Natl Acad Sci USA, 76, 5192–5196.

Ungewickell, E., Bennett, P.M., Calvert, R., Ohanian, V., and Gratzer, W.B. (1979) In vitro formation of a complex between cytoskeletal proteins of the human erythrocyte. Nature, 280, 811–814.

Ursitti, J.A. and Fowler, V.M. (1994) Immunolocalization of tropomodulin, tropomyosin and actin in spread human erythrocyte skeletons. J Cell Sci, 107, 1633–1639.

Ursitti, J.A., Kotula, L., DeSilva, T.M., Curtis, P.J., and Speicher, D.W. (1996) Mapping the human erythrocyte beta-spectrin dimer initiation site using recombinant peptides and correlation of its phasing with the alpha-actinin dimer site. J Biol Chem, 271, 6636–6644.

Vancompernolle, K., Gimona, M., Herzog, M., Van Damme, J., Vandekerckhove, J., and Small, V. (1990) Isolation and sequence of a tropomyosin-binding fragment of turkey gizzard calponin. FEBS Lett, 274, 146–150.

Virel, A. and Backman, L. (2004) Molecular evolution and structure of alpha-actinin. Mol Biol Evol, 21, 1024–1031.

Virel, A. and Backman, L. (2007) A comparative and phylogenetic analysis of the alpha-actinin rod domain. Mol Biol Evol, 24, 2254–2265.

Vogel, C., Berzuini, C., Bashton, M., Gough, J., and Teichmann, S.A. (2004) Supra-domains: evolutionary units larger than single protein domains. J Mol Biol, 336, 809–823.

Wang, X., Fukuda, K., Byeon, I.J., Velyvis, A., Wu, C., Gronenborn, A., and Qin, J. (2008) The structure of alpha-parvin CH_2-paxillin LD1 complex reveals a novel modular recognition for focal adhesion assembly. J Biol Chem, 283, 21113–21119.

Warren, D.T., Zhang, Q., Weissberg, P.L., and Shanahan, C.M. (2005) Nesprins: intracellular scaffolds that maintain cell architecture and coordinate cell function? Expert Rev Mol Med, 7, 1–15.

Way, M., Pope, B., Cross, R.A., Kendrickjones, J., and Weeds, A.G. (1992) Expression of the n-terminal domain of dystrophin in escherichia-coli and demonstration of binding to f-actin. FEBS Lett, 301, 243–245.

Weaver, D.C. and Marchesi, V.T. (1984) The structural basis of ankyrin function. I. Identification of two structural domains. J Biol Chem, 259, 6165–6169.

Weaver, D.C., Pasternack, G.R., and Marchesi, V.T. (1984) The structural basis of ankyrin function. II. Identification of two functional domains. J Biol Chem, 259, 6170–6175.

Winder, S.J., Hemmings, L., MacIver, S.K., Bolton, S.J., Tinsley, J.M., Davies, K.E., Critchley, D.R., and Kendrick-Jones, J. (1995) Utrophin actin binding domain: analysis of actin binding and cellular targeting. J Cell Sci, 108, 63–71.

Winkelmann, J.C., Chang, J.G., Tse, W.T., Scarpa, A.L., Marchesi, V.T., and Forget, B.G. (1990) Full-length sequence of the cDNA for human erythroid beta-spectrin. J Biol Chem, 265, 11827–11832.

Wu, J.Q., Bahler, J., and Pringle, J.R. (2001) Roles of a fimbrin and an alpha-actinin-like protein in fission yeast cell polarization and cytokinesis. Mol Biol Cell, 12, 1061–1077.

Wu, X., Kodama, A., and Fuchs, E. (2008) ACF7 regulates cytoskeletal-focal adhesion dynamics and migration and has ATPase activity. Cell, 135, 137–148.

Yabana, N. and Shibuya, M. (2002) Adaptor protein APS binds the NH_2-terminal autoinhibitory domain of guanine nucleotide exchange factor Vav3 and augments its activity. Oncogene, 21, 7720–7729.

Yan, Y., Winograd, E., Viel, A., Cronin, T., Harrison, S.C., and Branton, D. (1993) Crystal structure of the repetitive segments of spectrin. Science, 262, 2027–2030.

Ylanne, J., Scheffzek, K., Young, P., and Saraste, M. (2001) Crystal structure of the alpha-actinin rod reveals an extensive torsional twist. Structure (Camb), 9, 597–604.

Young, P. and Gautel, M. (2000) The interaction of titin and alpha-actinin is controlled by a phospholipid-regulated intramolecular pseudoligand mechanism. EMBO J, 19, 6331–6340.

Yu, H., Zhang, Y., Ye, L., and Jiang, W.G. (2011) The FERM family proteins in cancer invasion and metastasis. Front Biosci, 16, 1536–1550.

Zahnd, C., Kawe, M., Stumpp, M.T., de Pasquale, C., Tamaskovic, R., Nagy-Davidescu, G., Dreier, B., Schibli, R., Binz, H.K., Waibel, R. et al. (2010) Efficient tumor targeting with high-affinity designed ankyrin repeat proteins: effects of affinity and molecular size. Cancer Res, 70, 1595–1605.

Zhang, P., Talluri, S., Deng, H., Branton, D., and Wagner, G. (1995) Solution structure of the pleckstrin homology domain of *Drosophila* beta-spectrin. Structure, 3, 1185–1195.

11

UNUSUAL SPECIES DISTRIBUTION AND HORIZONTAL TRANSFER OF PEPTIDASES

NEIL D. RAWLINGS

Wellcome Trust Sanger Institute, Wellcome Trust Genome Campus, Hinxton, Cambridgeshire, UK

CHAPTER SUMMARY

Proteolytic enzymes release amino acids and peptides from proteins, and those that hydrolyze peptide bonds are known as peptidases. Over 200 families and over 40 clans of proteolytic enzymes have been identified. It had been noted previously that, within a family, prokaryote and eukaryote peptidase sequences were either very similar or so distantly related that the relationships were difficult to detect. An explanation put forward was that the distant relationships represented the very ancient divergence between bacteria and eukaryotes, whereas the more similar sequences were the result of horizontal gene transfers. Estimates of the divergence time also led to the suggestion that some peptidase genes had been introduced into eukaryotes at the time that mitochondria and chloroplasts were acquired by assimilation of endosymbiotic bacteria. At the time, supporting evidence for this hypothesis was expected to come from analyses of the genomes of Archezoa, eukaryotes that were derived from a pre-mitochondrial stock. However, genomes from species alleged to be Archezoans revealed that genes of mitochondrial proteins were present and that the organelles had been lost rather than never having been present.

Protein Families: Relating Protein Sequence, Structure, and Function, First Edition.
Edited by Christine Orengo and Alex Bateman.

This chapter presents an alternative method for detecting horizontal transfer of peptidase genes in bacteria, by examining the distribution of each peptidase family amongst the different phyla of bacteria. More than 2000 bacterial genomes have been sequenced and there are representatives from each bacterial phylum. Each peptidase family with an unusual distribution amongst the bacterial phyla is discussed.

11.1 INTRODUCTION

Proteolytic enzymes release amino acids and peptides from larger peptides and proteins by cleaving peptide bonds. Peptidases are by far the largest group of proteolytic enzymes, in which peptide bonds are cleaved by hydrolysis. The number of different peptidases in a proteome varies enormously from species to species. In a viral proteome there may just be one peptidase (and in some viruses none at all), whereas in the human proteome there are over 600. Peptidases are responsible for degrading proteins and peptides to amino acids so that new proteins can be built, by recycling existing proteins, destroying misfolded proteins, or degrading proteins acquired nutritionally. Peptidases also process proteins by removing targeting signals; by activating enzymes, neuropeptides and peptide hormones synthesized as precursors; releasing bioactive peptides from proteins; and deactivating these bioactive peptides, peptide hormones, neurotransmitters, and enzymes. There are many different types of peptidases utilizing different catalytic mechanisms, and the known sequences can be assembled into 215 different families. Some families are divided into subfamilies. This almost always happens because a sequence has been found that links together two groups of sequences that are divergent and had previously been considered separate families. The tertiary structure of a peptidase from one family might be related to that of a peptidase from another family (even though there is no significant sequence similarity) and peptidase families can be grouped into a higher taxon called a *clan* based on these structural similarities. There are at least 43 different clans of peptidases, each representing a unique origin for a group of peptidases (Rawlings and Barrett, 1993).

The number of sequences within a peptidase family can vary enormously. In some families (C27, C63, M5, S69, S75, U40) there is only one known sequence, whereas three families have over 10,000 sequences (S1, S9, and S33). There are now nearly 50 peptidase families with more than a 1000 members.

Similarly, the number of families within a clan also varies. Because a clan represents a unique fold, some clans contain singleton families. The clans with most families are MA (38 families), CA (26 families), PA (20 families), and PB (11 families).

Peptidases can be broadly divided into six groups depending on the nature of the nucleophile in the catalytic reaction. For three of these groups, the nucleophile is a thiol or hydroxyl group of an amino acid and these catalytic types are named after the residues involved: cysteine, serine, and threonine. For the other three catalytic types, the nucleophile is an activated water molecule and it is the mechanism of activation that is important. This is achieved either by binding to an aspartic or glutamic residue, in which case the catalytic type is aspartic or glutamic, or by binding to one or more metal cations, in which case the enzymes are known as *metallopeptidases*. In any one family, all the peptidase will be of the same catalytic type, so families are named with a single letter to reflect this (C for cysteine, S for serine, T for threonine, A for aspartic, G for glutamic, M for metallo, and U for unknown). However, this is not always reflected at the clan level where a clan may contain a mix of cysteine, serine, or threonine peptidase families, and so an extra letter "P" is used for some clan names. A family consists of an initial letter followed by a number (e.g., A1, C89) and a clan consists of two letters, the first of which indicates catalytic type (e.g., CA, PB, ST). Subfamilies are numbered alphabetically so that subfamilies in family S1 are S1A, S1B, etc. There are 5 clans and 16 families of aspartic peptidases, 8 clans and 72 families of cysteine peptidases, 2 clans and 2 families of glutamic peptidases, 14 clans and 62 families of metallopeptidases, 12 clans and 49 families of serine peptidases, 5 families of threonine peptidases, and 4 clans of mixed catalytic type (which contain all the threonine peptidase families). Thirty-four families cannot be assigned to clans because there are no known tertiary structures, and there are nine families in which the catalytic type is yet to be determined (Rawlings and Barrett, 1993).

The distribution of species with members of a peptidase family is also very varied (see Fig. 11.1). Eighty-two families are restricted to just one kingdom of life. Of these families, the vast majority are restricted to viruses (39 families) or bacteria (30 families). There are 10 families restricted to animals (C46, C67, S68, S71, S72, T6, U47, U60, U63, U71), 2 restricted to fungi (M77, S64), and 1 restricted to archaea (A5). There are no known families restricted just to protozoa or plants.

Families (or subfamilies) of peptidases are assembled around a sequence of a peptidase that has been well characterized biochemically, which is termed the *type example*. When comparing sequences to group them into peptidase families, it is important to consider only the subdomains bearing the active site residues (the "peptidase unit") and to discard any other domains. Usually, the range of the peptidase unit has already been determined. Homologs are found by searching a public protein sequence database (such as UniProt (2010) or the nonredundant database at NCBI (Sayers et al., 2011)) using the peptidase unit of each type example. The homologs found (those hits returned that are statistically significant with an Expect (E) value lower than 0.001) are assembled into the new family. Periodically, fresh searches are performed to find newly deposited homologs. An alignment is generated with multiple alignment using fast Fourier transform (MAFFT) (Katoh et al., 2002), which is used to identify catalytic residues in each sequence and to create a phylogenetic tree (using the UPGMA algorithm from QuickTree (Howe et al., 2002)). Outliers to the tree may be used to search for more distant homologs that were not returned as significant matches to the type example sequence.

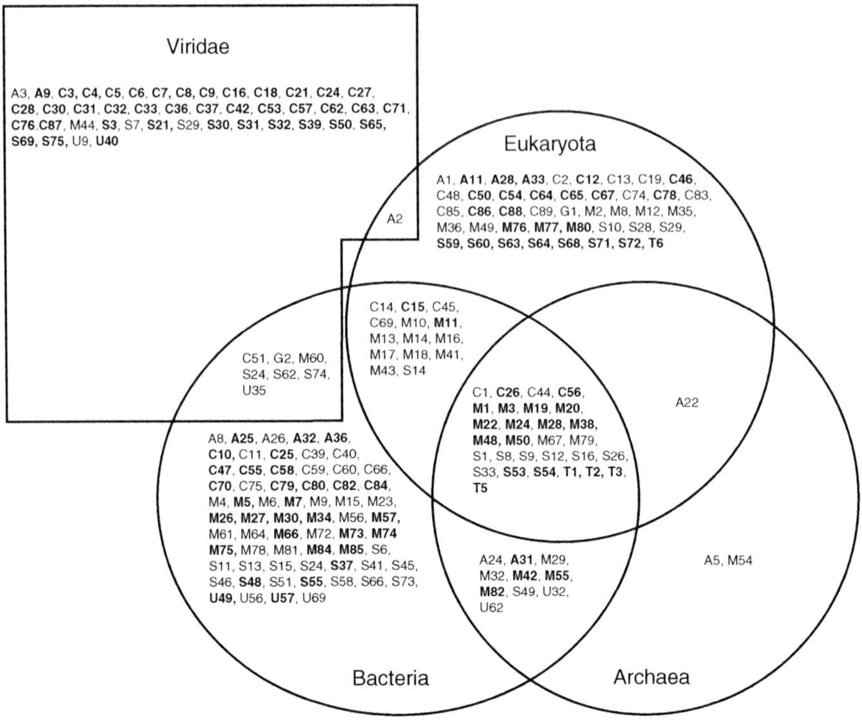

Figure 11.1 Distribution of peptidases families. Distribution is shown as a Venn diagram. Families with peptidases that are predominantly found in organism superkingdoms are listed. "Predominant" means the number of homologs exceeds half the number of completely sequenced genomes or that all the homologs are from only one superkingdom (shown in bold). The same principles are used where the circles overlap, with the additional condition that the family is also included if the number of homologs from each superkingdom is approximately equal. This condition caters for the instance where the number of homologs is much lower than the number of species with completely sequenced genomes in all the superkingdoms involved. Families found only in viruses are also shown within a rectangle, which overlaps with the circles for eukaryota and bacteria. Some families found in eukaryotes, bacteria, and archaea are shown in bold because homologs are also known from viruses.

When the families of peptidases were reviewed (Rawlings and Barrett, 1995), it was apparent from the alignments and trees that for some families the bacterial and eukaryote sequences were much more closely related than in other families. These sequence differences did not seem to reflect different mutation rates between peptidase families because the datable divergences (mainly within the vertebrates) were constant. In some families, such as M24 (methionyl aminopeptidase), there were both close and distant bacterial homologs of the eukaryote sequences. Our interpretation of these results, based on a similar observation from Young et al. (1978) who had noticed the phenomenon in family S1 (chymotrypsin), was that the divergence between some eukaryote and bacterial sequences coincided with the time of acquisition of the mitochondrion and chloroplast, which are believed to be derived from endosymbiont bacteria. The eukaryote genes were thus acquired by horizontal gene transfer from the proto-organelles. The absence of homologs in archaea and pre-mitochondrial protozoa would provide some supporting evidence for this hypothesis. Since then, the genomes of nearly 2000 species of cellular organisms have been completely sequenced, and the hypothesis deserves reexamination.

11.2 MECHANISMS OF HORIZONTAL GENE TRANSFER

Figure 11.2 shows peptidase families for groups of organisms in which few homologs are known. Families shown in bold have few homologs, all of which are

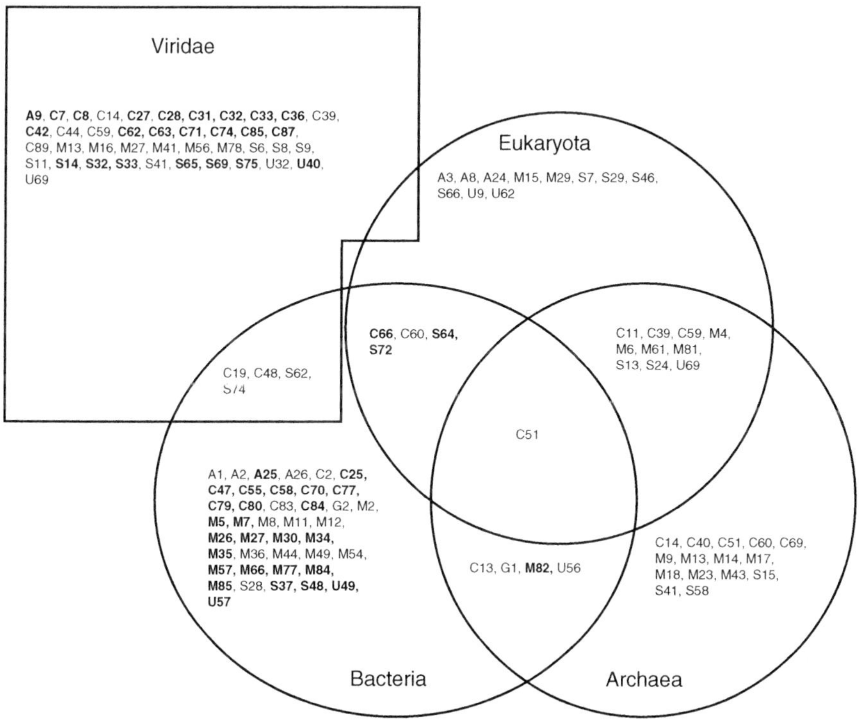

Figure 11.2 Distribution of peptidase families with few members. Distribution is shown as a Venn diagram. Only those families with a small number of homologs are shown. For eukaryotes and bacteria, a family is shown if the number of homologs is less than 10% of the number of species with genomes sequenced. For viruses and archaea, a family is shown if less than 10 homologs are known. Where all homologs in a family occur in one organism group, the family name is shown in bold.

in one (or two) organism groups. The rest represent families in which horizontal gene transfer or contamination is likely to have occurred.

Probable contaminants: Contamination of samples during sequencing is always a problem, and has been recognized as such for some unusual peptidase family members. The problem of contamination is much more likely to affect eukaryotes because assembling a completed genome is much more difficult because of multiple chromosomes, introns, and long stretches of intergenic DNA. Examples of likely bacterial contaminants in eukaryote genomes are given in Table 11.1.

Bacterial genomes can contain genes from lysogenic bacteriophages. These are infections rather than contamination, but the acquisition of the gene is more akin to contamination than horizontal gene transfer. Families of viral peptidases containing examples from infected bacteria are C51 (D-alanyl-glycyl peptidase, staphylococcal phage phi11), G2 (pre-neck appendage protein, *Bacillus* phage phi29), M60 (enhancin), S62 (PA peptidase, influenza A), S74 (K1F endosialidase CIMCD self-cleaving protein, *Escherichia coli* phage), and U35 (prohead peptidase, bacteriophage HK97).

Via viruses and retrotransposons: Transfer of a peptidase gene via a virus was first suggested by Hartley (Hartley, 1979) when discussing the origins of trypsin in *Streptomyces*. The occurrences of active peptidases in mammals that are similar but not identical to retropepsins (family A2) from RNA viruses lend support to the hypothesis that viruses can be intermediates for the transfer of genes (Fehrmann et al., 1997). It has been suggested, however, that the mammalian peptidases, often described as from "endogenous retroviruses," may simply be from defective retroviruses. This may also apply to retrotransposons, which are, in all probability, derived from retroviruses. The retrotransposon *Gypsy* from *Drosophila melanogaster* has been suggested to be a retrovirus (Kim et al., 1994). Several peptidases from retrotransposons have been characterized, including those from *Ulysses* (from *Drosophila virilis* (Volkov et al., 2004)) and *Ty3* (from *Saccharomyces cerevisiae* (Rinckel and Garfinkel, 1998)). Because RNA viruses have no error-correcting mechanisms, the rate of mutation is very high and some retrotransposon peptidase sequences are so different from the retrovirus equivalent that they are placed in a separate family, A11. Characterized peptidases from family A11 include those from the retrotransposons *Copia* (from *D. melanogaster* (Athauda et al., 2006)) and *Ty1* (from *S. cerevisiae* (Boeke, 2004)). The origin of retropepsins then becomes an issue. It had long been supposed that because structurally a retropepsin represents half a pepsin molecule and because dimerization is required for the peptidase to be active (Navia et al., 1989), either the ancestral retropepsin gene was derived from half an ancient pepsin gene, or the pepsin gene is derived from a duplication event involving a retropepsin gene (Doolittle et al., 1989). There are now two families of aspartic peptidases (families A32 and A33) from cellular organisms that, similarly to retropepsin, represent a half-pepsin structure and must dimerize to be active. The mouse skin aspartic protease (family A33; (Bernard et al., 2005)) and the PerP peptidase from the bacterium *Caulobacter crescentus* (family A32; (Chen et al., 2006)) have been

TABLE 11.1 Examples of Probable Bacterial Contamination in Eukaryote Genomes

Peptidase Family	Type Example	Organism	ProtID	Gene Name
A8	signal peptidase 2	*Populus trichocarpa*		Poptr1_1_597450
A24A	type 4 prepilin peptidase	*P. trichocarpa*		Poptr1_1_587453
C39	bacteriocin-processing peptidase	castor bean (*R. communis*)	EEF24206	
M4	thermolysin	castor bean (*R. communis*)	EEF24282	
M10B	serralysin	rice (*Oryza sativa*)	EEC80860	OsI_23479
M12	astacin	Sitka spruce	ABK23560	
M29	aminopeptidase T	tomato		Amp
		Physcomitrella patens	EDQ48191	
M56	BlaR1 peptidase	*P. trichocarpa*		Poptr1_1_596402
M61	glycyl aminopeptidase	castor bean (*R. communis*)	EEF25421	
M64	IgA peptidase	castor bean (*R. communis*)	EEF25385	
S13	D-Ala-D-Ala-peptidase C	*P. trichocarpa*	XP_002335028	
S24	LexA peptidase	rat		RvuZ
		Nematostella vectensis	EDO29243	v1g145237
		N. vectensis	EDO25984	v1g9678
		Hydra magnipapillata	XP_002164597	LOC100207154
S45	penicillin G acylase precursor	*P. trichocarpa*	XP_002339502	
S46	dipeptidyl-peptidase 7	castor bean (*R. communis*)	EEF23090	
S58	aminopeptidase DmpA	*P. trichocarpa*		Poptr1_1_584967
S66	murein tetrapeptide LD-carboxypeptidase	*P. patens*	EDQ62488	
		malaria mosquito		Ebip201
		pea aphid	XP_001948860	LOC100168405
U62	microcin-processing peptidase 1	*N. vectensis*	EDO26561	v1g224925
		Paulinella chromatophora	YP_002049464	PCC_0844
		P. chromatophora	ACB43253	PCC_0843

characterized. Unfortunately, the existence of these families does not solve the question of the origin of retropepsin but it probably does move the origin back to the time before the divergence of bacteria and eukaryotes.

There is other circumstantial evidence that viruses can acquire peptidase genes from their hosts. Peptidase family C1 (the papain family) is predominantly found in eukaryotes (there are homologs in bacteria and archaea). However, a homolog was discovered in a baculovirus (Rawlings et al., 1992), and the peptidase has now been characterized (Slack et al., 1995). Similarly, there is a viral homolog of subtilisin, a member of family S8, which is widely distributed in bacteria, archaean, and eukaryotes. The viral member is from ictalurid herpesvirus 1 (Davison, 1992), which infects the Channel catfish, but the peptidase has not been characterized.

Via plasmids: A plasmid is a double-stranded DNA molecule that is independent of the chromosome and can replicate independently. A plasmid is found in the cytoplasm and the DNA is described as "naked" because it is not encased in a protein coat as in viruses. Plasmids are found principally in bacteria, but there are examples from yeasts and archaea. Plasmids are implicated in horizontal gene transfer because plasmids themselves are transferred from one cell to another during conjugation, when chromosomal DNA is transferred. Transfer of the plasmid requires a special sex pilus, the proteins of which may be encoded by genes in the plasmid DNA. By encoding genes for antibiotic resistance or to enable the host to access different nutrients or to become pathogenic, plasmids can be advantageous to the host. Because most plasmids are specific to a particular host species, the plasmid genome is considered to be a component of the complete genome of that species. Some plasmids can infect multiple host species, and taxonomically these are considered separate entities. It is these "promiscuous plasmids" that are most likely to be involved in the horizontal transfer of genes between species (Lipps, 2008).

Via endosymbionts: The idea that eukaryote organelles such as the mitochondrion and the chloroplast were derived from endosymbiotic bacteria has been proposed several times (Schwartz and Dayhoff, 1978; Yang et al., 1985; Cavalier-Smith, 1987). Evidence for this hypothesis has come from the comparison of mitochondrial and bacterial genes, especially *Rickettsia* genome sequences (Andersson et al., 1998). It is known that genes can be transferred from the mitochondrion to the nucleus, presumably via a reverse transcriptase (Gellissen and Michaelis, 1987), and the mitochondria of some higher plants retain genes that have been transferred to the nucleus in other plants (Brennicke et al., 1993; Eriksson et al., 1993). The proposal that peptidase genes may have been transferred from a proto-organelle to the nucleus during early eukaryote evolution (Rawlings and Barrett, 1995) was based on the lack of homologs in archaea and some protozoa because it had been suggested that archaea were more closely related to eukaryotes than to bacteria (Cavalier-Smith, 1989) and that some extent protozoa may have arisen from ancient eukaryotes that predated the acquisition of mitochondria and chloroplasts, such as the Microsporidia *Encephalitozoon*, the diplomonad *Giardia*, and others known as *Archezoa*

(Cavalier-Smith, 1991). However, recent analyses of the completely sequenced genomes of protozoa that lack mitochondria have shown that homologs of mitochondrially encoded genes exist, implying that mitochondria have been lost in these organisms and not that they are derived from a pre-mitochondrial stock (Sogin, 1997a; Katinka et al., 2001; Roger et al., 1998). It has also been shown that *Trichomonas vaginalis* possesses bacterial-like heat-shock proteins, implying that an even more ancient endosymbiotic relationship with bacteria may have existed (Sogin, 1997b). *Entamoeba* and *Trichomonas* possess hydrogenosomes for the generation of energy, which have been proposed to be derived from mitochondria (Cavalier-Smith, 1991; Sogin, 1997b)—further evidence that eukaryotes derived from pre-mitochondrial stock simply do not exist today.

This means that the only evidence that can be used to determine if a family of peptidases might have originated in an endosymbiotic bacterium is its wide presence in bacteria and eukaryotes and absence in archaea.

11.3 DISTRIBUTION OF PEPTIDASES IN BACTERIA

By far the largest number of completely sequenced genomes is from bacteria. At the time of writing, the genomes of 912 bacterial species, 73 archaean species, 22 protists, 26 fungi, 20 animals, and 4 plants had been completely sequenced and an estimate at least of the total number of protein-coding genes in each genome is available. In order to better understand evolution of peptidase families and to distinguish horizontal from vertical gene transfer, it is essential to understand the distribution of those families in bacteria. Families that have examples in almost all bacterial phyla presumably arose in the universal ancestor of bacteria, and examples in archaea are most probably derived by vertical gene transfer. Families with members restricted to one bacterial phylum are likely to have arisen much later in bacterial evolution, and would not be expected to be found in archaea or eukaryotes as a result of vertical gene transfer. Families with members from some bacterial phyla but not others are likely to be the result of horizontal gene transfers between bacterial species.

The genomes of some bacterial species have been sequenced several times, to see how the genomes vary from one strain to another. There are 30 strains of *E. coli* where the genome has been sequenced and the number of protein-coding genes varies from only 4084 in strain BW2952 to 5516 from serovar O26:H11 strain 11368. Partly, this discrepancy is due to whether all the plasmids have been sequenced, but there are also major differences in the number of genes in the chromosome from one strain to another. The total number of different peptidases recognized in *E. coli* is 212, and there is a core set of 70 peptidases that are found in all 30 strains. However, there are 40 peptidases that are found in only 1 strain. Some of the peptidases found in a few strains are plasmid specific, especially members of peptidase families A24 (processing of pilus protein precursors essential for plasmid propagation and transmission), A26 (e.g., outer membrane protease 7), C39 (colicin V secretion ATP-binding protein CvaB), C59 (choloylglycine hydrolase), M20 (an *N*-acyl-L-amino acid amidohydrolase), M66 (StcE protein), M79, S6 (EspP gamma, EatA, and tsh proteins), S9 (Ypt2 protein), S12 (class C beta-lactamase CMY-18), S24 (ImpA UV protection protein), and S33 (SAT-1 protein). Chromosomally encoded peptidases that are restricted to very few strains are rare, and most frequently relate to genes derived from lysogenic bacteriophages, with peptidases families M79, S24, S49, and S73 being principally affected. These provide the bulk of the sequences that represent the remainder of the unique peptidases. These are little more than variants of the original viral protein, though in most cases the bacteriophage from which the gene is derived is unknown.

Not all *E. coli* peptidases are variants on an existing theme. Two peptidases predicted to be active (i.e., all the active site residues and/or metal ligands are conserved), known only from one or two strains, are not only unlike any other *E. coli* peptidase, but so distantly related to other peptidases in the same family that had they been biochemically characterized they would be candidate founders of new subfamilies. The sequences in question are homologs in families M41 (an FtsH peptidase homolog, hypothetical protein E2348C_3897 from *E. coli* O127:H6 str. E2348/69) and S8 (an unnamed subtilisin homolog, also from *E. coli* O127:H6 str. E2348/69 (CAS09358) and RN587/1 (EFZ72340)). The origin of these genes is unknown, but presumably they did not develop in these *E. coli* strains and are most likely derived by horizontal transfer from a group of bacteria, none of which have had their genomes sequenced.

There is only one member in peptidase family M5: mycolysin from *Streptomyces cacaoi*. The origin of this peptidase gene is unclear because as there are no homologs in other *Streptomyces* species, it is extremely unlikely that the gene evolved as a result of speciation. It is much more likely that the gene was horizontally transferred. Members of peptidase family M27 are restricted to a single genus, *Clostridium*, and the family includes toxins from the causative agents of tetanus (*Clostridium tetani*) and botulism (*Clostridium botulinum*). Other species of *Clostridium* do not have homologs, so again the origin of the ancestral gene is a mystery. On the other hand, homologs have now been found for the anthrax lethal factor from *Bacillus anthracis*, which had been the only member of family M34. These include closely related proteins in *Bacillus cereus* and *Paenibacillus larvae*. The product of the BC03BB108_C0180 gene in *B. cereus* is also a member of M34, but encoding a much shorter protein that is more distantly related to the anthrax toxin (36% sequence identity), implying a much more ancient origin for the ancestral gene. If the ancestral gene is a result of a horizontal transfer from an as yet unknown source, then the transfer happened long enough ago to allow anthrax lethal toxin and the BC03BB108_C0180 protein to diverge greatly.

Table 11.2 shows that there are peptidase homologs from bacteria in 145 peptidase families. These are homologs that are predicted to be active; in other words, the known or putative catalytic residues are conserved, though for some families, particularly those of unknown catalytic mechanism, active site

TABLE 11.2 Distribution of Peptidase Families Amongst Bacterial Phyla

	Acidobacteria	Aquificae	Bacteroidetes	Chlamydiae	Chlorobi	Chloroflexi	Chrysiogenetes	Cyanobacteria	Deferribacteres	Dictyoglomi	Fibrobacteres	Firmicutes	Fusobacteria	Gemmatimona-detes	Lentisphaerae	Nitrospirae	Planctomycetes	Proteobacteria	Spirochaetes	Synergistetes	Thermodesulfobacteria	Thermotogae	Verrucomicrobia	Xenobacteria	Viridae	Archaea	Eukaryota
Completed genomes	4	9	56	8	11	13	1	41	3	2	1	308	4	1	2	1	8	448	19	3	0	11	6	11		77	110
A1																		10									*2259*
A2																		1							159		*284*
A8	*5*	*11*	41	*9*	*12*	*16*	*3*	*45*	*3*	*2*	*1*	*313*	*4*	*2*	1	*3*	*9*	*554*	18	2	1	10	5	*12*			1
A24	*7*	*10*	1		7	12	2	*43*	*3*	*2*	*1*	*314*	3	*1*		*4*	*25*	*616*	3	*4*	1	*12*	4	4		76	1
A25			1									94															
A26													1					40	3								
A31	*5*	*13*	11		10	*24*	4	35	*6*	*2*		97		*1*		3	1	312		2	2		5			*108*	
A32			1														1	347									
A36			1									93															
C1	1	1	*82*		8	5		1				195	2				2	54	4	2		2		2	53	50	*2512*
C2			5			3		12				13					1	6									*703*
C10			***110***									2						3					1				
C11		3	32			10		2		1		34						11				14				9	6
C13			4															35								2	*445*
C14	*5*	1	30	1	8	*15*		*69*	*7*		*1*	49				7	6	219	10				2			7	*842*

(*Continued*)

TABLE 11.2 (*Continued*)

	Acidobacteria	Aquificae	Bacteroidetes	Chlamydiae	Chlorobi	Chloroflexi	Chrysiogenetes	Cyanobacteria	Deferribacteres	Dictyoglomi	Fibrobacteres	Firmicutes	Fusobacteria	Gemmatimona-detes	Lentisphaerae	Nitrospirae	Planctomycetes	Proteobacteria	Spirochaetes	Synergistetes	Thermodesulfobacteria	Thermotogae	Verrucomicrobia	Xenobacteria	Viridae	Archaea	Eukaryota
C15			2			2		9				137	4				4	98	4	2		6		7		24	94
C19			2	2														20							1		*1757*
C25			27		2	5					***1***						4	3									
C26	*17*	*37*	*126*	*11*	*50*	*57*	*9*	*97*	*15*	*13*	*5*	*1155*	*25*	*6*		*10*	*22*	*1618*	*48*	*15*	*5*	*38*	*16*	*53*		*313*	*355*
C39		1	56		2	2		62				183		2			2	206	3					3	1	10	1
C40	1	*7*	*110*	*9*	*20*	10		*46*	2	*2*	*1*	*652*	*9*	*2*		*1*	1	*677*	8	1	*1*	4	1	*20*	22	2	11
C44	*15*	*31*	*105*	*10*	*38*	*39*	*10*	*138*	*12*	*7*	*3*	*1000*	*9*	*4*		*14*	*31*	*1647*	*27*	*8*	*2*	*31*	*12*	*39*	9	282	*439*
C45			4			8		1		2		41			1		2	55	2	1				5			87
C47												7															
C48				8														12							3		*532*
C51			7									38	2					14						1	67	1	18
C55																		45									
C56	2		6			7		12	2			40			1		*14*	165	2		*1*		3	4		43	52
C58																		26									
C59			27	1				8				138	1				10	125		2					5	8	2
C60	4		2			32						641				1		31								2	
C66			1									8											1				1
C69			32		1							176	2					46	5	22		2				9	43
C70																		5									

C75												77															
C79																		15									
C80												5						36									
C82	2	6	10		2	*21*		38	1		*1*	*458*	2	*1*		2	2	*637*	3	2	1	4	*26*				
C83								14										25									80
C84			1					1				3						1									
G1												16						2								2	69
G2												28						2							10		
M1	*7*	1	*156*		5	7		32	1	*2*		*326*		*5*		*3*	*9*	*494*	4			1	3	2		53	*988*
M2	3							1				1						53									236
M3	*4*	2	*140*	*10*	1	*18*	*2*	*45*	2		*1*	*533*	*8*	*1*	1	*1*	*15*	*742*	*26*	*6*		*15*	*15*	*34*		29	*404*
M4		1	13	9		5		4				252					1	190	4					3		2	11
M5												1															
M6		1	12					1		4	3	134					1	65				1				3	11
M7												40															
M8								4				1		1			3	2						3			*334*
M9			3			5						124						103	2							8	
M10	2		7		3			45				82		1			5	207					1		19	20	*926*
M11												2						16									29
M12			41		2	3		1				5						27					2	4			*1963*
M13	15		71	1		1						160	2				1	157		1			3		1	10	*661*
M14	1		7		1	3		5			1	122					3	78	3					2		4	*1131*
M15	3		*57*		5	2		*96*	*3*		2	*455*	1	*1*			3	275	16	2		2	5	1	16		6
M16	*15*	*9*	*123*	*18*	*26*	*23*	*4*	*100*	*8*	2	2	*412*	*5*	*4*	*3*	*8*	*16*	*790*	*36*	1	*1*	*11*	*20*	*20*	1	8	*757*
M17	5	*11*	8	*10*	*12*	*14*	2	*49*	*3*		*1*	213	*5*	*1*	1	*4*	*9*	*797*	6	*5*			2			8	*245*
M18			4	1							1	260	*11*		1			68	*20*	*4*	1	*11*				1	*234*

(*Continued*)

TABLE 11.2 *(Continued)*

	Acidobacteria	Aquificae	Bacteroidetes	Chlamydiae	Chlorobi	Chloroflexi	Chrysiogenetes	Cyanobacteria	Deferribacteres	Dictyoglomi	Fibrobacteres	Firmicutes	Fusobacteria	Gemmatimona-detes	Lentisphaerae	Nitrospirae	Planctomycetes	Proteobacteria	Spirochaetes	Synergistetes	Thermodesulfobacteria	Thermotogae	Verrucomicrobia	Xenobacteria	Viridae	Archaea	Eukaryota
M19	7		48	8		10		3				175	1	3			7	161	5	4		2	4	10		30	171
M20	32	8	286	10	22	64	5	85	5	4	2	1738	42	11	2	1	39	2474	67	39	1	32	13	84		249	431
M22	5	12	64	10	12	17	2	49	3	2	1	373	6	1	1	3	9	524	19	4	1	12	8	14		100	323
M23	8	23	208		19	45	5	180	6	4	5	842	5	3		8		1133	101	10		21	2	47		2	60
M24	19	19	158	20	25	45	4	144	6	4	2	1022	9	5	2	7	27	1342	42	8	2	22	18	39		199	788
M26			1									49						1									
M27												15													2		
M28	40	3	217		3	3		25		2		205	2	11	1	2	27	357	4				4	5		35	644
M29	4		1	1		12			3		1	183	2				3	64	17			4	7	30		28	2
M30			1			4						9		3			1	8	6								
M32			5	4		8		23				93	3		1	1	7	170	7			5	4	16		48	19
M34												11															
M35												1						24									129
M36	1		20									16						26					3				128
M38	6	1	21			7		21		1		289	5	4		2		305	5	2		7	1	3		45	65
M41	8	11	56	10	24	22	2	173	5	2	1	468	5	1	2	5	18	653	23	4	2	13	11	31	1		563
M42	1		42		3	16				4		338	2	1			6	66	8	1		28	9	18		112	
M43			55					6				40					1	14								2	147

M44												1													30		
M48	18	18	90	6	24	23	5	97	5	4	1	324	12	3		11	7	1288	16	8	4	8	12	4		176	240
M49	1		35															13					1				145
M50	11	19	39	10	27	47	5	132	6	4	1	484	5	2		9	30	609	12	7	2	6	17	26		243	84
M54	3	4	4							2					1	1		10					1			67	29
M55						4		1				66		2	1			61	4	4		7	1	3		19	
M56	19	1	83					5				199	1	4			11	74					3	1	1		
M57			27															4									
M60												14						7					1		21		4
M61	4		37		3	6		39				2					3	197	3				2			7	1
M64	1		25									15						11									1
M66												1		1				24									
M67	1	8	4		11	2		17		2	2	65				3	6	100						8	13	91	571
M72		1	3					1				4					1	42									3
M73												77															
M74								1				1						202									
M75			17					8										155	2				1	1			
M78			3					1		1		65	6					13	1	1				1	3		
M79	10	4	52	7	16	31		84	4			614	10	1			11	228	20	2			10			80	178
M81			10			5						43					5	186	3				2			3	8
M82			1					1				17						4						1		8	
M85																		11									
S1	18	14	136	12	24	65	3	271	7	4	1	1442	6	3	3	14	99	1876	57	4	1	10	24	54	51	59	7038
S6													4					101							2		
S8	6	6	168	1	11	56	3	109	5		1	1117	12	1			25	786	9	1	11		18	114	2	126	1498
S9	88	7	695	20	28	74	4	240	3	9	2	1937	43	31	3	6	129	3133	43	6		56	39	66	3	213	1833

(*Continued*)

TABLE 11.2 (*Continued*)

	Acidobacteria	Aquificae	Bacteroidetes	Chlamydiae	Chlorobi	Chloroflexi	Chrysiogenetes	Cyanobacteria	Deferribacteres	Dictyoglomi	Fibrobacteres	Firmicutes	Fusobacteria	Gemmatimona-detes	Lentisphaerae	Nitrospirae	Planctomycetes	Proteobacteria	Spirochaetes	Synergistetes	Thermodesulfobacteria	Thermotogae	Verrucomicrobia	Xenobacteria	Viridae	Archaea	Eukaryota
S10	9		14	2		4		2				10					3	132		1				5			*1047*
S11	*5*	1	18	*8*	10	*14*	*1*	24				*819*	*7*	*1*	*1*	*2*	3	*948*	*15*		*1*		*8*	*17*	3		23
S12	*41*		*279*	1	*12*	*30*	*3*	37	1		*1*	*1032*		*18*	*1*	*2*	*38*	*1071*	*14*	2	*1*	3	*27*	*21*	36	31	*235*
S13	6	3	55		2			73			1	146		2			3	304				3	2	5		7	8
S14	*5*	*10*	*70*	*18*	*12*	*18*	2	*122*	*3*	2	2	*531*	*5*	*1*	*1*	*4*	*17*	*574*	*35*	*5*	*1*	*13*	*9*	*11*	5	2	*309*
S15	9		25	2		5		22				240		7			1	102	3					1		6	25
S16	*9*	*15*	*65*	*12*	7	*31*	2	3	*9*	2	2	*525*	*7*	*4*	*1*	*10*	3	*917*	*42*	*8*	*4*	*20*	5	*30*		*129*	*237*
S24	*5*	4	34	6	*13*	*14*	*6*	*60*	*4*	*2*	*1*	*385*	3	*1*	*3*	*4*	*9*	*990*	14	2	*1*	*12*	*6*	*13*	48	4	6
S26	*16*	*12*	*81*	*9*	*17*	*16*	*3*	*110*	*4*	2	*1*	*586*	*9*	2		*4*	*13*	*768*	*33*	*5*	*1*	10	*6*	*21*		*138*	*329*
S28																		1									*402*
S33	*31*	*10*	*209*	5	*87*	*102*	5	*350*	*4*	*4*	5	*2814*	*27*	*18*	2	*4*	*27*	*3783*	*51*	*4*		*18*	*12*	*74*	4	*149*	*1545*
S37												2															
S41	*7*	1	*146*	*12*	*37*	11		*53*	1			82	*4*		2		*13*	*533*	12			2	*8*	7	1	7	24
S45	*7*		32		1	11		13				101		*3*		*1*	3	272	3				1	*15*		21	17
S46	3		121					1									3	88					2				2
S48								**82**																			
S49	*4*	*21*	43	*11*	*19*	12	1	*73*	*7*	*4*	2	128	*8*	2	1	*6*	*11*	*904*	18	*3*	*3*	6	*15*	*14*	18	*128*	27
S51	1		33			2		25				50	1	3			8	83						2			27

S53	16		2									60						100					3			39	*141*
S54	*7*	8	*103*	3	*13*	16		*43*	*3*	*2*	2	*309*		*3*		*2*	*25*	441	23		*1*	*19*	5	*13*		*79*	*426*
S55			1									143															
S58	3		11			3						39		1		1		116	1	2		1	1	1		10	13
S62																		1							1		
S66	3		54	1	*13*			*68*			1	216	*6*	*1*	*3*	*2*		211	5		*1*		1	4		13	4
S73																		130							11		
S74												1						3							5		
T1	*7*	*10*	13		*13*	2	*2*	8	*4*			222		*1*		6	7	*515*	*20*	*5*	*1*	*13*	2	8		*218*	*1249*
T2	4		49			11		22				20		1		2	8	191						4		48	*247*
T3	*8*	4	44			*19*		*45*	1			183	5	*4*	2		*8*	*805*	6	*7*		2	4	*11*		30	*384*
T5	*5*	*10*	3		12	*18*	2	*47*	*4*	2	1	237	2	*1*	1	3	*8*	322	13		*1*	8	*8*	*11*		35	68
U32	2	7	*80*		*13*	*14*	*5*	*42*	*3*		2	*469*	*12*	*1*	*3*		*17*	*626*	*19*			1	*17*	9	2	*85*	45
U35			3									77				1	1	205	1	2					40		
U49																		3									
U56		6							1			41					1	27	3	3		7				6	
U57												78															
U62	*18*	*18*	38			11		*163*	*6*	*12*	2	176		*8*		*10*	*2*	*865*	16	*6*	2	*24*	2	*18*		*225*	3
U69	1		10	1				2				4	5				1	454							8	1	1

The first row shows the number of completely sequenced genomes (at the species level) for each group of organisms. Family names in bold represent families where homologs are known from bacteria, archaea, and eukaryotes. Family names in italic are where homologs are found in more than half of the bacterial phyla. Family names in bold italic have homologs in bacteria but not viruses, archaea, or eukaryotes. Where the number of homologs exceeds or equals the number of genomes, the number is in italics.

residues are unknown and all homologs are assumed to be active. Counts are for known homologs, which includes homologs from organisms that have not had the genome sequenced and all instances where paralogs exist in the same genome. Table 11.2 shows the number of homologs per bacterial phylum and for viruses, archaea, and eukaryotes. The first row shows the number of species with completely sequenced genomes in each bacterial phylum or superkingdom. Note that at the time of writing no genome had been completely sequenced for any member of the Thermodesulfobacteria.

There are 41 families with homologs in all 3 cellular superkingdoms of life. None are aspartic or glutamic peptidases, 6 are cysteine peptidases, 16 are metallopeptidases, 14 are serine peptidases, 3 are threonine peptidases, and 2 are of unknown catalytic mechanism. It might be presumed that these represent families found in the universal ancestor; however, for several of these families presence in one or more superkingdoms is limited. There are only 12 families with members present in most bacterial phyla, archaea, and eukaryotes, namely, families C26 (gamma-glutamyl hydrolase), C44 (amidophosphoribosyltransferase precursor), M20 (glutamate carboxypeptidase), M22 (O-sialoglycoprotein peptidase), M24 (methionyl aminopeptidase), M48 (Ste24 peptidase), S8 (subtilisin), S9 (prolyl oligopeptidase), S16 (Lon-A peptidase), S26 (signal peptidase), S33 (prolyl aminopeptidase), and S54 (rhomboid). These are all assumed to have been present in the universal ancestor and the sequences are derived by normal, vertical evolution. The remaining 29 families are candidates for horizontal transfer of genes.

Peptidase families with homologs from all bacterial phyla are very few: A8 (signal peptidase 2), M16 (pitrilysin), M20, M22, M24, M41 (FtsH peptidase), S1 (trypsin), S14 (peptidase Clp), S16, S24 (LexA peptidase), and family S26. Family S26 had been missing from the Lentisphaerae, but a homolog has now been found in *Victivallis vadensis* (ZP_06244436). A homolog from each family is presumed to have been present in the universal bacterial ancestor.

Peptidase families restricted to bacteria are often restricted to just one phylum. There are 13 families predominantly found only in Firmicutes: A25, A36, C47, C75, C84, M5, M7, M26, M34, M73, S37, S55, and U57. There are 12 families found mainly only in Proteobacteria: A26, A32, C55, C58, C70, C79, C80, M66, M74, M75, M85, and U49. In family A26 (omptin), there is a single homolog from the promiscuous plasmid F. Three families have their peptidases concentrated amongst members of the Bacteriodetes: C10, C25, and M57. The only peptidase family restricted to Cyanobacteria is S48. However, in very few of these families does the number of homologs exceed the number of completely sequenced genomes—the exception is S48. This implies that most of these families arose during evolution of the phylum. The remaining families of bacterial peptidases (C82 and M30) are more widely distributed, but only C82 is found in most of the bacteria with completely sequenced genomes. Thus family C82 probably arose during early bacterial evolution.

11.4 UNUSUAL OCCURRENCES IN PEPTIDASE FAMILIES

The following section gives examples of peptidase families where unusual distribution can be explained by horizontal gene transfer, though in most cases it is not possible to predict the mechanism. Unusually distributed peptidases are identified by protein identifiers or gene names in brackets. The direction of gene transfer can be inferred, usually from the superkingdom with most known homologs to that with fewer ones; these transfers are indicated in Fig. 11.3. Families present in the universal ancestor (see Fig. 11.1) will not be considered because it will not be possible to distinguish horizontal from vertical gene transfer. Similarly, families with homologs only in bacteria and archaea (M42, M55, M82, U56) or only in archaea and eukaryotes (A22) need not be considered further for the same reason. Families restricted to one superkingdom of organisms will also be excluded because horizontal gene transfer, if any has occurred, could also be explained by gene loss. In family M32, the species distribution does not predominate in any superkingdom of organisms, even though homologs are found in bacteria, archaea, and eukaryotes. Understanding the evolution here is difficult because multiple gene losses cloud the issue. Absence from a major subgroup of organisms can be explained by gene loss rather than by horizontal gene transfer in families S49 (from animals) and U32 (from fungi).

A1A (pepsin): There are only a handful of prokaryote homologs in subfamily A1A. This family includes vertebrate peptidases involved in the digestion of food proteins (pepsin, chymosin, and gastricsin) and turnover of phagocytosed

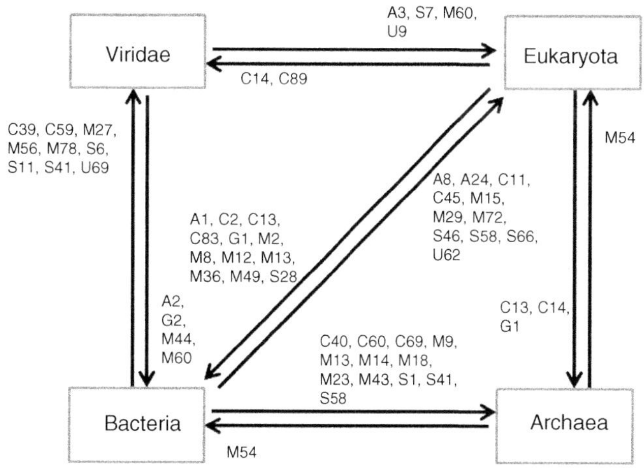

Figure 11.3 Presumed direction of horizontal gene transfers. The boxes indicate the four superkingdoms, and the arrows indicate the presumed direction of horizontal gene transfer. The names of peptidase families involved are shown near the arrowheads. Note than no transfers between viruses and archaeans are postulated.

and intracellular proteins in the lysosome (cathepsin D), as well as more specialized peptidases such as renin, which is important for processing the precursor of the peptide hormone angiotensin that controls blood pressure (Rawlings and Barrett, 2004a). Other homologs are known from protists, fungi, and plants. Most pepsin homologs are proteins that enter the secretory pathway and have an acidic pH optimum. Homologs from bacteria were detected in 2009 (Rawlings and Bateman, 2009) but have never been biochemically characterized. All of the bacteria homologs are from Proteobacteria, and all lack signal peptides, implying that they are cytoplasmic proteins. The active site residues are conserved, but an important structural motif is missing. It is not known if the homologs are inactive as peptidases because of this, or if this structural motif is unnecessary for activity in a cytoplasmic protein. Intriguingly, the bacterial homologs cluster into two groups. In one group, all the homologs are from marine bacteria (*Shewanella*, *Colwellia*, and *Kangiella*) and cluster with a sequence from the marine protist *Monosiga brevicollis*. In the second group (*Marinomonas* and *Sinorhizobium*), the pepsin homologs are derived from bacteria associated with symbiotic relationships with plants.

A2 (retropepsin): There is a single bacterial homolog in family A2, which is otherwise restricted to retroid viruses and animal retrotransposons. The bacterial homolog is from the *Wolbachia* endosymbiont of *Drosophila ananassae* (EAL58701), for which the genome has been completely sequenced, and has probably been acquired from the host fruit fly.

C1A (papain): Subfamily C1A is widely distributed amongst eukaryotes, bacteria, and archaea. It is also found in most kingdoms of eukaryotes with the exception of fungi. Only three fungal species possess homologs: the basidiomycete *Laccaria bicolor* (ProtID EDR10121), and the ascomycete *Chaetomium globosum* (CHGG_05418 gene) and *Podospora anserine* (ProtID CAP65700). This is surprising given that there are many fungal examples in subfamily C1B (bleomycin hydrolase). The most likely explanation is that the ancient gene duplication that gave rise to ancestors of both subfamilies occurred in an ancestral single-celled eukaryote but the gene for a C1A member was lost from the line that gave rise to fungi. Its presence in some modern day fungi must either be as contaminants or derived by horizontal gene transfer.

C1B (bleomycin hydrolase): This subfamily is found in eukaryotes and bacteria, but there are no known homologs in archaea, protozoa, or plants. Most of the 190 homologs are from species in the phyla Bacteroidetes and Firmicutes (there are just 4 homologs from the species in other bacterial phyla). The presence in bacteria could be explained by a lateral gene transfer from a eukaryote to a bacterium.

C2A (calpains): Calpains are large, multidomain, cytoplasmic endopeptidases from animals, plants, fungi, and protists. The peptidase domain is structurally related to that of papain, and the enzymes are calcium dependent and calcium controlled. Calpains are heterodimers consisting of a heavy chain carrying the peptidase domain and calcium-binding sites, plus a light chain with further calcium-binding sites (Rawlings and Barrett, 2004c). Distribution in fungi and protists is patchy with some completed genomes lacking homologs (e.g., *Schizosaccharomyces*, *Dictyostelium*, and *Encephalitozoon*). There are just over 40 bacterial homologs of calpains (and no known archaean homologs), from 7 phyla (Bacteroidetes, Chloroflexi, Cyanobacteria, Firmicutes, Planctomycetes, and Proteobacteria). The distribution is disjointed (many complete proteomes from organisms within these phyla lack homologs), which is indicative of either multiple horizontal gene transfers or multiple gene losses. All of the bacterial homologs are on the bacterial chromosomes and not on plasmids.

The bacterial homologs vary considerably in length. The shortest sequence is from *Segniliparus rugosus* (gene HMPREF9336_02911), which contains only 343 residues consisting of a predicted signal peptide and a peptidase domain. At the opposite extreme, a sequence from *Salinispora tropica* (gene Strop_0483) contains 3754 residues, with the peptidase domain located near the C-terminus (residues 3399–3629); this unusual protein is also predicted to have two transmembrane regions, one near each terminus. The peptidase domain is usually at the C-terminus, but in the homolog from *Granulibacter bethesdensis* (ABI62444) there is presumably at least one domain following the peptidase domain (this domain appears to be distantly related to the calcium-binding, secretion domain of serralysins, subfamily M10B). The homolog from *Acaryochloris marina* (gene AM1_6343) has a large insert within the peptidase domain between the catalytic Cys and His residues; this insert is unique to this protein. Clearly, the calpain-like domain is being utilized by bacteria in novel ways.

C11 (clostripain): Clostripain is a virulence factor from *Clostridium histolyticum* and a cysteine peptidase distantly related to papain (Rawlings and Barrett, 2004c). Several bacterial homologs are known, from eight phyla (Aquificae, Bacteroidetes, Chloroflexi, Cyanobacteria, Dictyoglomi, Firmicutes, Proteobacteria, Thermotogae) but the distribution is patchy; even some other *Clostridrium* species lack homologs. Homologs are also known from archaea (*Aciduliprofundum boonei* (EDY34731 and EDY35283), *Methanocorpusculum labreanum* (gene Mlab_1476), *Methanosarcina acetivorans* (gene MA4003), and *Methanospirillum hungatei* (ABD42499, ABD42500 and gene Mhun_2076)) and eukaryotes. The eukaryote homologs are all from single-celled organisms, include the apicomplexans *Cryptosporidium parvum* (EAK89763; a mammalian parasite which infects the intestinal tract), *Cryptosporidium muris* (XP_002140096), *Neospora caninum* (CBZ53884), and *Toxoplasma gondii* (EEE34214), and the algae *Chlamydomonas reinhardtii* (XP_001702001) and *Volvox carteri* (XP_002950572). The sequences from apicomplexans are all large, multidomain proteins; that from *Cryptosporidium* has 1607 residues and includes two epidermal growth factor-like domains and a notch-like domain besides the clostripain domain. All nonbacterial homologs are predicted to be active because the catalytic dyad is conserved.

C13 (legumain): Legumains are eukaryote cysteine peptidases distantly related to caspases. While caspases have a predominant preference for Asp in P1, legumains prefer Asn. The family also contains GPI transamidase, which not only cleaves a protein but also transfers glycosylphosphatidylinositide to the new C-terminus, which permits the protein to attach to a membrane (Rawlings and Barrett, 2004c). There are around 30 known bacterial homologs, the majority from Proteobacteria and 4 from members of the phylum Bacteroidetes (*Bacteroides cellulosilyticus* (gene BACCELL_01867), *Bacteroides helcogenes* (YP_004160604), and 2 homologs from *Prevotella ruminicola* (ADE82740 and ADE83205)). There is a homolog from the archaean *M. hungatei* (ABD40250).

C14A (caspase): Caspases are best known as the executor proteins that degrade proteins during apoptosis, and are known from multicellular organisms. However, there are homologs that perform other roles, such as caspase-1, which processes the precursor of the cytokine interleukin-1β (Rawlings and Barrett, 2004c). There is only one known prokaryote homolog, from the proteobacterium *Desulfovibrio salexigens* (ACS78236). The protein is uncharacterized, but has the catalytic dyad conserved and is presumed to be that of an active peptidase. This is the most divergent sequence in the family. Caspases are predominantly intracellular proteins; however, a signal peptide is predicted for the *Desulfovibrio salexigens* sequence.

There is also a homolog from the double-stranded DNA virus *Spodoptera frugiperda* ascovirus 1a (CAD24627), which has probably been acquired from the host moth. However, despite the fact that the active site His has been replaced by Asp, the viral caspases can induce apoptosis (Bideshi et al., 2005).

C14B (metacaspase): Meta- and paracaspases are predominantly plant and fungal peptidases. There are homologs from the archaeans *Methanosarcina barkeri* (gene Mbar_A1574), *Methanothermus fervidus* (YP_004004559), and *Thermococcus onnurineus* (YP_002307598).

C15 (pyroglutamyl-peptidase I): There are homologs in bacteria, archaea, and eukaryotes. Bacteria homologs are found mainly in Firmicutes and Proteobacteria, and homologs are known from only a third of archaean genomes. It is not clear if this distribution represents multiple horizontal gene transfers or losses.

C45 (acyl-coenzyme A:6-aminopenicillanic acid acyl-transferase precursor): Homologs of this family are present in bacteria and eukaryotes, but absent from archaea and protozoa. The most divergence sequences are from bacteria implying a horizontal gene transfer from a bacterium to a eukaryote, perhaps via the protomitochondrion with subsequent loss in protozoa.

*C51 (*D*-Ala-Gly-peptidase)*: The number of homologs from bacteria and viruses is similar, so it is unclear from where the peptidase family arose. There are 18 homologs in protozoa, mostly trypanosomes. The homolog from *Trypanosma cruzi* has been characterized as a trypanothione synthetase, catalyzing the formation of trypanothione from glutathione and spermidine (Oza et al., 2002); this is an unusual horizontal gene transfer in which the function of the protein has changed. There is a single homolog from the archaean *Methanosarcina mazei* (the product of the MM2478 gene), also presumably derived by horizontal gene transfer.

C69 (dipeptidase A): Homologs in this family are predominantly from bacteria; however, there are 45 homologs from eukaryotes (animals, fungi, plants, and protozoa) and 8 from archaea, presumably derived from horizontal gene transfers. Mammalian homologs are secernins, cytoplasmic proteins involved in the regulation of exocytosis from peritoneal mast cells (Way et al., 2002); it is not known if these are peptidases.

G1 (scytalidoglutamic peptidase): This family predominates in fungi, but two homologs are known from archaea (from *Acidilobus saccharovorans* (YP_003816089) and *Caldivirga maquilingensis* (ABW02092)) and 18 from bacteria, all bar 2 from the phylum Firmicutes.

M2 (angiotensin-converting enzyme): Most homologs are from animals, but there are over 50 homologs from Proteobacteria. The bacterial homologs have probably been acquired from a eukaryote by horizontal gene transfer.

M3A (thimet oligopeptidase): Thimet oligopeptidase and its homologs are intracellular peptidases that can only degrade short peptides or remove N-terminal mitochondrial targeting signals from proteins (Rawlings and Barrett, 2004b). Homologs are present in almost all eukaryote genomes (*Encephalitozoon* is an exception) and bacteria (where oligopeptidase A degrades fragments of signal peptides) but homologs in archaea are few. This makes subfamily M3A an ideal candidate for bacterial origin with a horizontal gene transfer to eukaryotes when the mitochondrion was acquired. The archaean homologs are all from the order Methanomicrobiales. This implies that another ancient, horizontal gene transfer has occurred between a bacterium and an archaean.

M3B (oligopeptidase F): Members of subfamily M3B are widely distributed in bacteria and archaea. However, there are a few eukaryote homologs, including examples from protozoa (the apicomplexan parasites *Toxoplasma gondii* (EEB03785 and EEE33876) and *Neospora caninum* (CBZ54222 and CBZ55576), plants (the green algae *Micromonas* (EEH60113 and XP_002501639) and *Ostreococcus* (XP_001417073 and the gene Ot03g05100), and the moss *Physcomitrella patens* (XP_001786868)) and animals (the sea-anemone *Nematostella vectensis* (XP_001617582)). All of the eukaryote homologs are predicted to be active enzymes with at least two domains, with the peptidase domain at the C-terminus. Each would appear to be derived from an individual horizontal gene transfer.

M4 (thermolysin): In a family otherwise restricted to bacteria, there are, however, several homologs from fungi and two from the archaean *M. acetivorans* (genes MA2075 and MA2529).

M11 (gametolysin): Members of family M11 are found in single-celled eukaryotes (such as *Volvox* and *Chlamydomonas*) and also in Proteobacteria and Firmicutes. This limited distribution is indicative of horizontal gene transfer, but because the number of homologs (29 from eukaryotes and 18 from bacteria) is so few it is not possible to determine the direction of transfer.

M12B (ADAMs): ADAM proteins are predominantly found in animals and fungi. There are about 50 bacterial homologs from several bacteria phyla. Homologs are not known from plants or protozoa, with the exception of four Microsporidia species (*Encephalitozoon cuniculi* (gene ECU06_0380),

Encephalitozoon intestinalis (XP_003072998), *Enterocytozoon bieneusi* (geneE-BI_27098), and *Nosema ceranae* (gene NCER_100968)). There has been much debate about the relationship between Microsporidia and fungi, the difference in ribosomal RNAs implying a great difference (Vossbrinck et al., 1987) but other studies suggesting that Microsporidia are fungi (Katinka et al., 2001). The presence of ADAM homologs supports the hypothesis that Microsporidia and Fungi are closely related.

M13 (neprilysin): Homologs in this family are found in eukaryotes, bacteria, and archaea. There is also a single viral homolog from *Acanthamoeba polyphaga* mimivirus (AAV50783), the gene for which is probably derived from the protozoan host. Eukaryote homologs are predominantly from animals and fungi, with very few homologs from plants, red algae, and protozoa. Bacterial homologs are widely distributed, but the sequences are more similar to one another than are the eukaryote homologs. This implies that the eukaryote homologs are more divergent and therefore older, and the bacterial homologs were derived from a lateral gene transfer from a eukaryote. The archaean homologs are very similar to the bacterial PgPepO oligopeptidase, and the genes were probably derived from a bacterium by a second lateral transfer. It is also possible that the ancestral gene was present in the ultimate ancestor and multiple losses have occurred.

M14 (carboxypeptidase A): This family has numerous, widely distributed homologs in eukaryotes and bacteria. There are only three active homologs in archaea, mostly from *Methanosarcina* species (CBH38710 and genes cpa and Meth1525), and all of them from the same subfamily (M14A). The paucity of archaean homologs makes family M14 a candidate to have been transferred from a proto-mitochondrion to a eukaryote genome, with the few archaean homologs derived from a bacterium by a second horizontal gene transfer. The *Methanosarcina* homologs are indeed closer to a bacterium (*Bacillus thuringiensis*, 60% sequence identity) than to eukaryote sequences, but the numbers of bacterial and eukaryote sequences are approximately equal, which makes determining the direction of the transfer more difficult. The most divergent sequences in the family are from bacteria, however.

*M15A (*D*-Ala-*D*-Ala-carboxypeptidase)*: A single archaean homolog is known from *M. mazei* (gene MM3062) adjacent to a gene encoding an S8B (kexin) homolog. This is indicative of a portion of a bacterial genome being incorporated into the archaean.

M16 (pitrilysin): Members of family M16 are found from species in all bacterial genomes and eukaryotes, but there are very few homologs known in archaea. The presence of this family in eukaryotes would therefore be a candidate for horizontal gene transfer at the time of the acquisition of the mitochondrion. Family M16 is divided into subfamilies, and the archaean homologs are found only in subfamily M16B (mitochondrial processing peptidase). Homologs are known from the Crenarchaeota *Aeropyrum pernix* gene APE0212, *Thermoproteus* species (YP_001794471 and YP_004338264), *Pyrobaculum* species (ABL89100, ABP50596, and YP_001056400, and gene PAE0368), *Vulcanisaeta* species (YP_003901456, YP_004245255, and YP_004245556), and *C. maquilingensis* (ABW02494), as well as *A. boonei* (YP_002578747; Euryarchaeota) and *Nanoarchaeum equitans* (gene NEQ023; Nanoarchaeota).

M17 (leucyl aminopeptidase): Family M17 is found mainly in eukaryotes and bacteria, with just eight homologs known in archaea making M17 another candidate for horizontal transfer of a gene from a proto-mitochondrion to a eukaryote genome. The majority of sequences are from bacteria, with examples in almost all phyla. The archaean sequences are more closely related to those from bacteria than eukaryotes, implying a second lateral transfer from bacteria to eukaryotes. However, there is an alternative hypothesis that the ancestral M17 gene was present in the universal ancestor and was subsequently lost from some archaean lines, most notably the Euryarchaeota from which there are no known homologs.

M18 (aminopeptidase I): This family is found in most eukaryotes but is patchily distributed in bacteria. There is only one archaean homolog, from *A. boonei* (YP_002578450), which is 55% identical to a sequence from the bacterium *Sebaldella termitidis*, implying a horizontal gene transfer from a bacterium to a eukaryote. This makes family M18 another candidate for a proto-mitochondrial origin in eukaryotes. This is supported by the fact that the most divergent sequences in the family are bacterial.

M41 (FtsH peptidase): Family M41 is found in all bacterial phyla and many eukaryotes, but is absent from archaea. Thus, family M41 is another candidate for horizontal transfer via the proto-mitochondrion.

M43 (cytophagalysin): There is only a single archaean homolog in this family, ulilysin from *M. acetivorans*, which has been characterized (Tallant et al., 2006). The sequence is 63% identical to the sequence of cytophagalysin (*B. thuringiensis*), making it most likely that a gene has been transferred from a bacterium to an archaean. Homologs are known from eukaryotes, but there are few homologs from plants and protozoa, and from just a few bacterial phyla, with most homologs from members of Bacteroidetes, Firmicutes, and Proteobacteria. The unusual distribution makes this family a candidate for horizontal gene transfer. The fact that the most divergent sequences are bacterial implies a transfer from a bacterium to a eukaryote, possibly via the proto-mitochondrion.

M54 (archaelysin): Homologs from this family are found in almost all archaean proteomes, but in only a few bacteria. Homologs are also found in animals, with a few homologs in fungi and protozoa, but none in plants. The implication is that the family arose in archaea and then genes were horizontally transferred to bacteria (the archaean and bacterial sequences are closely related) and vertically to eukaryotes, but were subsequently lost in plants.

M60 (enhancin): Family M60 consists of viral proteins that occur in bacterial genomes during the lysogenic phase of infection. There are also four homologs in the fungi *Aspergillus* (BAE55343 and EED56799) and *Uncinocarpus reesii* (gene UREG_04003), presumably also of viral origin.

M72 (peptidyl-Asp metallopeptidase): All the eukaryote homologs are from the brown alga *Ectocarpus siliculosus* (CBN76176, CBN78156, and CBN79239). These are sequences only distantly related to the other bacterial homologs in the family (the closest homolog from *Moritella* sp. PE36 is 44% identical). Bacterial members of the family have an N-terminal signal peptide followed by the peptidase domain and a large C-terminal domain. The *Ectocarpus* sequences show relationship only in the peptidase domain, which means they are unlikely to be simple contaminants.

S1A (trypsin): The trypsin subfamily contains sequences mostly from eukaryotes and bacteria. There is just one homolog from an archaean, the product of the gene Mhun_2415 from *M. hungatei*. This is a divergent sequence showing at most 36% identity to other members of the subfamily. The protein is likely to be secreted because a signal peptide is predicted. This is followed by the trypsin-like domain and a C-terminal region containing a PAN/apple domain, which is involved in protein–carbohydrate or protein–protein interactions (Tordai et al., 1999).

Homologs are also absent from plants, and only a few homologs are known from protozoa (10 proteins from *Naegleria gruberi*). There are 30 homologs from fungi, some of which have been characterized, but homologs are absent from *Saccharomyces* and *Schizosaccharomyces*. Clearly, the distribution of peptidases from this subfamily amongst organisms is very complex, probably involving multiple horizontal gene transfers and/or gene losses.

*S13 (*D*-Ala-*D*-Ala-peptidase C)*: Besides the homolog from *Populus trichocarpa*, which is almost certainly a bacterial contaminant (see Table 11.1), there are homologs from the protozoa *Paulinella chromatophora* (YP_002048804) and *Dictyostelium discoideum* (AAM44351).

S14 (peptidase Clp): Homologs in this family are found in all bacterial phyla and in many eukaryotes. However, only one homolog has been found in an archaean (the product of the Mvol_0834 gene from *Methanococcus voltae*), which may not be a peptidase because the third member of the catalytic triad has been replaced. This may be a bacterial contaminant; however, the closest known homolog to this sequence is only 31% identical. The absence of homologs in Archaea means that family S14 is an ideal candidate for being transferred to eukaryotes at the time of the acquisition of the mitochondrion.

S45 (penicillin G acylase precursor): Besides the homolog from *P. trichocarpa*, which is almost certainly a bacterial contaminant (see Table 11.1), there are a number of protozoan homologs.

S58 (aminopeptidase DmpA): Besides the homolog from *P. trichocarpa*, which is almost certainly a bacterial contaminant (see Table 11.1 there are several homologs from fungi (ascomycetes) and one from the protozoan *N. gruberi* (EFC37030). These were probably derived from horizontal gene transfers from a bacterium, but probably not via the proto-mitochondrion unless the gene had subsequently been lost from plants and animals. There are 12 homologs from archaea, species of *Pyrococcus*, *Thermococcus*, and *Sulfolobus*. The archaean sequences are closely related to bacterial sequences, implying the transfer of a gene from a bacterium to an archaean.

11.5 CONCLUSIONS

There are many examples among peptidase families of unusual occurrences in organisms. Some can be explained by contamination or infection, but the vast majority are best explained by horizontal gene transfers. Rawlings and Barrett (Rawlings and Barrett, 1995) proposed that some peptidase families may have been acquired by eukaryotes at the time of the development of the mitochondrion and chloroplast from endosymbiotic bacteria. Although many more genomes have now been completely sequenced and the absence of a gene proved beyond doubt, this hypothesis still holds true for several peptidase families (C45, M3, M14, M16, M17, M18, M41, M43, and S14). Just as in 1995, the majority of families are of metallopeptidases. The existence of present day eukaryotes derived from pre-mitochondrial stock has been called into question; proving this route of gene transfer is difficult because it could equally well be argued that no horizontal gene transfer has occurred but a homolog of each family was present in the universal ancestor and subsequently lost from Archaea. However, transfer of genes from the mitochondrion to the nucleus is known to occur, and the presence of some peptidase families in eukaryotes can be explained by one event whereas the absences in archaea require several gene loss events.

Horizontal gene transfer clearly provides the recipient with an evolutionary advantage, most likely by making a new food source available to the organism, which in the case of peptidases would be amino acids from otherwise undegradable proteins. The advantage of transferring a gene from the mitochondrion to the nucleus would be to place it under the control of the cell, which in the case of a peptidase gene would prevent unwanted and possibly destructive digestion of cellular proteins.

Determining whether a gene has been transferred horizontally or vertically depends very much on the distribution of known homologs, which for many taxonomic groups is poorly known. More sequence data from a wider variety of organisms will help resolve the evolutionary history of many of these peptidase families.

REFERENCES

Andersson, S.G., Zomorodipour, A., Andersson, J.O., Sicheritz-Ponten, T., Alsmark, U.C., Podowski, R.M., Naslund, A.K., Eriksson, A.S., Winkler, H.H., and Kurland, C.G. (1998) The genome sequence of *Rickettsia prowazekii* and the origin of mitochondria. Nature, 396, 133–140.

Athauda, S.B., Yoshioka, K., Shiba, T., and Takahashi, K. (2006) Isolation and characterization of recombinant *Drosophila* copia aspartic proteinase. Biochem J, 399, 535–542.

Bernard, D., Mehul, B., Thomas-Collignon, A., Delattre, C., Donovan, M., and Schmidt, R. (2005) Identification and characterization of a novel retroviral-like aspartic protease specifically expressed in human epidermis. J Invest Dermatol, 125, 278–287.

Bideshi, D.K., Tan, Y., Bigot, Y., and Federici, B.A. (2005) A viral caspase contributes to modified apoptosis for virus transmission. Genes Dev, 19, 1416–1421.

Boeke, J.D. (2004) Fungal, plant and animal retrotransposon elements, in Handbook of Proteolytic Enzymes (eds A.J. Barrett, N.D. Rawlings, and J.F. Woessner), Elsevier, London, pp. 190–195.

Brennicke, A., Grohmann, L., Hiesel, R., Knoop, V., and Schuster, W. (1993) The mitochondrial genome on its way to the nucleus: different stages of gene-transfer in higher-plants. FEBS Lett, 325, 140–145.

Cavalier-Smith, T. (1987) The simultaneous symbiotic origin of mitochondria, chloroplasts, and microbodies. Ann N Y Acad Sci, 503, 55–71.

Cavalier-Smith, T. (1989) Archaebacteria and Archezoa. Nature, 339, 100–101.

Cavalier-Smith, T. (1991) Archamoebae: the ancestral eukaryotes? Biosystems, 25, 25–38.

Chen, J.C., Hottes, A.K., McAdams, H.H., McGrath, P.T., Viollier, P.H., and Shapiro, L. (2006) Cytokinesis signals truncation of the PodJ polarity factor by a cell cycle-regulated protease. EMBO J, 25, 377–386.

Davison, A.J. (1992) Channel catfish virus: a new type of herpesvirus. Virology, 186, 9–14.

Doolittle, R.F., Feng, D.F., Johnson, M.S., and McClure, M.A. (1989) Origins and evolutionary relationships of retroviruses. Q Rev Biol, 64, 1–30.

Eriksson, A.C., Sjoeling, S., and Glaser, E. (1993) A general processing proteinase of spinach leaf mitochondria is associated with the bc1 complex of the respiratory chain, in Plant Mitochondria (eds A. Brennicke and U. Kueck), Wiley-VCH Verlag GmbH, Weinheim, pp. 299–306.

Fehrmann, F., Welker, R., and Krausslich, H.G. (1997) Intracisternal A-type particles express their proteinase in a separate reading frame by translational frameshifting, similar to D-type retroviruses. Virology, 235, 352–359.

Gellissen, G. and Michaelis, G. (1987) Gene transfer: mitochondria to nucleus. Ann N Y Acad Sci, 503, 391–401.

Hartley, B.S. (1979) Evolution of enzyme structure. Proc R Soc Ser B, 205, 443–452.

Howe, K., Bateman, A., and Durbin, R. (2002) QuickTree: building huge Neighbour-Joining trees of protein sequences. Bioinformatics, 18, 1546–1547.

Katinka, M.D., Duprat, S., Cornillot, E., Metenier, G., Thomarat, F., Prensier, G., Barbe, V., Peyretaillade, E., Brottier, P., Wincker, P., Delbac, F., El Alaoui, H., Peyret, P., Saurin, W., Gouy, M., Weissenbach, J., and Vivares, C.P. (2001) Genome sequence and gene compaction of the eukaryote parasite *Encephalitozoon cuniculi*. Nature, 414, 450–453.

Katoh, K., Misawa, K., Kuma, K., and Miyata, T. (2002) MAFFT: a novel method for rapid multiple sequence alignment based on fast Fourier transform. Nucleic Acids Res, 30, 3059–3066.

Kim, A., Terzian, C., Santamaria, P., Pelisson, A., Purd'homme, N., and Bucheton, A. (1994) Retroviruses in invertebrates: the gypsy retrotransposon is apparently an infectious retrovirus of *Drosophila melanogaster*. Proc Natl Acad Sci USA, 91, 1285–1289.

Lipps, G. (2008) Plasmids: Current Research and Future Trends. Horizon Scientific Press.

Oza, S.L., Tetaud, E., Ariyanayagam, M.R., Warnon, S.S., and Fairlamb, A.H. (2002) A single enzyme catalyses formation of Trypanothione from glutathione and spermidine in Trypanosoma cruzi. J Biol Chem, 277, 35853–35861.

Navia, M.A., Fitzgerald, P.M., McKeever, B.M., Leu, C.T., Heimbach, J.C., Herber, W.K., Sigal, I.S., Darke, P.L., and Springer, J.P. (1989) Three-dimensional structure of aspartyl protease from human immunodeficiency virus HIV-1. Nature, 337, 615–620.

Rawlings, N.D. and Barrett, A.J. (1993) Evolutionary families of peptidases. Biochem J, 290, 205–218.

Rawlings, N.D. and Barrett, A.J. (1995) Evolutionary families of metallopeptidases. Methods Enzymol, 248, 183–228.

Rawlings, N.D. and Barrett, A.J. (2004a) Introduction: aspartic peptidases and their clans, in Handbook of Proteolytic Enzymes (eds A.J. Barrett, N.D. Rawlings, and J.F. Woessner), London, Elsevier, pp. 3–12.

Rawlings, N.D. and Barrett, A.J. (2004b) Introduction: metallopeptidases and their clans, in Handbook of Proteolytic Enzymes (eds A.J. Barrett, N.D. Rawlings, and J.F. Woessner), Elsevier, London, pp. 231–268.

Rawlings, N.D. and Barrett, A.J. (2004c) Introduction: the clans and families of cysteine peptidases, in Handbook of Proteolytic Enzymes (eds A.J. Barrett, N.D. Rawlings, and J.F. Woessner), Elsevier, London, pp. 1051–1071.

Rawlings, N.D. and Bateman, A. (2009) Pepsin homologues in bacteria. BMC Genomics, 10, 437.

Rawlings, N.D., Pearl, L.H., and Buttle, D.J. (1992) The baculovirus *Autographa californica* nuclear polyhedrosis virus genome includes a papain-like sequence. Biol Chem Hoppe Seyler, 373, 1211–1215.

Rinckel, L.A. and Garfinkel, D.J. (1998) Fungal, plant and animal transposon elements, in Handbook of Proteolytic Enzymes (eds A.J. Barrett, N.D. Rawlings, and J.F. Woessner), Academic Press, London, pp. 956–960.

Roger, A.J., Svard, S.G., Tovar, J., Clark, C.G., Smith, M.W., Gillin, F.D., and Sogin, M.L. (1998) A mitochondrial-like chaperonin 60 gene in *Giardia lamblia*: Evidence that diplomonads once harbored an endosymbiont related to the progenitor of mitochondria. Proc Natl Acad Sci USA, 95, 229–234.

Sayers, E.W., Barrett, T., Benson, D.A., Bolton, E., Bryant, S.H., Canese, K., Chetvernin, V., Church, D.M., DiCuccio, M., Federhen, S., Feolo, M., Fingerman, I.M., Geer, L.Y., Helmberg, W., Kapustin, Y., Landsman, D., Lipman, D.J., Lu, Z., Madden, T.L., Madej, T., Maglott, D.R., Marchler-Bauer, A., Miller, V., Mizrachi, I., Ostell, J., Panchenko, A., Phan, L., Pruitt, K.D., Schuler, G.D., Sequeira, E., Sherry, S.T., Shumway, M., Sirotkin, K., Slotta, D., Souvorov, A., Starchenko, G., Tatusova, T.A., Wagner, L., Wang, Y., Wilbur, W.J., Yaschenko, E., and Ye, J. (2011) Database resources of the National Center for Biotechnology Information. Nucleic Acids Res, 39, D38–D51.

Schwartz, R.M. and Dayhoff, M.O. (1978) Origins of prokaryotes, eukaryotes, mitochondria and chloroplasts. Science, 199, 395–403.

Slack, J.M., Kuzio, J., and Faulkner, P. (1995) Characterization of V-cath, a cathepsin L-like proteinase expressed by the baculovirus *Autographa californica* multiple nuclear polyhedrosis virus. J Gen Virol, 76, 1091–1098.

Sogin, M. (1997a) History assignment: when was the mitochondrion founded? Curr Opin Genet Dev, 7, 792–799.

Sogin, M.L. (1997b) Organelle origins: energy-producing symbionts in early eukaryotes? Curr Biol, 7, R315–R317.

Tallant, C., Garcia-Castellanos, R., Seco, J., Baumann, U., and Gomis-Ruth, F.X. (2006) Molecular analysis of ulilysin, the structural prototype of a new family of metzincin metalloproteases. J Biol Chem, 281, 17920–17928.

The UniProt Consortium (2010) The Universal Protein Resource (UniProt) in 2010. Nucleic Acids Res, 38, D142–D148.

Tordai, H., Banyai, L., and Patthy, L. (1999) The PAN module: the N-terminal domains of plasminogen and hepatocyte growth factor are homologous with the apple domains of the prekallikrein family and with a novel domain found in numerous nematode proteins. FEBS Lett, 461, 63–67.

Volkov, D.A., Dergousova, N.I., and Rumsh, L.D. (2004) New aspartic proteinase of Ulysses retrotransposon from *Drosophila virilis*. Biochemistry (Mosc), 69, 697–701.

Vossbrinck, C.R., Maddox, J.V., Friedman, S., Debrunner-Vossbrinck, B.A., and Woese, C.R. (1987) Ribosomal RNA sequence suggests microsporidia are extremely ancient eukaryotes. Nature, 326, 411–414.

Way, G., Morrice, N., Smythe, C., and O'Sullivan, A.J. (2002) Purification and identification of secernin, a novel cytosolic protein that regulates exocytosis in mast cells. Mol Biol Cell, 13, 3344–3354.

Yang, D., Oyaizu, Y., Oyaizu, H., Olsen, G.J., and Woese, C.R. (1985) Mitochondrial origins. Proc Natl Acad Sci USA, 82, 4443–4447.

Young, C.L., Barker, W.C., Tomaselli, C.M., and Dayhoff, M.O. (1978) Serine proteases, in Atlas of Protein Sequence and Structure (ed M.O. Dayhoff), National Biomedical Research Foundation, Washington DC, pp. 73–93.

12

DEDUCING TRANSPORT PROTEIN EVOLUTION BASED ON SEQUENCE, STRUCTURE, AND FUNCTION

STEVEN T. WAKABAYASHI, MAKSIM A. SHLYKOV, UJJWAL KUMAR, VAMSEE S. REDDY, ANKUR MALHOTRA, ERIK L. CLARKE, JONATHAN S. CHEN, ROSTISLAV CASTILLO, RUSSELL DE LA MARE, ERIC I. SUN, AND MILTON H. SAIER JR.

Division of Biological Sciences, University of California, San Diego, La Jolla, CA, USA

CHAPTER SUMMARY

All of biology makes sense only in the context of evolution. This generalization applies to living organisms, metabolic pathways, and macromolecules including nucleic acids and proteins. Our research team has focused on the evolution of transmembrane transport proteins, a subset of the cellular proteins, comprising about 10% of the total. With few exceptions, transporters evolved from small amphipathic transmembrane peptides, which underwent duplication, triplication, and quadruplication as well as fusion and splicing, all at the gene level. In this review, we summarize many of the bioinformatic methodologies and results obtained that support the extensive and numerous pathways we and others have been able to elucidate using these tools. The Transporter Classification Database (TCDB; www.tcdb.org) has been an essential resource in these studies.

Protein Families: Relating Protein Sequence, Structure, and Function, First Edition.
Edited by Christine Orengo and Alex Bateman.

12.1 INTRODUCTION

About 10% of all functionally characterized genes across the three domains of life encode transport proteins, which mediate a range of cellular processes (Saier and Ren, 2006). First, many provide nutrients to cells and their compartments, while others are direct participants in the excretion of harmful metabolites, drugs, and toxins (Van Bambeke et al., 2003; Saier, 2007). Second, they control the uptake and efflux of ions, which are crucial for proper cell function, maintenance of membrane potentials, the activities of mineral cofactor-containing enzymes, and directional sensing for cell motility (Özkucur et al., 2011; Prakash et al., 2003; Schwarz et al., 2009). Third, transporters secrete lipids, carbohydrates, proteins, and nucleic acids. Macromolecular export is required for cell envelope assembly, extracellular molecular degradation, and the exchange of genetic material between organisms (Saier, 2000a). Fourth, they allow for cellular communication by transporting signaling molecules such as hormones, neurotransmitters, and pheromones (Lam et al., 2011; Rettner and Saier, 2010). Lastly, transport proteins allow entry of pore-forming/non-pore-forming toxins into cells and export of antiviral, antibiotic, and antifungal agents to allow killing of the target organism (Davies et al., 2002). Not surprisingly, therefore, the biological importance of transporters cannot be overstated.

Membrane transport proteins are classified in a large, online database called the *TCDB* (www.tcdb.org). Many of these transporters include proteins that have been shown to be evolutionarily related because of the similarities in their amino acid sequences. Most of these homologous proteins have structures similar to those of their evolutionarily linked family members, which rarely exhibit functions aside from transport. The primary exceptions are transport proteins that have evolved to become receptors, sometimes retaining, and sometimes losing their transport functions. Amino acid sequence data can yield information about the transmembrane β-strand or α-helical hydrophobic or amphipathic segments (transmembrane segments, TMSs). These proteins usually arose by multiplication of small peptide units with 1, 2, 3, or 4 TMSs (Saier, 2003a).

In this chapter, we shall discuss the ways in which the pathways of transport protein evolution are deduced. We first examine TCDB with respect to the system used for transporter classification (TC), revealing how information flow into the database is generated. Second, we discuss representative classes, subclasses, families, and superfamilies of these systems, presenting details of novel software developed for accurate phylogenetic analyses of distantly related homologs. Functional predictions become possible, especially for prokaryotes, by conducting genome context analyses using the SEED and RegPredict databases. Third, we consider methods for predicting transporter function. The superfamily tree (SFT) programs (Yen et al., 2009; 2010; Chen et al., 2011) allow depiction of distant relationships of families within superfamilies by construction of reliable phylogenetic trees based on BLAST bit scores rather than multiple alignments. Fourth, we shall consider the pathways taken for transport protein evolution. Fifth, we describe the pathways of transport protein evolution observed and recorded in

TCDB. Finally, we conclude with approaches used for conducting whole genome analyses of transport systems.

12.1.1 The Transporter Classification Database

TCDB (www.tcdb.org) (Saier et al., 2006, 2009) presents the comprehensive IUBMB (International Union of Biochemistry and Molecular Biology)-approved and adopted classification system for membrane transport proteins known as the *TC* system. The TC system, unlike The Enzyme Commission (EC) system for the classification of enzymes (Webb, 1992), incorporates both functional and phylogenetic details of the proteins. Class, family, and transport system descriptions, TC numbers, references, crystal structures (when available), and disease relationships are all included. Examples of over 600 families of transport proteins are provided. Each transport protein is given a unique TC number consisting of five components and is in the form of V.W.X.Y.Z. (Fig. 12.1). V (a number) corresponds to the transporter *class* (i.e., (1) channel, (2) carrier (porter), (3) primary active transporter (4) group translocator or (5) transmembrane electron flow carrier); W (a letter) corresponds to the transporter *subclass*, which in the case of primary active transporters, refers to the energy source used to drive transport; X (a number) corresponds to the transporter *family* (sometimes actually a superfamily); Y (a number) corresponds to the *subfamily* (or, in the case of a superfamily, the family), and Z corresponds to a specific *transporter* with its unique mechanism of action, polarity, and range of substrates.

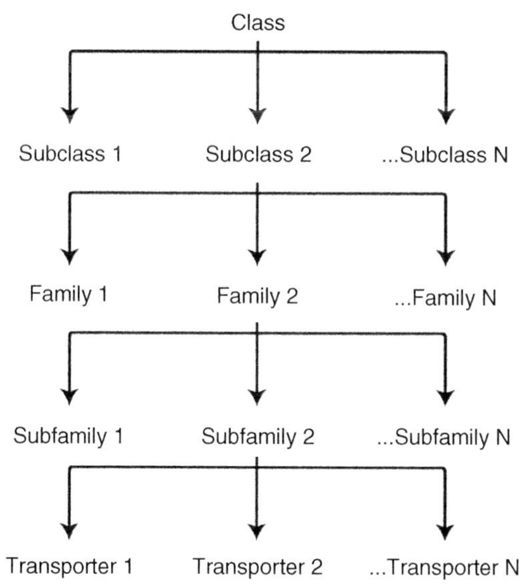

Figure 12.1 The TC scheme of transport system classification.

The following are classes of transport families included in TCDB:

1. Channels
2. Carriers
3. 1° Active porters
4. Group translocators
5. TM electron flow carriers
6. Open[1]
7. Open
8. Auxiliary transport proteins
9. Poorly characterized porters

Any two transport systems in the same subfamily of a transporter family that transport the same substrate(s) are given the same TC#, regardless of whether they are orthologs (e.g., arose in distinct organisms by speciation) or paralogs (e.g., arose within a single organism by gene duplication). However, more than one such system may be present in TCDB when different aspects of their structures, functions, and/or mechanisms are known. Superfamilies are either represented under a single TC#, or their family constituents are indicated by the use of hyperlinks. Sequenced homologs of unknown function are not usually assigned a TC# unless they represent a unique subfamily, are distant, unrepresented members of a subfamily, or are from an underrepresented organismal kingdom/domain. If multiple dissimilar subunits are present, they are numbered S1, S2, S3,. . . ,S*n*. Classification categories 6 and 7 are reserved for new, currently unidentified classes. Classes 8 and 9 are reserved for accessory transport proteins and incompletely characterized transporters, respectively. TCDB is hosted in the San Diego Supercomputing center (www.sdsc.edu) at the University of California, San Diego.

12.1.2 TCDB User Interface

12.1.2.1 Search TCDB gives a user many options to search, analyze, and browse transport proteins (Fig. 12.2). The user can view the superfamily taxonomy and search a protein by TC ID#, name, organismal source, associated diseases (if applicable), PubMed ID, title, authors of references, etc. One or more search terms can be entered to narrow a search utilizing the search box at the top of the web page.

The following details are returned with a protein search:

1. TC ID#
2. Reference
3. Accession number

[1]Open: no assignment has yet been made for classes 6 and 7.

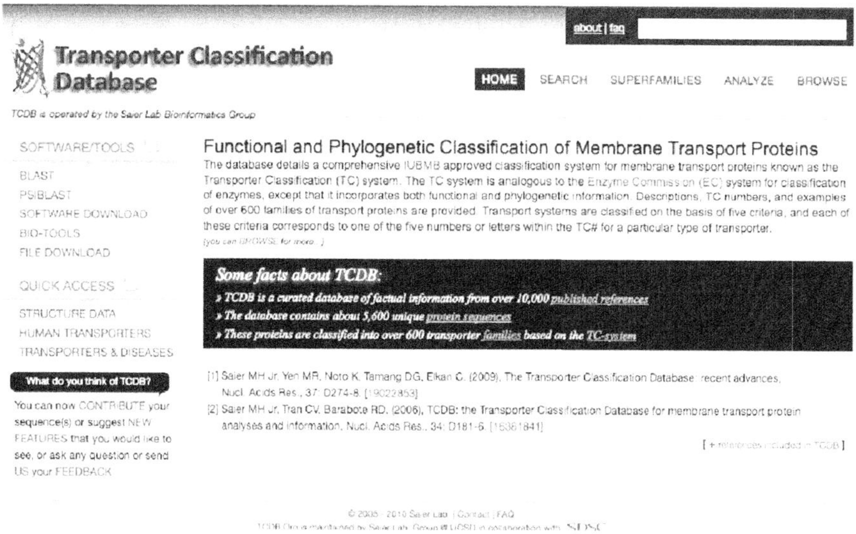

Figure 12.2 TCDB homepage (http://tcdb.org).

4. Protein name
5. Length
6. Molecular weight
7. Species
8. Predicted number of TMSs
9. Location/topology/orientation
10. Database of Interacting Proteins (DIP)
11. Protein Families (PFAM) reference
12. Gene ontology

The user is also given an option of either BLASTING/PSI-BLASTING the protein against the nonredundant National Center for Biotechnology Information (NCBI) or TCDB (accessed from the sidebar). Additional analysis options, such as predicting numbers of TMSs through hydropathy plots, are also available.

The TCDB web server hosts dedicated tools including the following:

1. Hydropathy analysis using the Kyte–Doolittle scale combined with hydrophobic moment and TMS predictions (WHAT program).
2. EMBOSS hydropathy analysis.
3. Hydrophobic moment calculation—the amphipathicity of a peptide measured for a specified angle of rotation per residue with EMBOSS Hmoment.
4. Identifying amphipathic TMSs by creating helical wheel plots that display a protein sequence looking down the axis of the α-helix with EMBOSS Pepwheel and helical wheel plot with Helixator.

5. TMS prediction (HMMTOP program).
6. Homolog search with BLAST and SSEARCH.
7. Pairwise sequence alignment using the Smith–Waterman or Needleman–Wunsch algorithm
8. Multiple sequence alignment using ClustalX
9. Average hydropathy, amphipathicity, and similarity on full or cut sequence alignments (AveHAS program)
10. Topological analyses for any class, subclass, superfamily, family, subfamily, cluster, or system (TMStats program)
11. Determining homology between distantly related protein sequences using a binary alignment tool (GSAT program)
12. Simplified program of obtaining distant homologs by removing redundant and abnormal sequences (Protocol1 program)
13. Comparing two FASTA files of different families from the Protocol1 program using Smith–Waterman algorithm to produce local alignment and GSAT scores, regions of homology, and global alignment of HSP (Protocol2 program)
14. Targeted Smith–Waterman Search tool used to find areas of homology between distantly related full length sequences (TSSearch program)

The web server also hosts other useful tools and scripts under the category of "Bio Tools" for analyzing transport proteins. These tools are available for download.

12.2 DATA INPUT FOR TCDB

Data can be input into TCDB in two ways: through built-in semiautomatic machine learning tools and through user contributions. "Spiders" periodically crawl through the Medline documents and collect the "keywords" appearing in a document's abstract or the set of keywords associated with the document, related to transport proteins. A "bag of words" is classified according to the abstracts of the scientific literature as related or nonrelated on the basis of the relevance of these keywords. The "related" literature is then manually reviewed for accuracy. This approach gives a recall approaching that of a human expert and considerably reduces the time required (Saier et al., 2009). The second "nonautomatic" approach gives an interface to researchers inputting newly characterized transporters into TCDB. This feature is available via "CONTRIBUTE your sequence(s)" link on the TCDB home page. A researcher can submit a protein sequence in FASTA format along with its other characteristics including protein name, gene name, mechanism of action, transport substrate(s), organismal source, reference(s), protein accession number, suggested TC ID#, subcellular location, etc. New proteins are added to TCDB after review (Saier et al., 2009).

12.2.1 Families and Superfamilies

Protein families are the primary phylogenetic unit of classification by which homologous proteins are organized. At the most basic level, protein families encompass a group of proteins with significant similarity and conserved general function (Dayhoff, 1974). It is widely accepted that sequence similarity between two proteins is proportional to their evolutionary relatedness. Owing to the accumulation of mutations (point substitutions, deletions, insertions, etc.), proteins tend to diverge in sequence over time, even though their functions may remain similar. The likelihood of mutations can be organized into protein evolutionary models represented by matrices, and these matrices (such as BLOSUM (Henikoff, 1992) or PAM (Dayhoff, 1978)) can then be used in algorithms that compare sequences to identify their evolutionary divergence. Because these point mutations arise seldomly, homologous sequences can be recognized across all domains of life (Dayhoff, 1974). Accurate phylogenetic trees reveal the relatedness of proteins within a family or superfamily.

Protein superfamilies can include multiple protein families. Members of a superfamily usually share a series of functional, mechanistic, and structural attributes. When Margaret Dayhoff described the first superfamilies in 1974 (Dayhoff, 1974), there were 619 protein sequences available. As of April 2011, there were over 72,000 three-dimensional (3-D) structures of proteins in the RCSB Protein Data Bank (PDB), and over 12 million protein sequences available in the NCBI databank. This exponential increase in the numbers of known protein sequences has aided in the dramatic increase in the numbers of families and superfamilies recognized. Here, we focus on representative families and superfamilies of transport proteins, which have characteristics particularly worth examining.

The individual TMSs most often found in transport proteins are composed of multiple α-helices or β-strands that, together, form the channels through which substrates pass. TMSs are detectable using hydropathy and amphipathicity plots of the primary sequence. The hydropathy plot of a transporter contains sharp spikes that generally indicate the presence of TMSs (Fig. 12.3a). An average hydropathy/amphipathy/similarity plot provides much more reliable results (Fig. 12.3b).

Because the TMSs play a critical role in the structure and orientation of a transport protein in the membrane, there is significant selective pressure against mutations that would alter or deprive a protein of a TMS. The loss of a single internal transmembrane segment would invert the remainder of the membrane protein and cause dramatic changes in function. Because of the relatively high selective pressure against alteration, these regions are relatively well conserved. Detection of homologous transmembrane domains has aided in the classification of hundreds of transporter families and many transport protein superfamilies. Superfamilies are presented in TCDB by hyperlinks to the individual families.

The major facilitator superfamily (MFS TC ID# 2.A.1) (Law et al., 2008; Marger and Saier, 1993; Pao et al., 1998; Saier et al., 1999) is the largest

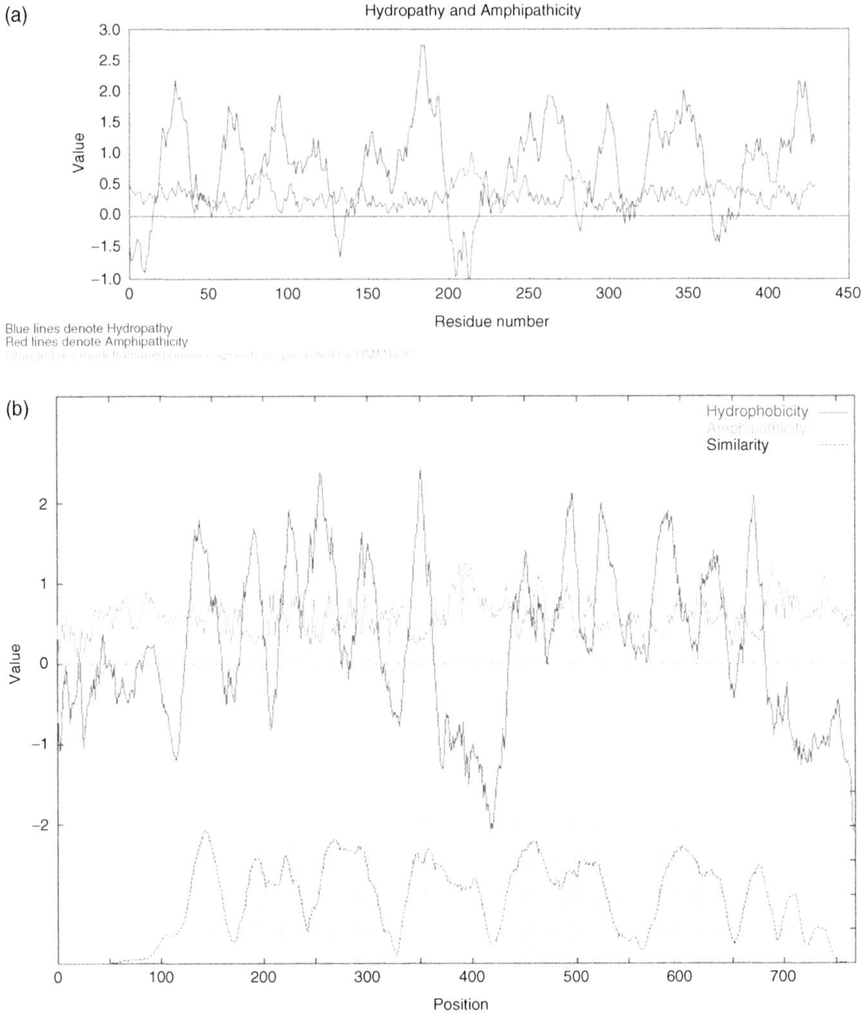

Figure 12.3 (a) Hydropathy and amphipathicity plots for an MFS carrier (Galactose : H^+ symporter) (WHAT) and (b) Hydropathy and amphipathicity and similarity plots for MFS porter proteins (AveHAS). (*See insert for color representation of the figure.*)

superfamily of secondary carriers currently recognized and is found in all extant domains of life. Remarkably, for such a large superfamily, its currently recognized members have a number of well-conserved topological attributes; most have 12 or 14 TMSs where the 12 TMS topology resulted from duplication of a basic 6 TMS element, and the extra 2 TMSs in the 14 TMS proteins are inserted in between the two repeat units.

We have recently demonstrated that the 6 TMSunit in MFS porters arose by duplication of a primordial 3 TMS unit (S. Meyers and M.H. Saier, unpublished results; see Fig. 12.3). Well-conserved similar hydrophilic motifs are found between the second and third TMSs as well as the eighth and ninth TMS in corresponding proteins of the two TMS repeat units. MFS porters are involved in the transport of small organic molecules such as nucleosides, amino acids, carbohydrates, vitamins, and organic ions. Inorganic ions are seldom transported. No current member is known to transport large organic macromolecules such as DNA, complex carbohydrates, or proteins. However, MFS-type lipid porters have been identified (Harvat et al., 2005). These systems exploit chemiosmotic gradients and are involved in uniport, symport, and antiport. The substrate-binding site is believed to orient sequentially toward the two sides of the membrane (Radestock and Forrest, 2011; Dang et al., 2010).

The ATP-binding cassette transporter superfamily (TC ID# 3.A.1) (Dean et al., 2001) is a functional superfamily consisting of three evolutionarily distinct types of porters with distinct transmembrane folds: ABC1, ABC2, and ABC3 (Wang et al., 2009). These three porter types have homologous cytoplasmic ATPases superimposed on them for the purpose of energy coupling. As its name implies, the ABC superfamily derives energy for transport by hydrolyzing ATP instead of using transmembrane ion or chemical gradients.

A typical ABC transporter is composed of two duplicated domains: the transmembrane domain and the nucleotide-binding domain. ABC transport proteins contain many substantial variations on that theme; for example, one ABC transporter, ABCC1 (P33527; TC# 3.A.1.208.8), contains a third transmembrane domain that may assist in the transport process. Entirely different folds are observed for ABC2 uptake transporters and ABC1 export systems (Aller et al., 2009; Dean, 2002; Jones et al., 2009; Oldham and Chen, 2011; Wang et al., 2009).

Several years ago, five families were recognized to be members of a superfamily designated as the amino acid/polyamine/organocation (APC) superfamily (Jack et al., 2000; Gao et al., 2010; Kowalczyk et al., 2011). All of the porters containing these families transported amino acids and their derivatives. With fairly well-represented structural motifs (such as two 5 TMS repeat elements), these proteins can exhibit a variety of typological types due to the gain or loss of TMSs. For example, members of the spore germination protein (SGP) family (TC# 2.A.3.9) have 10 putative TMSs, while members of the eukaryotic cationic amino acid transporter (CAT) family (TC# 2.A.3.3) have 14. SGP proteins have apparently lost transport activity and function as simple receptors (Cooper and Moir, 2011; Mongkolthanaruk et al., 2011). The 14 TMS topology has arisen twice in two different families of the APC superfamily, once in prokaryotes and once in eukaryotes. However, in both cases, the extra two TMSs are found C-terminal to the two repeat units (Chang et al., 2004 and unpublished observation).

The recently expanded APC superfamily now includes 10 families as initially suggested by 3-D comparisons of high resolution x-ray crystallography data (Fang et al., 2009; Gao et al., 2010, Kowalczyk et al., 2011). Iterative PSI-BLAST and the SFT program (Yen et al., 2009; Chen et al., 2011) were used to elucidate the relationships between members with reliable phylogenetic trees. The 10 constituent families in this superfamily show representation in different organismal types; some are found only in prokaryotes, others are found in bacteria and eukaryotes but not archaea, while some others are ubiquitous. Some families within the APC superfamily remain functionally uncharacterized. However, the five originally recognized amino-acid-transporting families are all more closely related to each other than they are to the more recently discovered families (Chang et al., 2004; Chen et al., 2011).

12.2.2 The SFT Programs

Phylogenetic programs help visualize the relationships of transport proteins and their constituent families belonging to a superfamily (sharing a common evolutionary ancestor; Chang et al., 2004). Many phylogenetic methods are available to estimate evolutionary distances and are based on different algorithms and assumptions. Prior to the development of the SFT programs (Yen et al., 2009; Yen et al., 2010; Chen et al., 2011), the available phylogenetic methods relied on multiple sequence alignments. However, when proteins within a superfamily are too distantly related to one another, they will not align properly. As a result, a reliable tree cannot be generated (Ogdenw and Rosenberg, 2006). The SFT programs use tens of thousands of BLAST bit scores instead of multiple alignments, thus avoiding the pitfalls often encountered when determining phylogeny for distantly related proteins (Yen et al., 2009; Yen et al., 2010; Chen et al., 2011).

To create a phylogenetic tree using the SFT1 program, a temporary database file containing proteins from TCDB is generated in FASTA format (Saier et al., 2006; Saier et al., 2009). The TC proteins are used to define the criteria for that superfamily and how that superfamily is later to be broken down into its respective families and/or subfamilies. The division of proteins into superfamilies, families, and subfamilies is conducted according to assignments within TCDB. In many cases, the SFT1 program proves valuable in allowing us to verify current TCDB classifications, as well as giving us the insight on how to improve TCDB assignments (Chen et al., 2011).

Using a program, *getNcbiSeq.pl*, the NCBI "nonredundant" database can be searched for proteins that are similar to each of the proteins within the temporary database file using *blastcl3* (Altschul et al., 1990; Altschul et al., 1997). The collected NCBI proteins are downloaded in FASTA format and sorted into separate database files representing families/subfamilies within the superfamily. Each database file generated contains NCBI proteins closely related to the defining TC protein to which that file relates, thus allowing better definition of the constituent proteins within each protein family or subfamily.

Using the newly generated protein database files, the program, *supertree.pl*, is used to define phylogenetic relationships among families/subfamilies using BLAST bit scores. This process begins with the random selection of several proteins from each protein database file. BLAST comparison values are generated and stored within matrices when all the randomly selected proteins within one database file are compared to those within another file using *blastall*. Thus, a mean value is generated for each interfamily/intersubfamily comparison. This process is repeated for 100 iterations, and the resultant matrices are used to generate 100 Fitch trees. Fitch trees are generated using BLAST bit score matrices as measures of genetic distance between protein families/subfamilies (Fitch and Margoliash, 1967). The 100 Fitch trees are then consolidated into a single consensus tree using the program Consense (http://evolution.genetics.washington.edu/phylip/doc/protpars.html). The resulting consensus tree is viewed as a radial phylogenetic tree using a version of the Tree View (TV) program (Zhai et al., 2002).

To generate a SFT2 tree, the protein database files are selectively combined, generated earlier by *getNcbiSeq.pl*, into new database files based on SFT1 and TCDB family classifications for that superfamily. These newly formed database files are then moved into a new directory, and the *supertree.pl*, Fitch, and Consense programs are used to create a new consensus phylogenetic SFT2 tree, revealing the phylogenetic relationships between subfamilies within the superfamily.

When analyzing SFT trees, small numbers adjacent to the branches represent the "bootstrap" values. These "bootstrap" values indicate the frequency with which certain phylogenetic branching and clustering patterns are observed within the 100 Fitch trees. They can thus be used as indications of the reliability of observed branching orders.

Using the TCDB, we can generate several types of phylogenetic trees. These approaches are exemplified using the MOP superfamily. Trees were generated using, for example, the ClustalX neighbor-joining or parsimony approach (Gascuel and Steel; 2006; Thompson et al., 1997). A neighbor-joining tree (Fig. 12.4a), an SFT1 tree of the proteins (Fig. 12.4b), and an SFT2 tree of the families (Fig.12.4c) were made. According to the ClustalX tree, members of the MATE family (TC# 2.A.66.1) are found in four distinct subclusters. The PST family (TC# 2.A.66.2) is scattered throughout the tree. The OLF family (TC# 2.A.66.3) has three members that cluster together while the fourth member is localized to an isolated branch. The two proteins from the MVF family (TC# 2.A.66.4) cluster loosely together. The three Ank family proteins (TC# 2.A.66.9) cluster together as do the two EPS-E homologs (TC# 2.A.66.8). The members of each remaining family cluster together, but separate from members of other families (Fig. 12.4a).

In comparison, the SFT1 tree correctly clusters all protein members according to their family designations within the MOP superfamily (Fig. 12.4b). Using the SFT2 tree (Fig. 12.4c), the phylogenetic relationships of the constituent families within the MOP superfamily are displayed. It is interesting to note that all prokaryotic proteins, except some in the ubiquitous MATE family, cluster together in the upper half of the tree, while all eukaryotic proteins, plus a few bacterial proteins as in the ubiquitous OLF family and one member of the Ank family cluster together in the lower half of the tree. The SFT2 tree suggests that poorly characterized MVF family proteins might be lipid-peptidoglycan flippases and/or precursor drug exporters similarly to members of the MATE family; the AgnG family might be capable of exporting poly- or oligosaccharides as do

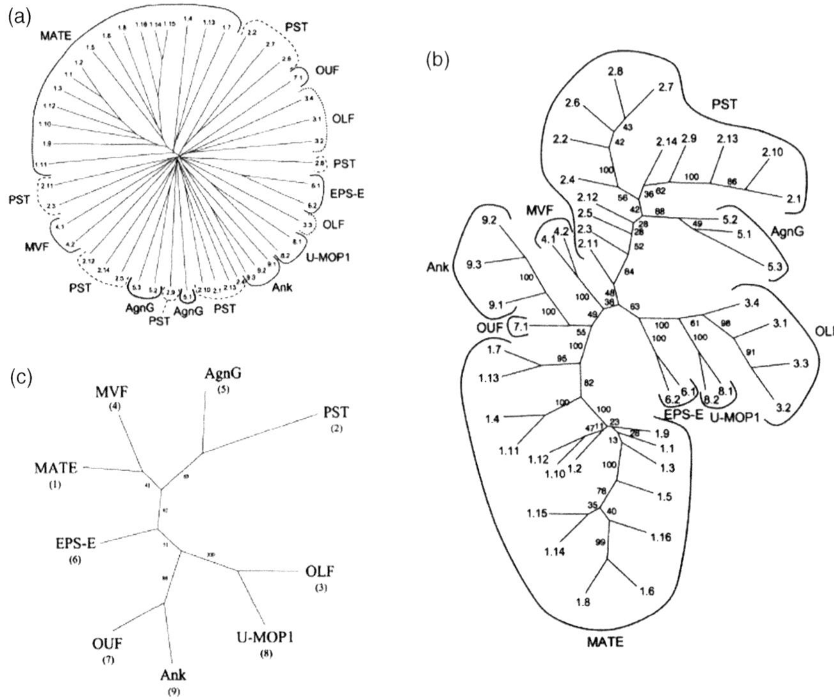

Figure 12.4 Phylogenetic (Fitch) trees for the MOP superfamily using the proteins in TCDB as of February 2011. Three different methods of tree construction were used: (a) ClustalX-based neighbor-joining method showing all MOP superfamily members, (b) the BLAST-derived SFT1-program, showing all MOP superfamily members, and (c) the SFT2-based program, showing all MOP superfamily families. In (a and b) numbers indicate the protein TC#'s (last two digits of the complete TC#). In (c), family abbreviations are presented with TC family numbers in parenthesis. In (b and c), small numbers adjacent to the branches present the "bootstrap" values, indicating the reliability of the branching order. See TCDB for protein and family identification.

PST family porters, and U-MOP1 proteins might export oligosaccharides as do OLF family transporters. These examples illustrate the superiority of the SFT programs over those based on multiple alignments when sequence divergence is extensive. When the sequences show greater conservation, the different methods agree (Yen et al., 2009; Chen et al., 2011). Figures 12.4a–c also reveal the predictive power made available by the use of these programs.

SFT1 and 2 use larger protein databases to help define each protein family and subfamily. These programs are more accurate in determining the correct phylogenetic relationships when compared to other phylogenetic programs and methods. The usage of protein databases coupled with the novel use of BLAST bit score comparison matrices allows the SFT programs to properly determine the phylogeny of superfamilies containing more evolutionarily divergent members

(Yen et al., 2009; Yen et al., 2010; Chen et al., 2011). Using other phylogenetic programs and methods, more distantly related proteins are usually either grouped together owing to long-branch clustering or excluded from clusters and given their own distinctive isolated branches (Felsenstein, 1978; Kolaczkowski and Thornton, 2009; Ogdenw and Rosenberg, 2006; Siddall and Whiting, 1999).

A minor drawback to using BLAST bit score comparisons for determining phylogeny is that the "bootstrap" values become less indicative of the reliability of observed clustering patterns for closely related proteins. However, BLAST bit scores allow the SFT programs to generate trees within a fraction of the time it would take to create phylogenetic trees of comparable reliability using other phylogenetic methods (Felsenstein, 2004; Liu et al., 2010).

The SFT programs have a wide range of applications for the study of any set of homologous protein or nucleic acid sequences. They should be useful for predicting and evaluating evolutionary, functional, mechanistic, and structural relationships (Saier, 1994).

12.3 FUNCTIONAL PREDICTIONS

Using bioinformatic tools, functions of gene products can be predicted. Two useful programs for these analyses are SEED and RegPredict (Overbeek et al., 2005; Novichkov et al., 2010). GENE (NCBI) and IMG (JGI) are also useful databases. In this section, we cover the principles that drive functional predictions and demonstrate its applicability.

SEED functions on the principle of subsystems; pathways fulfill a discreet physiological function and are defined on the basis of this function. User-defined boundaries determine the roles necessary to transform the substrate to the final products, and these limits define the role of the subsystem. It should be noted that multiple pathways may exist which produce the same final product. The subsystem, however, may substitute genes that are functionally if not structurally equivalent, providing the user with the flexibility to more easily investigate all genes that may function in multiple pathways within a single subsystem.

To propagate functional assignments, the user relies on the colocalization of genes within prokaryotic (and sometimes eukaryotic) operons. Colocalization is SEED's proxy for the coregulation resulting from the operon being controlled by an upstream regulator. Frequently, all genes within an operon play a role in the same pathway; it is evolutionarily favorable to regulate all necessary genes in the pathway together by placing them under the same promoter. Thus, if the functions of one or more genes are identified within an operon, one can use this information to effectively provide putative functional roles for the other genes. For example, if the user is given an operon including four genes, an ATPase, a metallochaperone, a regulator, and an arsenite efflux transporter, the user can surmise that the functionally uncharacterized genes probably encode proteins involved in arsenite resistance (Castillo and Saier, 2010). These predictions prove robust. Applying this principle, SEED allows the user to create vast networks of functionally interlinked genes within a subsystem.

Finding patterns within these pathways has led to unique discoveries. Taking the arsenic resistance subsystem as an example, patterns of colocalization yield numerous predictions, as we will enumerate. A novel arsenite efflux (ArsP) transporter family has been identified through frequent instances of colocalizing within arsenic resistance operons (Castillo and Saier, 2010). A previously uncharacterized link was found between ArsA homologs, ATPases energizing arsenite efflux via ArsB, and ACR3 homologs, which are arsenite transporters that are non-homologous to ArsB (Tisa and Rosen, 1990). It is believed that ACR3 can function with or without an ATPase, using the proton motive force (pmf) as an energy source when the ATPase is unavailable (Maciaszczyk-Dziubinska et al., 2011). The addition of the ArsA ATPase subunit to the pmf-dependent ArsB transporter couples ATP hydrolysis to arsenite efflux (Silver et al., 1993).

New roles for ArsA homologs have been proposed in carbon starvation and gas vesicle biogenesis. It is likely that ArsA homologs may energize peptide uptake via CstA, a putative peptide transporter involved in carbon stress relief. The role of ArsA homologs in gas vesicle biogenesis is not yet established, but it is possible that the ATPase activity energizes an aspect of the biogenesis of gas vesicles (Castillo and Saier, 2010).

By examining the phylogenetic trees of corresponding genes within operons, it has been shown that the cofunctioning proteins, ArsB, ArsP, ArsD, and CstA, coevolved with ArsA, and thus, they may physically interact. The exception is ACR3, which, despite colocalization, does not demonstrate coevolution. However, a lack of coevolution may result from the promiscuity of ArsA homologs and does not preclude physical interaction (Castillo and Saier, 2010; Paulsen et al., 1997).

P-type ATPases fall into subfamilies of known or unknown function. Even strong sequence identity (>70%) does not ensure a common function, so genomic context analyses must be applied. Ten prokaryotic functionally uncharacterized families (FUPAs) of P-type ATPases (TC# 3.A.3) exist for which phylogenetic similarity is too distant to place them firmly within a family of known function (Chan et al., 2010). FUPA27 and FUPA29 family members, for example, are found within operons containing at least three (and often up to five) members of a cbb3 type cytochrome c oxidase biogenesis (*cco*) pathway, strongly suggesting that these P-type ATPases are copper transporters, and implying that they provide the copper ions to these complexes (Hassani et al., 2010). They may lack the ability to transport Cu^{2+} across this membrane, but instead may translocate this ion halfway across the membrane to the active site of cytochrome oxidase in the interior of the membrane (R. De La Mare and M.H. Saier Jr., unpublished observation).

FUPA23 P-type ATPases are most closely related to Family 2 (Ca^{2+} P-type ATPases; see TCDB). Their genes colocalize in Actinobacteria with β-lactamase and RomA, a Zn^{2+}-dependent hydrolase of the β-lactamase fold, possibly implicating it in the transport of Zn^{2+} to metallo-β-lactamases, which require zinc for activity.

To acquire DNA regulatory sequences of previously unrecognized regulons, RegPredict can be used. This database allows identification of operons under coordinate control, and thus, completion of a regulon. The functions and interactions of these gene products can sometimes be accurately predicted through the use of SEED, BLAST searches, the conserved domain database (CDD) in NCBI, and phylogenetic analyses.

SEED predictions rely on gene colocalization. Although often an effective proxy, colocalization does not always capture the actual link between coregulated genes. As an example, the *Vibrio* hemin uptake operon, *hmu*, contains all genes necessary for hemin uptake, except for the outer membrane receptor *hmuR*. Missing this crucial component, Vibrios would be unable to recognize and transport hemin (Simpson et al., 2000). Utilizing RegPredict to analyze the regulatory sequences that precede the *hmu* operon, it became clear that *hmuR* is not missing; rather, it is located elsewhere in the *Vibrio* genome, regulated by the same sequence as the operon. This *hmu* network comprises a regulon, a group of coregulated genes and operons. Such identifiable DNA regulatory sequences are conserved nucleotide sequences that can usually be found within the first 250 base pairs upstream of the gene or operon that they regulate. In addition, a nearby operon, divergently transcribed from the *hmu* operon, appears to include hemin uptake genes and is probably regulated by the *hmu* operon palindromic regulatory sequence. Just as in an operon, all genes involved in the pathway are coregulated.

The FUPA27 and 29 families of the P-type ATPase superfamily mentioned above provide another example (Chan et al., 2010). Several organisms have more than one homolog of a given FUPA27 gene, each in its own cytochrome c oxidase biogenesis (*cco*) operon. RegPredict can be used to demonstrate that these are, in fact, under the control of similar promoters. In other cases, some of these "duplicate" operons appear incomplete. One of the more common situations involves an apparently missing histidine sensor kinase/response regulator pair. Many such pairs exist in a given organism, and without colocalization, matching them to the appropriate pathway can be difficult. RegPredict allows determination of which sensor kinase/response regulator pairs are coregulated and therefore probably function together. Thus, RegPredict fills gaps within the SEED subsystem network by predicting functions based on coregulation rather than colocalization. However, there exist inherent weaknesses within RegPredict analyses. Regulatory networks are most effective at predicting relationships when studying closely related genomes. Otherwise, sequences regulating the same functions within more distant organisms may be too divergent to link within the same network. For example, adequate conservation can be found within the flavobacterial phylum, but consistent sequence conservation within the spectrum of proteobacteria is not generally expected. Adequate conservation is often found within γ-proteobacteria, and excellent conservation is observed in the γ-proteobacterial order: Enterobacteriales.

12.3.1 Pathways of Transport Protein Evolution

Evolutionary studies have led to the finding that transport proteins usually arose through successive rounds of intragenic duplication and gene-fusion events (Rapp et al., 2007; Saier, 2000b; Shimizu et al., 2004). Moreover, enzymes and transport proteins are believed to have arisen independently of one another, with the proposed pathway for integral membrane transport protein evolution being peptide channels → channel proteins → secondary carriers → primary active transporters and group translocators (Saier, 2000b; 2003a; 2003b). In contrast to the situation with a unit of even-numbered TMSs, the duplication of a proteinaceous precursor unit with an odd number of TMSs leads to inversion of repeated protein topology (Saier, 2003a; 2003b). Topological control is imposed by the positive-inside rule, with lysine and arginine residues predominating in the connecting loops localized to the cytoplasmic side of the cell membrane (von Heijne, 1986). We next describe a few unique evolutionary cases and recent advances in the field of transport protein evolution.

12.3.2 The Drug/Metabolite Transporter Superfamily (TC# 2.A.7)

Members of the ubiquitous drug/metabolite transporter (DMT) superfamily fall into 26 recognized families, where each family tends to have a characteristic size, topological features, and function (Jack et al., 2001). For these families, the characteristic topologies are generally 4, 5, or 10 TMSs per polypeptide chain. The recent identification of 2 TMS protein members allowed us to propose that the 4 TMS topology arose by the duplication of a 2 TMS unit and not by the loss of a TMS from a 5 TMS unit (Lam et al., 2011). The 5 TMS topology was generated thereafter via a C-terminal fusion, and the 10 TMS topology arose by duplication of the 5 TMS unit (Lam et al., 2011). X-ray structural data for the 4 TMS EmrE protein within the DMT superfamily, as well as recent comparisons to the TSUP (TC#2.A.102) family support the conclusion of a common origin (Chen et al., 2007; Shlykov et al., 2012). These findings resolved earlier ambiguities in the evolution of the DMT family.

12.3.3 The Oligopeptide Transporter Family (TC# 2.A.67)

Members of the oligopeptide transporter (OPT) family of peptide and iron siderophore transporters are present in many prokaryotic phyla, but they are also represented in certain eukaryotic kingdoms (plants, fungi, and slime molds). Phylogenetic analyses suggested that slime molds acquired these proteins from fungi (Gomolplitinant and Saier, 2011). Limited functional analyses suggested that eukaryotic OPT family proteins segregate on a phylogenetic tree according to substrate type, some transporting peptides and others taking up iron-siderophores (Lubkowitz, 2006). In prokaryotes, however, this proposal does not appear to hold up (Gomolplitinant and Saier, 2011).

The prototypical OPT family 16 TMS topology, as well as a minor 17 TMS topology apparently arose from a 2 TMS precursor after three successive rounds of duplication as follows: 2 TMSs → 4 TMSs → 8 TMSs → 16 TMSs → 17 TMSs (Gomolplitinant and Saier, 2011). The 17 TMS proteins likely arose only once, being present exclusively in a single cluster on the OPT family phylogenetic tree. They may be the product of a gene-fusion event where the last C-terminal TMS was added to the 16 TMS transporter. The example of the OPT family exemplifies how far computer-based methods allow one to look into the evolutionary history of a transport protein family.

12.3.4 The Cation Diffusion Facilitator Family (TC# 2.A.4) and the Ca^{2+}-Release-Activated Ca^{2+} Channel Family (TC# 1.A.52)

It is usual that evolution through natural selection tends toward complexity, with relatively few exceptions (Adami et al., 2000; Barton and Zuidema, 2003; Soyer and Bonhoeffer, 2006). As an example, the endosymbiotic establishment of bacteria in early nucleus-containing cells, as well as subsequent gene loss and transfer, led to the evolution of mitochondria and chloroplasts, the power plants of eukaryotic cells (Adams and Palmer, 2003; Gray et al., 1999; Martin, 2010; Raven and Allen, 2003). These events eased energy restrictions and allowed for the evolution of higher forms of life.

In stark contrast to the norm, evidence has been presented leading to the conclusion that the ancient, ubiquitous, 6 TMS cation diffusion facilitator (CDF) carriers were the precursors of the 4 TMS Ca^{2+}-release-activated Ca^{2+} (CRAC) channels (CRAC-Cs) in animals (Matias et al., 2010). This conclusion was based on rigorous statistical analyses showing that TMSs 3-4 and 5-6 of CDF transporters are most closely related to TMSs 1-2 and 3-4, respectively, of CRAC channels. Therefore, eukaryotic CRAC channels most likely arose as a direct result of the loss of TMSs 1 and 2 from an ancestral 6 TMS precursor of the ubiquitous CDF transporters (Fig. 12.5). This appears to be a rare instance of "retroevolution" where a large and functionally more complex protein was the precursor of a structurally and functionally simpler protein.

12.4 SEMIAUTOMATED GENOME ANALYSIS

Because transport represents a primary mechanism of communication with the environment for all living organisms, it may be important to identify and then characterize the transporters encoded within the genome of an organism of interest. The quantity, types, and characteristics of the encoded transport proteins can reveal a wealth of information about the organism. In many cases, a unique life style correlates with the types of transport-protein-encoding genes present. For example, obligatory parasitic and symbiotic organisms usually have reduced genome sizes. The fish parasite, *Ichthyophthirius multifiliis* (ICH), encodes numerous members of the voltage-gated ion channel (VIC) superfamily, many of which lack voltage sensors. These attributes can vary substantially between even closely related organisms such as *Tetrahymena*

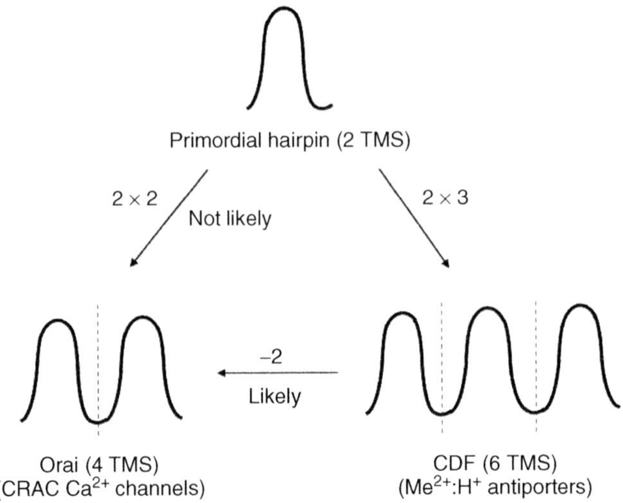

Figure 12.5 Proposed pathway for the evolutionary origin of 6 TMS carrier from a 2 TMS precursor followed by loss of TMSs 1 and 2 to give 4 TMS CRAC Ca^{2+} channels.

thermophila and *Paramecium tetraurelia*. One can use various bioinformatic methods to first find potential transporter proteins of an organism and then analyze and organize them. Steps are taken to remove errors, find trends, and identify unique proteins (Lorca et al., 2007; Paulsen et al., 2000; Tamang et al., 2009).

Genome analysis begins by first finding the most up-to-date proteome, in FASTA format, of the organism in question, often accomplished by doing a simple web search or using NCBI to generate a list of known proteins of the organism. An in-house program called *Gblast* is used to extract all potential transport proteins from the query proteome. Gblast works by simply blasting the contents of the proteome against all sequences in the TCDB(www.tcdb.org) (Saier et al., 2006; 2009). The results can be presented in an excel file, which displays the query proteins against the top hit proteins from TCDB as well as much relevant data. The primary initial goals are to identify the regions of homology between the query and hit proteins, particularly to determine if the homology is within the transmembrane domain(s). One can do this by determining what TMSs, if any, are shared by the hit and query proteins. This can be accomplished by using programs such as "WHAT" (Zhai and Saier, 2001) and "HMMTOP" (Tusnady and Simon, 1998), located in TCDB biotools (Saier et al., 2009).

It may be important to determine what domains are shared by the query and hit proteins. PSI-BLAST searches and the NCBI CDD are used to identify shared domains. The only homology shared between two proteins may, for example, be a CAP_ED domain (as revealed by CDD) that simply codes for a cAMP-binding domain. Such domains are common among proteins with a diversity of functions. Although their presence suggests a mode of regulation, it does not provide a guide to the primary function of a protein. In many cases, recognized domains cannot be identified because the domains present are too sequence divergent or have not yet been characterized and are therefore not in CDD.

12.5 CONCLUSIONS AND PERSPECTIVES

In this review, we have attempted to lead the interested reader through the development and utilization of useful software for studying the structures, functions, and evolutionary origins of integral membrane transport proteins. These proteins, almost without exception, did not derive from other classes of proteins such as structural proteins, enzymes, or regulatory proteins, but instead arose from small peptides by a series of intragenic duplication events. Few exceptions contradict this precept, although a few examples can be cited where a structural or catalytic protein has gained an alternative function by inserting into a membrane to form a hydrophilic ion-conducting transmembrane pore. Out of the over 600 families in the TCDB, only about 1% are believed to have evolved via such a pathway, and their study is still in its infancy. Examples include the (i) CLIC (TC# 1.A.12), (ii) Bcl-2 (TC# 1.A.21), (iii) Annexin (TC# 1.A.31), (iv) Hsp70 (TC# 1.A.33), (v) ICln (TC# 1.A.47), and (vi) Presenilin (TC# 1.A.54) families (see TCDB). It is interesting to note that all of these proteins form simple channels.

The evolutionary pathways discussed in this review provide valuable guidelines for structural, mechanistic, functional, pathological, and regulatory studies on these essential proteins. Phylogeny provides a valuable, but insufficient guide, but additional tools are available to enhance the value of a phylogenetic approach. Functional predictions can be made on the basis of genome context, particularly in prokaryotes, and they often extrapolate to homologs in eukaryotes. Using RegPredict, the regulated areas upstream of the encoding operons of overlapping physiological function can be identified.

It is often claimed that high resolution, 3-D structural analyses can predict homology with greater reliability than sequence-based bioinformatics. We would like to suggest that the two approaches taken together can lead us much further toward an understanding of evolutionary pathways than either one alone. In fact, we have been able to demonstrate homology between repeat elements and between distantly related proteins of differing topologies when 3-D structural similarities are minimal. We are thus coming to recognize that using sequence-based bioinformatics, evolutionary histories can often be traced much further back than is possible using 3-D structures. We have tentative evidence for the existence of transmembrane precursor peptides that may have provided the initial building blocks for the appearance of multiple currently recognized superfamilies. These may have developed in ways that allowed these precursors to generate dissimilar primordial repeat sequences that differed in numbers of TMSs, size, and sequence in spite of a (partial) common origin. Thus, a single small peptide precursor may have preceded the formation of structurally dissimilar families of proteins with different repeat units and 3-D structures.

We have previously discussed the hypothesis that the general routes for transport protein evolution probably involved the use of short oligomeric channel-forming hydrophobic/amphipathic peptides that underwent intragenic multiplication to yield longer, less oligomeric channels (Saier, 2003a, b). These then underwent mutations, allowing alternate configurations and promoting high affinity substrate recognition, thus creating secondary carriers. Finally, the generation of primary active transporters appears to have involved superimposition of catalytic energy-coupling proteins onto carriers (Saier, 2003a, b). The recently discovered energy-coupled factor (ECF) superfamily of the ABC functional superfamily, members of which can function either as secondary or primary active transporters, provides a prime example (Eitinger et al., 2011; Hebbeln et al., 2007; Rodionov et al., 2009). While exceptions are bound to exist, we believe that this generalized scheme will prove to account for the evolution of a vast majority of transporters.

Aspects of this postulate have been amply documented as a result of bioinformatic and 3-D structural elucidation. Bioinformatic analyses, as outlined here, have provided the bulk of the evidence while X-ray and multidimensional nuclear magnetic resonance (NMR) analyses have confirmed several essential aspects of the proposal. The next step is to elucidate the details of these processes and to identify the ancestral precursors. The approaches described here represent state-of-the-art technologies that are likely to provide these details. Further development of faster, more sensitive, and more comprehensive software will undoubtedly lead to further insights into the details of transport protein evolution.

ACKNOWLEDGMENT

This work was supported by PHS grant GM077402 from the National Institute of General Medical Sciences of the NIH.

REFERENCES

Adams, K.L. and Palmer, J.D. (2003) Evolution of mitochondrial gene content: gene loss and transfer to the nucleus. Mol Phylogenet Evol, 29, 380–395.

Adami, C., Ofria, C., and Collier, T.C. (2000) Evolution of biological complexity. Proc Natl Acad Sci USA, 97, 4463–4468.

Aller, S.G., Yu, J., Ward, A., Weng, Y., Chittaboina, S., Zhuo, R., Harrell, P.M., Trinh, Y.T., Zhang, Q., Urbatsch, I.L., and Chang, G. (2009) Structure of P-glycoprotein reveals a molecular basis for poly-specific drug binding. Science, 323, 1718–1722.

Altschul, S.F., Gish, W., Miller, W., Myers, E.W., and Lipman, D.J. (1990) Basic local alignment search tool. J Mol Biol, 215, 403–410.

Altschul, S.F., Madden, T.L., Schäffer, A.A., Zhang, J., Zhang, Z., Miller, W., and Lipman, D.J. (1997) Gapped BLAST and PSI-BLAST: a new generation of protein database search programs. Nucleic Acids Res, 25, 3389–3402.

Barton, N. and Zuidema, W. (2003) Evolution: the erratic path towards complexity. Curr Biol, 13, R649–R651.

Castillo, R. and Saier, M.H. Jr. (2010) Functional promiscuity of homologues of the bacterial ArsA ATPases. Int J Microbiol. doi: 10.1155/2010/187373

Chan, H., Babayan, V., Blyumin, E., Gandhi, C., Hak, K., Harake, D., Kumar, K., Lee, P., Li, T.T., Lo, T.C., Meyer, C.J., Stanford, S., Zamora, K.S., and Saier, M.H. Jr. (2010) The p-type ATPase superfamily. J Mol Microbiol Biotechnol, 19, 5–104.

Chang, A.B., Lin, R., Studley, W.K., Tran, C.V., and Saier, M.H. Jr. (2004) Phylogeny as a guide to structure and function of membrane transport proteins. Mol Membr Biol, 21, 171–181.

Chen, J.S., Reddy, V., Chen, J.H., Zheng, W.H., Shlykov, M.A., Yen, M.R., and Saier, M.H.J. (2011) Phylogenetic characterization of transport protein superfamilies: superiority of SFT programs over those based on multiple-alignments. J Mol Microbiol Biotechnol In Press.

Chen, Y.J., Pornillos, O., Lieu, S., Ma, C., Chen, A.P., and Chang, G. (2007) X-ray structure of EmrE supports dual topology model. Proc Natl Acad Sci USA, 104, 18999–19004.

Cooper, G.R. and Moir, A. (2011) Amino acid residues in the GerAB protein important in the function and assembly of the alanine spore germination receptor of Bacillus subtilis 168. J Bacteriol, 193, 2261–2267.

Dang, S., Sun, L., Huang, Y., Lu, F., Liu, Y., Gong, H., Wang, J., and Yan, N. (2010) Structure of a fucose transporter in an outward-open conformation. Nature, 467, 734–738.

Davies, R.L., Campbell, S., and Whittam, T.S. (2002) Mosaic structure and molecular evolution of the leukotoxin operon (lktCABD) in Mannheimi (Pasteurella) haemolytica, Mannheimia glucosidal, and Pasteurella trehalosi. J Bacteriol, 184, 266–277.

Dayhoff, M.O. (1974) Computer analysis of protein sequences. Fed Proc, 33, 2314–2316.

Dayhoff, M.O. (1978) A model of evolutionary change in proteins, in Atlas of Protein Sequence and Structure, National Biomedical Research Foundation, Washington, D.C.

Dean, M.C. (2002) The Human ATP-binding Cassette (ABC) Transporter Superfamily, National Library of Medicine, Bethesda, MD.

Dean, M.C., Hamon, Y., and Chimini, G. (2001) The human ATP-binding cassette (ABC) transporter superfamily. J Lipid Res, 42, 1007–1017.

Eitinger, T., Rodionov, D.A., Grote, M., and Schneider, E. (2011) Canonical and ECF-type ATP-binding cassette importers in prokaryotes: diversity in modular organization and cellular functions. FEMS Microbiol Rev, 35, 3–67.

Fang, Y., Jayaram, H., Shane, T., Kolmakova-Partensky, L., Wu, F., Williams, C., Xiong, Y., and Miller, C. (2009) Structure of a prokaryotic virtual proton pump at 3.2 A resolution. Nature, 460, 1040–1043.

Felsenstein, J. (1978) Cases in which parsimony and compatibility methods will be positively misleading. Syst Zool, 27, 401–410.

Felsenstein, J. (2004) Inferring Phylogenies, Sinauer Associates, Sunderland, MA.

Fitch, W.M. and Margoliash, E. (1967) Construction of phylogenetic trees. Science, 155, 279–284.

Gao, X., Zhou, L., Jiao, X., Lu, F., Yan, C., Zeng, X., Wang, J., and Shi, Y. (2010) Mechanism of substrate recognition and transport by an amino acid antiporter. Nature, 463, 828–832.

Gascuel, O. and Steel, M. (2006) Neighbor-joining revealed. Mol Biol Evol, 23, 1997–2000.

Gomolplitinant, K.M. and Saier, M.H. Jr. (2011) Evolution of the oligopeptide transporter family. J Membr Biol, 240, 89–110.

Gray, M.W., Burger, G., and Lang, B.F. (1999) Mitochondrial evolution. Science, 283, 1476–1481.

Harvat, E.M., Zhang, Y.M., Tran, C.V., Zhang, Z., Frank, M.W., Rock, C.O., and Saier, M.H. Jr. (2005) Lysophospholipid flipping across the *Escherichia coli* inner membrane catalyzed by a transporter (LplT) belonging to the major facilitator superfamily. J Biol Chem, 280, 12028–12034.

Hassani, B.K., Astier, C., Nitschke, W., and Ouchane, S. (2010) CtpA, a copper-translocating P-type ATPase involved in the biogenesis of multiple copper-requiring enzymes. J Biol Chem, 285, 19330–19337.

Hebbeln, P., Rodionov, D.A., Alfandega, A., and Eitinger, T. (2007) Biotin uptake in prokaryotes by solute transporters with an optional ATP-binding cassette-containing module. Proc Natl Acad Sci USA, 104, 2909–2914.

Henikoff, S. (1992) Amino acid substitution matrices from protein blocks. Proc Natl Acad Sci, 22, 10915–10919.

Jack, D.L., Paulsen, I.T., and Saier, M.H. Jr. (2000) The amino acid/polyamine/organocation (APC) superfamily of transporters specific for amino acids, polyamines and organocations. Microbiology, 146, 1797–1814.

Jack, D.L., Yang, N.M., and Saier, M.H. Jr. (2001) The drug/metabolite transporter superfamily. Eur J Biochem, 268, 3620–3639.

Jones, P.M., O'Mara, M.L., and George, A.M. (2009) ABC transporters: a riddle wrapped in a mystery inside an enigma. Trends Biochem Sci, 34, 520–531.

Kolaczkowski, B. and Thornton, J.W. (2009) Long-branch attraction bias and inconsistency in Bayesian phylogenetics. PLoS One, 4, e7891.

Kowalczyk, L., Ratera, M., Paladino, A., Bartoccioni, P., Errasti-Murugarren, E., Valencia, E., Portella, G., Bial, S., Zorzano, A., Fita, I., Orozco, M., Carpena, X., Vázquez-Ibar, J.L., and Palacín, M. (2011) Molecular basis of substrate-induced permeation by an amino acid antiporter. Proc Natl Acad Sci USA, 108, 3935–3940.

Lam, V.H., Lee, J., Silverio, A., Chan, H., Gomolplitinant, K.M., Povolotsky, T.L., Orlova, E., Sun, E.I., Welliver, C.H., and Saier, M.H. Jr. (2011) Pathways of transport protein evolution: recent advances. Biol Chem, 392, 5–12.

Law, C.J., Maloney, P.C., and Wang, D.N. (2008) Ins and outs of major facilitator superfamily antiporters. Annu Rev Microbiol, 62, 289–305.

Liu, K., Linder, C.R., and Warnow, T. (2010) Multiple sequence alignment: a major challenge to large-scale phylogenetics. PLoS Curr, 2RRN1198, 1–16.

Lorca, G.L., Barabote, R.D., Zlotopolski, V., Tran, C., Winnen, B., Hvorup, R.N., Stonestrom, A.J., Nguyen, E., Huang, L.W., Kim, D.S., and Saier, M.H. Jr. (2007) Transport capabilities of eleven gram-positive bacteria: comparative genomic analyses. Biochim Biophys Acta, 1768, 1342–1366.

Lubkowitz, M. (2006) The OPT family functions in long-distance peptide and metal transport in plants. Genet Eng, 27, 35–55.

Maciaszczyk-Dziubinska, E., Migocka, M., and Wysocki, R. (2011) Acr3p is a plasma membrane antiporter that catalyzes As(III)/H+ and Sb(III)/H+exchange in *Saccharomyces cerevisiae*. BBA - Biomembranes, 1808, 1855–1859.

Marger, M.D. and Saier, M.H. Jr. (1993) A major superfamily of transmembrane facilitators that catalyse uniport, symport and antiport. Trends Biochem Sci, 18, 13–20.

Martin, W. (2010) Evolutionary origins of metabolic compartmentalization in eukaryotes. Philos Trans R Soc Lond B Biol Sci, 365, 847–855.

Matias, M.G., Gomolplitinant, K.M., Tamang, D.G., and Saier, M.H. Jr. (2010) Animal Ca2+ release-activated Ca2+ (CRAC) channels appear to be homologous to and derived from the ubiquitous cation diffusion facilitators. BMC Res Notes 3, 158, 1–12.

Mongkolthanaruk, W., Cooper, G.R., Mawer, J.S., Allan, R.N., and Moir, A. (2011) Effect of amino acid substitutions in the GerAA protein on the function of the alanine-responsive germinant receptor of Bacillus subtilis spores. J Bacteriol, 193, 2268–2275.

Novichkov, P.S., Rodionov, D.A., Stavrovskaya, E.D., Novichkova, E.S., Kazakov, A.E., Gelfand, M.S., Arkin, A.P., Mironov, A.A., and Dubchak, I. (2010) RegPredict: an integrated system for regulon inference in prokaryotes by comparative genomics approach. Nucleic Acids Res, 38, W299–W307.

Ogdenw, T.H. and Rosenberg, M.S. (2006) Multiple sequence alignment accuracy and phylogenetic inference. Syst Biol, 55, 314–328.

Oldham, M.L. and Chen, J. (2011) Crystal structure of the maltose transporter in a pretranslocation intermediate state. Science (In Press).

Overbeek, R., Begley, T., Butler, R.M., Choudhuri, J.V., Chuang, H.Y., Cohoon, M., de Crécy-Lagard, V., Diaz, N., Disz, T., Edwards, R., Fonstein, M., Frank, E.D., Gerdes, S., Glass, E.M., Goesmann, A., Hanson, A., Iwata-Reuyl, D., Jensen, R., Jamshidi, N., Krause, L., Kubal, M., Larsen, N., Linke, B., McHardy, A.C., Meyer, F., Neuweger, H., Olsen, G., Olson, R., Osterman, A., Portnoy, V., Pusch, G.D., Rodionov, D.A., Rückert, C., Steiner, J., Stevens, R., Thiele, I., Vassieva, O., Ye, Y., Zagnitko, O., and Vonstein, V. (2005) The subsystems approach to genome annotation and its use in the project to annotate 1000 genomes. Nucleic Acids Res, 33, 5691–5702.

Özkucur, N., Perike, S., Sharma, P., and Funk, R.H.W. (2011) Persistent directional cell migration requires ion transport proteins as direction sensors and membrane potential differences in order to maintain directedness. BMC Cell Biol, 12, 4.

Pao, S.S., Paulsen, I.T., and Saier, M.H. Jr. (1998) Major facilitator superfamily. Microbiol Mol Biol Rev, 62, 1–34.

Paulsen, I.T., Park, J.H., Choi, P.S., and Saier, M.H. Jr. (1997) A family of gram-negative bacterial outer membrane factors that function in the export of proteins, carbohydrates, drugs and heavy metals from gram-negative bacteria. FEMS Microbiol Lett, 156, 1–8.

Paulsen, I.T., Nguyen, L., Sliwinski, M.K., Rabus, R., and Saier, M.H. Jr. (2000) Microbial genome analyses: comparative transport capabilities in eighteen prokaryotes. J Mol Biol, 301, 75–100.

Prakash, S., Cooper, G., Singhi, S., and Saier, M.H. Jr. (2003) The ion transporter superfamily. Biochem Biophys Acta, 1618, 79–92.

Radestock, S. and Forrest, L.R. (2011) The alternating-access mechanism of MFS transporters arises from inverted-topology repeats. J Mol Biol, 407, 698–715.

Rapp, M., Seppälä, S., Granseth, E., and von Heijne, G. (2007) Emulating membrane protein evolution by rational design. Science, 315, 1282–1284.

Raven, J.A. and Allen, J.F. (2003) Genomics and chloroplast evolution: what did cyanobacteria do for plants. Genome Biol, 4, 209–5.

Rettner, R.E. and Saier, M.H. Jr. (2010) The autoinducer-2 exporter superfamily. J Mol Microbiol Biotechnol, 18, 195–205.

Rodionov, D.A., Hebbeln, P., Eudes, A., ter Beek, J., Rodionova, I.A., Erkens, G.B., Slotboom, D.J., Gelfand, M.S., Osterman, A.L., Hanson, A.D., and Eitinger, T. (2009) A novel class of modular transporters for vitamins in prokaryotes. J Bacteriol, 191, 42–51.

Saier, M.H. Jr. (1994) Computer-aided analyses of transport protein sequences: gleaning evidence concerning function, structure, biogenesis, and evolution. Microbiol Rev, 58, 71–93.

Saier, M.H. Jr. (2000a) A functional-phylogenetic classification system for transmembrane solute transporters. Microbiol Mol Biol Rev, 64, 354–411.

Saier, M.H. Jr. (2000b) Vectorial metabolism and the evolution of transport systems. J Bacteriol, 182, 5029–5035.

Saier, M.H. Jr. (2003a) Tracing pathways of transport protein evolution. Mol Microbiol, 48, 1145–1156.

Saier, M.H. Jr. (2003b) Review on tracing pathways of transport protein evolution. Mol Microbiol, 48, 1145–1156.

Saier, M.H. Jr. (2007) Active transport in communication, protection and nutrition. J Mol Microbiol Biotechnol, 12, 161–164.

Saier, M.H. Jr. and Ren, Q. (2006) The bioinformatic study of transmembrane molecular transport. J Mol Microbiol Biotechnol, 11, 289–290.

Saier, M.H. Jr., Tran, C.V., and Barabote, R.D. (2006) TCDB: the Transporter Classification Database for membrane transport protein analyses and information. Nucleic Acids Res, 34, D181–D186.

Saier, M.H. Jr., Yen, M.R., Noto, K., Tamang, D.G., and Elkan, C. (2009) The Transporter Classification Database: recent advances. Nucleic Acids Res, 37, D274–D278.

Saier, M.H. Jr., Beatty, J.T., Goffeau, A., Harley, K.T., Heijne, W.H., Huang, S.C., Jack, D.L., Jähn, P.S., Lew, K., Liu, J., Pao, S.S., Paulsen, I.T., Tseng, T.T., and Virk, P.S. (1999) The major facilitator superfamily. J Mol Microbiol Biotechnol, 2, 257–279.

Schwarz, G., Mendel, R.R., and Ribbe, M.W. (2009) Molybdenum cofactors, enzymes and pathways. Nature, 460, 839–847.

Shimizu, T., Mitsuke, H., Noto, K., and Arai, M. (2004) Internal gene duplication in the evolution of prokaryotic transmembrane proteins. J Mol Biol, 339, 1–15.

Shlykov, M.A., Zheng, W.H., Chen, J.S., Saier, M.H. Jr. (2012) Bioinformatic characterization of the 4-Toluene Sulfonate Uptake Permease (TSUP) family of transmembrane proteins. Biochim Biophys Acta. 1818(3), 703–717. Doi: 10.1016/j.bbamem.2011.12.005. Epub 2011 Dec 13. PMID: 22192777.

Siddall, M.E. and Whiting, M.F. (1999) Long-branch abstractions. Cladistics, 15, 9–24.

Silver, S., Ji, G., Bröer, S., Dey, S., Dou, D., and Rosen, B.P. (1993) Orphan enzyme or patriarch of a new tribe: the arsenic resistance ATPase of bacterial plasmids. Mol Microbiol, 8, 637–642.

Simpson, W., Olczak, T., and Genco, C.A. (2000) Characterization and expression of HmuR, a TonB-dependent hemoglobin receptor of Porphyromonas gingivalis. J Bacteriol, 182, 5737–5748.

Soyer, O.S. and Bonhoeffer, S. (2006) Evolution of complexity in signaling pathways. Proc Natl Acad Sci USA, 103, 16337–16342.

Tamang, D.G., Rabus, R., Barabote, R.D., and Saier, M.H. Jr. (2009) Comprehensive analyses of transport proteins encoded within the genome of "Aromatoleum aromaticum" strain EbN1. J Membr Biol, 229, 53–90.

Tisa, L.S. and Rosen, B.P. (1990) Transport systems encoded by bacterial plasmids. J Bioenerg Biomembr, 22, 493–507.

Thompson, J.D., Gibson, T.J., Plewniak, F., Jeanmougin, F., and Higgins, D.G. (1997) The CLUSTAL_X windows interface: flexible strategies for multiple sequence alignment aided by quality analysis tools. Nucleic Acids Res, 25, 4876–4882.

Tusnady, G.E. and Simon, I. (1998) Principles governing amino acid composition of integral membrane proteins: application to topology prediction. J Mol Biol, 283, 489–506.

Van Bambeke, F., Glupczynski, Y., Plésiat, P., Pechère, J.C., and Tulkens, P.M. (2003) Antibiotic efflux pumps in prokaryotic cells: occurrence, impact on resistance and strategies for the future of antimicrobial therapy. J Antimicrob Chemother, 51, 1055–1065.

von Heijne, G. (1986) The distribution of positively charged residues in bacterial inner membrane proteins correlates with the trans-membrane topology. EMBO J, 5, 3021–3027.

Wang, B., Dukarevich, M., Sun, E.I., Yen, M.R., and Saier, M.H. Jr. (2009) Membrane porters of ATP-binding cassette transport system are polyphyletic. J Membr Biol, 231, 1–10.

Webb, E.C. (1992) Enzyme Nomenclature 1992. Recommendations of the Nomenclature Committee of the International Union of Biochemistry and Molecular Biology, Academic Press, New York.

Yen, M.R., Choi, J., and Saier, M.H. Jr. (2009) Bioinformatic analyses of transmembrane transport: novel software for deducing protein phylogeny, topology, and evolution. J Mol Microbiol Biotechnol, 17, 163–176.

Yen, M.R., Chen, J.S., Marquez, J.L., Sun, E.I., and Saier, M.H. (2010) Multidrug resistance: phylogenetic characterization of superfamilies of secondary carriers that include drug exporters. Methods Mol Biol, 637, 47–64.

Zhai, Y. and Saier, M.H. Jr. (2001) A web-based program (WHAT) for the simultaneous prediction of hydropathy, amphipathicity, secondary structure and transmembrane topology for a single protein sequence. J Mol Microbiol Biotechnol, 3, 501–502.

Zhai, Y., Tchieu, J., and Saier, M.H. Jr. (2002) A web-based Tree View (TV) program for the visualization of phylogenetic trees. J Mol Microbiol Biotechnol, 4, 69–70.

13

CRISPR-CAS SYSTEMS AND CAS PROTEIN FAMILIES

Kira S. Makarova

National Center for Biotechnology Information, National Library of Medicine, National Institutes of Health, Bethesda, MD, USA

Daniel H. Haft

J Craig Venter Institute, Rockville, MD, USA

Eugene V. Koonin

National Center for Biotechnology Information, National Library of Medicine, National Institutes of Health, Bethesda, MD, USA

CHAPTER SUMMARY

Clustered regularly interspaced short palindrome repeats (CRISPR) CRISPR-associated (Cas) modules are adaptive antivirus immunity systems that are present in most archaea and many bacteria. These systems function by incorporating fragments of alien genomes into specific genomic loci, transcribing the inserts and using the transcripts as guide RNAs to destroy the genome of the cognate virus or plasmid. This RNA interference-like immune response is mediated by numerous, highly diverse Cas proteins, several of which form the Cascade (CRISPR-associated complex for antiviral defense) complex involved in the processing of CRISPR loci transcripts and cleavage of the target DNA. Comparative analysis of the CRISPR-Cas modules led to the classification of

Protein Families: Relating Protein Sequence, Structure, and Function, First Edition.
Edited by Christine Orengo and Alex Bateman.

the CRISPR-Cas systems into three types (I, II, and III) that are characterized by distinct sets of *cas* genes. Classification of Cas proteins into families and superfamilies is a nontrivial task because of the fast evolution of many *cas* genes. Exhaustive sequence comparison aided by analysis of the available crystal structures led to the delineation of approximately 30 protein families, some of which can be further unified in several superfamilies. A more "splitting" approach implemented in the TIGRFAM protein family system currently counts approximately 80 Cas protein families (Selengut et al., 2007).

By far the most common domain in Cas proteins is the RNA recognition motif (RRM). The RRM domains show remarkable diversity within the CRISPR-Cas systems and in particular comprise the scaffold of the Cascade complex. In addition to the numerous RRM domains, the Cas proteins contain a distinct Superfamily II helicase domain, a predicted polymerase domain, and several diverse nuclease domains. Detailed comparative analysis of the Cas protein sequences and structures shed light on the deep relationships between Type I and Type III systems and allowed us to propose a simple evolutionary scenario for the origin of the CRISPR-Cas system. Moreover, the combination of experimental structural studies and comparative analysis provides for detailed models of the structures of the Cascade complexes from different CRISPR-Cas types, revealing remarkable architectural uniformity.

13.1 INTRODUCTION

The CRISPR-Cas is an adaptive immunity system that is present in most archaea and many bacteria, and functions on the "Lamarckian inheritance" principle. The CRISPR-Cas loci in prokaryote genomes consist of an array of direct, typically palindromic or imperfect dyad repeats known as CRISPR and unique spacers 35–40 base pairs in length that are located between the CRISPR repeats (Jansen et al., 2002). The CRISPR repeat arrays are usually associated with *cas* (CRISPR-associated) genes that encode proteins with a variety of predicted nucleic-acid-manipulating activities (Jansen et al., 2002; Makarova et al., 2002).

The discovery of the immune function of CRISPR-Cas makes for a veritable detective story. The results of the original comparative analysis of *cas* gene neighborhoods and Cas protein sequences and structures suggested the possibility that Cas proteins comprised a novel DNA repair system, given the presence among these proteins of several nucleases, a distinct helicase, and a polymerase (Makarova et al., 2002). A simultaneous, independent study has shown that in the archaeon *Archaeoglobus fulgidus*, the CRISPR cassettes are expressed yielding multiple small RNA species (Tang et al., 2002). However, the biological function(s) of the CRISPR cassettes remained an enigma, with a variety of unsubstantiated speculations having been proposed on potential regulatory functions of the small RNAs transcribed from CRISPR loci. The light of understanding shone on CRISPRs in 2005 when comparison of the spacer sequences to the growing sequence databases revealed that some of them were identical to fragments of viral (bacteriophage) genes (Bolotin et al., 2005; Mojica et al., 2005). This finding, combined with the predicted activities of the Cas proteins, prompted the hypothesis that CRISPR-Cas is a system of adaptive immunity that integrates short genomic segments of selfish elements (viruses or plasmids) into specific loci in prokaryotic genomes and then employs these inserts to abrogate the replication of the cognate agents via an RNAi-like mechanism (Makarova et al., 2006).

This prediction was promptly followed by experimental validation of the CRISPR immunity hypothesis (Barrangou et al., 2007), and subsequent experiments have elucidated many key details of the molecular mechanisms of the CRISPR-Cas systems. The current view distinguishes between three functional stages of CRISPR-Cas operation (Jore et al., 2011b; van der Oost et al., 2009). During the first stage, adaptation, short pieces of DNA (characteristic length of approximately 30 bp) homologous to virus or plasmid sequences (known as *protospacers*) are integrated into the CRISPR loci (Barrangou et al., 2007; Garneau et al., 2010; Sontheimer and Marraffini, 2010). The short (3 or 4 nucleotides) protospacer adjacent motifs (PAMs) located immediately downstream of the protospacer appear to determine the selection of the protospacer followed by integration into a preexisting CRISPR array (Marraffini and Sontheimer, 2010; Mojica et al., 2009). The second stage, expression and processing, involves transcription and cleavage of the long primary transcript of a CRISPR locus (pre-crRNA) that is processed into short crRNAs. This step is catalyzed by endoribonucleases encoded by the *cas* genes that operate as subunits of a large complex denoted Cascade. The Cascade has been isolated and characterized both from *Escherichia coli* (Brouns et al., 2008) and from the hyperthermophilic archaeon *Sulfolobus solfataricus* (Lintner et al., 2011), suggesting that the formation of this helical complex comprised of several Cas subunits is a conserved feature of CRISPR-Cas systems. However, in the archaeon *Pyrococcus furiosus*, a stand-alone enzyme, Cas6, has been reported to mediate pre-crRNA processing (Carte et al., 2008; Hale et al., 2009). At the third stage, interference, the alien nucleic acid (DNA or RNA) is targeted by a modified form of the Cascade complex containing a crRNA guide, and cleaved within or in the vicinity of the PAM sequence (Garneau et al., 2010; Lintner et al., 2011; Sontheimer and Marraffini, 2010; van der Oost et al., 2009). In several CRISPR-Cas systems, crRNAs have been shown to be complementary to either strand of the phage or plasmid, which is best compatible with DNA being the target (Barrangou et al., 2007; Haurwitz et al., 2010). Direct demonstration of DNA being the target of the CRISPR-Cas machinery has come from experiments in *Staphylococcus epidermidis*. In this case, insertion of a self-splicing intron into the protospacer sequence of the target gene rendered the respective plasmid resistant to the CRISPR-mediated immunity (Marraffini and Sontheimer, 2008). Recently, the *E. coli* Cascade complex containing crRNA has been shown to recognize the target DNA, with the specificity defined by the crRNA sequence, and to displace the noncomplementary strand in an energy-independent manner (Jore et al., 2011b). However, *in vitro* experiments with one of the CRISPR-Cas systems (Type IIIB, formerly known as *Cmr system* or repeat-associated mysterious protein, RAMP, module) from the archaeon *P. furiosus* showed that the crRNA rather targets the mRNA (Hale et al., 2009). These findings emphasize the remarkable mechanistic and functional diversity of the CRISPR-Cas systems.

The apparent functional diversity of CRISPR-Cas systems is paralleled by the equally notable diversity of Cas proteins: at least 45 distinct protein families have been identified in association with CRISPR loci in various bacterial and archaeal genomes (Haft et al., 2005). An analysis involving more sensitive methods of sequence comparison and additional evidence from genomic context has revealed distant homologous relationships between several of these families, paring down the number of distinct protein groups to approximately 25 and suggesting that additional Cas protein families might be linked subsequently thanks to the growth of genomic and structural data sets and further advances in computational analysis (Makarova et al., 2006).

The recently updated classification of CRISPR-Cas systems divides them into three distinct types (I, II, and III) (Makarova et al., 2011b). All these systems contain two universal genes: *cas1* encoding a metal-dependent DNAse with no apparent sequence specificity that could be involved in the integration of the alien DNA (spacer) into CRISPR cassettes (Marraffini and Sontheimer, 2009; Wiedenheft et al., 2009), and *cas2* encoding a metal-dependent endoribonuclease that also appears to be involved in the spacer acquisition stage (Beloglazova et al., 2008). Otherwise, the three types of CRISPR-Cas systems substantially differ in their sets of constituent genes, and each is characterized, respectively, by a unique signature gene. The signature genes for the three types are, respectively, *cas3* (a superfamily 2 helicase containing an N-terminal hydrophilic domain (HD) superfamily nuclease domain) (Sinkunas et al., 2011)), *cas9* (a large protein containing a predicted RuvC-like and HNH nuclease domains), and *cas10* (a protein containing a domain homologous to the Palm domain of nucleic acid polymerases and nucleotide cyclases) (Makarova et al., 2011b). Within these three types, CRISPR-Cas systems can be further classified into subtypes based on a number of considerations that include distinct signature genes along with the phylogeny of the universal *cas1* gene (Makarova et al., 2011b). The Cas proteins known as RAMPs are present in several copies in both type I and III systems. Some of the RAMPs have been shown to possess sequence- or structure-specific RNAse activity that is involved in the processing of pre-crRNA transcripts (Brouns et al., 2008; Carte et al., 2008; Haurwitz et al., 2010). The crystal structures of several RAMPs have been solved and indicate that they contain one or two domains that display distinct versions of the RRM, also known as the ferredoxin fold (Haurwitz et al., 2010; Lintner et al., 2011; Makarova et al., 2006; Sakamoto et al., 2009; Wang et al., 2011).

Given the progress of structural biology that has already yielded crystal structures of many Cas proteins and the ever-increasing, diversity of Cas protein sequences and genomic organizations of CRISPR-Cas loci due to rapid genome sequencing, it seems timely to try and put in order our current understanding of

Cas proteins. Here we present an overview of all proteins families related to the CRISPR-Cas systems, with particular emphasis on their characteristic domains and domain architectures, and briefly discuss the functional and evolutionary implications.

13.2 CAS PROTEIN FAMILIES

13.2.1 Cas1 and Cas2: Signature Cas Proteins Implicated in Spacer Acquisition

Two Cas proteins, Cas1 and Cas2, are present in all CRISPR/Cas systems that are predicted to be functionally active. These proteins are thought to function as the "information processing" subsystem of CRISPR-Cas that is involved in spacer integration (the adaptation stage). The predicted roles of Cas1 and Cas2 in spacer acquisition are in agreement with the observations that these proteins are not involved in the antiviral defense stage of the mechanism when a spacer is already present in the CRISPR array (Brouns et al., 2008; Hale et al., 2009). The *cas1* and *cas2* genes comprise the cores of the three distinct types of CRISPR/Cas systems (Makarova et al., 2011b). The putative nuclease/integrase Cas1 is the most conserved among all Cas proteins. This protein is widely used as a marker for detection in CRISPR-Cas systems in bacterial and archaeal genomes and for construction of phylogenetic trees that provide a framework for reconstruction of CRISPR-Cas system evolution. Based on the evolutionary conservation of several acidic residues and a histidine, Cas1 has been predicted to possess nuclease activity (Makarova et al., 2006). To date, two Cas1 proteins have been experimentally characterized and their respective structures have been solved (Babu et al., 2011; Wiedenheft et al., 2009). The Cas1 protein from *Pseudomonas aeruginosa* is a metal-dependent nuclease that cleaves ssDNA and dsDNA, generating 80-bp DNA fragments from dsDNA. The previously identified conserved residues line up a metal-binding pocket in the α-helical domain of a novel fold (Fig. 13.1). The catalytic domain is connected to the N-terminal, mostly β-stranded domain by a flexible linker (Fig. 13.1); Cas1 protein forms homodimers (Wiedenheft et al., 2009). Mutation of metal-ion-binding amino acid residues of Cas1 inhibits Cas1-catalyzed DNA degradation. The function of the N-terminal domain is not clear. Similar properties have been reported for the Cas1 protein (YgbT) of *E. coli* (Babu et al., 2011). In addition, nuclease activity of *E. coli* Cas1 against branched DNAs including Holliday junctions, replication forks, and 5′-flaps has been demonstrated (Babu et al., 2011).Furthermore, genome-wide screens have shown that YgbT physically and genetically interacts with key components of DNA repair systems such as *recB, recC*, and *ruvB*, suggesting a dual role for Cas1 protein in bacterial antivirus immunity and DNA repair (Babu et al., 2011).

Several conserved fusions of Cas1 with other protein domains have been detected; in all cases, the genes encoding the fused proteins belong to *cas* operons. The most common fusion is with the Cas4 protein, a PD-(D/E)XK superfamily

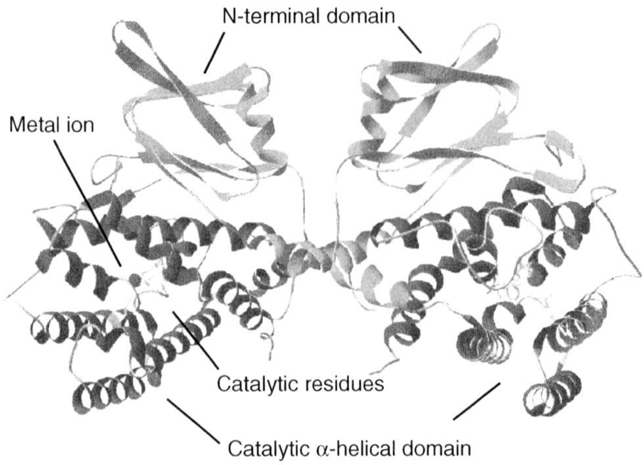

Figure 13.1 Cas1 structure and domain fusions. The cartoon shows the dimeric structure of Cas1. The catalytic residues and metal ions are circled and indicated with an arrow.

nuclease containing a C-terminal three-cysteine cluster. This fusion might indicate a role for Cas4 in spacer acquisition (van der Oost et al., 2009). Several fusions of Cas1 with reverse transcriptase (RT) similarly might be indicative of involvement of RT in the function of some CRISPR-Cas systems (Kojima and Kanehisa, 2008; Makarova et al., 2006). Furthermore, some RT appear to be involved in a distinct abortive infection mechanism of antivirus defense (Kojima and Kanehisa, 2008), suggesting the possibility that CRISPR-Cas systems and the abortive infection mechanism could be functionally linked.

The *cas2* gene is typically located immediately downstream of the *cas1* gene and encodes a small protein of approximately 100 amino acids; in Type I-F CRISPR-Cas systems, *cas2* is fused to the *cas3* gene. Based on the conservation of aspartate or asparagine located after the N-terminal β-strand, Cas2 protein has been predicted to possess nuclease activity (Makarova et al., 2006). There is statistically significant sequence similarity between Cas2 and the VapD component of one of the experimentally characterized toxin–antitoxin (TA) systems (Daines et al., 2004; Makarova et al., 2006). This suggests a functional link between CRISPR-Cas and TA systems, with the further implication that Cas2 is likely to be an endoribonuclease with an activity similar to that of interferases, the toxin components of numerous TA systems that cleave ribosome-associated mRNAs (Yamaguchi and Inouye, 2009). Several Cas2 proteins have been crystallized and studied biochemically (Beloglazova et al., 2008; Samai et al., 2010). These proteins adopt an RRM (ferredoxin) fold and form homodimers (Fig. 13.2). For Cas2 protein from *S. solfataricus* (Sso1404) the ribonuclease activity has been experimentally demonstrated. It has been shown that *in vitro* this protein cleaves the phosphodiester linkage on the 3′-side and generates 5′-phosphate- and 3′-hydroxyl-terminated oligonucleotides with in U-rich

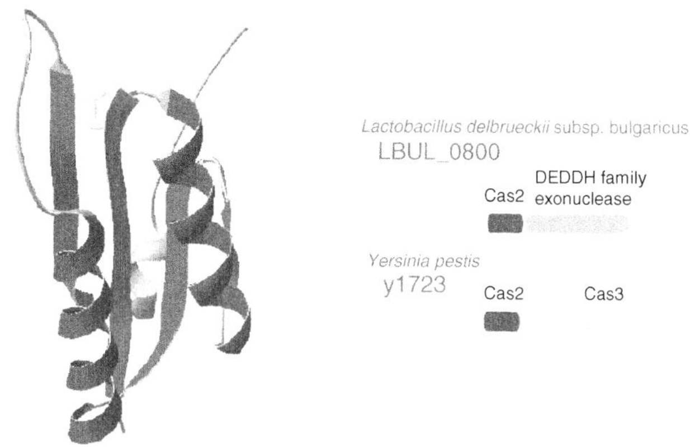

Figure 13.2 Cas2 structure and domain fusions. The cartoon shows the two RRM domain-containing Cas6 family proteins. An arrow points to the catalytic aspartate.

region preference. Alanine scanning revealed a number of residues that affect the ribonuclease activity including the predicted N-terminal catalytic aspartate (Beloglazova et al., 2008).However, for Cas2 from *Desulfovibrio vulgaris* neither nuclease activity nor ssRNA or ssDNA binding have been demonstrated despite the conservation of the N-terminal aspartate (Samai et al., 2010). Thus, it currently remains unclear whether Cas2 proteins from different organisms actually are so functionally diverse, or the differences in biochemical properties of Cas2 proteins are caused by unrecognized differences in isolation and assay conditions.

Several conserved fusions of Cas2 have been detected including the fusion to Cas3 in Type I-F systems (Makarova et al., 2006) and a fusion with a DEDDh family exonuclease in several genomes with a distinct Type I-E system version, mostly in Firmicutes.

13.2.2 The HD Domain is a Single-Stranded DNAse Required for Interference

The CRISPR-associated HD nuclease is a component of all Type I and Type III systems. In most Type I systems, the HD domain forms an N-terminal fusion with the Cas3 helicase but in some Type I-A systems it appears as a stand-alone gene (*cas3″*) in some. A few Type I-C systems (e.g., GSU0051) contain the HD domain as a C-terminal fusion with Cas3. Several organisms possess Type III CRISPR-Cas systems in which the HD domain is fused to the Cas10 protein. In some of these Cas10-HD fusions (e.g., TM1794), the HD domain shows a circular permutation so that the N-terminal metal-binding histidine is displaced to the extreme C-terminus (Makarova et al., 2002). However, the HD domain of Cas10d (Subtype I-D) does not show the circular permutation that makes it similar to the HD domain present in Cas3.

Several HD domains from different CRISPR-Cas systems have been studied experimentally, and two crystal structures have been resolved (PDB: 3S4L from *Methanocaldococcus jannaschii* and PDB: 3SKD from *Thermus thermophilus* HB8 (Mulepati and Bailey, 2011)). In particular, it has been demonstrated that, in addition to the helicase activity, Cas3 also shows ATP-independent nuclease activity mapped to the HD domain (Sinkunas et al., 2011) and that the HD domain itself possesses metal-dependent single-stranded DNA-endonuclease activity (Mulepati and Bailey, 2011; Sinkunas et al., 2011). Earlier, however, it has been reported that HD protein SSO2001 from *S. solfataricus* P2 CRISPR-Cas system I-A cleaves double-stranded oligonucleotides *in vitro* (Han and Krauss, 2009). In general, the reported properties of the HD domain are compatible with the hypothesis that Cas3 functions by cleaving the DNA region exposed by Cascade upon crRNA-guided target DNA binding.

13.2.3 Cascade-Associated Protein Families

At present, expression and processing is the best characterized stage of CRISPR-Cas-mediated immunity. It has been shown that the long primary transcript of a CRISPR locus (pre-crRNA) is processed into short crRNAs. Processing of pre-crRNA is catalyzed by endoribonucleases encoded by *cas* genes that operate either as a subunit of Cascade (CRISPR-associated complex for antiviral defense in *E. coli* (Brouns et al., 2008)) complex consisting of several Cas proteins, or as a stand-alone enzyme, for example, Cas6 in the archaeon *P. furiosus* (Carte et al., 2008; Hale et al., 2009). In the latter case, the formation of a multisubunit complex (denoted Cmr complex of Type III-B system) also has been observed (Hale et al., 2009). Recently, two additional Cas protein complexes have been characterized. The first one is the Csy complex associated with Type I-F system from *P. aeruginosa* (Wiedenheft et al., 2011), which also includes the CRISPR transcript processing endoribonuclease, Cas6f (Csy4), a homolog of Cas6 (Gesner et al., 2011; Haurwitz et al., 2010; Makarova et al., 2011b). The second complex is Cascade from *S. solfataricus*, which corresponds to CRISPR-Cas system Type I-A. Preliminary models of the architectures of these complexes are shown in Fig. 13.3. The general features of the Cascade complexes in Type I CRISPR-Cas systems are (i) multiple subunits of Cas7, apparently involved in binding crRNA; (ii) strong association between Cas7 and Cas5 proteins; (iii) a more loose association with Cas6; (iv) loose association between the large and small subunits (if the small subunit is present).

The crystal structures of several Cas6 homologs as well as the structure of the cleavage products produced by Cascade have been characterized. All Cas6 homologs adopt a double RRM fold (although in the case of Cas6f the second RRM fold is heavily distorted) and feature a conserved histidine located after the first β-strand of the N-terminal RRM, domain which is crucial for nuclease activity (Fig. 13.4). It seems that all other amino acids involved in catalysis are different in different Cas6 families. The cleavage of the pre-crRNA occurs within a CRISPR repeat at the 3′ side of the phosphodiester bond, generating a 5- end

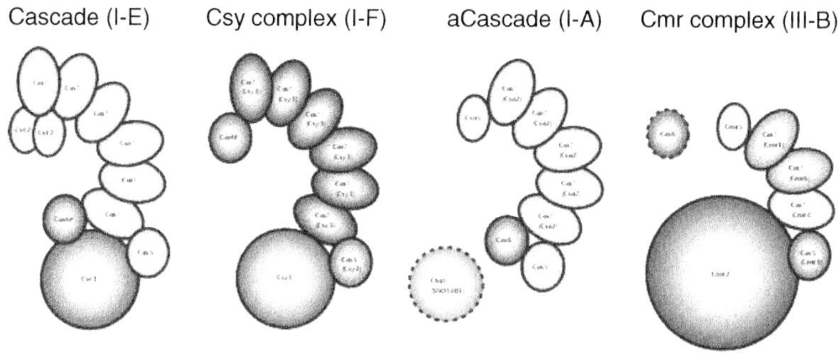

Figure 13.3 Cascade complexes models. The models for four characterized Cascade complexes include Cascade from *E. coli* (Brouns et al., 2008), Csy complex for the system Type I-F from *P. aeruginosa* (Wiedenheft et al., 2011), Cascade from *S. solfataricus* (Lintner et al., 2011), and Cmr complex from *P. furiosus* (Hale et al., 2009). For the first three complexes, the observed or inferred stoichiometry of subunits is reflected in the cartoons. The stably associated subunits are shown by solid circles and weakly associated subunits are shown by dashed circles. Three groups of RAMPs (Cas5, Cas6, and Cas7) are indicated along with the corresponding gene names.

hydroxyl group and a 2′, 3′ end cyclic phosphate group, and yields crRNA of approximately 60 nt size (for details see recent review; Jore et al., 2011a).

Recent data suggest that the same Cascade complex bound to a mature crRNA is involved in the interference stage by promoting R-loop formation to match a spacer within crRNA to the target ssDNA (Jore et al., 2011b).

In addition to Cas6, Cascade complexes of Type I systems typically include products of *cas7* and *cas5* genes, a large (typically, approximately 500 aa) protein and a small, mostly α-helical protein (Fig. 13.3). Along with Cas6, Cas5, and Cas7 proteins belong to the RAMP superfamily of RRM-containing proteins; at least four components of cmr complex (Cmr1, Cmr3, Cmr4, Cmr6) also belong to the same superfamily (Makarova et al., 2006).

13.3 THE THREE MAJOR GROUPS OF RAMPS

The first systematic sequence comparison of Cas proteins led to the identification of an extensive (super)family of diverse proteins that showed limited similarity to each other, centering on a glycine-rich loop. These proteins were denoted RAMPs (repair-associated mysterious proteins) given that Cas proteins were initially thought to represent a distinct repair system (Makarova et al. 2002). Subsequently, when the association of Cas proteins with CRISPR was realized, this superfamily was renamed repeat-associated mysterious proteins, with the acronym RAMP surviving. Comparison of the several resolved crystal structures of RAMPs led to the realization that they all contained distinct forms of the RRM domain (also often described as a ferredoxin-like fold). The latest

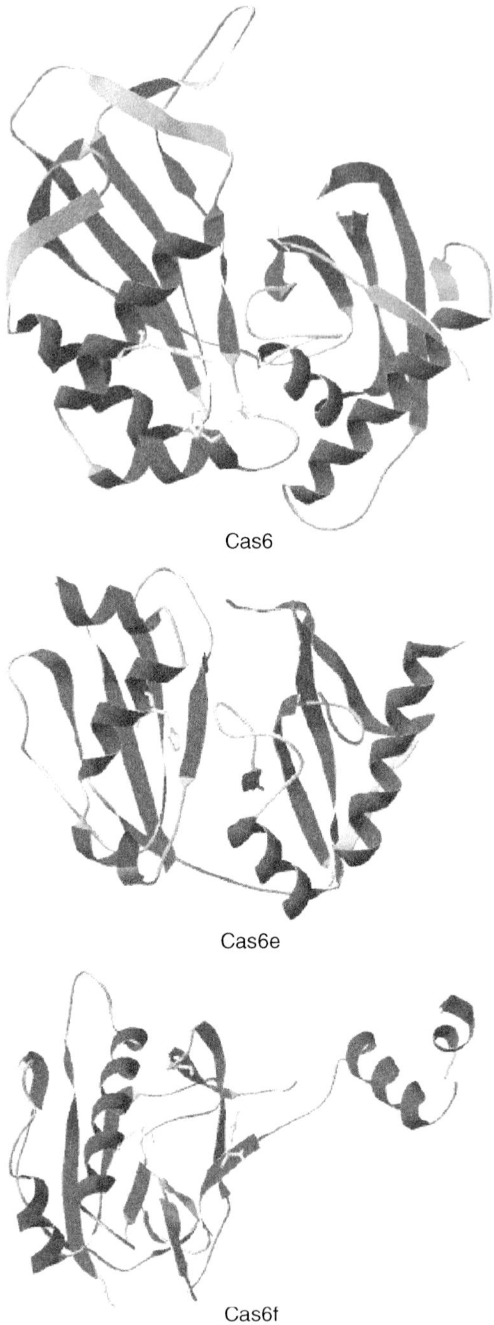

Figure 13.4 Cas6 structures. The cartoon shows the two RRM domain-containing Cas6 family proteins. The catalytic residues, including the N-terminal histidine, are indicated by arrows.

sequence and particularly structure comparisons allowed a tentative classification of the RAMPs into three major groups: Cas5, Cas6, and Cas7. Each of these groups includes many previously uncharacterized proteins now shown to adopt the RAMP structure.

The recent report of the crystal structure of Cas7 (Csa2) from the crenarchaeon *S. solfataricus* (system Type I-A) (Lintner et al., 2011) was an important breakthrough that provided for a comprehensive classification of the RAMPs. This structure clearly shows the presence of a single RRM domain that is structurally similar to the N-terminal RRM domain of Cas6 proteins. In addition, Cas7 contains four inserts within the RRM core and a C-terminal extension (Lintner et al., 2011). Independent sequence analysis showed clear similarity between the Cas7 family and other RAMP families including those from Type III CRISPR-Cas systems. Several sequence blocks that are conserved in all RAMP families include the core elements of the RRM domain and an insert containing a "hhG" motif and located immediately before the second β strand of the RRM fold (Fig. 13.5).

The demonstration that the Cas7 family belongs to the RAMP superfamily prompted detailed investigation of the relationships between the RAMPs. By combining the results of comparison of all available RAMP structures, secondary structure prediction, and sequence profile searches (using PSI-BLAST and or HHpred), the RAMP superfamily could be classified into three major groups: Cas5, Cas6, and Cas7 (Fig. 13.6).

The Cas5 group RAMPs (Cas5/COG1688, Cmr3/COG1769, Csm4/COG1567, Csy2, Csc1) were unified on the basis of sequence similarity that in most cases was identifiable by HHpred and the presence of a C-terminal domain downstream of the G-rich loop (Fig. 13.6). For some of these C-terminal domains, an RRM fold can be predicted. For example, in the Cmr3 subfamily (Subtype III-B), the predicted secondary structure elements of the C-terminal domain are compatible with the RRM structure. Moreover, this domain ends with a second G-rich loop whereas in the Csm4 subfamily from Type III-A CRISPR-Cas systems (the closest homolog of the Cmr3 family), this loop is almost completely degraded. The proteins of the Csx10 subfamily that is also related to Cmr3/Csm4 contain two predicted RRM domains followed by a readily identifiable G-rich loop. The Csx10 subfamily can be unequivocally linked to the Cas5 group and specifically to Cmr3 and Csm4 families (Makarova et al., 2011a). The remaining Cas5 proteins including Cas5 proper, Csy2, Csc1, and Csf3 contain a single N-terminal RRM domain that terminates with the G-rich loop and is followed by a distinct C-terminal β-meander domain. Thus, the large Cas5 group of RAMPs consists of two distinct subgroups, one of which contains two RRM domains and the other one contains only one RRM domain (Fig. 13.6). It remains uncertain as to which is the ancestral form, that is, whether the ancestor of the Cas5 group already contained two RRM domains and the C-terminal domain was lost or severely deteriorated in one of the subgroups, or the ancestral form possessed a single RRM domain that was duplicated in one of the subgroups.

(a)

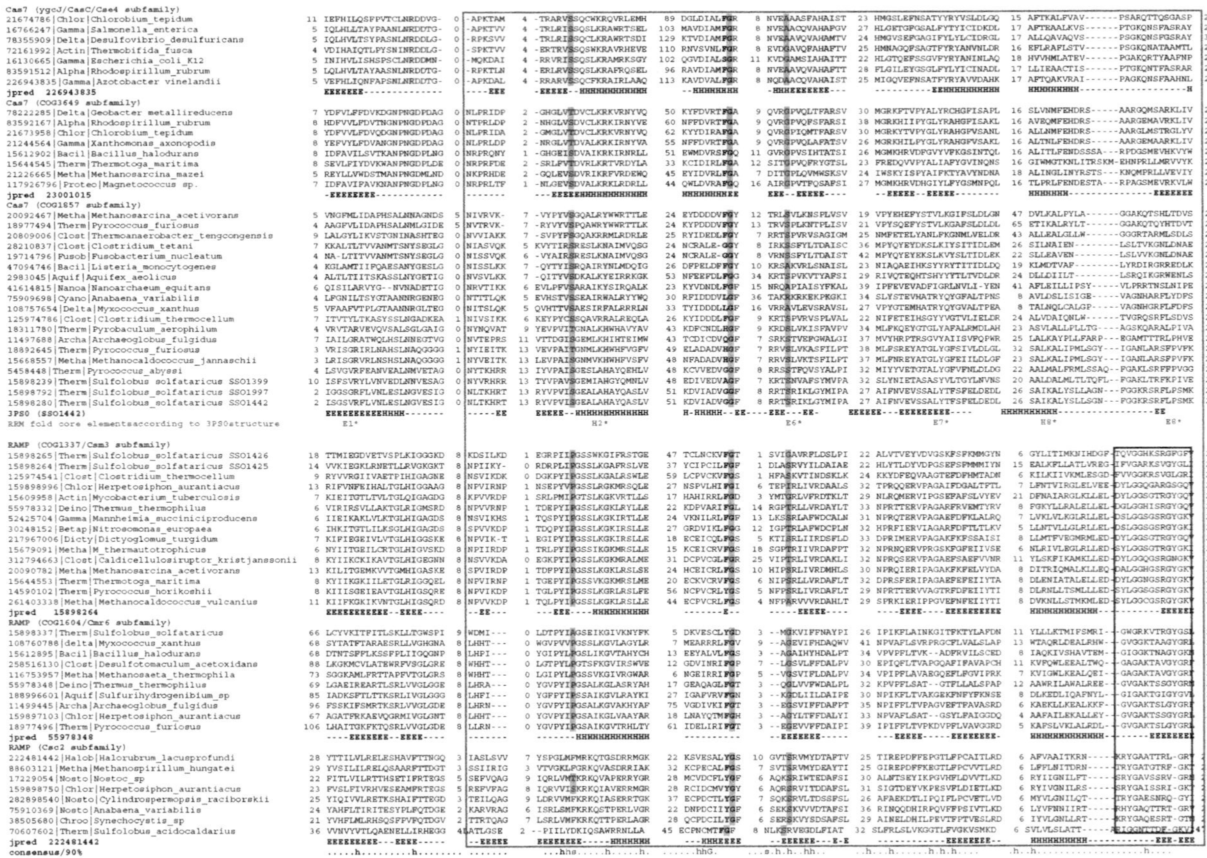

Figure 13.5 *(Continued)*

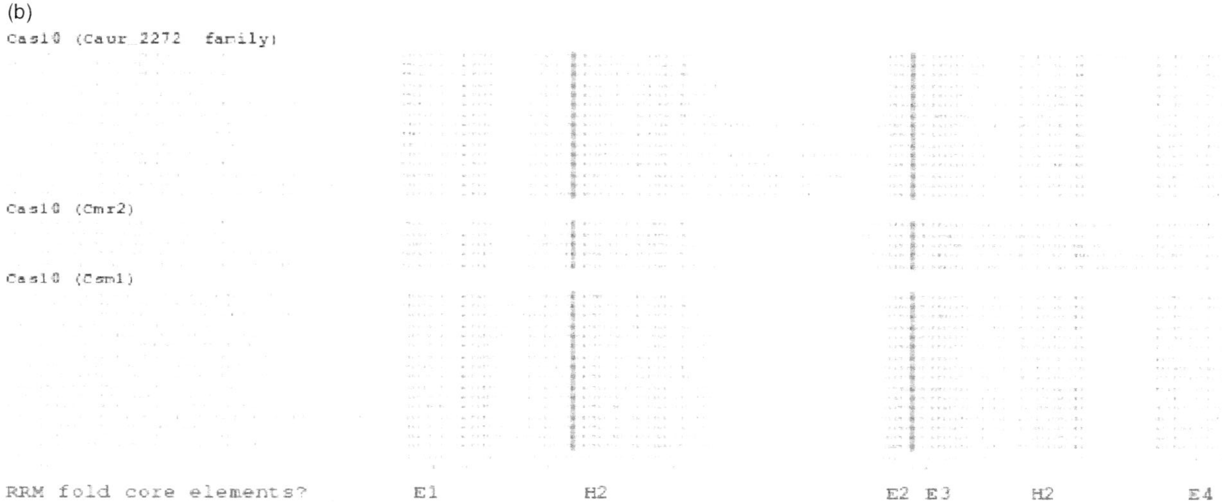

Figure 13.5 Multiple alignment of Cas7 subfamilies and related families of RAMPs. The multiple sequence alignment includes the conserved blocks identified by HHpred (red box), secondary structure predictions, and the secondary structure elements extracted from the crystal structure of the Cas7 from *S. solfataricus* (Lintner et al., 2011). Secondary structure prediction is shown as follows: "H" indicates α-helix, "E" indicates extended conformation (β-strand). The sequences are denoted by their GI numbers and species names. G-rich loop region of RAMPs is shown by the blue box. The positions of the first and the last residues of the aligned region in the corresponding protein are indicated for each sequence. The numbers within the alignment represent poorly conserved inserts that are not shown. The coloring is based on the consensus shown underneath the alignment; "h" indicates hydrophobic residues (WFYMLIVACTH), "p" indicates polar residues (EDKRNQHTS), and "s" indicates small residues (ACDGNPSTV). (*See insert for color representation of the figure.*)

The Cas6 group includes Cas6 proteins proper (COG1853/COG5551) as well as highly diverged homologs from the I-E (Cas6e) and I-F (Cas6f) CRISPR-Cas subtypes. This grouping is supported by the available structures and is compatible with the reported functions for the representatives of each family. Most of the Cas6 proteins encompass two well-defined RRM domains, which are connected by a "flange" in the extended conformation and contain a glycine-rich loop upstream of the last strand of the second RRM domain. Thus, the ancestor of the Cas6 group can be confidently inferred to have possessed two RRM domains. However, Cas6f proteins contain a typical N-terminal RRM domain and a distinct C-terminal domain that display certain topological features reminiscent of the RRM fold and contains a C-terminal G-rich loop but does not show significant sequence or structural similarity to any RRM domains (Fig. 13.4). This domain could be either a grossly distorted RRM or a distinct β-meander that convergently acquired the G-rich loop.

The Cas7 group includes Cas7 proper (COG1857) and a variety of RAMPs that are mostly associated with the Type III CRISPR-Cas systems. All these proteins contain a single RRM domain with additional elaborations as demonstrated by examination of the recently reported Cas7 structure (Fig. 13.6), sequence comparison, and secondary structure prediction. The Type III RAMP families (Csm3/COG1337 and Csm5/COG1332 in Subtype III-A; Cmr6/COG1604, Cmr4/COG1336, and Cmr1/ COG1367 in Subtype III-B; Csc2 in Subtype I-D and Csf2 from the system of unknown subtype) are more similar to each other than to Cas7 but share with Cas7 several conserved sequence motifs, the overall sequence similarity identifiable by HHpred (Soding et al., 2006), and the absence of the additional RRM domain after the G-rich loop (or its counterpart). The Csy3 subfamily is tentatively included in this group on the basis of the secondary structure prediction (no extension after the G-rich loop compatible with another RRM domain). Some members of the Cas7 group, such as Cmr1, contain a second predicted RRM domain. Furthermore, the RAMPs of the Cas7 group show a tendency for gene duplication at least in Type III CRISPR-Cas systems.

The only RAMP family that could not be confidently assigned to any of the three groups is Csf3: despite some weak sequence similarity to both Cas6 and Cas5 in the G-rich loop region, these proteins contain fewer predicted β-strands than Cas6 or Cas7 and no predicted RRM domain downstream of the G-rich loop; although the latter feature resembles the organization of Cas7, there is otherwise no similarity between these proteins.

The diversity and weak conservation of the sequences and structures of the RAMPs hamper the elucidation of the evolutionary relationships between the three major groups. Considering only the relationship between the domain architectures of the RAMPs, the most parsimonious evolutionary scenario would involve an ancestral RAMP with a single enzymatically active RRM domain, resembling Cas7, and a single duplication in the putative common ancestor of the Cas5 and Cas6 groups, with subsequent deterioration or displacement of the C-terminal RRM domains in several Cas5 and Cas6 lineages (Fig. 13.6).

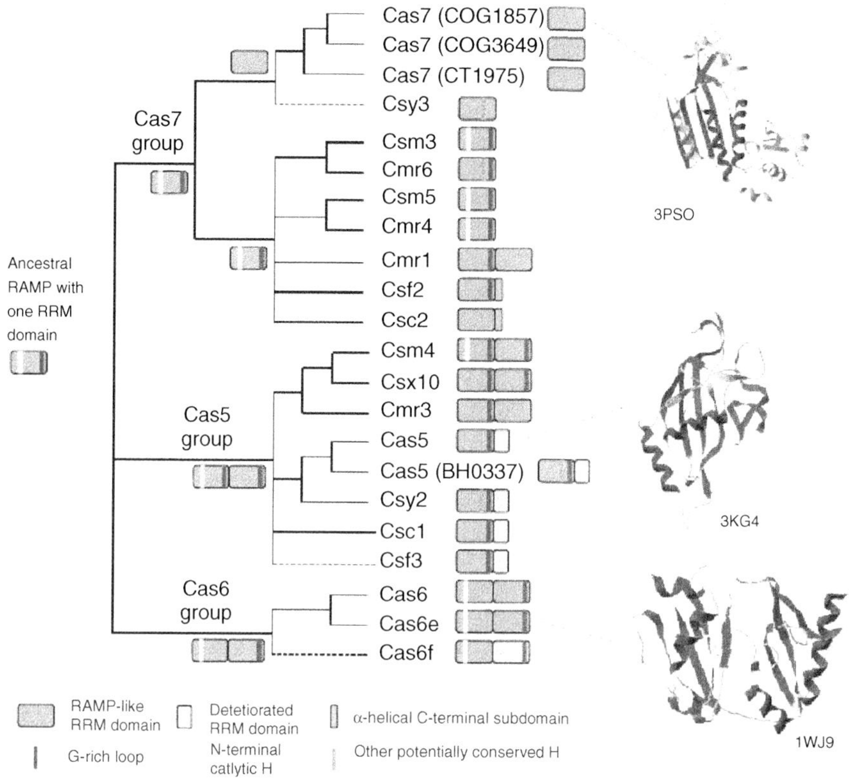

Figure 13.6 Classification of the RAMPs. The tree-like scheme of RAMP relationships is based on the sequence similarity, structural features, and neighborhood analysis described in the text, and should not be construed as a phylogenetic tree. Unresolved relationships are shown as multifurcations and tentative assignments are shown by broken lines. The catalytic activity of some of the RAMP proteins of the Cas5 and Cas7 groups involving the partially conserved histidines shown in the figure should be considered a tentative prediction. The structures for the RAMPs of the Cas5, Cas6, and Cas7 groups are shown. The RRM fold domains are depicted as in Fig. 13.4. (*See insert for color representation of the figure.*)

13.4 THE CHARACTERISTIC ARRANGEMENT OF RAMPS IN CRISPR-CAS OPERONS

Mapping the new classification of RAMPs described in the preceding section onto the operons of the Type I and Type III CRISPR-Cas systems reveals a common architectural pattern. Most subtypes of Type I CRISPR-Cas systems encode one RAMP of the Cas5, Cas6, and Cas7 groups each. Operons of type III CRISPR-Cas system are organized similarly except that they typically encode multiple Cas7 group RAMPs. Notably, Cas5 and a Cas7 usually are encoded in a pair of adjacent genes. Remarkably, the Cas5 and Cas7 orthologs in two distinct CRISPR-Cas systems belong to the stable core of the Cascade complex in extremely diverse organisms, *E. coli* (Type I-E) (Jore et al., 2011b; van der Oost et al., 2009) *and S. solfataricus* (Type I-A) (Lintner et al., 2011). Unclassified (U-type) CRISPR-Cas systems form operons that lack *cas5* but in which a cas7 (csf2) gene is located adjacent to the csf3 gene suggesting that Csf3 could be a truncated derivative of Cas5 performing an analogous function.

13.5 PUTATIVE HOMOLOGY AMONG THE LARGE AND SMALL SUBUNITS OF DIVERSE TYPE I AND TYPE III CRISPR-CAS SYSTEMS

Multiple lines of evidence suggest that large subunits contained in most of the CRISPR-Cas systems could be homologous to Cas10 proteins that contain a polymerase-like Palm domain and are predicted to be enzymatically active in Type III CRISPR-Cas systems but inactivated in Type I systems.

Among the large subunits of Type I CRISPR-Cas systems, sequence conservation has been demonstrated previously (Makarova et al., 2006) for several subfamilies of the Cas8 family (Cas8a1/Csa6, a subfamily of Subtype I-A; Cas8b/Csh1/Cst1, a subfamily of Subtype I-B; and Cas8c/Csd1, a subfamily of Subtype I-C). Using HHpred and PSI-BLAST, other subfamilies could be linked to the Cas8 family, in particular, Cmx1/Csx13/LA3191 associated with some diverged variants of Subtype I-C and Cas8a2 (Csa4/Csx9 subfamily) associated with some Subtype I-A systems (Makarova et al., 2011a). The large Cascade subunit of Subtype I-D shows similarity to the Zn-finger regions of the Cas8b/Cst1of I-B system and is, in addition, fused to an HD domain analogously to the Type III Cas10 proteins. The large subunits of subtypes I-E (Cse1) and I-F (Csy1) do not show any sequence similarity to each other (despite the fact that these systems are related by the Cas1 phylogeny and the *cas* gene sets) or to any Cas8 family proteins.

Type III CRISPR-Cas systems contain several subfamilies of Cas10 (Csm1, Cmr2, and Csx11 according to (Haft et al., 2005)) that have been denoted CRISPR polymerases because of the presence of a readily identifiable Palm/Cyclase domain (Anantharaman et al., 2010; Makarova et al., 2002; Pei and Grishin, 2001). The CRISPR polymerase consists of several domains, namely, the HD domain (ssDNAase), a distinct domain so far unique to this protein family; a Zn-finger domain; and the Palm domain, the signature domain of various polymerases and cyclases, which adopts a distinct RRM fold (Makarova et al., 2002). The Palm domain of CRISPR polymerases is more similar to the Palm domain of cyclases than to those of 3′-5′ DNA and RNA polymerases, and contains all typical secondary structure elements including four β-strands of the core RRM fold (Anantharaman et al., 2010). Many structures of Palm domain-containing polymerases from all domains of cellular life and numerous viruses have been solved and compared (Steitz and Yin, 2004). Most of these polymerases show a common arrangement of the core domains and the same modes of nucleic acid binding; the polymerases additionally contain a variety of editing nuclease domains and regulatory domains. The core domains (usually arranged in the same order from the N-terminus to the C-terminus) are the following: the "Fingers" domain that binds a nucleotide, the catalytic "Palm" domain that binds a single-stranded nucleic acid, and the "Thumb" domain that binds a double-stranded nucleic acid (Steitz and Yin, 2004). Despite this structural and mechanistic uniformity, only the Palm domains of these numerous polymerase families are clearly homologous (Iyer et al., 2008; Steitz and Yin, 2004). The most conserved feature of the Palm domains is the β-hairpin formed by strands 2 and 3 of the RRM fold (Aravind et al., 2002; Iyer et al., 2008). The thumb domain is usually enriched in α helices, some of which interact directly with the DNA or RNA duplex (Steitz and Yin, 2004).

Recently, we undertook a detailed analysis of multiple alignments and predicted secondary structures of the large subunits of Type I and Type III systems, Cas8 and Cas10 (predicted CRISPR polymerases), respectively (Makarova et al., 2011a). The C-terminal portion of the CRISPR polymerases (starting immediately after the Zn-binding treble clef domain) contains a region that consists mostly of α-helices and directly follows the fourth strand of the RRM fold of the Palm domain. This arrangement is consistent with the general structure of Palm-domain polymerases described above and suggests that this region of the CRISPR polymerases could be equivalent to the Thumb domain of other polymerases. Furthermore, given the compactness of the core Palm domain in these proteins, the region located after the HD nuclease domain and upstream of the Zn-binding domain might be the equivalent of the Fingers domain.

Most of the large subunits of different subtypes of Type I CRISPR-Cas systems contain a readily identifiable Zn-finger domain in the middle of the protein sequence (Makarova et al., 2006). If the large subunits are highly diverged, inactivated Palm-domain containing polymerases as proposed above, and the Zn-finger is equivalent to the treble-clef domain found in the CRISPR polymerase, one should expect that a domain containing several β-strands compatible with the general structure of the Palm-domain followed by an α-helical region would be located downstream of the Zn-finger. Indeed, in various subfamilies of Cas8, Cas10d, inactivated Cas10 (Csx11 subfamily), and Cse1, the same structural pattern is observed, namely, at least three predicted β-strands that might belong to a RRM fold, including the core β-hairpin, followed by an α-helical region (Makarova et al., 2011a). Because two other subfamilies (Csy1 and Cmx1) do not contain Zn-fingers, it is difficult to map the beginning of the putative Palm-domain within these sequences. However, sequence similarity between Cmx1 and Cas8 (see above) could be identified (Makarova et al., 2011a), and given that Cmx1 proteins possess an α-helical C-terminal region, it seems likely that Cmx1 is a diverged homolog of Cas8. The Csy1 protein might be homologous to Cse1 (the large subunit of the Subtype I-F system) given the overall similarity in the operon organization between the I-E and I-F systems and the clustering of these systems in the Cas1 phylogeny (Makarova et al., 2011b). Similar to Cse1, Csy1 also has an α-helical C-terminal domain and an N-terminal region with mixed

α-helices and β-strands. Although the pattern of the predicted secondary structure elements of Csy1 cannot be confidently aligned with either Cse1 or Cas8, the possibility that it contains a derived RRM-like fold cannot be ruled out. Most of the large subunits of Type I CRISPR-Cas systems containing Zn-fingers also possess an N-terminal region with mixed β-strands and α helices, which is compatible with the general organization of the region following the HD domain and preceding the Zn-finger in Cas10 subfamilies. Taken together, analysis of the general secondary structure features, the presence of the Zn-finger domain in many large subunits, the similar operon organization, and the experimentally demonstrated functional link to RAMPs and the Cascade complex (Brouns et al., 2008; Hale et al., 2009; Jore et al., 2011b) raise the possibility that all large subunits of CRISPR-Cas systems might be inactivated derivatives of the CRISPR polymerase (Fig. 13.7a). However, there is currently not enough evidence to rule out nonhomologous displacement of some large subunits or their individual domains.

Interestingly, the pattern of secondary structure elements in the putative Fingers domain in Cas10 and several large subunits, in particular Csx11, Cas8a2/Csa4, and Csc3, resembles the structures of the RRM domain of the RAMPs. Like the RRM core domain, many of the Fingers-like domains contain four predicted β-strands. Furthermore, the Fingers-like domains start with a β strand-α helix element and ends with a helix-β-strand element, which are the two most conserved structural patterns in the RAMPs. Thus, the Fingers domain of the large subunits might adopt an RRM fold.

In several families of the large subunits (Cas8a1, Cas8b, Cas8c, Cmx1, and Cas10d) of the I-A, B, C, and D system subtypes, the C-terminal region (the predicted Thumb domain) is longer than it is in Cas10 proteins (eight α helices compared tofour in Cas10). Interestingly, in these subtypes, the small Cascade subunit is missing in the CRISPR-Cas operons. Typically, the small subunit is an α-helical protein containing six α helices (structure is available for cmr5: AF1862, 2OEB and TTHB164, 2ZOP; see Fig. 13.8). The size and predicted structure of the small subunit appears to be compatible with the size and structure of the extra α helical region at the C-termini of the longest large subunits (Fig. 13.7a). The Csy1 protein, the Subtype I-F specific large subunit, contains eight predicted α helices at the C-terminus and four helices at the extreme N-terminus. Because none of the predicted RAMP proteins from this system contain extended α-helical regions compatible with the size of the small subunit, it is tempting to hypothesize that the small subunit might be "hidden" within Csy1.

The demonstration that at least some of the large subunits of Type I CRISPR-Cas systems are homologous to the CRISPR polymerase suggests that all these large proteins function and interact with DNA or RNA in a mode analogous to that of other Palm domain polymerases (Table 13.1). In particular, the Palm domain probably interacts with ssDNA whereas the analog of the Thumb interacts with dsDNA. Notably, evolutionarily conserved inactivated derivatives of Palm domain polymerases have been detected in archaea and eukaryotes although their functions remain uncharacterized (Rogozin et al., 2008; Tahirov et al., 2009). The small subunits of CRISPR-Cas systems might be responsible for the recognition

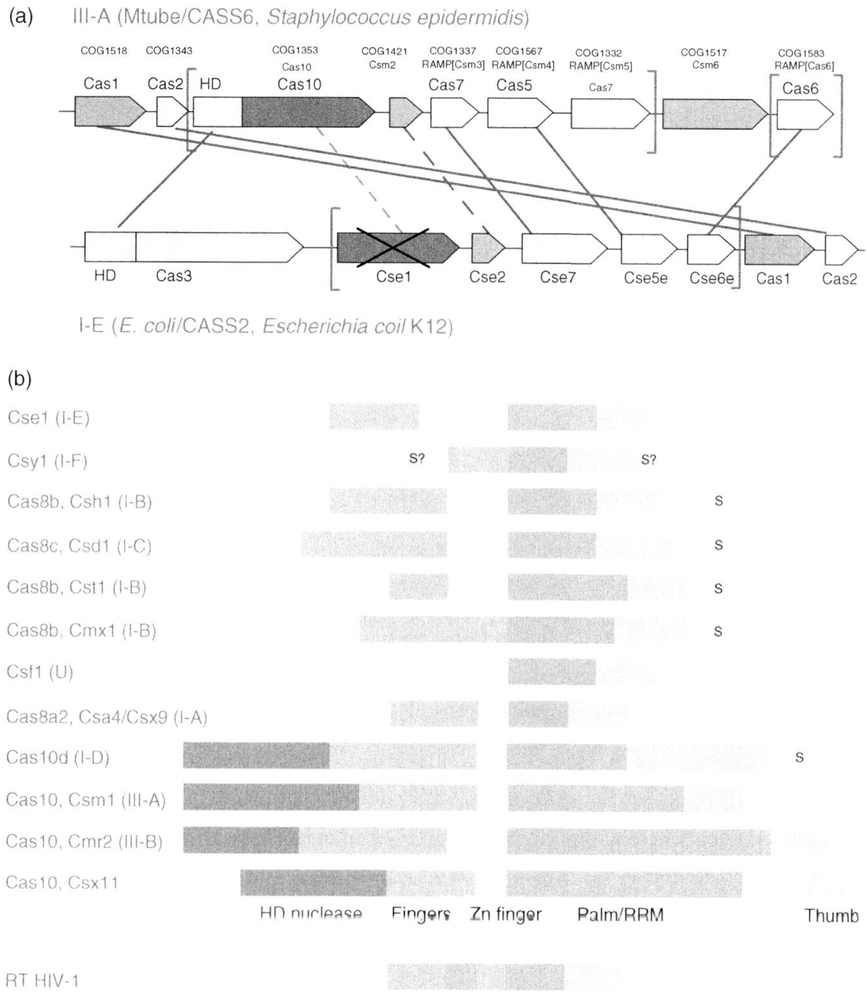

Figure 13.7 Gene content similarity between Type I-E and Type III-A systems and structural organizations of large subunits of different CRISPR-Cas systems of Type I and III. (a). Genes in the operons for I-E and III-A subtypes are shown by arrows with size roughly proportional to the size of the corresponding gene. Homologous genes are shown by the arrows of the same color or hashing. RAMPs are shown by pink or pink hashing. Solid lines connect genes for which homology can be confidently demonstrated, and dashed lines connect genes for which homology is inferred tentatively. The Cascade complex subunits are shown by square brackets. Two previously published domain annotations are included for comparison. (b) Domain organization of large subunits of different Type I III CRISPR-Cas systems. Domain size is roughly proportional to the corresponding sequence length. The letter "S" marks the regions that could be homologous to small subunits of Cascade complex encoded as separated genes in Type III systems, I-E subtype, and some systems of I-A subtype. (*See insert for color representation of the figure.*)

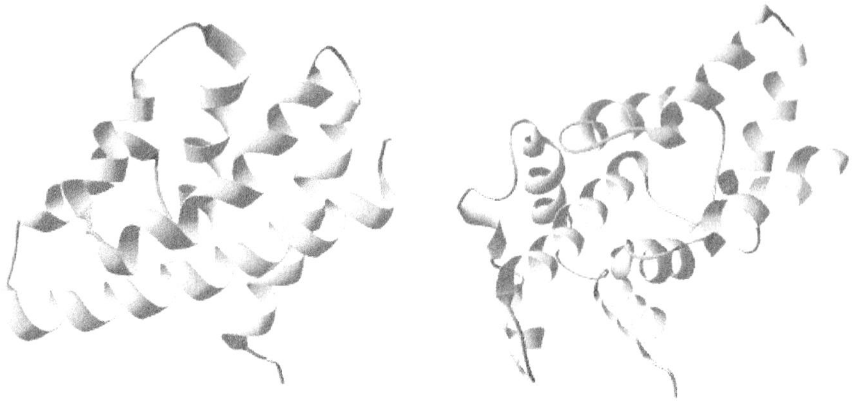

Figure 13.8 Structures and motifs of the small subunits of CRISPR-Cas systems. Two available structures for small subunits are shown. Conserved tryptophanes within the C-terminal α helix are shown in yellow. (*See insert for color representation of the figure.*)

of the PAM motif that is required for the selection and incorporation of new spacers (Table 13.1).

The conservation of the complete set of catalytic residues typical of Palm domain polymerases and cyclases implies that the Palm domain of Cas10 is enzymatically active but the nature of this activity remains unknown. There is no indication that a processive polymerase is involved at any stage of the CRISPR-Cas system functioning. The possibility remains that Cas10 is a nucleotidyltranferase or even a nucleotide cyclase, perhaps involved in crRNA modification. This type of activity is compatible with the activity of the tRNA(His) guanylyltransferase THG1 (Jackman and Phizicky, 2006), which belongs to the same clade of Palm domain proteins with Cas10 and the GGDEF diguanylate cyclases (Anantharaman et al., 2010) (see above). Another possibility is that Cas10 has a secondary role as a helicase in one or more stages of CRISPR/Cas functioning. A helicase activity dependent on the cleavage of the c-α bond in NTP during polymerization has been demonstrated for the bacteriophage T7 RNA polymerase (Steitz, 2004; Steitz and Yin, 2004; Yin and Steitz, 2004), which is a derivative of the Palm domain DNA polymerases (Iyer et al., 2008). Remarkably, all Type I CRISPR-Cas systems in which the large subunits are inactivated Cas10 homologs (i.e, those Cas10 homologs in which some or all of the predicted catalytic residues in the Palm domain are replaced) also include the Cas3 helicase, and conversely, all Type III systems that contain Cas10 proteins predicted to be active lack Cas3 (Makarova et al., 2011b). Thus, it seems likely that Cas3 compensates for the loss of the original enzymatic function of Cas10 in Type I CRISPR-Cas system whereas the inactivated derivative of Cas10 performs an accessory structural role in these systems. Notably, some Type U CRISPR-Cas systems that contain degraded versions of Cas10 and lack Cas3 include a DinG-like helicase

TABLE 13.1 Structures, Domain Architectures, and Functions of the Core Components of CRISPR-Cas Systems

Family	Biochemical/*in silico* Evidence	Structural Features	Prediction
Cas1	Metal-dependent deoxyribonuclease; (Han et al., 2009; Wiedenheft et al., 2009); deletion of Cas1 in *E. coli* results in increased sensitivity to DNA damage and impaired chromosomal segregation (Babu et al., 2011).	PDB: 3GOD, 3LFX, and 2YZS Unique fold with two domains: N-terminal β-stranded domain and catalytic C-terminal α-helical domain Fusions: Cas4, PD-D/ExK family nuclease and reverse transcriptase	Involved in integration of spacer DNA into CRISPR repeats
Cas2	RNAse specific to U-rich regions (Beloglazova et al., 2008)	PDB: 2IVY, 2I8E, and 3EXC RRM (ferredoxin) fold Fusions: Cas3 and DEDDh family exonuclease	Facilitates spacer selection and/or integration. Could be involved in further crRNA cleavage
Cas3 (helicase and HD domain)	Single-stranded DNA nuclease (HD domain) and ATP-dependent helicase (Sinkunas et al., 2011); required for interference (Brouns et al., 2008).		Cuts DNA during interference; promotes strand separation
Stand-alone HD nuclease	Metal-dependent deoxyribonuclease specific for double-stranded oligonucleotides (Han and Krauss, 2009).	PDB: 3S4L and 3SKD	Cuts DNA during interference

(*Continued*)

TABLE 13.1 (*Continued*)

Family	Biochemical/*in silico* Evidence	Structural Features	Prediction
Cas4		PD-D/ExK nuclease homolog with three-cysteine C-terminal cluster (Makarova et al., 2006)	Might be involved in spacer acquisition
Cas5	Subunit of Cascade complex (Brouns et al., 2008; Jore et al., 2011b).	PDB: 3KG4 RRM (ferredoxin) fold, RAMP superfamily	Might substitute for Cas6 if catalytically active. Otherwise might be involved in both interference and adaptation stages
Cas6	Metal-independent endoribonuclease that generates crRNAs, subunit of Cascade complex (Brouns et al., 2008; Carte et al., 2008; Hale et al., 2009; Haurwitz et al., 2010; Jore et al., 2011b).	PDB: 2XLJ, 1WJ9, and 3I4H Double RRM (ferredoxin) fold, RAMP superfamily	
Cas7	Subunit of Cascade complex (Brouns et al., 2008); present Cascade complex of I-E systems in six copies (Jore et al., 2011b) and in several copies in I-A systems (Lintner et al., 2011).	PDB: 3PS0 RRM (ferredoxin) fold with subdomains, RAMP superfamily	Implicated in interference; binds crRNA; if enzymatically active, might be involved in RNA-guided RNA cleavage
Cas8(large subunit)	Subunit of Cascade complex (Brouns et al., 2008).		Inactivated Cas10 polymerase-like protein, binds DNA, interacts with HD domain and a RAMP carrying crRNA; could be involved in both interference and spacer selection stages

Cas10 (large subunit, CRISPR polymerase)	Subunit of Cascade (Cmr) complex (Hale et al., 2009);	Homologous to Palm domain polymerases and cyclases; Fusion: HD nuclease domain	Same as Cas8, but fused to HD and thus cuts ssDNA; might be involved in strand separation
Small subunit	Small, mostly α helical protein, subunit of Cascade complex (Brouns et al., 2008; Hale et al., 2009); present in Cascade complex of I-E systems in two copies (Jore et al., 2011b)	PDB: 2ZCA (Cse2) and 2ZOP, 2OEB (Cmr5); Both families have a unique fold with α helical structure	Specifically binds DNA; might recognize PAM
Cas9	In Type II CRISPR-Cas systems, Cas9 is sufficient both to generate crRNA and to cleave the target DNA (Barrangou et al., 2007; Garneau et al., 2010); the mutations of both RuvC and HNH domains of Cas9 of *Streptococcus thermophilus* abolish phage interference (Sapranauskas et al., 2011).	Contains a predicted RuvC-like (RNAse H fold) and HNH nuclease domains	May be considered a functional analog of CASCADE and HD nuclease; generates crRNAs, probably via RuvC domain; cleaves target DNA, probably via HNH domain; binds both crRNA and DNA

(see below), in further support of the possibility that a helicase activity required for the CRISPR-Cas function can be supplied by different, in some cases, unrelated proteins.

13.5.1 Type II CRISPR-Cas Systems and Homologs of Cas9

The signature protein of the Type II CRISPR-Cas systems II, Cas9, does not show any detectable similarity to any proteins in Type I and Type III systems. It appears that Cas9 is sufficient both to generate crRNA and to cleave the target DNA (Barrangou et al., 2007; Garneau et al., 2010; Makarova et al., 2011b). The large Cas9 protein (~1000 amino acids) contains two predicted nuclease domains, namely, the HNH (McrA-like) nuclease domain that is located in the middle of the protein and the RuvC-like nuclease domain (RNAse H fold) that contains all the characteristic catalytic motifs of the RuvC nuclease family (Aravind et al., 2000; Makarova et al., 2011a; Makarova et al., 2006) and hence is predicted to be enzymatically active, but contains a long (~450 amino acids) insert including the HNH nuclease domain (Fig. 13.9).

The roles of the two predicted nuclease domains of Cas9 in the function of the Type II CRISPR remain unclear. However, the insertion of the HNH nuclease domain into the RNAse H fold domain suggests that the two nuclease activities are closely coupled and that their active sites are proximally located. The HNH nuclease domain, which is common in restriction enzymes and possesses DNA-endonuclease activity (Jakubauskas et al., 2007; Kleanthous et al., 1999), might be responsible for the target cleavage. Conversely, the RuvC-like RNAseH fold domain might be involved in CRISPR transcript processing. The mutations of the

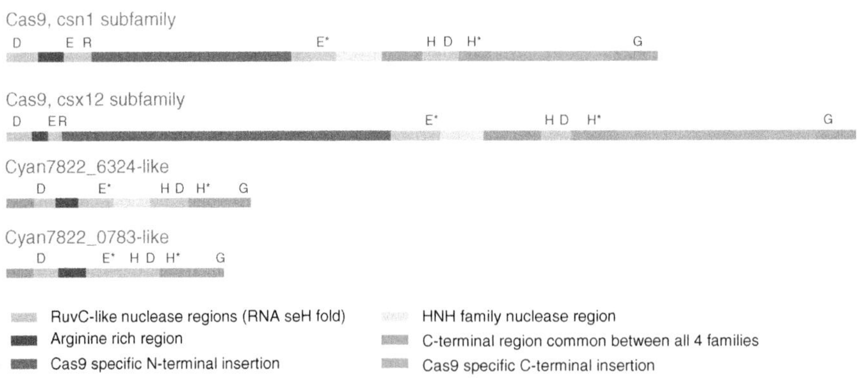

Figure 13.9 Structural organization of Cas9 proteins and their homologs. Homologous regions are shown by the same color. Distinct sequence motifs are denoted by the corresponding conserved amino acid residues above the respective domains (when the same conserved amino acid occurs in different motifs, one is marked by an asterisk to avoid confusion). (*See insert for color representation of the figure.*)

predicted catalytic amino acids of both RuvC and HNH domains of Cas9 of *Streptococcus thermophilus* abolish phage interference (Sapranauskas et al., 2011).

Several PSI-BLAST searches using different Cas9 sequences as queries detected similarity to a large family of prokaryotic proteins containing both RuvC-like and HNH-nuclease domain. This family could be divided into at least two subfamilies by domain architecture (Fig. 13.9). Analysis of the genomic context of the genes encoding these Cas9 homologs did not reveal any stable associations, and there are no CRISPR repeats in the vicinity of any of these genes. Hence, the function of these proteins remains obscure. An intriguing possibility is that they might represent a novel system of RNA-guided DNA interference involved in antivirus defense, which in some respects could be analogous to the prokaryotic Argonaute proteins (Makarova et al., 2009b). Some of these proteins form large lineage-specific paralogous families (e.g., 49 genes in *Ktedonobacter racemifer* or 17 genes in *Microcoleus chthonoplastes*). These expansions of closely related paralogs in the same genome suggest that at least this subset of the family could represent novel mobile elements. The *cas9* gene might have been co-opted by the CRISPR/Cas system from such mobile elements with the concomitant loss of typical CRISPR/Cas components, such as RAMPs and CRISPR polymerases resulting in the emergence of the distinctive Type II gene neighborhoods. The emergence of Cas9 involved two distinct insertions, namely, a mostly α-helical insert near the middle of the protein sequence and a mostly β-stranded region near the C-terminus (Fig. 13.9). These large inserts did not show sequence similarity to any other proteins but, given the close functional similarity between Type II and Type I/III CRISPR-Cas, it cannot be ruled out that the inserts originate from CRISPR-Cas components.

13.5.2 Unusual CRISPR-Cas Systems and Cas Protein Families

An unusual CRISPR-Cas system has been recently identified in several bacterial genomes, for example, *Acidithiobacillus ferrooxidans* ATCC 23270 (operon AFE_1037-AFE_1040) (denoted Type U as it did not contain signature genes of any of the three CRISPR-Cas types) (Makarova et al., 2011b). This system contains neither the two ubiquitous core *cas* genes, *cas1* or *cas2*, nor any other signature genes of the three CRISPR-Cas types. The *A. ferrooxidans* system consists of four genes denoted *csf1*, *csf2*, *csf3*, and *csf4*. The Csf2 protein is a Cas7 group RAMP closely related to the Csm3 subfamily. Csf3 is yet another diverged RAMP protein that might be functionally analogous to the Cas5 group (Fig. 13.6). Csf1 is a Zn-finger containing protein that shows a weak sequence similarity with the Zn-finger sequence of Cas10 proteins of the Crm2 family, and the predicted secondary structure of Csf1 is compatible with the treble clef fold (Makarova et al., 2011a). The secondary structure prediction for these proteins generally shows the same pattern as in the large Cascade subunits discussed above, namely, several β-strands (some of them forming a potential hairpin) and several α-helices at the C-terminus. Taken together, these observations suggest that Csf1 could be a highly divergent, inactivated and N-terminally truncated Cas10-like polymerase derivative lacking the N-terminal Fingers domain. The fourth gene in this system, *csf4*, is usually located on the complementary DNA strand in the divergent orientation and encodes a DinG family helicase (White, 2009). Homologs of Csf1, Csf2, and Csf3 were identified in several Actinobacteria in a somewhat different genomic context (e.g., pREL1_0084-pREL1_0087 *Rhodococcus erythropolis*). There is no gene encoding a DinG-like helicase in the neighborhood. A gene encoding a small, largely α-helical protein with conserved positively charged and aromatic amino acids in several positions follows the *csf1* gene. This arrangement resembles the large and small Cascade subunits of the I-E and III-A subtypes. All these loci are located on plasmids. There are no CRISPR repeats detected on these plasmids and, in many cases, in other partitions of the respective genomes either (see the CRISPRdb database (Grissa et al., 2007)). Thus, this variant of the Type U CRISPR-Cas system might be a mobile Cascade-like module functioning in a completely different context, not associated with CRISPR repeats and other Cas proteins.

A few CRISPR-Cas systems that could be readily classified into established subtypes or at least types based on signature genes contain, in addition, unusual protein families, domain fusions and/or operon rearrangements (Fig. 13.10). For example, a distinct Subtype I-C system variant possesses a number of specific features, in particular, fused *cas1* and *cas4* genes and two extremely divergent RAMPs (Fig. 13.10a). One of these distinct RAMPs is a homolog of Cas7 group RAMPs (GSU0053), and the other one is an apparent fusion of Cas5 and Cas6 group RAMPs (GSU0054). The ancestral version of these systems could be similar to that present in *Methanosarcina barkeri*, with a probable homolog of Cas8 (inferred Cas8 family protein with characteristic α-helical domain at C-terminus, which could also include fusion to the small subunit). Several CRISPR-Cas systems (e.g., in *Geobacter sulfurreducens*) encode a protein that appears to be a deteriorated version of Cas8 as suggested by the presence of an α-helical C-terminal domain and the location in the operon. In several other genomes, for example, *Bifidobacterium animalis*, there are no traces of a Cas8-like subunit. The additional gene in this operon (Csb3 family) resembles RAMPs of the Cas6 family by secondary structure prediction and several motifs; however, this protein also contains a C-terminal extension resembling the α-helical region present in Cas8 family proteins. The variant of the Subtype I-F system in *Photobacterium profundum* contains three genes that are clearly orthologous to Cas1, Cas2/Cas3 fusion, and Cas6f of the I-F system, respectively; however, two additional genes in this system encode proteins (PBPRB1993 and PBPRB1992) that show no detectable sequence similarity to any known protein families (Fig. 13.10b). By length and the position in the operon, these proteins resemble Csy2 and Csy3, respectively. The predicted secondary structures of these proteins are also compatible with the RAMP structure but not with that of the Cas8 family (no α-helical domain). Thus, these proteins might belong to the Cas5 and Cas7 groups, respectively. The *cas8* (large subunit) gene is absent in this system, which is predicted to be active on the basis of the presence of a large array of CRISPR repeats in the genome.

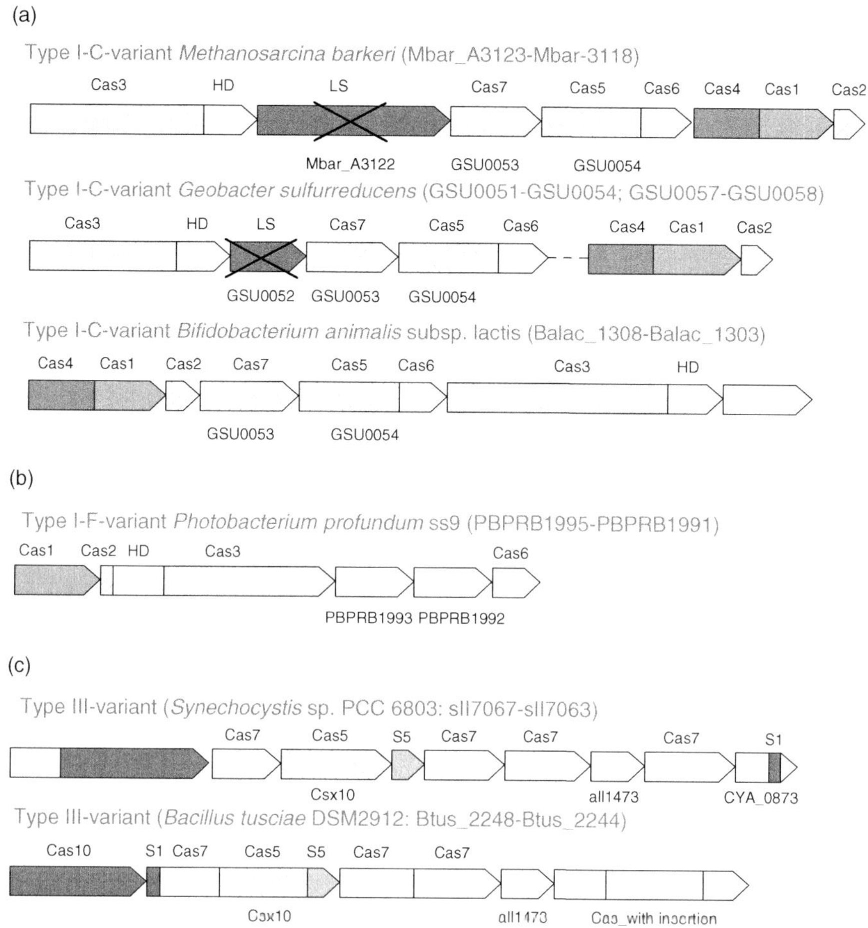

Figure 13.10 Unusual CRISPR-Cas systems. (a) Type I-C-variants with GSU0054 (or GSU0053) signature gene. (b) Type I-F-variant. (c) Type III-variant. (*See insert for color representation of the figure.*)

Some variants of the Subtype III-B system encompass the signature Csx10 family, which belongs to the Cas5 group of RAMPs (Fig. 13.10c). Another feature of this system is the presence of a protein of all1473 family, which does not show any similarity to known Cas protein families but the predicted secondary structure resembles that of the RAMPs. These systems also contain the ribosomal protein S1 domain (the OB fold that forms two distinct fusions). In some systems (e.g., in *Bacillus tusciae*), several additional fusions occurred, mostly between adjacent genes in the operon (Fig. 13.10c). The Cas10 homolog in the latter systems lost the HD domain but retained all catalytic residues of the Palm domain.

Comparative analysis of these unusual variants of CRISPR-Cas system architectures could shed additional light on CRISPR-Cas evolution as discussed in the next section.

13.5.3 An Evolutionary Scenario for the Origin of CRISPR-Cas Systems

Taken together, the findings described here allow us to propose a simple scenario for the origin and the major stages in the evolution of the CRISPR-Cas system (Fig. 13.11). The primary observations that contribute to this reconstruction of CRISPR-Cas evolution are

1. the demonstration that Cas7 proteins represent a distinct group of RAMPs;
2. classification of all RAMPs into three major groups, Cas5, Cas6, and Cas7;
3. the unification, even if more tentative, of Cas8 and Cas10 in the CRISPR polymerase family (large subunits of CRISPR-Cas systems);
4. the tentative unification of small, Csm2-like subunits.

Taking into account these newly discovered unifying connections between the Cas proteins, comparison of the gene compositions and operon organizations of the three major types and 12 subtypes of CRISPR-Cas systems allows us to reconstruct the ancestral forms with some confidence.

The ancestral functional CRISPR-Cas system probably resembled Subtype III-A and consisted of six or seven genes, namely, the two universal *cas* genes, *cas1* and *cas2* ("information processing" subsystem involved in the adaptation phase) along with four or five additional genes that comprised the "executive" subsystem (Cascade complex) involved in crRNA processing and interference. The "executive" module included the large subunit (Cas10/Cas8, or the CRISPR polymerase), the small subunit (an α-helical protein or domain enriched in positively charged and aromatic amino acids), and two or three RAMPs (of the Cas5, Cas6, and Cas7 groups). Given that Cas5 and Cas6 are structurally similar and considering that Cas5 probably substitutes for Cas6 in Subtype I-C, the ancestral system could have contained only one protein representing these two families. Most of the ancestral components are retained in many extant CRISPR-Cas subtypes, in particular, the Type III systems that show relatively little variation. In the most parsimonious scenario, only a few evolutionary events suffice to explain the emergence of Type I and Type III systems with their subtypes (Fig. 13.11)

The key events that apparently gave rise to Type I CRISPR-Cas systems include the acquisition of the helicase Cas3 and the PD-D/ExK family nuclease Cas4; inactivation of the Palm domain of Cas10 protein that yielded Cas8; and fission of HD domain and Cas10 followed by fusion of HD domain with the Cas3-like helicase. The preservation of 6–7 ancestral components in most of the Type I and Type III CRISPR-Cas systems suggests tight structural and functional links among these proteins. However, a degree of independence between the "informational" and "executive" modules has been reported previously (Haft

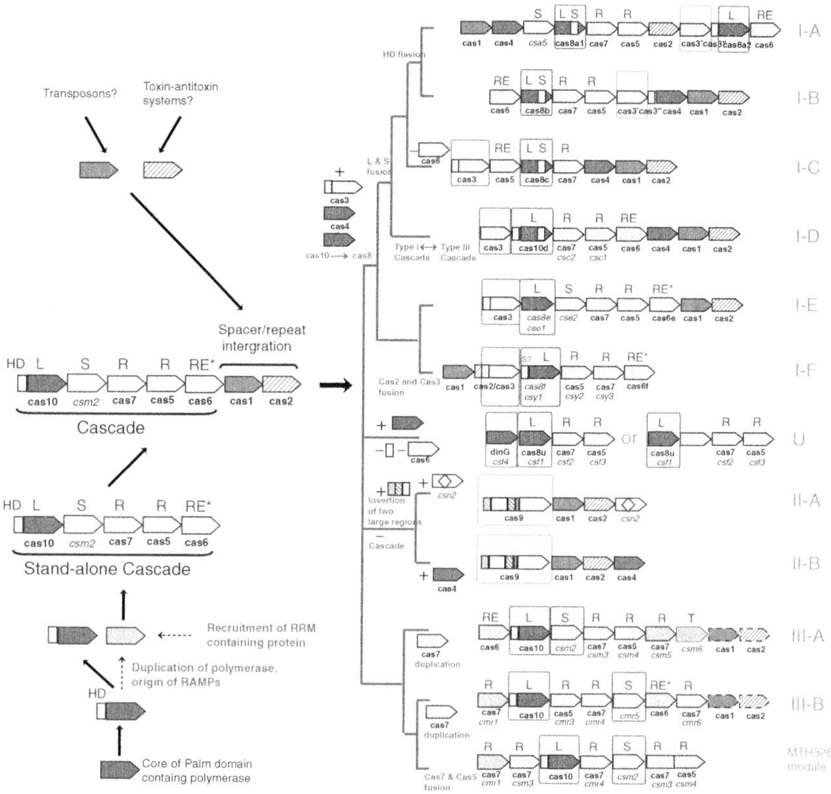

Figure 13.11 Evolutionary scenario for the origin of CRISPR-Cas systems. Homologous genes are color-coded and identified by a family name (names follow the classification from (Makarova et al., 2011b)). Names in bold are proposed systematic names including those proposed in this work; "legacy names" are in regular font. The signature genes for CRISPR-Cas types are shown within green boxes, and for subtypes within red boxes. The bold letters above the genes show major categories of Cas proteins: L, large Cascade subunit; S, small Cascade subunit; R, RAMP Cascade subunit; RE, RAMP family RNase involved in crRNA processing (experimentally characterized nucleases shown be asterisks); T, transcriptional regulator. Genes coding for inactivated (putative) polymerases are indicated by crosses. Major evolutionary events are shown in the corresponding branches. Broken lines denote alternative evolutionary scenarios for the origin of RAMPs. (*See insert for color representation of the figure.*)

et al., 2005; Makarova et al., 2006; Makarova et al., 2011b). In particular, Type III "executive" modules (Cascades) are often encoded separately (not in proximity to *cas1* and *cas2* genes) and often occur in a genome along with Type I and/or Type II CRISPR-Cas systems. Furthermore, Cas1 sequences from Type III systems are not monophyletic in the phylogenetic tree (Makarova et al., 2011b), suggesting that Type III "executive" modules have combined with diverse "informational modules" on multiple occasions. This is a likely evolutionary scenario for Subtype I-D in which the Cascade complex (especially Csc2, a RAMP of the Cas7 group) resembles the Type III counterpart rather than other Type I Cascades. Interestingly, the HD domain in this subtype is associated with the large subunit (Cas10d) rather than with Cas3, again similarly to Type III rather than to other Type I systems. However, the HD domain of Subtype I-D systems does not show the circular permutation that is characteristic of the HD domain fused with Cas10 in Type III systems. Thus, in this case, the similarity of domain architectures seems to be convergent, that is, the HD domain in Subtype I-D systems probably was translocated from *cas3* to inactivated *cas10* (or fused with the latter if the ancestral form was a stand-alone HD domain).

There are currently no archaeal or bacterial genomes that would possess the "information processing" module but not the "executive" module of the CRISPR-Cas system. Although involvement of Cas1 in various repair processes has been suggested by recent experiments (Babu et al., 2011), this tight linkage indicates that the primary functions of Cas1 and Cas2 depends on the Cascade complex (the "executive module"). In contrast, "Cascade only" systems (Type U) that are not associated with CRISPR arrays have been identified, suggesting the intriguing possibility that some variants of Cascade might function as an independent defense system, without relying on Cas1, Cas2, or CRISPR arrays for the acquisition of spacers. Although the source of RNA guides for such a system is unclear, an interesting possibility is that this version of Cascade might recognize alien DNA molecules and process nascent alien mRNA to generate RNA guides; such mechanism would be directly analogous to the siRNA branch of the eukaryotic RNA interference systems (Csorba et al., 2009). From the evolutionary perspective, stand-alone Cascades could be one of the antecedents of CRISPR-Cas systems.

The ancestor of the CRISPR polymerase (Cas10) could have originated from an ancient Palm domain polymerase, such as RT. On the basis of a number of derived shared characters, the CRISPR polymerase has been classified as a member of a distinct group of Palm domain proteins that also includes Thg1-type 3′- 5′ nucleic acid polymerases and adenylate and diguanylate cyclases (Anantharaman et al., 2010). The association with the HD domain probably goes deep into the evolutionary past given that HD family hydrolases are also commonly associated with GGDEF family diguanylate cyclases (Anantharaman et al., 2010; Galperin, 2006). The ancestral function of the CRISPR polymerase that was probably associated with the HD hydrolase domain could potentially involve a distinct form of signal transduction, a role in repair and/or in antivirus defense. The latter possibility seems attractive given the tight association of this protein with the CRISPR-Cas systems.

Genomic islands, in which viral defense, mobile elements, and stress response genes, such as TA systems, are often present together, are likely to be "melting pots" for the emergence of new functional systems through recombination, duplication, and HGT (Kawano et al., 2007; Makarova et al., 2011c; Makarova et al., 2009b). It appears likely that the CRISPR-Cas systems evolved in such genomic environments, by combination of distinct mobile elements. The origin of RAMPs remains an enigma: these highly diverged RRM-domain proteins possess shared derived characters that are strongly suggestive of their monophyly (such as the presence of a glycine-rich loop and a conserved histidine implicated in catalysis in numerous RAMPs) but show no significant similarity to any other proteins. An intriguing possibility is that there is a direct evolutionary connection between the CRISPR polymerase and the RAMPs given that the cores of all these proteins consist of RRM domains. The first RAMP proteins could have emerged by duplication of an inactivated polymerase followed by rapid evolution that involved the emergence of the endoribonuclease catalytic center. The ancestral RAMP might have resembled Cas7 proteins that contain a single RRM domain with structural embellishments along with a Zn-finger domain (in a subset of the Cas7 proteins), and so resemble polymerases in their domain architecture. Furthermore, several CRISPR-Cas systems apparently remain functional despite having a highly degraded form of the large subunit (Type U system) or lacking the large subunit altogether in some variants of Subtype I-C and Subtype I-F (Fig. 13.10b), suggesting that RAMPs might be able to substitute for the function of large subunits. The Cas6 and Cas5 groups of RAMPs could have subsequently evolved from the Cas7-like RAMPs. This scenario seems plausible considering that RAMP duplications, including tandem duplications and fusions, are often present in CRISPR-Cas loci, especially among the Type III systems in which Cas7 group RAMPs are particularly prone to duplication. Interestingly, in both Type I Cascade complexes that have been characterized in detail, those from *E. coli* and *S. solfataricus* Jore, 2011 #246;Lintner, 2011 #260}, the Cas7 subunit is present in multiple copies. In Type III Cascades, these homo-oligomers appear to be replaced by hetero-oligomers made of paralogous Cas7 proteins. Furthermore, lineage-specific (and hence relatively recent) inactivation of the CRISPR polymerase (Cas10) was detected in some Type III MTH326-like systems (Fig. 13.11). All these observations attest to the dynamic character of the evolution of CRISPR-Cas systems and seem to lend plausibility to the route of evolution from the CRISPR polymerase to the RAMP-based Cascade complexes (Fig. 13.11). Nevertheless, this scenario remains speculative given the absence of specific similarity between the RAMPs and CRISPR polymerases. An alternative involving recruitment of a different RRM-domain protein as the ancestral RAMP gene cannot be ruled out; identification of this potential ancestor of the RAMPs is an extremely difficult task given the high sequence divergence of these proteins, which implies rapid evolution.

The CRISPR polymerase and the entire ancestral, Subtype III-A-like CRISPR-Cas system most likely evolved in hyperthermophilic archaea. Indeed, this system, in particular the *cas10* gene, is present in a substantial majority of archaea and can be inferred to have been present in the last archaeal common ancestor (LACA) (Makarova et al., 2007) although a later origin with subsequent dissemination via HGT cannot be strictly ruled out. By contrast, Type III CRISPR-Cas systems are much less common in bacteria and often contain variants of Cas10 that are predicted to be inactivated (Makarova et al., 2011b). Similarly to most antiviral defense systems, CRISPR-Cas is prone to HGT and could have rapidly spread among bacteria. Notably, many thermophilic bacteria possess Type III systems, which might have started the dissemination of CRISPR-Cas among bacteria. The active Cas10 could be particularly beneficial in thermal environments, in agreement with the previous observations that identified Cas10 as a prominent genomic determinant of the thermophilic life style (Makarova et al., 2002; Makarova et al., 2003).

The close association between Cas1 and Cas2 is more difficult to explain in terms of function or evolution. Given that Cas1 is a DNAse with a Holliday junction resolvase-like activity (Babu et al., 2011; Wiedenheft et al., 2009), the prediction is that this protein functions as a recombinase and integrase at the spacer acquisition stage. These activities are typical of transposable elements, so the origin of Cas1 from a transposon appears likely. The endoribonuclease Cas2 might have evolved from another class of equally widespread mobile elements, namely, TA systems. Cas2 is yet another RRM-domain-containing component of CRISPR-Cas systems that is homologous to VapDHi, the toxin in the two-component TA system vapDHi/VapX (Daines et al., 2004; Makarova et al., 2011a; Makarova et al., 2006). It remains unclear whether Cas1 and Cas2 ever formed a distinct two-gene unit or have independently joined the evolving CRISPR-Cas system.

Type II CRISPR-Cas systems are the only group for which the origin of Cascade complex components could not be confidently inferred. Nevertheless, experimental data suggests that Type II systems in general function similarly to the Cascade complexes of Type I and Type III systems (Garneau et al., 2010). Of the three types of CRISPR-Cas systems, the Type II systems have changed the most compared to the inferred ancestral form during which the genes encoding the subunits of the ancestral Cascade complex as well as the large (polymerase) and small subunits appear to have been replaced by a single large, multidomain protein, Cas9, which contains two unrelated nuclease domains (Fig. 13.9) and appears to be responsible for both the CRISPR transcript processing and interference.

13.5.4 Regulatory and Auxiliary Components of CRISPR-Cas Systems

Regulation of the CRISPR-Cas systems is poorly understood. Genomic context analysis suggests that a distinct family of Cas proteins represented by COG1517 (Table 13.2) includes regulators of expression of CRISPR-Cas systems (Makarova et al., 2006). The structures of several proteins in this family have been solved, and a DNA-binding winged HTH domain, likely involved in transcriptional regulation, has been identified in all of these (Lintner et al., 2010)(Fig. 13.12). The core domain of COG1517 proteins is a unique variation of the dinucleotide-binding domain (Rossmann-like fold), which is implicated in the binding of an unidentified ligand(s) (Lintner et al., 2010). In some cases, additional domains are fused to the core two-domain protein. In the case of VC1899 from *Vibrio cholerae* (PDB: 1xmx), the additional domain is a PD-D/ExK superfamily endonuclease. This particular subfamily of the COG1517 family is not strongly associated with

TABLE 13.2 Regulatory and Auxiliary Components of CRISPR-Cas Systems

Current Name	CRISPR-Cas System Type or Subtype	Structure (PDB Code)	Other Family Names	Representatives	Comment and References
cmr7	III-B	2X5Q		SSO1986	A component of cmr complex in *S. solfataricus;* has a unique fold. Specific to Sulfolobales
csn2	II-A	3S5U	SPy1049-like	SPy1049	dsDNA-binding protein forming a tetrameric ring; inactivated ATPase homolog
	II-U			stu0660	Distant homolog of *csn2*
csm6	III-A	2WTE	COG1517	SSO1445, APE2256	HTH-type transcriptional regulator; often fused to COG1517-like domain
csx16	III-U		*VVA1548*	VVA1548	~100 aa protein; often seen in proximity to COG1517
csx3	III-U			AF1864	~100 aa domain, in some cases fused to COG1517 family domains
csx1	III-U	1XMX, 2I71	COG1517, COG4006; *csa3, csx2, NE0113*, DxTHG motif *TIGR02710*	PF1127, MJ1666, TM1812, NE0113	Some are fused to HTH domain (see COG1517/HTH); some proteins have the domain duplication; some have a fusion with HTH and PD-D/ExK family nuclease domain; domain appears to have a Rossmann-like fold
csx15	III-U		TTE2665	TTE2665	~130 aa protein, no prediction; some are fused to AAA ATPase domain
			COG2378	sll7009	HTH-type transcriptional regulator, containing an additional C-terminal ligand-binding domain

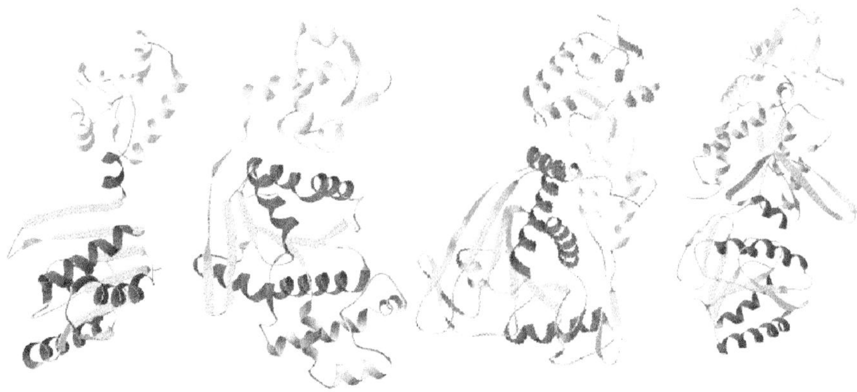

Figure 13.12 COG1517 domain-containing proteins structures. The cartoon shows structures of proteins that belong to the COG1517 superfamily.

CRISPR-Cas systems and is represented also in genomes lacking CRISPR-Cas, suggesting that this protein can perform an independent role, most likely in antiviral defense, given the presence of the nuclease domain (Makarova et al., 2006; Makarova et al., 2011b). The role of other additional domains (e.g., PDB: 2I71) and stand-alone genes often found in the vicinity of COG1517 (Table 13.2) genes is not yet understood.

Genes encoding HTH domains of other families are also found in CRISPR-Cas neighborhoods. One such family is COG2378, which also contains at least two domains. The expression of CRISPR-Cas systems appears to be regulated not only by specific regulators but also, in some organisms, by global or stress response regulators. For example, for *E. coli* and *Salmonella enterica*, the involvement of Lrp, LeuO, and H-NS has been demonstrated (Medina-Aparicio et al., 2011; Pougach et al., 2010; Pul et al., 2010; Westra et al., 2010).

Csn2 protein, a component of Type II-A CRISPR-Cas systems (Nam et al., 2011), was hypothesized to be an analog of Cas4 homolog, PD-D/ExK superfamily endonuclease, which is located in the same place in the operon on closely related system Type II-B (van der Oost et al., 2009). The crystal structure and biochemical analysis show that Csn2 protein is a dsDNA-binding protein that forms a tetrameric ring (Nam et al., 2011); thus, it is unlikely that the function of Csn2 is related to that of Cas4. Oligomerization of Csn2 is regulated by Ca^{2+} ions (Nam et al., 2011). Csn2 consists of an $\alpha\beta$ and an α-helical domain and the helical domain binds Ca^{2+} ions (Nam et al., 2011). Neither of these domains could be assigned to any known fold. Given that there are no indications that Csn2 is a nuclease, it appears most likely that this protein is an auxiliary component of Type II-A systems but not any other CRISPR-Cas systems. The same reasoning seems to apply to Cas4 in Type II-B system. Indeed, there are multiple examples of the Type II system consisting of only three genes, *cas1*, *cas2*, and *cas9* (e.g., NMA0629-NMA0631 from *Neisseria meningitidis* Z2491). Some Type II systems contain a distant homolog of Csn2 that belong to stu0660 family (Table 13.2). Although the exact role of Csn2 proteins remains unclear, the fact that the homologs of these proteins are not found outside Type II CRISPR-Cas systems, a unique involvement of Csn2 in the function of these systems appears likely.

13.6 CONCLUSIONS

The CRISPR-Cas systems are extremely variable in their gene composition, and most of the *cas* genes evolve fast compared to other genes in prokaryotes. Owing to this rapid evolution, comparative analyses of the Cas protein sequences and structures present a history of progressive detection of increasingly subtle relationship leading to unification of protein families previously thought to be unrelated (Haft et al., 2005; Makarova et al., 2002; Makarova et al., 2006; Makarova et al., 2011b). At this stage, it has become clear that Cas proteins can be classified into no more than a dozen major superfamilies including the Cas1–Cas10 proteins, another group of small subunits (perhaps to be denoted Cas11) and additionally a few regulatory protein families such as Csm6 (Table 13.1).

The central structural unit of the CRISPR-Cas systems is the RRM domain that is present in numerous Cas proteins in a striking variety of structurally and functionally distinct forms. The RRM domains reach extreme diversity of enzymatically active and inactivated versions in the RAMP superfamily and in addition are also present in Cas10 (CRISPR polymerase) and Cas2. Given the extensive diversification of RRM domains within the CRISPR-Cas systems, it appears likely that additional, barely recognizable RRM domains exist in poorly characterized Cas proteins such as the large subunits of Cas3-containing systems that might be inactivated derivatives of the CRISPR polymerase.

The comparative analyses of Cas protein sequences and structures and genomic organizations of the CRISPR-Cas loci have introduced considerable order to the seemingly intractable diversity of CRISPR-Cas systems. Three major types and 10 subtypes of CRISPR-Cas systems have been delineated, each with its own signature gene(s). Moreover, these comparative analyses suggested a simple scenario for the origin and evolution of the CRISPR-Cas machinery in hyperthermophilic archaea. Under this scenario, the CRISPR-Cas systems originated as a large protein that combined the polymerase and HD hydrolase domain and might have functioned as a stand-alone antivirus defense system. The next step of evolution might have involved duplication of the RRM portion of the polymerase followed by inactivation that yielded the ancestral, Cas7-like RAMP or recruitment of a distinct RRM-domain protein that became the ancestral RAMP. Regardless of the exact origin of the ancestral RAMP, it has undergone a series of additional duplications and rapid diversification that yielded the stand-alone Cascade complex. The formation of the ancestral CRISPR-Cas system was then completed through the unification of Cascade with Cas1 and Cas2. The central theme of this scenario is the origin of the components of the CRISPR-Cas system from different classes of mobile elements. Other prokaryotic defense systems such as restriction–modification (Fukuda et al., 2008; Kobayashi, 2001) and TA systems (Makarova et al., 2009a; Van Melderen and Saavedra De Bast, 2009) also comprise of such elements, indicating a major trend in the relationships between prokaryotes, viruses that infect them, other classes of selfish elements, and defense mechanisms.

Most of the *cas* genes evolve fast, apparently because of the arms race between the adaptive immunity systems and viruses, resulting in extreme sequence divergence (Takeuchi et al., 2012). Therefore, identification of domains in Cas proteins and classification of these proteins into families and superfamilies is a nontrivial task that requires detailed comparative analysis of the sequences and structures of these proteins using the most sensitive available methods. Extensive comparison of the CRISPR-Cas modules led to the classification of the CRISPR-Cas systems into three types (I, II, and III) that are characterized by distinct signature of *cas* genes and patterns of operon architecture. The Cas proteins have been classified into approximately 30 families, some of which have been further unified in several superfamilies. The CRISPR-Cas system is centered around the RRM) domain. The remarkably diverse RRM domains comprise the core structure of the numerous RAMP proteins, which form the scaffold of the Cascade complex, Cas2 protein, and the CRISPR-associated polymerase/cyclase protein. In addition to the numerous RRM domains, the Cas proteins contain a distinct Superfamily II helicase domain and several diverse nuclease domains. Detailed comparative analysis of the sequences and structures of Cas proteins yielded a simple scenario for the origin of the three types of CRISPR-Cas systems.

ACKNOWLEDGMENT

KSM and EVK are supported by intramural funds of the US Department of Health and Human Services (to the National Library of Medicine, NIH).

REFERENCES

Anantharaman, V., Iyer, L.M., and Aravind, L. (2010) Presence of a classical RRM-fold palm domain in Thg1-type 3′- 5′nucleic acid polymerases and the origin of the GGDEF and CRISPR polymerase domains. Biol Direct, 5, 43.

Aravind, L., Makarova, K.S., and Koonin, E.V. (2000) SURVEY AND SUMMARY: holliday junction resolvases and related nucleases: identification of new families, phyletic distribution and evolutionary trajectories. Nucleic Acids Res, 28, 3417–3432.

Aravind, L., Mazumder, R., Vasudevan, S., and Koonin, E.V. (2002) Trends in protein evolution inferred from sequence and structure analysis. Curr Opin Struct Biol, 12, 392–399.

Babu, M., Beloglazova, N., Flick, R., Graham, C., Skarina, T., Nocek, B., Gagarinova, A., Pogoutse, O., Brown, G., Binkowski, A. et al. (2011) A dual function of the CRISPR-Cas system in bacterial antivirus immunity and DNA repair. Mol Microbiol, 79, 484–502.

Barrangou, R., Fremaux, C., Deveau, H., Richards, M., Boyaval, P., Moineau, S., Romero, D.A., and Horvath, P. (2007) CRISPR provides acquired resistance against viruses in prokaryotes. Science, 315, 1709–1712.

Beloglazova, N., Brown, G., Zimmerman, M.D., Proudfoot, M., Makarova, K.S., Kudritska, M., Kochinyan, S., Wang, S., Chruszcz, M., Minor, W. et al. (2008) A novel family of sequence-specific endoribonucleases associated with the clustered regularly interspaced short palindromic repeats. J Biol Chem, 283, 20361–20371.

Bolotin, A., Quinquis, B., Sorokin, A., and Ehrlich, S.D. (2005) Clustered regularly interspaced short palindrome repeats (CRISPRs) have spacers of extrachromosomal origin. Microbiology, 151, 2551–2561.

Brouns, S.J., Jore, M.M., Lundgren, M., Westra, E.R., Slijkhuis, R.J., Snijders, A.P., Dickman, M.J., Makarova, K.S., Koonin, E.V., and van der Oost, J. (2008) Small CRISPR RNAs guide antiviral defense in prokaryotes. Science, 321, 960–964.

Carte, J., Wang, R., Li, H., Terns, R.M., and Terns, M.P. (2008) Cas6 is an endoribonuclease that generates guide RNAs for invader defense in prokaryotes. Genes Dev, 22, 3489–3496.

Csorba, T., Pantaleo, V., and Burgyan, J. (2009) RNA silencing: an antiviral mechanism. Adv Virus Res, 75, 35–71.

Daines, D.A., Jarisch, J., and Smith, A.L. (2004) Identification and characterization of a nontypeable Haemophilus influenzae putative toxin-antitoxin locus. BMC Microbiol, 4, 30.

Fukuda, E., Kaminska, K.H., Bujnicki, J.M., and Kobayashi, I. (2008) Cell death upon epigenetic genome methylation: a novel function of methyl-specific deoxyribonucleases. Genome Biol, 9, R163.

Galperin, M.Y. (2006) Structural classification of bacterial response regulators: diversity of output domains and domain combinations. J Bacteriol, 188, 4169–4182.

Garneau, J.E., Dupuis, M.E., Villion, M., Romero, D.A., Barrangou, R., Boyaval, P., Fremaux, C., Horvath, P., Magadan, A.H., and Moineau, S. (2010) The CRISPR/Cas bacterial immune system cleaves bacteriophage and plasmid DNA. Nature, 468, 67–71.

Gesner, E.M., Schellenberg, M.J., Garside, E.L., George, M.M., and Macmillan, A.M. (2011) Recognition and maturation of effector RNAs in a CRISPR interference pathway. Nat Struct Mol Biol, 18, 688–692.

Grissa, I., Vergnaud, G., and Pourcel, C. (2007) The CRISPRdb database and tools to display CRISPRs and to generate dictionaries of spacers and repeats. BMC Bioinformatics, 8, 172.

Haft, D.H., Selengut, J., Mongodin, E.F., and Nelson, K.E. (2005) A guild of 45 CRISPR-associated (Cas) protein families and multiple CRISPR/Cas subtypes exist in prokaryotic genomes. PLoS Comput Biol, 1, e60.

Hale, C.R., Zhao, P., Olson, S., Duff, M.O., Graveley, B.R., Wells, L., Terns, R.M., and Terns, M.P. (2009) RNA-guided RNA cleavage by a CRISPR RNA-Cas protein complex. Cell, 139, 945–956.

Han, D. and Krauss, G. (2009) Characterization of the endonuclease SSO2001 from Sulfolobus solfataricus P2. FEBS Lett, 583, 771–776.

Han, D., Lehmann, K., and Krauss, G. (2009) SSO1450–a CAS1 protein from Sulfolobus solfataricus P2 with high affinity for RNA and DNA. FEBS Lett, 583, 1928–1932.

Haurwitz, R.E., Jinek, M., Wiedenheft, B., Zhou, K., and Doudna, J.A. (2010) Sequence- and structure-specific RNA processing by a CRISPR endonuclease. Science, 329, 1355–1358.

Iyer, L.M., Abhiman, S., and Aravind, L. (2008) A new family of polymerases related to superfamily A DNA polymerases and T7-like DNA-dependent RNA polymerases. Biol Direct, 3, 39.

Jackman, J.E. and Phizicky, E.M. (2006) tRNAHis guanylyltransferase adds G-1 to the 5′ end of tRNAHis by recognition of the anticodon, one of several features unexpectedly shared with tRNA synthetases. RNA, 12, 1007–1014.

Jakubauskas, A., Giedriene, J., Bujnicki, J.M., and Janulaitis, A. (2007) Identification of a single HNH active site in type IIS restriction endonuclease Eco31I. J Mol Biol, 370, 157–169.

Jansen, R., Embden, J.D., Gaastra, W., and Schouls, L.M. (2002) Identification of genes that are associated with DNA repeats in prokaryotes. Mol Microbiol, 43, 1565–1575.

Jore, M.M., Brouns, S.J., and van der Oost, J. (2011a) RNA in defense: CRISPRs protect prokaryotes against mobile genetic elements. Cold Spring Harb Perspect Biol, 4, a003657.

Jore, M.M., Lundgren, M., van Duijn, E., Bultema, J.B., Westra, E.R., Waghmare, S.P., Wiedenheft, B., Pul, U., Wurm, R., Wagner, R. et al. (2011b) Structural basis for CRISPR RNA-guided DNA recognition by Cascade. Nat Struct Mol Biol, 18, 529–536.

Kawano, M., Aravind, L., and Storz, G. (2007) An antisense RNA controls synthesis of an SOS-induced toxin evolved from an antitoxin. Mol Microbiol, 64, 738–754.

Kleanthous, C., Kuhlmann, U.C., Pommer, A.J., Ferguson, N., Radford, S.E., Moore, G.R., James, R., and Hemmings, A.M. (1999) Structural and mechanistic basis of immunity toward endonuclease colicins. Nat Struct Biol, 6, 243–252.

Kobayashi, I. (2001) Behavior of restriction-modification systems as selfish mobile elements and their impact on genome evolution. Nucleic Acids Res, 29, 3742–3756.

Kojima, K.K. and Kanehisa, M. (2008) Systematic survey for novel types of prokaryotic retroelements based on gene neighborhood and protein architecture. Mol Biol Evol, 25, 1395–1404.

Lintner, N.G., Frankel, K.A., Tsutakawa, S.E., Alsbury, D.L., Copie, V., Young, M.J., Tainer, J.A., and Lawrence, C.M. (2010) The structure of the CRISPR-sssociated protein Csa3 provides insight into the regulation of the CRISPR/Cas system. J Mol Biol, 405, 939–955.

Lintner, N.G., Kerou, M., Brumfield, S.K., Graham, S., Liu, H., Naismith, J.H., Sdano, M., Peng, N., She, Q., Copie, V. et al. (2011) Structural and functional characterization of an archaeal CASCADE complex for CRISPR-mediated viral defense. J Biol Chem, 286, 21643–21656.

Makarova, K.S., Aravind, L., Grishin, N.V., Rogozin, I.B., and Koonin, E.V. (2002) A DNA repair system specific for thermophilic Archaea and bacteria predicted by genomic context analysis. Nucleic Acids Res, 30, 482–496.

Makarova, K.S., Aravind, L., Wolf, Y.I., and Koonin, E.V. (2011a) Unification of Cas protein families and a simple scenario for the origin and evolution of CRISPR-Cas systems. Biol Direct, 6, 38.

Makarova, K.S., Grishin, N.V., Shabalina, S.A., Wolf, Y.I., and Koonin, E.V. (2006) A putative RNA-interference-based immune system in prokaryotes: computational analysis of the predicted enzymatic machinery, functional analogies with eukaryotic RNAi, and hypothetical mechanisms of action. Biol Direct, 1, 7.

Makarova, K.S., Haft, D.H., Barrangou, R., Brouns, S.J.J., Charpentier, E., Horvath, P., Moineau, S., Mojica, F.J.M., Wolf, Y.I., Yakunin, A.F. et al. (2011b) Evolution and classification of the CRISPR/Cas systems. Nat Rev Microbiol, 9, 467–477.

Makarova, K.S., Sorokin, A.V., Novichkov, P.S., Wolf, Y.I., and Koonin, E.V. (2007) Clusters of orthologous genes for 41 archaeal genomes and implications for evolutionary genomics of archaea. Biol Direct, 2, 33.

Makarova, K.S., Wolf, Y.I., and Koonin, E.V. (2003) Potential genomic determinants of hyperthermophily. Trends Genet, 19, 172–176.

Makarova, K.S., Wolf, Y.I., and Koonin, E.V. (2009a) Comprehensive comparative-genomic analysis of type 2 toxin-antitoxin systems and related mobile stress response systems in prokaryotes. Biol Direct, 4, 19.

Makarova, K.S., Wolf, Y.I., Snir, S., and Koonin, E.V. (2011c) Defense islands in bacterial and archaeal genomes and prediction of novel defense systems. J Bacteriolin press.

Makarova, K.S., Wolf, Y.I., van der Oost, J., and Koonin, E.V. (2009b) Prokaryotic homologs of Argonaute proteins are predicted to function as key components of a novel system of defense against mobile genetic elements. Biol Direct, 4, 29.

Marraffini, L.A. and Sontheimer, E.J. (2008) CRISPR interference limits horizontal gene transfer in staphylococci by targeting DNA. Science, 322, 1843–1845.

Marraffini, L.A. and Sontheimer, E.J. (2009) Invasive DNA, chopped and in the CRISPR. Structure, 17, 786–788.

Marraffini, L.A. and Sontheimer, E.J. (2010) Self versus non-self discrimination during CRISPR RNA-directed immunity. Nature, 463, 568–571.

Medina-Aparicio, L., Rebollar-Flores, J.E., Gallego-Hernandez, A.L., Vazquez, A., Olvera, L., Gutierrez-Rios, R.M., Calva, E., and Hernandez-Lucas, I. (2011) The CRISPR/Cas immune system is an operon regulated by LeuO, H-NS, and leucine-responsive regulatory protein in Salmonella enterica serovar Typhi. J Bacteriol, 193, 2396–2407.

Mojica, F.J., Diez-Villasenor, C., Garcia-Martinez, J., and Almendros, C. (2009) Short motif sequences determine the targets of the prokaryotic CRISPR defence system. Microbiology, 155, 733–740.

Mojica, F.J., Diez-Villasenor, C., Garcia-Martinez, J., and Soria, E. (2005) Intervening sequences of regularly spaced prokaryotic repeats derive from foreign genetic elements. J Mol Evol, 60, 174–182.

Mulepati, S. and Bailey, S. (2011) Structural and biochemical analysis of the nuclease domain of the clustered regularly interspaced short palindromic repeat (CRISPR) associated protein 3(CAS3). J Biol Chem, 286, 31896–31903.

Nam, K.H., Kurinov, I., and Ke, A. (2011) Crystal structure of clustered regularly interspaced short palindromic repeats (CRISPR)-associated Csn2 protein revealed Ca^{2+} dependent double-stranded DNA-binding activity. J Biol Chem, 286, 30759–30768.

Pei, J. and Grishin, N.V. (2001) GGDEF domain is homologous to adenylyl cyclase. Proteins, 42, 210–216.

Pougach, K., Semenova, E., Bogdanova, E., Datsenko, K.A., Djordjevic, M., Wanner, B.L., and Severinov, K. (2010) Transcription, processing and function of CRISPR cassettes in Escherichia coli. Mol Microbiol, 77, 1367–1379.

Pul, U., Wurm, R., Arslan, Z., Geissen, R., Hofmann, N., and Wagner, R. (2010) Identification and characterization of E. coli CRISPR-cas promoters and their silencing by H-NS. Mol Microbiol, 75, 1495–1512.

Rogozin, I.B., Makarova, K.S., Pavlov, Y.I., and Koonin, E.V. (2008) A highly conserved family of inactivated archaeal B family DNA polymerases. Biol Direct, 3, 32.

Sakamoto, K., Agari, Y., Agari, K., Yokoyama, S., Kuramitsu, S., and Shinkai, A. (2009) X-ray crystal structure of a CRISPR-associated RAMP module [corrected] Cmr5 protein [corrected] from Thermus thermophilus HB8. Proteins, 75, 528–532.

Samai, P., Smith, P., and Shuman, S. (2010) Structure of a CRISPR-associated protein Cas2 from Desulfovibrio vulgaris. Acta Crystallogr Sect F Struct Biol Cryst Commun, 66, 1552–1556.

Sapranauskas, R., Gasiunas, G., Fremaux, C., Barrangou, R., Horvath, P., and Siksnys, V. (2011) The Streptococcus thermophilus CRISPR/Cas system provides immunity in Escherichia coli. Nucleic Acids Resin press.

Selengut, J.D., Haft, D.H., Davidsen, T., Ganapathy, A., Gwinn-Giglio, M., Nelson, W.C., Richter, A.R., and White, O. (2007) TIGRFAMs and Genome Properties: tools for the assignment of molecular function and biological process in prokaryotic genomes. Nucleic Acids Res, 35, D260–D264.

Sinkunas, T., Gasiunas, G., Fremaux, C., Barrangou, R., Horvath, P., and Siksnys, V. (2011) Cas3 is a single-stranded DNA nuclease and ATP-dependent helicase in the CRISPR/Cas immune system. EMBO J, 30, 1335–1342.

Soding, J., Remmert, M., Biegert, A., and Lupas, A.N. (2006) HHsenser: exhaustive transitive profile search using HMM-HMM comparison. Nucleic Acids Res, 34, W374–W378.

Sontheimer, E.J. and Marraffini, L.A. (2010) Microbiology: slicer for DNA. Nature, 468, 45–46.

Steitz, T.A. (2004) The structural basis of the transition from initiation to elongation phases of transcription, as well as translocation and strand separation, by T7 RNA polymerase. Curr Opin Struct Biol, 14, 4–9.

Steitz, T.A. and Yin, Y.W. (2004) Accuracy, lesion bypass, strand displacement and translocation by DNA polymerases. Philos Trans R Soc Lond B Biol Sci, 359, 17–23.

Tahirov, T.H., Makarova, K.S., Rogozin, I.B., Pavlov, Y.I., and Koonin, E.V. (2009) Evolution of DNA polymerases: an inactivated polymerase-exonuclease module in Pol epsilon and a chimeric origin of eukaryotic polymerases from two classes of archaeal ancestors. Biol Direct, 4, 11.

Takeuchi, N., Wolf, Y.I., Makarova, K.S., and Koonin, E.V. (2012) Nature and intensity of selection pressure on CRISPR-associated genes. J Bacteriol, 194, 1216–1225.

Tang, T.H., Bachellerie, J.P., Rozhdestvensky, T., Bortolin, M.L., Huber, H., Drungowski, M., Elge, T., Brosius, J., and Huttenhofer, A. (2002) Identification of 86 candidates for small non-messenger RNAs from the archaeon Archaeoglobus fulgidus. Proc Natl Acad Sci USA, 99, 7536–7541.

van der Oost, J., Jore, M.M., Westra, E.R., Lundgren, M., and Brouns, S.J. (2009) CRISPR-based adaptive and heritable immunity in prokaryotes. Trends Biochem Sci, 34, 401–407.

Van Melderen, L. and Saavedra De Bast, M. (2009) Bacterial toxin-antitoxin systems: more than selfish entities? PLoS Genet, 5, e1000437.

Wang, R., Preamplume, G., Terns, M.P., Terns, R.M., and Li, H. (2011) Interaction of the Cas6 riboendonuclease with CRISPR RNAs: recognition and cleavage. Structure, 19, 257–264.

Westra, E.R., Pul, U., Heidrich, N., Jore, M.M., Lundgren, M., Stratmann, T., Wurm, R., Raine, A., Mescher, M., Van Heereveld, L. et al. (2010) H-NS-mediated repression of CRISPR-based immunity in Escherichia coli K12 can be relieved by the transcription activator LeuO. Mol Microbiol, 77, 1380–1393.

White, M.F. (2009) Structure, function and evolution of the XPD family of iron-sulfur-containing 5′–>3′ DNA helicases. Biochem Soc Trans, 37, 547–551.

Wiedenheft, B., van Duijn, E., Bultema, J., Waghmare, S., Zhou, K., Barendregt, A., Westphal, W., Heck, A., Boekema, E., Dickman, M. et al. (2011) RNA-guided complex from a bacterial immune system enhances target recognition through seed sequence interactions. Proc Natl Acad Sci USA, 108, 10092–10097.

Wiedenheft, B., Zhou, K., Jinek, M., Coyle, S.M., Ma, W., and Doudna, J.A. (2009) Structural basis for DNase activity of a conserved protein implicated in CRISPR-mediated genome defense. Structure, 17, 904–912.

Yamaguchi, Y. and Inouye, M. (2009) mRNA interferases, sequence-specific endoribonucleases from the toxin-antitoxin systems. Prog Mol Biol Transl Sci, 85, 467–500.

Yin, Y.W. and Steitz, T.A. (2004) The structural mechanism of translocation and helicase activity in T7 RNA polymerase. Cell, 116, 393–404.

14

FAMILIES OF SEQUENCE-SPECIFIC DNA-BINDING DOMAINS IN TRANSCRIPTION FACTORS ACROSS THE TREE OF LIFE

VARODOM CHAROENSAWAN

Department of Biochemistry, Faculty of Science, Mahidol University, Bangkok, Thailand; Integrative Computational BioScience (ICBS) Center, Mahidol University, Nakhon Pathom, Thailand

SARAH TEICHMANN

EMBL-European Bioinformatics Institute; Wellcome Trust Sanger Institute, Wellcome Trust Genome Campus, Hinxton, Cambridge, UK

CHAPTER SUMMARY

Control of gene expression by transcription factors (TFs) is essential to determine morphological diversity and complexity in all organisms. Sequence-specific DNA-binding TFs are important for genetic regulation because they directly interpret the genetic code through protein–DNA contacts via DNA-binding domains (DBDs). Here, we survey the phylogenetic distributions of DBD families in bacteria, archaea, and eukaryotes.

14.1 INTRODUCTION

Regulation of gene expression is one of the most widely studied areas in modern genetics, beginning with the elegant regulatory mechanism of the *lac* operon

Protein Families: Relating Protein Sequence, Structure, and Function, First Edition.
Edited by Christine Orengo and Alex Bateman.

in *Escherichia coli*, introduced by Jacob and Monod (Jacob and Monod, 1961). Since then, numerous studies have shown that regulation of gene expression is essential to determining physiological and morphological diversity in different species, (e.g., Carroll, 2000; Levine and Tjian, 2003; McClintock, 1956; Struhl, 1999).

On top of genetic variation in DNA sequence, distinctive regulatory mechanisms and their components are employed to control differential expression of genes in different living organisms. To illustrate the point, even though vertebrates are usually considered to be more evolutionarily advanced than invertebrates in terms of diversity of cell types (Vogel and Chothia, 2006), the number of protein-coding genes in vertebrate genomes is only approximately twice that of invertebrates (Levine and Tjian, 2003). This underlines the tremendous influence of the regulation of gene expression in different phylogenetic subsets, in addition to the diversity in the genome template itself.

Transcription is a crucial step of genetic regulation because the genetic information is directly extracted from the DNA template. Sequence-specific DNA-binding TFs are key players in this process because they specifically recognize and directly bind to their cognate sites on DNA. The central role of TFs is illustrated by the many "master regulators" that have been characterized in development and differentiation, as well as reprogramming of fibroblasts into embryonic stem cells, by a few TFs including Oct3/4, Sox2, c-Myc, and Klf4, also known as the *Yamanaka factors* (Takahashi and Yamanaka, 2006).

Because of their function, DBDs are important components of sequence-specific TFs, which mediate the specificity of the TF–DNA interaction. As a result, DBDs are often used to represent TF families (Latchman, 1997; Luscombe et al., 2000), which is appropriate from functional, structural, as well as evolutionary points of view. Characterized TFs belong to many different DBD families that have distinctive amino acid sequences, structural motifs, and thus importantly bind to different regulatory sequences on DNA. Despite their key role, TF and DBD repertoires were reviewed only twice over the past decade from a global perspective across protein families and phylogenetic groups to the best of our knowledge. The first review (Luscombe et al., 2000) surveyed and summarized all known DBDs from a structural perspective, and the second (Charoensawan et al., 2010a) was based on a perspective of genomic repertoire of TFs in different species.

In this chapter, we first demonstrate the current understanding and trends on TF identification and classification in genomes according to DBD families, by summarizing key publications of genome-wide studies of TFs. We also survey TF resources currently available and discuss the criteria and techniques used to attain TF catalogs. To provide an insight into the evolution of TFs in conjunction with the rest of the genes in genomes, we recapitulate the known power law relationship between the number of TFs and the total number of genes in genomes in different organisms across the tree of life (covering Bacteria, Archaea, and Eukaryota superkingdoms). Finally, we provide a comprehensive overview of conserved and lineage-specific DBD families, as well as review interesting domain architectures found in TFs in certain taxonomic groups, using network representation.

14.2 GENOMIC REPERTOIRES OF TFs BASED ON DBD FAMILIES

A great number of TF studies have provided valuable insight into the conserved and specific TFs in different organisms. Initially, most publications focused on prokaryotic genomes owing to the abundant genomic data for these organisms. As an illustration, since the first whole-genome sequence was completed in 1995 for the pathogenic bacterium *Haemophilus influenzae* (Fleischmann et al., 1995), the number of completely sequenced prokaryotic genomes has been increasing exponentially, with many hundreds of bacterial and archaeal genomes now available (Koonin and Wolf, 2008).

Most genome-wide studies of TFs employed various computational methods to search for homologs of DBDs, which were frequently used to represent TF families (more details on TF annotation methods are discussed in later sections). One of the earlier analyses on the repertoire of TF families was focused on four completely sequenced archaeal genomes: *Methanococcus jannaschii*, *Methanobacterium thermoautotrophicum*, *Archaeoglobus fulgidus*, and *Pyrococcus horikoshii* (Aravind and Koonin, 1999). Using sequence profile methods in conjunction with protein structure information, the authors showed that the majority of archaeal DBDs had helix-turn-helix (HTH) structures similar to bacterial HTH domains. This is in contrast to the core archaeal transcriptional machinery, such as basal TFs and RNA polymerases, which are closely related to the eukaryotic systems. More recent articles recapitulated this finding (Coulson et al., 2007; Geiduschek and Ouhammouch, 2005; Peeters and Charlier, 2010).

Other key literature on genome-wide TF analysis includes an earlier study addressing the distribution of 75 TF families across 90 prokaryotes based on the well-characterized set of TFs in *E. coli* K12 (Perez-Rueda et al., 2004). In this study, the reference TFs were derived from a single bacterial species (i.e., *E. coli*), while in a later study (Minezaki et al., 2005), the reference DBD families were collected from eight different archaea and bacteria, with additional DBDs from the Pfam database (Finn et al., 2010) (read more on the Pfam classification in Chapter 2). The authors describe the classification of TFs from 154 complete prokaryotic genomes into 52 TF families. Although the two studies applied different criteria to construct the reference TFs, both confirmed the predominance of HTH DBDs in prokaryotes. Other studies focused on more restricted groups of prokaryotes, such as Brune et al. (2005) and Moreno-Campuzano et al. (2006), which provided TF repertoires for firmicutes and corynebacteria, respectively.

Apart from those on prokaryotic species, numerous studies have been conducted on eukaryotic genomes, although the number of completely sequenced eukaryotic genomes increases more slowly than that of prokaryotic genomes. This is likely due to the larger genome sizes on average, and a large amount of repetitive and noncoding elements. Among the key publications, Riechmann et al.

surveyed specific TF families occurring in four eukaryotic genomes: *Arabidopsis thaliana, Caenorhabditis elegans, Drosophila melanogaster*, and *Saccharomyces cerevisiae* (Riechmann et al., 2000). They pointed out that while DBD families are shared across all three major eukaryotic kingdoms, that is, Metazoa (animal), Fungi, and Viridiplantae (plant), domain combinations of DBDs plus other domains in TFs are often highly kingdom specific. Another analysis (Coulson and Ouzounis, 2003) was performed on six genomes (four genomes as in Riechmann et al., 2000, with the addition of *Homo sapiens*, and *Schizosaccharomyces pombe*). The authors found that each eukaryotic kingdom possesses a number of kingdom-specific TF families in addition to the shared families. Some of the phylogenetically restricted families may be responsible for regulating kingdom-specific processes.

Most other studies focused on particular phylogenetic eukaryotic subgroups. In plants, it was found that the TF families are more diverse than in fungi and animals and also that family expansion and duplication rates are greater in plants (Shiu et al., 2005). This suggests a more rapid and frequent adaptation to selective pressure, perhaps because plants do not have the mobility to avoid stress stimuli. In a study on 62 fungal species, Shelest (Shelest, 2008) found 37 TF families, only 6 ($\sim$16%) of which are fungal specific.

Messina and colleagues (Messina et al., 2004) were among the first to compile a catalog of animal TFs by focusing on human. They used a hidden Markov model (HMM) search based on seed sequences from TFs in TRANSFAC (Matys et al., 2006), InterPro (Mulder et al., 2007), and Flybase (Drysdale, 2008) to define an initial set of human TFs, followed by manual curation. More recently, Vaquerizas and coworkers (Vaquerizas et al., 2009) compiled another human TF repertoire, using annotated DBDs from the InterPro database (Mulder et al., 2007) as a starting point. They then manually curated, added other probable TFs from sources such as TRANSFAC (Matys et al., 2006) DBD (Wilson et al., 2008) and Gene Ontology (Ashburner et al., 2000), and finally analyzed expression patterns and evolutionary conservation of TFs. Beyond human TFs, the mouse TFs are of central interest as a favorite mammalian model organism. There have been several attempts to catalog mouse TFs, including the TFCat database (Fulton et al., 2009), TFdb database (Kanamori et al., 2004), and a list the International Regulome Consortium (IRC), an initiative to characterize the transcription regulatory network in mammalian cells, have been putting together [see review in (Kondro, 2004)].

Protists, such as apicomplexans and ciliates, are eukaryotic unicellular species that are phylogenetically distant from animals, fungi, and plants. Iyer and colleagues (Iyer et al., 2008) were the first to survey the repertoires of TFs and chromatin proteins in parasitic protists. They demonstrated lineage-specific TF families, which could be observed even between close sister lineages. The diversification of TFs was most likely attained via rapid gene loss and lineage-specific expansions. (Read more on eukaryotic TFs and chromatin proteins in Chapter 6).

14.3 TF ANNOTATION RESOURCES CURRENTLY AVAILABLE

Being aware of the importance of TFs on genetic regulation, the community has put a great amount of effort into compiling the systematic collection and classification of annotated TFs in different genomes. These resources are either focused around the DBD families of TFs or include this as an aspect of the database. In addition to the studies discussed in the previous section, we will now survey the TF resources publicly available to date and summarize them in Table 14.1. The TF resources are categorized using two different criteria: (i) categories based on the superkingdoms that the organisms for which the TF annotations are provided belong to; (ii) categories based on the annotation methods used to compile TF catalogs, for instance, whether they are manually or computationally annotated. Note that databases that have ceased to be developed or not been updated since 2004 are not discussed here.

14.3.1 Categories of TF Resources Based on Superkingdoms

14.3.1.1 Prokaryotic TF Resources The continuous development of sequencing technology has allowed a large number of genomes to be sequenced and assembled within a short period, especially for prokaryotic genomes, which are typically smaller and contain fewer repeats and noncoding DNA fractions compared to eukaryotes. This enormously facilitates TF annotation and has resulted in a large number of TF databases across a wide range of prokaryotic lineages.

Early databases tended to serve as integrated resources on transcription of specific model organisms, rather than focusing on TF annotation *per se*. One of the earliest developed and most well-recognized databases is RegulonDB (Gama-Castro et al., 2011). The database provides high quality information relating to the transcriptional regulatory in the gram-negative bacterium *E. coli* K12, including a literature curated list of TFs and operon structures. RegulonDB is regularly updated and many new features have been added since the first release in 1999. In the latest version (RegulonDB 7.0), the authors have integrated signal transduction with transcription as part of large regulatory units known as *genetic sensory response* (Gensor units). Analogous to RegulonDB (Sierro et al., 2008), the DBTBS database provides TFs as well as other transcriptional regulation-related information such as recognition sequences and target genes for the gram-positive bacterium *Bacillus subtilis*.

Apart from RegulonDB and DBTBS, most other prokaryotic TF resources provide TF catalogs for multiple species. The key databases include GTOP_TF (Minezaki et al., 2005), which contains TF repertoires in more than 150 species across archaea and bacteria. In addition, proTF is another database that has recently been developed. Apart from providing an annotated list of specific TFs from over 800 prokaryotic species, the web interface also allows the users to perform comparative genomics analyses to investigate evolutionary relationship between TF families, through multiple sequence alignment and constructing phylogenetic tree. The focus of PRODORIC (Grote et al., 2009) and RegTransBase

TABLE 14.1 TF Resources Currently Available[a]

Database	Annotation	Organism	Link
GTOP_TF	A/M	Prokaryotes	http://spock.genes.nig.ac.jp/~gtop_tf/index2.html/
ProTF	A/M	Prokaryotes	http://centre.bioinformatics.zj.cn/proTF/
PRODORIC	M	Prokaryotes	http://prodoric.tu-bs.de
RegTransBase	A/M	Prokaryotes	http://regtransbase.lbl.gov
ArchaeaTF	A/M	Archaea	http://bioinformatics.zj.cn/archaeatf/
CoryneRegNet	A/M	Corynebacteria	http://www.coryneregnet.de/
cTFbase	A/M	Cyanobacteria	http://cegwz.com/
DBTBS	M	*Bacillus subtilis*	http://dbtbs.hgc.jp/
RegulonDB	M	*Escherichia coli* K-12	http://regulondb.ccg.unam.mx/
TRANSFAC	M	Eukaryotes	http://www.gene-regulation.com/pub/databases.html/
JASPAR	A/M	Eukaryotes	http://jaspar.cgb.ki.se/
ITFP	A	Mammals	http://itfp.biosino.org/itfp/
TFCat	M	Mammals	http://www.tfcat.ca/
FTFD	A/M	Fungi	http://ftfd.snu.ac.kr/
PlanTAPDB	A/M	Plants	http://www.cosmoss.org/bm/plantapdb/
PlantTFDB	A/M	Plants	http://planttfdb.cbi.pku.edu.cn/
PlnTFDB	A/M	Plants	http://plntfdb.bio.uni-potsdam.de/v2.0/
RARTF	A/M	*Arabidopsis thaliana* (thale cress)	http://rarge.gsc.riken.jp/rartf/
SoyDB	A/M	*Glycine max* (soybean)	http://casp.rnet.missouri.edu/soydb/
wDBTF	A/M	*Triticum aestivum T* (wheat)	http://wwwappli.nantes.inra.fr:8180/wDBFT/
TOBFAC	A/M	*Nicotiana tabacum* (Tobacco)	http://compsysbio.achs.virginia.edu/tobfac/
FlyTF	M	*Drosophila melanogaster* (fruit fly)	http://flytf.org/
EDGEdb	A/M	*Caenorhabditis elegans* (worm)	http://edgedb.umassmed.edu/
DBD	A	Cellular organisms	http://www.transcriptionfactor.org/

[a]*Source*: Adapted from Table 1 in Charoensawan et al. (2010a).
The databases can be divided into three categories: (i) Prokaryotic TF databases; (ii) Eukaryotic TF databases; (iii) the DBD databases. The databases that have ceased to be developed or not been updated since 2004 are excluded. Annotation methods are indicated as A, automated; m, manually curated.

(Kazakov et al., 2007) is genetic regulation and regulatory sites, but they also provide lists of annotated TFs. The two databases cover approximately 700 and 130 prokaryotic species, respectively.

Other resources concentrate on more specific prokaryotic lineages. These include CoryneRegNet (Baumbach, 2007), covering TFs and gene-regulatory networks of eight corynebacteria, two mycobacteria and *E. coli* K12, and ArchaeaTF (Wu et al., 2008), dedicated to archaea. Archaea are less well studied than the two other superkingdoms, Bacteria and Eukaryota. By providing TF predictions for 37 archaeal species, the resource is a platform for investigating their genetic regulation mechanisms. cTFbase (Wu et al., 2007) is devoted to TF prediction in cyanobacteria. These are an ancient group of gram-negative bacteria that reside in diverse environments and are believed to be the origin of chloroplasts.

14.3.1.2 Eukaryotic TF Resources TRANSFAC (Matys et al., 2006) is one of the earliest established eukaryotic TF resources and is mainly focused on the TF binding sites of experimentally characterized sequence-specific TFs. TRANSFAC is available in two versions. The public version can be accessed freely by academic and nonprofit users. However, this version is more than 5 years old and thus out of date. The professional version is regularly updated and contains additional information such as ChIP-chip and ChIP-seq fragments and miRNA data, but requires a license to access. In contrast, JASPAR (Portales-Casamar et al., 2010) is constructed for a similar purpose to TRANSFAC but is a completely open-access resource. It also contains more family-specific information, such as the structural classes of TFs that share binding properties.

Other databases are more specific to eukaryotic kingdoms or lineages. There are several databases for specific animals, including FlyTF (Pfreundt et al., 2010) for *D. melanogaster* (fruit fly); EDGEdb (Barrasa et al., 2007) for *C. elegans* (roundworm); TFdb (Kanamori et al., 2004) for (mouse); ITFP (Zheng et al., 2008) for *H. sapiens, Mus musculus*, and *Rattus norvegicus* (human, mouse, and rat); and TFCat (Fulton et al., 2009) for *H. sapiens* and *M. musculus*. For fungi, to the best of our knowledge, there is only one resource dedicated to fungal TF annotation, FTFD (Park et al., 2008). The database provides TF predictions for over 70 fungal species.

Recently, there has been a burgeoning of plant genome sequences owing to their importance as model organisms (e.g., thale cress), food (e.g., rice, potato, wheat), and renewable energy sources (e.g., corn, sugarcane, oil palm, grebe algae). Along with this, several groups have focused on plant TF annotation and developed a number of databases that provide plant TF collections across different plant species, including PlanTAPDB (Richardt et al., 2007), PlnTFDB (Perez-Rodriguez et al., 2010), and PlantTFDB (Zhang et al., 2011). The last two databases have recently been updated and now cover as many as 19 and 49 plant species, respectively. In addition, there are databases for specific plant species such as RARTF (Iida et al., 2005) and AtTFDB (Davuluri et al., 2003) for *A. thaliana* (thale cress), SoyDB (Wang et al., 2010) for *Glycine max* (soybean), wDBTF (Romeuf et al., 2010) for *Triticum aestivum* (wheat), and TOBFAC (Rushton et al., 2008) for *Nicotiana tabacum* (tobacco).

14.3.1.3 DBD: A Pangenomic TF Prediction Database Centered Around DNA-Binding Domain Families Of the currently available TF resources, the DBD TF database (Kummerfeld and Teichmann, 2006; Wilson et al., 2008) is unique because it is based on a collection of DBD families as diagnostic tools for TF annotation across genomes. This means it is the only resource that covers organisms from all three superkingdoms of life (Bacteria, Archaea, and Eukaryota), currently roughly 1,000 cellular organisms. The database automatically predicts sequence-specific TFs based on the presence of protein domains that are homologous to the reference DBD HMMs, taken from two existing HMM libraries: Pfam (sequence-based) (Finn et al., 2010) and SUPERFAMILY (structure-based) (de Lima Morais et al., 2011). This means other transcription-associated proteins including basal TFs, chromatin-remodeling proteins or sigma factors in prokaryotes are excluded. The reference DBD list was manually curated and undergoes occasional refinement. Since TFs in all species from diverse phylogenetic lineages are predicted by one consistent platform, the DBD database serves as an appropriate resource for exploring the phylogenetic distribution of TF families across the tree of life (Charoensawan et al., 2010a, b).

14.3.1.4 Other Relevant Resources In addition to the TF databases discussed above, there are other resources that concentrate on different aspects but also contain useful information on annotated TFs. For instance, numerous databases and web tools were intended to provide information on TF–DNA interactions and binding site predictions, but some TF annotation can also be found. This type of resources include MATCH (Kel et al., 2003), MAPPER (Marinescu et al., 2005), CORE_TF (Hestand et al., 2008), and CompleteMOTIFs (Kuttippurathu et al., 2011) to name a few. General resources for specific genomes such as FlyBase (Tweedie et al., 2009) and Flymine (Lyne et al., 2007) for fruit flies, Mouse Genome Database (MGD) (Blake et al., 2011) for mouse, and Saccharomyces Genome Database (SGD) (Engel et al., 2010) for *S. cerevisiae* usually contain literature-curated gene annotations, as well as lists of known TFs. Apart from DNA-binding TFs, TcoF-DB (Schaefer et al., 2011) provides a list of human transcription cofactors, the proteins that interact with specific TFs to form transcription initiation complex, but do not directly interact with DNA.

14.3.2 Categories of TF Resources Based on Annotation Methods

14.3.2.1 Manual Literature Curation In terms of the methods used to identify and compile TF catalogs, a variety of criteria and techniques have been used by different groups, ranging from manual literature curation to fully automated using computers (Table 14.1). The first-generation TF databases (e.g.,

RegulonDB, DBTBS, and TRANSFAC) mainly relied on literature curation. There are more recent resources that also contain compilations of manually curated TFs (e.g., PRODORIC, FlyTF, and TFCat). Although some of these databases used computationally predicted TFs from other sources as a starting point, experimental evidence is required for each TF annotation. Thus, these manually curated databases serve as valuable resources for high confidence TFs in specific organisms, but usually involve considerable amount of manpower to construct and maintain them.

14.3.2.2 Sequence-Based Searches of DBD Families Plus Manual Curation

Automatic approaches became more popular as the number of completely sequenced genomes increased rapidly. Computational methods allow rapid annotation of newly sequenced genomes. Initially, pairwise alignment sequence search methods such as BLAST (Altschul et al., 1990) were widely used to detect homologs of known DBDs, and thus predict novel TFs. The initial searches were often followed by literature curation and benchmarking against other studies or resources such as Gene Ontology (Ashburner et al., 2000) or InterPro (Mulder et al., 2007). However, due to the superior sensitivity and specificity in remote homolog detection of the profile-based searches (to be discussed in the next section), this approach became less popular and most of the databases using this approach have now ceased to be developed further. The examples of these second-generation TF databases are BacTregulators (Martinez-Bueno et al., 2004) for bacteria, TrSDB (Hermoso et al., 2004) for eukaryotes, and TFdb (Kanamori et al., 2004) for mouse.

14.3.2.3 Profile-Based Searches of DBD Families Plus Manual Curation

At present, most of the regularly updated TF databases use profile-based searches such as hidden Markov models (HMM) [see review in (Eddy, 1998)] and PSI-BLAST (Altschul et al., 1997) to detect even distantly related members of protein families. The profile-based approaches have a number of advantages, including better sensitivity in recognizing remote homologs of DBDs and automatic classification into families of related TFs (Kummerfeld and Teichmann, 2006). Examples of these third-generation TF databases are GTOP_TF, RATTF, DBD, PlanTAPDB, PlantTFDB, PlnTFDB, cTFbase, and ArchaeaTF. Reference libraries of protein families can either be taken directly from existing projects including Pfam (Finn et al., 2010), SUPERFAMILY (de Lima Morais et al., 2011), PROSITE (Hulo et al., 2008), or compiled from scratch. For most databases, post-automatic manual refinement is performed after the initial computational searches to improve the specificity of the predicted TF lists. False negatives and true positives are normally added and filtered using a variety of post-processing steps, according to experimentally verified evidence.

14.4 GENOMIC REPERTOIRES OF TFs AND THEIR FAMILIES ACROSS THE TREE OF LIFE

14.4.1 TF Abundance in Prokaryotes and Eukaryotes Follows Power Law Increase with Number of Genes

Sequence-specific TFs are a crucial class of proteins because they directly bind to DNA and thus regulate differential expression of genes. Evidently, organisms require a greater proportion of TFs for gene expression regulation as their physiological complexity increases. Previous studies have shown that TFs, among other classes of proteins including proteins that involve cell adhesion, extracellular processes, and signal transduction, are highly and disproportionally expanded in animals, especially vertebrates, compared with unicellular eukaryotes (Tordai et al., 2005; Vogel and Chothia, 2006). In addition, several studies have observed a power law increase of numbers of TFs, with respect to the total numbers of genes in both prokaryotic and eukaryotic genomes (Charoensawan et al., 2010a; Cordero and Hogeweg, 2007; Maslov et al., 2009; Molina and van Nimwegen, 2009; Ranea et al., 2004; Ranea et al., 2005; van Nimwegen, 2003), whereas TF numbers in prokaryotes seem to increase faster with gene numbers than in eukaryotes. Here, we revisit this interesting finding and discuss what could be the implications of this phenomenon.

To demonstrate the relationship between TF abundance and the total number of genes in genomes, we first extracted predicted DNA-binding TFs from the DBD database (Wilson et al., 2008). In this comparative genomic study of TFs, using annotated TFs from the DBD database has a major advantage because TFs in all species from diverse phylogenetic lineages were annotated using one consistent platform. Thus, there is no bias from combining TFs from different resources, which might have been annotated using different methods and criteria. Figure 14.1a shows the total numbers of TFs in the DBD database, *versus* the total number of genes in 449 nonredundant bacterial (shown as blue dots), 49 archaeal (green), and 185 eukaryotic (red) genomes. Note that sigma factors in bacteria and other nonsequence-specific TFs were not included in this data set.

When TF abundance is plotted against the number of genes on a log–log scale, the relationship is linear, which implies a power law relationship between the two variables. In bacteria, as the number of genes becomes larger, the TF expansion strictly follows a power law increase with a gradient of 1.98 (blue trend line in Fig. 14.1a, with $R^2 = 0.87$). The value of this gradient, which is known as *a power law exponent*, close to 2 implies a quadratic increase between TF number and gene number on a normal scale. In other words, a bacterial genome that is twice the size of another one would require approximately fourfold more TFs than the smaller genome to regulate those additional genes. Earlier studies found similar relationships even within individual bacterial phyla (Charoensawan et al., 2010a; Molina and van Nimwegen, 2009). In eukaryotic genomes, there is a similar exponential TF expansion but with a lower exponent (gradient = 1.23, red trend line in Fig. 14.1a) and fitting quality ($R^2 = 0.61$). Nonetheless, exponents

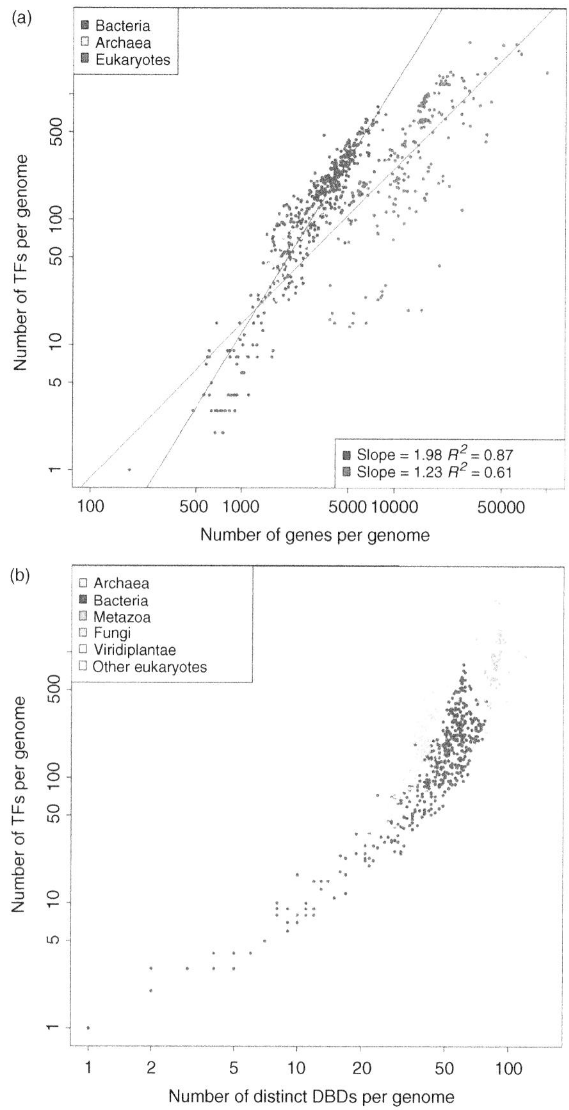

Figure 14.1 (a) Number of TFs against number of genes per genome in bacterial (blue), archaeal (green), and eukaryotic (red) genomes. Each colored dot represents a genome. According to the linear model fit on a log–log scale, TF expansion in bacteria strictly follows a power law increase, with an exponent close to quadratic (power law exponent = 1.98 and $R^2 = 0.87$), whereas TF increase in eukaryotes has a lower exponent as well as degree of correlation (exponent = 1.23 and $R^2 = 0.61$). (b) Number of TFs against number of distinct DBD families (Pfam and SCOP families) per genome. The number of TFs per DBD family rapidly increases in larger genomes in both prokaryotic and eukaryotic species. These multiple TFs belonging to the same DBD families have arisen from series of gene duplication events. Gray dots represent eukaryotic species not belonging to Metazoa, Fungi, and Viridiplantae, such as apicomplexans and euglenozoa. (*See insert for color representation of the figure.*)

greater than one in both superkingdoms mean that the TF repertoire expands faster than linearly for every gene added to the genome in general. We did not perform linear regression on archaeal genomes because only a small number of genomes were available, but noted that archaeal TF seemed to increase in a manner similarly to bacteria. In the next section, we review the probable implications of this observed power law relationship proposed by earlier studies, as well as possible explanations of why TF number in bacteria increases faster than that in eukaryotes.

14.4.2 Why Do Bacterial TF Repertoires Increase at a Higher Rate Than Eukaryotic TF Repertoires?

Although several groups pointed out the power law relationship between TF and gene numbers, to the best of our knowledge, very few actually discussed the possible implications of this observation. From the point of view of metabolic network evolution, Maslov and coworkers (Maslov et al., 2009) have proposed the following explanation. In order to explore and survive in a new environment, an organism requires a new set of TFs to regulate enzymes that allow it to adapt to the new growth conditions and environmental stresses. At the same time, some of the established metabolic enzymes (called the *toolbox* by the authors) can be reused. This may be the explanation for why the number of metabolic reactions or "tasks" as well as TFs required to regulate them increases faster than linearly with the number of genes encoding enzymes and the total number of genes in the genome.

So why do bacterial TFs increase faster than eukaryotic TFs? An obvious explanation is the more complex mechanisms for gene expression control in eukaryotes, compared to the bacterial systems. For instance, eukaryotic genomes contain a greater fraction of noncoding DNA, which has several important roles. First, small interfering RNAs that provide an additional layer of gene regulation at the posttranscriptional level are often encoded in these regions. Secondly, noncoding regions harbor cis-regulatory sequences with more complex binding site architectures than in prokaryotes (Ahnert et al., 2008; Harafuji et al., 2002; Mattick, 2007). In addition to the large regions of noncoding DNA in eukaryotes versus bacteria, it is also worth mentioning a difference in chromatin architecture: eukaryotic DNA is packaged into chromatin repressed in the transcriptional ground state and the promoter is only accessible in the presence of chromatin-remodeling proteins (Struhl, 1999). In multicellular eukaryotes, tissue-specific regulatory circuits are another way to utilize the same transcriptional elements to temporally and spatially control gene expression (Furlong et al., 2001; Odom et al., 2004). These are a few examples of differences between bacteria and eukaryotes, which might enhance eukaryotic gene regulation much beyond the context of TFs and their target genes.

On the other hand, the slow increase in eukaryotic TF numbers might simply be an artifact of less well-characterized TF and DBD families in some eukaryotic lineages, especially plants and fungi. We have shown that the average fraction of genes predicted as TFs in genomes (i.e., TF content) is highest in animals, where approximately 4.7% of genes in genomes are TFs (Charoensawan et al., 2010a). In fungi and plants, the proportion of TFs is between 2% and 2.5% of all genes, which is even lower than the average of 4.2% in bacteria. In unicellular obligate parasites, such as apicomplexa and euglenozoa, for example, *Plasmodium falciparum* (malaria apicomplexan) and *Trypanosoma brucei* (sleeping sickness euglenozoa), this TF content is even lower. Their entire gene repertoires contain less than 0.5% TFs on average. This is most likely because the current understanding on eukaryotic gene regulation is biased toward the study of animal model organisms and because the DBD repertoires in different eukaryotic kingdoms are highly lineage specific (Charoensawan et al., 2010b).

Another interesting implication of the rapid TF increase in bacteria was put forward by Ranea and colleagues (Ranea et al., 2005). In the context of microeconomics, they speculated that bacterial genomes might have saturated their genome size, given the limits of their transcriptional mechanisms. The sharper TF increase with number of genes in bacteria might be linked to the absence of more sophisticated transcriptional mechanisms. This could hinder the organisms from having a larger genome in the sense that a further increase in number of genes would be "economically" ineffective: the extra cost to regulate a gene (that is, additional TFs required) would become prohibitively expensive.

14.4.3 Numbers of DBD Families and Domain Duplication

We have discussed the increase in the number of TFs with the total number of genes. How are these TFs distributed according to their family membership? In Fig. 14.1b, the number of TFs is plotted against the number of distinct DBD families (Pfam and SCOP families classified as DBDs) of each genome on a log–log scale. In small and reduced genomes, the increase in TFs closely follows the increase in families (e.g., in small bacteria, ~50 TFs belong to ~50 DBD families). In larger genomes, there is extensive gene and domain duplication, so that over 1000 TFs belong to just over 100 DBD families in the largest eukaryotic genomes. This result corresponds to previous findings, showing that multiple TFs belonging to the same DBD family have arisen through multiple gene duplication events, followed by protein sequence divergence, and protein domain recombination (Madan Babu and Teichmann, 2003; Perez-Rueda et al., 2004). It is also worth noting that metazoans (pink dots) have the largest diversity in terms of number of DBD families, even in genomes with comparable numbers of TFs in fungi (orange) and plants (yellow), for instance.

14.5 PHYLOGENETIC DISTRIBUTION OF DBD FAMILIES IS HIGHLY LINEAGE-SPECIFIC

Having mentioned the differences in TF repertoires and families between lineages above, we now concentrate on TF and DBD families that are conserved across diverse phylogenetic groups or are specific to certain lineages. As mentioned earlier, sequence-specific TFs contain DBDs, the evolutionary units that mediate the specificity of the TF–DNA interaction. Consequently, the domain-based analysis of TFs is most appropriate for studying TF family conservation. Previous publications on DBD families tended to focus on specific superkingdoms or phylogenetic groups (e.g., Aravind and Koonin, 1999; Perez-Rueda et al., 2004; Riechmann et al., 2000), while here we take a more global view across all superkingdoms.

To systematically access the conservation of DBD families in diverse species across the tree of life, we employed our earlier published automated method for inferring the origin of protein domains, called *taxonomic limits* (Charoensawan et al., 2010b). In brief, DBD family occurrence from the DBD database (Wilson et al., 2008) was combined with taxonomic information from the NCBI taxonomy tree (Benson et al., 2009). In all the species where the DBD family of interest was detected, we traced down the phylogenetic tree to find their last common ancestors. The most frequent last common ancestor was then identified as the taxonomic limit of that particular DBD family.

Figure 14.2 demonstrates the fractions of taxonomic limits at the major taxonomic lineages, when our algorithm was applied to all Pfam families (release 18.0), and to 131 Pfam families classified as DBDs. According to Fig. 14.2, approximately a third (~33%) of all Pfam families have "cellular organisms" as taxonomic limits (i.e., shared by more than one superkingdom). In contrast, only approximately 15% of Pfam DBD families have "cellular organisms" as their taxonomic limits, whereas most of DBDs are specific to superkingdoms, kingdoms (for eukaryotes), or phyla (for bacteria). This evidently suggests that the repertoires of DBD families are more lineage-specific than proteins of other functions.

Within the next few subsections, we discuss in detail interesting DBD families that are phylogenetically conserved, or are specific to particular taxonomic groups (summarized in Tables 14.2 and 14.3), as well as the biological processes they are implicated in. In addition to the taxonomic limit, we calculated the so-called *conservation density*: this is the fraction of species containing the DBD over the total number of species below the taxonomic limit. The logic here is that DBD families that emerged from the same speciation event should be detectable in most of the children species (i.e., conservation density close to 1). In contrast, DBDs present sporadically in taxonomically distant lineages (i.e., conservation density much less than 1) are likely to have been disseminated through horizontal gene transfer or to have gone through multiple gene loss events.

14.6 FEW DBD FAMILIES ARE CONSERVED IN MULTIPLE SUPERKINGDOMS

Out of 131 Pfam families classified as DBDs, our algorithm identified that only 19 of them or approximately 15% had "cellular organisms" as their taxonomic limit. This result is well in line with an earlier study (Wunderlich and Mirny, 2009) that used criteria for TF identification that are different to ours. The authors

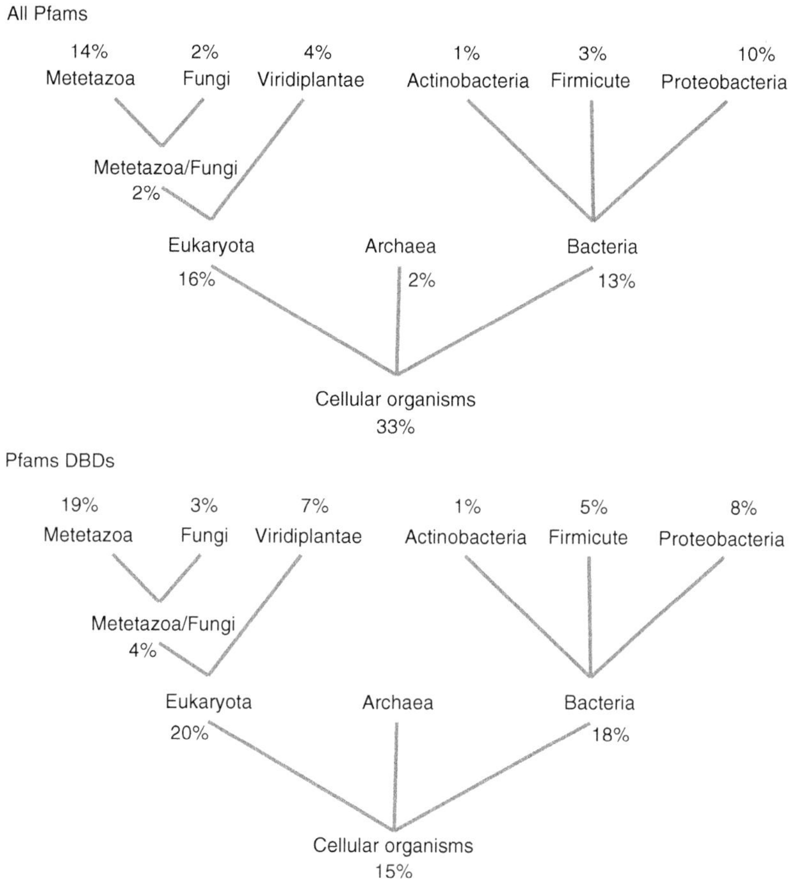

Figure 14.2 The percentages of all Pfam and Pfam DBD families having taxonomic limits assigned to the major taxonomic nodes on simplified taxonomic tree. The number of DBDs families having "cellular organisms" as taxonomic limits (~15%) is greater than those of all Pfam families (~33%).

also reached a similar conclusion that bacteria and eukaryotes share very few DBDs. Interestingly, out of these DBDs having "cellular organisms" as their taxonomic limits, only three (~2% of all Pfam DBDs) are shared by all the three superkingdoms: Archaea, Bacteria, and Eukaryota. These three universally shared DBD families are CSD (cold-shock domain) (Jones et al., 1996), HTH_psq (the pipsqueak protein) (Lehmann et al., 1998), and HTH_3 (Ren et al., 2010). On the other hand, the majority of these families (11 out of 19) are shared by the two prokaryotic kingdoms, Archaea and Bacteria, but not Eukaryota. This result corresponds with the results previously proposed by several groups that archaeal sequence-specific TFs were more similar to those in bacteria, than in eukaryotes, which is discussed in greater detail in the next section.

TABLE 14.2 Examples of Pfam DBD Families Specific to the Bacterial Superkingdom or Nested Bacterial Taxa

Pfam DBD Family	Taxonomic Limit	Conservation Density
FUR	Bacteria	0.88
GerE	Bacteria	0.78
GntR	Bacteria	0.79
HTH_8	Bacteria	0.75
LacI	Bacteria	0.60
LexA_DNA_bind	Bacteria	0.68
MerR	Bacteria	0.86
Pencillinase_R	Bacteria	0.31
Trans_reg_C	Bacteria	0.93
HTH_CodY	Bacteria	0.13
Trp_repressor	Bacteria	0.23
WhiB	Actinomycetales	1.00
CodY	Firmicutes	0.68
CtsR	Firmicutes	0.74
CoiA	Bacilli	0.94
ComK	Bacillales	0.91
FlhC	Proteobacteria	0.22
FlhD	Proteobacteria	0.18
SirB	Proteobacteria	0.30
ROS_MUCR	Proteobacteria	0.22
Crl	Gammaproteobacteria	0.27
MetJ	Gammaproteobacteria	0.55

Taxonomic limits and conservation density are provided.

14.7 PROKARYOTIC DBD REPERTOIRES

14.7.1 DBD Families Shared by Archaea and Bacteria

Across prokaryotes, by far the most common DBD structure is the HTH motif. The majority of prokaryotic DBD families belong to the winged helix structural class (Aravind and Koonin, 1999; Minezaki et al., 2005) and are classified into the HTH clan according to Pfam.

Archaea are thought to be phylogenetically closer to the Eukaryota superkingdom. It is worth noting that the core components of transcription machinery in archaea, such as RNA polymerases and basal TFs, are more closely related to eukaryota than bacteria. However, the archaeal sequence-specific TFs and also the DBD family repertoire are more closely related to bacteria (Aravind and Koonin, 1999; Coulson et al., 2007; Perez-Rueda et al., 2004). The common DBD families could have been established and maintained through multiple horizontal gene transfers, or may belong to an ancestral core preceding the divergence of the two superkingdoms.

Some of the DBDs shared by the two prokaryotic superkingdoms are iron-dependent repressor (Schiering et al., 1995), HTH_5 (arsenic resistance) (Cook

TABLE 14.3 Examples of Pfam DBD Families Specific to the Eukaryotic Superkingdom or Nested Eukaryotic Taxa

Pfam DBD Family	Taxonomic Limit	Conservation Density
CBFB_NFYA	Eukaryota	0.81
GATA	Eukaryota	0.88
HLH	Eukaryota	0.86
HSF_DNA-bind	Eukaryota	0.90
Homeobox	Eukaryota	0.89
SRF-TF	Eukaryota	0.88
E2F_TDP	Eukaryota	0.59
WRKY	Eukaryota	0.13
Tub	Eukaryota	0.61
bZIP_1	Eukaryota	0.91
bZIP_2	Eukaryota	0.90
zf-C5HC2	Eukaryota	0.49
CP2	Fungi/Metazoa group	0.65
Fork_head	Fungi/Metazoa group	1.00
TEA	Fungi/Metazoa group	0.79
CUT	Bilateria	0.99
Ets	Bilateria	1.00
T-box	Bilateria	1.00
PAX	Bilateria	1.00
Pou	Bilateria	1.00
TF_AP-2	Bilateria	0.99
zf-C2HC	Bilateria	1.00
zf-C4	Bilateria	1.00
BESS	Coelomata	0.48
Basic	Coelomata	0.97
GCM	Coelomata	0.95
P53	Coelomata	0.97
RHD	Coelomata	1.00
Runt	Coelomata	1.00
STAT_bind	Coelomata	1.00
Churchill	Chordata	0.81
Fez1	Chordata	0.91
IRF	Chordata	1.00
Myc_N	Chordata	1.00
APSES	Fungi	0.98
MAT_Alpha1	Ascomycota	0.56
Copper-fist	Dikarya	0.85
B3	Viridiplantae	0.81
SBP	Viridiplantae	0.81
zf-Dof	Viridiplantae	0.91
EIN3	Embryophyta	1.00
TCP	Embryophyta	1.00
NAM	Embryophyta	1.00
YABBY	Magnoliophyta	1.00

Taxonomic limits and conservation density are provided.

et al., 1998), MarR (antibiotic resistance) (Alekshun et al., 2001), NikR (nickel responsive regulator) (Schreiter et al., 2003), PadR (phenolic acid metabolism) (Gury et al., 2004), and TrmB (trehalose/maltose sugar-specific regulator) (Lee et al., 2003). These DBD families shared by Archaea and Bacteria regulate specific genes required for adaptation to the environmental stresses except for HTH_1 (LysR family), which participates in amino acid synthesis (Schell, 1993).

14.7.2 DBD Families Shared by all Bacteria Phyla

Out of all 131 Pfam DBD families in this study, 42 (~32%) are specific to bacteria or one of its phyla (Fig. 14.2). Intriguingly, approximately 18% of all Pfam DBDs or more than half of the DBDs specific to the bacterial branch are commonly found across different bacterial lineages. Most of these DBDs are detected in the vast majority of species within the Bacteria superkingdom (conservation densities >0.60). Some of them participate in basic carbon source metabolism such as LacI (Nguyen and Saier, 1995) and GntR (Hillerich and Westpheling, 2006), and others participate in more specific biological processes such as FUR (ferric uptake regulator) (Escolar et al., 1998), MerR (mercury resistance) (Helmann et al., 1989), LexA repressor (DNA repair system) (Campoy et al., 2002), GerE (Lux family, quorum sensing) (Miller and Bassler, 2001), and HTH_8 (Fis family, virulence gene expression) (Goldberg et al., 2001). Conservation densities close to 1 mean the DBDs are observed in nearly all species under the taxonomic limit, as in the case of FUR and GntR (conservation densities of 0.88 and 0.79, respectively), and it can be inferred that the DBDs are most likely inherited from the last common ancestor of all bacterial species.

In contrast, some of the DBD families that are involved in gene regulation under specific environmental stresses have relatively low conservation densities, despite being found in diverse bacterial lineages. This suggests they might have been disseminated through horizontal gene transfer. Examples of DBD families in this group are Pencillinase_R (penicillinase repressor) (Wittman et al., 1993), and Trp_repressor (tryptophan repressor protein) (Otwinowski et al., 1988). Table 14.2 provides examples of DBD families conserved throughout bacterial genomes or are specific to certain bacterial lineages, together with their conservation densities.

14.7.3 Bacterial Phylum-Specific DBD Families

Despite both being gram-positive bacteria, Actinobacteria and Firmicutes do not share any DBD family beyond the ones that are conserved across all bacteria. Since most of the Actinobacteria are filamentous, it makes sense that this bacterial phylum has a DBD WhiB (Kormanec et al., 1998) specific to mycelium formation regulation. A number of DBDs, which control expression of genes in different pathways, are specific to the Firmicutes: CodY GAF-like domain (Guedon et al., 2001), ComK protein (Nanamiya et al., 2003), and Firmicute transcriptional repressor of class III stress genes (CtsR) (Nair et al., 2000), for instance. The sporulation initiation factor (Spo0A) is, however, the most interesting of all.

This family reflects the lifestyle of many Firmicutes that reproduce by forming spores in undesirable conditions (Zhao et al., 2002).

Owing to the greater number of completely sequenced genomes available, Proteobacteria possess more phylum-specific DBDs than any other bacterial lineage (~8% of all DBD families). The DBD families that fall into this category include Crl (fibronectin binding activators) (Arnqvist et al., 1992), ROS/MUCR (virulence region repressor) (Martin et al., 2000), and Met repressor (MetJ, methionine synthesis) (Wu et al., 1993). The FlhC and FlhD TFs have been shown to be global regulators involved in many cellular processes as well as flagella transcriptional activators (Pruss et al., 2001). They are only present in gram-negative Proteobacteria but not in gram-positive Firmicutes and Actinobacteria. These DBD families may be restricted phylogenetically because of their link to the four-support-ring flagella in gram-negative bacteria, as opposed to the two-support-ring flagella in gram-positive bacteria.

14.8 EUKARYOTIC DBD REPERTOIRES

14.8.1 DBD Families Shared by all Eukaryotic Superkingdoms

A relatively smaller proportion of eukaryotic DBDs are ubiquitously present across the eukaryotic kingdoms, compared with a larger fraction of bacterial DBDs detected in all bacterial phyla. Out of all 131 DBD families, 69 (~53%) have the Eukaryota superkingdom or one of its kingdoms as their taxonomic limits (Fig. 14.2). However, approximately 16% of all Pfam DBDs or only approximately 38% of the 69 eukaryotic specific DBDs are common in all eukaryotic kingdoms.

There are a small number of DBD families shared by all three major kingdoms, Metazoa, Fungi, and Viridiplantae. These include most of the zinc finger families (Klug, 1999; Laity et al., 2001), the helix-loop-helix family (HLH) (Littlewood and Evan, 1995) and the basic leucine zipper family (bZIPs) (Amoutzias et al., 2007). It may seem surprising that the Homeobox family, famous for its role in morphogenesis and animal body development (Gehring, 1992; Pavlopoulos and Akam, 2007), is also found throughout eukaryotic organisms, including fungi and plants. A list of examples of DBD families having Eukaryota or one of its nested taxa as their taxonomic limits can be found in Table 14.3.

14.8.2 Eukaryotic Kingdom-Specific DBD Families

The Metazoa (animal) kingdom possesses a considerably larger DBD repertoire than the Fungi and Viridiplantae (plant) kingdoms (Charoensawan et al., 2010b). This reflects the greater morphological complexity and number of body structures in animals, as well as a potential bias toward the study of animal model organisms. The majority of DBDs found in metazoans are present in both vertebrates and invertebrates, for example, STAT (signal transduction) (Kisseleva et al., 2002), T-box (body plan and organogenesis) (Wilson and Conlon, 2002), p53 (cell cycle arrest and apoptosis) (Sutcliffe et al., 2003), and Runt (developmental processes) (Kagoshima et al., 1993). Other families such as interferon regulator factor (IRF, regulation of immunity) (Miyamoto et al., 1988), Churchill (neural development) (Akai and Storey, 2003), and the oncogene Myc family (Facchini and Penn, 1998), are present in vertebrates only. Some of these families are directly linked to the more elaborate immune and nervous systems in vertebrates.

Being phylogenetically closer to Metazoa, Fungi share more DBD families with animals than with plants (Charoensawan et al., 2010b; Riechmann et al., 2000). The DBD families that are common to metazoans and fungi but completely absent in plants include CP2 (developmental processes) (Rodda et al., 2001), Fork_head (embryogenesis) (Hacker et al., 1992), and NDT80/PhoG (cell cycle) (Montano et al., 2002). As pointed out previously (Shelest, 2008), there are a number of DBDs found in Fungi but not in other eukaryotes. Interestingly, not all fungal-specific DBDs are restricted to fungal-specific processes. To illustrate the point, the DBD of Mlu1-box-binding protein MBP1 is mainly involved in regulation of the cell cycle (Machado et al., 1997). Zn2/Cys6 (Zn cluster) has many regulatory roles such as in sugar and amino acid metabolism, cell cycle, as well as drug resistance (MacPherson et al., 2006). In contrast, the Copper-fist domain participates in copper utilization and stress response processes (Keller et al., 2005). MAT_alpha1 and APSES regulate fungal-specific processes as they activate mating-type specific genes (Sengupta and Cochran, 1991) and regulate yeast–hyphal transitions (Wang and Szaniszlo, 2007), respectively. There are a few exceptional cases of bacterial DBDs found in many fungi, such as HTH_AraC (Saviola et al., 1998) and FMN (Flavin mononucleotide) binding domain (Honjo et al., 1990). These families are known to be involved in sporulation regulation and sugar uptake in bacteria, but their role in fungi remains unclear.

Apart from the majority of DBD families found in plants, which are also shared with animals and fungi, there is also a small set of DBD families specific to plants, including AP2/GCC-box-binding domain (activation of defence genes) (Ecker, 1995), SBP (flowering development) (Yamasaki et al., 2004), and WRKY (pathogen defence and biosynthesis of secondary metabolites) (Eulgem et al., 2000). In addition, we observe that a number of DBD families found in the Streptophyta phylum (land plants) are absent in Chlorophyta (green algae). These families are discussed in the next section.

14.8.3 From Uni- to Multicellular Eukaryotes: Additional DBD Families Emerge

Apart from the three major kingdoms, the DBD database also provides TF predictions for many unicellular eukaryotes. Among the unicellular eukaryotic species available, *Monosiga brevicollis* is one of the most interesting organisms as it is the only well-annotated representative of choanoflagellates, the closest known relatives of metazoans (King et al., 2008). Previous studies on the organism have concentrated on its signal transduction mechanisms and found that the species contained a considerable amount of signaling components in common with animals (Manning et al., 2008; Pincus et al., 2008).

Besides the more elaborate signaling machineries, uni- to multicellular transitions also require a greater number of components that contribute to the more complex genetic regulatory networks in functionally diverse cell types (Rokas, 2008). One possible way to enhance regulation capacity is by recruiting novel sets of TFs. By investigating the *M. brevicollis* genome, we observe not only DBDs common to the Fungi/metazoa group such as Homeobox, HLH, Fork_head, and bZIPs but also many metazoan-specific DBDs not found in fungi. Among the animal-specific DBDs, there are families that regulate animal-specific processes such as STAT (signal transduction), p53 (apoptosis), Tub (nervous system development) (Carroll et al., 2004), as well as those involved in more general pathways such as E2F/DP (cell cycle) (Wu et al., 1995) and CSD (low temperature response) (Jones et al., 1996).

A shift of lifestyle from single to multicellular organisms also occurred in the Viridiplantae kingdom, in parallel to the Fungi/metazoa lineage. In a similar manner to choanoflagellates, a unicellular green alga *Chlamydomonas* was used to demonstrate the presence/absence of DBD families in unicellular viridiplantae (Riano-Pachon et al., 2008). Our results confirm a number of DBDs shared by Streptophyta (land plants) and Chlorophyta (green algae) previously proposed including AP2, SBP, WRKY, and zf-Dof domains. Apart from NAM (NAC, apical meristem development) (Souer et al., 1996), we observe additional families that are involved in processes more specific to higher plants and absent from Chlorophyta altogether. Examples of these DBDs are SRF (cell proliferation and differentiation) (Pellegrini et al., 1995), YABBY (abaxial polarity in lateral organs) (Bowman et al., 2002), EIN3 (ethylene-mediated responses) (Chang and Shockey, 1999), and TCP (multiple developmental control pathways) (Navaud et al., 2007).

14.9 PROTEIN FAMILIES COMBINE TO FORM COMPLEX TF DOMAIN ARCHITECTURES

Though DBDs can be considered as the core components of TFs, some TFs also contain other protein domains, which we term *partner domains*. DBDs and partner domains have been combined together during the course of evolution, giving rise to complex domain architectures in the protein chains. A network-style representation has been successfully used to demonstrate domain combinations and architectures of multidomain proteins as, for example, in Charoensawan et al. (2010b) and Wilson et al. (2007). In this final section, we touch upon this domain network representation, and discuss the most common partner domains and domain architectures observed in TFs.

14.9.1 Protein Domain Network Representation in a Nutshell

Figure 14.3 demonstrates how domain architecture information of multidomain proteins can be transformed into a network representation. DBD families are represented as circular nodes and partner domains as rectangles in protein chains and as squares in the network representation. Adjacent domains in a polypeptide chain

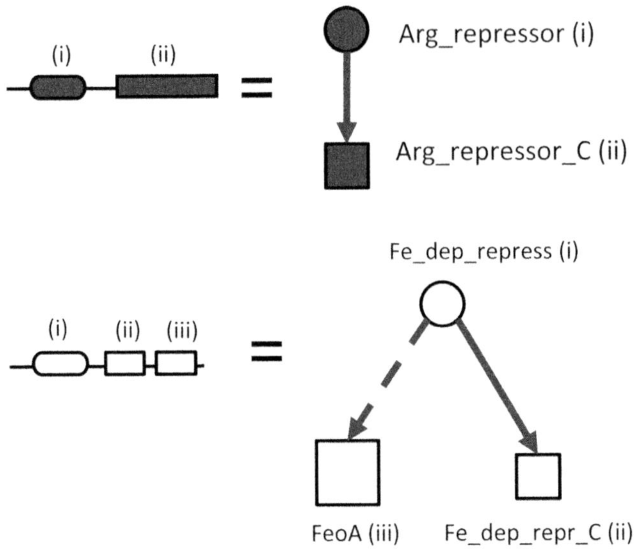

Figure 14.3 Examples of network representation of bacterial TF architectures. DBDs are shown as oblongs in protein chains and as circular nodes in our network representation. Partner domains are shown as rectangles in protein chains and as squares in the network representation. DBDs, for example, Arg_repress and Fe_dep_repress, and their adjacent partner domains, for example, Arg_repressor_C and Fe_dep_repr_C, are linked by solid arrows, pointing in N- to C-terminal orientation. Dotted arrows connect DBDs and partner domains that occur in the same TF chain but are not adjacent to DBDs, for example, FeoA. Numbers on the top of each domain indicate its order from N- to C- termini. Node sizes and arrow thickness are proportional to the abundance of domains and their combination, respectively. Shaded nodes and arrows indicate phylum-specific domain occurrence and domain combination, obtained from the taxonomic limit method described in the text (e.g., arrow linking Arg_repressor to Arg_repressor_C indicates the combination is common to all bacteria). An unshaded (white) node means the DBD is also shared with other superkingdoms, for example, Fe_dep_repress and their partner domains are also shared by Archaea.

are connected by solid arrows pointing in N- to C-terminal direction. Domains in the same polypeptide chain but not immediately adjacent are connected by dotted lines. We indicate the phylogenetic distribution of a family by colors of nodes and arrows (see Fig. 14.3 legend for more detail). The sizes of nodes and arrow thicknesses are proportional to the genomic abundance of TFs containing the protein domains and combinations in all the organisms used in this study, respectively.

As an example, the DBD Arg_repressor (argenine repressor) family is usually found toward the N-termini of TFs, and often co-occurs with the partner domain Arg_repressor_C. Consequently, a circular node representing Arg_repressor is joined to a square representing Arg_repressor_C by a solid arrow pointing from Arg_repressor to Arg_repressor_C. The Arg_repressor DBD, its partner domain Arg_repressor_C, and their domain combination are observed exclusively in bacteria. So they are all labeled in blue, the color code for bacteria used throughout this chapter.

In the case of the DBD Fe_dep_repress, it is regularly present toward the N-termini of TFs, followed by two partner domains, Fe_dep_repr_C, and FeoA, in this restricted order. Similar to the previous example, Fe_dep_repress is joined to Fe_dep_repr_C by a solid arrow because they are immediately adjacent. In contrast, Fe_dep_repress is joined to FeoA by a dotted arrow because they are not adjacent, but are still present in the same TF. The Fe_dep_repress DBD and its two partner domains are found in both bacteria and archaea, so these nodes are labeled in white (shared by more than one superkingdom). On the other hand, their domain combinations are found only in bacteria, so the arrows are labeled in blue. For the TFs that only contain one DBD with no partner domain present (single-domain TF), we simply represent this using an unconnected circular node.

14.9.2 Partner Domains of Bacterial DBDs are Usually Involved in Small Molecule Binding

Figure 14.4 demonstrates the domain architectures most commonly found in bacterial TFs. Almost all bacterial DBD families that combine with partner domains in TFs are shared by all bacteria (blue circles). In contrast, most of phylum-specific DBDs occur in single-domain TFs without any partner domains (16 out of 19, bottom of the figure), as in the case of FhlC described above. This may be because these lineage-specific domains emerged relatively recently, and did not have sufficient time to combine with other domains to form more elaborate architectures.

Regarding the partner domains found in bacterial TFs, earlier studies (Madan Babu and Teichmann, 2003; Martinez-Antonio et al., 2006) showed that a large fraction of partner domains in bacterial TFs participate in two-component signaling systems or small molecule binding. These systems enable bacteria to sense and respond to a wide range of environmental stressors in various growth conditions (Stock et al., 2000). Notable partner domains found in bacteria include the Response_reg (response regulator receiver) domain, which precedes many DBDs such as GerE (luxR family), HTH_8 (Fis family), and Trans_reg_C. HTH_1 (LysR family), the most abundant DBD in prokaryotes, is always located upstream of the LysR substrate-binding domain. Crp (cAMP receptor protein), being one of the most important bacterial master regulators (Martinez-Antonio et al., 2008), usually occurs after cNMP-binding (cyclic nucleotide-binding) domain. Other partner domains that have small molecule binding functions include IclR, FCD, UTRA, and CBS.

HTH_AraC is one exceptional case of prokaryotic DBDs because it can participate in multiple alternative architectures. The orange border around the

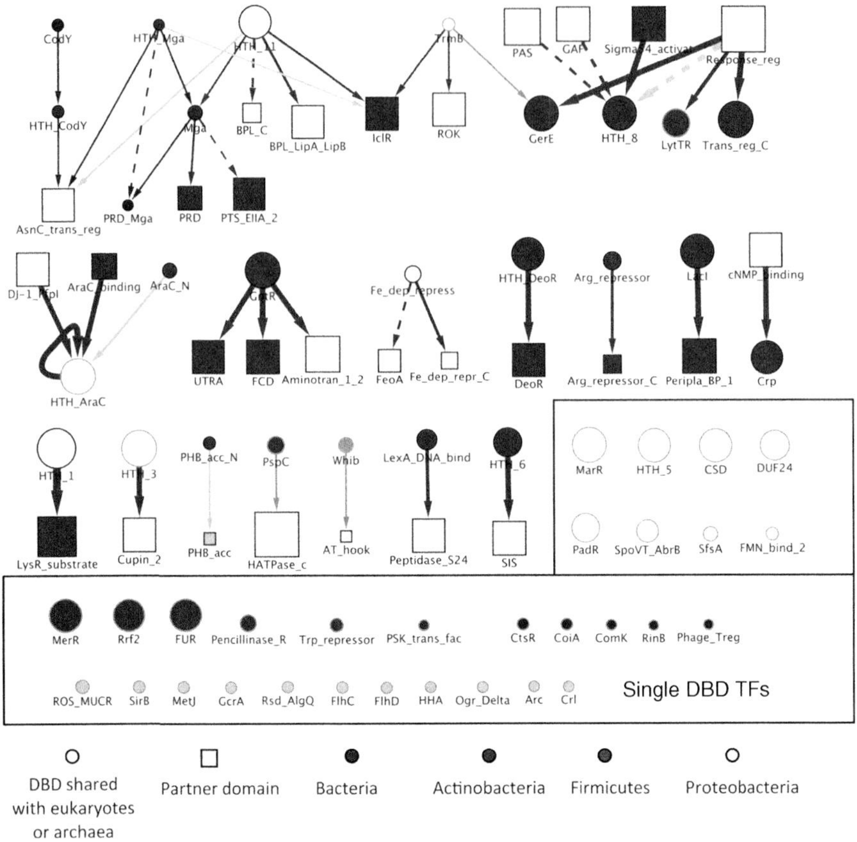

Figure 14.4 Network representing common bacterial TF architectures. Circular nodes correspond to DBDs and squares represent partner domains. Neighboring domains are represented by solid arrows in N- to C- termini orientation. Dotted arrows represent partner domains which occur in the same TF chain but are not adjacent to DBDs. Node sizes and arrow thickness are proportional to the abundance of domains and their combination, respectively. Colored nodes and arrows indicate phylum-specific domain occurrence and domain combination, respectively. Color codes are aurora blue for Bacteria, purple for Actinobacteria, navy for Firmicutes, and cyan for Proteobacteria. White means the DBD is also shared with other superkingdoms (Eukaryota and/or Archaea). DBDs that occur alone as single-domain TFs in more than 25% of all their architectural patterns have orange borders. (*See insert for color representation of the figure.*)

HTH_AraC node indicates its frequent presence in single-domain TFs (more than 25% of all TFs containing HTH_AraC) (Fig. 14.4). The self-looping arrow suggests it is the only bacterial DBD capable of occurring more than once within the same TF chain. Moreover, the DBD also occurs toward the C-terminus of a sugar-binding domain AraC-binding, protease DJ01/PfpI, and another DBD AraC_N.

14.9.3 DBD Repeats in Eukaryotic TFs Permit Increased Diversity of Motif Recognition

We next discuss the most common domain architectures of eukaryotic TFs (Fig. 14.5). Several DBD families occur repeatedly several times in a single TF chain, represented by self-looping arrows. These DBD repeats frequently occur in the zinc finger families and other DBDs such as CUT, E2F/DP, and Tea. Interestingly, AP2, B3, and WRKY are the only DBD families that exhibit repeats exclusively in plants (yellow self-looping arrows), while the AP2 and WRKY DBDs themselves are not plant specific. The function of repeated DBDs in eukaryotic TFs is most likely to increase the specificity and diversity of sequence recognition in the DNA. To illustrate this point, each zinc finger C2H2 domain, the most abundant eukaryotic DBD, recognizes only a short nucleotide sequence of three consecutive base pairs (Mandel-Gutfreund and Margalit, 1998). The DBD, however, may occur consecutively up to 30 times in a single TF (Itzkovitz et al., 2006). This results in an increased number of possible DNA-binding sequences from a limited number of DBDs. Importantly, DBDs of different families rarely occur together within the same TFs, possibly to minimize potential cross-binding errors in multi-DBD TFs (Itzkovitz et al., 2006).

Besides employing DBD repeats, eukaryotic TFs also combine DBDs and partner domains in numerous different arrangements. For instance, STAT TFs have one of the most conserved TF structures. The STAT_bind DBD is often preceded by STAT_int and STAT_alpha domains, which are both involved in protein–protein interaction mediation, and followed by an SH2 domain, which acts as a switch to receptor recognition control (Chen et al., 1998). In NF-kB, a TF that regulates the immune response to infection, the RHD (Rel homology domain) DBD is usually followed by an immunoglobulin-like domain TIG, a domain that mediates dimerization (Muller et al., 1995). The nuclear receptors are one of the largest eukaryotic TF classes. They always contain the zf-C4 DBD that occurs upstream to Hormone_rep, a partner domain that binds different ligands and alters the TF conformation, thus affecting DNA-binding specificity (Edwards, 2000).

Lineage-specific linkages between DBDs and their partner domains can be seen throughout the domain network. For instance, Homeobox mostly occurs by itself in a single domain TF but is also observed to be preceded by three other DBDs exclusively in animals: Pou, Pax, and CUT. In plants, Homeobox co-occurs with another DBD called *DDT* in a small number of TFs. Another example of lineage-specific architectures is the Tub family. In animals, the DBD frequently occurs downstream to SOCS_box, but in plants it co-occurs with F-box only. This information on lineage-specific domain combinations is useful for an understanding of both TF function and evolution (Apic et al., 2001; Vogel et al., 2004).

Similarly to DBD sharing, different superkingdoms share limited numbers of partner domains. In contrast to prokaryotes, partner domains in eukaryotic TFs have more diverse functions; the most common one is to mediate protein–protein interaction, including with histones and chromatin factors, and dimerization (Amoutzias et al., 2008). Examples of these domains include

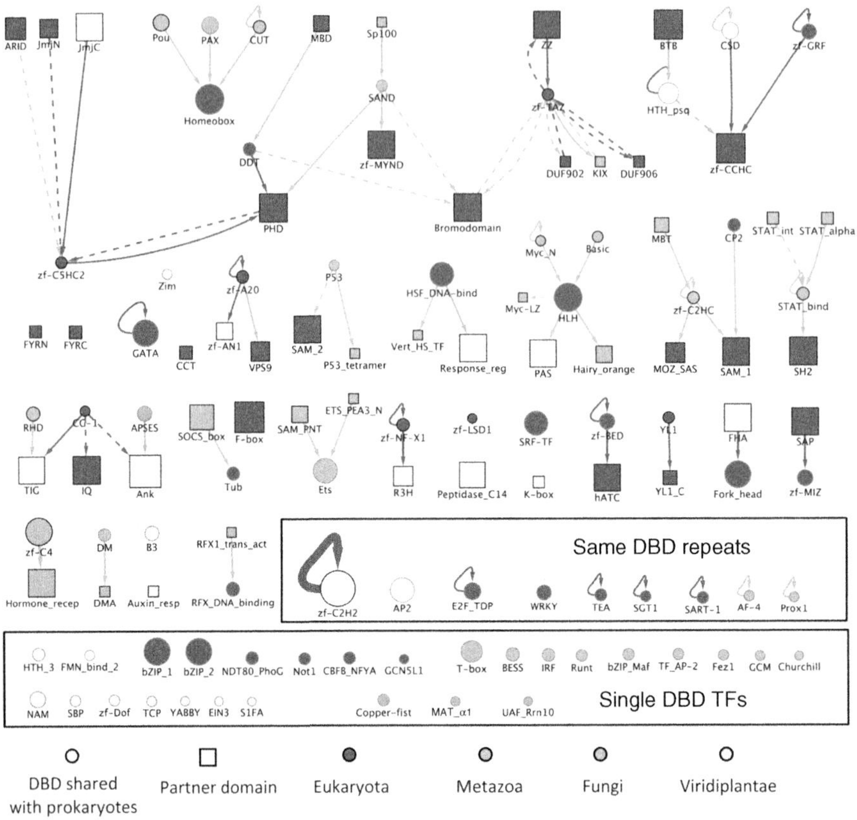

Figure 14.5 A network representing common eukaryotic TF architectures . Color codes are red for Eukaryota, pink for Metazoa, orange for Fungi, and yellow for Viridiplantae. DBDs that occur alone as single-domain TFs in more than 25% of all their architectural patterns have green borders. *Source*: From Figure 2c in Charoensawan et al. (2010b). *(See insert for color representation of the figure.)*

BTB, Bromodomain, SAM (sterile alpha motif), ANK, and hATC. In addition, eukaryotic TFs usually contain a high fraction of intrinsically disordered (ID) regions (Bell et al., 2002; Minezaki et al., 2006). These protein segments are naturally unfolded and unstructured but may serve as flexible linkers that aid protein interactivity (Dyson and Wright, 2005). Both partner domains involved in protein interaction and long ID regions together promote the formation of composite protein elements, a crucial step toward combinatorial regulation.

14.10 GENOME-WIDE STUDIES OF TFs: WHAT HAVE WE LEARNED?

The DBD is a key functional component of a sequence-specific TF because it directly binds to DNA and importantly, defines the specificity of genetic regulation. The different TF repertoires in different phylogenetic groups influence their divergent physiology.

Several groups have surveyed genome-wide occurrence of TFs and DBDs in different genomes, but most studies tended to be restricted to phylogenetic lineages. A number of databases have been established to provide TF collections to the scientific community. These databases serve as valuable resources to better understand the genetic regulation in diverse phylogenetic groups, through identifying the conserved and lineage-specific TF families. Of all the databases available to date, the DBD database (Wilson et al., 2008) is unique in the sense that it provides TF predictions for all completely sequenced genomes from species across the tree of life using a single platform. Consequently, the database is ideal for reviewing the large-scale phylogenetic distribution of all known TF families.

In this chapter, we have reviewed the genomic distributions of TFs and DBD repertoires in diverse organisms, based on the lists of manually curated DBDs and computationally predicted TFs from the DBD database. This genome-wide study of TFs and DBDs will greatly benefit the community by providing new insights into the evolution of this important class of proteins, in conjunction with the rest of other genes in genomes. Moreover, a better understanding of lineage-specific TFs and DBDs will aid functional annotation based on computational analysis, as well as facilitate experimental design. An obvious experimental design where definition of TF repertoires is important are high-throughput studies of transcriptional regulators and TF–DNA interactions, such as systematic evolution of ligands by exponential enrichment (SELEX) (Hallikas and Taipale, 2006), protein-binding microarray (Mukherjee et al., 2004), ChIP-chip (Horak and Snyder, 2002), ChIP-seq (Park, 2009), and bacterial (Meng and Wolfe, 2006) and yeast (Deplancke et al., 2004) one-hybrid systems.

Numerous studies have shown that TF abundance follows a power law increase with the number of genes. The additional regulatory mechanisms unique to eukaryotes, such as combinatorial complexity of TFs through availability of binding sites in large noncoding regions, chromatin packaging, and alternative splicing, are employed to enhance the genetic regulation in eukaryotic cells. These mechanisms together possibly enable the development of multicellular organisms. With their multicellularity, eukaryotes can further leverage a limited TF repertoire through differential expression of TFs in different cell types and also at different times.

We have also reviewed a limited conservation of DBD families between prokaryotes and eukaryotes. Lineage-specific DBD repertoires can also be seen clearly at the eukaryotic kingdom level. In contrast, more than half of prokaryotic DBD families are detected in diverse lineages and are thus less specific to a particular prokaryotic lineages. Our earlier published taxonomic limit algorithm allows us to obtain the taxonomic limits for all DBD families and their partner domains, and thus approximate the origin of the protein families on the NCBI taxonomy tree.

Together with the lineage-specific DBD families, the unique combination of DBDs and their partner domains in TFs can also be used as a signature for the genetic regulatory circuits in different species groups. Other work (Apic et al., 2001; Vogel et al., 2004) has demonstrated over-represented domain architectures in general in protein sequences. Similar analyses of TFs will undoubtedly bring about an improved understanding of evolution and function of the transcription machinery. In addition, knowledge of the phylogenetic distribution of DBD families and their domain combinations can contribute to improvement of methods for remote homology detection as in Coin et al. (2003, 2004). This would advance the discovery of yet more new TFs in genomes.

REFERENCES

Ahnert, S.E., Fink, T.M., and Zinovyev, A. (2008) How much non-coding DNA do eukaryotes require? J Theor Biol, 252, 587–592.

Akai, J. and Storey, K. (2003) Brain or brawn: how FGF signaling gives us both. Cell, 115, 510–512.

Alekshun, M.N., Levy, S.B., Mealy, T.R., Seaton, B.A., and Head, J.F. (2001) The crystal structure of MarR, a regulator of multiple antibiotic resistance, at 2.3 A resolution. Nat Struct Biol, 8, 710–714.

Altschul, S.F., Gish, W., Miller, W., Myers, E.W., and Lipman, D.J. (1990) Basic local alignment search tool. J Mol Biol, 215, 403–410.

Altschul, S.F., Madden, T.L., Schaffer, A.A., Zhang, J., Zhang, Z., Miller, W., and Lipman, D.J. (1997) Gapped BLAST and PSI-BLAST: a new generation of protein database search programs. Nucleic Acids Res, 25, 3389–3402.

Amoutzias, G.D., Robertson, D.L., Van de Peer, Y., and Oliver, S.G. (2008) Choose your partners: dimerization in eukaryotic transcription factors. Trends Biochem Sci, 33, 220–229.

Amoutzias, G.D., Veron, A.S., Weiner, J. 3rd, Robinson-Rechavi, M., Bornberg-Bauer, E., Oliver, S.G., and Robertson, D.L. (2007) One billion years of bZIP transcription factor evolution: conservation and change in dimerization and DNA-binding site specificity. Mol Biol Evol, 24, 827–835.

Apic, G., Gough, J., and Teichmann, S.A. (2001) Domain combinations in archaeal, eubacterial and eukaryotic proteomes. J Mol Biol, 310, 311–325.

Aravind, L. and Koonin, E.V. (1999) DNA-binding proteins and evolution of transcription regulation in the archaea. Nucleic Acids Res, 27, 4658–4670.

Arnqvist, A., Olsen, A., Pfeifer, J., Russell, D.G., and Normark, S. (1992) The Crl protein activates cryptic genes for curli formation and fibronectin binding in Escherichia coli HB101. Mol Microbiol, 6, 2443–2452.

Ashburner, M., Ball, C.A., Blake, J.A., Botstein, D., Butler, H., Cherry, J.M., Davis, A.P., Dolinski, K., Dwight, S.S., Eppig, J.T. et al. (2000) Gene ontology: tool for the unification of biology. The Gene Ontology Consortium. Nat Genet, 25, 25–29.

Barrasa, M.I., Vaglio, P., Cavasino, F., Jacotot, L., and Walhout, A.J. (2007) EDGEdb: a transcription factor-DNA interaction database for the analysis of C. elegans differential gene expression. BMC Genomics, 8, 21.

Baumbach, J. (2007) CoryneRegNet 4.0—a reference database for corynebacterial gene regulatory networks. BMC Bioinformatics, 8, 429.

Bell, S., Klein, C., Muller, L., Hansen, S., and Buchner, J. (2002) p53 contains large unstructured regions in its native state. J Mol Biol, 322, 917–927.

Benson, D.A., Karsch-Mizrachi, I., Lipman, D.J., Ostell, J., and Sayers, E.W. (2009) GenBank. Nucleic Acids Res, 37, D26–D31.

Blake, J.A., Bult, C.J., Kadin, J.A., Richardson, J.E., and Eppig, J.T. (2011) The Mouse Genome Database (MGD): premier model organism resource for mammalian genomics and genetics. Nucleic Acids Res, 39, D842–D848.

Bowman, J.L., Eshed, Y., and Baum, S.F. (2002) Establishment of polarity in angiosperm lateral organs. Trends Genet, 18, 134–141.

Brune, I., Brinkrolf, K., Kalinowski, J., Puhler, A., and Tauch, A. (2005) The individual and common repertoire of DNA-binding transcriptional regulators of Corynebacterium glutamicum, Corynebacterium efficiens, Corynebacterium diphtheriae and Corynebacterium jeikeium deduced from the complete genome sequences. BMC Genomics, 6, 86.

Campoy, S., Mazon, G., Fernandez de Henestrosa, A.R., Llagostera, M., Monteiro, P.B., and Barbe, J. (2002). A new regulatory DNA motif of the gamma subclass Proteobacteria: identification of the LexA protein binding site of the plant pathogen Xylella fastidiosa. Microbiology, 148 (Pt 11), 3583–3597.

Carroll, K., Gomez, C., and Shapiro, L. (2004) Tubby proteins: the plot thickens. Nat Rev Mol Cell Biol, 5, 55–63.

Carroll, S.B. (2000) Endless forms: the evolution of gene regulation and morphological diversity. Cell, 101, 577–580.

Chang, C. and Shockey, J.A. (1999) The ethylene-response pathway: signal perception to gene regulation. Curr Opin Plant Biol, 2, 352–358.

Charoensawan, V., Wilson, D., and Teichmann, S.A. (2010a) Genomic repertoires of DNA-binding transcription factors across the tree of life. Nucleic Acids Res, 38, 7364–7377.

Charoensawan, V., Wilson, D., and Teichmann, S.A. (2010b) Lineage-specific expansion of DNA-binding transcription factor families. Trends Genet, 26, 388–393.

Chen, X., Vinkemeier, U., Zhao, Y., Jeruzalmi, D., Darnell, J.E. Jr., and Kuriyan, J. (1998) Crystal structure of a tyrosine phosphorylated STAT-1 dimer bound to DNA. Cell, 93, 827–839.

Coin, L., Bateman, A., and Durbin, R. (2003) Enhanced protein domain discovery by using language modeling techniques from speech recognition. Proc Natl Acad Sci USA, 100, 4516–4520.

Coin, L., Bateman, A., and Durbin, R. (2004) Enhanced protein domain discovery using taxonomy. BMC Bioinformatics, 5, 56.

Cook, W.J., Kar, S.R., Taylor, K.B., and Hall, L.M. (1998) Crystal structure of the cyanobacterial metallothionein repressor SmtB: a model for metalloregulatory proteins. J Mol Biol, 275, 337–346.

Cordero, O.X. and Hogeweg, P. (2007) Large changes in regulome size herald the main prokaryotic lineages. Trends Genet, 23, 488–493.

Coulson, R.M. and Ouzounis, C.A. (2003) The phylogenetic diversity of eukaryotic transcription. Nucleic Acids Res, 31, 653–660.

Coulson, R.M., Touboul, N., and Ouzounis, C.A. (2007) Lineage-specific partitions in archaeal transcription. Archaea, 2, 117–125.

Davuluri, R.V., Sun, H., Palaniswamy, S.K., Matthews, N., Molina, C., Kurtz, M., and Grotewold, E. (2003) AGRIS: Arabidopsis gene regulatory information server, an information resource of Arabidopsis cis-regulatory elements and transcription factors. BMC Bioinformatics, 4, 25.

de Lima Morais, D.A., Fang, H., Rackham, O.J., Wilson, D., Pethica, R., Chothia, C., and Gough, J. (2011) SUPERFAMILY 1.75 including a domain-centric gene ontology method. Nucleic Acids Res, 39, D427–D434.

Deplancke, B., Dupuy, D., Vidal, M., and Walhout, A.J. (2004) A gateway-compatible yeast one-hybrid system. Genome Res, 14, 2093–2101.

Drysdale, R. (2008) FlyBase : a database for the Drosophila research community. Methods Mol Biol, 420, 45–59.

Dyson, H.J. and Wright, P.E. (2005) Intrinsically unstructured proteins and their functions. Nat Rev Mol Cell Biol, 6, 197–208.

Ecker, J.R. (1995) The ethylene signal transduction pathway in plants. Science, 268, 667–675.

Eddy, S.R. (1998) Profile hidden Markov models. Bioinformatics, 14, 755–763.

Edwards, D.P. (2000) The role of coactivators and corepressors in the biology and mechanism of action of steroid hormone receptors. J Mammary Gland Biol Neoplasia, 5, 307–324.

Engel, S.R., Balakrishnan, R., Binkley, G., Christie, K.R., Costanzo, M.C., Dwight, S.S., Fisk, D.G., Hirschman, J.E., Hitz, B.C., Hong, E.L. et al. (2010) Saccharomyces Genome Database provides mutant phenotype data. Nucleic Acids Res, 38, D433–D436.

Escolar, L., Perez-Martin, J., and de Lorenzo, V. (1998). Binding of the fur (ferric uptake regulator) repressor of *Escherichia coli* to arrays of the GATAAT sequence. J Mol Biol, 283, 537–547.

Eulgem, T., Rushton, P.J., Robatzek, S., and Somssich, I.E. (2000) The WRKY superfamily of plant transcription factors. Trends Plant Sci, 5, 199–206.

Facchini, L.M. and Penn, L.Z. (1998) The molecular role of Myc in growth and transformation: recent discoveries lead to new insights. FASEB J, 12, 633–651.

Finn, R.D., Mistry, J., Tate, J., Coggill, P., Heger, A., Pollington, J.E., Gavin, O.L., Gunasekaran, P., Ceric, G., Forslund, K. et al. (2010) The Pfam protein families database. Nucleic Acids Res, 38, D211–D222.

Fleischmann, R.D., Adams, M.D., White, O., Clayton, R.A., Kirkness, E.F., Kerlavage, A.R., Bult, C.J., Tomb, J.F., Dougherty, B.A., Merrick, J.M. et al. (1995) Whole-genome random sequencing and assembly of Haemophilus influenzae Rd. Science, 269, 496–512.

Fulton, D.L., Sundararajan, S., Badis, G., Hughes, T.R., Wasserman, W.W., Roach, J.C., and Sladek, R. (2009) TFCat: the curated catalog of mouse and human transcription factors. Genome Biol, 10, R29.

Furlong, E.E., Andersen, E.C., Null, B., White, K.P., and Scott, M.P. (2001) Patterns of gene expression during Drosophila mesoderm development. Science, 293, 1629–1633.

Gama-Castro, S., Salgado, H., Peralta-Gil, M., Santos-Zavaleta, A., Muniz-Rascado, L., Solano-Lira, H., Jimenez-Jacinto, V., Weiss, V., Garcia-Sotelo, J.S., Lopez-Fuentes, A. et al. (2011) RegulonDB version 7.0: transcriptional regulation of Escherichia coli K-12 integrated within genetic sensory response units (Gensor Units). Nucleic Acids Res, 39, D98–D105.

Gehring, W.J. (1992) The homeobox in perspective. Trends Biochem Sci, 17, 277–280.

Geiduschek, E.P. and Ouhammouch, M. (2005) Archaeal transcription and its regulators. Mol Microbiol, 56, 1397–1407.

Goldberg, M.D., Johnson, M., Hinton, J.C., and Williams, P.H. (2001) Role of the nucleoid-associated protein Fis in the regulation of virulence properties of enteropathogenic *Escherichia coli*. Mol Microbiol, 41, 549–559.

Grote, A., Klein, J., Retter, I., Haddad, I., Behling, S., Bunk, B., Biegler, I., Yarmolinetz, S., Jahn, D., and Munch, R. (2009) PRODORIC (release 2009): a database and tool platform for the analysis of gene regulation in prokaryotes. Nucleic Acids Res, 37, D61–D65.

Guedon, E., Serror, P., Ehrlich, S.D., Renault, P., and Delorme, C. (2001) Pleiotropic transcriptional repressor CodY senses the intracellular pool of branched-chain amino acids in Lactococcus lactis. Mol Microbiol, 40, 1227–1239.

Gury, J., Barthelmebs, L., Tran, N.P., Divies, C., and Cavin, J.F. (2004) Cloning, deletion, and characterization of PadR, the transcriptional repressor of the phenolic acid decarboxylase-encoding padA gene of Lactobacillus plantarum. Appl Environ Microbiol, 70, 2146–2153.

Hacker, U., Grossniklaus, U., Gehring, W.J., and Jackle, H. (1992) Developmentally regulated Drosophila gene family encoding the fork head domain. Proc Natl Acad Sci USA, 89, 8754–8758.

Hallikas, O. and Taipale, J. (2006) High-throughput assay for determining specificity and affinity of protein-DNA binding interactions. Nat Protoc, 1, 215–222.

Harafuji, N., Keys, D.N., and Levine, M. (2002) Genome-wide identification of tissue-specific enhancers in the Ciona tadpole. Proc Natl Acad Sci USA, 99, 6802–6805.

Helmann, J.D., Wang, Y., Mahler, I., and Walsh, C.T. (1989) Homologous metalloregulatory proteins from both gram-positive and gram-negative bacteria control transcription of mercury resistance operons. J Bacteriol, 171(1), 222–229.

Hermoso, A., Aguilar, D., Aviles, F.X., and Querol, E. (2004) TrSDB: a proteome database of transcription factors. Nucleic Acids Res, 32, D171–D173.

Hestand, M.S., van Galen, M., Villerius, M.P., van Ommen, G.J., den Dunnen, J.T., and t Hoen, P.A. (2008) CORE_TF: a user-friendly interface to identify evolutionary conserved transcription factor binding sites in sets of co-regulated genes. BMC Bioinformatics, 9, 495.

Hillerich, B. and Westpheling, J. (2006) A new GntR family transcriptional regulator in streptomyces coelicolor is required for morphogenesis and antibiotic production and controls transcription of an ABC transporter in response to carbon source. J Bacteriol, 188, 7477–7487.

Honjo, M., Nakayama, A., Fukazawa, K., Kawamura, K., Ando, K., Hori, M., and Furutani, Y. (1990) A novel Bacillus subtilis gene involved in negative control of sporulation and degradative-enzyme production. J Bacteriol, 172, 1783–1790.

Horak, C.E. and Snyder, M. (2002) ChIP-chip: a genomic approach for identifying transcription factor binding sites. Methods Enzymol, 350, 469–483.

Hulo, N., Bairoch, A., Bulliard, V., Cerutti, L., Cuche, B.A., de Castro, E., Lachaize, C., Langendijk-Genevaux, P.S., and Sigrist, C.J. (2008) The 20 years of PROSITE. Nucleic Acids Res, 36, D245–D249.

Iida, K., Seki, M., Sakurai, T., Satou, M., Akiyama, K., Toyoda, T., Konagaya, A., and Shinozaki, K. (2005) RARTF: database and tools for complete sets of Arabidopsis transcription factors. DNA Res, 12, 247–256.

Itzkovitz, S., Tlusty, T., and Alon, U. (2006) Coding limits on the number of transcription factors. BMC Genomics, 7, 239.

Iyer, L.M., Anantharaman, V., Wolf, M.Y., and Aravind, L. (2008) Comparative genomics of transcription factors and chromatin proteins in parasitic protists and other eukaryotes. Int J Parasitol, 38, 1–31.

Jacob, F. and Monod, J. (1961) Genetic regulatory mechanisms in the synthesis of proteins. J Mol Biol, 3, 318–356.

Jones, P.G., Mitta, M., Kim, Y., Jiang, W., and Inouye, M. (1996) Cold shock induces a major ribosomal-associated protein that unwinds double-stranded RNA in Escherichia coli. Proc Natl Acad Sci USA, 93, 76–80.

Kagoshima, H., Shigesada, K., Satake, M., Ito, Y., Miyoshi, H., Ohki, M., Pepling, M., and Gergen, P. (1993) The Runt domain identifies a new family of heteromeric transcriptional regulators. Trends Genet, 9, 338–341.

Kanamori, M., Konno, H., Osato, N., Kawai, J., Hayashizaki, Y., and Suzuki, H. (2004) A genome-wide and nonredundant mouse transcription factor database. Biochem Biophys Res Commun, 322, 787–793.

Kazakov, A.E., Cipriano, M.J., Novichkov, P.S., Minovitsky, S., Vinogradov, D.V., Arkin, A., Mironov, A.A., Gelfand, M.S., and Dubchak, I. (2007) RegTransBase–a database of regulatory sequences and interactions in a wide range of prokaryotic genomes. Nucleic Acids Res, 35, D407–D412.

Kel, A.E., Gossling, E., Reuter, I., Cheremushkin, E., Kel-Margoulis, O.V., and Wingender, E. (2003) MATCH: a tool for searching transcription factor binding sites in DNA sequences. Nucleic Acids Res, 31, 3576–3579.

Keller, G., Bird, A., and Winge, D.R. (2005) Independent metalloregulation of Ace1 and Mac1 in Saccharomyces cerevisiae. Eukaryot Cell, 4, 1863–1871.

King, N., Westbrook, M.J., Young, S.L., Kuo, A., Abedin, M., Chapman, J., Fairclough, S., Hellsten, U., Isogai, Y., Letunic, I. et al. (2008) The genome of the choanoflagellate Monosiga brevicollis and the origin of metazoans. Nature, 451, 783–788.

Kisseleva, T., Bhattacharya, S., Braunstein, J., and Schindler, C.W. (2002) Signaling through the JAK/STAT pathway, recent advances and future challenges. Gene, 285, 1–24.

Klug, A. (1999) Zinc finger peptides for the regulation of gene expression. J Mol Biol, 293, 215–218.

Kondro, W. (2004) Molecular biology. Consortium tackles mouse regulome. Science, 304, 942.

Koonin, E.V. and Wolf, Y.I. (2008) Genomics of bacteria and archaea: the emerging dynamic view of the prokaryotic world. Nucleic Acids Res, 36, 6688–6719.

Kormanec, J., Sevcikova, B., Sprusansky, O., Benada, O., Kofronova, O., Novakova, R., Rezuchova, B., Potuckova, L., and Homerova, D. (1998) The Streptomyces aureofaciens homologue of the whiB gene is essential for sporulation; its expression correlates with the developmental stage. Folia Microbiol (Praha), 43, 605–612.

Kummerfeld, S.K. and Teichmann, S.A. (2006) DBD: a transcription factor prediction database. Nucleic Acids Res, 34, D74–D81.

Kuttippurathu, L., Hsing, M., Liu, Y., Schmidt, B., Maskell, D.L., Lee, K., He, A., Pu, W.T., and Kong, S.W. (2011) CompleteMOTIFs: DNA motif discovery platform for transcription factor binding experiments. Bioinformatics, 27, 715–717.

Laity, J.H., Lee, B.M., and Wright, P.E. (2001) Zinc finger proteins: new insights into structural and functional diversity. Curr Opin Struct Biol, 11, 39–46.

Latchman, D.S. (1997). Transcription factors: an overview. Int J Biochem Cell Biol, 29, 1305–1312.

Lee, S.J., Engelmann, A., Horlacher, R., Qu, Q., Vierke, G., Hebbeln, C., Thomm, M., and Boos, W. (2003) TrmB, a sugar-specific transcriptional regulator of the trehalose/maltose ABC transporter from the hyperthermophilic archaeon Thermococcus litoralis. J Biol Chem, 278, 983–990.

Lehmann, M., Siegmund, T., Lintermann, K.G., and Korge, G. (1998) The pipsqueak protein of Drosophila melanogaster binds to GAGA sequences through a novel DNA-binding domain. J Biol Chem, 273, 28504–28509.

Levine, M. and Tjian, R. (2003) Transcription regulation and animal diversity. Nature, 424, 147–151.

Littlewood, T.D. and Evan, G.I. (1995) Transcription factors 2: helix-loop-helix. Protein Profile, 2, 621–702.

Luscombe, N.M., Austin, S.E., Berman, H.M., and Thornton, J.M. (2000) An overview of the structures of protein-DNA complexes. Genome Biol, 1, REVIEWS001.

Lyne, R., Smith, R., Rutherford, K., Wakeling, M., Varley, A., Guillier, F., Janssens, H., Ji, W., McLaren, P., North, P. et al. (2007) FlyMine: an integrated database for Drosophila and Anopheles genomics. Genome Biol, 8, R129.

Machado, A.K., Morgan, B.A., and Merrill, G.F. (1997) Thioredoxin reductase-dependent inhibition of MCB cell cycle box activity in Saccharomyces cerevisiae. J Biol Chem, 272, 17045–17054.

MacPherson, S., Larochelle, M., and Turcotte, B. (2006) A fungal family of transcriptional regulators: the zinc cluster proteins. Microbiol Mol Biol Rev, 70, 583–604.

Madan Babu, M. and Teichmann, S.A. (2003) Evolution of transcription factors and the gene regulatory network in Escherichia coli. Nucleic Acids Res, 31, 1234–1244.

Mandel-Gutfreund, Y. and Margalit, H. (1998) Quantitative parameters for amino acid–base interaction: implications for prediction of protein-DNA binding sites. Nucleic Acids Res, 26, 2306–2312.

Manning, G., Young, S.L., Miller, W.T., and Zhai, Y. (2008) The protist, Monosiga brevicollis, has a tyrosine kinase signaling network more elaborate and diverse than found in any known metazoan. Proc Natl Acad Sci USA, 105, 9674–9679.

Marinescu, V.D., Kohane, I.S., and Riva, A. (2005) MAPPER: a search engine for the computational identification of putative transcription factor binding sites in multiple genomes. BMC Bioinformatics, 6, 79.

Martin, M., Lloret, J., Sanchez-Contreras, M., Bonilla, I., and Rivilla, R. (2000) MucR is necessary for galactoglucan production in Sinorhizobium meliloti EFB1. Mol Plant Microbe Interact, 13, 129–135.

Martinez-Antonio, A., Janga, S.C., Salgado, H., and Collado-Vides, J. (2006) Internal-sensing machinery directs the activity of the regulatory network in Escherichia coli. Trends Microbiol, 14, 22–27.

Martinez-Antonio, A., Janga, S.C., and Thieffry, D. (2008) Functional organisation of Escherichia coli transcriptional regulatory network. J Mol Biol, 381, 238–247.

Martinez-Bueno, M., Molina-Henares, A.J., Pareja, E., Ramos, J.L., and Tobes, R. (2004) BacTregulators: a database of transcriptional regulators in bacteria and archaea. Bioinformatics, 20, 2787–2791.

Maslov, S., Krishna, S., Pang, T.Y., and Sneppen, K. (2009) Toolbox model of evolution of prokaryotic metabolic networks and their regulation. Proc Natl Acad Sci USA, 106, 9743–9748.

Mattick, J.S. (2007) A new paradigm for developmental biology. J Exp Biol, 210, 1526–1547.

Matys, V., Kel-Margoulis, O.V., Fricke, E., Liebich, I., Land, S., Barre-Dirrie, A., Reuter, I., Chekmenev, D., Krull, M., Hornischer, K. et al. (2006) TRANSFAC and its module TRANSCompel: transcriptional gene regulation in eukaryotes. Nucleic Acids Res, 34, D108–D110.

McClintock, B. (1956) Controlling elements and the gene. Cold Spring Harb Symp Quant Biol, 21, 197–216.

Meng, X. and Wolfe, S.A. (2006) Identifying DNA sequences recognized by a transcription factor using a bacterial one-hybrid system. Nat Protoc, 1, 30–45.

Messina, D.N., Glasscock, J., Gish, W., and Lovett, M. (2004) An ORFeome-based analysis of human transcription factor genes and the construction of a microarray to interrogate their expression. Genome Res, 14(10B), 2041–2047.

Miller, M.B. and Bassler, B.L. (2001). Quorum sensing in bacteria. Annu Rev Microbiol, 55, 165–199.

Minezaki, Y., Homma, K., Kinjo, A.R., and Nishikawa, K. (2006) Human transcription factors contain a high fraction of intrinsically disordered regions essential for transcriptional regulation. J Mol Biol, 359, 1137–1149.

Minezaki, Y., Homma, K., and Nishikawa, K. (2005) Genome-wide survey of transcription factors in prokaryotes reveals many bacteria-specific families not found in archaea. DNA Res, 12, 269–280.

Miyamoto, M., Fujita, T., Kimura, Y., Maruyama, M., Harada, H., Sudo, Y., Miyata, T., and Taniguchi, T. (1988) Regulated expression of a gene encoding a nuclear factor, IRF-1, that specifically binds to IFN-beta gene regulatory elements. Cell, 54, 903–913.

Molina, N. and van Nimwegen, E. (2009) Scaling laws in functional genome content across prokaryotic clades and lifestyles. Trends Genet, 25, 243–247.

Montano, S.P., Pierce, M., Cote, M.L., Vershon, A.K., and Georgiadis, M.M. (2002) Crystallographic studies of a novel DNA-binding domain from the yeast transcriptional activator Ndt80. Acta Crystallogr D Biol Crystallogr, 58, 2127–2130.

Moreno-Campuzano, S., Janga, S.C., and Perez-Rueda, E. (2006) Identification and analysis of DNA-binding transcription factors in Bacillus subtilis and other Firmicutes–a genomic approach. BMC Genomics, 7, 147.

Mukherjee, S., Berger, M.F., Jona, G., Wang, X.S., Muzzey, D., Snyder, M., Young, R.A., and Bulyk, M.L. (2004) Rapid analysis of the DNA-binding specificities of transcription factors with DNA microarrays. Nat Genet, 36, 1331–1339.

Mulder, N.J., Apweiler, R., Attwood, T.K., Bairoch, A., Bateman, A., Binns, D., Bork, P., Buillard, V., Cerutti, L., Copley, R. et al. (2007) New developments in the InterPro database. Nucleic Acids Res, 35, D224–D228.

Muller, C.W., Rey, F.A., Sodeoka, M., Verdine, G.L., and Harrison, S.C. (1995) Structure of the NF-kappa B p50 homodimer bound to DNA. Nature, 373, 311–317.

Nair, S., Derre, I., Msadek, T., Gaillot, O., and Berche, P. (2000) CtsR controls class III heat shock gene expression in the human pathogen Listeria monocytogenes. Mol Microbiol, 35, 800–811.

Nanamiya, H., Shiomi, E., Ogura, M., Tanaka, T., Asai, K., and Kawamura, F. (2003) Involvement of ClpX protein in the post-transcriptional regulation of a competence specific transcription factor, ComK protein, of Bacillus subtilis. J Biochem, 133, 295–302.

Navaud, O., Dabos, P., Carnus, E., Tremousaygue, D., and Herve, C. (2007) TCP transcription factors predate the emergence of land plants. J Mol Evol, 65, 23–33.

Nguyen, C.C. and Saier, M.H., Jr. (1995). Phylogenetic, structural and functional analyses of the LacI-GalR family of bacterial transcription factors. FEBS Lett, 377, 98–102.

Odom, D.T., Zizlsperger, N., Gordon, D.B., Bell, G.W., Rinaldi, N.J., Murray, H.L., Volkert, T.L., Schreiber, J., Rolfe, P.A., Gifford, D.K. et al. (2004) Control of pancreas and liver gene expression by HNF transcription factors. Science, 303, 1378–1381.

Otwinowski, Z., Schevitz, R.W., Zhang, R.G., Lawson, C.L., Joachimiak, A., Marmorstein, R.Q., Luisi, B.F., and Sigler, P.B. (1988) Crystal structure of trp repressor/operator complex at atomic resolution. Nature, 335, 321–329.

Park, J., Jang, S., Kim, S., Kong, S., Choi, J., Ahn, K., Kim, J., Lee, S., Park, B., Jung, K. et al. (2008) FTFD: an informatics pipeline supporting phylogenomic analysis of fungal transcription factors. Bioinformatics, 24, 1024–1025.

Park, P.J. (2009) ChIP-seq: advantages and challenges of a maturing technology. Nat Rev Genet, 10, 669–680.

Pavlopoulos, A. and Akam, M. (2007) Hox go omics: insights from Drosophila into Hox gene targets. Genome Biol, 8, 208.

Peeters, E. and Charlier, D. (2010) The Lrp family of transcription regulators in archaea. Archaea, 2010, 750457.

Pellegrini, L., Tan, S., and Richmond, T.J. (1995) Structure of serum response factor core bound to DNA. Nature, 376, 490–498.

Perez-Rodriguez, P., Riano-Pachon, D.M., Correa, L.G., Rensing, S.A., Kersten, B., and Mueller-Roeber, B. (2010) PlnTFDB: updated content and new features of the plant transcription factor database. Nucleic Acids Res, 38, D822–D827.

Perez-Rueda, E., Collado-Vides, J., and Segovia, L. (2004) Phylogenetic distribution of DNA binding transcription factors in bacteria and archaea. Comput Biol Chem, 28, 341–350.

Pfreundt, U., James, D.P., Tweedie, S., Wilson, D., Teichmann, S.A., and Adryan, B. (2010) FlyTF: improved annotation and enhanced functionality of the Drosophila transcription factor database. Nucleic Acids Res, 38, D443–D447.

Pincus, D., Letunic, I., Bork, P., and Lim, W.A. (2008) Evolution of the phosphotyrosine signaling machinery in premetazoan lineages. Proc Natl Acad Sci USA, 105, 9680–9684.

Portales-Casamar, E., Thongjuea, S., Kwon, A.T., Arenillas, D., Zhao, X., Valen, E., Yusuf, D., Lenhard, B., Wasserman, W.W., and Sandelin, A. (2010) JASPAR 2010: the greatly expanded open-access database of transcription factor binding profiles. Nucleic Acids Res, 38, D105–D110.

Pruss, B.M., Liu, X., Hendrickson, W., and Matsumura, P. (2001) FlhD/FlhC-regulated promoters analyzed by gene array and lacZ gene fusions. FEMS Microbiol Lett, 197, 91–97.

Ranea, J.A., Buchan, D.W., Thornton, J.M., and Orengo, C.A. (2004) Evolution of protein superfamilies and bacterial genome size. J Mol Biol, 336, 871–887.

Ranea, J.A., Grant, A., Thornton, J.M., and Orengo, C.A. (2005) Microeconomic principles explain an optimal genome size in bacteria. Trends Genet, 21, 21–25.

Ren, J., Sainsbury, S., Nettleship, J.E., Saunders, N.J., and Owens, R.J. (2010) The crystal structure of NGO0477 from Neisseria gonorrhoeae reveals a novel protein fold incorporating a helix-turn-helix motif. Proteins, 78, 1798–1802.

Riano-Pachon, D.M., Correa, L.G., Trejos-Espinosa, R., and Mueller-Roeber, B. (2008) Green transcription factors: a chlamydomonas overview. Genetics, 179, 31–39.

Richardt, S., Lang, D., Reski, R., Frank, W., and Rensing, S.A. (2007) PlanTAPDB, a phylogeny-based resource of plant transcription-associated proteins. Plant Physiol, 143, 1452–1466.

Riechmann, J.L., Heard, J., Martin, G., Reuber, L., Jiang, C., Keddie, J., Adam, L., Pineda, O., Ratcliffe, O.J., Samaha, R.R. et al. (2000) Arabidopsis transcription factors: genome-wide comparative analysis among eukaryotes. Science, 290, 2105–2110.

Rodda, S., Sharma, S., Scherer, M., Chapman, G., and Rathjen, P. (2001) CRTR-1, a developmentally regulated transcriptional repressor related to the CP2 family of transcription factors. J Biol Chem, 276, 3324–3332.

Rokas, A. (2008) The molecular origins of multicellular transitions. Curr Opin Genet Dev, 18, 472–478.

Romeuf, I., Tessier, D., Dardevet, M., Branlard, G., Charmet, G., and Ravel, C. (2010) wDBTF: an integrated database resource for studying wheat transcription factor families. BMC Genomics, 11, 185.

Rushton, P.J., Bokowiec, M.T., Laudeman, T.W., Brannock, J.F., Chen, X., and Timko, M.P. (2008) TOBFAC: the database of tobacco transcription factors. BMC Bioinformatics, 9, 53.

Saviola, B., Seabold, R., and Schleif, R.F. (1998) Arm-domain interactions in AraC. J Mol Biol, 278, 539–548.

Schaefer, U., Schmeier, S., and Bajic, V.B. (2011) TcoF-DB: dragon database for human transcription co-factors and transcription factor interacting proteins. Nucleic Acids Res, 39, D106–D110.

Schell, M.A. (1993) Molecular biology of the LysR family of transcriptional regulators. Annu Rev Microbiol, 47, 597–626.

Schiering, N., Tao, X., Zeng, H., Murphy, J.R., Petsko, G.A., and Ringe, D. (1995) Structures of the apo- and the metal ion-activated forms of the diphtheria tox repressor from Corynebacterium diphtheriae. Proc Natl Acad Sci USA, 92, 9843–9850.

Schreiter, E.R., Sintchak, M.D., Guo, Y., Chivers, P.T., Sauer, R.T., and Drennan, C.L. (2003) Crystal structure of the nickel-responsive transcription factor NikR. Nat Struct Biol, 10, 794–799.

Sengupta, P. and Cochran, B.H. (1991) MAT alpha 1 can mediate gene activation by a-mating factor. Genes Dev, 5, 1924–1934.

Shelest, E. (2008) Transcription factors in fungi. FEMS Microbiol Lett, 286, 145–151.

Shiu, S.H., Shih, M.C., and Li, W.H. (2005) Transcription factor families have much higher expansion rates in plants than in animals. Plant Physiol, 139, 18–26.

Sierro, N., Makita, Y., de Hoon, M., and Nakai, K. (2008) DBTBS: a database of transcriptional regulation in Bacillus subtilis containing upstream intergenic conservation information. Nucleic Acids Res, 36, D93–D96.

Souer, E., van Houwelingen, A., Kloos, D., Mol, J., and Koes, R. (1996) The no apical meristem gene of Petunia is required for pattern formation in embryos and flowers and is expressed at meristem and primordia boundaries. Cell, 85, 159–170.

Stock, A.M., Robinson, V.L., and Goudreau, P.N. (2000) Two-component signal transduction. Annu Rev Biochem, 69, 183–215.

Struhl, K. (1999) Fundamentally different logic of gene regulation in eukaryotes and prokaryotes. Cell, 98, 1–4.

Sutcliffe, J.E., Korenjak, M., and Brehm, A. (2003) Tumour suppressors–a fly's perspective. Eur J Cancer, 39, 1355–1362.

Takahashi, K. and Yamanaka, S. (2006) Induction of pluripotent stem cells from mouse embryonic and adult fibroblast cultures by defined factors. Cell, 126, 663–676.

Tordai, H., Nagy, A., Farkas, K., Banyai, L., and Patthy, L. (2005) Modules, multidomain proteins and organismic complexity. FEBS J, 272, 5064–5078.

Tweedie, S., Ashburner, M., Falls, K., Leyland, P., McQuilton, P., Marygold, S., Millburn, G., Osumi-Sutherland, D., Schroeder, A., Seal, R. et al. (2009) FlyBase: enhancing Drosophila Gene Ontology annotations. Nucleic Acids Res, 37, D555–D559.

Vaquerizas, J.M., Kummerfeld, S.K., Teichmann, S.A., and Luscombe, N.M. (2009). A census of human transcription factors: function, expression and evolution. Nat Rev Genet, 10, 252–263.

van Nimwegen, E. (2003) Scaling laws in the functional content of genomes. Trends Genet, 19, 479–484.

Vogel, C., Berzuini, C., Bashton, M., Gough, J., and Teichmann, S.A. (2004) Supra-domains: evolutionary units larger than single protein domains. J Mol Biol, 336, 809–823.

Vogel, C. and Chothia, C. (2006) Protein family expansions and biological complexity. PLoS Comput Biol, 2, e48.

Wang, Q. and Szaniszlo, P.J. (2007) WdStuAp, an APSES transcription factor, is a regulator of yeast-hyphal transitions in Wangiella (Exophiala) dermatitidis. Eukaryot Cell, 6, 1595–1605.

Wang, Z., Libault, M., Joshi, T., Valliyodan, B., Nguyen, H.T., Xu, D., Stacey, G., and Cheng, J. (2010) SoyDB: a knowledge database of soybean transcription factors. BMC Plant Biol, 10, 14.

Wilson, D., Charoensawan, V., Kummerfeld, S.K., and Teichmann, S.A. (2008) DBD–taxonomically broad transcription factor predictions: new content and functionality. Nucleic Acids Res, 36, D88–D92.

Wilson, D., Madera, M., Vogel, C., Chothia, C., and Gough, J. (2007) The SUPERFAMILY database in 2007: families and functions. Nucleic Acids Res, 35, D308–D313.

Wilson, V. and Conlon, F.L. (2002) The T-box family. Genome Biol, 3, REVIEWS3008.

Wittman, V., Lin, H.C., Wong, H.C. (1993). Functional domains of the penicillinase repressor of *Bacillus licheniformis*. J Bacteriol, 175, 7383–7390.

Wu, C.L., Zukerberg, L.R., Ngwu, C., Harlow, E., and Lees, J.A. (1995) In vivo association of E2F and DP family proteins. Mol Cell Biol, 15, 2536–2546.

Wu, J., Wang, S., Bai, J., Shi, L., Li, D., Xu, Z., Niu, Y., Lu, J., and Bao, Q. (2008) ArchaeaTF: an integrated database of putative transcription factors in Archaea. Genomics, 91, 102–107.

Wu, J., Zhao, F., Wang, S., Deng, G., Wang, J., Bai, J., Lu, J., Qu, J., and Bao, Q. (2007) cTFbase: a database for comparative genomics of transcription factors in cyanobacteria. BMC Genomics, 8, 104.

Wu, W.F., Urbanowski, M.L., and Stauffer, G.V. (1993) MetJ-mediated regulation of the Salmonella typhimurium metE and metR genes occurs through a common operator region. FEMS Microbiol Lett, 108, 145–150.

Wunderlich, Z. and Mirny, L.A. (2009) Different gene regulation strategies revealed by analysis of binding motifs. Trends Genet, 25, 434–440.

Yamasaki, K., Kigawa, T., Inoue, M., Tateno, M., Yamasaki, T., Yabuki, T., Aoki, M., Seki, E., Matsuda, T., Nunokawa, E. et al. (2004) A novel zinc-binding motif revealed by solution structures of DNA-binding domains of Arabidopsis SBP-family transcription factors. J Mol Biol, 337, 49–63.

Zhang, H., Jin, J., Tang, L., Zhao, Y., Gu, X., Gao, G., and Luo, J. (2011) PlantTFDB 2.0: update and improvement of the comprehensive plant transcription factor database. Nucleic Acids Res, 39, D1114–D1117.

Zhao, H., Msadek, T., Zapf, J., Madhusudan , Hoch, J.A., and Varughese, K.I. (2002) DNA complexed structure of the key transcription factor initiating development in sporulating bacteria. Structure, 10, 1041–1050.

Zheng, G., Tu, K., Yang, Q., Xiong, Y., Wei, C., Xie, L., Zhu, Y., and Li, Y. (2008) ITFP: an integrated platform of mammalian transcription factors. Bioinformatics, 24, 2416–2417.

15

EVOLUTION OF EUKARYOTIC CHROMATIN PROTEINS AND TRANSCRIPTION FACTORS

L. Aravind, Vivek Anantharaman, Saraswathi Abhiman, and Lakshminarayan M. Iyer

National Center for Biotechnology Information, National Library of Medicine, National Institutes of Health, Bethesda, MD, USA

CHAPTER SUMMARY

Comparative genomics of eukaryotes has profoundly impacted our understanding of the regulatory systems involved in transcription and chromatin dynamics. While certain components of these systems are highly conserved, others are rapidly evolving, thereby presenting a possible explanation for the observed variety in the regulatory adaptations of eukaryotes. The absolute numbers of specific transcription factors (TFs) and chromatin proteins (CPs) are positively correlated with proteome size in eukaryotes. However, individual types of specific TFs show an enormous diversity across different eukaryotic lineages. Dominant families of specific TFs differ even between sister lineages, and have been shaped by gene loss and lineage-specific expansions (LSEs). Recognition of this principle has helped in identifying previously unknown, dominant specific TFs of several non–model eukaryotes. Comparative analysis of known and predicted CPs allows reconstruction of the early evolutionary history of histone and DNA modification, nucleosome assembly, and chromatin-remodeling systems. Many key catalytic-binding,

Protein Families: Relating Protein Sequence, Structure, and Function, First Edition.
Edited by Christine Orengo and Alex Bateman.

peptide-binding and DNA-binding domains (DBDs) in these systems have homologs in bacteria, but they were put together into distinctive domain architectures and regulatory complexes typical of CPs only in eukaryotes. In the case of histone methylases, histone demethylases, and SWI2/SNF2 ATPases, proliferation of paralogous families followed by acquisition of novel domain architectures played a major role in producing a diverse set of enzymes that create and respond to an epigenetic code of modified histones. Eukaryotic DNA methylases in particular appear to have emerged via multiple independent transfers from bacteria. Similarly, the Tet/JBP family of DNA-modifying enzymes appears to have been acquired by the eukaryotes from bacteriophages. Even though key histone modifications are universal to eukaryotes, domain architectures of proteins binding posttranslationally modified histones vary considerably across eukaryotes. This indicates that any epigenetic information stored in them might be "interpreted" differently in different lineages. In the course of eukaryotic evolution a strong interlock developed between CPs and two distinct forms of RNA-based regulation, namely, the RNA interference (RNAi) system and the system of long noncoding RNAs. The complexity of domain architectures of CPs appears to have increased in several lineages in the course of eukaryotic evolution and may have had a role in the origin of multicellularity and cell differentiation.

15.1 INTRODUCTION

15.1.1 Uniqueness of the Organization of Eukaryotic Chromatin and Transcription Apparatus

The three superkingdoms of life utilize rather distinct strategies for packaging genomic DNA. Though most bacteria utilize members of the IHF/HU family as their primary DNA-packaging protein, certain bacteria, such as chlamydiae, have specialized DNA-packaging proteins of the HC1/HC2 family that establish condensed chromatin typical of certain stages in their life cycle (Dorman, 2009; Pedersen et al., 1996; Perara et al., 1992). Archaea show a notable diversity of DNA-packaging proteins that include members of the Alba, MC1, Sac7/Cren7/Sso7, and histone fold families (Aravind et al., 2003; Sandman and Reeve, 2005). Histone fold proteins, which they share with eukaryotes, are primarily observed only in two of the great divisions of archaea, namely, the euryarchaea and the thaumarchaea. These histone-like proteins appear to represent a packaging strategy that was a precursor of the eukaryotic system. Currently known archaeal nucleosomes contain a single or a pair of distinct histone subunits, assembling into a tetramer that wraps approximately 80 bp of DNA around it (comparable to the eukaryotic histone H3-H4 tetrasomes) (Reeve et al., 2004). The origin of eukaryotes was accompanied by a dramatic development of this ancestral histone template. First, there was proliferation followed by divergence, resulting in four distinct histones (H2A, H2B, H3, and H4) that are conserved throughout eukarya (Reeve et al., 2004). Second, these histones assembled into an octamer, as opposed to the archaeal tetrasome, and wrapped nearly twice as much DNA (~146 bp) (Reeve et al., 2004). Third, the eukaryotic histones acquired low complexity extensions to the N-terminus and/or C-terminus of the globular DNA-binding histone fold that are enriched in positively charged residues. These extensions are known as the histone tails and provide additional surfaces that neutralize the negative charges of the DNA backbone (Aravind et al., 2006; Reeve et al., 2004).

In eukaryotes, the histone octamer-based packaging was also accompanied by several major structural innovations in chromosomal organization (Aravind et al., 2006; Denhardt et al., 2005). Even in the common ancestor of all extant eukaryotes a transition was made from the predominantly circular chromosomes of prokaryotes to multiple linear ones with ends capped by telomeres, which were synthesized by a reverse transcriptase (telomerase) using an RNA template. Further, the chromosomes were separated from the rest of the cell by a membrane bilayer, resulting in the quintessential feature of the eukaryotes, the nucleus (Mans et al., 2004). This decoupled cytoplasmic translation from nuclear transcription and marked a major departure from the prokaryotic situation. This also appears to have relaxed the constraints on the eukaryotic genes, allowing them to be colonized by introns, as translation of mRNA was no longer coupled to its transcription (Anantharaman et al., 2007; Aravind et al., 2006; Denhardt et al., 2005; Iyer et al., 2008b; Mans et al., 2004). The emergence of introns favored the origin of a new set of large protein complexes: the spliceosomal complexes that associated with transcribed genes and acted on the intron-containing primary transcripts (Anantharaman et al., 2002). The decoupling of translation from transcription with the origin of the nucleus also appears to have favored the emergence of a distinct subnuclear organelle, the nucleolus, where the ribosomal proteins could be combined with freshly synthesized rRNAs to generate functional ribosomal subunits (Staub et al., 2004). This further appears to have resulted in the origin of an entirely new cytological feature, the nuclear pore, to facilitate trafficking of RNA and proteins between the nucleus and the cytoplasm (Denhardt et al., 2005; Mans et al., 2004). Hence, the landscape of eukaryotic chromatin diverged considerably from that of prokaryotes, with spliceosomal, rRNA processomal, and telomerase ribonucleoprotein complexes adding to the protein and nucleic acid mass of the chromosomes, beyond the genome and the histone octamers.

The basal or general transcription apparatus of eukaryotes and archaea share several unique features. These include (i) structure of the RNA polymerase core subunits (the subunits equivalent to the bacterial β' and β, which are the catalytic subunits and RPB3 and RPB11 that are cognates of the α subunits); (ii) specific accessory RNA polymerase subunits (e.g., RPB10); (iii) proteins constituting the basal transcription initiation apparatus (general or global TFs), such as TATA box-binding protein (TBP), TFIIB, TFIIE, and MBF (Conaway and Conaway, 2004; Latchman, 2005; Reeve, 2003). On the other hand, certain components of the eukaryotic transcription elongation complex, such as the Spt6p-type RNA-binding proteins, are shared with bacteria rather than archaea (Anantharaman et al., 2002). Thus, eukaryotic systems appear chimeric in origin—archaea-like elements dominate the core basal or general transcription apparatus, while bacteria-like elements supply some additional factors of the basal transcription apparatus, in particular, the elongation apparatus (Aravind et al., 2005; Best et al., 2004; Conaway and Conaway, 2004; Dacks and Doolittle, 2001; Iyer et al., 2008b; Reeve, 2003). However, the eukaryotes display a unique innovation of a protein complex, the mediator, which is critical for interactions between the basal transcription apparatus and specific TFs (Conaway and Conaway, 2011). Similar to the two prokaryotic superkingdoms, several eukaryotes possess specific TFs that are required for the transcriptional regulation of particular sets of genes (Iyer et al., 2008b). In both prokaryotic superkingdoms, majority of specific TFs are members of a relatively small group of protein families containing the helix-turn-helix (HTH) DBD (Aravind et al., 2005; Iyer and Aravind, 2011). Certain families of eukaryote-specific TFs, such as the homeodomain and Myb domain proteins, also bind DNA via the HTH domain (Gehring et al., 1994). However, almost all eukaryotic HTH containing specific TFs do not belong to any of the prokaryotic HTH families, and are only very distantly related to them in sequence (Aravind et al., 2005; Babu et al., 2004). In addition, eukaryotes possess numerous large families of specific TFs containing an astonishing array of DBDs that span the entire spectrum of protein folds (Iyer et al., 2008b). This deployment of specific TFs with an immense structural diversity of DBDs and the interaction with the basal TFs via a mediator are dramatic departures *vis-à-vis* the prokaryotic superkingdoms. Thus, not only the organization of the chromatin but also specific innovations in the transcription apparatus set eukaryotes apart from the two prokaryotic superkingdoms.

15.1.2 Eukaryotic Chromatin is Characterized by Modification and Dynamics of Resident Proteins and DNA

In terms of protein structure, the origin of eukaryotes was characterized by an expansion of low complexity sequences in proteins (Aravind et al., 2006; Babu et al., 2011; Liu et al., 2002). These form nonglobular segments of proteins, which typically exist as disordered or unstructured random coils and tend to be enriched in a single or few amino acids. In addition to histone tails, such low complexity regions are also abundant in eukaryotic nuclear proteins such as TFs and spliceosomal proteins (e.g., RGG and SR repeats), and might play roles in low-specificity interactions with proteins and nucleic acids (Anantharaman et al., 2002; Babu et al., 2011).

Furthermore, low complexity regions offered a niche for the diversification of a veritable ecosystem of enzymes that catalyze addition of covalent modifications to the amino acid side chains or the N- and C-termini of polypeptides (Allis et al., 2007; Kouzarides, 2007; Rouleau et al., 2004; van Dijk et al., 2008; Walsh et al., 2005). There also arose a corresponding array of enzymes that catalyzed removal of such covalent modifications to restore the unmodified state.

These modifications span a dramatic range in terms of molecular weight and biochemical diversity. The simplest of these are low molecular weight adducts (methyl, phosphate, and acetyl groups). Slightly higher molecular weight modifications include mono-ADP ribosylation, biotinylation, glycosylation, and spermidinylation. The largest modifications involve the addition of whole biopolymers such as branched or linear polyADP-ribose (up to 200 ADP-ribose units), homopolymeric peptides such as polyglutamate or polyglycine (up to 20 amino acids), and polypeptides of the ubiquitin family such as ubiquitin (Ub) and Sumo (Allis et al., 2007; Fujiki et al., 2009; Kouzarides, 2007; Rouleau et al., 2004; van Dijk et al., 2008; Walsh et al., 2005). In addition to these covalent adducts, there are covalent modifications that directly alter the amino acid side chain. These include citrullination, which results from the deimidation of the guanidino group of arginine (releasing an ammonium ion), and ornithination, which results from the hydrolysis of the guanidino group (releasing urea) (Thompson and Fast, 2006). Other direct modifications are hydroxylations of the side chains of proline, lysine, and asparagine, which generate the corresponding hydroxy amino acids (Aravind and Koonin, 2001; Elkins et al., 2003; Iyer et al., 2010; Webby et al., 2009).

All of these modifications can affect both the surface electrostatics and the net size of the modified polypeptide and sterically affect its interactions with nucleic acids and proteins. For example, the acetylation of lysines can reduce the net positive charge, phosphorylation and polyglutamylation can increase the net negative charge, and ubiquitination and poly-ADP-ribosylation can drastically alter the size of the polypeptide (Allis et al., 2007; Fujiki et al., 2009; Kouzarides, 2007; Rouleau et al., 2004; van Dijk et al., 2008; Walsh et al., 2005). In addition, many of these modifications carry epigenetic information, commonly termed the *histone code* (Allis et al., 2007). Introduction of these modifications by specific enzymes can be seen as a coding step, in which extragenetic information is "written" into the histones and other CPs to be transmitted through subsequent cell divisions (Allis et al., 2007; Kouzarides, 2007). Discrimination between modified and unmodified peptides by specific peptide-binding domains, which might then recruit other chromatin remodeling or modifying activities to chromatin, can be conceptualized as the "interpretation" of the epigenetic code (Allis et al., 2007; Jenuwein and Allis, 2001; Kouzarides, 2007). Finally, the removal of these marks by other enzymes can be conceived as "resetting" or erasure of the epigenetic information and usually accompanies major differentiation events or transitions such as post-zygotic development in multicellular organisms (Allis et al., 2007; Hajkova et al., 2010; Mayer et al., 2000). These protein-based marks also functionally interact with both DNA modifications and the RNAi system to comprise the complete complex of epigenetic coding in eukaryotes (Allis et al., 2007; Grewal, 2010). The most prevalent DNA modifications in eukaryotes are methylations of cytosine at the C5 position (5mC) and the adenine at the N6 position (Goll and Bestor, 2005; Iyer et al., 2011a). More recently, it has become clear that the 5mC is not a terminal product and is further modified to 5 hydroxymethylcytosine (5hmC), formylcytosine, and carboxycytosine through successive oxygenation catalyzed by the Tet enzymes (He et al., 2011; Tahiliani et al., 2009). In certain eukaryotes, such as trypanosomes, a similar modification with epigenetic consequences, namely, conversion of thymine to form a novel base (base J), has been reported and is catalyzed by the related JBP enzymes (Borst and Sabatini, 2008). A combination of computational analysis of protein sequences and structures, with experimental investigations, have identified most of the major enzyme classes involved in the generation and erasure of epigenetic marks on proteins as well as the domains that discriminate among them. Another aspect of chromatin remodeling in eukaryotes is the use of multiple distinct types of conserved ATP-dependent engines that alter chromatin structure both on a chromosomal scale and locally (Hirano, 2005; Martens and Winston, 2003; Mohrmann and Verrijzer, 2005; Uhlmann and Hopfner, 2006). These include the ATPases that translocate to different epigenetic marks and locally displace nucleosomes from particular regions of the DNA and alter its accessibility to other CPs. Other ATPases act to cause condensation of whole or partial sections of a chromosome or help in bringing together distantly located regions of chromosomes with concomitant looping or movement of DNA.

15.1.3 General Evolutionary Considerations Relevant to Eukaryotic CPs and Transcription Regulators

Phylogenetic investigations have shown that model organisms represent only a small portion of the vast eukaryotic tree, with much of the bewildering phylogenetic diversity being found in the microbial eukaryotes or "protists" (Cavalier-Smith, 2009; Moon-van der Staay et al., 2001). Hence, a proper understanding of the diversity of eukaryotic transcription regulation and chromatin dynamics across phylogenetically distant lineages will be critical in any attempt to understand eukaryotic evolution. A major boost for these studies has come from the recent large-scale genome sequencing efforts that have generated complete or near-complete genome sequences of several "protists," which are either agents of major parasitic diseases or key players in worldwide ecosystems. In this chapter, we use this treasure-trove of data to present an overview of the key aspects of the transcription-regulatory and chromatin-remodeling apparatus in eukaryotes from a domain-centric viewpoint. Despite the availability of genome-scale data, reconstruction of eukaryotic phylogeny has not been straightforward. Some principal problems confound the determination of higher order relationships amongst eukaryotes: (i) *Rampant gene loss*. This is common throughout the fungal kingdom and is especially pronounced in the microsporidian lineage (Aravind et al., 2000; Peyretaillade et al., 2011). *Entamoeba* amongst amoebozoans, *Cryptosporidium* amongst apicomplexans, and *Giardia* amongst basal eukaryotes also display extreme gene loss relative to their sister lineages. (ii) *Rapid lineage-specific divergence*. Gene loss also spurs concomitant rapid sequence divergence of the proteins that have been retained on account of release from selective constraints due to lost interacting partners (Aravind et al., 2000). (iii) *Lateral gene transfer*. This is a major factor in several eukaryotic lineages such as chromists (stramenopiles) and apicomplexans, which have emerged via secondary or tertiary endosymbiosis involving engulfment of other eukaryotic cells from the plant lineage (Bhattacharya et al., 2004; Bowler et al., 2008). As a result their proteins show chimeric affinities to either those of the original lineage or to those of the endosymbiont's lineage. In addition to these issues, there are controversies concerning the rooting of the eukaryotic tree (Arisue et al., 2005; Dacks and Doolittle, 2001; Walsh and Doolittle, 2005). This has resulted in considerable problems with respect to reconstructing the nature of the last eukaryotic common ancestor (LECA) (Aravind et al., 2006; Cavalier-Smith, 2009; Dacks and Doolittle, 2001). Nevertheless, multiple independent recent studies using large multiprotein datasets and algorithms to correct for differential evolutionary rates have been robustly reproducing several higher order groupings among eukaryotes (Bapteste et al., 2002; Cavalier-Smith, 2009; Iyer et al., 2008b; James et al., 2006). Thus, the reconstructions presented in this chapter must be viewed in light of the above-stated caveats—it might be considered a reasonable general picture for eukaryotes though certain specific conclusions might change with availability of new data from hitherto unsampled organisms.

The most parsimonious reconstruction considering a conservative phylogenetic scenario suggests that the LECA was a flagellated motile organism that already possessed a distinctly larger gene complement (at least ~10,000 genes) than its prokaryotic precursors (Anantharaman et al., 2007; Aravind et al., 2006; Iyer et al., 2008b). This complement coded numerous families of proteins with multiple paralogous members and several novel regulatory systems, including CPs and TFs, with no direct prokaryotic equivalents. Availability of complete genome sequences also allows us to estimate the gross differences in the effects of natural selection on completely conserved orthologous proteins belonging to different functional categories. Examination of residues evolving at different rates in individual functional classes reveals certain interesting features (Fig. 15.1a). The machinery related to protein stability, namely, chaperones and proteasomal subunits, comprise one of the most conserved groups of eukaryotic proteins with the majority of their residues evolving slowly. In contrast, nuclear proteins, especially those related to transcription and chromatin structure and dynamics, display a bimodality of evolutionary rates—a subset of the residues belong to the most slowly evolving category amongst all eukaryotic proteins, whereas another subset is rapidly evolving. Specifically, all core histones that comprise the nucleosomal octamer and parts of the RNA-polymerase catalytic subunits belong to the most slowly evolving categories (Fig. 15.1a). However, there are other parts of the same RNA-polymerase subunits that exhibit amongst the most rapid evolutionary rates of all the universally conserved orthologous proteins. An apparently bimodal pattern of evolutionary rates is also observed amongst proteins comprising the replication apparatus. These observations suggest that while a subset or parts of chromosomal proteins have settled into highly conserved roles since the LECA, the remainder or remaining parts are rapidly diverging, indicating lineage-specific adaptations in these proteins (Fig. 15.1a).

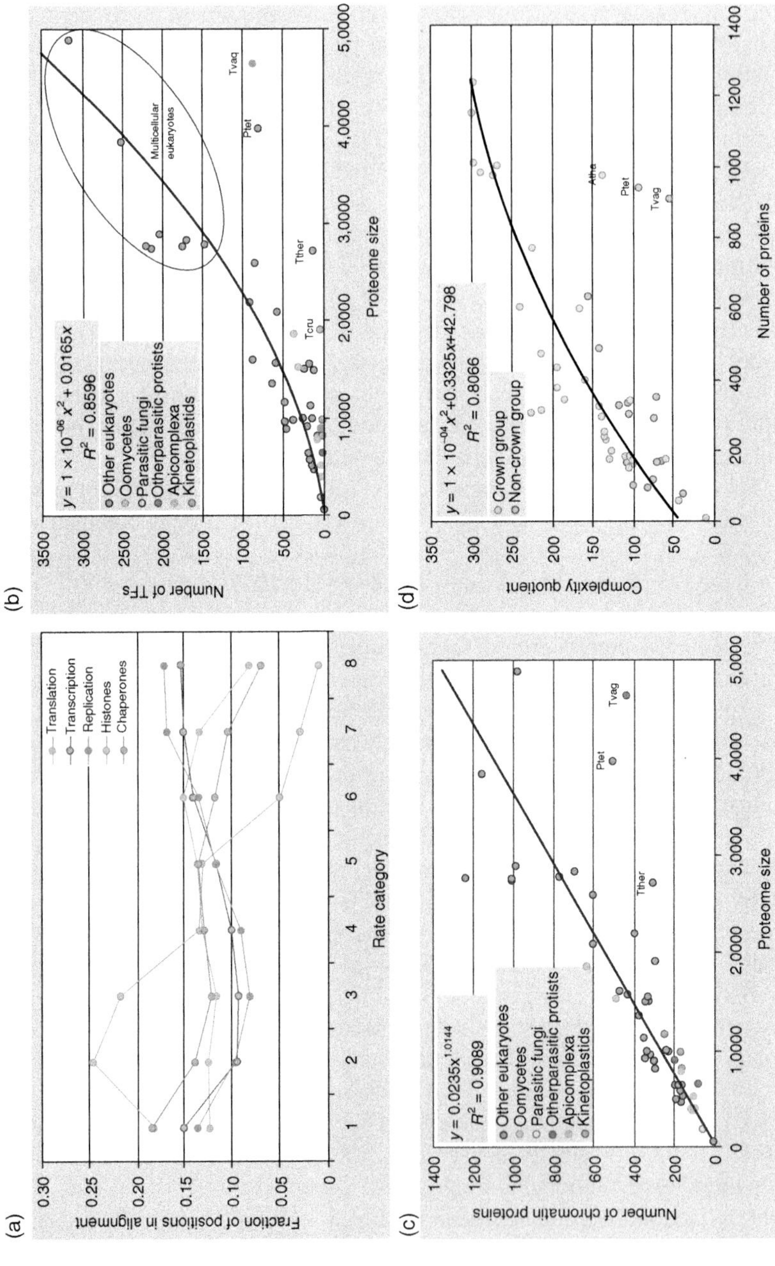
(a)
Translation
Transcription
Replication
Histones
Chaperones
Fraction of positions in alignment
Rate category
(b)
y = 1 × 10^-06 x^2 + 0.0165x
R^2 = 0.8596
Other eukaryotes
Oomycetes
Parasitic fungi
Otherparasitic protists
Apicomplexa
Kinetoplastids
Multicellular eukaryotes
Tvaq
Ptet
Tther
Tcru
Number of TFs
Proteome size
(c)
y = 0.0235x^1.0144
R^2 = 0.9089
Other eukaryotes
Oomycetes
Parasitic fungi
Otherparasitic protists
Apicomplexa
Kinetoplastids
Ptet
Tvag
Tther
Number of chromatin proteins
Proteome size
(d)
y = 1 × 10^-04 x^2+0.3325x+42.798
R^2 = 0.8066
Crown group
Non-crown group
Atha
Ptet
Tvag
Complexity quotient
Number of proteins

Figure 15.1 Differences in rate categories in different functional classes, scaling of TFs and CPs, and complexity quotient plots. (a) Among-site rate variation for different functional classes of eukaryotic proteins. These were calculated using multiple alignments of highly conserved proteins that are present in all eukaryotes in each functional category shown in the graph. The total numbers of positions in each category were—translation: 6,357; transcription: 2,275; replication: 5,436; histones: 381; chaperones: 5,154. A Whelan and Goldman (WAG) substitution matrix was used for the calculation. The fraction of the positions in each rate category is plotted for each functional class—the categories on the left evolve slower than those on the right. Note that the distribution for transcription and replication proteins is U-shaped, indicating an over-representation of extremes—slowest evolving and fastest evolving positions. (b) Scaling of TFs with proteome size. The names of organisms used for the plot and their abbreviations are indicated below. Organisms with a significantly lower-than-expected fraction of CPs are labeled. (c) Scaling of CPs with proteome size. The organisms are the same as in (b). Organisms with a lower-than-expected fraction of CPs are marked. (d) Complexity quotient plot for CPs. The "complexity quotient" for an organism is defined as the product of the number of different types of domains that co-occur in signaling proteins, and the average number of domains detected in these proteins. The complexity quotient is plotted against the total number of CPs in a given organism. A polynomial curve fitting the general trend of the majority of organisms is shown. Crown group members are shown in orange and the non-crown group members are in green. Some organisms with much lower complexity than those along the general trend are marked. Each protein has at least a single known or predicted domain with a chromatin/transcription-related function. A total of 363 domains were considered, among which 121 were domains specifically found in chromatin and TFs, and the rest were other domains with wider distributions encompassing other functional systems. The organisms included in all these plots and their species abbreviations are as follows: *Aspergillus fumigatus*, Afum; *Candida glabrata*, Cgla; *Debaryomyces hansenii*, Dhan; *Ashbya gossypii*, Egos; *Ectocarpus siliculosus*, Esil; *Gibberella zeae*, Gzea; *Kluyveromyces lactis*, Klac; *Neurospora crassa*, Ncra; *Saccharomyces cerevisiae*, Scer; *Schizosaccharomyces pombe*, Spom; *Yarrowia lipolytica*, Ylip; *Cryptococcus neoformans*, Cneo; *Ustilago maydis*, Umay; *Encephalitozoon cuniculi*, Ecun; *Anopheles gambiae*, Agam; *Apis mellifera*, Amel; *Branchiostoma floridae*, Bflo; *Caenorhabditis elegans*, Cele; *Ciona intestinalis*, Cint; *Danio rerio*, Drer; *Drosophila melanogaster*, Dmel; *Homo sapiens*, Hsap; *Mus musculus*, Mmus; *Pan troglodytes*, Ptro; *Rattus norvegicus*, Rnor; *Strongylocentrotus purpuratus*, Spur; *Tetraodon nigroviridis*, Tnig; *Tribolium castaneum*, Tcas; *Monosiga brevicollis*, Mbre; *Nematostella vectensis*, Nvec; *Entamoeba histolytica*, Ehis; *Dictyostelium discoideum*, Ddis; *Chlamydomonas reinhardtii*, Crei; *Ostreococcus tauri*, Otau; *Arabidopsis thaliana*, Atha; *Phaeodactylum tricornutum*, Ptri; *Phytophthora sojae*, Psoj; *Phytophthora ramorum*, Pram; *Thalassiosira pseudonana*, Tpse; *Tetrahymena thermophila*, Tthe; *Paramecium Tetraurelia*, PTet; *Toxoplasma gondii*, Tgon; *Theileria parva*, Tpar; *Theileria annulata*, Tann; *Cryptosporidium parvum*, Cpar; *Plasmodium falciparum*, Pfal; *Trypanosoma cruzi*, Tcru; *Trypanosoma brucei*, Tbru; *Leishmania major*, Lmaj; *Naegleria gruberi*, Ngru; *Giardia lamblia*, Glam; *Trichomonas vaginalis*, Tvag; *Guillardia theta*, Gthe. Additional species abbreviations used in various figures include: *Albugo laibachii*, Alai; *Chlorella variabilis*, Cvar; *Phytophthora infestans*, Pinf. Crown group eukaryotes include animals, plants, fungi, and amoebozoa. (*See insert for color representation of the figure.*)

15.2 EUKARYOTIC CPs AND TFs

15.2.1 Demographic Patterns in the Distribution of TFs and CPs

Generation of sensitive position-specific score matrices and hidden Markov models for conserved domains found in TFs (typically their DBD) and CPs allows their exhaustive and systematic detection across all complete eukaryotic proteomes (Babu et al., 2004; Finn et al., 2006; Iyer et al., 2008b). As a result, reasonably robust counts or demography of potential TFs and CPs encoded by a given organism can be obtained. These indicate positive correlations between the number of CPs or TFs coded by an organism and its proteome size (Fig. 15.1b and c). These trends are best approximated by linear or mildly nonlinear fits (weak quadratic fit for TFs or weak power-law in chromatin factors) suggesting that, in general, there is a proportional increase in the number of TFs for an increasing number of protein-coding genes. The trend observed in TFs is in contrast to that seen in prokaryotes wherein a fit to a much stronger power-law trend is observed (Aravind et al., 2005; Babu et al., 2004). In prokaryotes, there are few dedicated CPs and their number does not vary significantly with proteome size (Dorman, 2009). This suggests that eukaryotes might optimize their transcription-regulatory potential by increasing the numbers of both TFs and chromosomal proteins as their gene numbers increase. Parasites belonging to fungal, apicomplexan, and stramenopile lineages show greater or lesser degrees of gene loss in comparison with their free-living sister clades, but typically counts of their TFs and CPs do not deviate to a large extent from the general trend observed across eukaryotes. Hence, despite a degree of genomic reduction, the overall regulatory input per protein-coding gene in these parasites is roughly comparable with other eukaryotes.

Significant exceptions to the general eukaryotic trend in TFs are observed in trypanosomes, while *Trichomonas vaginalis* and ciliates displayed significant deviations in counts of both their TFs and CPs (Fig. 15.1b and c). Notably, trypanosomes appear to have hardly any of the known families of TFs. This might mean that they either possess unique TFs that are unrelated to any previously characterized versions or perhaps, they use a radically distinct mode of regulation (see below). In *T. vaginalis* and ciliates the absolute counts of TFs and CPs exceed those seen in other parasites or free-living protists. However, their proteome size is similar to that of multicellular animals and plants, and as result, they have relatively fewer TFs and CPs for their proteome sizes compared to multicellular forms (Fig. 15.1b and c). This might reflect different parallel causes: (i) Multicellular forms show both temporal transcriptional changes during development and spatially differentiated cell types with diverse gene-expression states (Aravind et al., 2009; Meyerowitz, 1999, 2002; Rokas, 2008). In contrast, *T. vaginalis* and ciliates only show simple temporal cell state transitions during development and lack equivalents of multiple differentiated cell fates (Carlton et al., 2007). Consequently, lower normalized counts of TFs in these organisms might reflect differences in the amount of transcriptional control required to regulate similarly sized genomes in the unicellular context (*T. vaginalis* or ciliates) as opposed to multicellular forms with differentiation. (ii) These protists also show tremendous genetic redundancy with several closely related or near-identical gene copies that, rather than being differentially regulated, might merely provide higher effective concentrations of particular gene products. The gene counts, especially in *T. vaginalis*, are also exaggerated by numerous transposable elements of diverse types.

15.3 DIVERSITY OF EUKARYOTIC-SPECIFIC TFs

15.3.1 LSEs Characterize the Evolutionary History of Specific TFs

The largest families of TFs in a given genome differ widely across major eukaryotic lineages: for example, in multicellular plants TFs with the MADS, VP1, and Apetala2 (AP2) DBDs are most prevalent, whereas in animals TFs containing homeodomains and C2H2 ZnFs are dominant, and in fungi and the heterolobosean amoeba *Naegleria* the C6-binuclear ZnFs are dominant (Fig. 15.2) (Iyer et al., 2008b; Lespinet et al., 2002). It is notable that this lineage-specific diversity of specific TFs exists despite a fairly strong global trend in TF demography across eukaryotes (Fig. 15.1b). This points to a general constraint in terms of the number of TFs required to regulate a proteome of a given size, even though there appears to be no major constraint on the actual type of TF being deployed (i.e. their evolutionary origin). A corollary is that different superfamilies of TFs have independently expanded in each major lineage to convergently produce overall counts corresponding to that dictated by the general constraint (Figs 15.1b and 15.2). In practical terms, this often makes their prediction in poorly studied eukaryotic lineages a difficult task. This was poignantly illustrated by the apicomplexans, where multiple studies had initially failed to recover *bona fide* specific TFs (Gardner et al., 2002; Templeton et al., 2004). However, sensitive sequence profile analysis revealed a major lineage-specific expanded family of proteins (ApiAP2 family) with one or more copies of the AP2 DBD, similar to those found in plant AP2 TFs, to be present in all studied apicomplexan clades from *Cryptosporidium* to *Plasmodium* (Balaji et al., 2005) (Fig. 15.2). Further analysis of expression of the ApiAP2 genes in the course of the *Plasmodium* developmental cycle showed that they clustered into specific co-expression guilds that notably corresponded to the major development stages, namely, the ring, trophozoite, early schizont, and schizogony/merozoite. Analysis of physical interactions of ApiAP2 proteins based on published large-scale protein interaction data revealed homo- and heterodimeric interaction with other ApiAP2 proteins, as well as interaction with various CPs such as the GCN5 histone acetyltransferase, CHD1 and Rad5/16-type SWI2/SNF2 ATPases, and the HMG1 ortholog (*Plasmodium* MAL8P1.72) (Iyer et al., 2008b). These observations suggested that the ApiAP2 proteins are indeed the predominant specific TFs of apicomplexans, and are likely to function similarly to their counterparts from model organisms by recruiting histone-modifying and chromatin-remodeling factors to their target sites. This prediction from computational analysis has subsequently been confirmed by numerous follow-up experimental studies (Painter et al., 2011).

Similar expansions of uncharacterized TFs are seen in other eukaryotic lineages, such as *T. vaginalis*. Transcription initiation in this organism is primarily dependent on the protein IBP39, which binds the initiator element (Inr) by means of a specialized winged HTH (wHTH) domain, termed the *IBD*, and recruits the RNA polymerase via its C-terminal tail (Lau et al., 2006; Schumacher et al., 2003). In the IBD, the recognition helix of the wHTH binds the major groove of DNA, while a positively charged loop from a bi-helical hairpin from the distinctive N-terminal extension of the IBD contacts the adjacent minor groove (Schumacher et al., 2003). This novel DBD, while containing an ancient protein fold, has no close relatives in other currently studied organisms. Sequence profile searches revealed that the IBD defines a novel family of lineage-specific TFs comprising of at least 100 proteins in the *T. vaginalis* proteome, containing single IBDs and organization congruent to IBP39 (Iyer et al., 2008b) (Fig. 15.2). This suggests that the IBD indeed defines a lineage-specific DBD that is utilized by a large family of specific TFs in this organism. Sequence divergence in the recognition helix as well as the N-terminal positively charged loop across the IBD family suggests that different versions of the domain have potentially specialized to contact a range of target sites, other than the *T. vaginalis* INR. Likewise, the MYB domain shows enormous LSEs in multicellular plants, the free-living ciliate *Paramecium* and phylogenetically distant eukaryotes such as *T. vaginalis*, *Entamoeba histolytica*, and *Naegleria gruberi* (Iyer et al., 2008b). The expanded MYB proteins are likely to constitute the predominant specific TFs in *E. histolytica* (Fig. 15.2). Other examples of major independent LSEs of TFs observed both in diverse parasites and free-living protist groups include the bZIP domain in *Phytophthora* and *Paramecium*, and the heat-shock factor (HSF) in most stramenopiles and *Paramecium* and CXC domain TFs in *Paramecium* (Iyer et al., 2008b) (Fig. 15.2). While the C2H2 Zn-finger (ZnF) is prevalent in most eukaryotic lineages, its rise in each lineage appears to be a result of independent LSEs (Fig. 15.2) (Lander et al., 2001).

A potential corollary to these observations is that the lineage-specific variety in specific TFs might provide the regulatory basis for the emergence of enormous biodiversity in terms of structure, life styles and life cycles across the eukaryotic evolutionary tree (Aravind et al., 2009). One of the classical examples of LSEs is that of the family of TFs with the POZ (also called BTB) domain (Aravind and Koonin, 1999a). These TFs have an N-terminal POZ domain combined with a C-terminal DBD that is usually a C2H2 ZnF. Both vertebrates and insects have large numbers of these TFs (over 50 paralogs per genome). However, phylogenetic analysis reveals that these expansions happened independently in the insect and vertebrate lineages after they separated from their common ancestor—the vertebrate POZ domain TFs group with each other to the exclusion of the insect versions, and similarly the insect versions group with themselves to the exclusion of the vertebrate forms. Developmental genetics in model systems have

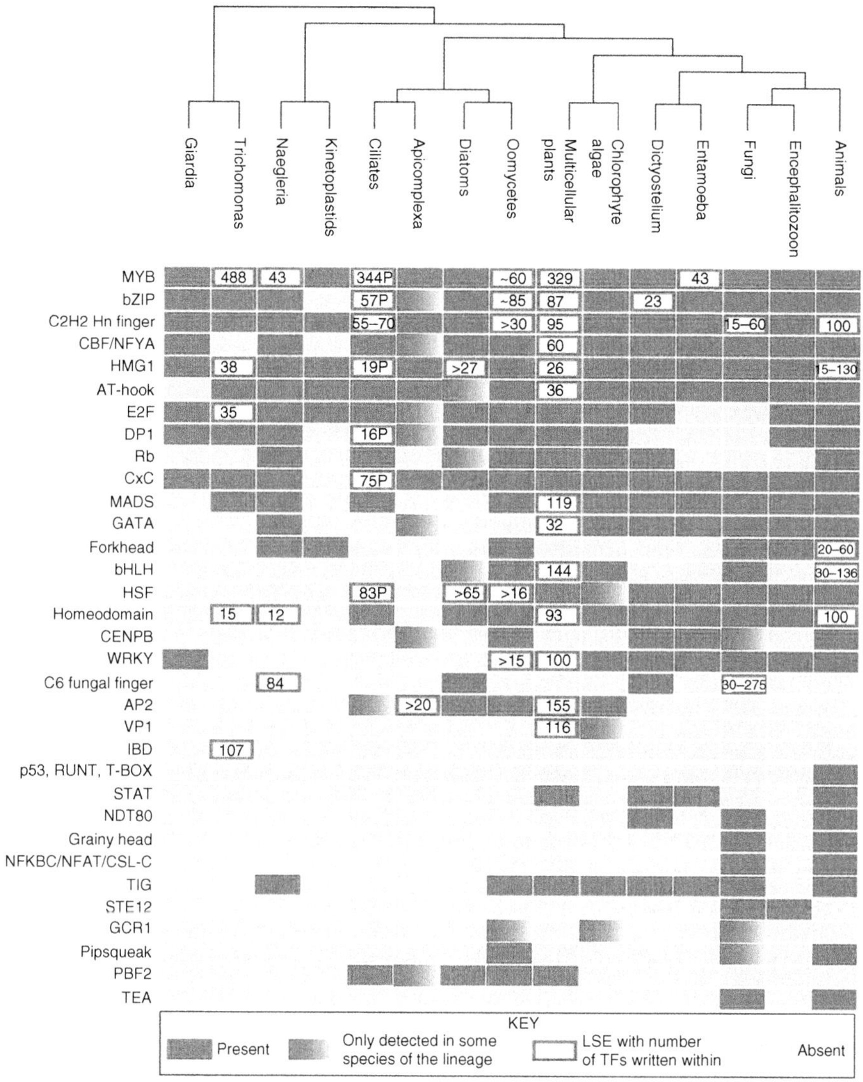

Figure 15.2 Lineage-specific expansions (LSEs) and phyletic distributions of specific transcription factors (TFs). Only those specific TFs that are widely present across eukaryotes and have LSEs or notable sporadic phyletic patterns are shown. The distribution of the TFs across eukaryotic species is shown below the eukaryotic tree. The key below the distribution gives the notations used to describe presence, absence, or LSEs in TFs. A "P" or a "Ps" next to the number of TFs in the ciliate and oomycete columns represents LSE in *Paramecium* and *Phytophthora sojae*, respectively.

shown that many key developmental processes are regulated by TFs belonging to lineage-specific clades (e.g., p53 in multicellular animals) or LSEs (e.g., POZ domain TFs) (Lespinet et al., 2002; Rutkowski et al., 2010; Spokony and Restifo, 2007). Gene products of the LSE of POZ domain TFs have been recovered as regulating a diverse range of developmental decisions in *Drosophila melanogaster* at the transcription level in contexts such as axonal path finding (*Lola*), morphological diversity of neuronal dendrites (*abrupt*), specification of neurons that determine sexual orientation (*Fruitless*), specification of cell fates in the eye (*Tramtrack*), the development of distinctive external genitalia (*ken* and *barbie*), epithelial morphogenesis (*ribbon*), and early oogenesis (*Pipsqueak*), to name just a few representatives (Tweedie et al., 2009). This functional "colonization" of a large number of disparate functions after the LSE suggests that this phenomenon might have had a major role in driving the diversification of morphology and adaptations in multicellular forms (Aravind et al., 2009). These LSEs also imply that the transcription regulatory networks might undergo massive reorganization and rewiring with the emergence of new lineages. This is consistent with studies on Tnets, which indicate that hubs are routinely displaced by new TFs or that hubs are lost and new hubs emerge in their place (Babu et al., 2004).

15.3.2 The DNA-Binding Domains of Eukaryotic Specific TFs and Their Evolutionary Origins

A survey of DBDs found in specific TFs shows that there are at least 50 distinct superfamilies spanning all structural classes, with some of those present in almost all eukaryotic lineages studied to date (Fig. 15.2). This latter group contains at least seven distinct DBDs, namely, the Basic-zipper (bZIP), C2H2 ZnF, HMG box, AT-hook, MYB, CBF/NFYA, and E2F/DP1 DBDs. These, along with DBDs of general TFs such as TBP, TFIIB, TFIIE, and MBF, which are shared with archaea, and the BRIGHT/ARID which emerged in eukaryotes, comprise the set of DBDs in TFs that can be confidently traced to the LECA. Although the majority of DBDs in the ancient set shared with archaea contain a HTH fold domain, amongst the early eukaryotic innovations only the BRIGHT domain possesses this fold (Aravind et al., 2005). This suggests that recruitment of a structurally diverse set of DBDs, not just limited to the HTH domain, had already begun early in the evolution of eukaryotes (Aravind et al., 2005; Iyer et al., 2008b). The wide distribution of specific TFs with several other DBDs, such as the MADS, GATA, and Forkhead (FKH) domains, in early branching eukaryotes also suggests a relatively ancient origin for these proteins in eukaryotic evolution (Fig. 15.2). Another major round of innovation of TFs, with new DBDs such as the CENPB, HSF, and bHLH domains, appears to have happened prior to divergence of the clade including ophisthokonts, amoebozoans, plants, and the chromalveolates (Iyer et al., 2008b). Finally, there were extensive innovations of several other DBDs within the crown group, defined herein as the assemblage of eukaryotes uniting ophisthokonts, amoebozoans, and plants—for example, the fast-evolving p53-like DBDs displaying a cytochrome f-like fold (Rutkowski et al., 2010). The earliest representatives of this fold were present in the common ancestor of the ophisthokonts, amoebozoans, and plants and were typified by the DBD of the STAT proteins (Fig. 15.2) (Hoey and Schindler, 1998; Kawata, 2011). The p53-like fold subsequently appears to have diversified greatly in animals and fungi giving rise to at least five distinct families (p53, RUNT, T-BOX, Grainyhead/CP2, ND80, and NFκB/NFAT/CSL N-terminal domain); of these p53 proper emerged early in the history of multicellular animals.

Irrespective of their point of origin, individual eukaryote-specific TFs show highly variable demographic patterns. For example, the AP2 domain has been independently expanded in both multicellular plants and apicomplexa but is present in very low numbers in its respective immediate sister groups, namely, the chlorophyte algae (*Chlamydomonas* and *Ostreococcus*) and ciliates (Balaji et al., 2005). Several families of TFs are shared by animals and plants or amoebozoans to the exclusion of the fungi (Iyer et al., 2008b) (Fig. 15.2). However, phylogenetic analysis strongly supports the exclusive grouping of animals and fungi, suggesting loss in the latter (Fig. 15.2). A striking case is furnished by the dimeric E2F and DP1 TFs, which are present in animals, amoebozoans, plants, chromalveolates, and basal eukaryotes such as *Trichomonas* and *Giardia*, while being absent in all fungal lineages except the basal microsporidian clade (Templeton et al., 2004). This pattern is highly suggestive of secondary loss of this ancient TF in the other fungi after their separation from microsporidians. In contrast, some TFs such as PBF2/TIF1, exclusively shared by plants and chromalveolates, might have been acquired by the latter during endosymbiotic association with the plant lineage (Oakley et al., 2007; Yakisich et al., 2006). A specific version of the WRKY TF is shared by plants, the plant parasite *Phytophthora* (shows a notable expansion of over 20 copies), and *Giardia*. The C6 finger, which was believed to be exclusive to the fungal lineage, has now been found in *Dictyostelium*, the stramenopile alga *Thalassiosira*, and *Naegleria* (Fig. 15.2) (Iyer et al., 2008b). The sporadic phyletic patterns of the WRKY and C6 domains in the protists are possibly the consequence of lateral transfer between the different eukaryotic lineages after their divergence (Babu et al., 2006). Thus, both gene losses and lateral transfers appear to contribute to the sporadic phyletic patterns of eukaryotic TF superfamilies. Differentiating between these alternatives is not always straightforward with the current state of the data. For example, the multiple lineages of the homeodomain superfamily are found in ophisthokont, amoebozoan, plant, and stramenopile lineages (Burglin, 2003; Gehring et al., 1994). Of these different lineages of homeodomains only the atypical TALE family of homeodomains (Burglin, 1997) is found sporadically in ciliates, stramenopiles, *Naegleria*, and *Trichomonas*. This could either imply that the TALE class represents the most ancient version of the homeodomain (with gene losses in many lineages) or that it was introduced through lateral transfer from one of the lineages with more diversified homeodomain complements.

This leads to the question regarding the ultimate origin of eukaryotic TFs. Several families, such as the BRIGHT, homeo, POU, paired, HSF, IBD, TEA, FKH, and pipsqueak domains contain the HTH fold, albeit only distantly related to that seen in prokaryotic TFs (Aravind et al., 2005). Hence, they could have potentially emerged through rapid diversification of older HTH domains inherited from prokaryotes. Certain other non-HTH DBDs, such as the C2H2 Znf and the AP2 domain might have also been acquired from some of the less frequent families of prokaryotic specific TFs (Iyer and Aravind, 2011). DBDs of TFs displaying the TIG domain-type immunoglobulin fold and the cytochrome f-like fold can be unified with the homologs of these two domains found in bacterial cell surface adhesins (Aravind and Koonin, 1999b). Thus, it is possible that the ultimate origin of the eukaryotic TFs with these DBDs lies among bacterial surface proteins (Iyer et al., 2008b). Several specific TFs appear to have been derived from DBDs of transposases and other selfish elements (Babu et al., 2006; Balaji et al., 2005; Smit and Riggs, 1996). Examples of major eukaryotic DBDs that appear to have a transposon origin are the WRKY, AP2, VP1, paired, pipsqueak, CENPBP, APSES, BED-finger, and GCR1 domains. While bacteria contain a few MYB domain TFs, the eukaryotic versions appear to be closer to MYB domain proteins encoded by a newly identified group of prokaryotic transposons (Iyer and Aravind, 2011). Thus, the eukaryotic MYB domain TFs might also have a transposon provenance. Typically, inactive mobile elements that have lost the catalytic activity of their transposase domain but retain their DBD appear to be "recycled" as new TFs. On the other hand, PBF2/TIF1 appears to have been acquired first by plants from a bacterial plasmid, where a cognate DNA-binding protein plays a role in plasmid segregation or replication (Oakley et al., 2007). Finally, as in the case of many other functional classes, eukaryotes have innovated TFs with DBDs containing entirely new folds. These are mostly α-helical or metal-chelation-supported structures, consistent with the greater "ease" with which such structures might be innovated *de novo* (Aravind et al. 2006).

15.4 THE NATURAL HISTORY AND EVOLUTION OF MAJOR FUNCTIONAL TYPES OF CPs

Advances in genomics and sequence analysis of the eukaryotic CPs allow us to answer certain questions more robustly: (i) What are the major functional and structural categories of CP domains? (ii) What was the complement of CP domains and CPs functioning in LECA? (iii) What were the lineage-specific innovations in eukaryotic CPs? (iv) What implications do differences in complements of CPs have for the epigenetic regulation (e.g., generation and "interpretation" of the histone code)? However, answering these questions is contingent on a reasonable definition of CPs and the domain found in them.

15.4.1 Definition and Characterization of CP Domains

It is impossible to precisely compartmentalize the disparate regulatory and structural complexes in chromatin from the complexes primarily participating in essential housekeeping processes such as replication, recombination, DNA repair, and transcription. Nevertheless, herein we adopt a restricted definition for CPs by focusing chiefly on "regulatory" and structural components. These regulatory components chiefly include enzymes catalyzing modifications of both regulatory and structural proteins including histones. These enzymes typically function in conjunction with energy-driven chromatin-remodeling enzymes. Some modifications attract other proteins that bind to modified peptides, and, in several cases, such interactions with histones have been termed *reading* of the histone code (Allis et al., 2007; Jenuwein and Allis, 2001; Peterson and Laniel, 2004). Domains mediating discrimination of these modifications as well as the covalent modifications of bases in DNA are also included in our definition of CP domains. The distinctness of this set of proteins being defined here as CPs is primarily supported by the observation that they are mostly comprised of a relatively small set of conserved protein domains (about 70–80), the majority of which are found largely in eukaryotic CPs (Table 15.1). These domains can be placed, in large part, under two broad biochemical categories: (i) non-catalytic interaction or adaptor domains and (ii) enzymatic regulatory domains. The former category can again be further subdivided into DNA-binding and protein–protein interaction domains (see Table 15.1 for summary). We first briefly discuss the DBDs and then consider the remaining domains.

15.4.2 DNA-Binding Domains in CPs

The most basic DNA–protein interaction in eukaryotic chromatin is mediated by the four core histones that are universally conserved in all eukaryotes (Allis et al., 2007; Allis et al., 2006; Woodcock, 2006). In addition to the core histones, there are other homologous histone fold proteins, namely, the smaller TAFs (TATA-binding protein associated factors) and general TFs such as NFYB and NFYC that appear to form octamer-like structures in the context of transcription initiation complexes (Gangloff et al., 2001). The four core histones, NFYB, NFYC, and at least three of the TAFs with a histone fold (TAF6, TAF8, and TAF12) had diverged from each other by the time of the LECA (Iyer et al., 2008b). Interestingly, these TAFs and the slightly later derived paralog TAF9 were independently, repeatedly lost in most or all apicomplexans and all kinetoplastids. The four core nucleosomal histones often show variants that have been shown in model systems to specify "specialized chromatin" in the regions where they are deposited on DNA. For example, centromere-specific histone H3 (cenH3) is critical for the assembly of the kinetochore complexes (Dalal et al., 2007). The kinetoplastids, on the other hand, contain rapidly evolving histones such as H4, which might indicate adaptive evolution (Lukes and Maslov, 2000). The histone variant H2A.Z is strongly anticorrelated with DNA cytosine methylation across eukaryotes (Zilberman et al., 2008). More generally, H2A.Z is correlated with active chromatin and the prevention of the spread of repressive heterochromatin into euchromatic regions, even in eukaryotes with no DNA methylation such as *Saccharomyces cerevisiae* (Mizuguchi et al., 2004). However, it is absent in most basal eukaryotes such as *Trichomonas*

TABLE 15.1 **Selected Domains Commonly Found in Chromatin Proteins**

Domain	Structure	Comments
Enzymatic domains		
Acetyltransferases (GNAT)	$\alpha + \beta$ fold with six core strands	No particular universally conserved active site residues but a structurally conserved acetyl coA binding loop
RPD3/HDAC-like deacetylases	Haloacid dehalogenase class of Rossmanoid folds	Chelates active metal using two conserved aspartates and one histidine. Most prokaryotic versions act on small molecule metabolites
Sir2-like deacetylases	Classical 6-stranded dehydrogenase-type Rossmann fold with a Zn-ribbon insert	Contains a specific active site with a conserved histidine, which is required for the NAD-dependent deacetylation
MACRO domain	Derived α/β fold with N-terminal β-hairpin in core sheet	There are at least eight independent transfers of this domain from prokaryotes and are probably involved in several distinct hydrolytic reactions involving NAD derivatives and ADP-ribose. For example, the POA1 proteins are cyclic phosphodiesterases that break down ADP-ribose $1'',2''$-cyclic phosphate during tRNA splicing
SET-like methylases	β-clip fold	The internal duplication in the β-clip fold creates two distinct active site pockets, one for AdoMet binding and the other for the peptide
Rossmann fold protein methyltransferases	Classical 7-stranded Rossmann fold	This domain is found in PRMT-like histone arginine methyltransferases where it is combined with a C-terminal avidin-like all β strand domain. The DOT1–like methylases have an extended element followed by a helix leading to the core Rossmann fold
Jumonji-related (JOR/JmjC) domain	Double-stranded β-helix fold	The active site consists of two histidines and an asparate that chelate an active metal, typically iron. They use a 2-oxoglutarate cofactor
LSD1-like demethylase	Classical dehydrogenase-type Rossmann fold	This enzyme is also believed to catalyze demethylation by an oxidative process but utilizes the classical flavin moiety as many other classical Rossmann fold enzymes

SWI2/SNF2 ATPase	Superfamily-II helicase type P-loop ATPase. Tandem duplication of two P-loop fold domains	These ATPases share with ERCC4 and ERCC3 a trihelical unit after the first strand of the second P-loop domain. The second and third helices are contiguous and are interrupted by a helix-breaking loop. The SWI2/SNF2 ATPases have a conserved histidine between the second and third helix that distinguishes them from the other closely related members of SF-II
MORC ATPase	-Gyrase B subunit- Hsp90 fold Histidine kinase-MutL domain fused to a S5-like domain	The S5-like domain provides a conserved lysine that acts a stabilizer of the hypercharged state of the phosphate group during hydrolysis
SMC ATPases	ABC superfamily of P-loop ATPases with a massive coiled-coil insert within the ATPase fold	SMC proteins are distinguished from all other members of the coiled-coil insert-containing ABC ATPases by the presence of a distinctive hinge domain
DNA methylase	Classical 7-stranded Rossmann fold	The DNA methylases, which act on adenine, tend to have a conserved [ND]PP[FY] motif. Those that act on cytosines are related to the RNA cytosine methylases but have a distinct catalytic cysteine
DNA base Hydroxylase/dioxygenase domain	Double-stranded β-helix fold	Found in the Tet/JBP family of enzymes that oxidatively modify bases in DNA. Distant homologs of AlkB. The animal Tet proteins contain a large cysteine-rich insert within the DSBH fold
DNA-binding domains		
Histone fold	Trihelical fold with long central helix	At least nine distinct members of this fold were present in LECA, including the core nucleosomal histones
Histone H1	Winged HTH domain	Possibly derived from the forkhead domain
HMG box	Simple trihelical fold	A eukaryote-specific DNA-binding domain, with at least a single representative in LECA, which might have functioned as a chromosome structural protein. Expansions of this domain are found in *Trichomonas* and diatoms suggesting that these might be deployed as TFs

(*Continued*)

TABLE 15.1 (*Continued*)

Domain	Structure	Comments
AT-hook	Flap-like element with projecting basic residues	A small motif that binds the DNA minor groove. Many versions show preference for AT-rich DNA and are often found as an accessory domain coupled with a larger globular DBD
CXXC	Binuclear ZnF with 8-metal chelating cysteines	The fold shows a duplication of a core CXXCXXCX(n) unit with the second unit inserted into the first. Versions in several alga are mono-CXXC domains with just a single unit
CXC	A trinuclear Zn cluster	Three extended segments bear rows of cysteines that cooperatively chelate Zn. *Paramecium* shows a lineage-specific expansion of CXC domain proteins that might function as specific TFs in this organism
BRIGHT (ARID)	Tetrahelical HTH domain	Shows a preference for AT-rich DNA. The ancestral version traceable to LECA might have been a core component of the chromatin-remodeling complex containing the brahma ortholog
JBP1C	Multihelical HTH domain	Found only in euglenozoans and specifically binds DNA containing the base J
SAND (KDWK)	SH3-like β-barrel	Most versions contain a conserved KDWK motif that forms part of the DNA-binding motif. Currently known only from animal and plant lineages
TAM (methylated DNA-binding domain—MBD)	Simple fold with three strands and helix	Found in animals, plants, and stramenopiles. Apparently lost in fungi and amoebozoans
SAD (SRA)	PUA/ASCH domain–like β-barrel fold	Methylated DNA-binding domain with conserved N-terminal histidine and C-terminal YDG signature. Of bacterial origin and fused to McrA-type HNH (Endonuclease VII) endonucleases in them
HIRAN	All β-fold	Typically fused to SWI2/SNF2 ATPases in eukaryotes. Found as a stand-alone domain in bacteria in conserved operons encoding a range of phage replication enzymes

PARP finger	Single Zn coordinated by three cysteines and histidine	Prototyped by the Zn-finger found in crown group polyADP-ribose polymerases. Appears to be a specialized nicked and damaged DNA sensing domain
RAD18 finger	Single Zn coordinated by three cysteines and histidine	Prototyped by the Zn-finger found in RAD18p and some Y-family DNA polymerases and SNM1-like nucleases
Ku	7-stranded β-barrel	Contains an extended insert in the β-barrel fold that encircles DNA. Related to the so-called SPOC domain found in the histone deacetylase complex proteins such as SHARP
Helix-extension-helix fold	Trihelical domain with a characteristic extended region between the second and third helix	Two superfamilies, namely, the SAP and LEM domains, contain this fold and are involved in the distinctive function of binding nuclear envelope associated DNA or tethering chromosomes to the nuclear membrane. The version traceable to LECA, in Src1p orthologs, appears to be the precursor of the SAP and LEM domains. Eukaryotic versions might be derived from bacteriophages
HARE-HTH	Winged HTH domain	Predicted to bind DNA and possibly discriminate modified bases. Found in eukaryotes and bacterial restriction enzymes and firmicute RNA polymerase delta subunit
Peptide-binding domains		
Bromo domain	Left-handed tetrahelical bundle	Contains an unusually structured loop between helix 1 and helix 2, which is critical for recognition of the acetylated peptide
SRI domain	Trihelical domain	Typically found at the C-termini of proteins with SET, RING, LSD1, RecQ helicase, and splicing factor domains. It links different catalytic and modifying activities to the gene bodies via association with the phosphorylated RNA polymerase C-terminal domain
Chromo (includes AGENET, MBT)	SH3-like β barrel	Some versions (e.g., in HP1) exhibit a truncated SH3-like barrel with loss of the N-terminal β-hairpin of the barrel and contain an extended C-terminal helix; binds methylated peptides

(*Continued*)

TABLE 15.1 ***(Continued)***

Domain	Structure	Comments
TUDOR	SH3-like β barrel	Some versions are found in RNA associated proteins of splicing complexes; they bind dimethylated arginines
BMB (PWWP)	SH3-like β barrel	This version of the SH3 fold is related to the TUDOR domain; binds methylated peptides
Aux-RF domain	SH3-like β barrel	Predicted methylated peptide-binding domain. Found in the plant auxin-response transcription factor and shows a massive LSE in plants
BAM/BAH	SH3-like β barrel	Contains an extensive elaboration with additional helical and β-stranded inserts
SANTA	β-rich structure	Usually found N-terminal to the SANT domain in crown group and heteroloboseans
PHD finger	Treble-clef fold with binuclear Zn-chelation sites	Apparently entirely absent in *Entamoeba*. ZfCW/PHD-X is a degenerate version of the PHD finger, which has lost the first Zn-chelation site. Binds methylated peptides, certain acetylated peptides, and unmodified lysines in peptides
ZZ-finger	Treble-clef fold with binuclear Zn-chelation sites	Earliest versions traceable to LECA are present in ADA2 orthologs
SJA (Set JOR associated domains)	$\alpha + \beta$ Similar fold as the POLO domain fused to kinases	Erroneously classified as two distinct domains FYRN and FYRC in domain databases. Found associated with SET and JOR domains. Might recruit both histone methylases and demethylases to target peptides
SWIB	Duplication of a core β-α-β-α-β unit with a swapping of the terminal strands between the two units. The helices form a bundle.	Stand-alone version traceable to LECA is a part of the SWI2/SNF2 chromatin-remodeling complex. The oomycete *Phytophothora* has an LSE of this domain. SWIB co-occurs with the SET domain in several bacteria

BRCT	α/β Rossmanoid topology	Domain of bacterial origin in LECA. Several eukaryotic versions bind phosphorylated peptides in the context of DNA repair
Other chromatin-associated domains		
KRI motif	Single α-helix	This helical motif is found fused to the SET methylases, the HhH-DNA glycosylases, and KRAB-type zinc finger proteins. Might play a role in recognition of epigenetic modification in histones like methylation
HUN	α-helix followed by an extended region	The helix has a conserved periodic pattern of acidic residues, which might be required for interaction with histone tails. Found in histone chaperones like HPC2 and ubinuclein
Scm3	α-helix followed by β-hairpin	CenH3 chaperone. Fused to diverse C-terminal DNA-binding domains in different fungi/metazoa
EP2	α-helical	Solo versions of this domain are seen in early branching eukaryotes like kinetoplastids and heteroloboseans and in *Tetrahymena* characterized by a stretch of basic conserved residues. Mostly associated with the EP1 domain.
SWIRM	Tetrahelical HTH similar to BRIGHT	The versions traceable to LECA (e.g., orthologs of SWI3p) are a part of a conserved remodeling complex containing a SWI2/SNF2 ATPase orthologous to Brahma
RFD module	Comprised of a tetrahelical HTH and circularly permuted Sm domain	RFD module either as a stand-alone protein or fused to PHD, chromo, bromo, or DNA methylase (DNMT1) domains. Might recruit different repressive factors to chromatin
Kleisins	Winged helix-turn-helix domain	Helps SMC ATPases in forming a ring around DNA
HSA	α-helical domain	Several positively charged residues are present, suggestive of a nucleic acid binding role. Earliest version is seen in the SWR1-like SWI2/SNF2 helicases

(*Continued*)

TABLE 15.1 (Continued)

Domain	Structure	Comments
SAM	α-helical bundle with core bi-helical hairpins	Known chromatin-associated versions are primarily found in the crown group and might mediate interactions with RNA
MYND finger	Metal chelating structure	A potential peptide-binding domain recruiting modifying activities to chromatin. Found associated in SET domains of the SKM-BOP2 family. Also found fused to aminopeptidases
DDT	Trihelical domain	Found in crown group and chromalveolates. Has a characteristic basic residue in the last helix and is usually N-terminal to a PHD finger. It may form a specialized peptide interaction unit along with the neighboring PHD finger
ELM2	α-helical domain	Usually found N-terminal to a MYB/SANT or PHD finger. Found in crown group, chromalveolates and heteroloboseans. Might form an extended peptide interaction interface with the adjacent MYB/SANT domain
BrC	α-helical domain	Found C-terminal to the bromodomain and TAF14
HARP	$\alpha + \beta$ domain	Found in a single or two copies N-terminal to certain HepA-like SWI2/SNF2 helicases. Found fused to certain DNA repair nucleases in bacteria
HORMA	$\alpha + \beta$ domain	A common domain found in mitotic and meiotic spindle assembly proteins
YEATS	β-sandwich	Binds histones H3 and H4. Structurally related to the ASF1 domain
ASF1	β-sandwich	Histone chaperone that binds H3 and H4

(a parabasalid) and *Giardia* (a diplomonad), which appear to lack the main cytosine methylase DNMT1 (see below). Hence, it apparently arose coevally with 5C DNA methylation (Iyer et al., 2011a). Histone H1, which binds internucleosomal linkers, is found in ophisthokonts, amoebozoans, plants, stramenopiles, and *Naegleria* (Iyer et al., 2008b). Its distribution is suggestive of an origin in the crown group from the more widespread paralogous FKH domain, followed by lateral transfers to stramenopiles during endosymbiosis with the plant lineage and independently to *Naegleria*.

DBDs of CPs such as the HMG box, CXC domains, BRIGHT, SAND (KDWK), C2H2 ZnFs, and the AT-hook motif are shared with specific TFs. However, excluding C2H2 ZnFs, these DBDs are predominantly found in CPs and, unlike in TFs, they are typically found in the context of multidomain proteins in the CPs. The HMG box and AT-hook proteins can mediate bending of the helical axis of DNA and play an important role in altering chromosomal structure (Aravind and Landsman, 1998; Travers, 2000). Others such as the HIRAN domain appear to specifically recruit chromatin-remodeling activities to damaged DNA (Iyer et al., 2006a). The Ku DNA-binding proteins (Table 15.1) bind matrix attachment regions of chromosomes, are part of the telomere-binding complex, and are associated with the perinuclear localization of telomeres. The ancestral Ku protein appears to have been acquired by the eukaryotes after the divergence of the two basal eukaryotic lineages, the parabasalids and diplomonads, from alphaproteobacteria (Krishna and Aravind, 2010). In the latter, they are coded by a mobile DNA repair operon with two paralogous copies of the Ku protein. These appear to have diverged in eukaryotes as the two paralogous subunits, Ku70 and Ku80, which were vertically inherited in eukaryotes since then. Interestingly, Ku was lost independently in all studied apicomplexan lineages, with the exception of *Toxoplasma*.

At least three major types of domains are involved in the discrimination of epigenetic modification in DNA (Iyer et al., 2011a). The methylated DNA-binding domain (MBD) was first observed in various methylated CpG-binding proteins (Bellacosa et al., 1999; Cross et al., 1997). Sequence profile analysis showed that a related domain was also found in a number of other proteins such as the mammalian BAZ2A/B (TTF-IIP5) and SETDB2, several *Caenorhabditis elegans* proteins, such as Flt-1, and *Drosophila* Toutatis, in which it was not originally recognized (Aravind and Landsman, 1998). These versions of the domain, while clearly related to the 5mCpG binding-MBDs, did not contain all the conserved residues required for 5mCpG binding. Hence, it became clear that not all versions of this domain are likely to bind 5mCpG-containing DNA and the more inclusive superfamily of these domains was accordingly named *TAM* (after TTF-IIP5, ARBP, and MeCP2) (Aravind and Landsman, 1998). Despite this suggestion more accurately reflecting the natural history of this domain, the term MBD has unfortunately been used indiscriminately in the literature. The TAM/MBD domain shows a rather distinctive phyletic pattern, being found in animals, plants, and stramenopiles (Iyer et al., 2011a). The complete loss of this domain in fungi is intriguing because several fungi display noticeable amounts of CpG methylation

(Zemach et al., 2010). Methylation patterns in fungi suggest that the ancestral fungus is likely to have possessed transposon and repeat element methylation, but not the gene body methylation observed in both animals and plants (Zemach et al., 2010). Hence, we speculate that the loss of gene body methylation in the common ancestor of most extant fungi might be correlated with the loss of the TAM/MBD.

The SET-associated *Deinococcus* endonuclease domain (SAD)/SET and RING finger associated (SRA) domain was first identified in Np65, certain plant SET domain histone methylases, and a Deinococcus McrA-like REase, and was accordingly named the SET-associated *Deinococcus* endonuclease domain (SAD) (Makarova et al., 2000). The same domain was subsequently given names such as YDG after an eponymous motif found in a subset of these domains and SRA by other workers (Citterio et al., 2004). A number of studies on the eukaryotic SAD/SRA domains have shown that they bind hemimethylated CpG dinucleotides and also other 5mC-containing dinucleotides (Bostick et al., 2007; Hashimoto et al., 2008; Johnson et al., 2007; Sharif et al., 2007; Woo et al., 2007). The SAD/SRA domain adopts a β-barrel fold, with a core of eight strands similar to that seen in the PUA and ASCH domains that bind different types of RNA (Iyer et al., 2011a; Iyer et al., 2006b). In bacteria, the SAD/SRA domain is usually fused to or found in an operon with either of two distinct restriction endonucleases of the EndoVII/HNH-fold or a domain of the classical restriction endonuclease fold. Eukaryotes appear to have acquired the SAD/SRA domain through a single lateral transfer from such a restriction system. In eukaryotes, the domain is found in animals, fungi, plants, stramenopiles, apicomplexans, and heteroloboseans (Iyer et al., 2011a). Given its wider phyletic spread in eukaryotes than TAM/MBD and its clear bacterial antecedents, SAD/SRA appears to have been the first dedicated 5mC-recognizing domain to have been acquired and recruited by the eukaryotes rather early in their evolution.

The CXXC domain was originally identified in the vertebrate MeCP2, in the N-terminal region of the vertebrate SET domain histone methyltransferase MLL1, and in the animal DNMT1 (Jorgensen et al., 2004; Ono et al., 2002). Subsequent studies have shown that, unlike versions of the TAM/MBD and SAD/SRA domains, it primarily recognizes unmethylated CpG nucleotides and thus plays a role complementary to theirs in discriminating epigenetic marks (Allen et al., 2006; Blackledge et al., 2010; Cierpicki et al., 2010; Tate et al., 2009, 2010). The CXXC domain is characterized by eight conserved cysteines, whose arrangement includes multiple CXXC motifs that give the domain its name. Analysis of its sequence and structure showed that the classical CXXC domain is comprised of a peculiar internal duplication, in which the second unit is inserted into the first one (Iyer et al., 2011a). Each of these units, the mono-CXXC domain, is characterized by four conserved cysteines displaying a signature of the form CXXCXXCX(n)C that together chelate a Zn^{2+} ion. The NMR structures of the bi-CXXC domain–DNA structure complex reveal that the two CXXC units form a crescent-shaped clasp around both grooves of the DNA bearing the target CpG dinucleotide (Allen et al., 2006; Tate et al., 2009). Sequence and structure comparisons show that the mono-CXXC domain is homologous to the structural Zn-binding domain of the medium-chain dehydrogenases/reductases (MDRs), which is inserted into the β-barrel GroES-like domain of the latter enzymes (Tate et al., 2009). As the version of the domain found in MDRs is present across the three superkingdoms of life, it is likely to represent the ancestral form. The DNA-binding properties of the CXXC domain appear to be a later innovation on the core scaffold offered by the MDR Zn-binding domain. In eukaryotes, the CXXC domain is found only in stramenopiles, plants, and metazoans.

More recently, a fourth domain, the HARE-HTH, has been identified in diverse eukaryotic CPs (Aravind and Iyer, 2012) such as the polycomb complex ASXL proteins (Table 15.1), orthologs of *Drosophila* Additional sex combs and the algal Demeter-type DNA demethylating glycosylases. In prokaryotes, the domain is found fused to diverse restriction endonucleases and DNA glycosylases, the RNA polymerase delta subunit of gram-positive bacteria and certain bacterial proteins that combine features of the RNA polymerase α-subunit and sigma factors. In ASXL proteins, this domain is fused to a central ASXH domain that appears to mediate interactions between chromatin-linked protein complexes and TFs and UCH37-like deubiquitinases via its conserved LXXLL motif (Aravind and Iyer, 2012). This domain adopts the winged helix-turn-helix fold (wHTH) and is predicted to bind DNA. In particular, its repeated association with different types of restriction endonuclease domains, which are responsive to DNA methylation of their target sequences and genetic interactions of the ASXL genes with the Tet genes suggest that the HARE-HTH domain could be a sensor of the state of epigenetic modifications such as methyladenine, 5-methylcytosine, and its oxidized derivatives.

15.4.3 Evolutionary History of Histone Acetylation and ADP-Ribosylation-Based Regulatory Systems

The catalytic domain of most histone lysine acetyltransferases (HATs) belong to the ancient superfamily of *N*-acetyltransferases typified by the GCN5 (also called *GNAT acetyltransferases*; Table 15.1) (Dutnall et al., 1998; Neuwald and Landsman, 1997). Some acetylase domains, such as those of CBP and the fungal-specific Rtt109p family had been originally reported as being unrelated to the other GNAT enzymes (Driscoll et al., 2007). However, sequence and structure analysis have revealed that they contain the classical GNAT fold, but have greatly diverged in sequence (Iyer et al., 2011a; Liu et al., 2008). These examples underscore the dramatic sequence divergence, which occurs in this superfamily of enzymes, without altering the catalytic capabilities of the divergent versions. At least 14 distinct families with the GNAT fold appear to be dedicated acetylases that are specialized in performing numerous specific roles in eukaryotic chromatin (Fig. 15.3). Conservatively, four can be traced back to the LECA. These are multidomain proteins fused to peptide-binding domains such as bromo (Gcn5) and chromo (Esa1) or other catalytic domains such as an ATPase domain

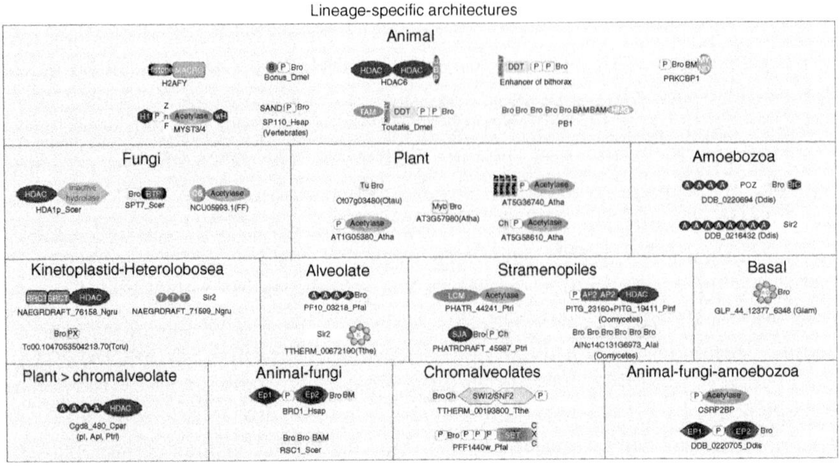

Figure 15.3 Ancient and lineage-specific domain architectures in acetylation-based regulatory systems. Evolution of acetylation-based systems are shown using various domain architectures that evolved either at different early stages in the evolution of eukaryotes or more recently in different lineages. The number of ancient conserved acetylases, deacetylases, and acetyl-peptide-detecting adaptors that were present in the different temporal epochs are shown on the right. Architectures are denoted by their gene name and species abbreviations, separated by underscores. Proteins from *Phytophthora sojae* are represented with fakegis and the sequence can be accessed from the supplementary material. If an architecture is restricted to a subset of species or lineages in a group then the species or lineage abbreviations in which they are present are listed in brackets below the architecture. Domain architectures of well-known proteins are only denoted by the protein names. For species abbreviations consult Fig. 15.1. Abbreviations of lineages include Amoe, Amoebozoans; Api, Apicomplexans; Chl, Chlorophytes; Chy, Chytrids; Cil, Ciliates; FF, Filamentous fungi; Kin, Kinetoplastids; Myceto, Mycetozoa; Pl, Plants; Stram, Stramenopiles. Domains are denoted by their standard names and abbreviations. For a comprehensive list of domain names and functions refer to Table 15.1. A typical domain abbreviations include: 5C-MTase, 5C DNA cytosine methyltransferase; A, Ankyrin repeat; B, B-box; BM, BMB/PWWP; BrC, Brd2/TAF14 C-terminal domain; UBP: Bro, Bromo; C2C2, a distinct treble-clef domain with a C2C2 signature; C6, C6 fungal finger; Ch, Chromo; Deam, Nucleotide deaminase; ECH, Enoyl-coA hydratase; FB, Fbox; H, HARP domain; Ing1N, Ing1-like N-terminal domain; JN, JOR/JmjC N-terminal domain; K, Kelch repeats; LCM, Leucine carboxymethyltransferase; MYND, MYND finger; N6-MTase, DNA adenine methyltransferase; OB nuclease, Staphylococcal nuclease-like domain of the OB fold; OTU, OTU-like thiol protease; P, PHD finger; PARPf, Zinc-chelating finger associated with poly ADP-ribose polymerases; PX, PHD-X/ZfCW; RAD16f, Zinc-chelating finger found in all RAD16 proteins; RAD18, Zinc-chelating finger associated with RAD18; R, RING finger; T, Treble-clef domain; TF2S2, the second domain of the TFIIS-like proteins; SnoC, Stawberry notch C-terminal domain; T, TPR repeat; TopC, Zinc ribbon found at the C-terminii of Topoisomerases; Tu, Tudor; WD, WD repeats; wH, winged HTH; Ubhyd, Ubiquitin carboxy-terminal hydrolase of the papain-like thiol protease fold, X, uncharacterized globular domain; Z, ZZ-finger. (*See insert for color representation of the figure.*)

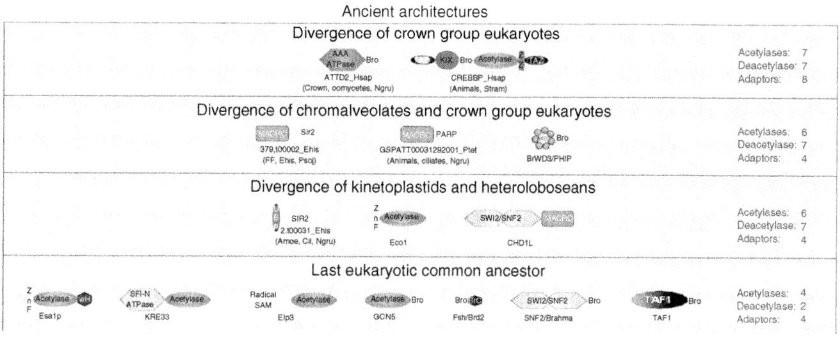

Figure 15.3 (*Continued*)

related to the N-terminal domain of the superfamily-I helicase module (Kre33) and a radical SAM (*S*-adenosylmethionine) enzyme domain (Elp3) (Fig. 15.3). Gcn5 is critical for acetylation of multiple lysines on histone H3 (e.g., K9, K14, K18, and K36) in connection with transcriptional activation by specific TFs (Dutnall et al., 1998). In contrast, Esa1 appears to have a primarily negative regulatory role by facilitating transcriptional silencing via acetylation of multiple positions in histone H4 (K5, K8, K12, and K16) (Doyon et al., 2004; Durant and Pugh, 2006). Elp3 is required for transcription elongation and acetylates both histone H3 (K14) and H4 (K8) (Winkler et al., 2002). Elp3 has additional roles in tRNA wobble uridine modifications that appear to involve its conserved radical SAM domain (Huang et al., 2005). Although it was recently implicated in DNA demethylation (Okada et al., 2010) the weight of the available evidence argues against a direct catalytic role in this process (Iyer et al., 2011a). Kre33 appears to regulate assembly of chromatin-associated ribonucleoprotein complexes and probably also in ribosomal assembly (Oeffinger et al., 2007; Ossareh-Nazari et al., 2010). This reconstruction suggests that in the LECA two distinct acetylases, one targeting H3 (Gcn5-like) and the other H4 (Esa1-like), had emerged, which were respectively associated with transcriptional activation and silencing. A third, which targeted both these histones simultaneously, appears to have been used in a distinct context of transcription elongation. The Elp3p and Kre33p acetylases are shared by eukaryotes and archaea, suggesting an inheritance from the archaeal precursor, whereas Esa1p and Gcn5p orthologs appear to be innovations specific to eukaryotes, which were derived through rapid divergence from a preexisting version of the fold (Iyer et al., 2008b).

Of the remaining families of HATs, the Eco1 orthologs (which acetylate chromosome cohesion proteins such as the cohesin Smc3 and Mcd1 (Heidinger-Pauli et al., 2009)) emerged prior to the branching-off of the heteroloboseans and kinetoplastids from other eukaryotes. Others such as Hat1p, CBP/P300, CSRP2BP, and some paralogs of the Esa1p, which form the MYST clade of HATs, emerged in the crown group or the common ancestor of the crown group and chromalveolates (Iyer et al., 2008b) (Fig. 15.3). Of these, the CBP/p300 acetylases, which were secondarily lost in fungi, acquired promiscuous activity for multiple lysines on all four histones (Fig. 15.3). Several HAT clades are restricted to a particular lineage. For example, fungi appear to have at least four lineage-specific families (orthologs of Spt10p, Hpa2p, Rtt109p, and *Neurospora* NCU05993.1), while plants have a lineage-specific family of their own with fusion of the acetylase domain with PHD fingers, TAM/MBD, chromo, or AT-hook motifs (Fig. 15.3). Amongst protists, an unusual lineage-specific representative is seen in *Phytophthora* and related stramenopiles, where the acetylase domain is fused to a carboxymethyltransferase domain (Fig. 15.3). It is possible that these enzymes might carry out a second covalent protein modification, perhaps of acidic side chains. The affinities of most of the lineage-specific versions suggest that they were acquired repeatedly by eukaryotes from the diverse bacterial GNAT clades (Iyer et al., 2008b).

Histone deacetylases belong to two structurally distinct superfamilies, namely, the RPD3/HDAC superfamily and the Sir2 superfamily (Blander and Guarente, 2004; Leipe and Landsman, 1997). Both of these are universally present in eukaryotes. Prokaryotic members of both superfamilies appear to have played predominantly metabolic roles, as opposed to a regulatory role in chromatin: representatives of the RPD3/HDAC superfamily appear to have a role in acetoin metabolism, whereas those of the Sir2 superfamily are involved in regulating acyl CoA biosynthesis and nicotinamide metabolism (Blander and Guarente, 2004; Leipe and Landsman, 1997). While both these families display a Rossmannoid α/β fold, they appear to have emerged independently within distinct radiation of this fold (Burroughs et al., 2006). The RPD3 superfamily belongs to the 2-aspartate superclass of Rossmannoid folds (include the haloacid dehalogenase superfamily) and uses metal-dependent catalysis (Burroughs et al., 2006). The Sir2 superfamily is closer to the classical Rossmann fold enzymes, and similarly to them uses an NAD cofactor (Avalos et al., 2004; Blander and Guarente, 2004). At least one deacetylase of the HDAC/Rpd3 superfamily was present in the LECA and appears to have been derived from bacterial acetoin-hydrolyzing enzymes (Fig. 15.3). There have been several lineage-specific innovations within this superfamily amongst eukaryotes. In line with its expansion of HATs, *T. vaginalis* also shows an expansion of HDAC deacetylases, while kinetoplastids show a unique family typified by LmjF21.1870 from *Leishmania* (Iyer et al., 2008b). The chromalveolate clade, including the apicomplexans *Cryptosporidium* and *Toxoplasma gondii*, has a distinctive version of HDAC that contains N-terminal ankyrin repeats, which is shared with plants (Fig. 15.3). Fungal-specific HDA1p deacetylases combine the HDAC domain with a C-terminal inactive α/β hydrolase domain that might be utilized for specific peptide interactions. The parasites *Phytophthora* and *Naegleria* possess lineage-specific architectures that, respectively, combine the HDAC domain with AP2 and PHD finger domains and the DNA repair-related phosphopeptide-binding BRCT domain (Fig. 15.3).

At least one member of the Sir2 superfamily deacetylases, the classical Sir2, can be traced back to the common ancestor of eukaryotes and archaea (Iyer et al., 2008b). All other major families appear to have been acquired from bacteria much later in eukaryotic evolution: Sirtuin 4, 5, and 6 appear to have been independently acquired prior to the divergence of heteroloboseans and kinetoplastids from other eukaryotic lineages. Another sporadic lineage of Sir2-like proteins typified by *Cryptosporidium* cgd7_2030 is present in gut parasites such as *Giardia* and *Cryptosporidium* and could have been laterally exchanged between them (Iyer et al., 2008b). Like the HDAC superfamily, members of this family show parallel domain fusions in various protists (Fig. 15.3): *Dictyostelium* and *Tetrahymena* show fusions to Tetratricopeptide and ankyrin repeats. A Sir2 deacetylase from ciliates, amoebozoans, and *Naegleria* contains a fusion to the ubiquitin-binding ZnF domain which, interestingly, parallels a similar fusion of the ubiquitin-binding ZnF domain to a HDAC deacetylase in animal HDAC6 enzymes (Fig. 15.3) (Pandey et al., 2007). These fusions point to several unique interactions being used to recruit enzymes containing deacetylase domains of either superfamily to specific contexts.

Members of the Sir2 superfamily have also been shown to carry out NAD-dependent mono-ADP ribosylation of proteins and generate ADP-ribose as a by-product of the deacetylation reaction (Avalos et al., 2004; Frye, 1999). Versions of the Macro domain, prototyped by the vertebrate macrohistone 2A, have been shown to bind *O*-acetyl-ADP-ribose or hydrolyze ADP-ribose-1″-phosphate (Aravind, 2001; Karras et al., 2005; Shull et al., 2005). In *E. histolytica*, certain fungi, and *Phytophthora*, the Sir2 domain is fused to the Macro domain (Fig. 15.3). Versions of the Macro domain are also found in other CPs, for instance, fused to the SWI2/SNF2 ATPase module (Iyer et al., 2008b). These occurrences suggest that the *O*-acetyl-ADP-ribose generated by Sir2 action might elicit additional regulatory roles in chromatin dynamics (Karras et al., 2005). It is possible that the Macro domain might recognize mono-ADP-ribosylated proteins and catalyze the removal of this modification. This is supported by their fusion to classical protein ADP-ribosyl transferases in animals. By binding or hydrolyzing *O*-acetyl-ADP-ribose, it might elicit a regulatory effect on Sir2 action by potentially favoring the forward (deacetylation) reaction by removing ADP-ribose. A representative of the Macro domain appears to have been acquired from bacteria prior to the LECA itself. It is possible that these versions have a role in RNA metabolism rather than chromatin dynamics (de Souza and Aravind, 2012). Versions involved in chromatin dynamics appear to represent independent transfers from bacteria on multiple occasions in evolution (Iyer et al., 2008b). Another domain, which might sense NAD-derived metabolites in chromatin is the inactive Nudix domain found in proteins such as DBC1 and CARP1 (Anantharaman and Aravind, 2008; Kim et al., 2008; Zhao et al., 2008). One of the largest histone modifications is the addition of branched or linear polyADP-ribose (up to 200 ADP-ribose units) that targets the acidic side chain of glutamate. At least one core histone and probably linker histones in several organisms are targets of poly-ADP-ribosylation (Rouleau et al., 2004). The enzyme catalyzing this reaction polyADP-ribose polymerase (PARP) appears to have been derived from a bacterial toxin prior to the LECA (de Souza and Aravind, 2012). In plants, the PARP enzymes are fused to the DNA-binding SAP domain that is likely to tether the catalytic domain to chromosome scaffold attachment regions (Aravind and Koonin, 2000b).

Acetylated peptides on histone H3 and H4 are chiefly recognized by the tetrahelical bromo domain that appears to be a unique eukaryotic innovation (Dhalluin et al., 1999; Kouzarides, 2007). Their role in recognizing the acetyl marks on other histones is poorly understood. Bromo domains are found in all eukaryotes and have at least four representatives in the LECA (Fig. 15.3). Two ancient and highly conserved versions of the bromo domain are fused to enzymatic domains (Fig. 15.3). The presence of a bromo domain in the basal transcription factor TAF1, which goes back to the LECA, indicates an ancestral role for this modification (potentially catalyzed by GCN5) in the context of transcription initiation. Another ancestral bromo domain is represented by orthologs of the *Drosophila* Fsh protein that interacts with acetylated H4. These proteins appear to interact with the TFIID transcription initiation complex, and probably recognize acetylation by Esa1p orthologs. It combines one to two bromo domains with another conserved C-terminal α-helical domain, the BrC domain, also found in TAF14 (Table 15.1). In *T. vaginalis*, consistent with the LSE of acetylases and deacetylases, this version shows an extraordinary expansion with at least 100 representatives (Fig. 15.3). Recently, some acetyl marks, such as H3K13ac and H4S acetylated N-terminus, have been shown to be recognized by tandem dyads of the PHD, such as those seen in DPF3b. Such PHD dyads cannot be traced back to LECA, but are found in *Naegleria*, suggesting that acetyllysine recognition by PHD fingers probably emerged somewhat later in eukaryotic evolution.

15.4.4 Natural History of Histone-Methylation-Based Regulation: The Methylases

Protein methylases have evolved among two structurally unrelated folds. The first group of protein methylases belongs to the classical methyltransferase superfamily along with numerous other methylases, and possesses the Rossmann fold (Fauman et al., 1999). The second group of currently known protein methylases, the SET domain superfamily, contains the β-clip fold (Iyer and Aravind, 2004). Among the classical Rossmann fold-type methylases are several distinct protein methylase families, and two of these methylate histones and other CPs (Dlakic, 2001; Lee and Stallcup, 2009). In the SET domain superfamily, most characterized eukaryotic versions act on histones and other CPs, but a few also methylate cytoplasmic proteins such as cytochrome C, Rubisco, ribosomal proteins, (L42, L12, and L23) and the translation initiation factor eIF1A (Demirci et al., 2008; Porras-Yakushi et al., 2007; Trievel et al., 2002).

Eukaryotic CP methylases with the Rossmann fold belong to two distinct families, namely, the PRMT and the Dot1 families. The PRMT family comprises dedicated arginine methyltransferases in eukaryotes that methylate specific arginines in histones and ribonucleoproteins (Chen et al., 2002; Frankel and Clarke, 2000; Lee and Stallcup, 2009; Strahl et al., 2001; Yue et al., 2007). They are characterized by a distinctive all-β domain with an avidin-like fold C-terminal

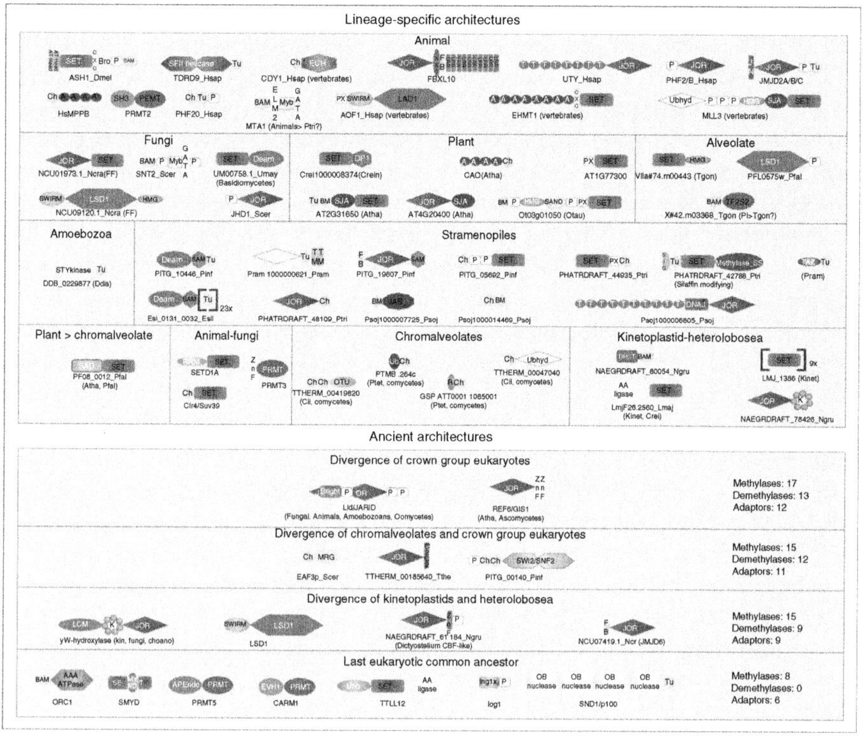

Figure 15.4 Ancient and lineage-specific domain architectures in the methylation-dependent regulatory systems. Evolution of methylation-based regulation is shown using various domain architectures that evolved either at different early stages in the evolution of eukaryotes or more recently in different lineages. The number of ancient conserved protein methylases, demethylases, and methylated-peptide-detecting adaptors that were present in the different temporal epochs are shown on the right. The scheme of labeling domain architectures, species, and lineages abbreviations is as in Fig. 15.3. (*See insert for color representation of the figure.*)

to the core Rossmann fold (Aravind et al., 2011). Phylogenetic analysis suggests that the PRMT family had already diversified in the LECA, as indicated by the presence of multiple members of this family in the early branching eukaryote *T. vaginalis* (Fig. 15.4). Among the eukaryotic PRMTs, the primary split in the tree separates the PRMT5 clade from the rest of the PRMTs such as those of the CARM1 and PRMT1/PRMT3 clades (Aravind et al., 2011). Prior to the LECA, the PRMT family had already radiated into at least one version that catalyzed symmetric arginine dimethylation and two distinct versions that catalyzed asymmetric dimethylation. Concomitant with their radiation, the PRMT proteins also acquired several distinct N-terminal domains (Fig. 15.4): The PRMT5 clade acquired an N-terminal catalytically inactive TIM-barrel domain related to the AP-endonucleases. The CARM1-clade acquired an N-terminal EVH1-like domain that adopts a PH-like β-barrel fold. In animals, the PRMT2 clade acquired an N-terminal SH3 domain that recognizes proline-rich peptides. The PRMT3 clade acquired a C2H2 zinc finger domain in the common ancestor of the animals and the fungi. The acquisition of these N-terminal domains appears to have played an important role in the emergence of distinct interactions and functional partners within the PRMT family. Outside of eukaryotes, members of the PRMT family can be seen in several proteobacteria and members of the chlamydia lineages (Aravind et al., 2011). These bacterial versions are components of systems that are involved in the synthesis of secondary metabolites such as antibiotics, toxins, and siderophores from modified amino acids or peptides (Aravind et al., 2011). The PRMT family probably emerged first in bacteria in the context of the diversification of secondary metabolite pathways and was subsequently transferred to eukaryotes.

Members of the Dot1 family rather exclusively modify the H3K79, and processively methylate it to give rise to mono-, di-, and tri-methylated forms (Dlakic, 2001; Feng et al., 2002). With the exception of the basal eukaryotes *Trichomonas* and *Giardia*, Dot1 orthologs are present in all other major eukaryotic lineages (Fig. 15.4). However, within those lineages there are certain notable instances of gene loss—while the basal plant lineages such as the chlorophyte algae and lycopodiophytes have one or more Dot1 paralogs, they have been completely lost in the crown group land plants such as angiosperms (Aravind et al., 2011). In the caenorhabditiform nematodes there has been a notable lineage-specific expansion of Dot1, with at least five paralogs in *C. elegans*. Phylogenetic analysis also suggests that the precursor of Dot1A and Dot1B in trypanosomes appears to have been acquired via lateral transfer from the animal lineage (Aravind et al., 2011). Outside of eukaryotes, the closest homologs of Dot1 are seen in bacteria (Aravind et al., 2011). Some of these bacterial forms, such as those seen in *Legionella*, myxobacteria, and *Protochlamydia*, are particular close to the eukaryotic forms and are further nested within a large prokaryotic radiation of methylases that are involved in the synthesis of secondary metabolites such as mycolic acids in mycobacteria (including the mycolic acid cyclopropane synthases) and polyether antibiotics such as nigericin (e.g., NigE of *Streptomyces* sp. DSM4137) (Boissier et al., 2006; Harvey et al., 2007). Thus, similar to the PRMT family, it appears that the Dot1 family arose in bacteria as part of the radiation of methylases involved in generating diversity among secondary metabolites and was subsequently transferred to eukaryotes (Harvey et al., 2007). Interestingly, of the bacterial versions closest to the eukaryotic forms, some are encoded by intracellular pathogens (e.g., *Legionella*) or endosymbionts (*Protochlamydia*). These versions have signal peptides that are likely to allow their secretion into the eukaryotic host cells. Hence, it would be of interest to investigate if they play a role in regulating host behavior via histone methylation comparable to the endogenous Dot1 (Aravind et al., 2011).

The SET domains are the primary methylases responsible for mono-, di-, and tri-methylation of lysines in the tails of the histones H3 and H4 (Aravind et al., 2011; Feng et al., 2002; Jacobs et al., 2002; Milne et al., 2002; Nakayama et al., 2001). In addition, they methylate a number of other CPs such as the DNA methylase (DNMT1) and the basal transcription factor TAF10 (Couture et al., 2006; Esteve et al., 2011). Of the currently known methylations catalyzed by SET domains, four are on lysines in the N-terminal tail of H3, namely, K4, K9, K27, and K36. Under regular conditions, H4 contains a single lysine target for methylation by SET domains, K20 (Allis et al., 2007; Kouzarides, 2007). However, recently it has been claimed that under stress conditions K5, K8, and K12 can be methylated by the Set5-type SET domains, which are associated with the chromatin-modifying COMPASS and NuA4 complexes (Green et al., 2012). An as yet poorly understood modification, H2BK5me1, associated with transcriptionally active chromatin, might also be catalyzed by a SET domain enzyme. The positions corresponding to H3K4, H3K9, H3K27, H3K36, and H4K20 are highly conserved throughout eukaryotes, and are confidently inferred as being lysine even in the LECA (Aravind et al., 2011). Thus, the primary histone sites modified by the SET domain proteins were available in the ancestral eukaryote. The SET domain displays the peculiar β-clip fold, which is additionally found in the SAF domain, the dUTPases, the molybdenum cofactor biosynthesis protein MoeA C-terminal domain, urease $\alpha\beta$ subunits, and some bacteriophage structural proteins (Iyer and Aravind, 2004). It is likely that the catalytic SET domain evolved from an ancestral sugar-binding version of the β-clip fold similar to the extant SAF domains. All eukaryotic genomes sequenced to date include genes for at least one SET domain protein, suggesting that a minimal histone methylation activity by these enzymes is essential for eukaryotic life. The basal eukaryotes *Giardia* and *Trichomonas* contain 5 and 10 SET domain proteins, respectively (Aravind et al., 2011). Of these, at least five distinct versions can be interpreted as being present in the LECA, including members of the trithorax, Ash1/NSD1, Su(var)3-9, SMYD (Skm-Bop2), and TTLL12 type SET domain clades. The TTLL12, which combines an N-terminal SET domain with a C-terminal ATP-grasp peptide ligase domain that probably catalyzes polyglutamylation, is the most poorly studied of these ancient lineages of SET domains (Fig. 15.4). The above observations imply that the ancestral eukaryote had a well-developed histone and CP methylation system.

Several additional clades of SET domains appear to have arisen relatively late and are restricted to particular eukaryotic lineages. The lineage archetyped by SETD1A and SETD1B branched off from the Trithorax clade in the common ancestor of plants, animals, and fungi, and the SET domain was combined with an RNA-recognition motif (RRM) domain in the precursor of the latter two lineages (Fig. 15.4). Some distinct lineages of SET domains emerged even within the terminal branches of the tree of life. A striking instance of this is the fusion of a Su(var)3-9 clade SET domain with mariner-type transposase domain in the anthropoid primates resulting in the SETmar protein that functions as both an endonuclease and a H3K36 methylase (Miskey et al., 2007; Shaheen et al., 2010). Kinetoplastids are remarkable in showing an LSE of 25 SET domain proteins, which is quite significant as it might explain certain important features of transcription regulation in kinetoplastids (Aravind et al., 2011). Kinetoplastids have a paucity of conventional TFs, and have large arrays of closely spaced genes that are cotranscribed into a polycistronic transcript by RNA polymerase II (Siegel et al., 2009). The transcription start sites of each of these gene arrays is marked by a specific type of nucleosome that is enriched in the H4K10ac mark, and its generation appears to depend on H3K4me3 marks (Mandava et al., 2008; Siegel et al., 2009). It is possible that the different expanded SET domains of kinetoplastids act as potential transcription regulators that introduce H3K4me3 and other methylated histone marks at specific start sites to activate transcription in a condition-specific manner. Another extraordinary expansion of SET domains with 14–20 members per haploid genome is observed in diatoms such as *Thalassiosira*, *Phaeodactylum*, and *Fragillariopsis* (Aravind et al., 2011). These differ from all other eukaryotic SET domain families in bearing N-terminal signal peptides, implying that they are secreted proteins. Several members of this expansion show fusions to a spermidine synthase domain, which is a member of the Rossmann fold methylase superfamily. This fusion suggests that in addition to lysine methylation they might catalyze other modifications of the lysine side chain such as addition of an amine adduct analogous to the formation of spermidine from putrescine. These are predicted to be involved in the modifications of the cell-surface silaffin proteins of the diatoms. The "epigenetic" code of modified residues on silaffin is a key determinant for the process of biomineralization by which the silica shells of diatoms are assembled (Kroger et al., 2002; Sumper et al., 2007). The diversity in the complement of these secreted SET domain proteins might be a key determinant for morphological diversity of the silicaceous shells of diatoms.

The SET domain often co-occurs in several parasitic as well as free-living bacteria with a protein containing the SWIB domain (Murata et al., 2007; Stephens et al., 1998). In certain bacteria, SWIB and SET domain genes are even in the same operon. In addition, the SWIB domain is fused to topoisomerase-I in organisms such as *Chlamydia*, and topoisomerase-III in several β-proteobacteria (Aravind et al., 2011). These observations suggest that at least a subset of the bacterial SET domains are likely to constitute a chromatin-remodeling and -modifying complex comprised of the SET, SWIB, and a topoisomerase protein. Alternatively, it has been proposed that some versions from parasitic bacteria might target host proteins (Pennini et al., 2010). In addition, certain eukaryotic DNA viruses of nucleocytoplasmic large DNA virus clade (e.g., PBCV) have acquired SET domains that could have been transferred either from the host or bacterial endosymbionts that cohabit the same host cells (Manzur et al., 2003; Manzur and Zhou, 2005).

15.4.5 Domains that Preempt or Reverse Peptide Methylation in Chromatin

Protein arginine deiminases (PADIs) use water to catalyze the release of ammonia from the arginine guanidino group, yielding citrulline (Thompson and Fast, 2006). Citrullination is an important histone modification, which antagonizes arginine methylation by PRMT family methylases, by preemptive removal of the methyl-accepting sites (Cuthbert et al., 2004; Wang et al., 2004). The catalytic domain of the PADIs contains the unique pentein fold, which is comprised of five sub-domains, each with a β-β-α-β secondary structure progression, arranged in a pentamerically symmetric manner around a central aperture (Shirai et al., 2001). This fold is found in a superfamily of enzymes that operate on the guanidino group of arginine and its derivatives, succinylarginine dihydrolase, dimethylarginine dimethylaminohydrolase, and L-arginine: inosamine-phosphate amidinotransferase. Analysis of phyletic patterns of the PADI family suggests that the classical members are restricted to animals (Aravind et al., 2011). Hence, it is possible that CP citrullination by these enzymes is a late innovation that emerged only in the metazoa. However, other members of the pentein fold might be active in other eukaryotic lineages in a similar capacity. The LSD1-like lysine demethylases combine a Rossmann fold oxidoreductase domain to the SWIRM domain that is found in several distinct chromatin-associated proteins and primarily demethylates H3K4me1 and H3K4me2 (Aravind and Iyer, 2002; Forneris et al., 2008). The LSD1 demethylases are found in all major eukaryotic lineages, with the exception of the basal eukaryotes *Giardia* and *Trichomonas* (Fig. 15.4). LSD1 also shows multiple fusions to different DBDs: in fungi it is fused to a C-terminal DNA-binding HMG domain, while in the chlorophyte algae *Micromonas* and *Chlorella* it is fused, respectively, to an N-terminal BRIGHT (ARID) and a VP1 (B3) domain in place of the SWIRM domain (Fig. 15.4). These fusions might recruit LSD1 to specific DNA sequences or structures in these organisms. In plants, certain LSD1 paralogs are also fused to the SRI domain (Kizer et al., 2005; Vojnic et al., 2006) that recruits proteins to the RNA polymerase C-terminal domain, suggesting the possibility of cotranslational demethylation (Kizer et al., 2005; Vojnic et al., 2006). Outside eukaryotes, LSD1-like enzymes are found in several bacterial lineages as part of systems involved in synthesis of peptide secondary metabolites with modified amino acids. Thus, similar to some of the other components of the methylation system, LSD1-like enzymes might have their ultimate origin among bacterial secondary metabolism systems (Aravind et al., 2011).

The Jumonji-related (JOR) or JmjC enzymes are 2-oxoglutarate-dependent dioxygenases that catalyze the peptide demethylation of CP (Klose et al., 2006; Tsukada et al., 2006). Unlike the LSD1-demethylases, these enzymes can demethylate mono-, di- and tri-methylated lysines and perhaps the different forms of methylated arginines (Chang et al., 2007; Klose et al., 2006; Tsukada et al., 2006). These enzymes are by far the most prevalent demethylases active in eukaryotic chromatin. Some members of this family also catalyze the hydroxylation of asparagine in proteins such as the hypoxia-induced factor-$\alpha\beta$ and of the modified base wybutosine in tRNA (the yW-hydroxylase) (Iyer et al., 2010; Webby et al., 2009). JOR/JmjC domains belong to a specific clade containing the double-stranded β-helix fold, which also includes the cNMP-binding domains, domains traditionally labeled as the cupins, and classical 2-oxoglutarate-iron-dependent dioxygenases. The JOR/JmjC enzymes insert one oxygen atom into 2-oxoglutarate, converting it to succinate + CO_2, and the second one into the demethylation substrate via hydroxylation. Unlike the SET domain methylases, JOR/JmjC domains are absent from the earliest branching eukaryotes such as *Giardia* and *Trichomonas*, and also from parasites such as the microsporidian *Encephalitozoon* and *Entamoeba*. A comprehensive phylogenetic analysis of the JOR/JmjC domains revealed the presence of at least 17 distinct lineages in eukaryotes, divided into three major clades: histone demethylase-like, FIH1/yW-hydroxylase-like, and MINA/No66-like (Aravind et al., 2011; Iyer et al., 2010). The first of these clades includes all well-established histone demethylases in addition to several poorly studied lineages. The FIH1/yW-hydroxylase-like clade includes the FIH1, yW-hydroxylase, JMJD5, JMJD7, HSPBAP1, and *S. pombe* Jmj4-like lineages (Dann et al., 2002). The MINA/No66-like is a unitary clade with no further differentiation of lineages within it. A notable innovation of multiple lineages within the histone demethylase-like clade appears to have occurred in the common ancestor of animals, fungi, and plants. Certain versions, such as Jumonji itself, appear to have lost their catalytic residues. Hence, these forms might function as regulatory proteins that bind the methylated sites on histones, but do not catalyze demethylation (Aravind et al., 2011). The major clades of JOR/JmjC appear to have diversified in bacteria, followed by a transfer of at least one member from each of the three clades to the eukaryotes, prior to the divergence of the heterolobosean-kinetoplastid clade and the remaining eukaryotes (Fig. 15.4). The bacterial representatives are coded by gene clusters involved in biosynthesis of several secondary metabolites, such as pyoverdine-like peptide-derived siderophores and possible peptide antibiotics. In this respect, they mirror the PRMT and Dot1-like methylases and the LSD1-type oxidoreductases (see above), but not the SET domains, which show no apparent connections to secondary metabolism in bacteria.

15.4.6 Domains Sensing Methylation Status of Peptides in Chromatin

The discrimination of the methylation states of modified peptides in CPs is mediated by a number of structurally diverse domains (Yap and Zhou, 2010): (i) Simple globular domains that chiefly include the chromo-like domains with the SH3 fold and possibly the catalytically inactive version of JOR/JmjC domain (Brehm et al., 2004; Maurer-Stroh et al., 2003). (ii) Metal-chelation-supported domains that include versions of the treble-clef fold typified by the PHD finger domain and its structural derivatives (Aasland et al., 1995; Chakravarty et al., 2009; He et al., 2010; Yap and Zhou, 2010). (iii) Superstructure-forming repeats that include versions of the WD40 and ankyrin repeats that bind methylated histones (Brent and Marmorstein, 2008; Wysocka et al., 2005). An interesting unifying theme, which has been frequently observed in several structurally distinct peptide-binding domains, is the use of a cage of aromatic residues to sense the methylated lysine (Brent and Marmorstein, 2008). The chromo-like domains include the classical chromo domain, tudor, BMB(PWWP), MBT, Agenet, and

BAM/BAH domains (Maurer-Stroh et al., 2003). Of these, the tudor, Agenet, and MBT domains are likely to be unified into a higher order clade within which the boundaries are not clear at all. Hence, all these domains might be referred to simply as tudor-like domains. This assemblage of tudor-like domains also includes the domains typified by the SAGA-associated factor 29, the "domains of unknown function" DUF3590 and DUF1325 defined by the PFAM database (Finn et al., 2006), the *S. cerevisiae* RAD9 RAD53BP domain, and the so-called tudor-knot domain found in the histone acetyltransferase Esa1 (Aravind et al., 2011; Shimojo et al., 2008). Sequence profile analysis revealed that a conserved domain typified by the central conserved region in the plant auxin-response transcription factor, the Aux-RF domain (PFAM: PF06507) is yet another previously unknown chromo-like domain that is distantly related to the tudor-like assemblage (Aravind et al., 2011). The strands forming the surface of the barrel are deployed to bind DNA or RNA in several representatives of the chromo-like fold.

By far, the chromo-like domains constitute the most versatile class of methylated histone-binding proteins (Flanagan et al., 2005; Kim et al., 2010). Currently available evidence suggests that the classical chromo domain can recognize methylation at H3K4, H3K9, H3K27, and H3K36 (Kouzarides, 2007). Members of the tudor-like assemblage target methylated forms of H3K4, H3K9, and H4K20 (Kouzarides, 2007). In eukaryotes, the explosive radiation of the chromo-like family appears to have occurred independently in several lineages (Fig. 15.4). In particular, multiple such radiations are seen within the tudor-like domains: the so-called plant Agenet domains represent one such example (Maurer-Stroh et al., 2003). Another such expansion, typified by proteins with multiple tandem tudor domains is specifically seen in stramenopiles (Fig. 15.4) (Iyer et al., 2011b). It would be of considerable interest to investigate if the large tandem arrays of tudor domains in these proteins play a role in higher order organization of CP complexes by acting as a scaffold for multiple methylation sites on different proteins. The initial radiation of the chromo-like domains in eukaryotes might coincide with the expansion of protein methylases that happened prior to the LECA. The classical SH3 and chromo-like domains had diverged from each other in bacteria, and were independently acquired by eukaryotes from bacterial precursors. Bacterial homologs of the SH3 and chromo-like domains are found in secreted or periplasmic proteins associated with peptidoglycan, along with the bacterial precursor of another eukaryotic peptide-binding domain called the *Slap homology domain 1* (SHD1). This suggests that these domains probably acquired their peptide-recognition functions in the context of peptides found on the bacterial cell surface (Aravind et al., 2011; Iyer et al., 2008b).

The PHD finger contains a binuclear version of the treble-clef fold that chelates two Zn^{2+} ions. PHD fingers can bind methylated peptides (H3K4me2/3 and H3K9me2/3), acetylated peptides (H3K14ac and H4S acetylated N-terminus), and non-acetylated peptides(He et al., 2010; Iwase et al., 2007; Karagianni et al., 2008; Lan et al., 2007; Mansfield et al., 2011; Miller et al., 2010; Pasillas et al., 2010; Pena et al., 2006). Versions involved in methylated peptide recognition appear to constitute a numerically dominant group of PHD fingers, suggesting that they are one of the primary domains involved in the recognition of methylated H3K4. H3K4me2/H3K4me3 or unmethylated H3K4 peptides stack in an antiparallel manner with the first strand of the core finger region of the PHD finger, in the extended configuration, via strand-like hydrogen bonding (Fig. 15.5). Currently, the PHD finger is not known to occur outside of eukaryotes. At least a single copy of the PHD finger is inferred as being present

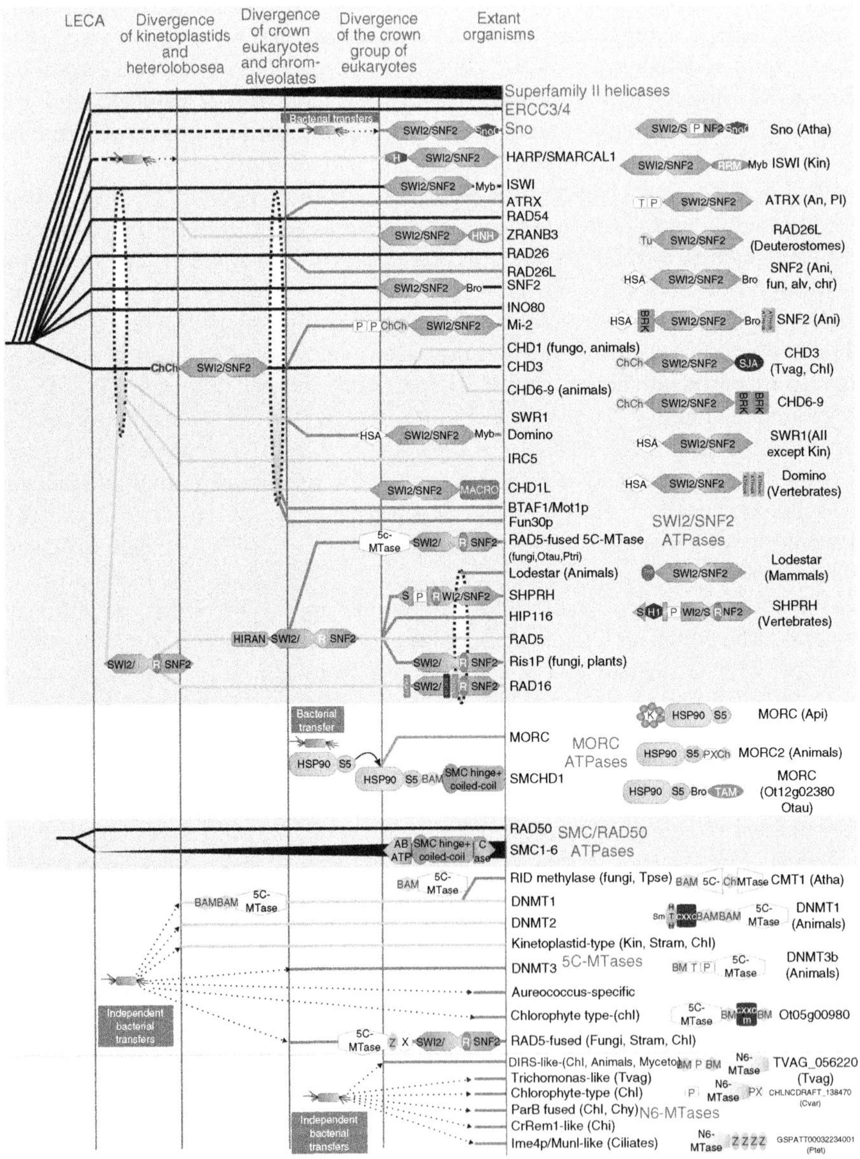

in the LECA, making it a candidate for a potential early eukaryotic innovation. Its origin could be linked to the pre-LECA expansion of protein methylation that resulted in the emergence of H3K4 methylation. The recognition of H3K9me2/3 appears to be a secondary innovation that emerged later, independently in certain PHD domains (Mansfield et al., 2011). The PHD-X/ZF-CW is a structurally degenerate version of the PHD domain that has lost the N-terminal flap of the treble-clef domain and the associated Zn-chelating site (He et al., 2010). However, it retains the H3-binding interface and binds H3K4me3 in a manner similar to the conventional PHD domains. The PHD-X/ZF-CW domain shows a remarkable expansion in the haptophyte alga *Emiliania huxleyi* (Aravind et al., 2011). The ASXL proteins of the metazoan polycomb complexes contain an unusual C-terminal PHD finger with structural modifications, such as loss of the C-terminal bilobed brace, which holds the H3 N-terminal peptide. These modifications might allow it to recognize internal modified lysines, making it the mediator of previously unexpected interactions in chromatin.

15.4.7 Evolution of Chromatin Remodeling and Assembling Systems

Enzymes mediating the dynamics of eukaryotic chromatin on local and global scales typically do so by utilizing the free energy of NTP hydrolysis. Not surprisingly, most of these enzymes contain motor domains of the P-loop NTPase fold (Table 15.1), two major classes of which are the SWI2/SNF2 ATPases and the SMC ATPases (Durr and Hopfner, 2006; Hirano, 2005, 2006; Martens and Winston, 2003; Mohrmann and Verrijzer, 2005; Uhlmann and Hopfner, 2006). SWI2/SNF2 ATPases are primarily involved in local chromatin-remodeling events by affecting nucleosome positioning and assembly.

Figure 15.5 Evolution of ATP-dependent remodeling and DNA methylation systems. The evolutionary history and interfamilial relationships of four different remodeling ATPases, Sno ATPases, SWI2/SNF2 ATPases, MORC ATPases, and SMC ATPases are shown in addition to 5C DNA cytosine methylases and the N6A adenine DNA methylases. Horizontal lines represent temporal epochs that correspond to the major transitions of eukaryote evolution; the last eukaryotic common ancestor, the divergence of kinetoplastids and heteroloboseans, the divergence of the chromalveolates and crown group eukaryotes, and the divergence of crown group eukaryotes. Solid lines show the maximum depth to which a particular family can be traced. Solid triangles are used to group together multiple families. The ellipses encompass all potential families from which a new family with a limited phyletic distribution could have emerged. Domain architectures common to all members are shown along the line depicting the family. Domain architectures limited to a few members of the family are shown on the right with their phyletic distribution or species abbreviations in brackets. Phyletic distribution of families with a limited distribution is shown next to the family name. For a full expansion of species and lineage abbreviations, please refer to Fig. 15.1 and Fig. 15.3, respectively. For a correct expansion of atypical domain names, refer to Fig. 15.3 legend. (*See insert for color representation of the figure.*)

They are usually core subunits of large functional complexes that include other chromatin-modifying activities, such as DNA-modifying enzymes and histone-acetylating, -methylating, or -ubiquitinating complexes (Durr and Hopfner, 2006; Martens and Winston, 2003; Mohrmann and Verrijzer, 2005). SWI2/SNF2 ATPases have their origins in bacteriophage replication systems and restriction-modification systems found in the prokaryotic superkingdoms (Iyer et al., 2008b; Iyer et al., 2006a). They appear to have been recruited from such a source in the earliest stages of eukaryotic evolution and expanded to give rise to at least six representatives by the time of the LECA (Fig. 15.5). A comparable count of these ATPases is found in the degraded genomes of *Giardia* and *Encephalitozoon* and includes most versions traceable to the LECA. Thus, this ancient set of SWI2/SNF2 ATPases is likely to comprise the most essential group of chromatin-remodeling enzymes required by any eukaryote. Domain architectures of these predicted ancestral versions show that the ATPase module was already fused to different peptide-binding domains such as chromo, bromo, and MYB (SANT), which allowed them to specifically interact with modified or unmodified nucleosomes (Fig. 15.5).

Prior to the divergence of the kinetoplastid-heterolobosean lineages from other eukaryotes the number of SWI2/SNF2 ATPases had increased to at least 13 representatives, and at least 19–20 representatives can be extrapolated to the common ancestor of chromalveolates and the crown group (Fig. 15.5). By the time of the former radiation, new architectures combining the SWI2/SNF2 ATPase module with different DBDs, a HNH (endonuclease VII-like) nuclease domain, a MACRO domain, and the RING finger had occurred (Iyer et al., 2008b). This implies that their functional roles were expanding, with the new versions sensing and repairing DNA damage or performing additional protein modifications through ubiquitination or sumoylation. In subsequent radiations of SWI2/SNF2 ATPases, several lineage-specific architectures appear to have arisen. Examples of these include convergent fusions to PHD fingers in apicomplexans and the crown group, and fusions to different DNA-modifying enzyme domains in kinetoplastids and fungi (Fig. 15.5, see below). In light of these associations with DNA metabolism, it remains to be seen if SWI2/SNF2 ATPases act as conventional DNA helicases, similarly to other Superfamily-II helicases in mediating chromatin remodeling. A striking lineage-specific expansion of a SWI2/SNF2 ATPase fused to the SJA (often wrongly annotated as two distinct domains Fyr-N/C) domain, which is predicted to bind peptides and potentially discriminate their phosphorylation state, is encountered in the parasitic protist, *T. vaginalis* (Iyer et al., 2008b). A distinctive version of the SWI2/SNF2 ATPase, typified by the *Drosophila* protein Strawberry notch, appears to have independently laterally transferred from bacteria or bacteriophages to the crown group eukaryotes, but was lost in amebozoans and fungi (Fig. 15.5).

In functional terms, the characterized chromatin-remodeling SWI2/SNF2 ATPases of eukaryotes can be divided into three broad classes: (i) Those utilizing actin-like proteins. This class might be further divided into those that associate with the Reptin/pontin AAA+ ATPases, that is, the Ino80-like class and those that associate with SWIRM domain containing subunits, that is, the Brahma-like class; (ii) the CHD/MI-2 like remodelers; (iii) the ISWI remodelers. All these classes can be traced to the LECA (Iyer et al., 2008b). Of these, the Brahma-like remodelers may slide or eject nucleosomes from chromatin. The Ino80-like remodelers include versions that facilitate exchange of canonical nucleosomes with those containing H2A.Z, promoting transcriptional activation by facilitating transcription start site exposure. The CHD/MI-2-like remodelers also tend to slide or eject nucleosomes in both repressive and activating contexts. The ISWI-like remodelers are unique in regulating nucleosome spacing—they might either optimize (e.g., the ACF and CHRAC) it to facilitate repression or randomize it to facilitate transcriptional activation. Prior experiments had demonstrated that the accessory subunits of the ISWI-like remodelers have a role in sensing linker DNA and thereby possibly regulating nucleosomal spacing (He et al., 2008). However, it remained unknown as to how exactly this was achieved. Recent sequence analysis studies showed all the large accessory subunits of ISWI ATPases across eukaryotes, such as hACF1, WSTF, RSF1, TIP5, WCRF180, BPTF, yeast Itc1, Ioc3, and Esc8, and the plant HB1 and MBD9 as having a common conserved module (Aravind and Iyer, 2012). This module is largely α- helical and is characterized by four conserved motifs. The first of these motifs maps to the previously identified DDT motif; the remaining three motifs are termed the *WHIM* motifs 1–3 (Aravind and Iyer, 2012). This module is predicted to interact with the C-terminus of ISWI ATPases, which are characterized by a HAND, SANT, and SLIDE domain (Yamada et al., 2011). Together with the C-terminal modules of the ISWI ATPases, the module formed by the DDT and WHIM motifs interacts with nucleosomal linker DNA and constitutes a protein ruler that measures out the spacing between two adjacent nucleosomes in a dinucleosome (Aravind and Iyer, 2012). The DDT–WHIM proteins can be traced back to LUCA, suggesting that reorganization of nucleosomal spacing was an ancestral feature of eukaryotes.

SMC-like ATPases belong to the ABC superfamily, and contain a coiled coil domain and a hinge domain inserted within the P-loop ATPase domain (Hirano, 2005, 2006; Uhlmann and Hopfner, 2006). The hinge domain, which is a defining feature of the SMC proteins, has been proposed to form a distinct DNA-binding site in these proteins (Griese et al., 2010; Haering et al., 2002). Working as dimers along with other accessory proteins such as kleisins, they are primarily responsible for the large-scale organizational dynamics of chromatin, including chromosome condensation. By the time of the LECA, the SMC ATPases had proliferated into at least six distinct versions, along with the more distantly related form Rad50 that lacks a hinge domain (Fig. 15.5). These six SMC ATPases have been vertically conserved in practically all eukaryotes, with apparent loss of SMC5 and SMC6 in kinetoplastids and ciliates. Another catalytic module found in CPs is prototyped by MORC (Microrchidia protein) domain, which is a unique version of the gyrase, histidine kinase, and MutL (GHKL) domain combined with a C-terminal S5 domains (Iyer et al., 2008a). Together, they constitute a catalytically active ATPase module. The eukaryotic MORCs and their prokaryotic counterparts belong to a larger radiation of several families of GHKL proteins (the paraMORCs) that are widespread in prokaryotes (Iyer et al., 2008a). These prokaryotic versions primarily function in restriction–modification systems, in conjunction with diverse superfamily-II DNA helicases and endonucleases (Iyer et al., 2008a). These GHKL modules, along with their homologs in MutL and topoisomerase ATPase modules catalyze structural reorganization of protein complexes and concomitant DNA superstructure manipulations, often in conjunction with fused or stand-alone nuclease domains. While the functions of the eukaryotic MORCs remains insufficiently understood, by analogy to their prokaryotic counterparts, it appears likely that they mediate long-distance interactions in chromatin by DNA superstructure manipulation (Iyer et al., 2008a). Their domain architectures suggest that such interactions might occur in response to the sensing of epigenetic signals such as histone and DNA methylation. In eukaryotes, there are two distinct lineages of MORC proteins (Fig. 15.5). One of those is, interestingly, fused to the hinge and coiled coil domains found in SMC ATPases and a BAM/BAH domain, which might recognize methylated histones (Fig. 15.5). These latter proteins might effectively function as analogs of SMC ATPases, with the MORC domain playing a role equivalent to the ABC ATPase domain of the former enzymes. Apicomplexans have a unique version of the MORC ATPase fused to kelch-type β-propellers (Fig. 15.5).

15.4.8 Other CP Modifications, Potential Histone Tail Interaction Domains, and Histone Chaperones

Another key covalent modification of CPs is the conjugation of ubiquitin (Ub) and other related modifiers (Ubls; e.g., Nedd8 and SUMO). This process involves a triad of ligases that transfers the Ub/Ubl to target proteins (typically to the epsilon NH_2 group of lysines) in a three-step reaction (Schwartz and Ciechanover, 2009; Schwartz and Hochstrasser, 2003). The substrate specificity for the transfer mainly lies in the third enzyme, the E3, which typically contains a RING finger domain. Several RING finger proteins are exclusive residents of eukaryotic chromatin: the PML family of SUMO-specific E3s (Park et al., 2007), the RING finger-containing Rad5/Rad8 family of SWI2/SNF2 ATPases (Gangavarapu et al., 2006), Rag1 (Grazini et al., 2010), the RNF20/RNF40 RING finger proteins (Pavri et al., 2006), and the posterior sex combs (PSC) family of proteins of the Polycomb group (Gearhart et al., 2006) that combine a RING finger with a C-terminal Ub-like domain. The latter family, which is conserved in both the crown group and alveolates, monoubiquitinates H2A (K120) (Gearhart et al., 2006). RNF20/RNF40 proteins are E3s that modify H2B (K123) (Pavri et al., 2006). The RING finger domain of the Rag1 V-D-J recombinase functions as an E3 for histone H3 as part of the process during immunoglobulin diversification. The ubiquitin-binding UBA domain and the "little finger"-type Ub-binding Zn-ribbon are found in several CPs and might perform the function of recognizing proteins ubiquitinated by the diverse ubiquitin ligases operating in chromatin (Aravind and Iyer, 2012; Iyer et al., 2011a). An unusual set of proteins in *Trichomonas* combine MYB domains with Ub-binding UBA domains, suggesting that they might interact with ubiquitinated chromosomal proteins (Iyer et al., 2008b).

The presence of dedicated enzymes for the removal of Ub modifications from histones and other nuclear proteins is indicated by the deubiquitinating enzymes that combine the JAB peptidase domain with the SWIRM domain in animals and *Dictyostelium* (MYSM1 which acts on H2AK120) (Aravind and Iyer, 2002; Zhu et al., 2007). Chromatin-specific deubiquitinating activity targeting H2AK119-monoUb is catalyzed by the BAP1 deubiquitinase of the UCH37 family of deubiquitinating enzymes. In addition to the ASXH domain (see above), this deubiquitinase is also recruited by the HCF-1 transcriptional activator (Yu et al., 2010). HCF-1 is a large CP that is found only in animals, which is comprised of Kelch repeats that form a β-propeller as in the case of the V-D-J recombinase subunit RAG2 (Aravind and Koonin, 1999b) combined to a C-terminal fibronectin type-III (FNIII) domain. In addition to recruiting the histone deubiquitinase BAP1, this β-propeller domain has autoproteolytic properties possibly dependent on conserved histidine and cysteine residues (Capotosti et al., 2011). Given that kelch repeats as well as the FNIII domain are found in several sugar/polysaccharide-binding enzymes and proteins (e.g., galactose oxidase and sialidases) (Aravind and Koonin, 1999b), it appears that they might form sugar-binding sites. In this context, it is of interest to note that HCF-1 physically interact with *O*-linked-*N*-acetylglucosaminyl transferase (Capotosti et al., 2011), an enzyme that modifies serines and threonines in proteins with sugars. Given the relationship of the FNIII domain in this protein to cell-surface adhesion proteins, it is conceivable that HCF-1 has evolved partly from such a precursor. Beyond ubiquitin-dependent protein degradation, other types of proteolytic events have also been reported in eukaryotic chromatin, primarily from metazoan model systems. One of these is the proteolytic cleavage of the histone methyltransferase trithorax and other CPs such as HCF1-1 and TFIIA by the NTN hydrolase taspase-1 (Capotosti et al., 2007; Zhou et al., 2006). Phylogenetic analysis suggests that the taspases appear to have emerged prior to the radiation of the eukaryotic crown group from a bacterial L-asparaginase-like precursor. Interestingly, histone-modifying kinases do not appear to show any notable fusions to other chromatin-specific peptide-binding domains, and are drawn from several ancient families of eukaryotic protein kinases (Kouzarides, 2007). Of these, the H3T3 phosphorylating haspin kinases are an unusual group of kinases that are characterized by unusual sequence and structure features in the BN-terminal domain of their catalytic module (Higgins, 2010).

In addition to well-characterized modified histone-interacting domains, there are numerous less studied potential peptide interaction domains in eukaryotic CPs that might also play analogous roles (Table 15.1). Several versions of the MYB domain found in CPs (often termed *SANT domains*) bind histone tails rather than DNA (Boyer et al., 2002; de la Cruz et al., 2005; Mo et al., 2005; Yang et al., 2006). This appears to represent a eukaryote-specific functional shift in the ancient DNA-binding HTH fold toward peptide interaction. Contextual information from domain architecture suggests that domains such as the ELM2, SANTA, SJA, and EP1/EP2 interact with histone tails, other CPs, or in recruiting other activities to the nucleosome (Table 15.1; Figs 15.6 and 15.7). The SJA domain adopts a fold similar to the Polo-box domain, which has been implicated in binding phosphoserine/phosphothreonine-containing peptides (Elia et al., 2003; Garcia-Alai et al., 2010). Hence, by comparison, it is possible that

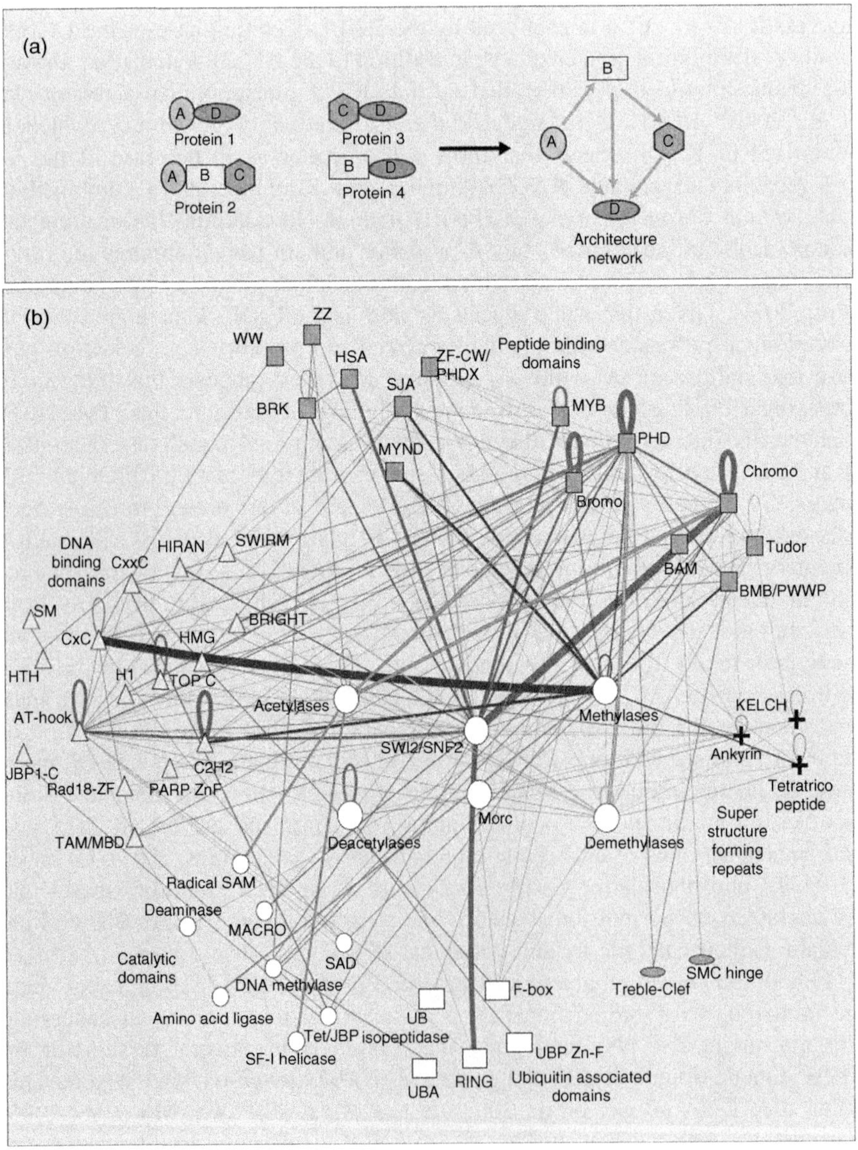

the SJA/FYR domain is involved in phosphopeptide recognition. One version of another peptide-binding domain, the SWIB domain, recruits ubiquitinating activities via the fused E3-ligase RING finger domain to TFs such as p53. The other stand-alone SWIB domain proteins (e.g., human BAF60a/b) are part of a chromatin-remodeling coactivator complex, BAF to which they recruit several unrelated TFs (Hsiao et al., 2003; Ito et al., 2001; Oh et al., 2008).

Four unrelated ancient families of histone-binding domains, namely, the HUN (HPC2-ubinuclein), nucleoplasmin/HD2, ASF1, and NAP1, appear to be primarily involved in the chaperoning and assembly of histones. The HUN domain has a characteristic pattern of conserved acidic residues that are likely to be critical for histone tail-binding (Balaji et al., 2009). The HUN domain proteins, such as HPC2 and ubinuclein, are probably involved in recruitment of the ancient conserved histone-loading Hir-Asf1 complex to different chromatin reorganization processes. The HD2 domain related to nucleoplasmin was originally claimed to be a histone deacetylase, but appears more likely to be a histone-binding domain (Aravind and Koonin, 1998). One version of the nucleoplasmin/HD2 domain contains a fusion to a peptidyl prolyl isomerase domain of the FKBP family. Orthologs of this protein are seen in several eukaryotes including *Giardia* and might play a role in the folding and assembly of histones by facilitating conformational isomerization of proline (Kuzuhara and Horikoshi, 2004; Winkler et al., 2000). The presence of the nucleoplasmin/HD2 and ASF1 domains in all eukaryotes, including early-branching forms such as *Giardia* and *Trichomonas*, points to the presence of at least two distinct histone chaperones in LECA (Iyer et al., 2008b). The ASF1 domain adopts the β-sandwich fold, a structure also seen in another CP domain, the YEATS domain, which, similarly to ASF1, also binds histones H3 and H4. In diatoms the YEATS domain is fused to a glutamine

Figure 15.6 Network representations of the domain architectures of eukaryotic chromatin proteins. (a) A hypothetical example showing how domain architecture networks are constructed. A, B, C, and D are globular domains that occur in a range of combinations. These are combined into an architectural network where the globular domains are nodes and the edges reflect their physical connectivity. (b) The domain architecture network for eukaryotic chromatin proteins with a focus on the primary catalytic regulatory systems, namely, acetylation, methylation, and ATP-dependent chromatin remodeling. Included within acetylases, deacetylases, methylases, and demethylases are all enzymes known or predicted to catalyze the respective activity, irrespective of the superfamily to which they belong. The links made by demethylase domains are shown in aquamarine, those by acetylases in red, by SWI2/SNF2 ATPases in purple, and by MORC ATPases in orange. Different functional categories of domains and their labels are colored in the same way and spatially grouped together. The thickness of the edges is approximately proportional to the relative frequency with which linkages between two domains reoccur in distinct polypeptides in all eukaryotes. The graphs were rendered using PAJEK (http://vlado.fmf.uni-lj.si/pub/networks/pajek/). (*See insert for color representation of the figure.*)

synthetase-like NH2-COOH ligase (e.g., gi: 223997528 from *Thalassiosira*), which points to the possibility of a novel histone modification in these organisms in the form of ligation of peptides to amino acid side chains. NAP1 is absent in the basal eukaryotic taxa, but appears to have emerged before the divergence of *Naegleria* and kinetoplastids from other eukaryotes. In contrast, histone chaperones of the Chz1p family and the Scm3 family have a more restricted distribution, being present only in animals and fungi (Luk et al., 2007; Zhou et al., 2011). The Scm3 domain is critical for the assembly of cenH3 into centromeric chromatin and is remarkable for its lineage-specific fusions to a diverse array of DBDs (Aravind et al., 2007). These fusions are likely to be required for recognizing centromeric DNA sequences and are important determinants of the form of the centromeres—that is, regional versus point centromere. Assembly of histone octamers using multiple chaperones appears to be an ancestral feature of eukaryotes, distinguishing them from archaea, and might be correlated with the origin of low complexity tails.

15.4.9 Natural History of Epigenetic DNA Modification Enzymes

The most frequent DNA modification in all the three superkingdoms of life is the methylation of cytosine at the fifth position of the pyrimidine ring (5mC) (Gommers-Ampt and Borst, 1995). The next most frequent is the methylation of adenine on the NH_2 group attached to the sixth position of the purine ring (N6mA),, which is fairly common in prokaryotes and certain eukaryotic lineages (Roberts, 1980; Roberts et al., 2003; Roberts et al., 2010). The catalytic deamination of cytosine in DNA has thus far only been confirmed in vertebrates, although the detection of divergent deaminase domains related to the DNA deaminases points to possibly more widespred distribution of this modification (Iyer et al., 2011a). Other modifications include (i) the formation of β-D-glucosyl-hydroxymethyluracil (base J) from thymine in euglenozoans, including the parasites *Trypanosoma* and *Leishmania* (Borst and Sabatini, 2008); (ii) a related modification, namely, 5hmC, first observed in the DNA of caudate phages (Kriaucionis and Heintz, 2009; Tahiliani et al., 2009). It has more recently been shown to occur in animals and is predicted to occur more widely across eukaryotes; (iii) N6-carbamoylmethyl adenines (called *Momylation* after the Mom enzyme of phage Mu that catalyzes this modification) are predicted in a subset of diatoms and haptophyte algae (Iyer et al., 2009b; Kaminska and Bujnicki, 2008).

15.4.10 DNA Methylation in Eukaryotes

In evolutionary terms, all DNA methylases belong to a large monophyletic assemblage containing the classical Rossmann fold, which is unified by the presence of a characteristic large loop immediately C-terminal to the core strand 4 (Fauman et al., 1999; Iyer et al., 2011a). Most members of this assemblage methylate bases in nucleic acids, or amino acid side chains of nucleoproteins. N6mA is relatively uncommon in most eukaryotes, but has been positively identified in several lineages of ciliates, chlorophyte algae, and dinoflagellates, where it constitutes 0.5–10% of the adenines in the genome (Gommers-Ampt and Borst, 1995). Of these, several distinct versions appear to be specified by different types of mobile elements. For example, N6A methylase domains are encoded by a distinct family of retroposons whose archetypal member is the *Dictyostelium* DIRS-1 element that has widely disseminated across eukaryotes and expanded in several distantly related organisms (Poulter and Goodwin, 2005). However, these might only be active in the genomes of certain chlorophyte algae (Iyer et al., 2011a). Other potential eukaryotic N6A methylases appear to be cellular enzymes with a role in chromatin organization. One of these, found across chlorophyte algae, is a multidomain protein with the N6A methylase domain fused to one or more N-terminal BMB/PWWP and C-terminal PHD-X/ZF-CW domains (Fig. 15.5). In addition, they often contain PHD finger domains N-terminal to the methylase domain (Iyer et al., 2011a). In ciliates, there is a distinctive version of the Ime4-like family, which is fused to four N-terminal ZZ ZnFs, a domain also found in CPs such as Ada2 and CBP/p300 (Iyer et al., 2011a). Given that all ciliates studied to date show substantial N6mA in DNA, and have no other candidate methylases to catalyze this reaction, we suggest that these ZZ-domain containing methylases indeed perform this function. In addition, orthologous methylases of this ciliate version are found in the heterolobosean amoeboflagellate *Naegleria* and the rhodophyte alga *Cyanidioschyzon*, suggesting a possibly wider distribution for this form of adenine methylation across eukaryotes (Iyer et al., 2011a).

5mC has been observed in the genomes of a wide range of eukaryotes, albeit with patchy phyletic patterns (Gommers-Ampt and Borst, 1995; Zemach et al., 2010). DNMT1 is the primary maintenance methylase that reestablishes the methylation marks at CpG sites on both strands of the duplex after replication (Jeltsch, 2006). Ascomycete fungi possess a second distinct methylase related to DNMT1, exemplified by RID (repeat-induced point mutation defective) from *Neurospora* and Masc1 from *Ascobolus*. These methylases are implicated in a related set of phenomena: repeat-induced point mutation (RIP) in *Neurospora* and methylation induced premeiotically (MIP) in *Ascobolus* (Freitag et al., 2002; Malagnac et al., 1997). Plants possess a second group of 5C DNA methylases related to DNMT1, the chromomethylases (CMTs), which are characterized by the distinctive insertion of a chromo domain into the methylase domain (Fig. 15.5) (Bartee et al., 2001; Cao et al., 2003; Cao and Jacobsen, 2002; Henikoff and Comai, 1998). In the multicellular plant *Arabidopsis*, one of the CMTs (CMT3) is involved in the methylation of CpNpG rather than CpG, and is a critical player in the RNA-directed DNA methylation process observed in plants. These methylases appear to have diverged from a single precursor within eukaryotes (Iyer et al., 2011a). DNMT1 was acquired early in eukaryotic evolution, prior to the divergence of the heteroloboseans, followed by multiple losses in lineages such as kinetoplastids, alveolates, stramenopiles, and amoebozoans. The ancestral architecture of DNMT1 can be reconstructed as comprising a methylase module (including the catalytic domain and the C-terminal DNA-binding module) fused to the N-terminal RFD module and two BAM/BAH domains (Fig. 15.5).

Structural analysis of the RFD module reveals two distinct globular domains, an N-terminal circularly permuted version of the Sm domain and a C-terminal HTH domain of the four-helical variety (Iyer et al., 2011a). The DNMT3 methylase clade was primarily characterized as the *de novo* methylase required for the reestablishment of the methylation patterns after they have been erased by demethylation (Jeltsch, 2006; Kaneda et al., 2004; Kato et al., 2007; Okano et al., 1999). DNMT3 is found primarily in the animal, plant, and stramenopile lineages, indicating that it has been entirely lost in the fungal and amoebozoan lineages. In land plants, one of the paralogous versions of DNMT3 underwent a circular permutation within the methylase module, which resulted in strand 5 of the Rossmann fold domain coming to the N-terminus of the entire methylase module (Fig. 15.5) (Henderson et al., 2010). In metazoans, the methylase module is fused at the N-terminus to the BMB/PWWP domain followed by a multinuclear Zn-chelating module shared with the SWI2/SNF2 ATPase ATRX1, referred to as the *ADD* module (Argentaro et al., 2007; Otani et al., 2009). The ADD module is comprised of an N-terminal mononuclear treble-clef domain and a C-terminal PHD finger domain. In contrast, in land plants, the methylase module is fused to three N-terminal UBA domains that are known to bind ubiquitin (Fig. 15.5).

The DNMT2 methylases have been at the center of controversy over whether they function as either DNA or RNA methylases or both (Goll et al., 2006; Kuhlmann et al., 2005; Phalke et al., 2009; Schaefer and Lyko, 2010). DNMT2 is the most widely distributed DNA MTase clade in eukaryotes, being present in the animal lineage, fungi, amoebozoans, the plant lineage, stramenopiles, apicomplexans, and the heterolobosean *Naegleria* (Fig. 15.2). Thus, it appears to have been acquired early in eukaryotic evolution and has been vertically inherited ever since. Unlike DNMT1 and DNMT3, DNMT2 shows simple domain architecture with no fusions to other CP domains in eukaryotes. A distinct group of cytosine methylases in eukaryotes are typified by a fusion to an ATPase module of the RAD5-clade of SWI2/SNF2 (Fig. 15.5) (Iyer et al., 2011a). These methylases are found in both ascomycete and basidiomycete fungi, chlorophyte algae, and stramenopiles. While they are likely to have been present in the common ancestor of most of the above groups, they have been frequently lost in several members. However, their overall distribution in eukaryotes is best interpreted as a consequence of lateral gene transfers occurring early in the evolution of these groups. They differ from most other methylases in that the methylase module is part of a large multidomain architecture with other enzymatic domains in the same polypeptide. The methylase module is fused at the C-terminus to a distinctive domain with a treble-clef fold related to the ZZ-domain (Ponting et al., 1996), followed by an uncharacterized globular domain, which in turn is followed by a C-terminal SWI2/SNF2 ATPase module with a RING finger domain inserted into it (Fig. 15.5). Kinetoplastids encode a conserved 5C DNMT1 typified by *Leishmania* LmjF25.1200 and *Trypanosoma brucei* TbDMT. In addition, representatives of this methylase family are found in several stramenopiles and the chlorophyte alga *Micromonas* (Fig. 15.5). Recently, it was demonstrated that TbDMT methylates cytosine at the retroposon insertion hotspots and variable surface antigen gene (VSG) loci in the *T. brucei* genome (Militello et al., 2008). The chlorophyte-type 5C DNA methylases are exclusively found in chlorophyte algae (Iyer et al., 2011a). Their methylase module is fused to two C-terminal BMB/PWWP domains that sandwich a distinct divergent CXXC domain (Fig. 15.5). Iridoviruses, such as the Lymphocystis disease virus, specify a distinct cytosine DNMT1 that is defined by a characteristic small CTDBM, which contains three conserved cysteines and a histidine that might stabilize the domain through chelation of a cation (Iyer et al., 2011a). This virally coded cytosine methylase could catalyze methylation of the viral genome (Doerfler, 2008) and could aid in both evasion of host foreign DNA surveillance systems and epigenetic regulation of viral gene expression.

15.4.11 DNA Demethylation and Further Modifications of Methylated DNA

DNA demethylation at 5mC is a critical process across eukaryotes. Nevertheless, the phenomenon is not well understood in terms of biochemistry or possible mechanisms. While a number of distinct enzymes and mechanisms have been proposed for the catalysis of demethylation, several of these appear either unlikely or dubious (Ooi and Bestor, 2008). The most notable candidates for demethylation are the DNA base glycosylases. The direct action of DNA 5mC glycosylases in demethylation has been demonstrated in plants, and is catalyzed by the Demeter-like family of glycosylases (Gehring et al., 2006; Morales-Ruiz et al., 2006). In vertebrates, MBD4 has been shown to be a bifunctional DNA glycosylase that removes 5mC in addition to G/T mismatches, generating a strand break (Hendrich et al., 1999; Kim et al., 2009; Zhu et al., 2000). The unrelated thymine DNA glycosylase (Tdg) may also possess this activity (Yoon et al., 2003). These enzymes have also been invoked in demethylation via a deamination step mediated by AID/APOBEC-like deaminases prior to the action of the DNA glycosylases (Rai et al., 2008) and also via the oxidation of methylcytosine by the Tet enzymes (He et al., 2011). The catalytic domains of all currently known DNA glycosylases belong to four structurally unrelated folds, two of which contain members that have currently been implicated in DNA demethylation (Iyer et al., 2011a). The first of these, the uracil DNA glycosylase (UDG) superfamily, typified by human Tdg and *Escherichia coli* Mug and Ung, contains an α/β domain with a central β-sheet formed by four conserved strands (Aravind and Koonin, 2000a). These enzymes are strictly monofunctional and only catalyze the removal of the base from the nucleotide (Krokan et al., 1997; Slupphaug et al., 1996). The second superfamily of DNA glycosylases implicated in demethylation (HhH-Glycosylase) is typified by the catalytic domains of Mbd4, Demeter, and their bacterial counterparts such as *E. coli* MutY and endonuclease III (Nth)(Krokan et al., 1997). This catalytic domain is comprised of four copies of the helix-hairpin-helix (HhH) motif, which also occurs independently as a DBD in diverse DNA repair proteins and the bacterial RNA polymerase α-subunit (Doherty et al., 1996). A third distinct fold of DNA glycosylases, typified by *E. coli* endonuclease VIII and vertebrate Neil1/2/3, has currently not been implicated in

DNA demethylation (Zhang et al., 2005). Nevertheless, versions of this superfamily from chlorophyte algae show fusions to the DNA-binding SAP domain, which specifically functions in tethering various DNA modification and repair activities to regions of chromatin (such as SARs/MARs). In light of this, a role in DNA demethylation or related epigenetic DNA modifications cannot be ruled out for this class of DNA glycosylases in certain eukaryotes (Iyer et al., 2011a).

The Mbd4-like clade is one the most widely distributed of the HhH-Glycosylase clades implicated in DNA demethylation. Mbd4 orthologs are known from animals, fungi, plants, and certain stramenopiles (Iyer et al., 2011a). The phyletic pattern of Mbd4 in eukaryotes usually shows a strong correlation with notable levels of genomic 5mC and has been repeatedly lost in many of the lineages with low levels of or no genomic methylation. In animals, basal members of the plant lineage (chlorophyte algae), and diatoms, MBD4 is fused to a TAM/MBD domain. This fusion suggests that the ancestral version of the MBD family probably directly translocated to sites enriched in methylated CpG by means of its TAM/MBD domain. However, this domain has been lost in the land plants and fungi (Iyer et al., 2011a). In land plants the MBD4 ortholog contains a long N-terminal extension with one to six copies of a short peptide repeat with a consensus motif [VI]SPxh (where x is any amino acid and h is a hydrophobic residue). Though the function of these repeats is currently unclear, it is possible that these repeats are the sites of posttranscriptional modification that regulates these enzymes. Chlorophyte algae possess a second paralog of MBD4 that contains, in place of the TAM/MBD domain, a distinct module known as the KRI motif, which is found in diverse eukaryotic CPs (Birtle and Ponting, 2006). On the basis of the analysis of KRI motif architectures, we predict that it is likely to have a role in recognizing epigenetic modification of histones, in particular, histone methylation. The Demeter-like clade has clear relatives within the vast radiation of bacterial Nth-like HhH-DNA glycosylases (Iyer et al., 2011a). Given that these bacterial cognates are found in the cyanobacteria and that the Demeter-like clade is restricted to plants and stramenopiles, it is possible that its ancestor was first acquired during the cyanobacterial endosymbiosis that gave rise to the plant lineage.

Until recently, it was thought that 5mC is a terminal DNA modification whose only further fate is removal by demethylation during the erasure of epigenetic marks. Studies in euglenozoans, such as the human parasites *Trypanosoma* and *Leishmania*, revealed the presence of two enzymes, JBP1 and JBP2, which catalyzed the hydroxylation of the methyl group in thymine, forming hydroxymethyl thymine (Borst and Sabatini, 2008). This base is further modified by glycosylation of the hydroxyl group resulting in the base "J." Sequence analysis of the JBP hydroxylase domains revealed that they were members of a distinctive family of 2-oxoglutarate and Fe^{2+}-dependent dioxygenases (2OGFeDOs), whose previously undetected representatives were found in several organisms (Iyer et al., 2009b). In particular, these studies showed that the metazoan Tet proteins (Tet1, Tet2, and Tet3 oncogenes in humans) are members of this family of 2OGFeDOs. Given that their domain architecture closely parallels that of the metazoan DNMT1, with an N-terminal DNA-binding CXXC domain combined to a C-terminal catalytic domain, it was proposed that they would act on 5mC and hydroxylate it to form 5hmC (Iyer et al., 2009b). Follow-up experimental studies showed that indeed the Tet proteins were 2OGFeDOs that generated 5hmC *in situ* from the 5mC in DNA. With the discovery of the catalytic activity of Tet proteins, it became clear that this further modification of 5mC is indeed a biologically relevant modification with profound significance as a novel epigenetic mark (Tahiliani et al., 2009). Studies are only just beginning to reveal the regulatory potential of this modification. In addition to a role in reprogramming of epigenetic marks on DNA, this modification might have a role to play as an intermediate in DNA demethylation by interacting with base excision repair, base deamination catalyzed by the AID/APOBEC family of cytosine deaminases, or base removal by DNA glycosylase (Guo et al., 2011; He et al., 2011; Iqbal et al., 2011; Ko et al., 2010). Indeed, it has been recently shown that the Tet enzymes catalyze further oxidation of 5hmC to generate formyl- and carboxycytosine, which are intermediates in a demethylation pathway(Iyer et al., 2009b). It has been demonstrated that the action the DNA glycosylase Tdg is responsible for the final demethylation step by removing carboxycytosine from DNA (He et al., 2011).

The catalytic domain of the Tet/JBP family displays a double-stranded β-helix fold (DSBH). This is characteristic of a vast class of 2OGFeDOs that catalyze dioxygenase reactions on a wide range of substrates, including peptides, nucleic acids, and small molecules (Iyer et al., 2010; Iyer et al., 2009b). The Tet/JBP family of enzymes is widely, albeit sporadically, distributed across the tree of life. All eukaryotic versions appear to have been derived via lateral transfers of bacteriophage versions on more than one occasion. In eukaryotes, the Tet/JBP proteins have diversified into five distinct subfamilies. Of these, the subfamily that is prototyped by the Tet proteins is restricted to metazoa and is strictly correlated with presence of DNA cytosine methylation. This subfamily is distinguished by the remarkable insertion of a cysteine-rich domain into the N-terminal region of the catalytic 2OGFeDO domains just upstream of the HxD motif (Iyer et al., 2009b). The next major Tet/JBP subfamily, the transposon-associated subfamily, is currently known from chlorophyte algae such as *Chlamydomonas* and *Volvox*, and mushrooms. These transposons are particularly expanded in the mushrooms with at least 40–60 copies in the genomes of *Coprinopsis* and *Laccaria*. These transposons appear to be located predominantly in the subtelomeric regions and might regulate their own gene expression or mobility via hydroxylation of methylcytosine (Iyer et al., 2009b). Of the remaining subfamilies, the JBP subfamily is currently only known from euglenozoans. These versions occur either fused to an SWI2/SNF2 ATPase module (JBP2) or fused to the J-base containing DBD characterized JBP1C domain (Iyer et al., 2009b). The fourth subfamily is currently only known from the heterolobosean amoeboflagellate *Naegleria*, at least one member of which is fused to a C-terminal chromo domain. Given the inference of the presence of 5mC (see above) in *Naegleria*, it is possible that these proteins generate 5hmC similarly to their homologs in other eukaryotes. The fifth subfamily is currently known from chlorophyte algae and stramenopiles. One version of this family is fused to an N-terminal TAM/MBD domain, suggesting that it is likely to recognize DNA with 5mC and modify the base to 5hmC (Fig. 15.3). However, the domain architectures of the remaining members of this subfamily are characterized by fusions to various RNA-binding or RNA-modifying enzymatic domains. It is likely that they generate a range of lineage-specific hmC or hT modifications in tRNAs and other small RNAs in these lineages (Iyer et al., 2009b).

Most bacterial cognates of each of the major eukaryotic cellular 5C and N6A methylases have primarily radiated as a part of the restriction–modification (R-M) systems of bacteria (Iyer et al., 2011a). Thus, selective pressures, which favor diversification of R-M systems, appear to have driven the evolution of a great variety of DNA methylases that were then repeatedly acquired by eukaryotes. However, the same epigenetic codes utilized by the R-M systems appear to have been deployed in the very distinct context of chromatin dynamics in eukaryotes. Indeed, several other components of R-M systems and other selfish elements have been acquired in parallel to the methylases and utilized in different facets of eukaryotic chromatin dynamics. The most prominent of these include chromatin-remodeling enzymes such as SWI2/SNF2 ATPases and MORCs, the Tet/JBP-like DNA base hydroxylases, DBDs such as HARE-HTH, SAD/SRA and HIRAN, and DNA repair enzymes such as the VRR-NUCs. Thus, it can be said that the bacterial mobile selfish systems have served as the "development labs" for not just the DNA methylases but also other key players in eukaryotic chromatin and nucleic-acid-modifying functions.

15.5 INTERACTIONS BETWEEN RNA-BASED REGULATORY SYSTEMS AND CHROMATIN FACTORS

Comparative genomic analysis of fungi predicted a functional link between the siRNA/miRNA biogenesis pathway and several CPs (Aravind et al., 2000). Accumulating recent experimental evidence has confirmed this and points to a functional link between RNA-based regulatory systems, including posttranscriptional gene silencing or RNAi and chromatin-level regulatory events (Allis et al., 2007; Grewal, 2010; Grewal and Rice, 2004). Studies in a number of eukaryotes have revealed a role for siRNAs in directing DNA methylation and heterochromatin formation (Chan et al., 2006; Grewal and Rice, 2004). RNAi-like systems have also been implicated in epigenetic phenomena such as paramutation in plants and meiotic silencing by unpaired DNA in *Neurospora* (Alleman et al., 2006; Shiu et al., 2001; Shiu et al., 2006). In ciliates, a similar small RNA-based pathway has been implicated in addition to histone H3 methylation and heterochromatin formation in the rearrangement and elimination of DNA sequences during the development of the macronucleus (Juranek et al., 2005; Mochizuki et al., 2002). The core molecular apparatus involved in RNA-based gene silencing can be traced back to the LECA (Muljo et al., 2010; Suzuki and Miyazono, 2011). On the basis of the currently available data, the ancestral apparatus can be reconstructed as primarily comprising of (i) a processing RNAse (Dicer), which generates small RNAs from a longer precursor through endonucleolytic cleavage (Braun et al., 2010; Malone et al., 2005; Pontes et al., 2006). The ancestral Dicer family of RNAses contained tandem catalytic domains of the RNAseIII superfamily linked to additional RNA-binding domains; (ii) an effector nuclease that uses the small RNAs generated by the former enzyme as guides to target transcripts and other RNAs. This nuclease belongs to the Piwi–Argonaute (Ago) family of the RNAseH superfamily (Braun et al., 2010; Li et al., 2006; Shi et al., 2004); (iii) additionally, in several eukaryotes, the siRNAs generated from transcripts might be amplified to generate further copies by means of an RNA-dependent RNA polymerase (RdRp) (Alleman et al., 2006; Iyer et al., 2003; Shiu et al., 2006). However, this enzyme is exclusive to the siRNA branch of the RNAi system and mammals appear to be missing RdRp orthologs.

All these enzymes, comprising the core of the ancestral eukaryotic silencing system, have bacterial and archaeal antecedents (Anantharaman et al., 2002; Iyer et al., 2003; Muljo et al., 2010; Suzuki and Miyazono, 2011). However, they appear to have come together as a single coherent functional system only in the eukaryotes. Within eukaryotes, the system has considerably diversified in terms of the domain architecture and the recruitment of additional components. For example, Dicers from early-branching eukaryotic lineages such as *Trichomonas* and *Giardia* contain two RNAseIII catalytic domains linked to an N-terminal PAZ domain and a divergent C-terminal dsRNA-binding domain (dsRBD) (Muljo et al., 2010). Further duplications and domain accretions to this core appear to have happened by the time of the divergence of the heterolobosean lineage, resulting in versions of Dicer with fusions to two more N-terminal modules, namely, the superfamily-II helicase and an additional dsRBD. Since that point on, most eukaryotic lineages possessed two versions of the Dicer family, one with an associated helicase domain and another without (Muljo et al., 2010). In animals and amoebozoans, a representative of the latter type, with two RNAseIII domains and a C-terminal dsRBD, seems to have been fixed as a conserved orthologous lineage of proteins prototyped by Drosha (Suzuki and Miyazono, 2011). The Piwi–Ago family similarly appears to have differentiated extensively in eukaryotes. The Piwi subgroup probably represents the ancestral versions of these proteins in eukaryotes, which in the crown group (animals, fungi, amoebozoans, and plants) spawned a new subgroup, the Ago proteins. The Ago subgroup underwent further LSEs independently in the plants and animals. In some lineages, the RNAseH-fold nuclease domain of Ago underwent apparent inactivation—three of the four mammalian Ago proteins have lost catalytic activity (Miyoshi et al., 2010; Muljo et al., 2010). The silencing action of the small RNAs is mainly mediated by certain members of Piwi-Argonaute family, which might localize to chromatin to specifically degrade transcripts right at the source (Grewal, 2010). The RNAi system was repeatedly lost, either partially or entirely, in several eukaryotes. Vertebrate apicomplexan parasites, with the exception of the *Toxoplasma* lineage, have lost both the Piwi nuclease and the RdRp, suggesting that they are unlikely to possess a *bona fide* RNAi system (Ullu et al., 2004). In *Toxoplasma*, this system generates miRNAs comparable to those seen in other organisms (Braun et al., 2010). Some parasites such as kinetoplastids and *Trichomonas* appear to have lost the RdRp but retain Piwi nucleases, and as a consequence display certain branches of the RNAi system (Iyer et al., 2008b; Shi et al., 2004; Ullu et al., 2004). While the RNAi system precedes the origin of 5mC DNA methylation in eukaryotes, it appears to have evolved a strong connection with this system at some point just prior to the radiation of the eukaryotic crown group (Iyer et al., 2011a). The evidence from ciliates suggests interaction of the RNAi system with the histone methylation apparatus and HP1-like chromo domain proteins might have emerged even earlier in eukaryotes (Allis et al., 2007; Aravind et al., 2011; Juranek et al., 2005).

Several studies in crown group eukaryotes have implicated large noncoding RNAs in heterochromatin formation and chromosome dosage compensation (Kanduri, 2011; Nagano and Fraser, 2011; Sanchez-Elsner et al., 2006; Tsai et al., 2010). Some chromo-like domains have been shown to interact with these RNAs (Bernstein et al., 2006; Brehm et al., 2004; Shimojo et al., 2008). Similarly, SAM domain proteins of the polycomb complex in animals have also been shown to interact with large RNAs in chromatin (Zhang et al., 2004). These suggest that there might be other RNA-based pathways, distinct from RNAi pathways, which might have a direct role in chromatin level regulation. Certain large noncoding RNA molecules such as HOTAIR, Air, Kcnq1ot1, Tsix, RepA, and the trithorax RNAs are incorporated as subunits of protein methylation complexes (Kanduri, 2011; Nagano and Fraser, 2011; Sanchez-Elsner et al., 2006; Tsai et al., 2010). HOTAIR in particular appears to function as a nucleic acid scaffold that links the SET domain histone methylase EZH2 with the demethylase LSD1 (Tsai et al., 2010). Several other long noncoding RNAs interact with the polycomb repressive complexes PRC1 and PRC2 in animals and appear to play a critical role in guiding these complexes to specific genomic loci (El Messaoudi-Aubert et al., 2010; Khalil et al., 2009). The weight of the evidence suggests that these large noncoding RNAs directly bind SET domain proteins such as Ash1, G9a, and EZH2 in chromatin complexes (Sanchez-Elsner et al., 2006; Zhao et al., 2010), suggesting that interactions between SET domain proteins and such long noncoding RNAs might be a more general feature of chromatin preserved throughout eukaryotic evolution. The RNA helicase Mov10 has been shown to be a component of the PRC1 complex that might interact with these RNAs (El Messaoudi-Aubert et al., 2010). Mov10/armitage-like RNA helicases have also been implicated in posttranscriptional gene regulation via the RNAi system. They are characterized by the presence of two domains N-terminal to the superfamily-I RNA helicase module, namely, a novel β-sandwich fold domain related to the ASH family of Ig-like domains (Anantharaman et al., 2007) and an SH3-fold β-barrel distantly related to the tudor-like domains. These domains might play a key role in recruiting the Mov10/armitage-like helicases to chromatin, with the latter domain probably binding the noncoding RNAs. Given that Mov10-like helicases are detectable in *Naegleria*, stramenopiles and all crown group eukaryotes, it is possible that such regulation emerged relatively early in eukaryotic evolution. This "world" of noncoding RNA-based chromatin-modifying processes in different eukaryotes is an exploration poised to yield rich dividends in the near future.

15.6 DOMAIN ARCHITECTURES OF CPs

15.6.1 Syntactical Features in Domain Architectures of CPs: Nature of Interactions Between Different Regulatory Systems

Domain architectures of CPs reveal certain strong "syntactical" patterns (Figs 15.6 and 15.7). For example, histone methylase and acetylase domains almost never co-occur in the same polypeptide in any eukaryote. Similarly, demethylases and deacetylases rarely, if ever, co-occur with each other or, respectively, with methylases and acetylases (Fig. 15.6). This suggests that acetylation and methylation are relatively stable modifications, and that in general their removal is not closely coupled in time with another remodification (Allis et al., 2007; Iyer et al., 2008b; Kouzarides, 2007). This is consistent with methylation and acetylation being epigenetic markers and being independent but potentially complementary in action. Two of the four acetyltransferases that can be traced to the LECA are closely associated with the core transcription apparatus (GCN5, Elp3 families). Hence, the earliest roles of acetylation were probably in the context of modulating histone–DNA interaction to facilitate transcription. On the other hand, methylation appears to have emerged in a more general context of establishing particular chromatin states, such as condensed chromatin or heterochromatin, by altering histone properties (Filion et al., 2010; Roudier et al., 2011). Methylases eventually developed greater domain architectural diversity than acetylases in eukaryotic evolution (Figs 15.4 and 15.7). Similarly, histone demethylases show a clearly greater architectural complexity than deacetylases (Figs 15.4 and 15.6). These patterns could suggest that methylases and demethylases might have evolved a greater selectivity for the specific contexts (e.g., other co-occuring modifications) of their target residues or respond to a larger range of inputs sensed by the fused domains (Kouzarides, 2007).

Acetylases and methylases show preferential associations with certain peptide-binding domains—acetylases most frequently combine with bromo domains, and methylases with PHD fingers (Fig. 15.6). Given the binding preferences of these peptide-binding domains, it is possible that, respectively, discriminating the existing methylation or acetylation state of histones might be an important functional feature of some versions of these enzymes, especially in the context of maintaining an epigenetic mark (Collins and Cheng, 2010; de la Cruz et al., 2005; Pasillas et al., 2010). Conversely, some methylases are also fused

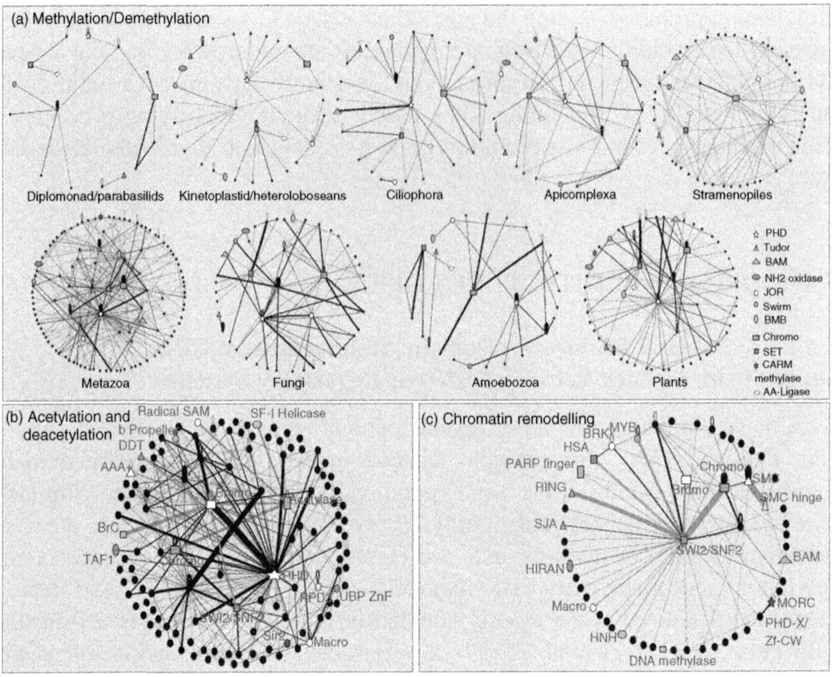

Figure 15.7 Domain architecture networks of proteins involved in protein methylation, acetylation, and ATP-dependent chromatin remodeling. (a) Domain architecture networks of proteins known or predicted to be involved in the chromatin protein methylation system are shown for representative eukaryotes. The proteins belonging to the methylation system include all proteins containing methylase, demethylase, and methylated peptide-binding domains. Their connections with each other and all other domains occurring in their respective polypeptides proteins are shown. Certain key domains of the system are marked with colored shapes as indicated in the right panel of the figure. Note the increasing architectural complexity as indicated by the increasing density of the network over eukaryotic evolution, especially in several crown group lineages. (b) The domain architecture network for the chromatin protein acetylation-based system across all eukaryotes. This set includes proteins containing acetylase, deacetylase, ADP-ribose metabolite-binding, and acetylated-peptide-binding domains. The architecture network was constructed as illustrated in Fig. 15.6a and for the methylation system, except that it includes all eukaryotes. Several key chromatin protein domains have colored shapes and are labeled. Red edges denote domain connections that can be traced back to the last eukaryotic common ancestor, green shows those emerging prior to the divergence of the kinetoplastid-heterolobosean clade, and cyan connections can be traced back to the common ancestor of the crown group and chromalveolates. Note the proliferation of lineage-specific architectures in the course of eukaryotic evolution. (c) A network similar to Fig. 15.7b for the ATP-dependent chromatin-remodeling system across all eukaryotes. This includes all proteins containing SWI2/SNF2, MORC, and SMC domains. Various notable domains are colored and labeled. Certain edges have been colored on the basis of their point of origin as described above. The thickness of the edges is approximately proportional to the frequency with which linkages between two domains appear in multiple polypeptides (thickness is relative within a given figure). (*See insert for color representation of the figure.*)

to acetylated-peptide-binding domains and some acetylases are fused to methylated peptide-binding domains (Figs 15.6 and 15.7), suggesting that a degree of cross-talk or interdependence developed between these processes in the course of evolution. Similarly, evidence from domain architectures suggests that both systems interact to a certain degree with the ubiquitin system; such associations began emerging in the chromalveolate and crown group clades. Peptide-binding domains recognizing different forms of histone modifications might also be combined with each other in the same polypeptide (Figs 15.4, 15.5, and 15.7). Often, such architectures have arisen in a lineage-specific manner (Figs 15.3 and 15.4). For example, *Phytophthora* shows proteins with six tandem bromo domains and serial bromo, PHD finger, and chromo domains, trypanosomes possess a protein with bromo and ZF-CW(PHD-X) domains, and *Giardia* possesses a protein combining the bromo domain and a WD-type β-propeller (Figs 15.3 and 15.4). In particular, the trypanosome versions might represent a lineage-specific adaptation for coordinating histone methylation with histone acetylation to activate transcription of polycistronic mRNAs (Siegel et al., 2009) (see above). This suggests that while histone modifications might be universal in eukaryotes, their "interpretation" by peptide-binding adaptors shows numerous lineage-specific differences. SWI2/SNF2 ATPases have been shown to work with different histone-modifying enzymes in eukaryotic model systems. However, their domain architectures across eukaryotes show that there are no known fusions between these ATPases and histone acetylase or methylase domains (or the corresponding de-modifying enzymes) (Fig. 15.6). Hence, though their actions are cooperative, they are not closely coupled mechanistically. However, SWI2/SNF2 ATPases are combined with Ub-conjugating E3 domains in the same polypeptide, suggesting possible coupled action between these activities.

15.6.2 Relationship Between Phylogeny, Organizational Complexity, and Domain Architectures of CPs

Domain architectures can be depicted as an ordered graph or a network, in which domains form the nodes, and their linkages with other domains within a given polypeptide (adjacent co-occurrence in polypeptide) are depicted as edges connecting nodes (Fig. 15.6) (Anantharaman et al., 2007). These domain architecture networks have proven to be useful in assessing the complexity of domain architectures. Complexity of domain architectures of proteins in a given functional system can also be independently assessed using the complexity quotient that measures both the variety and the number of domains in those proteins (Fig. 15.1d) (Burroughs et al., 2007). Anecdotal studies had indicated that domain architectural complexity correlated with increased organizational complexity of the organism—that is, emergence of multicellularity and increased cellular differentiation (Gibson and Spring, 1998; Lander et al., 2001). In functional terms, greater domain architectural complexity of CPs would imply a greater variety and number of interactions made by those with proteins, nucleic acids, and small molecules.

Domain architecture networks show a trend of increasing domain architectural complexity in CPs in the course of eukaryotic evolution (Fig. 15.7). Diplomonads and parabasalids have the least complex domain architectures. The *Naegleria*-kinetoplastid clade, apicomplexans, and ciliates have higher architectural complexity than these, and chromists have even higher values (Iyer et al., 2008b). However, the highest architectural complexity is observed in certain crown group clades, and amongst those, the animals are unparalleled in the complexity of their domain architecture networks (Fig. 15.7). When the complexity quotient of CPs is plotted against the total number of predicted CPs encoded by an organism, we observe a steady positively correlated rise in these values (Fig. 15.1d). In many cases this increase in architectural complexity occurs via "domain accretion" or fusion of new domains around an ancient orthologous core of the polypeptide (Koonin et al., 2000). This tendency is particularly prominent in histone methylases and SWI2/SNF2 ATPases (Figs 15.4, 15.5, and 15.7). Despite having large absolute numbers of CPs, ciliates and *Trichomonas* tend to have much lower architectural complexity (Fig. 15.1d). Mere increase in proteome size without increase in architectural complexity of CPs, as seen in ciliates and *T. vaginalis*, might be sufficient to achieve the required levels of regulation in a unicellular context. In contrast, the high complexity of animal proteins points to a possible relationship between architectural complexity and the number of CPs, and emergence of numerous differentiated cell types (Figs 15.1d and 15.7) (Aravind et al., 2009). Excluding *Naegleria* and *Trichomonas*, other protist parasites such as apicomplexans, kinetoplastids, and diplomonads have relatively fewer and architecturally less complex CPs (Figs 15.2d and 15.7).

In general, the observed architectures and phyletic patterns are consistent with the phylogenetic relationships among eukaryotes, albeit obscured by extensive losses in several parasites (Iyer et al., 2008b). Certain clades are strongly supported by shared architectures and phyletic patterns: (i) the animal-fungi clade; (ii) a crown group clade uniting plants, amoebozoans, fungi, choanoflagellates, nucleariids, and animals; (iii) apicomplexans, alveolates and, to a certain extent, the chromalveolate clade; and (iv) a clade comprised of all eukaryotes, excluding the diplomonad and parabasalid lineages (tree in Fig. 15.2). These points appear to coincide with notable innovations amongst CPs and TFs. Plants and stramenopiles exclusively share several TFs or CP domain architectures, compared with plants and alveolates. This is particularly intriguing given that the secondary endosymbiotic event is believed to have occurred in the common ancestor of the chromalveolate lineage (Bhattacharya et al., 2004). This might either imply selective loss of more plant-derived genes in both parasitic apicomplexans and free-living ciliates or a more recent tertiary endosymbiotic event in the ancestor of stramenopiles that delivered a new load of plant-derived genes or evidence for independent plastid acquisition (Baurain et al., 2010). It is also conceivable that the plant-derived TFs and CPs contributed to the rise of organizational complexity and multicellularity observed in stramenopiles.

15.7 GENERAL CONSIDERATIONS AND CONCLUSIONS

As seen from the above discussion, it has now become feasible to develop a reasonable reconstruction of various transcription- and chromatin-related regulatory systems in the LECA and their subsequent evolution. Though the core histones are of archaeal origin, several key nucleic-acid-associated players in chromatin and eukaryotic transcription regulation were possibly derived from bacterial mobile elements or selfish elements, such as prophages, plasmid, and restriction–modification systems (Babu et al., 2006; Balaji et al., 2005; Iyer et al., 2008b). These include the SWI2/SNF2 ATPases, the MORC ATPases, the HEH domain that helps in tethering chromosomes to the nuclear membrane, and the RDRP of the RNAi system (Iyer et al., 2008a; Iyer et al., 2006a; Iyer et al., 2003; Mans et al., 2004). In the case of peptide-modifying enzymes, the bacterial secondary metabolism systems involved in the production of non-ribosomally synthesized peptide metabolites appear to have made a major contribution. These systems gave rise to several protein methylases, demethylases, and amino acid ligases (e.g., polyglutamylases) (Aravind et al., 2011; Iyer et al., 2009a).

An important feature that defined the origin of eukaryotes was an early spurt of drastic evolutionary innovation (Aravind et al., 2006; Best et al., 2004; Dacks and Doolittle, 2001; Lopez-Garcia and Moreira, 1999). It appears that the melding of the archaeal and bacterial inheritances were key in generating both the raw material and new "biochemical niches" that favored such innovation leading to the distinctive eukaryotic systems. This appears to have happened between the point of emergence of the first eukaryotic progenitor and the LECA from which all extant eukaryotes have emerged. Hence, it is likely that the key factor that initiated this phenomenon was the primary endosymbiosis itself, which gave rise to the eukaryotic cell (Aravind et al., 2006). In general terms, the main innovations with respect to nuclear regulatory systems in this early phase were (i) multiple rounds of duplication giving rise to various paralogous protein families, which diversified into distinct functional niches (e.g., SWI2/SNF2 ATPases); (ii) apparent "invention" of new α-helical domains (e.g., the bromodomain) and diversification of metal-chelation supported structures (e.g., the PHD finger), leading to whole new sets of protein–protein interactions; (iii) emergence of proteins with long nonglobular or low complexity stretches accreted to the ancient globular domains (e.g., tails of eukaryotic histones) allowed for a greater degree of regulation of proteins through a variety of posttranslational modifications (Babu et al., 2011; Liu et al., 2002); (iv) origin of nucleocytoplasmic compartmentalization accompanied by diversification of several families of ancient domains into versions with specific cytoplasmic or nuclear roles (Mans et al., 2004). In the later phases of eukaryotic evolution, CPs, particularly enzymes, were marked by the phenomenon of "domain accretion" (Koonin et al., 2000). Typically, this phenomenon involved "complexification" of domain architectures via addition of multiple interaction domains around a core catalytic domain. This trend is strong in multicellular eukaryotes and raises the possibility that it might have had a role in the emergence of the multiple spatially and temporally differentiated cellular states typical of these organisms (Aravind et al., 2009; Gerhart and Kirschner, 1997; Meyerowitz, 2002).

In conclusion, the study of the CPs offers an opportunity to tackle a classical evolutionary problem (Kirschner and Gerhart, 2005; Maynard Smith and Szathmáry, 1995). The eukaryotic chromatin system is by all accounts a novelty that was largely absent in that form in the two prokaryotic superkingdoms. Hence, it presents a challenge to the evolutionary theory to explain this "major transition" in life in terms of the previously known evolutionary mechanisms (Maynard Smith and Szathmáry, 1995). Yet, careful sequence analysis reveals that most of the ingredients for the innovations marking the chromatin system were not derived in eukaryotes but were largely inherited from prokaryotes (mostly bacteria). Indeed, much of the original diversity of the components emerged from the intense selective pressures of the arms race that characterize the interactions of bacterial selfish elements, R-M systems, and antibiotic/siderophore biosynthetic systems. The eukaryotes merely deployed these components in entirely unprecedented ways, which was made possible by their unique cellular architecture emerging from the endosymbiotic process.

ACKNOWLEDGMENTS

The authors are supported by the intramural program of the National Center for Biotechnology Information. As the field under consideration is vast and extremely active, there are an enormous number of primary papers. We apologize to all colleagues whose important contributions could not be cited to keep the article within reasonable length. Supplementary information comprising a comprehensive collection of Genbank identifiers for all CPs and TFs included in this study is available at: ftp://ftp.ncbi.nih.gov/pub/aravind/chromatin/.

REFERENCES

Aasland, R., Gibson, T.J., and Stewart, A.F. (1995) The PHD finger: implications for chromatin-mediated transcriptional regulation. Trends Biochem Sci, 20, 56–59.

Alleman, M., Sidorenko, L., McGinnis, K., Seshadri, V., Dorweiler, J.E., White, J., Sikkink, K., and Chandler, V.L. (2006) An RNA-dependent RNA polymerase is required for paramutation in maize. Nature, 442, 295–298.

Allen, M.D., Grummitt, C.G., Hilcenko, C., Min, S.Y., Tonkin, L.M., Johnson, C.M., Freund, S.M., Bycroft, M., and Warren, A.J. (2006) Solution structure of the nonmethyl-CpG-binding CXXC domain of the leukaemia-associated MLL histone methyltransferase. EMBO J, 25, 4503–4512.

Allis, C.D., Jenuwein, T., and Reinberg, D. (2007) Epigenetics, Cold Spring Harbor Laboratory Press, Cold Spring Harbor, NY.

Allis, C.D., Jenuwein, T., Reinberg, D., and Caparros, M. (2006) Epigenetics, Cold Spring Harbor Laboratory Press, Cold Spring Harbor, NY.

Anantharaman, V. and Aravind, L. (2008) Analysis of DBC1 and its homologs suggests a potential mechanism for regulation of sirtuin domain deacetylases by NAD metabolites. Cell Cycle, 7, 1467–1472.

Anantharaman, V., Iyer, L.M., and Aravind, L. (2007) Comparative genomics of protists: new insights on evolution of eukaryotic signal transduction and gene regulation. Annu Rev Microbiol, 61, 453–475.

Anantharaman, V., Koonin, E.V., and Aravind, L. (2002) Comparative genomics and evolution of proteins involved in RNA metabolism. Nucleic Acids Res, 30, 1427–1464.

Aravind, L. (2001) The WWE domain: a common interaction module in protein ubiquitination and ADP ribosylation. Trends Biochem Sci, 26, 273–275.

Aravind, L., Abhiman, S., and Iyer, L.M. (2011) Natural history of the eukaryotic chromatin protein methylation system. Prog Mol Biol Transl Sci, 101, 105–176.

Aravind, L., Anantharaman, V., Balaji, S., Babu, M.M., and Iyer, L.M. (2005) The many faces of the helix-turn-helix domain: transcription regulation and beyond. FEMS Microbiol Rev, 29, 231–262.

Aravind, L., Anantharaman, V., and Venancio, T.M. (2009) Apprehending multicellularity: regulatory networks, genomics, and evolution. Birth Defects Res C Embryo Today, 87, 143–164.

Aravind, L. and Iyer, L.M. (2002) The SWIRM domain: a conserved module found in chromosomal proteins points to novel chromatin-modifying activities. Genome Biol, 3RESEARCH0039.

Aravind, L. and Iyer, L.M. (2012) The HARE-HTH and associated domains: novel modules in the coordination of epigenetic DNA and protein modifications. Cell Cycle, 11, 119–131.

Aravind, L., Iyer, L.M., and Anantharaman, V. (2003) The two faces of Alba: the evolutionary connection between proteins participating in chromatin structure and RNA metabolism. Genome Biol, 4, R64.

Aravind, L., Iyer, L.M., and Koonin, E.V. (2006) Comparative genomics and structural biology of the molecular innovations of eukaryotes. Curr Opin Struct Biol, 16, 409–419.

Aravind, L., Iyer, L.M., and Wu, C. (2007) Domain architectures of the Scm3p protein provide insights into centromere function and evolution. Cell Cycle, 6, 2511–2515.

Aravind, L. and Koonin, E.V. (1998) Second family of histone deacetylases. Science, 280, 1167a.

Aravind, L. and Koonin, E.V. (1999a) Fold prediction and evolutionary analysis of the POZ domain: structural and evolutionary relationship with the potassium channel tetramerization domain. J Mol Biol, 285, 1353–1361.

Aravind, L. and Koonin, E.V. (1999b) Gleaning non-trivial structural, functional and evolutionary information about proteins by iterative database searches. J Mol Biol, 287, 1023–1040.

Aravind, L. and Koonin, E.V. (2000a) The alpha/beta fold uracil DNA glycosylases: a common origin with diverse fates. Genome Biol, 1RESEARCH0007.

Aravind, L. and Koonin, E.V. (2000b) SAP—a putative DNA-binding motif involved in chromosomal organization. Trends Biochem Sci, 25, 112–114.

Aravind, L. and Koonin, E.V. (2001) The DNA-repair protein AlkB, EGL-9, and leprecan define new families of 2-oxoglutarate- and iron-dependent dioxygenases. Genome Biol, 2RESEARCH0007.

Aravind, L. and Landsman, D. (1998) AT-hook motifs identified in a wide variety of DNA-binding proteins. Nucleic Acids Res, 26, 4413–4421.

Aravind, L., Watanabe, H., Lipman, D.J., and Koonin, E.V. (2000) Lineage-specific loss and divergence of functionally linked genes in eukaryotes. Proc Natl Acad Sci USA, 97, 11319–11324.

Argentaro, A., Yang, J.C., Chapman, L., Kowalczyk, M.S., Gibbons, R.J., Higgs, D.R., Neuhaus, D., and Rhodes, D. (2007) Structural consequences of disease-causing mutations in the ATRX-DNMT3-DNMT3L (ADD) domain of the chromatin-associated protein ATRX. Proc Natl Acad Sci USA, 104, 11939–11944.

Arisue, N., Hasegawa, M., and Hashimoto, T. (2005) Root of the Eukaryota tree as inferred from combined maximum likelihood analyses of multiple molecular sequence data. Mol Biol Evol, 22, 409–420.

Avalos, J.L., Boeke, J.D., and Wolberger, C. (2004) Structural basis for the mechanism and regulation of Sir2 enzymes. Mol Cell, 13, 639–648.

Babu, M.M., Iyer, L.M., Balaji, S., and Aravind, L. (2006) The natural history of the WRKY-GCM1 zinc fingers and the relationship between transcription factors and transposons. Nucleic Acids Res, 34, 6505–6520.

Babu, M.M., Luscombe, N.M., Aravind, L., Gerstein, M., and Teichmann, S.A. (2004) Structure and evolution of transcriptional regulatory networks. Curr Opin Struct Biol, 14, 283–291.

Babu, M.M., van der Lee, R., de Groot, N.S., and Gsponer, J. (2011) Intrinsically disordered proteins: regulation and disease, Current Opinion in Structural Biology.

Balaji, S., Babu, M.M., Iyer, L.M., and Aravind, L. (2005) Discovery of the principal specific transcription factors of Apicomplexa and their implication for the evolution of the AP2-integrase DNA binding domains. Nucleic Acids Res, 33, 3994–4006.

Balaji, S., Iyer, L.M., and Aravind, L. (2009) HPC2 and ubinuclein define a novel family of histone chaperones conserved throughout eukaryotes. Mol Biosyst, 5, 269–275.

Bapteste, E., Brinkmann, H., Lee, J.A., Moore, D.V., Sensen, C.W., Gordon, P., Durufle, L., Gaasterland, T., Lopez, P., Muller, M. et al. (2002) The analysis of 100 genes supports the grouping of three highly divergent amoebae: Dictyostelium, Entamoeba, and Mastigamoeba. Proc Natl Acad Sci USA, 99, 1414–1419.

Bartee, L., Malagnac, F., and Bender, J. (2001) Arabidopsis cmt3 chromomethylase mutations block non-CG methylation and silencing of an endogenous gene. Genes Dev, 15, 1753–1758.

Baurain, D., Brinkmann, H., Petersen, J., Rodriguez-Ezpeleta, N., Stechmann, A., Demoulin, V., Roger, A.J., Burger, G., Lang, B.F., and Philippe, H. (2010) Phylogenomic evidence for separate acquisition of plastids in cryptophytes, haptophytes, and stramenopiles. Mol Biol Evol, 27, 1698–1709.

Bellacosa, A., Cicchillitti, L., Schepis, F., Riccio, A., Yeung, A.T., Matsumoto, Y., Golemis, E.A., Genuardi, M., and Neri, G. (1999) MED1, a novel human methyl-CpG-binding endonuclease, interacts with DNA mismatch repair protein MLH1. Proc Natl Acad Sci USA, 96, 3969–3974.

Bernstein, E., Duncan, E.M., Masui, O., Gil, J., Heard, E., and Allis, C.D. (2006) Mouse polycomb proteins bind differentially to methylated histone H3 and RNA and are enriched in facultative heterochromatin. Mol Cell Biol, 26, 2560–2569.

Best, A.A., Morrison, H.G., McArthur, A.G., Sogin, M.L., and Olsen, G.J. (2004) Evolution of eukaryotic transcription: insights from the genome of Giardia lamblia. Genome Res, 14, 1537–1547.

Bhattacharya, D., Yoon, H.S., and Hackett, J.D. (2004) Photosynthetic eukaryotes unite: endosymbiosis connects the dots. Bioessays, 26, 50–60.

Birtle, Z. and Ponting, C.P. (2006) Meisetz and the birth of the KRAB motif. Bioinformatics, 22, 2841–2845.

Blackledge, N.P., Zhou, J.C., Tolstorukov, M.Y., Farcas, A.M., Park, P.J., and Klose, R.J. (2010) CpG islands recruit a histone H3 lysine 36 demethylase. Mol Cell, 38, 179–190.

Blander, G. and Guarente, L. (2004) The Sir2 family of protein deacetylases. Annu Rev Biochem, 73, 417–435.

Boissier, F., Bardou, F., Guillet, V., Uttenweiler-Joseph, S., Daffe, M., Quemard, A., and Mourey, L. (2006) Further insight into S-adenosylmethionine-dependent methyltransferases: structural characterization of Hma, an enzyme essential for the biosynthesis of oxygenated mycolic acids in Mycobacterium tuberculosis. J Biol Chem, 281, 4434–4445.

Borst, P. and Sabatini, R. (2008) Base J: discovery, biosynthesis, and possible functions. Annu Rev Microbiol, 62, 235–251.

Bostick, M., Kim, J.K., Esteve, P.O., Clark, A., Pradhan, S., and Jacobsen, S.E. (2007) UHRF1 plays a role in maintaining DNA methylation in mammalian cells. Science, 317, 1760–1764.

Bowler, C., Allen, A.E., Badger, J.H., Grimwood, J., Jabbari, K., Kuo, A., Maheswari, U., Martens, C., Maumus, F., Otillar, R.P. et al. (2008) The Phaeodactylum genome reveals the evolutionary history of diatom genomes. Nature, 456, 239–244.

Boyer, L.A., Langer, M.R., Crowley, K.A., Tan, S., Denu, J.M., and Peterson, C.L. (2002) Essential role for the SANT domain in the functioning of multiple chromatin remodeling enzymes. Mol Cell, 10, 935–942.

Braun, L., Cannella, D., Ortet, P., Barakat, M., Sautel, C.F., Kieffer, S., Garin, J., Bastien, O., Voinnet, O., and Hakimi, M.A. (2010) A complex small RNA repertoire is generated by a plant/fungal-like machinery and effected by a metazoan-like Argonaute in the single-cell human parasite Toxoplasma gondii. PLoS Pathog, 6, e1000920.

Brehm, A., Tufteland, K.R., Aasland, R., and Becker, P.B. (2004) The many colours of chromodomains. Bioessays, 26, 133–140.

Brent, M.M. and Marmorstein, R. (2008) Ankyrin for methylated lysines. Nat Struct Mol Biol, 15, 221–222.

Burglin, T.R. (1997) Analysis of TALE superclass homeobox genes (MEIS, PBC, KNOX, Iroquois, TGIF) reveals a novel domain conserved between plants and animals. Nucleic Acids Res, 25, 4173–4180.

Burglin, T.R. (2003) The homeobox genes of Encephalitozoon cuniculi (Microsporidia) reveal a putative mating-type locus. Dev Genes Evol, 213, 50–52.

Burroughs, A.M., Allen, K.N., Dunaway-Mariano, D., and Aravind, L. (2006) Evolutionary genomics of the HAD superfamily: understanding the structural adaptations and catalytic diversity in a superfamily of phosphoesterases and allied enzymes. J Mol Biol, 361, 1003–1034.

Burroughs, A.M., Balaji, S., Iyer, L.M., and Aravind, L. (2007) Small but versatile: the extraordinary functional and structural diversity of the beta-grasp fold. Biol Direct, 2, 18.

Cao, X., Aufsatz, W., Zilberman, D., Mette, M.F., Huang, M.S., Matzke, M., and Jacobsen, S.E. (2003) Role of the DRM and CMT3 methyltransferases in RNA-directed DNA methylation. Curr Biol, 13, 2212–2217.

Cao, X. and Jacobsen, S.E. (2002) Locus-specific control of asymmetric and CpNpG methylation by the DRM and CMT3 methyltransferase genes. Proc Natl Acad Sci USA, 99(Suppl 4), 16491–16498.

Capotosti, F., Guernier, S., Lammers, F., Waridel, P., Cai, Y., Jin, J., Conaway, J.W., Conaway, R.C., and Herr, W. (2011) O-GlcNAc transferase catalyzes site-specific proteolysis of HCF-1. Cell, 144, 376–388.

Capotosti, F., Hsieh, J.J., and Herr, W. (2007) Species selectivity of mixed-lineage leukemia/trithorax and HCF proteolytic maturation pathways. Mol Cell Biol, 27, 7063–7072.

Carlton, J.M., Hirt, R.P., Silva, J.C., Delcher, A.L., Schatz, M., Zhao, Q., Wortman, J.R., Bidwell, S.L., Alsmark, U.C., Besteiro, S. et al. (2007) Draft genome sequence of the sexually transmitted pathogen Trichomonas vaginalis. Science, 315, 207–212.

Cavalier-Smith, T. (2009) Megaphylogeny, cell body plans, adaptive zones: causes and timing of eukaryote basal radiations. J Eukaryot Microbiol, 56, 26–33.

Chakravarty, S., Zeng, L., and Zhou, M.M. (2009) Structure and site-specific recognition of histone H3 by the PHD finger of human autoimmune regulator. Structure, 17, 670–679.

Chan, S.W., Henderson, I.R., Zhang, X., Shah, G., Chien, J.S., and Jacobsen, S.E. (2006) RNAi, DRD1, and histone methylation actively target developmentally important non-CG DNA methylation in arabidopsis. PLoS Genet, 2, e83.

Chang, B., Chen, Y., Zhao, Y., and Bruick, R.K. (2007) JMJD6 is a histone arginine demethylase. Science, 318, 444–447.

Chen, S.L., Loffler, K.A., Chen, D., Stallcup, M.R., and Muscat, G.E. (2002) The coactivator-associated arginine methyltransferase is necessary for muscle differentiation: CARM1 coactivates myocyte enhancer factor-2. J Biol Chem, 277, 4324–4333.

Cierpicki, T., Risner, L.E., Grembecka, J., Lukasik, S.M., Popovic, R., Omonkowska, M., Shultis, D.D., Zeleznik-Le, N.J., and Bushweller, J.H. (2010) Structure of the MLL CXXC domain-DNA complex and its functional role in MLL-AF9 leukemia. Nat Struct Mol Biol, 17, 62–68.

Citterio, E., Papait, R., Nicassio, F., Vecchi, M., Gomiero, P., Mantovani, R., Di Fiore, P.P., and Bonapace, I.M. (2004) Np95 is a histone-binding protein endowed with ubiquitin ligase activity. Mol Cell Biol, 24, 2526–2535.

Collins, R. and Cheng, X. (2010) A case study in cross-talk: the histone lysine methyltransferases G9a and GLP. Nucleic Acids Res, 38, 3503–3511.

Conaway, R.C. and Conaway, J.W. (2004) Proteins in Eukaryotic Transcription, Vol. 67, Academic Press, San Diego.

Conaway, R.C. and Conaway, J.W. (2011) Function and regulation of the mediator complex. Curr Opin Genet Dev, 21, 225–230.

Couture, J.F., Collazo, E., Hauk, G., and Trievel, R.C. (2006) Structural basis for the methylation site specificity of SET7/9. Nat Struct Mol Biol, 13, 140–146.

Cross, S.H., Meehan, R.R., Nan, X., and Bird, A. (1997) A component of the transcriptional repressor MeCP1 shares a motif with DNA methyltransferase and HRX proteins. Nat Genet, 16, 256–259.

Cuthbert, G.L., Daujat, S., Snowden, A.W., Erdjument-Bromage, H., Hagiwara, T., Yamada, M., Schneider, R., Gregory, P.D., Tempst, P., Bannister, A.J. et al. (2004) Histone deimination antagonizes arginine methylation. Cell, 118, 545–553.

Dacks, J.B. and Doolittle, W.F. (2001) Reconstructing/deconstructing the earliest eukaryotes: how comparative genomics can help. Cell, 107, 419–425.

Dalal, Y., Furuyama, T., Vermaak, D., and Henikoff, S. (2007) Structure, dynamics, and evolution of centromeric nucleosomes. Proc Natl Acad Sci USA, 104, 15974–15981.

Dann, C.E. 3rd, Bruick, R.K., and Deisenhofer, J. (2002) Structure of factor-inhibiting hypoxia-inducible factor 1: an asparaginyl hydroxylase involved in the hypoxic response pathway. Proc Natl Acad Sci USA, 99, 15351–15356.

de la Cruz, X., Lois, S., Sanchez-Molina, S., and Martinez-Balbas, M.A. (2005) Do protein motifs read the histone code? Bioessays, 27, 164–175.

de Souza, R.F. and Aravind, L. (2012) Identification of novel components of NAD-utilizing metabolic pathways and prediction of their biochemical functions, Mol Biosyst.

Demirci, H., Gregory, S.T., Dahlberg, A.E., and Jogl, G. (2008) Multiple-site trimethylation of ribosomal protein L11 by the PrmA methyltransferase. Structure, 16, 1059–1066.

Denhardt, D.T., Chaly, N., and Walden, D.B. (2005) The eukaryotic nucleus: a thematic issue. Bioessays, 9, 43.

Dhalluin, C., Carlson, J.E., Zeng, L., He, C., Aggarwal, A.K., and Zhou, M.M. (1999) Structure and ligand of a histone acetyltransferase bromodomain. Nature, 399, 491–496.

Dlakic, M. (2001) Chromatin silencing protein and pachytene checkpoint regulator Dot1p has a methyltransferase fold. Trends Biochem Sci, 26, 405–407.

Doerfler, W. (2008) In pursuit of the first recognized epigenetic signal–DNA methylation: a 1976 to 2008 synopsis. Epigenetics, 3, 125–133.

Doherty, A.J., Serpell, L.C., and Ponting, C.P. (1996) The helix-hairpin-helix DNA-binding motif: a structural basis for non-sequence-specific recognition of DNA. Nucleic Acids Res, 24, 2488–2497.

Dorman, C.J. (2009) Nucleoid-associated proteins and bacterial physiology. Adv Appl Microbiol, 67, 47–64.

Doyon, Y., Selleck, W., Lane, W.S., Tan, S., and Cote, J. (2004) Structural and functional conservation of the NuA4 histone acetyltransferase complex from yeast to humans. Mol Cell Biol, 24, 1884–1896.

Driscoll, R., Hudson, A., and Jackson, S.P. (2007) Yeast Rtt109 promotes genome stability by acetylating histone H3 on lysine 56. Science, 315, 649–652.

Durant, M. and Pugh, B.F. (2006) Genome-wide relationships between TAF1 and histone acetyltransferases in *Saccharomyces cerevisiae*. Mol Cell Biol, 26, 2791–2802.

Durr, H. and Hopfner, K.P. (2006) Structure-function analysis of SWI2/SNF2 enzymes. Methods Enzymol, 409, 375–388.

Dutnall, R.N., Tafrov, S.T., Sternglanz, R., and Ramakrishnan, V. (1998) Structure of the yeast histone acetyltransferase Hat1: insights into substrate specificity and implications for the Gcn5-related N-acetyltransferase superfamily. Cold Spring Harb Symp Quant Biol, 63, 501–507.

El Messaoudi-Aubert, S., Nicholls, J., Maertens, G.N., Brookes, S., Bernstein, E., and Peters, G. (2010) Role for the MOV10 RNA helicase in polycomb-mediated repression of the INK4a tumor suppressor. Nat Struct Mol Biol, 17, 862–868.

Elia, A.E., Rellos, P., Haire, L.F., Chao, J.W., Ivins, F.J., Hoepker, K., Mohammad, D., Cantley, L.C., Smerdon, S.J., and Yaffe, M.B. (2003) The molecular basis for phosphodependent substrate targeting and regulation of Plks by the Polo-box domain. Cell, 115, 83–95.

Elkins, J.M., Hewitson, K.S., McNeill, L.A., Seibel, J.F., Schlemminger, I., Pugh, C.W., Ratcliffe, P.J., and Schofield, C.J. (2003) Structure of factor-inhibiting hypoxia-inducible factor (HIF) reveals mechanism of oxidative modification of HIF-1 alpha. J Biol Chem, 278, 1802–1806.

Esteve, P.O., Chang, Y., Samaranayake, M., Upadhyay, A.K., Horton, J.R., Feehery, G.R., Cheng, X., and Pradhan, S. (2011) A methylation and phosphorylation switch between an adjacent lysine and serine determines human DNMT1 stability. Nat Struct Mol Biol, 18, 42–48.

Fauman, E.B., Blumenthal, R.M., and Cheng, X. (1999) Structure and evolution of AdoMet-dependent methyltransferases, in S-Adenosylmethionine-Dependent Methyltransferases: Structures and Functions (eds X. Cheng and R.M. Blumenthal), River Edge World Scientific, pp. 1–54.

Feng, Q., Wang, H., Ng, H.H., Erdjument-Bromage, H., Tempst, P., Struhl, K., and Zhang, Y. (2002) Methylation of H3-lysine 79 is mediated by a new family of HMTases without a SET domain. Curr Biol, 12, 1052–1058.

Filion, G.J., van Bemmel, J.G., Braunschweig, U., Talhout, W., Kind, J., Ward, L.D., Brugman, W., de Castro, I.J., Kerkhoven, R.M., Bussemaker, H.J. et al. (2010) Systematic protein location mapping reveals five principal chromatin types in Drosophila cells. Cell, 143, 212–224.

Finn, R.D., Mistry, J., Schuster-Bockler, B., Griffiths-Jones, S., Hollich, V., Lassmann, T., Moxon, S., Marshall, M., Khanna, A., Durbin, R. et al. (2006) Pfam: clans, web tools and services. Nucleic Acids Res, 34, D247–251.

Flanagan, J.F., Mi, L.Z., Chruszcz, M., Cymborowski, M., Clines, K.L., Kim, Y., Minor, W., Rastinejad, F., and Khorasanizadeh, S. (2005) Double chromodomains cooperate to recognize the methylated histone H3 tail. Nature, 438, 1181–1185.

Forneris, F., Binda, C., Battaglioli, E., and Mattevi, A. (2008) LSD1: oxidative chemistry for multifaceted functions in chromatin regulation. Trends Biochem Sci, 33, 181–189.

Frankel, A. and Clarke, S. (2000) PRMT3 is a distinct member of the protein arginine N-methyltransferase family. Conferral of substrate specificity by a zinc-finger domain. J Biol Chem, 275, 32974–32982.

Freitag, M., Williams, R.L., Kothe, G.O., and Selker, E.U. (2002) A cytosine methyltransferase homologue is essential for repeat-induced point mutation in Neurospora crassa. Proc Natl Acad Sci USA, 99, 8802–8807.

Frye, R.A. (1999) Characterization of five human cDNAs with homology to the yeast SIR2 gene: Sir2-like proteins (sirtuins) metabolize NAD and may have protein ADP-ribosyltransferase activity. Biochem Biophys Res Commun, 260, 273–279.

Fujiki, R., Chikanishi, T., Hashiba, W., Ito, H., Takada, I., Roeder, R.G., Kitagawa, H., and Kato, S. (2009) GlcNAcylation of a histone methyltransferase in retinoic-acid-induced granulopoiesis. Nature, 459, 455–459.

Gangavarapu, V., Haracska, L., Unk, I., Johnson, R.E., Prakash, S., and Prakash, L. (2006) Mms2-Ubc13-dependent and -independent roles of Rad5 ubiquitin ligase in postreplication repair and translesion DNA synthesis in Saccharomyces cerevisiae. Mol Cell Biol, 26, 7783–7790.

Gangloff, Y.G., Romier, C., Thuault, S., Werten, S., and Davidson, I. (2001) The histone fold is a key structural motif of transcription factor TFIID. Trends Biochem Sci, 26, 250–257.

Garcia-Alai, M.M., Allen, M.D., Joerger, A.C., and Bycroft, M. (2010) The structure of the FYR domain of transforming growth factor beta regulator 1. Protein Sci, 19, 1432–1438.

Gardner, M.J., Hall, N., Fung, E., White, O., Berriman, M., Hyman, R.W., Carlton, J.M., Pain, A., Nelson, K.E., Bowman, S. et al. (2002) Genome sequence of the human malaria parasite Plasmodium falciparum. Nature, 419, 498–511.

Gearhart, M.D., Corcoran, C.M., Wamstad, J.A., and Bardwell, V.J. (2006) Polycomb group and SCF ubiquitin ligases are found in a novel BCOR complex that is recruited to BCL6 targets. Mol Cell Biol, 26, 6880–6889.

Gehring, M., Huh, J.H., Hsieh, T.F., Penterman, J., Choi, Y., Harada, J.J., Goldberg, R.B., and Fischer, R.L. (2006) DEMETER DNA glycosylase establishes MEDEA polycomb gene self-imprinting by allele-specific demethylation. Cell, 124, 495–506.

Gehring, W.J., Affolter, M., and Burglin, T. (1994) Homeodomain proteins. Annu Rev Biochem, 63, 487–526.

Gerhart, J. and Kirschner, M. (1997) Cells, Embryos, and Evolution: Toward a Cellular and Developmental Understanding of Phenotypic Variation and Evolutionary Adaptability, Malden, Mass, Blackwell Science.

Gibson, T.J. and Spring, J. (1998) Genetic redundancy in vertebrates: polyploidy and persistence of genes encoding multidomain proteins. Trends Genet, 14, 46–49discussion 49–50.

Goll, M.G. and Bestor, T.H. (2005) Eukaryotic cytosine methyltransferases. Annu Rev Biochem, 74, 481–514.

Goll, M.G., Kirpekar, F., Maggert, K.A., Yoder, J.A., Hsieh, C.L., Zhang, X., Golic, K.G., Jacobsen, S.E., and Bestor, T.H. (2006) Methylation of tRNAAsp by the DNA methyltransferase homolog Dnmt2. Science, 311, 395–398.

Gommers-Ampt, J.H. and Borst, P. (1995) Hypermodified bases in DNA. FASEB J, 9, 1034–1042.

Grazini, U., Zanardi, F., Citterio, E., Casola, S., Goding, C.R., and McBlane, F. (2010) The RING domain of RAG1 ubiquitylates histone H3: a novel activity in chromatin-mediated regulation of V(D)J joining. Mol Cell, 37, 282–293.

Green, E.M., Mas, G., Young, N.L., Garcia, B.A., and Gozani, O. (2012) Methylation of H4 lysines 5, 8 and 12 by yeast Set5 calibrates chromatin stress responses. Nat Struct Mol Biol, 19, 361–363.

Grewal, S.I. (2010) RNAi-dependent formation of heterochromatin and its diverse functions. Curr Opin Genet Dev, 20, 134–141.

Grewal, S.I. and Rice, J.C. (2004) Regulation of heterochromatin by histone methylation and small RNAs. Curr Opin Cell Biol, 16, 230–238.

Griese, J.J., Witte, G., and Hopfner, K.P. (2010) Structure and DNA binding activity of the mouse condensin hinge domain highlight common and diverse features of SMC proteins. Nucleic Acids Res, 38, 3454–3465.

Guo, J.U., Su, Y., Zhong, C., Ming, G.L., and Song, H. (2011) Hydroxylation of 5-methylcytosine by TET1 promotes active DNA demethylation in the adult brain. Cell, 145, 423–434.

Haering, C.H., Lowe, J., Hochwagen, A., and Nasmyth, K. (2002) Molecular architecture of SMC proteins and the yeast cohesin complex. Mol Cell, 9, 773–788.

Hajkova, P., Jeffries, S.J., Lee, C., Miller, N., Jackson, S.P., and Surani, M.A. (2010) Genome-wide reprogramming in the mouse germ line entails the base excision repair pathway. Science, 329, 78–82.

Harvey, B.M., Mironenko, T., Sun, Y., Hong, H., Deng, Z., Leadlay, P.F., Weissman, K.J., and Haydock, S.F. (2007) Insights into polyether biosynthesis from analysis of the nigericin biosynthetic gene cluster in Streptomyces sp. DSM4137. Chem Biol, 14, 703–714.

Hashimoto, H., Horton, J.R., Zhang, X., Bostick, M., Jacobsen, S.E., and Cheng, X. (2008) The SRA domain of UHRF1 flips 5-methylcytosine out of the DNA helix. Nature, 455, 826–829.

He, F., Umehara, T., Saito, K., Harada, T., Watanabe, S., Yabuki, T., Kigawa, T., Takahashi, M., Kuwasako, K., Tsuda, K. et al. (2010) Structural insight into the zinc finger CW domain as a histone modification reader. Structure, 18, 1127–1139.

He, X., Fan, H.Y., Garlick, J.D., and Kingston, R.E. (2008) Diverse regulation of SNF2h chromatin remodeling by noncatalytic subunits. Biochemistry, 47, 7025–7033.

He, Y.F., Li, B.Z., Li, Z., Liu, P., Wang, Y., Tang, Q., Ding, J., Jia, Y., Chen, Z., Li, L. et al. (2011) Tet-mediated formation of 5-carboxylcytosine and its excision by TDG in mammalian DNA. Science, 333, 1303–1307.

Heidinger-Pauli, J.M., Unal, E., and Koshland, D. (2009) Distinct targets of the Eco1 acetyltransferase modulate cohesion in S phase and in response to DNA damage. Mol Cell, 34, 311–321.

Henderson, I.R., Deleris, A., Wong, W., Zhong, X., Chin, H.G., Horwitz, G.A., Kelly, K.A., Pradhan, S., and Jacobsen, S.E. (2010) The de novo cytosine methyltransferase DRM2 requires intact UBA domains and a catalytically mutated paralog DRM3 during RNA-directed DNA methylation in arabidopsis thaliana. PLoS Genet, 6, e1001182.

Hendrich, B., Hardeland, U., Ng, H.H., Jiricny, J., and Bird, A. (1999) The thymine glycosylase MBD4 can bind to the product of deamination at methylated CpG sites. Nature, 401, 301–304.

Henikoff, S. and Comai, L. (1998) A DNA methyltransferase homolog with a chromodomain exists in multiple polymorphic forms in Arabidopsis. Genetics, 149, 307–318.

Higgins, J.M. (2010) Haspin: a newly discovered regulator of mitotic chromosome behavior. Chromosoma, 119, 137–147.

Hirano, T. (2005) SMC proteins and chromosome mechanics: from bacteria to humans. Philos Trans R Soc Lond B Biol Sci, 360, 507–514.

Hirano, T. (2006) At the heart of the chromosome: SMC proteins in action. Nat Rev Mol Cell Biol, 7, 311–322.

Hoey, T. and Schindler, U. (1998) STAT structure and function in signaling. Curr Opin Genet Dev, 8, 582–587.

Hsiao, P.W., Fryer, C.J., Trotter, K.W., Wang, W., and Archer, T.K. (2003) BAF60a mediates critical interactions between nuclear receptors and the BRG1 chromatin-remodeling complex for transactivation. Mol Cell Biol, 23, 6210–6220.

Huang, B., Johansson, M.J., and Bystrom, A.S. (2005) An early step in wobble uridine tRNA modification requires the Elongator complex. RNA, 11, 424–436.

Iqbal, K., Jin, S.G., Pfeifer, G.P., and Szabo, P.E. (2011) Reprogramming of the paternal genome upon fertilization involves genome-wide oxidation of 5-methylcytosine. Proc Natl Acad Sci USA, 108, 3642–3647.

Ito, T., Yamauchi, M., Nishina, M., Yamamichi, N., Mizutani, T., Ui, M., Murakami, M., and Iba, H. (2001) Identification of SWI.SNF complex subunit BAF60a as a determinant of the transactivation potential of Fos/Jun dimers. J Biol Chem, 276, 2852–2857.

Iwase, S., Lan, F., Bayliss, P., de la Torre-Ubieta, L., Huarte, M., Qi, H.H., Whetstine, J.R., Bonni, A., Roberts, T.M., and Shi, Y. (2007) The X-linked mental retardation gene SMCX/JARID1C defines a family of histone H3 lysine 4 demethylases. Cell, 128, 1077–1088.

Iyer, L.M., Abhiman, S., and Aravind, L. (2008a) MutL homologs in restriction-modification systems and the origin of eukaryotic MORC ATPases. Biol Direct, 3, 8.

Iyer, L.M., Abhiman, S., and Aravind, L. (2011a) Natural history of eukaryotic DNA methylation systems. Prog Mol Biol Transl Sci, 101, 25–104.

Iyer, L.M., Abhiman, S., de Souza, R.F., and Aravind, L. (2010) Origin and evolution of peptide-modifying dioxygenases and identification of the wybutosine hydroxylase/hydroperoxidase. Nucleic Acids Res, 38, 5261–5279.

Iyer, L.M., Abhiman, S., Maxwell Burroughs, A., and Aravind, L. (2009a) Amidoligases with ATP-grasp, glutamine synthetase-like and acetyltransferase-like domains: synthesis of novel metabolites and peptide modifications of proteins. Mol Biosyst, 5, 1636–1660.

Iyer, L.M., Anantharaman, V., Wolf, M.Y., and Aravind, L. (2008b) Comparative genomics of transcription factors and chromatin proteins in parasitic protists and other eukaryotes. Int J Parasitol, 38, 1–31.

Iyer, L.M. and Aravind, L. (2004) The emergence of catalytic and structural diversity within the beta-clip fold. Proteins, 55, 977–991.

Iyer, L.M. and Aravind, L. (2011) Insights from the architecture of the bacterial transcription apparatus. J Struct Biol, 179, 299–319.

Iyer, L.M., Babu, M.M., and Aravind, L. (2006a) The HIRAN domain and recruitment of chromatin remodeling and repair activities to damaged DNA. Cell Cycle, 5, 775–782.

Iyer, L.M., Burroughs, A.M., and Aravind, L. (2006b) The ASCH superfamily: novel domains with a fold related to the PUA domain and a potential role in RNA metabolism. Bioinformatics, 22, 257–263.

Iyer, L.M., Koonin, E.V., and Aravind, L. (2003) Evolutionary connection between the catalytic subunits of DNA-dependent RNA polymerases and eukaryotic RNA-dependent RNA polymerases and the origin of RNA polymerases. BMC Struct Biol, 3, 1.

Iyer, L.M., Tahiliani, M., Rao, A., and Aravind, L. (2009b) Prediction of novel families of enzymes involved in oxidative and other complex modifications of bases in nucleic acids. Cell Cycle, 8, 1698–1710.

Iyer, L.M., Zhang, D., Rogozin, I.B., and Aravind, L. (2011b) Evolution of the deaminase fold and multiple origins of eukaryotic editing and mutagenic nucleic acid deaminases from bacterial toxin systems. Nucleic Acids Res, 39, 9473–9497.

Jacobs, S.A., Harp, J.M., Devarakonda, S., Kim, Y., Rastinejad, F., and Khorasanizadeh, S. (2002) The active site of the SET domain is constructed on a knot. Nat Struct Biol, 9, 833–838.

James, T.Y., Kauff, F., Schoch, C.L., Matheny, P.B., Hofstetter, V., Cox, C.J., Celio, G., Gueidan, C., Fraker, E., Miadlikowska, J. et al. (2006) Reconstructing the early evolution of Fungi using a six-gene phylogeny. Nature, 443, 818–822.

Jeltsch, A. (2006) Molecular enzymology of mammalian DNA methyltransferases. Curr Top Microbiol Immunol, 301, 203–225.

Jenuwein, T. and Allis, C.D. (2001) Translating the histone code. Science, 293, 1074–1080.

Johnson, L.M., Bostick, M., Zhang, X., Kraft, E., Henderson, I., Callis, J., and Jacobsen, S.E. (2007) The SRA methyl-cytosine-binding domain links DNA and histone methylation. Curr Biol, 17, 379–384.

Jorgensen, H.F., Ben-Porath, I., and Bird, A.P. (2004) Mbd1 is recruited to both methylated and nonmethylated CpGs via distinct DNA binding domains. Mol Cell Biol, 24, 3387–3395.

Juranek, S.A., Rupprecht, S., Postberg, J., and Lipps, H.J. (2005) snRNA and heterochromatin formation are involved in DNA excision during macronuclear development in stichotrichous ciliates. Eukaryot Cell, 4, 1934–1941.

Kaminska, K.H. and Bujnicki, J.M. (2008) Bacteriophage Mu Mom protein responsible for DNA modification is a new member of the acyltransferase superfamily. Cell Cycle, 7, 120–121.

Kanduri, C. (2011) Kcnq1ot1: A chromatin regulatory RNA. Semin Cell Dev Biol, 22, 343–350.

Kaneda, M., Okano, M., Hata, K., Sado, T., Tsujimoto, N., Li, E., and Sasaki, H. (2004) Essential role for de novo DNA methyltransferase Dnmt3a in paternal and maternal imprinting. Nature, 429, 900–903.

Karagianni, P., Amazit, L., Qin, J., and Wong, J. (2008) ICBP90, a novel methyl K9 H3 binding protein linking protein ubiquitination with heterochromatin formation. Mol Cell Biol, 28, 705–717.

Karras, G.I., Kustatscher, G., Buhecha, H.R., Allen, M.D., Pugieux, C., Sait, F., Bycroft, M., and Ladurner, A.G. (2005) The macro domain is an ADP-ribose binding module. EMBO J, 24, 1911–1920.

Kato, Y., Kaneda, M., Hata, K., Kumaki, K., Hisano, M., Kohara, Y., Okano, M., Li, E., Nozaki, M., and Sasaki, H. (2007) Role of the Dnmt3 family in de novo methylation of imprinted and repetitive sequences during male germ cell development in the mouse. Hum Mol Genet, 16, 2272–2280.

Kawata, T. (2011) STAT signaling in Dictyostelium development. Dev Growth Differ, 53, 548–557.

Khalil, A.M., Guttman, M., Huarte, M., Garber, M., Raj, A., Rivea Morales, D., Thomas, K., Presser, A., Bernstein, B.E., van Oudenaarden, A. et al. (2009) Many human large intergenic noncoding RNAs associate with chromatin-modifying complexes and affect gene expression. Proc Natl Acad Sci USA, 106, 11667–11672.

Kim, D., Blus, B.J., Chandra, V., Huang, P., Rastinejad, F., and Khorasanizadeh, S. (2010) Corecognition of DNA and a methylated histone tail by the MSL3 chromodomain. Nat Struct Mol Biol, 17, 1027–1029.

Kim, J.E., Chen, J., and Lou, Z. (2008) DBC1 is a negative regulator of SIRT1. Nature, 451, 583–586.

Kim, M.S., Kondo, T., Takada, I., Youn, M.Y., Yamamoto, Y., Takahashi, S., Matsumoto, T., Fujiyama, S., Shirode, Y., Yamaoka, I. et al. (2009) DNA demethylation in hormone-induced transcriptional derepression. Nature, 461, 1007–1012.

Kirschner, M. and Gerhart, J. (2005) The Plausibility of Life: Resolving Darwin's Dilemma, Yale University Press, New Haven.

Kizer, K.O., Phatnani, H.P., Shibata, Y., Hall, H., Greenleaf, A.L., and Strahl, B.D. (2005) A novel domain in Set2 mediates RNA polymerase II interaction and couples histone H3 K36 methylation with transcript elongation. Mol Cell Biol, 25, 3305–3316.

Klose, R.J., Kallin, E.M., and Zhang, Y. (2006) JmjC-domain-containing proteins and histone demethylation. Nat Rev Genet, 7, 715–727.

Ko, M., Huang, Y., Jankowska, A.M., Pape, U.J., Tahiliani, M., Bandukwala, H.S., An, J., Lamperti, E.D., Koh, K.P., Ganetzky, R. et al. (2010) Impaired hydroxylation of 5-methylcytosine in myeloid cancers with mutant TET2. Nature, 468, 839–843.

Koonin, E.V., Aravind, L., and Kondrashov, A.S. (2000) The impact of comparative genomics on our understanding of evolution. Cell, 101, 573–576.

Kouzarides, T. (2007) Chromatin modifications and their function. Cell, 128, 693–705.

Kriaucionis, S. and Heintz, N. (2009) The nuclear DNA base 5-hydroxymethylcytosine is present in Purkinje neurons and the brain. Science, 324, 929–930.

Krishna, S.S. and Aravind, L. (2010) The bridge-region of the Ku superfamily is an atypical zinc ribbon domain. J Struct Biol, 172, 294–299.

Kroger, N., Lorenz, S., Brunner, E., and Sumper, M. (2002) Self-assembly of highly phosphorylated silaffins and their function in biosilica morphogenesis. Science, 298, 584–586.

Krokan, H.E., Standal, R., and Slupphaug, G. (1997) DNA glycosylases in the base excision repair of DNA. Biochem J, 325(Pt 1), 1–16.

Kuhlmann, M., Borisova, B.E., Kaller, M., Larsson, P., Stach, D., Na, J., Eichinger, L., Lyko, F., Ambros, V., Soderbom, F. et al. (2005) Silencing of retrotransposons in Dictyostelium by DNA methylation and RNAi. Nucleic Acids Res, 33, 6405–6417.

Kuzuhara, T. and Horikoshi, M. (2004) A nuclear FK506-binding protein is a histone chaperone regulating rDNA silencing. Nat Struct Mol Biol, 11, 275–283.

Lan, F., Collins, R.E., De Cegli, R., Alpatov, R., Horton, J.R., Shi, X., Gozani, O., Cheng, X., and Shi, Y. (2007) Recognition of unmethylated histone H3 lysine 4 links BHC80 to LSD1-mediated gene repression. Nature, 448, 718–722.

Lander, E.S., Linton, L.M., Birren, B., Nusbaum, C., Zody, M.C., Baldwin, J., Devon, K., Dewar, K., Doyle, M., FitzHugh, W. et al. (2001) Initial sequencing and analysis of the human genome. Nature, 409, 860–921.

Latchman, D. (2005) Gene Regulation, 5th edn, Taylor & Francis, New York.

Lau, A.O., Smith, A.J., Brown, M.T., and Johnson, P.J. (2006) Trichomonas vaginalis initiator binding protein (IBP39) and RNA polymerase II large subunit carboxy terminal domain interaction. Mol Biochem Parasitol, 150, 56–62.

Lee, Y.H. and Stallcup, M.R. (2009) Minireview: protein arginine methylation of nonhistone proteins in transcriptional regulation. Mol Endocrinol, 23, 425–433.

Leipe, D.D. and Landsman, D. (1997) Histone deacetylases, acetoin utilization proteins and acetylpolyamine amidohydrolases are members of an ancient protein superfamily. Nucleic Acids Res, 25, 3693–3697.

Lespinet, O., Wolf, Y.I., Koonin, E.V., and Aravind, L. (2002) The role of lineage-specific gene family expansion in the evolution of eukaryotes. Genome Res, 12, 1048–1059.

Li, C.F., Pontes, O., El-Shami, M., Henderson, I.R., Bernatavichute, Y.V., Chan, S.W., Lagrange, T., Pikaard, C.S., and Jacobsen, S.E. (2006) An ARGONAUTE4-containing nuclear processing center colocalized with Cajal bodies in Arabidopsis thaliana. Cell, 126, 93–106.

Liu, J., Tan, H., and Rost, B. (2002) Loopy proteins appear conserved in evolution. J Mol Biol, 322, 53–64.

Liu, X., Wang, L., Zhao, K., Thompson, P.R., Hwang, Y., Marmorstein, R., and Cole, P.A. (2008) The structural basis of protein acetylation by the p300/CBP transcriptional coactivator. Nature, 451, 846–850.

Lopez-Garcia, P. and Moreira, D. (1999) Metabolic symbiosis at the origin of eukaryotes. Trends Biochem Sci, 24, 88–93.

Luk, E., Vu, N.D., Patteson, K., Mizuguchi, G., Wu, W.H., Ranjan, A., Backus, J., Sen, S., Lewis, M., Bai, Y. et al. (2007) Chz1, a nuclear chaperone for histone H2AZ. Mol Cell, 25, 357–368.

Lukes, J. and Maslov, D.A. (2000) Unexpectedly high variability of the histone H4 gene in Leishmania. Parasitol Res, 86, 259–261.

Makarova, K.S., Aravind, L., Daly, M.J., and Koonin, E.V. (2000) Specific expansion of protein families in the radioresistant bacterium Deinococcus radiodurans. Genetica, 108, 25–34.

Malagnac, F., Wendel, B., Goyon, C., Faugeron, G., Zickler, D., Rossignol, J.L., Noyer-Weidner, M., Vollmayr, P., Trautner, T.A., and Walter, J. (1997) A gene essential for de novo methylation and development in Ascobolus reveals a novel type of eukaryotic DNA methyltransferase structure. Cell, 91, 281–290.

Malone, C.D., Anderson, A.M., Motl, J.A., Rexer, C.H., and Chalker, D.L. (2005) Germ line transcripts are processed by a Dicer-like protein that is essential for developmentally programmed genome rearrangements of Tetrahymena thermophila. Mol Cell Biol, 25, 9151–9164.

Mandava, V., Janzen, C.J., and Cross, G.A. (2008) Trypanosome H2Bv replaces H2B in nucleosomes enriched for H3 K4 and K76 trimethylation. Biochem Biophys Res Commun, 368, 846–851.

Mans, B.J., Anantharaman, V., Aravind, L., and Koonin, E.V. (2004) Comparative genomics, evolution and origins of the nuclear envelope and nuclear pore complex. Cell Cycle, 3, 1612–1637.

Mansfield, R.E., Musselman, C.A., Kwan, A.H., Oliver, S.S., Garske, A.L., Davrazou, F., Denu, J.M., Kutateladze, T.G., and Mackay, J.P. (2011) Plant homeodomain (PHD) fingers of CHD4 are histone H3-binding modules with preference for unmodified H3K4 and methylated H3K9. J Biol Chem, 286, 11779–11791.

Manzur, K.L., Farooq, A., Zeng, L., Plotnikova, O., Koch, A.W., Sachchidanand, and Zhou, M.M. (2003) A dimeric viral SET domain methyltransferase specific to Lys27 of histone H3. Nat Struct Biol, 10, 187–196.

Manzur, K.L. and Zhou, M.M. (2005) An archaeal SET domain protein exhibits distinct lysine methyltransferase activity towards DNA-associated protein MC1-alpha. FEBS Lett, 579, 3859–3865.

Martens, J.A. and Winston, F. (2003) Recent advances in understanding chromatin remodeling by Swi/Snf complexes. Curr Opin Genet Dev, 13, 136–142.

Maurer-Stroh, S., Dickens, N.J., Hughes-Davies, L., Kouzarides, T., Eisenhaber, F., and Ponting, C.P. (2003) The Tudor domain 'Royal Family': Tudor, plant Agenet, Chromo, PWWP and MBT domains. Trends Biochem Sci, 28, 69–74.

Mayer, W., Niveleau, A., Walter, J., Fundele, R., and Haaf, T. (2000) Demethylation of the zygotic paternal genome. Nature, 403, 501–502.

Maynard Smith, J. and Szathmáry, E. (1995) The Major Transitions in Evolution, W.H. Freeman Spektrum, Oxford, NY.

Meyerowitz, E.M. (1999) Plants, animals and the logic of development. Trends Cell Biol, 9, M65–68.

Meyerowitz, E.M. (2002) Plants compared to animals: the broadest comparative study of development. Science, 295, 1482–1485.

Militello, K.T., Wang, P., Jayakar, S.K., Pietrasik, R.L., Dupont, C.D., Dodd, K., King, A.M., and Valenti, P.R. (2008) African trypanosomes contain 5-methylcytosine in nuclear DNA. Eukaryot Cell, 7, 2012–2016.

Miller, T.C., Rutherford, T.J., Johnson, C.M., Fiedler, M., and Bienz, M. (2010) Allosteric remodelling of the histone H3 binding pocket in the Pygo2 PHD finger triggered by its binding to the B9L/BCL9 co-factor. J Mol Biol, 401, 969–984.

Milne, T.A., Briggs, S.D., Brock, H.W., Martin, M.E., Gibbs, D., Allis, C.D., and Hess, J.L. (2002) MLL targets SET domain methyltransferase activity to Hox gene promoters. Mol Cell, 10, 1107–1117.

Miskey, C., Papp, B., Mates, L., Sinzelle, L., Keller, H., Izsvak, Z., and Ivics, Z. (2007) The ancient mariner sails again: transposition of the human Hsmar1 element by a reconstructed transposase and activities of the SETMAR protein on transposon ends. Mol Cell Biol, 27, 4589–4600.

Miyoshi, K., Miyoshi, T., and Siomi, H. (2010) Many ways to generate microRNA-like small RNAs: non-canonical pathways for microRNA production. Mol Genet Genomics, 284, 95–103.

Mizuguchi, G., Shen, X., Landry, J., Wu, W.H., Sen, S., and Wu, C. (2004) ATP-driven exchange of histone H2AZ variant catalyzed by SWR1 chromatin remodeling complex. Science, 303, 343–348.

Mo, X., Kowenz-Leutz, E., Laumonnier, Y., Xu, H., and Leutz, A. (2005) Histone H3 tail positioning and acetylation by the c-Myb but not the v-Myb DNA-binding SANT domain. Genes Dev, 19, 2447–2457.

Mochizuki, K., Fine, N.A., Fujisawa, T., and Gorovsky, M.A. (2002) Analysis of a piwi-related gene implicates small RNAs in genome rearrangement in tetrahymena. Cell, 110, 689–699.

Mohrmann, L. and Verrijzer, C.P. (2005) Composition and functional specificity of SWI2/SNF2 class chromatin remodeling complexes. Biochim Biophys Acta, 1681, 59–73.

Moon-van der Staay, S.Y., De Wachter, R., and Vaulot, D. (2001) Oceanic 18S rDNA sequences from picoplankton reveal unsuspected eukaryotic diversity. Nature, 409, 607–610.

Morales-Ruiz, T., Ortega-Galisteo, A.P., Ponferrada-Marin, M.I., Martinez-Macias, M.I., Ariza, R.R., and Roldan-Arjona, T. (2006) DEMETER and REPRESSOR OF SILENCING 1 encode 5-methylcytosine DNA glycosylases. Proc Natl Acad Sci USA, 103, 6853–6858.

Muljo, S.A., Kanellopoulou, C., and Aravind, L. (2010) MicroRNA targeting in mammalian genomes: genes and mechanisms. Wiley Interdiscip Rev Syst Biol Med, 2, 148–161.

Murata, M., Azuma, Y., Miura, K., Rahman, M.A., Matsutani, M., Aoyama, M., Suzuki, H., Sugi, K., and Shirai, M. (2007) Chlamydial SET domain protein functions as a histone methyltransferase. Microbiology, 153, 585–592.

Nagano, T. and Fraser, P. (2011) No-nonsense functions for long noncoding RNAs. Cell, 145, 178–181.

Nakayama, J., Rice, J.C., Strahl, B.D., Allis, C.D., and Grewal, S.I. (2001) Role of histone H3 lysine 9 methylation in epigenetic control of heterochromatin assembly. Science, 292, 110–113.

Neuwald, A.F. and Landsman, D. (1997) GCN5-related histone N-acetyltransferases belong to a diverse superfamily that includes the yeast SPT10 protein. Trends Biochem Sci, 22, 154–155.

Oakley, M.S., Kumar, S., Anantharaman, V., Zheng, H., Mahajan, B., Haynes, J.D., Moch, J.K., Fairhurst, R., McCutchan, T.F., and Aravind, L. (2007) Molecular factors and biochemical pathways induced by febrile temperature in intraerythrocytic plasmodium falciparum parasites. Infect Immun, 75, 2012–2025.

Oeffinger, M., Wei, K.E., Rogers, R., DeGrasse, J.A., Chait, B.T., Aitchison, J.D., and Rout, M.P. (2007) Comprehensive analysis of diverse ribonucleoprotein complexes. Nat Methods, 4, 951–956.

Oh, J., Sohn, D.H., Ko, M., Chung, H., Jeon, S.H., and Seong, R.H. (2008) BAF60a interacts with p53 to recruit the SWI/SNF complex. J Biol Chem, 283, 11924–11934.

Okada, Y., Yamagata, K., Hong, K., Wakayama, T., and Zhang, Y. (2010) A role for the elongator complex in zygotic paternal genome demethylation. Nature, 463, 554–558.

Okano, M., Bell, D.W., Haber, D.A., and Li, E. (1999) DNA methyltransferases Dnmt3a and Dnmt3b are essential for de novo methylation and mammalian development. Cell, 99, 247–257.

Ono, R., Taki, T., Taketani, T., Taniwaki, M., Kobayashi, H., and Hayashi, Y. (2002) LCX, leukemia-associated protein with a CXXC domain, is fused to MLL in acute myeloid leukemia with trilineage dysplasia having t(10;11)(q22;q23). Cancer Res, 62, 4075–4080.

Ooi, S.K. and Bestor, T.H. (2008) The colorful history of active DNA demethylation. Cell, 133, 1145–1148.

Ossareh-Nazari, B., Bonizec, M., Cohen, M., Dokudovskaya, S., Delalande, F., Schaeffer, C., Van Dorsselaer, A., and Dargemont, C. (2010) Cdc48 and Ufd3, new partners of the ubiquitin protease Ubp3, are required for ribophagy. EMBO Rep, 11, 548–554.

Otani, J., Nankumo, T., Arita, K., Inamoto, S., Ariyoshi, M., and Shirakawa, M. (2009) Structural basis for recognition of H3K4 methylation status by the DNA methyltransferase 3A ATRX-DNMT3-DNMT3L domain. EMBO Rep, 10, 1235–1241.

Painter, H.J., Campbell, T.L., and Llinas, M. (2011) The Apicomplexan AP2 family: integral factors regulating Plasmodium development. Mol Biochem Parasitol, 176, 1–7.

Pandey, U.B., Nie, Z., Batlevi, Y., McCray, B.A., Ritson, G.P., Nedelsky, N.B., Schwartz, S.L., DiProspero, N.A., Knight, M.A., Schuldiner, O. et al. (2007) HDAC6 rescues neurodegeneration and provides an essential link between autophagy and the UPS. Nature, 447, 859–863.

Park, S.W., Hu, X., Gupta, P., Lin, Y.P., Ha, S.G., and Wei, L.N. (2007) SUMOylation of Tr2 orphan receptor involves Pml and fine-tunes Oct4 expression in stem cells. Nat Struct Mol Biol, 14, 68–75.

Pasillas, M.P., Shah, M., and Kamps, M.P. (2010) Nsd1 PHD domains bind methylated H3K4 and H3K9 using interactions disrupted by point mutations in human Sotos syndrome. Hum Mutat, 32, 292–298.

Pavri, R., Zhu, B., Li, G., Trojer, P., Mandal, S., Shilatifard, A., and Reinberg, D. (2006) Histone H2B monoubiquitination functions cooperatively with FACT to regulate elongation by RNA polymerase II. Cell, 125, 703–717.

Pedersen, L.B., Birkelund, S., and Christiansen, G. (1996) Purification of recombinant Chlamydia trachomatis histone H1-like protein Hc2, and comparative functional analysis of Hc2 and Hc1. Mol Microbiol, 20, 295–311.

Pena, P.V., Davrazou, F., Shi, X., Walter, K.L., Verkhusha, V.V., Gozani, O., Zhao, R., and Kutateladze, T.G. (2006) Molecular mechanism of histone H3K4me3 recognition by plant homeodomain of ING2. Nature, 442, 100–103.

Pennini, M.E., Perrinet, S., Dautry-Varsat, A., and Subtil, A. (2010) Histone methylation by NUE, a novel nuclear effector of the intracellular pathogen Chlamydia trachomatis. PLoS Pathog, 6, e1000995.

Perara, E., Ganem, D., and Engel, J.N. (1992) A developmentally regulated chlamydial gene with apparent homology to eukaryotic histone H1. Proc Natl Acad Sci USA, 89, 2125–2129.

Peterson, C.L. and Laniel, M.A. (2004) Histones and histone modifications. Curr Biol, 14, R546–551.

Peyretaillade, E., El Alaoui, H., Diogon, M., Polonais, V., Parisot, N., Biron, D.G., Peyret, P., and Delbac, F. (2011) Extreme reduction and compaction of microsporidian genomes. Res Microbiol, 162, 598–606.

Phalke, S., Nickel, O., Walluscheck, D., Hortig, F., Onorati, M.C., and Reuter, G. (2009) Retrotransposon silencing and telomere integrity in somatic cells of Drosophila depends on the cytosine-5 methyltransferase DNMT2. Nat Genet, 41, 696–702.

Pontes, O., Li, C.F., Nunes, P.C., Haag, J., Ream, T., Vitins, A., Jacobsen, S.E., and Pikaard, C.S. (2006) The Arabidopsis chromatin-modifying nuclear siRNA pathway involves a nucleolar RNA processing center. Cell, 126, 79–92.

Ponting, C.P., Blake, D.J., Davies, K.E., Kendrick-Jones, J., and Winder, S.J. (1996) ZZ and TAZ: new putative zinc fingers in dystrophin and other proteins. Trends Biochem Sci, 21, 11–13.

Porras-Yakushi, T.R., Whitelegge, J.P., and Clarke, S. (2007) Yeast ribosomal/cytochrome c SET domain methyltransferase subfamily: identification of Rpl23ab methylation sites and recognition motifs. J Biol Chem, 282, 12368–12376.

Poulter, R.T. and Goodwin, T.J. (2005) DIRS-1 and the other tyrosine recombinase retrotransposons. Cytogenet Genome Res, 110, 575–588.

Rai, K., Huggins, I.J., James, S.R., Karpf, A.R., Jones, D.A., and Cairns, B.R. (2008) DNA demethylation in zebrafish involves the coupling of a deaminase, a glycosylase, and gadd45. Cell, 135, 1201–1212.

Reeve, J.N. (2003) Archaeal chromatin and transcription. Mol Microbiol, 48, 587–598.

Reeve, J.N., Bailey, K.A., Li, W.T., Marc, F., Sandman, K., and Soares, D.J. (2004) Archaeal histones: structures, stability and DNA binding. Biochem Soc Trans, 32, 227–230.

Roberts, R.J. (1980) Restriction and modification enzymes and their recognition sequences. Gene, 8, 329–343.

Roberts, R.J., Belfort, M., Bestor, T., Bhagwat, A.S., Bickle, T.A., Bitinaite, J., Blumenthal, R.M., Degtyarev, S., Dryden, D.T., Dybvig, K. et al. (2003) A nomenclature for restriction enzymes, DNA methyltransferases, homing endonucleases and their genes. Nucleic Acids Res, 31, 1805–1812.

Roberts, R.J., Vincze, T., Posfai, J., and Macelis, D. (2010) REBASE–a database for DNA restriction and modification: enzymes, genes and genomes. Nucleic Acids Res, 38, D234–236.

Rokas, A. (2008) The molecular origins of multicellular transitions. Curr Opin Genet Dev, 18, 472–478.

Roudier, F., Ahmed, I., Berard, C., Sarazin, A., Mary-Huard, T., Cortijo, S., Bouyer, D., Caillieux, E., Duvernois-Berthet, E., Al-Shikhley, L. et al. (2011) Integrative epigenomic mapping defines four main chromatin states in Arabidopsis. EMBO J, 30, 1928–1938.

Rouleau, M., Aubin, R.A., and Poirier, G.G. (2004) Poly(ADP-ribosyl)ated chromatin domains: access granted. J Cell Sci, 117, 815–825.

Rutkowski, R., Hofmann, K., and Gartner, A. (2010) Phylogeny and function of the invertebrate p53 superfamily. Cold Spring Harb Perspect Biol, 2, a001131.

Sanchez-Elsner, T., Gou, D., Kremmer, E., and Sauer, F. (2006) Noncoding RNAs of trithorax response elements recruit Drosophila Ash1 to Ultrabithorax. Science, 311, 1118–1123.

Sandman, K. and Reeve, J.N. (2005) Archaeal chromatin proteins: different structures but common function? Curr Opin Microbiol, 8, 656–661.

Schaefer, M. and Lyko, F. (2010) Solving the Dnmt2 enigma. Chromosoma, 119, 35–40.

Schumacher, M.A., Lau, A.O., and Johnson, P.J. (2003) Structural basis of core promoter recognition in a primitive eukaryote. Cell, 115, 413–424.

Schwartz, A.L. and Ciechanover, A. (2009) Targeting proteins for destruction by the ubiquitin system: implications for human pathobiology. Annu Rev Pharmacol Toxicol, 49, 73–96.

Schwartz, D.C. and Hochstrasser, M. (2003) A superfamily of protein tags: ubiquitin, SUMO and related modifiers. Trends Biochem Sci, 28, 321–328.

Shaheen, M., Williamson, E., Nickoloff, J., Lee, S.H., and Hromas, R. (2010) Metnase/SETMAR: a domesticated primate transposase that enhances DNA repair, replication, and decatenation. Genetica, 138, 559–566.

Sharif, J., Muto, M., Takebayashi, S., Suetake, I., Iwamatsu, A., Endo, T.A., Shinga, J., Mizutani-Koseki, Y., Toyoda, T., Okamura, K. et al. (2007) The SRA protein Np95 mediates epigenetic inheritance by recruiting Dnmt1 to methylated DNA. Nature, 450, 908–912.

Shi, H., Chamond, N., Tschudi, C., and Ullu, E. (2004) Selection and characterization of RNA interference-deficient trypanosomes impaired in target mRNA degradation. Eukaryot Cell, 3, 1445–1453.

Shimojo, H., Sano, N., Moriwaki, Y., Okuda, M., Horikoshi, M., and Nishimura, Y. (2008) Novel structural and functional mode of a knot essential for RNA binding activity of the Esa1 presumed chromodomain. J Mol Biol, 378, 987–1001.

Shirai, H., Blundell, T.L., and Mizuguchi, K. (2001) A novel superfamily of enzymes that catalyze the modification of guanidino groups. Trends Biochem Sci, 26, 465–468.

Shiu, P.K., Raju, N.B., Zickler, D., and Metzenberg, R.L. (2001) Meiotic silencing by unpaired DNA. Cell, 107, 905–916.

Shiu, P.K., Zickler, D., Raju, N.B., Ruprich-Robert, G., and Metzenberg, R.L. (2006) SAD-2 is required for meiotic silencing by unpaired DNA and perinuclear localization of SAD-1 RNA-directed RNA polymerase. Proc Natl Acad Sci USA, 103, 2243–2248.

Shull, N.P., Spinelli, S.L., and Phizicky, E.M. (2005) A highly specific phosphatase that acts on ADP-ribose 1′′-phosphate, a metabolite of tRNA splicing in Saccharomyces cerevisiae. Nucleic Acids Res, 33, 650–660.

Siegel, T.N., Hekstra, D.R., Kemp, L.E., Figueiredo, L.M., Lowell, J.E., Fenyo, D., Wang, X., Dewell, S., and Cross, G.A. (2009) Four histone variants mark the boundaries of polycistronic transcription units in *Trypanosoma brucei*. Genes Dev, 23, 1063–1076.

Slupphaug, G., Mol, C.D., Kavli, B., Arvai, A.S., Krokan, H.E., and Tainer, J.A. (1996) A nucleotide-flipping mechanism from the structure of human uracil-DNA glycosylase bound to DNA. Nature, 384, 87–92.

Smit, A.F. and Riggs, A.D. (1996) Tiggers and DNA transposon fossils in the human genome. Proc Natl Acad Sci USA, 93, 1443–1448.

Spokony, R.F. and Restifo, L.L. (2007) Anciently duplicated Broad Complex exons have distinct temporal functions during tissue morphogenesis. Dev Genes Evol, 217, 499–513.

Staub, E., Fiziev, P., Rosenthal, A., and Hinzmann, B. (2004) Insights into the evolution of the nucleolus by an analysis of its protein domain repertoire. Bioessays, 26, 567–581.

Stephens, R.S., Kalman, S., Lammel, C., Fan, J., Marathe, R., Aravind, L., Mitchell, W., Olinger, L., Tatusov, R.L., Zhao, Q. et al. (1998) Genome sequence of an obligate intracellular pathogen of humans: Chlamydia trachomatis. Science, 282, 754–759.

Strahl, B.D., Briggs, S.D., Brame, C.J., Caldwell, J.A., Koh, S.S., Ma, H., Cook, R.G., Shabanowitz, J., Hunt, D.F., Stallcup, M.R. et al. (2001) Methylation of histone H4 at arginine 3 occurs in vivo and is mediated by the nuclear receptor coactivator PRMT1. Curr Biol, 11, 996–1000.

Sumper, M., Hett, R., Lehmann, G., and Wenzl, S. (2007) A code for lysine modifications of a silica biomineralizing silaffin protein. Angew Chem Int Ed Engl, 46, 8405–8408.

Suzuki, H.I. and Miyazono, K. (2011) Emerging complexity of microRNA generation cascades. J Biochem, 149, 15–25.

Tahiliani, M., Koh, K.P., Shen, Y., Pastor, W.A., Bandukwala, H., Brudno, Y., Agarwal, S., Iyer, L.M., Liu, D.R., Aravind, L., and Rao, A. (2009) Conversion of 5-methylcytosine to 5-hydroxymethylcytosine in mammalian DNA by MLL partner TET1. Science, 324, 930–935.

Tate, C.M., Lee, J.H., and Skalnik, D.G. (2009) CXXC finger protein 1 contains redundant functional domains that support embryonic stem cell cytosine methylation, histone methylation, and differentiation. Mol Cell Biol, 29, 3817–3831.

Tate, C.M., Lee, J.H., and Skalnik, D.G. (2010) CXXC finger protein 1 restricts the Setd1A histone H3K4 methyltransferase complex to euchromatin. FEBS J, 277, 210–223.

Templeton, T.J., Iyer, L.M., Anantharaman, V., Enomoto, S., Abrahante, J.E., Subramanian, G.M., Hoffman, S.L., Abrahamsen, M.S., and Aravind, L. (2004) Comparative analysis of apicomplexa and genomic diversity in eukaryotes. Genome Res, 14, 1686–1695.

Thompson, P.R. and Fast, W. (2006) Histone citrullination by protein arginine deiminase: is arginine methylation a green light or a roadblock? ACS Chem Biol, 1, 433–441.

Travers, A. (2000) Recognition of distorted DNA structures by HMG domains. Curr Opin Struct Biol, 10, 102–109.

Trievel, R.C., Beach, B.M., Dirk, L.M., Houtz, R.L., and Hurley, J.H. (2002) Structure and catalytic mechanism of a SET domain protein methyltransferase. Cell, 111, 91–103.

Tsai, M.C., Manor, O., Wan, Y., Mosammaparast, N., Wang, J.K., Lan, F., Shi, Y., Segal, E., and Chang, H.Y. (2010) Long noncoding RNA as modular scaffold of histone modification complexes. Science, 329, 689–693.

Tsukada, Y., Fang, J., Erdjument-Bromage, H., Warren, M.E., Borchers, C.H., Tempst, P., and Zhang, Y. (2006) Histone demethylation by a family of JmjC domain-containing proteins. Nature, 439, 811–816.

Tweedie, S., Ashburner, M., Falls, K., Leyland, P., McQuilton, P., Marygold, S., Millburn, G., Osumi-Sutherland, D., Schroeder, A., Seal, R. et al. (2009) FlyBase: enhancing Drosophila Gene Ontology annotations. Nucleic Acids Res, 37, D555–559.

Uhlmann, F. and Hopfner, K.P. (2006) Chromosome biology: the crux of the ring. Curr Biol, 16, R102–105.

Ullu, E., Tschudi, C., and Chakraborty, T. (2004) RNA interference in protozoan parasites. Cell Microbiol, 6, 509–519.

van Dijk, J., Miro, J., Strub, J.M., Lacroix, B., van Dorsselaer, A., Edde, B., and Janke, C. (2008) Polyglutamylation is a post-translational modification with a broad range of substrates. J Biol Chem, 283, 3915–3922.

Vojnic, E., Simon, B., Strahl, B.D., Sattler, M., and Cramer, P. (2006) Structure and carboxyl-terminal domain (CTD) binding of the Set2 SRI domain that couples histone H3 Lys36 methylation to transcription. J Biol Chem, 281, 13–15.

Walsh, C.T., Garneau-Tsodikova, S., and Gatto, G.J. (2005) Protein posttranslational modifications: the chemistry of proteome diversifications. Angew Chem, 2005, 7342–7372.

Walsh, D.A. and Doolittle, W.F. (2005) The real 'domains' of life. Curr Biol, 15, R237–240.

Wang, Y., Wysocka, J., Sayegh, J., Lee, Y.H., Perlin, J.R., Leonelli, L., Sonbuchner, L.S., McDonald, C.H., Cook, R.G., Dou, Y. et al. (2004) Human PAD4 regulates histone arginine methylation levels via demethylimination. Science, 306, 279–283.

Webby, C.J., Wolf, A., Gromak, N., Dreger, M., Kramer, H., Kessler, B., Nielsen, M.L., Schmitz, C., Butler, D.S., Yates, J.R. 3rd et al. (2009) Jmjd6 catalyses lysyl-hydroxylation of U2AF65, a protein associated with RNA splicing. Science, 325, 90–93.

Winkler, G.S., Kristjuhan, A., Erdjument-Bromage, H., Tempst, P., and Svejstrup, J.Q. (2002) Elongator is a histone H3 and H4 acetyltransferase important for normal histone acetylation levels in vivo. Proc Natl Acad Sci USA, 99, 3517–3522.

Winkler, K.E., Swenson, K.I., Kornbluth, S., and Means, A.R. (2000) Requirement of the prolyl isomerase Pin1 for the replication checkpoint. Science, 287, 1644–1647.

Woo, H.R., Pontes, O., Pikaard, C.S., and Richards, E.J. (2007) VIM1, a methylcytosine-binding protein required for centromeric heterochromatinization. Genes Dev, 21, 267–277.

Woodcock, C.L. (2006) Chromatin architecture. Curr Opin Struct Biol, 16, 213–220.

Wysocka, J., Swigut, T., Milne, T.A., Dou, Y., Zhang, X., Burlingame, A.L., Roeder, R.G., Brivanlou, A.H., and Allis, C.D. (2005) WDR5 associates with histone H3 methylated at K4 and is essential for H3 K4 methylation and vertebrate development. Cell, 121, 859–872.

Yakisich, J.S., Sandoval, P.Y., Morrison, T.L., and Kapler, G.M. (2006) TIF1 activates the intra-S-phase checkpoint response in the diploid micronucleus and amitotic polyploid macronucleus of Tetrahymena. Mol Biol Cell, 17, 5185–5197.

Yamada, K., Frouws, T.D., Angst, B., Fitzgerald, D.J., DeLuca, C., Schimmele, K., Sargent, D.F., and Richmond, T.J. (2011) Structure and mechanism of the chromatin remodelling factor ISW1a. Nature, 472, 448–453.

Yang, M., Gocke, C.B., Luo, X., Borek, D., Tomchick, D.R., Machius, M., Otwinowski, Z., and Yu, H. (2006) Structural basis for CoREST-dependent demethylation of nucleosomes by the human LSD1 histone demethylase. Mol Cell, 23, 377–387.

Yap, K.L. and Zhou, M.M. (2010) Keeping it in the family: diverse histone recognition by conserved structural folds. Crit Rev Biochem Mol Biol, 45, 488–505.

Yoon, J.H., Iwai, S., O'Connor, T.R., and Pfeifer, G.P. (2003) Human thymine DNA glycosylase (TDG) and methyl-CpG-binding protein 4 (MBD4) excise thymine glycol (Tg) from a Tg:G mispair. Nucleic Acids Res, 31, 5399–5404.

Yu, H., Mashtalir, N., Daou, S., Hammond-Martel, I., Ross, J., Sui, G., Hart, G.W., Rauscher, F.J. 3rd, Drobetsky, E., Milot, E. et al. (2010) The ubiquitin carboxyl hydrolase BAP1 forms a ternary complex with YY1 and HCF-1 and is a critical regulator of gene expression. Mol Cell Biol, 30, 5071–5085.

Yue, W.W., Hassler, M., Roe, S.M., Thompson-Vale, V., and Pearl, L.H. (2007) Insights into histone code syntax from structural and biochemical studies of CARM1 methyltransferase. EMBO J, 26, 4402–4412.

Zemach, A., McDaniel, I.E., Silva, P., and Zilberman, D. (2010) Genome-wide evolutionary analysis of eukaryotic DNA methylation. Science, 328, 916–919.

Zhang, H., Christoforou, A., Aravind, L., Emmons, S.W., van den Heuvel, S., and Haber, D.A. (2004) The C. elegans Polycomb gene SOP-2 encodes an RNA binding protein. Mol Cell, 14, 841–847.

Zhang, Q.M., Yonekura, S., Takao, M., Yasui, A., Sugiyama, H., and Yonei, S. (2005) DNA glycosylase activities for thymine residues oxidized in the methyl group are functions of the hNEIL1 and hNTH1 enzymes in human cells. DNA Repair (Amst), 4, 71–79.

Zhao, J., Ohsumi, T.K., Kung, J.T., Ogawa, Y., Grau, D.J., Sarma, K., Song, J.J., Kingston, R.E., Borowsky, M., and Lee, J.T. (2010) Genome-wide identification of polycomb-associated RNAs by RIP-seq. Mol Cell, 40, 939–953.

Zhao, W., Kruse, J.P., Tang, Y., Jung, S.Y., Qin, J., and Gu, W. (2008) Negative regulation of the deacetylase SIRT1 by DBC1. Nature, 451, 587–590.

Zhou, H., Spicuglia, S., Hsieh, J.J., Mitsiou, D.J., Hoiby, T., Veenstra, G.J., Korsmeyer, S.J., and Stunnenberg, H.G. (2006) Uncleaved TFIIA is a substrate for taspase 1 and active in transcription. Mol Cell Biol, 26, 2728–2735.

Zhou, Z., Feng, H., Zhou, B.R., Ghirlando, R., Hu, K., Zwolak, A., Miller Jenkins, L.M., Xiao, H., Tjandra, N., Wu, C. et al. (2011) Structural basis for recognition of centromere histone variant CenH3 by the chaperone Scm3. Nature, 472, 234–237.

Zhu, B., Zheng, Y., Angliker, H., Schwarz, S., Thiry, S., Siegmann, M., and Jost, J.P. (2000) 5-Methylcytosine DNA glycosylase activity is also present in the human MBD4 (G/T mismatch glycosylase) and in a related avian sequence. Nucleic Acids Res, 28, 4157–4165.

Zhu, P., Zhou, W., Wang, J., Puc, J., Ohgi, K.A., Erdjument-Bromage, H., Tempst, P., Glass, C.K., and Rosenfeld, M.G. (2007) A histone H2A deubiquitinase complex coordinating histone acetylation and H1 dissociation in transcriptional regulation. Mol Cell, 27, 609–621.

Zilberman, D., Coleman-Derr, D., Ballinger, T., and Henikoff, S. (2008) Histone H2A.Z and DNA methylation are mutually antagonistic chromatin marks. Nature, 456, 125–129.

INDEX

A1A (pepsin), 303–304
ab-initio methods for protein identification, 43–44
Acaryochloris marina, 305
acetylases, 477
acetylated peptides, 452
ACF7/MACF1, 244
Acidilobus saccharovorans, 307
Acidithiobacillus ferrooxidans ATCC 23270, 365
Aciduliprofundum boonei, 305
α-actinins, 247
  actinin-associated lim protein, 251
  C-terminal regions, 257
  functional distinction between spectrin and, 253
  ligands, 251
  as a protein binding module, 250–252
  proteolytic analysis of, 254
  as a spacer module, 250
acyl-CoA binding protein, 265
acylhomoserine lactone-dependent quorum-sensing systems, 142
adenosine diphosphate ADP-ribosyl cyclase-like families, 53–54
adipocyte plasma membrane-associated protein (APMAP), 144
ADP-ribosylation-based regulatory systems, 447–452
ADP-ribosyl transferases, 451
Aeropyrum pernix gene APE0212, 308
alignment masking, 115–116
amino acid protein sequences, 71
amino acid residues, 199
aminoacyl-tRNA synthetases, 160–161
  connective polypeptide in, 185
  domain–domain interfaces, 187
  multidomain architectures (MDAs), 174
  structural clusters, 179
aminoacyl-tRNA synthetases (AATRS), 162, 165–166, 170, 173
Amphitrite ornata, 223
analysis tools
  analysis platform, 84–85
  comparative genomics, 88–89
  domain families, 88
  generating additional forms of data, 89–90
  sequence alignments, 85–88
ancestral eukaryotic silencing system, 475
ancestral functional CRISPR-Cas system, 368–370
androglobin, 220

Protein Families: Relating Protein Sequence, Structure, and Function, First Edition.
Edited by Christine Orengo and Alex Bateman.

Ank family proteins, 325
ankyrin
  binding functions, 261–263
  carbon dioxide transport, 261
  cross-links proteins, 263
  genes, 259
  obscurin/titin-binding-related domain (OTBD), 259
  overview, 258–260
  repeats as nanosprings, 262–263
  spliced modules, 259
  structure and prevalence, 261
  trimeric transmembrane complex, 262
ankyrin-based complex, 242
annotation transfer, 103
A24 peptidase family, 293
A26 peptidase family, 293
AP2/GCC-box-binding domain, 402
Arabidopsis thaliana, 119, 386, 389
Archaeoglobus fulgidus, 342
Archaeplastida, 212, 219
Archezoa, 292
A2 (retropepsin), 304
Arg 40 C6, 230
Arg 92 FG4, 230
arginyl-tRNA synthetase, 46
Arg-tRNA synthetase, 39
ArsA homologs, 328
arsenite efflux (ArsP) transporter family, 328
arylesterase, 136–139
arylesterase-like subgroup, 141–143, 145
Asn synthetase B, 39
Asp138, 141
Asp229, 149
aspartic acid, 201
Asp 94 G1, 230
Asp–Tyr hydrogen bond, 230
ATP-pyrophosphatases (ATP-PPASE), 163, 171
automated protocol of functional family identification, 172
Automatic Domain Decomposition Algorithm (ADDA)
  block model, 13
  domain composition, 13–15
  domain decomposition, 13–16
  efficiency, 19
  footprint of a sequence, following a, 17
  homologous overextension, 15
  implementation of, 17–18
  limitations, 19
  multidomain proteins, 13
  objective of, 19
  overview, 13
  parameters, 17
  spurious alignments, 15
  structure of multiple alignments, 14

Bacillus anthracis, 294
Bacillus subtilis, 387
Bacillus tusciae, 367
bacterial contamination in eukaryote genomes, 290–291
Bacteroides cellulosilyticus, 306
Bacteroides helcogenes, 306
BC03BB108_C0180 protein, 294
Berkeley PHOG (for PhyloFacts Orthology Group) database, 106–107
Berkeley Phylogenomics Group, 107
beta-naphthyl acetate, 144
bidirectional nearest-neighbor sequences, 9
Bifidobacterium animalis, 366
binding sites, 182
bioinformatics, 99–100
  boundaries of structural domains, identifying, 100
  prediction of catalytic residues, 100
bioinformatics algorithms, development of, 78–80
2,3-Bisphosphoglycerate (BPG), 225–226
BLAST, 53, 103, 113, 316
  Cas9 sequences, 365
  E-values, 101
  "leave-one-out" benchmark, 75
  similarity networks, assessment of, 136
Bohr effect, 227
Bron–Kerbosch algorithm, 51
Burkholderia sp. H160, 152

C. maquilingensis, 309
C2A (calpains), 304–305
C14A (caspase), 306
C45 (acyl-coenzyme A:6-aminopenicillanic acid acyl-transferase precursor), 306
Caenorhabditis elegans, 223, 386, 445
Caldivirga maquilingensis (ABW02092), 307
calmodulin-like domain, 257–258
calponin homology domains
  CH domains, 255–256
  discovery, 254–255
  functional interactions and regulation, 256–257
  prevalence, 255–256
C44 (amidophosphoribosyltransferase precursor), 302
CAMPS 2.0, 41
CAMSAP family, 256
C1A (papain), 304

Ca^{2+}-release-activated Ca^{2+} (CRAC) channels (CRAC-Cs), 331
Cascade complexes, 348–349
Cas8 family proteins, 366
Cas6f proteins, 354
cas2 gene, 346
cas3 gene, 344
cas9 gene, 344
cas10 gene, 344
Cas10-HD fusions, 347
Cas6 homologs, 348
Cas9 homologs, 364–365
  of *Streptococcus thermophilus,* 365
caspases, 306
Cas1 protein, 345
Cas proteins
  antiviral defense, 348
  Cas1 and Cas2, 345–347
  HD domain of, 347–348
Cas10 proteins, 365
catalytic residues, 88
Catalytic Site Atlas (CSA), 88, 118, 182
CATH
  NAD binding domains, 193
CATH–HMMscan algorithm, 46
CATHEDRAL algorithm, 45–46, 51
CATH PSI-blast, 73
cation diffusion facilitator (CDF), 331
Caulobacter crescentus, 290
C51 bacteria, 290
C1B (bleomycin hydrolase), 304
C14B (metacaspase), 306
CBP/p300 acetylases, 449
C11 (clostripain), 305
C51 (d-Ala-Gly-peptidase), 306
C69 (dipeptidase A), 306–307
cellular localization, 99
C26 (gamma-glutamyl hydrolase), 302
Chaetomium globosum, 304
chimeric proteins (FHbs), 217
Chlamydomonas, 403
Chlamydomonas reinhardtii, 305
choanoflagellates, 247
CHOP algorithm, 10
chromatin proteins (CPs)
  covalent modification of, 464–468
  definition and characterization of domains, 436–437
  distribution of, 430–431
  DNA-binding domains in, 436–445
  domain architectures of, 477–480
  domains sensing methylation status of peptides, 458–461
  functional type, 436–447
  interactions between RNA-based regulatory systems and, 474–477
  organizational complexity, 479–480
  remodeling and assembling systems, 461–464
  reverse peptide methylation in, 456–458
  "syntactical" patterns, 477–479
chymotrypsin, 289
clans, 286
class, architecture, topology, homology (CATH), 29, 38
  approach to recognize folds, 49–50
  automated protocols, 57
  classifications of domains, 160
  comparison with SCOP, 59–61
  domain and residue coverage, 80–81
  domain architectures for Follistatin-related protein 4, 78
  domain boundary assignment, 44
  fold recognition in, 49–50
  hierarchies of population, 58, 60
  hierarchy, 42
  history, 73
  machine learning protocol (CATHsvm), 54
  multiple structural alignments in, 51
  websites, 63
classification protocols
  domain boundary recognition algorithms, 52–53
  fold divergence in homologous superfamilies, 55–56
  fold group identification, 54–55
  subclassification of homologous superfamilies, 56–58
  superfamily identification, 53–54
C13 (legumain), 305–306
Clostridium, 294
Clostridium difficile proteome, 31
Clostridium histolyticum, 305
CLUSS algorithm, 11
ClustalW, 115
ClustalX tree, 325
cluster of orthologous groups (COGs), 9
Clusters of Orthologous Groups (COGs), 104
ClusTr, 104
ClusTr algorithm, 10
Cnidaria, 248
CoA-binding N-terminal domain, 193
Cocoa genome, domain of, 90
COCO-CL, 105
coenzyme A biosynthesis pathways, 174
COG1517 proteins, 372
colocalization of genes, 327
ComK protein, 400
common ancestry, 101

community annotation, 32
comparative genomics tools, 88–89
COMPARER algorithm, 40
comparison of protein structures, approaches used in, 48
COMPASS method, 168
computational assignment of protein structure, 4
connective polypeptide, 185
conservation density, 396
Conserved Domain Database (CDD), 26, 28, 41
CORA algorithm, 51
CoryneRegNet, 389
C39 peptidase family, 294
C59 peptidase family, 294
C15 (pyroglutamyl-peptidase I), 306
CRISPR-Cas systems
- *A. ferrooxidans* system, 365
- ancestral functional, 368
- classification of, 344
- definition, 342
- functional diversity of, 344
- immune function of, 342–343
- origin of, 368–372
- plasmid resistance of, 343
- RAMP arrangement in, 355–356
- regulatory and auxiliary components of, 372–373
- subunits, 356–375
- types of, 344, 351, 364–368, 372

CRISPR polymerase, 371
Critical Assessment for Protein Structure Prediction (CASP), 108
Critical Assessment of protein Function Annotation (CAFA) experiment, 58
Critical Assessment of Structure Prediction (CASP6), 71
cryptochromes/DNA photolyases (C-DNAP), 163, 166–167, 171, 180
- multidomain architectures (MDAs), 174
- structural diversity, 179

Cryptosporidium, 305
Cryptosporidium muris, 305
Cryptosporidium parvum, 305
C-terminal embellishments, 183
cTFbase, 389
Cupredoxin superfamily, 76
cyanobacterium *Oscillatoria* sp. PCC 6506, 247
7-cyano-7-deazaguanine, 169, 173
7-cyano-7-deazaguanine synthase, 169
2-cyano-6-hydroxybenzothiazole, 140
cysteine residue (C1167), 253
cysteines, 19
cytoglobin, 218
cytophagalysin *(B. thuringiensis),* 309
Cytoscape, 136
Cytoscape-Web, 88

DALI algorithm, 40, 48–49, 51
Dali Domain Dictionary (DDD), 38
DBConsensus algorithm, 46
DBTBS, 391
D-cysteine, 140
dehydrogenases, 193
designed ankyrin-repeat proteins (darpins), 262
Desulfovibrio salexigens, 306
Desulfovibrio vulgaris, 347
Detective algorithm, 44, 46
Dictyostelium, 305
Dictyostelium discoideum (AAM44351), 310
dicyclopentyl phosphoramidate inhibitor, 149
diisopropylfluorophosphatase (DFPase), 151
- aspartic acid and, 141
- superposition of the active site, 149

diisopropyl fluorophosphate (DFP), 141
dimer–tetramer interconversion, 241
DinG family, 366
diphosphoglycerate (DPG), 225
dissociation equilibrium equation, 224
Distributed Annotation System (DAS), 83
divergent proteins, 129
DNA-binding domain families, 390
- bacterial phylum-specific, 400–401
- in bacteria phyla, 400
- conservation of, 396–398
- eukaryotic repertoires, 401–403
- phylogenetic groups, 395–396
- prokaryote repertoires, 398–401
- shared by archaea and bacteria, 398–400
- in unicellular eukaryotic species, 402–403

DNA demethylation, 471–474
DNA methylation in eukaryotes, 468–471
Dobzhansky's proclamation, 100
Dollo parsimony, 92
domain architecture, 77, 113
domain boundaries, 39
domain chaining, 6
domain cutting, 8
domain–domain interfaces of HUP superfamily, 187
domain families, 4
- analysis tools for, 88
- classification of domains, 4–5, 8
- identification of, 4. *see also* sequence space graph

DomainFinder3, 77
DomainFinder process, 81
Domain Maker (DOMAK) algorithm, 44, 46
domain prediction resources, 52

domain predictions in completed genomes, 39
domain problem, 74
domain shuffling, 103
domain structure classifications
  class, architecture, topology, homology (CATH), 38
  COMPASS, 38
  Dali Domain Dictionary (DDD), 38
  3DEE, 38
  structural classification of proteins (SCOP), 38
Drosophila ananassae, 304
Drosophila bristles, 263
Drosophila Fsh protein, 452
Drosophila melanogaster, 290, 386, 389
Drosophila virilis, 290
drug/metabolite transporter (DMT) superfamily, 330
drug-responsive protein 35 (Drp35), 141
duplication of domains, 44
D-xylonate, 152
dystonin/BPAG1, 244
dystrophin, 253–254

EC, 56
Eco1 orthologs, 449
Ectocarpus sequences, 310
Ectocarpus siliculosus (CBN76176, CBN78156, and CBN79239), 309
eggNOG, 104
electron transfer flavoproteins (ETFs), 160, 162, 165, 170
  multidomain architectures (MDAs), 174
Encephalitozoon, 292, 305, 307
Encephalitozoon cuniculi, 307
Encephalitozoon intestinalis, 308
end-binding (EB) proteins, 256
endogenous retroviruses, 290
endosymbiotic bacteria, 292
endothelial nitric oxide synthase, 251
EnsemblCompara tool, 116, 160
Ensembl Genomes, 73, 81
Entamoeba, 293
Enterocytozoon bieneusi, 308
entropy-based Scorecons algorithm, 58
Enzyme Classification (EC), 85
enzyme domain (Elp3), 449
epigenetic DNA modification enzymes, 468
error-prone annotation protocol, 103
errors per query, 82
erythrocyte ankyrin-R (ANK1), 261
erythrocyte membranes, 241
erythrocytes, 252–254
erythrocyte spectrin tetramers, 241
Escherichia coli, 106, 343, 345, 384
  peptidases, 293–294
Escherichia coli flavohemoglobin (HMP), 223
Eukaryotic Clusters of Orthologous Groups KOGs, 31
eukaryotic cytoskeleton, 240
  spectrin-type triple helical repeat domain, 243
eukaryotic evolution
  ADP-ribosylation-based regulatory systems, 447–452
  distribution of TFs and CPs, 430–431
  DNA-binding domains of, 434–436
  DNA demethylation, 471–474
  DNA methylation, 468–471
  DNA methylation in, 468–471
  eukaryotic chromatin, 422–426
  funtional type CPs, 436–447
  higher order relationships amongst eukaryotes, 426
  histone lysine acetyltransferases (HATs), 447–452
  histone-methylation-based regulation, 452–456
  lateral gene transfer, 426–427
  rampant gene loss, 426
  rapid lineage-specific divergence, 426
  specific TFs, 431–434
Eukaryotic genomes, 80
eukaryotic genomes, 106
eukaryotic linear motifs (ELM), 89
eukaryotic TF resources, 389–390
European Bioinformatics Institute (EBI), 40
e-values, 82
EVEREST algorithm, 11
evolutionary reconstruction, 100
ezrin, radixin and moesin (ERM proteins), 266–267

F-actin, 256
false discovery rate (FDR), 82
families, definition of, 41
family annotation using Pfam database, 32–34
family unification, 16
FASTA format, 107
FERM domain, 264–266
  of ezrin, radixin and moesin (ERM proteins), 266–267
  phosphorylation, effects of, 266–267
  4.1R, 265–267
field programmable gate arrays (FPGAs), 78
fimbrin, 255
flavohemoglobin family, 217–218, 220
flavohemoglobin (SDFgbs), 217
FlowerPower algorithm, 109–110, 113
FMN (Flavin mononucleotide) binding domain, 402

focal-adhesion kinases, 265
fold divergence in homologous superfamilies, 55–56
fold group clusters, 41
fold group identification, 54–55
fold recognition in CATH, 49–50
FORCE algorithm, 11
formate/glycerate dehydrogenases, 193
full-length globins, 210–212
  helix interfaces, 210–211
  mutations at interfaces, 212
  residue–residue contacts in, 212
  ridges-into-grooves model, 210–211
  sequence–structure relationships in, 212
functional annotation errors, 103
functional boundaries, 151
functional conservation, 5
functional families (FunFams) protocol, 58
functional groups, 161
  classification of HUP domains in, 162–165
  comparison of structural clusters with, 179
  embellishments in, 181. *see also* structural embellishment
  function diversity and evolution, 173–174
  Gene3D HUP domains, 168–173
  multidomain architectures (MDAs), 174–175
  structural diversity within, 178–180
functionally diverse enzyme superfamilies (SFs), 128–132
fungi, 212, 219
FUPA27 family, 328
FUPA29 family, 328
FUPA27 gene, 329
FUPA23 P-type ATPases, 328–329
FUR (ferric uptake regulator), 400

G2 bacteria, 290
Gblast, 332
GeMMA functional family, 168–173
GenBank, 145
Gene3D, 29, 52, 58, 74
  applications, 71–72
  coverage and domain statistics, 84
  cut-off in, 82
  data access, 83
  domain annotations using, 77
  domain architectures for Follistatin-related protein 4, 78
  dynamic programming for HMM scoring, 78
  e-values, 82
  genome assemblies, 80–81
  graphical means of comparing genomes, 89
  high throughput technology, 70–71
  history, 72–73
  HMMER3, 75
  HUP superfamily domains, 168–173
  phylogenetic profiling of domains, 79
  precalculated annotations, distribution of, 83
  recognition and display of structural domain assignments, 85
  structural information, 71
  superfamily comparisons, 90–91
  technology, 73–74
gene duplication events, 100
gene ontology (GO) resources, 53, 57, 109, 161
  molecular functions, 58
GENERAGE algorithm, 10
genetic sensory response, 387
GenomeRNAi, 85
genome sequencing projects, 4
GenTHREADER, 52, 71
Geobacter sulfurreducens, 366
getNcbiSeq.pl, 324–325
Giardia, 292, 445, 451, 479
globin-coupled sensor (GCS) family, 218–219
globins, 38
  amino acid sequence, 208
  amino acid side chains in, 213
  B10 and E7, position of, 221
  B–E and G–H packings, 216
  C-helices in, 216
  folding pattern, 208
  full-length, 210–212
  functions of, 220
  heme group in, 213
  hexacoordinate, 213–214
  hydrogen-bonding residues, 222
  model of evolution, 220
  monomers, 210
  monomer structure, 210–216
  multimeric states of, 216–217
  peroxidase activity, 223
  phenotypic differences in, 223
  phycocyanins and, 214–216
  rates of O_2 dissociation, 221–222
  secondary structure of, 208
  species distribution in, 217–220
  studies, 208–210
  truncated, 212–213
globular (ordered) protein sequence, 80
Glu309, 143, 147, 149
  mutagenesis of, 147
glycerol-3-phosphate cytidylyltransferase, 39
glycerol-3-phosphate cytidylyltransferase, HUP domain embellishments of, 183
Glycine max, 389
glycosylphosphatidylinositide, 306

GNAT acetyltransferases, 447
GO Consortium, 112
Golgi alpha-mannosidase II, 89
G-protein coupled receptors, 115
Granulibacter bethesdensis, 305
graph-clustering algorithms, 104
GRATH algorithm, 45, 51
Group I, II, and III proteins, 220
grouping of proteins into families, 4
G1 (scytalidoglutamic peptidase), 307
GSDH protein, 149

haemophilus influenza genome, 73
H2A.Z, 437
β_{Heavy}-spectrin, 243
hemoglobin, 217
  allosteric changes in, 224–231
  crystal structures of the R and T states of, 230–231
  Hill plot of, 225–226
  ligated forms, 227
  mechanism of allosteric change in, 222
  mutants of, 233
  oxy and deoxy structures, 230
  oxygen affinity of oxy form of, 228
  Patterson analysis of, 210
  quaternary states, 230
  tense and relaxed states, 227–228
  tertiary structural changes, 229
  tetramer, 229
hemoglobins, 208
  heme groups, 208
Her2-specific darpins, 262
hexacoordinate globins, 213–214
hidden Markov models (HMMs), 4, 25, 43, 74, 136, 160, 193, 247
  Cupredoxin superfamily, 76
  graphical representation, 76
hierarchies of population, 58, 60
  vs structural continuum, 61–62
High-quality Automated and Manual Annotation of microbial Proteomes (HAMAP), 40
high signature proteins, 160
high throughput technology, 70–71, 77–78
Hill plot method, 224–225
histone-like proteins, 422
histone lysine acetyltransferases (HATs), 447–452
histone-methylation-based regulation, 452–456
HMMER software, 27
  HMMER2, 30
  HMMER3, 30
HMM formalism for proteins, 30
hmu operon, 329
homologous superfamilies, subclassification of, 56–58
homologs, 4, 41, 101, 306, 310
  clustering of, 113–115
  sequence identity with proteins, 38
homology
  functional prediction by, 103
  sequence identity and, 101
homology detection as multidomain proteins, 6
HOMSTRAD, 41
horizontal gene transfer, mechanism of, 289–293
  problem of contamination, 290–291
  via endosymbionts, 292–293
  via plasmids, 292
  via viruses and retrotransposons, 290–292
human Lamin-B receptor, phylogenetic analysis of, 114, 119
human neuroglobin, 218
HUP superfamily, 160
  classification of, 162–165
  domain–domain interfaces, 187
  protein–protein interfaces, 184–187
  structural diversity of, 175–180
  structural embellishments in, 180–187
hybrid pair-wise/profile method, 79
hydrolytic reactions, 143
hydrolytic SSL proteins, 152

αI- and βI-spectrin polypeptides, 252
Ichthyophthirius multifiliis, 331
ictalurid herpesvirus 1, 292
identification of domains
  ab-initio methods, 43–44
  of "attractor" motifs, 49
  automatic structure-based approaches, 43
  fold divergence in homologous superfamilies, 55–56
  fold group, 54–55
  homology-based methods, 44–46
  intra-residue comparison method, 48–49
  midnight zone, 47
  in multidomain proteins, 43
  multiple structural alignments, 51–52
  sequence-based automated methods, 45–46
  structural similarity, 50–51
  structure-based approaches, 46–48
  subclassification of homologous superfamilies, 56–58
  superfamily, 53–54
  twilight zone, 47, 101
Ile/Leu/Val-tRNA synthetases, 175
immunoglobulins, 38
inparalog, 101
InParanoid, 104

InParanoid database, 106
InterPro domain database, 28, 32, 40, 73, 193
InterProScan software, 34, 83
intra-residue comparison method, 48–49
INTREPID, 118
INTREPID functional site identification algorithm, 111

Jackhmmer program, 30–32, 74, 113
automated SAM iterative "target" procedure, 74–75
HMMER3-based, 77
Janus tyrosine kinases, 265

Kre33p acetylases, 449
Ktedonobacter racemifer, 365
Kyoto Encyclopedia of Genes and Genomes (KEGG), 57, 85, 109

Laccaria bicolor, 304
lactate dehydrogenase, 193
Lamarckian inheritance principle, 342
L-ascorbic acid, 140
last archaeal common ancestor (LACA), 371
last eukaryotic common ancestor (LECA), 427, 451, 477
latent class analysis (LCA), 12
"leave-one-out" benchmark, 75
leghemoglobin (LegHb), 221–222
legumains, 305–306
leucine-rich-repeat (LRR) regions, 119
LexA repressor, 400
limited proteolysis, 241
lim protein, 251
lipids, 252
LOFT program, 105
low-density lipoprotein repeat class B (Ldl_recept_b), 136–138, 149
luciferin-regenerating enzyme (LRE), 140
lysine, 201
lysosome (cathepsin D), 304

MACF1/ACF7, 254
machine learning methods, 12
MACHOS algorithm, 11
MACiE, 198, 200
M15A (D-Ala-d-Ala-carboxypeptidase), 308
major royal jelly protein (MRJP), 136, 149
M18 (aminopeptidase I), 309
mammalian red blood cells, 240
M2 (angiotensin-converting enzyme), 307
M54 (archaelysin), 309
marine bacteria, 304
Markov clustering, 106
Markov cluster algorithm (MCL), 9–10
M3A (thimet oligopeptidase), 307
maximum likelihood phylogenetics, 92
M60 bacteria, 290
M12B (ADAMs), 307
M3B (oligopeptidase F), 307
M14 (carboxypeptidase A), 308
M43 (cytophagalysin), 309
M60 (enhancin), 309
Mermis nigrescens, 223
metagenomics, 128
metal-binding ligands, 149
metal-coordinating active site residues, 145
metal-coordinating ligands, 141
of DFPase, 151
of *V. vinifera* protein, 152–153
metal-dependent SSL proteins, 152
metallopeptidases, 287
MetaPhOrs, 112
metazoa, 248
Methanocaldococcus jannaschii, 348
Methanocorpusculum labreanum, 305
Methanosarcina acetivorans, 305, 309
Methanosarcina barkeri, 306, 366
Methanosarcina homologs, 308
Methanosarcina mazei, 306
Methanosarcina species, 308
Methanospirillum hungatei, 305–306
Methanothermus fervidus, 306
methylases, 477
M41 (FtsH peptidase), 309
M11 (gametolysin), 307
M20 (glutamate carboxypeptidase), 302
Microcoleus chthonoplastes, 365
Microsporidia species, 307–308
misannotations, 103
mitochondria, 293
M17 (leucyl aminopeptidase), 309
Mlu1-box-binding protein (MBP1), 402
M24 (methionyl aminopeptidase), 288, 302
M13 (neprilysin), 308
molecular biology, 4
molecular shock absorber, 254
mono-ADP-ribosylated proteins, 451
Monod–Wyman–Changeux model of allosteric change, 208
Monosiga brevicollis, 247, 304, 402
Monte-Carlo simulation, 10
Moritella sp. PE36, 310
M22 (O-sialoglycoprotein peptidase), 302
Mov10/armitage-like RNA helicases, 476
M5 peptidase family, 294
M20 peptidase family, 294
M27 peptidase family, 294
M34 peptidase family, 294

M41 peptidase family, 294
M66 peptidase family, 294
M79 peptidase family, 294
M72 (peptidyl-Asp metallopeptidase), 309
M16 (pitrilysin), 308–309
M48 (Ste24 peptidase), 302
M4 (thermolysin), 307
multidomain architecture (MDA)
  associated functions for human sequences, 202
  NAD binding domain, 193–195
multidomain architecture of proteins, 5, 74, 100
multidomain architectures (MDAs), 174–175
multidomain protein families, 39
MultiParanoid database, 106
multiple alignment using fast Fourier transform (MAFFT), 26, 115–116, 287
multiple sequence alignment (MSA), 100
  construction and alignment masking, 115–116
  phylogenetic tree estimation, 116
  program, 111
multiple structural alignments, 51–52
  HMM–HMM comparisons, 57
MUSCLE, 115
MUSCLE sequence, 26
mutants of hemoglobin, 232–233
myoglobin, 72, 217, 220
  Hill plot of, 225–226

N. gruberi (EFC37030), 310
NAD biosynthesis pathways, 174
NAD Rossmann-like binding domain superfamily, 192
  atomic structures of, 192–193
  catalytic residues, 198
  in CATH, 193
  C–H bond formation/cleavage, 201
  course of evolution, 196–197
  C-terminal extension, 204
  exchanges in EC classes, 196
  functional changes, 197
  multidomain architecture (MDA), 193–195
  network of hydrogen bonds, 193
  secondary structure topology of, 192
  subfunctionalization, 195–196
NAD synthetase (N-type ATP-pyrophosphatases functional group), 174
Naegleria, 445, 450–451, 477, 480
Naegleria gruberi, 310
Nanoarchaeum equitans, 309
National Center for Biotechnology Information, 28, 324
National Center for Biotechnology Information (NCBI), 92
  RefSeq, 82
Ncd80 complex, 257
N-deoxyribosyltransferase family, 53–54
neo-functionalization, 101
Neospora caninum, 305
nesprins, 244
neuroglobin, 218
next-generation sequencing machines, 78
NHL family, 149
nicotinamide adenine dinucleotide phosphate (NADP), 192
  role in enzyme reactions, 198
nonmetazoan eukaryotes, 217
nonnucleated mammalian erythrocytes, 252
nonsynonymous mutations, 100
Nosema ceranae, 308
NOTUNG tool, 116
N6P superfamily, 128–129
  boundaries, 148–151
  catalytic mechanisms, 149
  comparison of the SSs with, 146
  criterion for inclusion of proteins, 146
  definition of, 138–139
  online classification of, 129
  phytases and, 151
  sequences in, 129
  structural alignments between neuraminidases and, 151
N-terminal embellishments, 185
N-type ATP-pyrophosphatases, 166, 169, 173, 180
  C-terminal embellishments in, 187
  domain–domain interfaces, 187
Nucleic Acids Research, 26
nucleotide sequences, 85
nucleotide-triphosphates, 173
nucleotidyltransferases (NT), 163–164, 167, 171, 173
  multidomain architectures (MDAs), 174

O-acetyl-ADP-ribose or hydrolyze ADP-ribose-1″-phosphate, 451
oligopeptide transporter (OPT) family, 330–331
Online Mendelian Inheritance in Man (OMIM), 85
Online Mendelian Inheritance in Man (OMIM)™, 232
organophosphatase activity, 142
Orthologous MAtrix (OMA), 104
orthologs, 101–102, 106
  pufferfish, 119
  tree analysis, 116
orthology databases, 106–107
orthology identification methods, 104

orthology identification methods (*Continued*)
  as basis for functional annotation, 118
  challenges, 104–105
  evaluation of, 105
  functional inferences to, 118
  graph-based methods, 104
  phylogenetic tree-based, 104
  phylogeny-based prediction of relationships, 112
OrthoMCL, 104
OrthoMCL algorithm, 106
Oryza sativa nonsymbiotic hemoglobin, 214
Oryza sativa protein Q851L1_ORYSJ, 119
outparalog, 101
over-annotation, 130
oxidoreductases, 197
oxyluciferin, 140

P. trichocarpa, 310
Paenibacillus larvae, 294
pair-wise sequences
  comparison, 73–74
  search protocols, 53, 74
pantetheine-phosphate adenyltransferase, 46
PANTHER (Protein Analysis Through Evolutionary Relationships), 112
pantoate-β-alanine ligase, 39
pantothenate synthetases (PS), 164, 167, 171, 173
  embellishments in, 185
  structural diversity, 179
paralogs, 101–102, 106
Paramecium caudatum, 210
Paramecium tetraurelia, 332
paraoxonases, 142–143
partner domains, 403
Paulinella chromatophora (YP_002048804), 310
paxillin LD1, 257
PDBSprotEC, 132
PD40 family, 149
peptidase family C1, 292
peptidases, 286
  in bacteria, 293–302
  distribution of, 288
  families (or subfamilies) of, 286
  family C1 (the papain family), 292
  unusual occurences in, 303–310
Periodic Table of Coiled Coils (CC+ database), 41
PerP peptidase, 290
Pfam-A profile-HMMs, 26–27
Pfam database, 26, 396
  alignments, 31–32
  curated entries, 26
  DBD families, 396
  domain of ATP-synthase alpha and beta chains, 31
  Domains of Unknown Function (DUFs), 32
  family annotation, 32–34
  flow of work in building a, 26
  FULL alignments, 27
  human genome coverage, 80
  leucine-rich-repeat (LRR) regions, 119
  Pfam-A profile-HMMs, 26–27
  philosophy of, 28–30
  release 26.0, 33
  screenshot of, 33
  SEED alignment, 26
  sequence classification, future, 34–35
  sourcing new families, 31–32
  spectrin, 247
phenyl acetate, 144
5′-phosphate- and 3′-hydroxyl-terminated oligonucleotides, 346
phosphatidylcholine, 252
phosphatidylethanolamine (PE), 252
phosphatidylserine (PS), 252
phosphoadenosine phosphosulfate reductases, 174
phosphoadenylyl sulfate reductases
  embellishments in, 185
phosphoadenylyl-sulfate reductases (PAPSR), 164, 168, 171–172
phosphopantetheine adenylyltransferase, 39
Photobacterium profundum, 366
phycocyanins, 214–216
  C-helices in, 216
  helices in, 216
  interactions of residues, 216
PhyloBuilder, 111
PhyloFacts Orthology Group (PHOG) database, 102, 105–106, 118
  orthology relationships, prediction of, 109
PhyloFacts Phylogenomic Encyclopedia
  collection of primary and derived data, 108
  concept of protein family "books," 108
  construction pipeline, 109
  PhyloFacts 3.0, 113
  phylogenetic trees for protein functional and structural domains, 109, 115
  PhyloScope display of a protein family, 117
  subfamily classification in, 111
phylogenetic orthology prediction protocols, 102, 104–105
  PhylomeDB, 112
phylogenetic tree estimation, 116
phylogenomic databases, 107

domain-based phylogenetic reconstruction, problems with, 119–120
specific issues in, 113–118
PhylomeDB, 105, 113
Phylo-Tuner, 79
Phytophthora, 450
Pictet–Spengler reaction, 143
PlanTAPDB, 389
plant receptor-like proteins (RLPs), 119
PlantTFDB, 389
plasmids, 292
Plasmodium falciparum, 395
plectin, 244
PlnTFDB, 389
P-loop hydrolase superfamily, 55
p-nitrophenyl acetate, 152
Podospora anserine, 304
polyADP-ribose polymerase (PARP), 451
Populus trichocarpa, 310
positional homology, 115
position-specific scoring matrices (PSSMs), 4
PP-ATPase, 160
Prevotella ruminicola, 306
PRINTS, 40
PRINTS database, 26
ProbCons, 115
ProClust algorithm, 10
PRODOM algorithm, 9
PRODORIC, 387
profile-based clustering approach, 79–80
profile-HMM software, 27
profile–profile method (PRC), 75
Pro 95 G2, 230
prokaryotes, 212, 219
prokaryotic TF resources, 387–389
PROMALS program, 150
promiscuous domains, 120
β-propeller proteins, 151
boundaries, 148–151
"4-pro-S" hydride, 201
PROSITE, 26
PROSITE profiles, 26, 28, 31
protein arginine deiminases (PADIs), 456
Protein Data Bank (PDB), 10, 38, 82, 255, 265
domain boundaries, 43
protein domain, definition of
evolutionary history and, 40
structure-based, 39
protein family collections, 28
protein–protein interfaces of HUP superfamily, 184–187
protein 4.1R, 263–265
protein sequence clustering, 7
algorithms, 9–11
domain classification and, 8
hierarchical clustering, 8
single-linkage clustering, 8
protein sequence families, 5
protein targets, selection of, 151–153
protein unfolding units (PUU) algorithm, 43–44, 46
Proteobacteria, 304
proteolytic enzymes, 286
ProtoMap algorithm, 9
ProtoNet, 104
ProtoNet algorithm, 10
protozoans, 212, 219
pseudogenes, 101
Pseudomonas aeruginosa, 142, 345
PSI-BLAST algorithm, 30
P-type ATPases, 328
pufferfish *(Tetraodon nigroviridis),* 119
Pyrobaculum species, 308
Pyrococcus furiosus, 343, 348

quality assessment, 12–13

RAP tool, 116
Ras GTPase-activating protein (rng2), 255
receptor-like protein kinases (RLKs), 119
red cell cytoskeleton, 242
red cell survival, 240
Reference Genome Project, 112
reference species tree of sequenced genomes, 92
RefSeq, 160
RegPredict, 327–329
regucalcin, 140
RegulonDB, 387, 391
repeat-associated mysterious protein (RAMP), 344
arrangement in CRISPR-Cas systems, 355–356
Cas3 group, 368
Cas4 group, 368
Cas5 group, 351, 367
Cas6 group, 350, 354
Cas7 group, 354, 365
Cas8 group, 366–367
classification, 355
Csf3, 354
Csy3 subfamily, 354
domain architectures of, 354
major groups of, 349–355
sequences and structures of, 354
residue-based structure comparison, 49–50

retrograde pathway evolution, 173
retrotransposons, 290
Rex, 193
Rfam database of RNA families, 33
Rickettsia genome sequences, 292
rigid body superposition method, 50
RIO tool, 116
root mean squared deviation (RMSD), 50, 179
Rossmann domain, 30
Rossmann fold, 160
RPD3/HDAC superfamily, 450
RuvC-like RNAseH fold domain, 364

S. pombe, 255
Saccharomyces, 310
Saccharomyces cerevisiae, 290, 386, 437
Saccharomyces Genome Database (SGD), 390
Salinispora tropica, 305
SAM
  iterative procedure, 77
  "target" procedures, 77
S58 (aminopeptidase DmpA), 310
SAM (*S*-adenosylmethionine) enzyme domain (Elp3), 449
SATCHMO-JS, 115
SATCHMO-JS simultaneous tree construction, 111
S1A (trypsin), 310
S62 bacteria, 290
S74 bacteria, 290
Schizosaccharomyces, 305, 310
Schizosaccharomyces pombe Ain1 protein, 247
SCI-PHY algorithm, 110
S12 (class C beta-lactamase CMY-18), 294
SCOP, 29
  classifications of domains, 160
  domain and residue coverage, 80–81
  domain architectures for Follistatin-related protein 4, 78
  history, 73
  "leave-one-out" benchmark, 75
Scorecons, 88
SCPS algorithm, 10
S13 (d-Ala-d-Ala-peptidase C), 310
Sebaldella termitidis, 309
secologanin, 147
secologanin bound, 149
secondary structure orientations, 50
SEED alignment, 26
  resources for, 26, 28–29
SEED predictions, 327–329
Segniliparus rugosus (gene HMPREF9336_02911), 305
senescence marker protein-30/gluconolactonase/luciferin-regenerating enzyme-like (SGL) subgroup, 139–141
  DFPase, 146–147
  gene context for an uncharacterized, 153
  metal-coordinating ligands common to, 144
senescence marker protein-30 (SMP-30), 140
sequence alignments, 85–88
sequence-based automated methods, 45–46
sequence databases, 5
sequence homology, transitivity of, 74
sequence identity, 101
sequence-profile comparison, 73
sequence similarity network, 144–145
sequence similarity networks, 151
  of arylesterase-like subgroup, 142
  of six-bladed β-propeller proteins, 150
  of SSL group, 151
sequence space graph, 6–9
  algorithms, 10
  mutational paths, 6
  for protein sequence clustering, 7–9
  putative instances of domain chaining and, 8–9
  p-values, 6
  spurious similarities, 6
  structure comparisons using, 6
sequence-specific DNA-binding TFs, 384
Sequential Structure Alignment Program (SSAP), 40
  Cβ atoms, structure of, 49
  domain structures in CATH, 49
  fold recognition, 49
  strategies for aligning equivalent regions, 51
  using SISYPHUS, 51–52
seven-bladed β-propeller proteins, 151
SFT programs, 324–327
short-chain dehydrogenase/reductases (SDR)
  in humans, 203–204
  motif of specific residues for, 201
  reactions undertaken by, 201
  Rossmann fold, 201
short-chain dehydrogenases, 193
sickle-cell anemia, 208, 232
S24 (ImpA UV protection protein), 294
Simple Modular Architecture Research Tool (SMART), 28, 31, 40, 250, 261
Sir2 superfamily deacetylases, 450
Sirtuin 4, 5, and 6, 450
SISYPHUS algorithm, 51
six-bladed β-propeller fold proteins, 136–138
six-bladed β-propeller proteins, 149
  sequence similarity networks of, 150

S16 (Lon-A peptidase), 302
SMART, 26
Smith–Waterman (SSEARCH), 53
sparse taxon sampling, 113
SPC algorithm, 10
special clustering of protein sequence (SCPS), 9
spectraplakins, 244, 250, 254
spectrin, 241
  calponin homology domains, 244
  CH domain HMM (PF00307), 247
  C-terminal regions of α-spectrins, 257
  domain origin in, 245–248
  EF-hand domains, 247
  functional distinction between α-actinins, 253
  genome duplications, 244
  βI-, 253–254
  4.1 proteins and, 263–265
  spectrin-actin-binding (SAB) domain, 256, 264
  superfamilies, 244–245
  triple helical bundle of, 248–250
β-spectrin, 243
spectrin–actin interaction, 241
S45 (penicillin G acylase precursor), 310
S14 (peptidase Clp), 310
sperm whale myoglobin, 210–211, 213–214
sperm whale *(Physeter macrocephalus, a.k.a. Physeter catodon)* myoglobin, 208
spirochetes, 246–247
Spodoptera frugiperda ascovirus 1a (CAD24627), 306
sporulation initiation factor (Spo0A), 400
S33 (prolyl aminopeptidase), 302
S9 (prolyl oligopeptidase), 302
S54 (rhomboid), 302
S33 (SAT-1 protein), 294
S26 (signal peptidase), 302
S8 (subtilisin), 302
standard functional annotation protocol, 103–104
Staphylococcus epidermidis, 343
STAT (signal transduction), 403
Streptomyces cacaoi, 294
strictosidine synthase-like (SSL) subgroup, 143
  diisopropylfluorophosphatase (DFPase), 146
  functional predictions and prediction of misannotation, 143–146
  phylogenetic analysis of, 146–147
  sequence similarity network for, 144, 153
strictosidine synthase (SS), 136, 138–139, 143–149, 151–153
Structural Alignment of Multiple Proteins (STAMP) algorithm, 40
structural classification of proteins (SCOP), 12, 38
  comparison with CATH, 59–61
  domain boundary assignment, 44
  hierarchies of population, 58, 60
  multiple structural alignments in, 51–52
  websites, 63
structural embellishment
  domain–domain interfaces, 187
  functional role of, 181–183
  in ligand-binding sites, 183–184
  in protein–protein interaction, 186
  in shaping catalytic sites, 183–184
structural phylogenomics, 100, 113
structural similarity, measuring
  in CATH, 51
  random score distribution analysis, 51
  rigid body superposition method, 50
structure-based domain classifications
  application, 71
  commonly-used resources, 40
structured query language (SQL), 92
Structure–Function Linkage Database (SFLD), 41, 128–129, 133
  criterion for inclusion of proteins, 146
  paraoxonases, 141–143
  sequence- and structure-similarity networks, 133–136
subclassification of homologous superfamilies, 56–58
sub-functionalization, 101
Sulfolobus solfataricus, 343, 348, 351, 356, 371
superfamilies, 5
  actin-binding proteins, 244
  definition of, 41
  drug/metabolite transporter (DMT), 330
  HDAC/Rpd3, 450
  homologous, subclassification of, 56–58
  population account of, 58
  RPD3/HDAC, 450
  spectrin, 244–245
  structural diversity in, 176–178
  TolB C-terminal domain, 149
  voltage-gated ion channel (VIC), 331
superfamily characters, 51
SUPERFAMILY database, 29
  annotation projects and, 72
  applications, 71–72
  automated SAM iterative "target" procedure, 74–75
  Cocoa genome, domain of, 90
  coverage and domain statistics, 84
  cut-off in, 82
  data access, 83
  domain annotations using, 77
  domain architectures for Follistatin-related protein 4, 78

SUPERFAMILY database (*Continued*)
dynamic programming for HMM scoring, 78
e-values, 82
genome assemblies, 80–81
high throughput technology, 70–71
history, 72–73
HMM-based server of, 73, 78
HMMER3 version, 78
"leave-one-out" benchmark, 75
phylogenetic pattern of a domain architecture, 80
phylogenetic profiling of domains, 79–80
precalculated annotations, distribution of, 83
recognition and display of structural domain assignments, 85
reference species tree of sequenced genomes, 92
structural information, 71
technology, 73–74
superfamily identification, 53–54
criteria for, 53
enzyme commission (EC) resources, 53
fold variations, 77
gene ontology (GO) resources, 53
HMM–HMM-based methods, 53
sequence-based protocols, 54
structure comparison methods, 53
superfamily phylogenies, 117
superfamily tree (SFT) programs, 316
superfolds, 58
super-orthologs, 101, 105
super paramagnetic clustering (SPC), 9
SwissProt, 118
SYSTERS, 104
SYSTERS algorithm, 9

TAFs (TATA-binding protein associated factors), 437
taxonomic limits, 395
TaxViz image, 88
TCDB database, 316–318
data input for, 320–327
FASTA format genome analysis, 332
functional predictions using, 327–331
Gblast program, 332
phylogenetic trees, 325
semiautomated genome analysis, 331–333
SFT programs, 324–327
user interface, 318–320
T-Coffee, 115
Tetrahymena thermophila, 331–332
thalassaemias, 232–233
Thermococcus onnurineus, 306
thermodesulfobacteria, 302
thermoplasmas, 246
Thermoproteus species, 308
Thermus thermophilus, 348
Thiamine Diphosphate Dependent Enzyme Engineering Database (TEED), 41
thresholded similarity networks, 134–135
TIGRFAMs, 26, 28, 31
TMS EmrE protein, 330
Toxoplasma gondii, 305, 450
transcription factors (TFs)
annotation resources, 387–391
based on DBD families, 385–386
DBD sharing, 407
DNA-binding domain families and, 390
domain architectures, 403–408
in eukaryotic genomes, 386
genome-wide studies of, 385–386, 408–410
genomic repertoires of, 392–395
intrinsically disordered (ID) regions, 408
manual literature curation, 390–391
parasitic protists and, 386
phylogenetic groups, 395–396
profile-based searches, 391
regulatory mechanisms, 409
resources, 387–390
sequence-based searches, 391
sequence-specific DNA-binding, 384
TRANSFAC, 386, 391
transitive annotation, 103
transmembrane (TM) helix, 119
transporter classification (TC), 316–318
transport protein evolution, 330
tree reconciliation tools, 116
TribeMCL algorithm, 10
Trichomonas vaginalis, 247, 293, 452, 480
triosephosphate isomerase (TIM) barrel enzymes, 173
triple helices of spectrin, 248–250
as the basis for enzyme structures, 254
tRNA(Ile)-lysidine synthases, 173
tRNA-processing HUP domain enzymes, 173
tRNA specific 2-thiouridylases (TRMU), 164, 167, 171, 173
tRNA sulfurtransferases, 173
tRNA 2-thiocytidine biosynthesis protein, 173
tropomodulin, 241
tropomyosin, 241
truncated globins, 212–213, 219–220
helices, 213
Trypanosma cruzi, 306
Trypanosoma brucei, 395
tryptamine, 147
tryptamine bound, 149
twilight zone, 47, 101

Type II CRISPR-Cas systems, 364–365, 372
Type III-A CRISPR-Cas systems, 351
Type III RAMP families, 354
Type U CRISPR-Cas system, 366
Tyr 37 C3, 230
tyrosine, 198
tyrosine-dependent oxidoreductases, 193

U35 bacteria, 290
ubiquitin-binding ZnF domain, 451
ultraparalogs, 101, 116
unclassified domain (UNC), 84
Uniprot Consortium, 160
UniProtKB sequence database, 27, 29, 31, 45, 73, 75, 112
  protein sets, 81
  substrate identifier, 204
  Tree of Life, 109
UniProtKB/TrEMBL protein database, 128
UniRef100, 145, 150
universal stress protein A, 160
universal stress proteins, 174
universal stress proteins A (USPA), 164–165, 168, 172
unusual Superfamilies, 88

Valyl-tRNA synthetases, 172–173
vertebrate globins, 217
vertebrate hemoglobin, 220
vertebrate homologs, 218
Vibrio cholerae, 372
Vibrio genome, 329
Victivallis vadensis, 302
vinculin, 251
viral peptidases, 290
virginiamycin B lyases, 151
visual identification of domains, 11
Vitis vinifera, 152–153
Vitreoscilla Hb, 222
voltage-gated ion channel (VIC) superfamily, 331
Volvox carteri, 305
Vulcanisaeta species, 308–309

Wikipedia, 33

X-Prolyl dipeptidyl aminopeptidase PepX, 255
X-ray crystallographic techniques, 38
xylono-1,4 lactone, 152

Yamanaka factors, 384
yeast arginyl-tRNA synthetase, 181

Z-disc, 258
Zu5 domain, 259

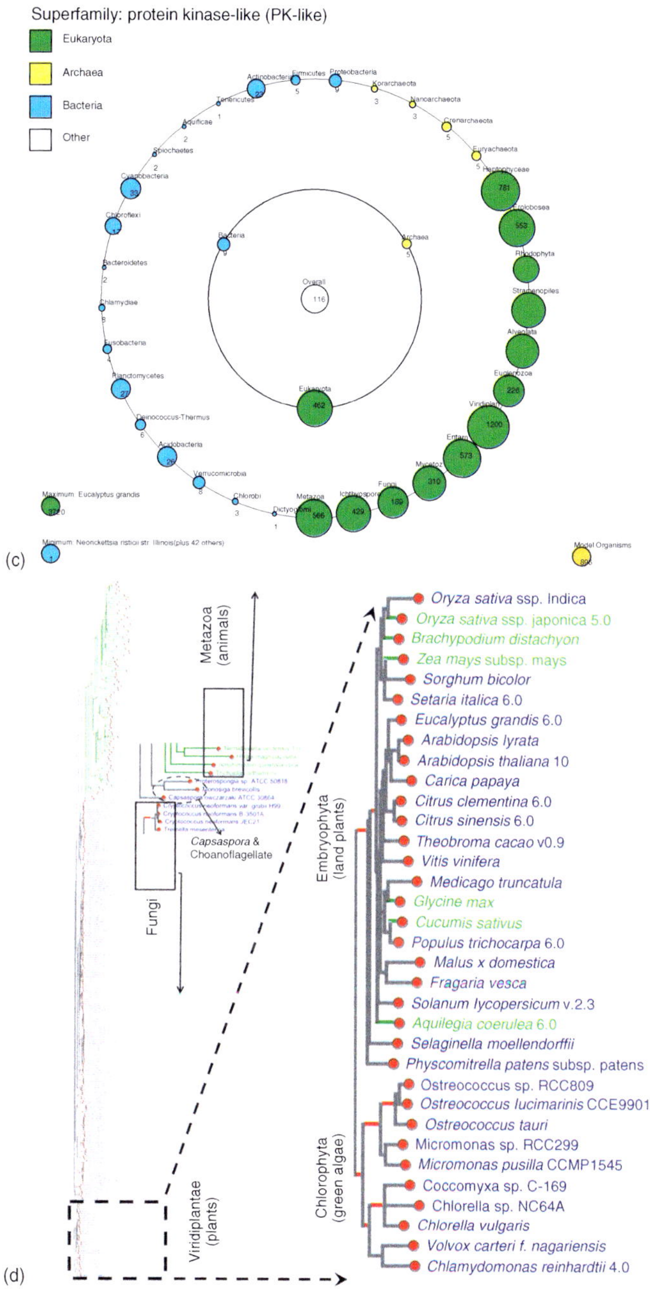

Figure 4.6c,d See full caption on page 87.

Protein Families: Relating Protein Sequence, Structure, and Function, First Edition.
Edited by Christine Orengo and Alex Bateman.

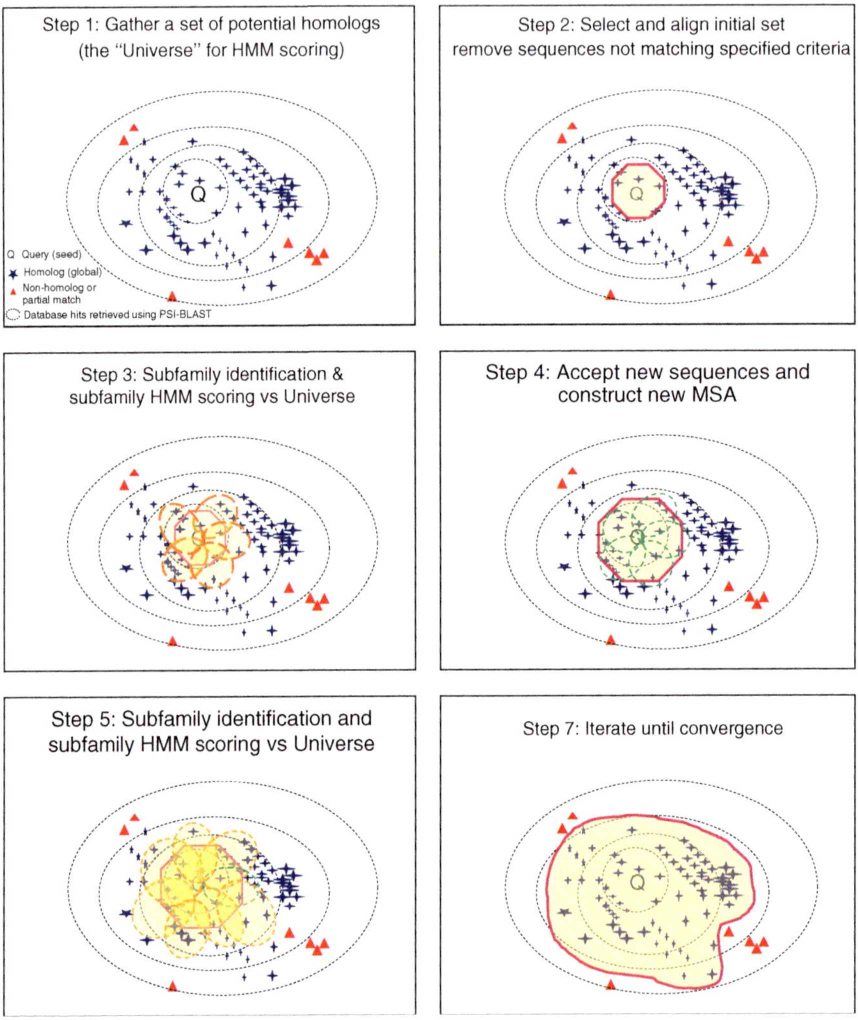

Figure 5.3 See full caption on page 110.

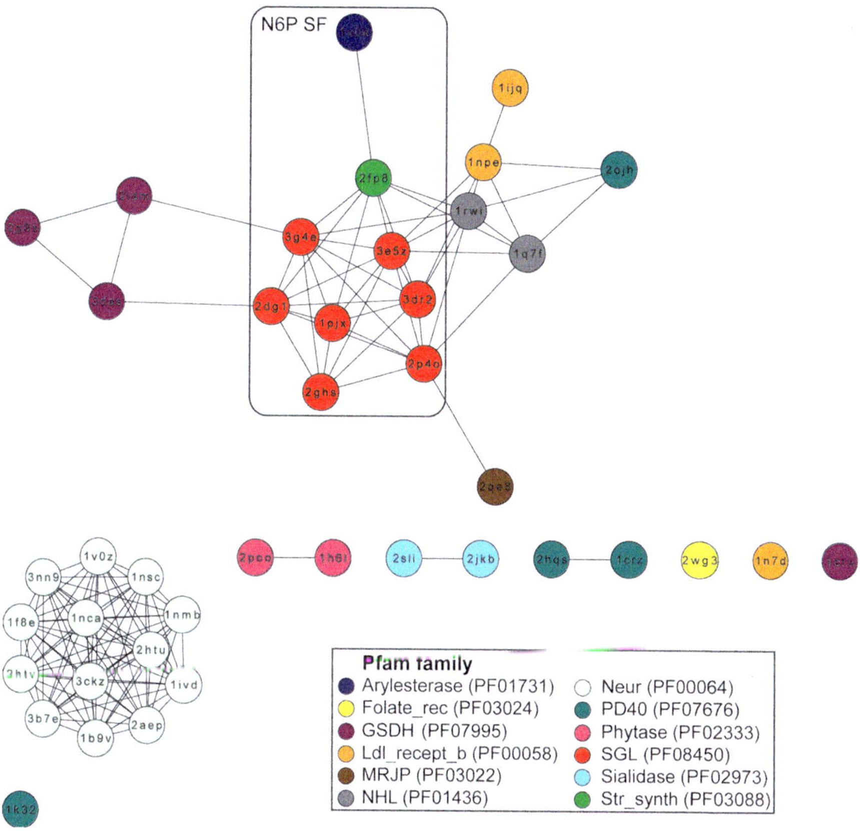

Figure 6.4 See full caption on page 137.

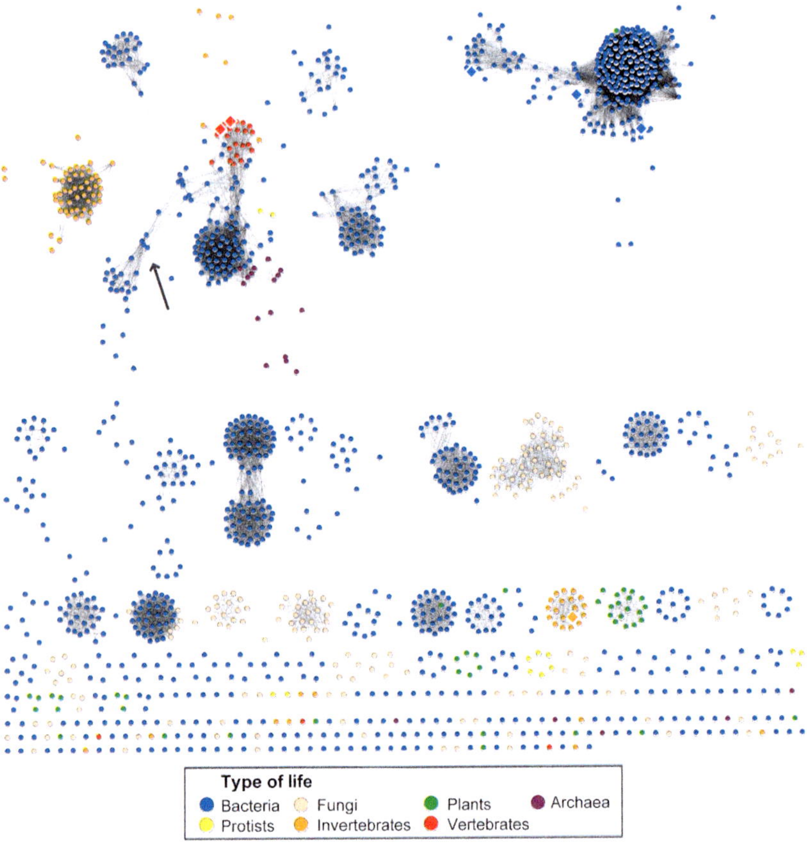

Figure 6.6 See full caption on page 140.

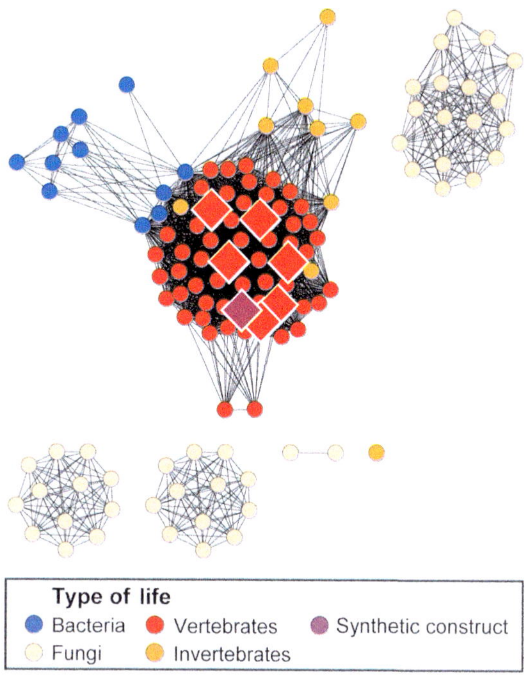

Figure 6.7 See full caption on page 142.

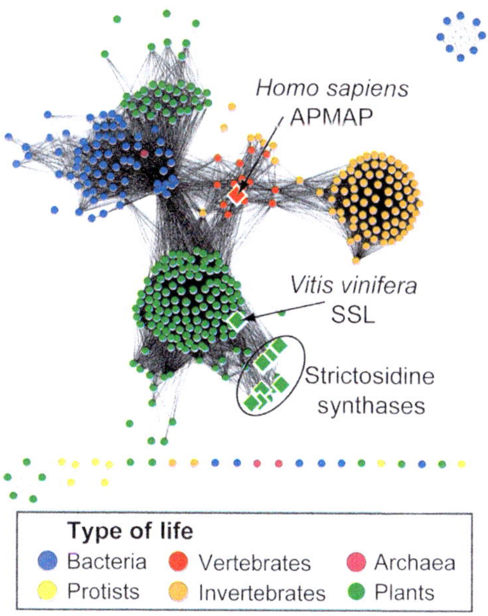

Figure 6.8 See full caption on page 144.

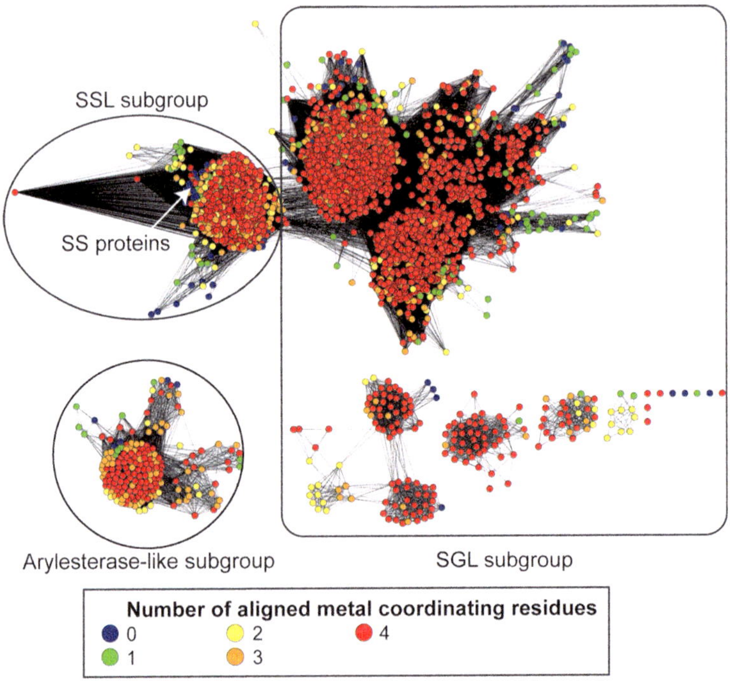

Figure 6.9 See full caption on page 145.

(a)

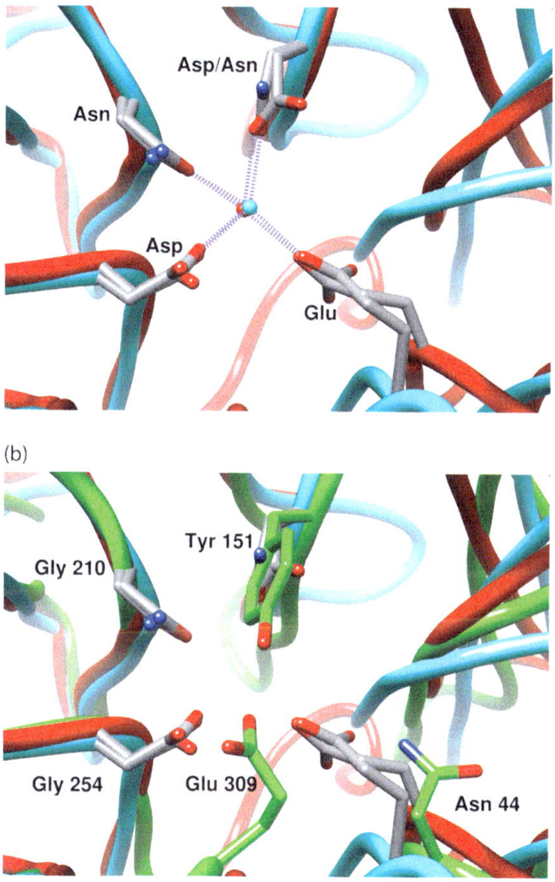

Figure 6.10 See full caption on page 147.

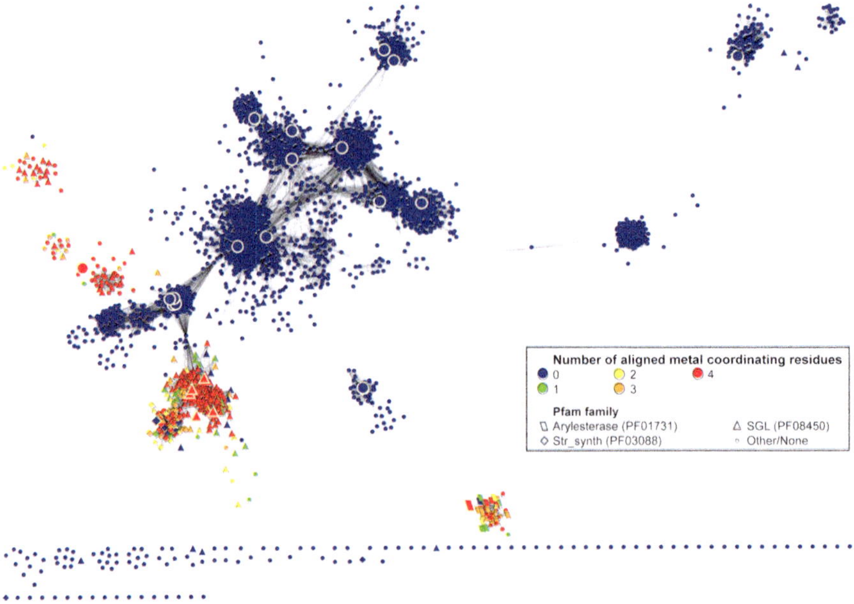

Figure 6.12 See full caption on page 150.

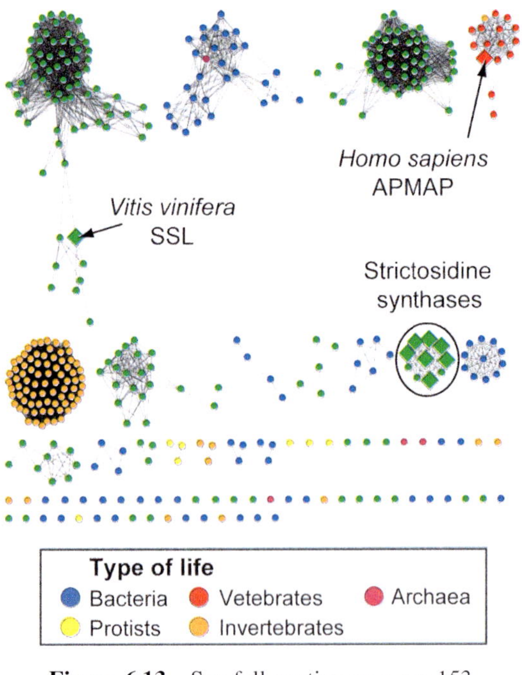

Figure 6.13 See full caption on page 153.

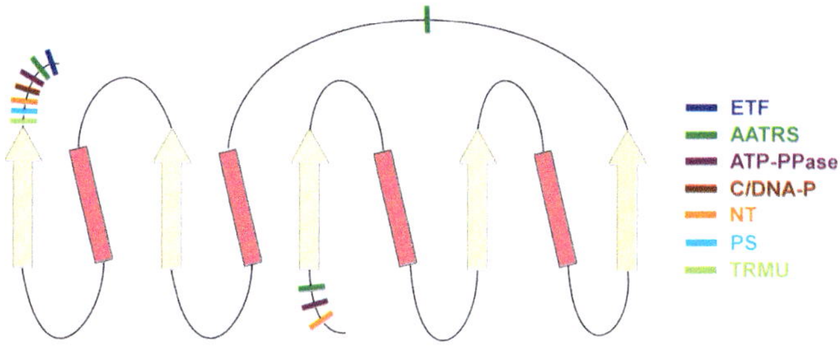

Figure 7.2 See full caption on page 175.

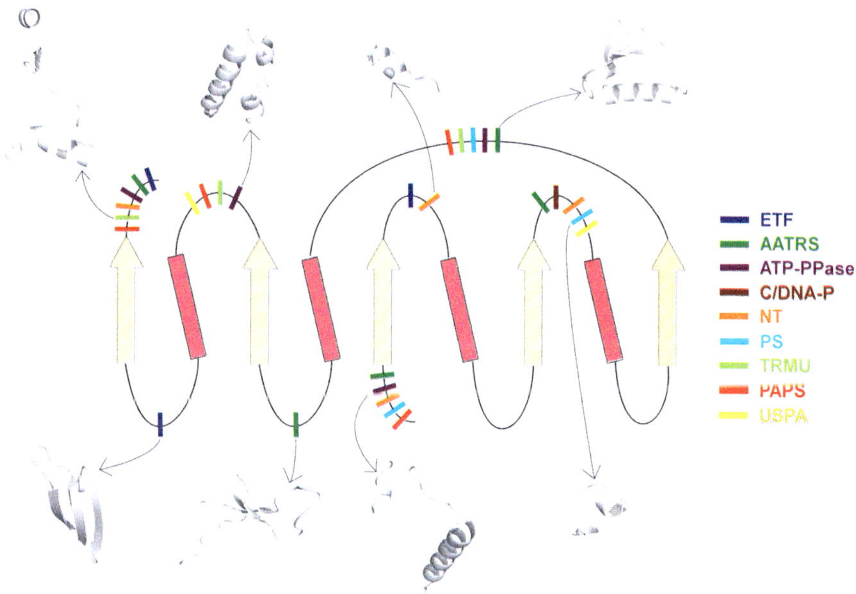

Figure 7.8 See full caption on page 183.

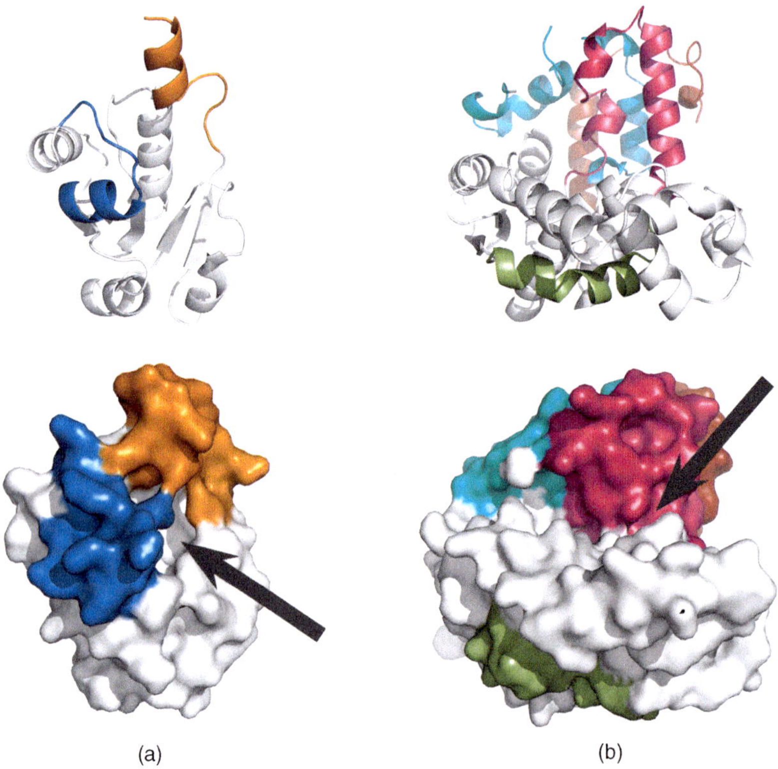

Figure 7.9 See full caption on page 184.

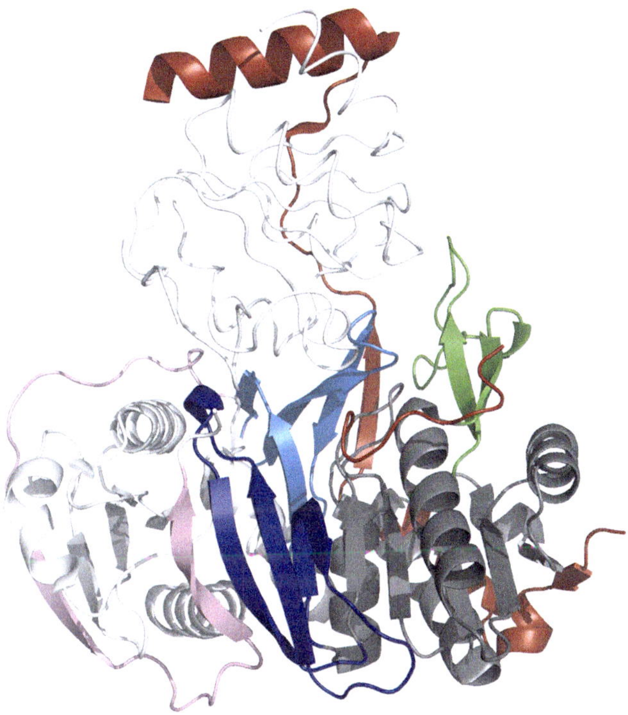

Figure 7.10 See full caption on page 185.

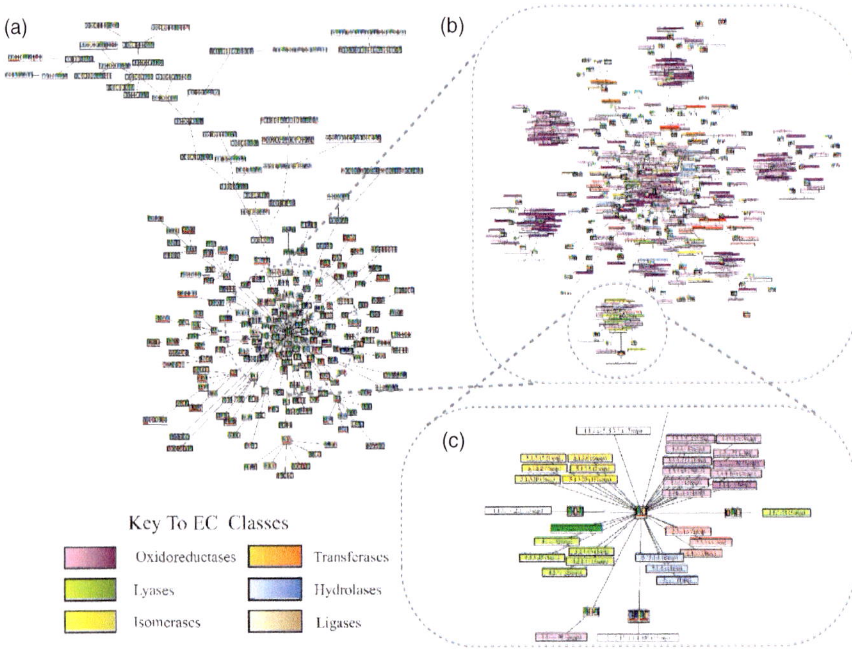

Figure 8.2 See full caption on page 194.

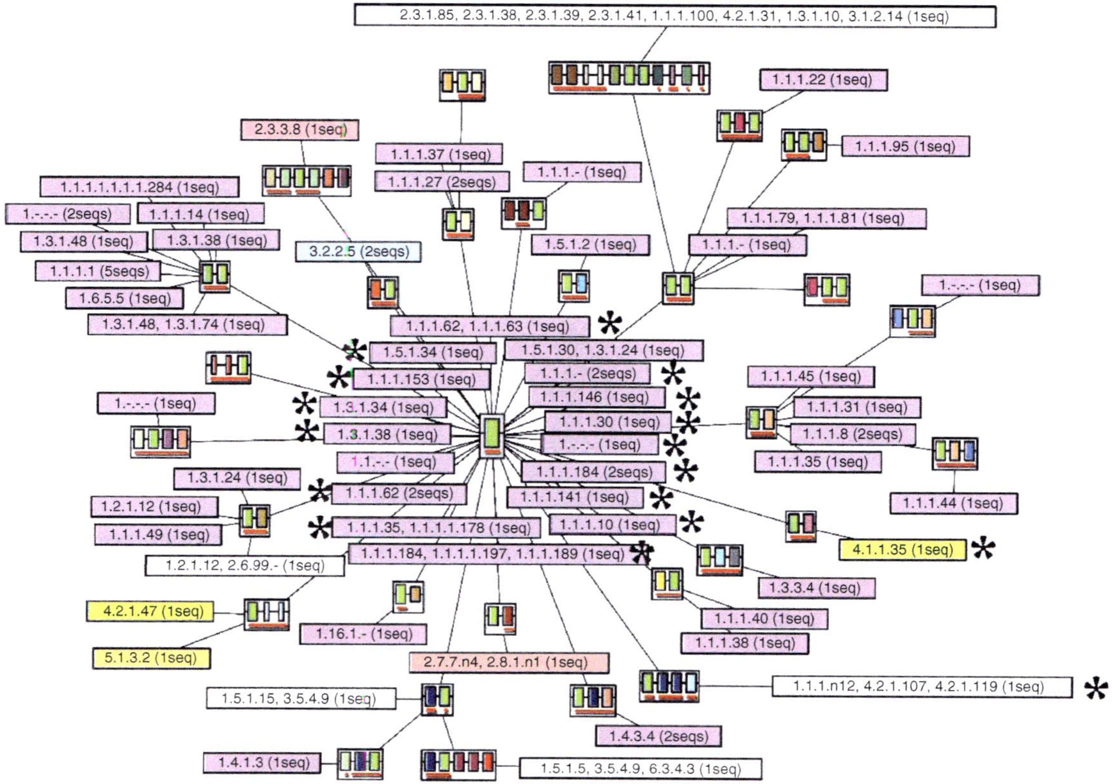

Figure 8.7 See full caption on page 202.

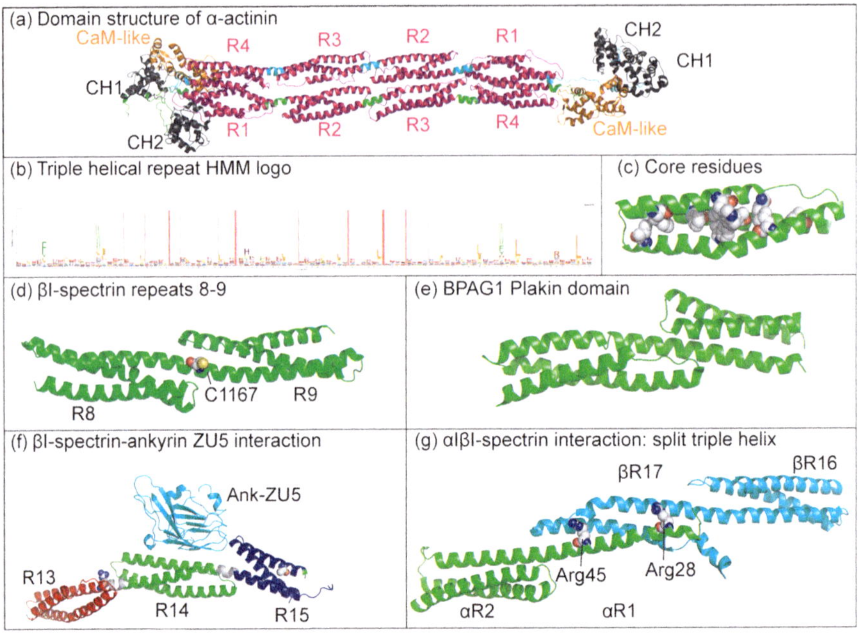

Figure 10.3 See full caption on page 249.

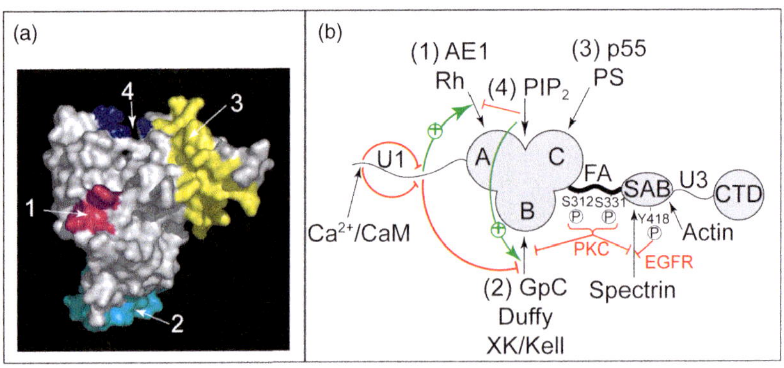

Figure 10.5 See full caption on page 264.

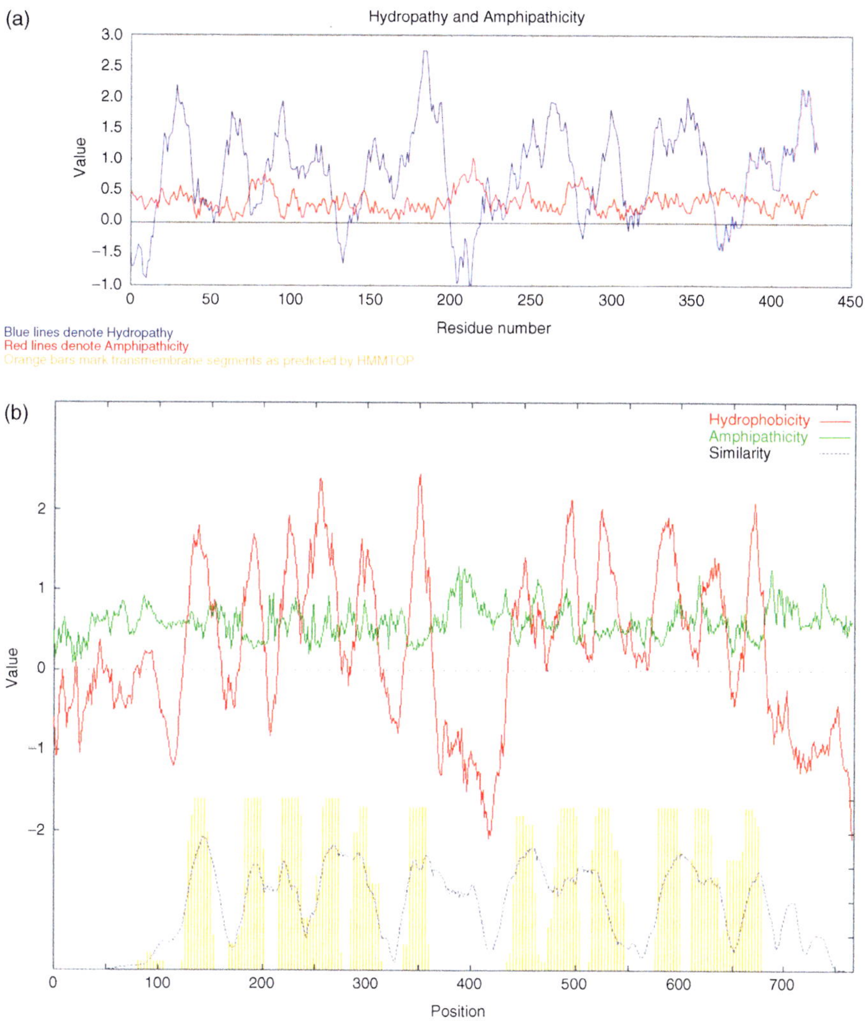

Figure 12.3 See full caption on page 322.

Figure 13.5a See full caption on page 352.

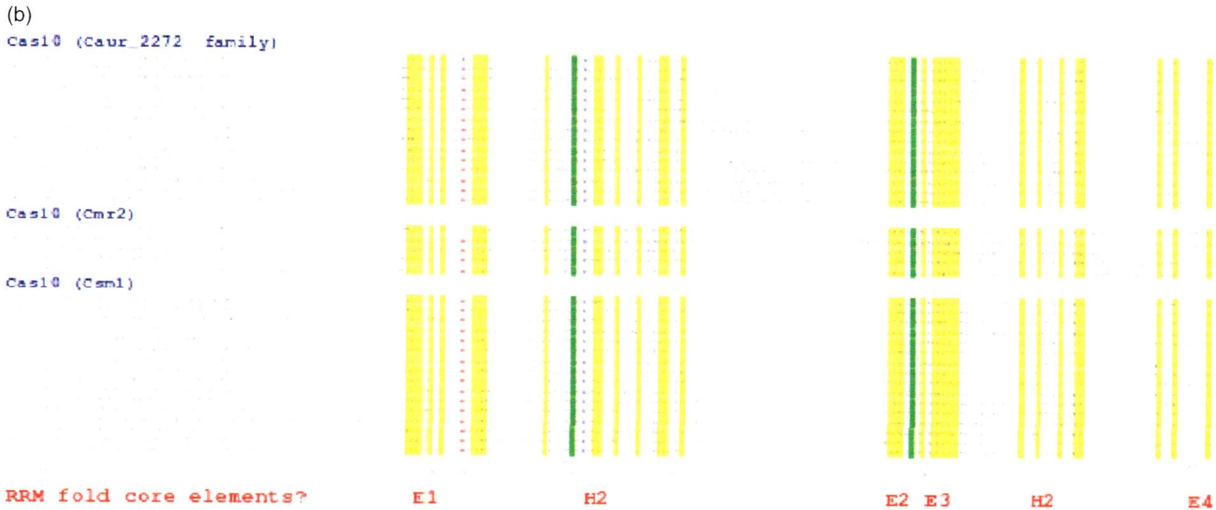

Figure 13.5b See full caption on page 353.

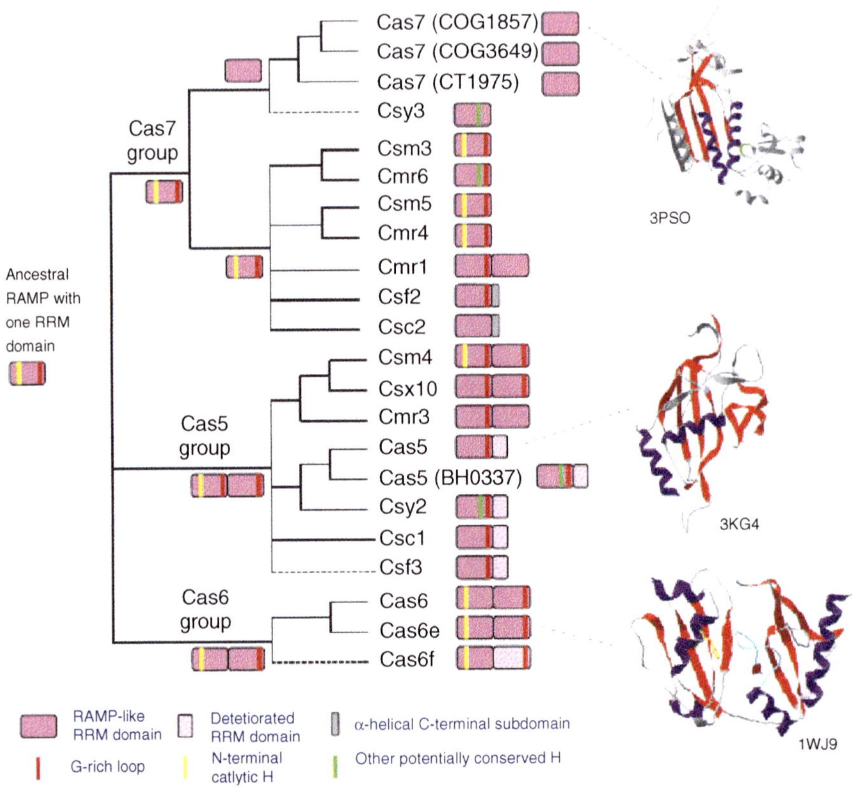

Figure 13.6 See full caption on page 355.

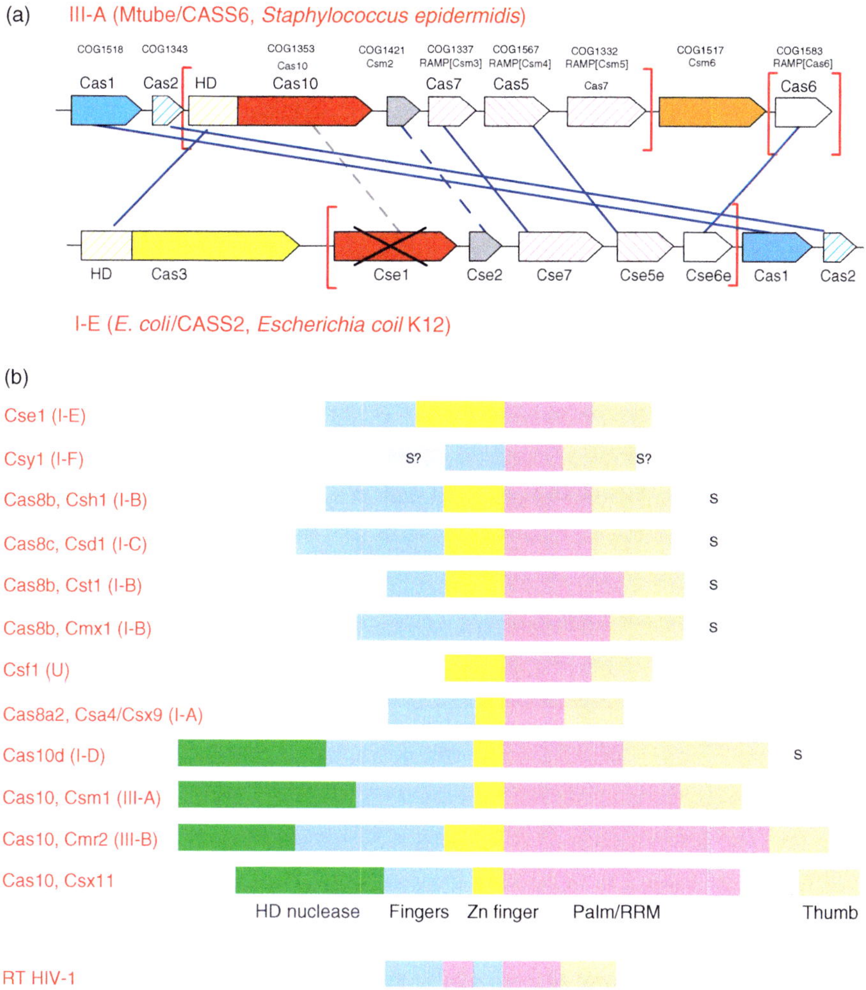

Figure 13.7 See full caption on page 359.

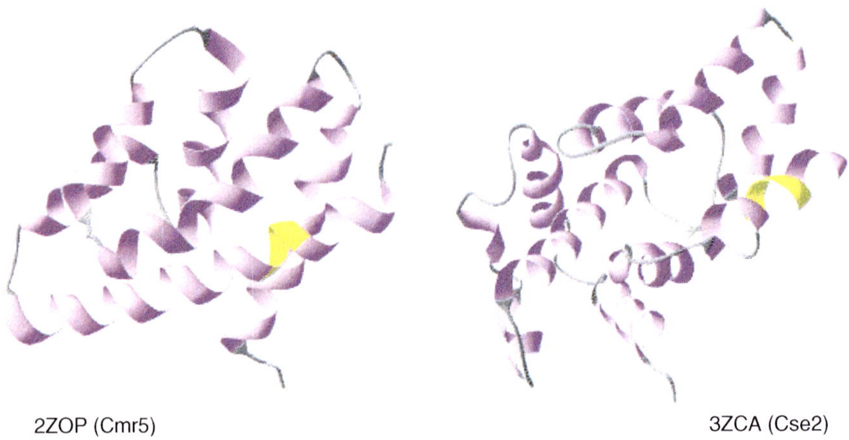

Figure 13.8 See full caption on page 360.

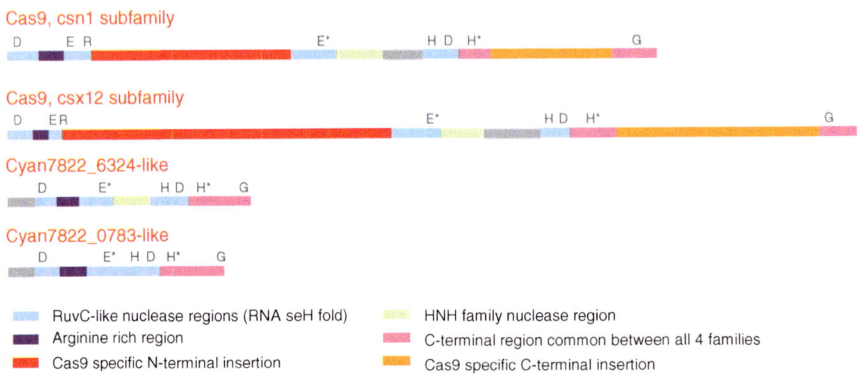

Figure 13.9 See full caption on page 364.

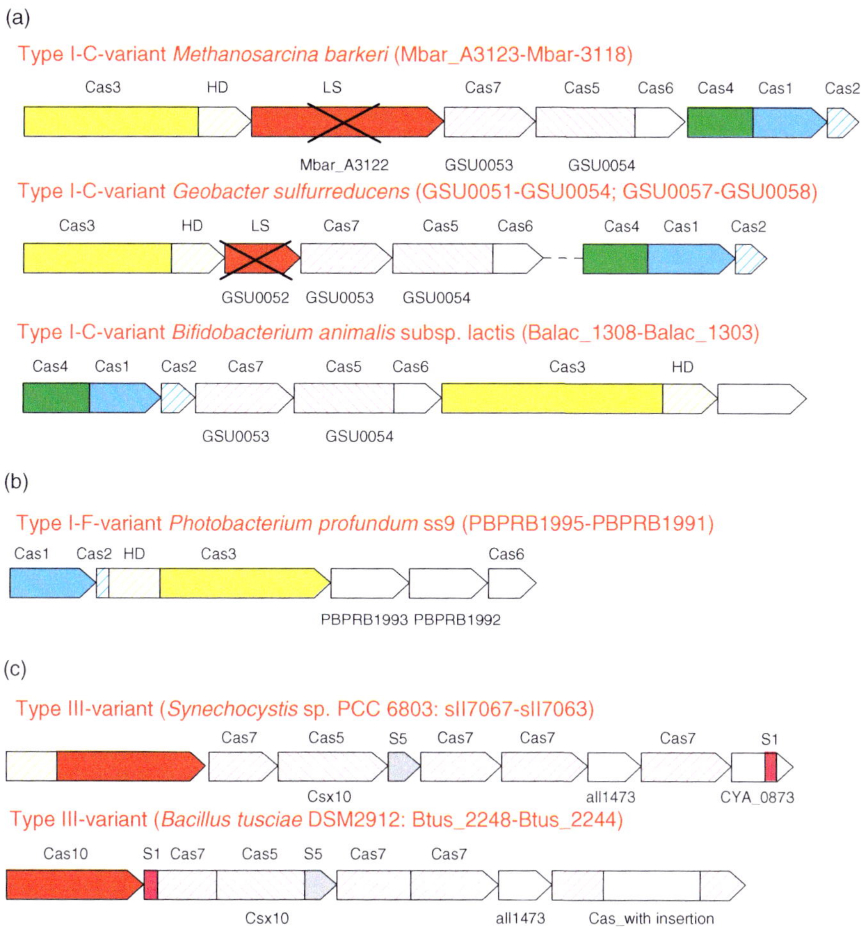

Figure 13.10 See full caption on page 367.

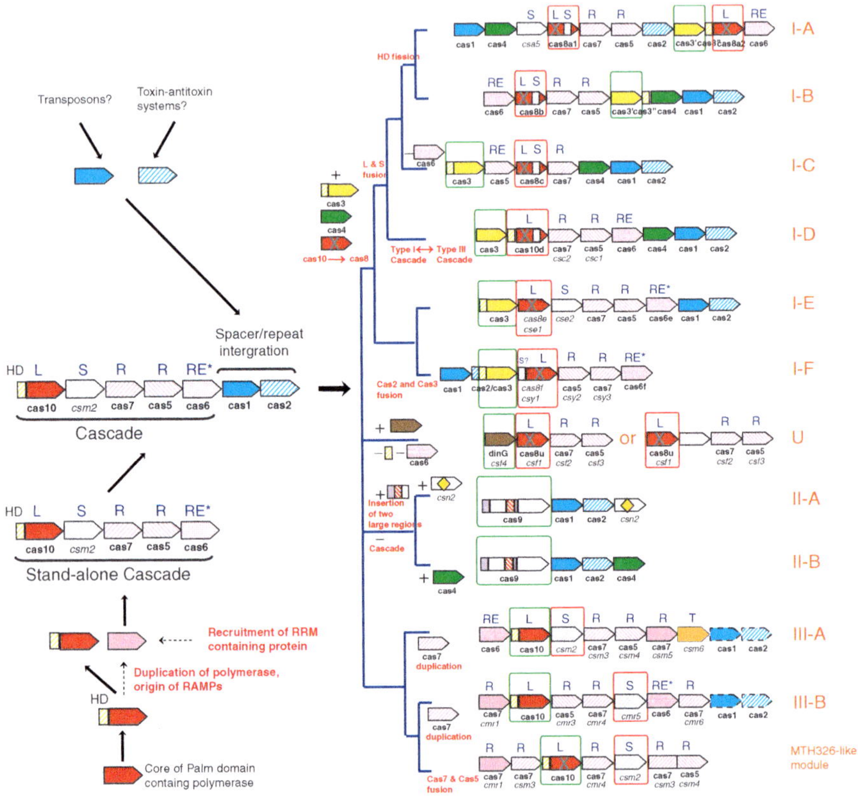

Figure 13.11 See full caption on page 369.

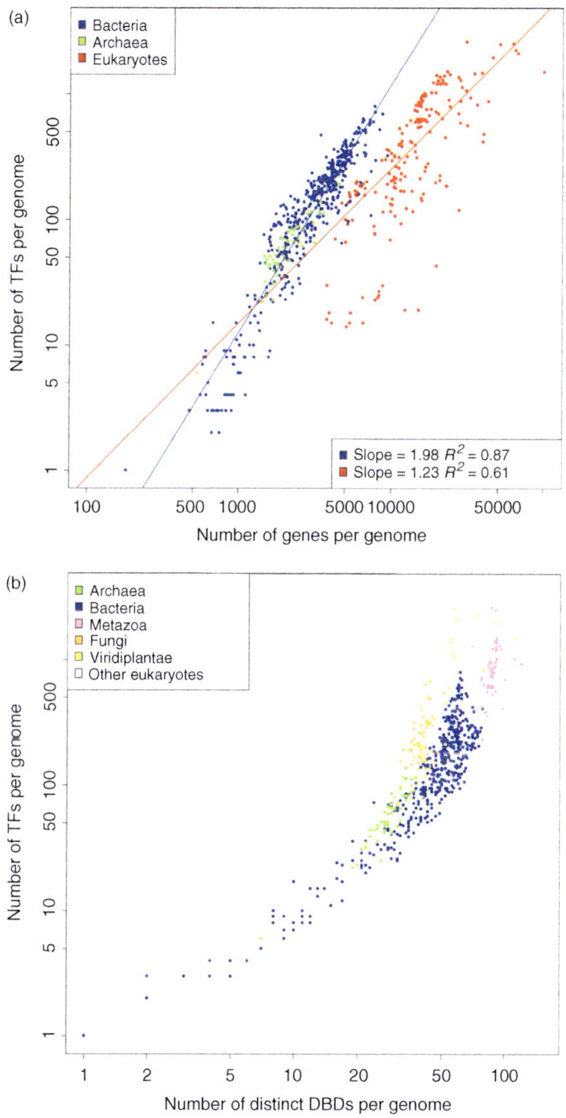

Figure 14.1 See full caption on page 393.

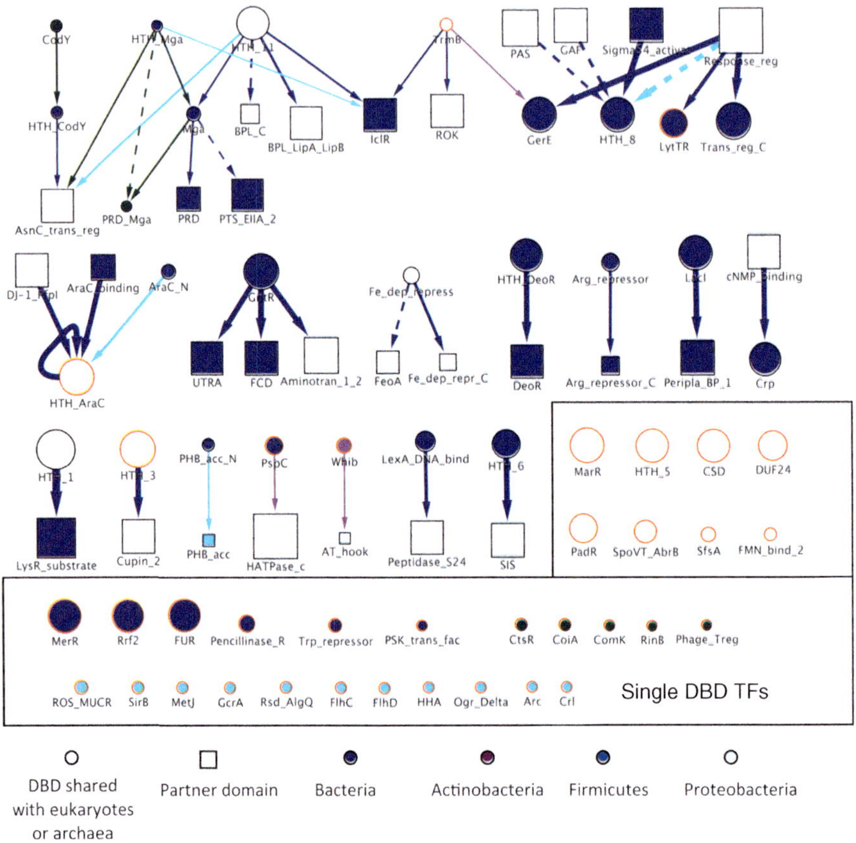

Figure 14.4 See full caption on page 406.

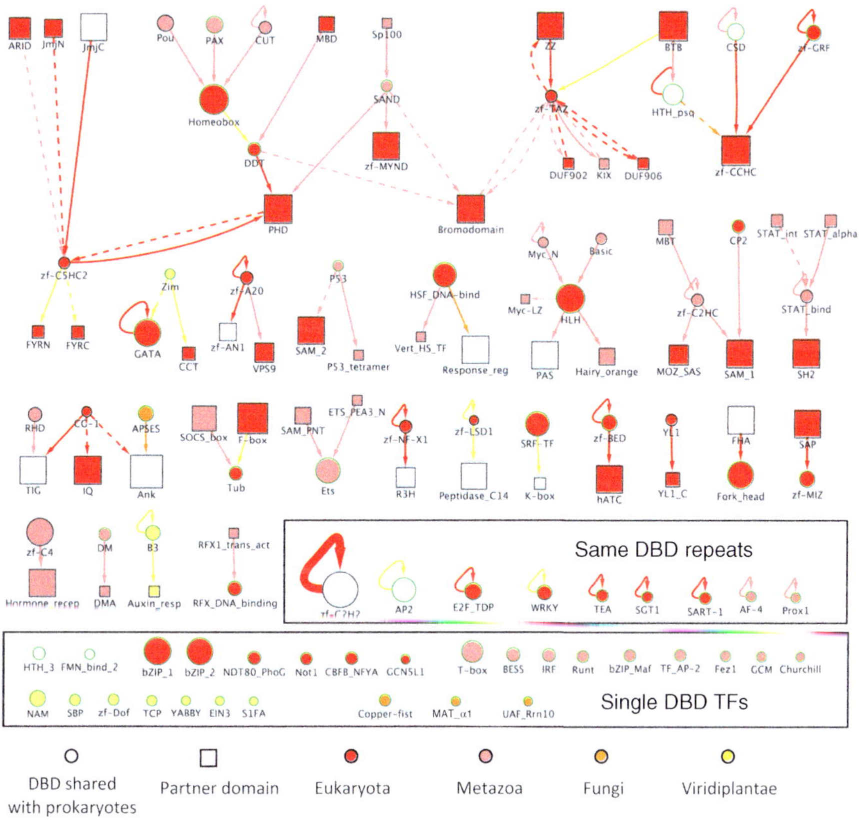

Figure 14.5 See full caption on page 408.

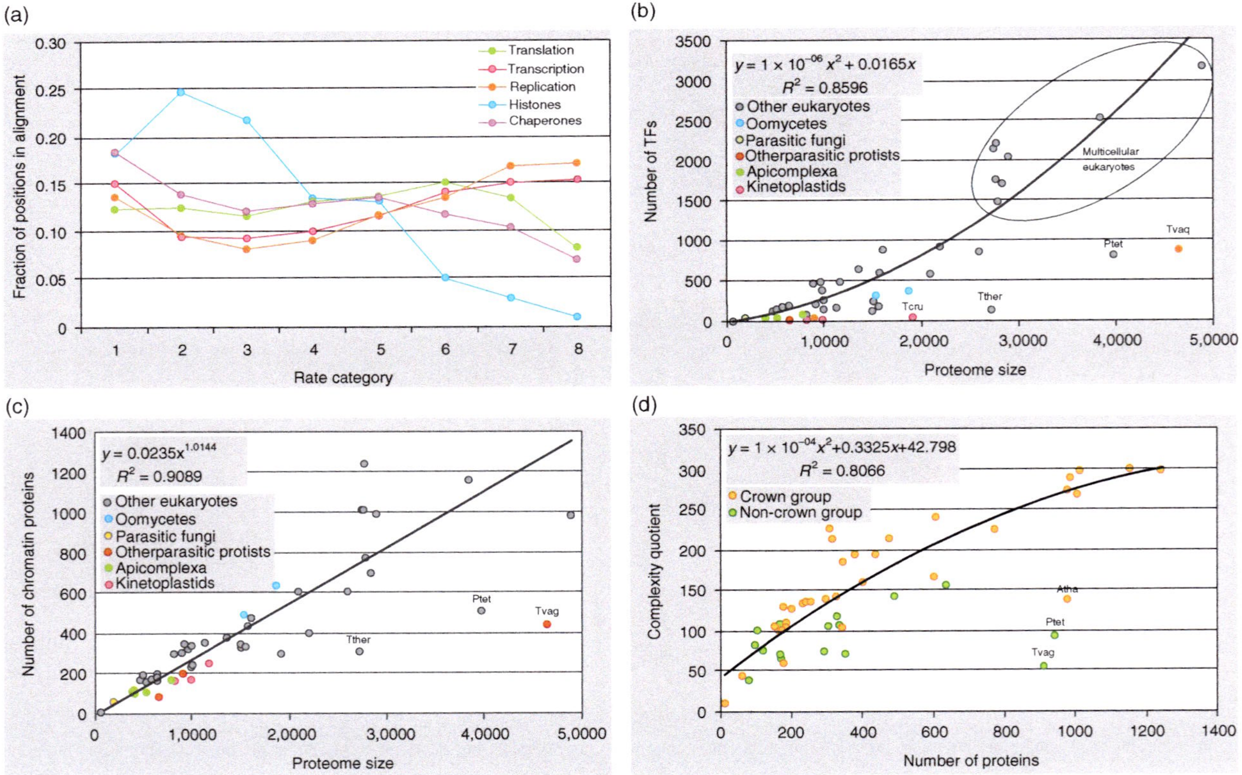

Figure 15.1 See full caption on page 428.

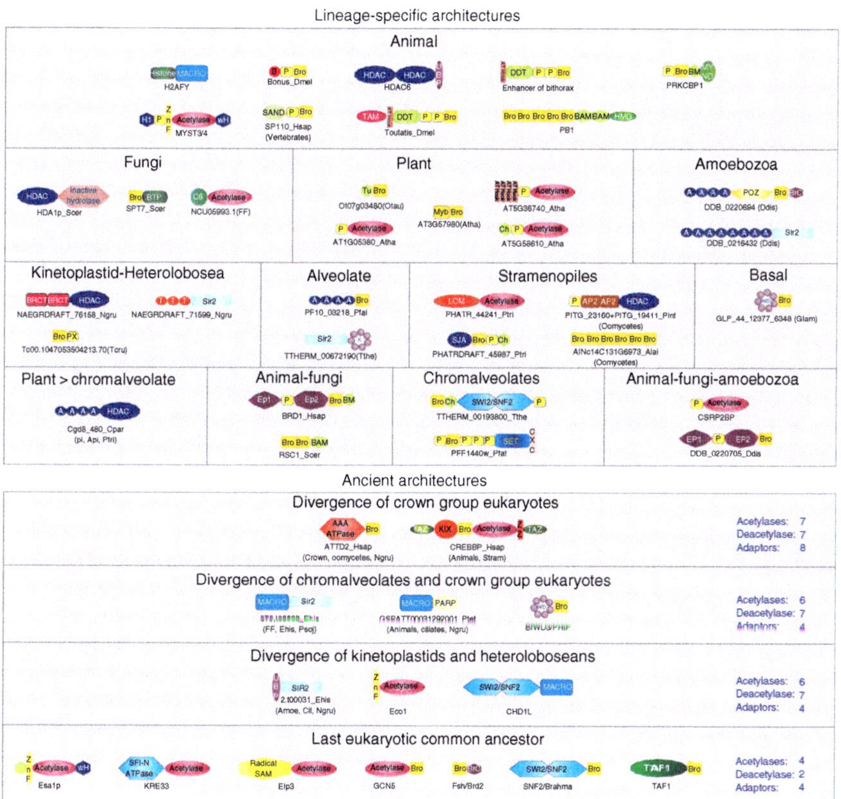

Figure 15.3 See full caption on page 448.

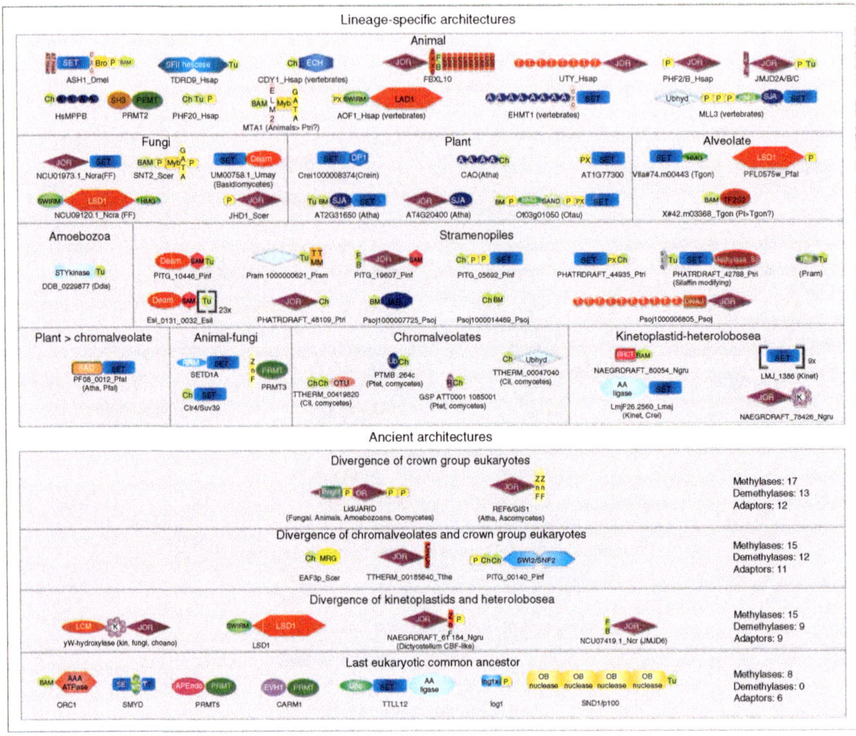

Figure 15.4 See full caption on page 453.

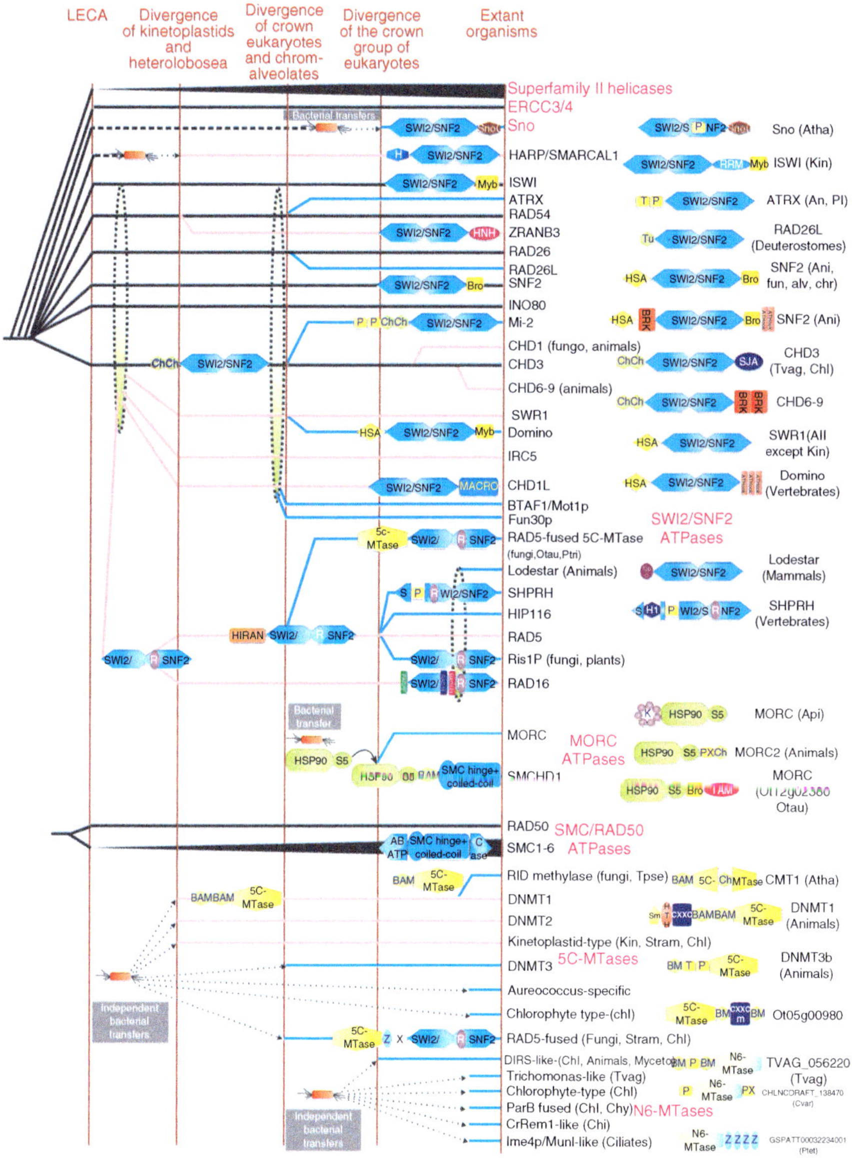

Figure 15.5 See full caption on page 460.

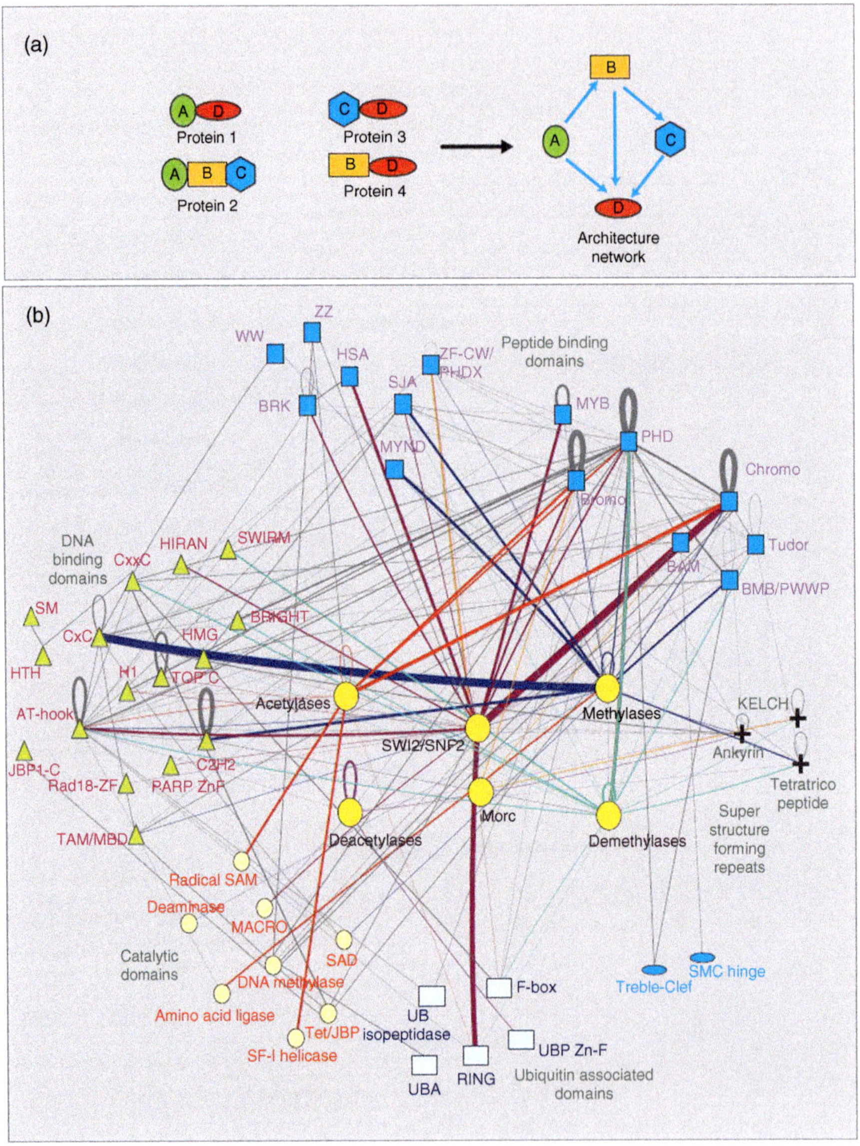

Figure 15.6 See full caption on page 466.

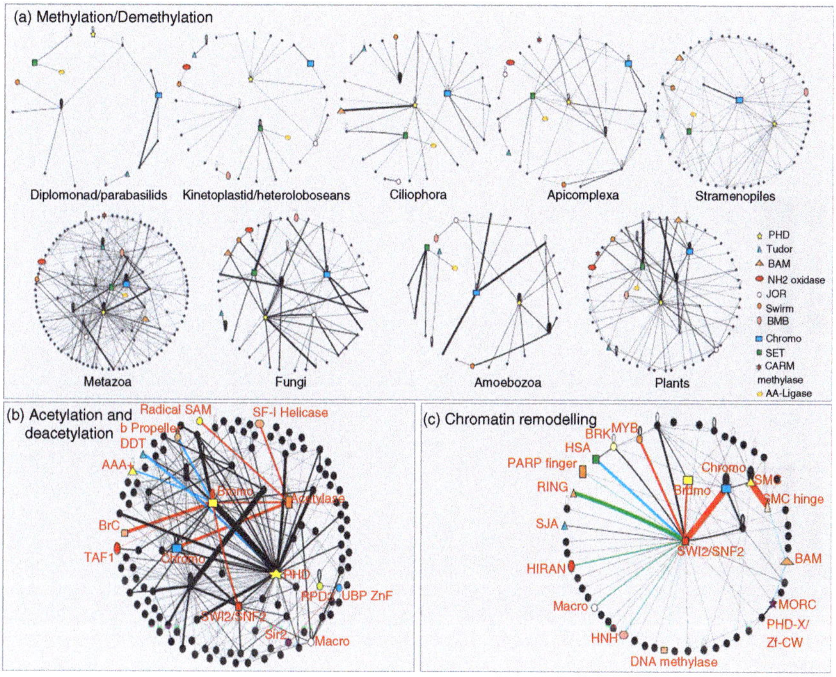

Figure 15.7 See full caption on page 478.

Printed and bound by CPI Group (UK) Ltd, Croydon, CR0 4YY
16/09/2026
14963416-0001

Low-Voltage
SOI CMOS VLSI
Devices and Circuits